Principles of

INFORMATION SYSTEMS

A Managerial Approach

Fourth Edition

Ralph M. Stair
Florida State University

George W. Reynolds
The University of Cincinnati

COURSE
TECHNOLOGY

ONE MAIN STREET, CAMBRIDGE, MA 02142

an International Thomson Publishing company I(T)P®

Cambridge • Albany • Bonn • Boston • Cincinnati • London • Madrid • Melbourne • Mexico City
New York • Paris • San Francisco • Singapore • Tokyo • Toronto • Washington

Senior Editor: Jennifer Normandin
Associate Publisher: Kristen Duerr
Product Manager: Cheryl Ouellette
Project Management: Elm Street Publishing Services, Inc.
Editorial Assistant: Amanda Young
Marketing Manager: Susan Ogar
Text Design: Elm Street Publishing Services, Inc.
Cover Design: Efrat Reis
Composition House: GEX, Inc.
Photo Researcher: Abby Reip

© 1999 by Course Technology—I(T)P®

For more information contact:

Course Technology
One Main Street
Cambridge, MA 02142

International Thomson Editores
Seneca, 53
Colonia Polanco
11560 Mexico D.F. Mexico

ITP Europe
Berkshire House 168-173
High Holborn
London WCIV 7AA
England

ITP GmbH
Königswinterer Strasse 418
53227 Bonn
Germany

Nelson ITP, Australia
102 Dodds Street
South Melbourne, 3205
Victoria, Australia

ITP Asia
60 Albert Street, #15-01
Albert Complex
Singapore 189969

ITP Nelson Canada
1120 Birchmount Road
Scarborough, Ontario
Canada M1K 5G4

ITP Japan
Hirakawacho Kyowa Building, 3F
2-2-1 Hirakawacho
Chiyoda-ku, Tokyo 102
Japan

Library of Congress Cataloging-in-Publication Data
Stair, Ralph M.
   Principles of information systems: a managerial approach/Ralph M. Stair, George W. Reynolds.—4th ed.
     p.   cm.
   Includes bibliographical references and index.
   ISBN 0-7600-1079-X
1. Management information systems.   I. Reynolds, George Walter. II. Title. III. Title: Information systems.
T58.6.S72 1999
658.4'038—dc21   98-47464
                               CIP

Trademarks
Course Technology and the open book logo are registered trademarks of Course Technology.
I(T)P® The ITP logo is a registered trademark of International Thomson Publishing.
Microsoft, Windows 95, and Windows 98 are registered trademarks of Microsoft Corporation.
Some of the product names and company names used in this book have been used for identification purposes only and may be trademarks or registered trademarks of their manufacturers and sellers.

Disclaimer
Course Technology reserves the right to revise this publication and make changes from time to time in its content without notice.

ISBN: 0-7600-1079-X

Printed in the United States of America

For Lila and Leslie
  —RMS

To Ginnie, Tammy, Kim, Kelly, and Kristy
  —GWR

Preface

Principles of Information Systems: A Managerial Approach, fourth edition, continues the tradition, principles, and approach of the previous editions. Our primary objective is to develop the best information systems text and accompanying materials for the first computer course required of all business students. Through surveys, questionnaires, focus groups, and feedback that we have received from current and past adopters, as well as others who teach in the field, we have been able to develop the highest quality set of teaching materials available.

Because information systems are key to business functions today, the information systems (IS) discipline deserves its own course on a par with those of other academic lines. This is even truer now than it was with the first edition's publication. This book stands proudly at the beginning of the IS curriculum and remains unchallenged in its position as the only IS principles text offering the basic IS ideas and concepts that every business student must learn to be successful. In the past, instructors of the introductory course faced a dilemma. On one hand, experience in business organizations allows students to grasp the complexities underlying important IS concepts. For this reason, many delayed presenting these concepts until students completed a large portion of the core business requirements. On the other hand, delaying the presentation of IS concepts until students have matured within the business curriculum often forces the one or two required introductory IS courses to focus only on microcomputer software tools and, at best, merely to introduce computer concepts.

This text has been written specifically for the principles course in the IS curriculum. It represents an exciting alternative to texts used in the introductory IS course in the past. *Principles of Information Systems: A Managerial Approach*, fourth edition, treats the appropriate computer and IS concepts together with a strong managerial emphasis.

APPROACH OF THE TEXT

Principles of Information Systems: A Managerial Approach, fourth edition, offers the traditional coverage of computer concepts material, but it places the material within a highly structured framework of overall IS functionality. The text stresses principles of IS, which are brought together and presented in a way that is directly accessible. These fundamental ideas are not buried within historical detail or technical jargon. In addition, this book offers an overview of the entire IS discipline as well as a solid preparation for further study in advanced IS courses. It will serve both general business students and those who will become IS professionals. In particular, this book provides a solid groundwork from which to build advanced courses in such areas as systems development, programming, database management, Internet deployment, and decision support.

The overall vision, framework, and pedagogy that made the previous editions so popular have been retained in the fourth edition, offering a number of benefits to students. We continue to present IS concepts with a managerial emphasis. While much of the fundamental vision of this market-leading text remains unchanged, the fourth edition more clearly highlights established principles and draws out new ones that have emerged as a result of corporate and technological change. This text:

- Begins with a solid grounding in the principles of systems theory.
- Utilizes a problem-solving framework throughout and expands the scope of IS problem identification to include all activities, especially those at the strategic level of the organization.
- Covers the latest technologies, including connectivity and electronic commerce, in a business setting. Boxes, vignettes, examples, and cases are current, exciting, and relevant to today's businesses.
- Includes appropriate coverage of computer concepts, stressing characteristics of information systems relevant to aspiring decision makers.
- Presents the tenets, rules, guidelines—the principles—of information systems with which every business student must be knowledgeable.
- Stresses a single, all-encompassing concept: The right information, if it is delivered to the right person, in the right fashion, and at the right time, can improve and ensure organizational effectiveness and efficiency.
- Shows the value of the discipline as an attractive field of specialization and emphasizes the role of the IS professional as a change agent and manager who exercises special skills from a generalist perspective.
- Emphasizes the promise of integrated information systems in providing vastly superior organizational efficiencies.
- Shows that information systems are often intimately intertwined with value-added business processes.
- Presents IS objectives as supportive of, yet subordinate to, broader organizational goals.

IS Principles First, Where They Belong

Exposing students to fundamental IS concepts provides a service to students who do not later return to the discipline for advanced courses. Since most functional areas in business rely on information systems, an understanding of IS principles helps students in other course work. In addition, introducing students to the principles of information systems helps future functional area managers avoid mishaps that often result in unfortunate consequences. Furthermore, presenting IS concepts at the introductory level creates interest among general business students who will later choose information systems as a field of concentration.

Current Examples, Boxes, Cases, and References

We take great pride in including the most recent examples, boxes, cases, and references throughout the text. Some were developed at the last possible moment, just a few months before the publication of the book. Our adopters have come to expect the best and most recent material. We have done everything we can to meet or exceed these expectations. Every effort was made to include the newest, freshest, and most relevant examples, boxes, cases, and references possible.

Author Team

Ralph Stair and George Reynolds team up again for the fourth edition. Together, they have over fifty years of academic and industrial experience. Ralph Stair brings years of writing, teaching, and academic experience. He has written over twenty books and a large number of articles while at Florida State University. George Reynolds brings a wealth of computer and industrial experience to the project, with over thirty years experience working in government, institutional, and commercial IS organizations. He has also authored eight texts and has been an adjunct professor at the University of Cincinnati, teaching the introductory IS course. The Stair and Reynolds team brings a solid conceptual foundation along with practical IS experience to students.

GOALS OF THIS TEXT

This text has three main goals:

1. To present a core of IS principles with which every business student should be familiar and to offer a survey of the IS discipline that will enable all business students to understand the relationship of advanced courses to the curriculum as a whole
2. To present the changing role of the IS professional
3. To show the value of the discipline as an attractive field of specialization

These goals help students, regardless of major, understand and use fundamental information systems principles so that they will efficiently and effectively function as future business employees and managers. Because *Principles of Information Systems: A Managerial Approach*, fourth edition, is written for all business majors, we believe it is important to present not only a realistic perspective on IS in business but also to provide students with the skills they can use to be effective leaders in their companies.

IS Principles

Information systems are critical to the success of every business. In the past, advances in IS technology and applications have moved faster than the business curriculum. Books and courses dealing with important IS concepts have fallen too late in the curriculum. Introductory texts have tended to focus more on the descriptive (What have business managers been doing with information systems?) rather than on the prescriptive (What should business managers do with information systems to succeed?).

This text offers the traditional coverage of computer concepts but stresses the broad framework to provide students solid grounding in business uses of technology. The book, although comprehensive, does not attempt to cover every aspect of the IS discipline. Instead, it offers an essential core of guiding IS principles for students to use as they face the career challenges ahead. From the opening vignettes to the end-of-chapter material, each chapter emphasizes these fundamental IS principles. IS principles sections at the end of each chapter further reinforce important concepts and principles.

Information systems technology has outpaced academic guidelines directing how information technology should best be employed. The driving force of the technology has captured the attention of most authors of introductory IS textbooks. The first course in information systems has historically concentrated on discussing the components of an information system (primarily the technological components) and perhaps to some degree how information systems are developed. Yet, the first course has often neglected the important aspects of *why* and *how* information systems should be used to meet organizational goals. As a result, general business graduates have been thrust into work environments in which information systems were critical to their success, but the graduates were unprepared to interact with these systems. Even those with some understanding of IS technology have floundered; an understanding of the technology alone does not enable one to apply it successfully.

In addition to serving general business students, this book offers an overview of the entire IS discipline and gives a solid preparation for further study in advanced IS courses. It gives students who will become IS professionals a solid foundation to face the rapidly changing role of the profession and the IS discipline.

Changing Role of the IS Professional

As business and the IS discipline have changed, so too has the role of the IS professional. Once considered a dedicated specialist, the IS professional now is often an internal consultant to all functional areas, knowledgeable about their needs, and competent in bringing the power of information systems to bear throughout the business. The IS professional must exercise a broad perspective, encompassing the entire organization and often going beyond it.

The scope of responsibilities of an IS professional today ranges not only throughout the organization but also throughout the entire interconnected network of suppliers, customers, competitors, and other entities, no matter where they may be located. This broad scope offers IS professionals a new challenge: how to help the organization survive in a highly interconnected, highly competitive international environment. In accepting that challenge, the IS professional plays a pivotal role in shaping the business itself and ensuring its success. To survive, businesses must now strive for ultimate customer satisfaction and loyalty through ever-improving product and service quality. The IS professional assumes critical responsibility in determining the organization's approach to quality performance and therefore plays an important role in the ongoing survival of the organization. This new duality in the role of the IS employee—a professional who exercises specialist's skills with a generalist's perspective—is reflected throughout the book.

IS as a Field for Further Study

The IS field is exciting, challenging, and rewarding. It is important to show the value of the IS discipline as an attractive field of study for the average business student. The need to draw bright and interested students into the IS discipline is part of our ongoing responsibility. The IS graduate is no longer a technical recluse. Increasingly, we are seeing the brightest and most talented students enter the IS field. IS graduates at many schools are among the highest paid of all business graduates. Throughout this text, the many challenges and opportunities available to IS professionals are highlighted and emphasized. The "Making a Difference" supplemental interest boxes strengthen this theme by showing how IS professionals and organizations have used information systems to achieve their goals. Students are shown that the IS discipline is not only rewarding but fun!

CHANGES IN THE FOURTH EDITION

Like the previous editions, the fourth edition retains the focus on IS principles and strives to be the most current text on the market. We are excited about a number of changes to the text, particularly those that were made in response to feedback on how the course is now being taught. Some of the highlights follow:

- *Themes for the Fourth Edition.* This text looks to the future with the latest content and pedagogy. We stress *commerce* and *connectivity* as major themes for the fourth edition. Placing IS concepts in a business context has always set us apart from general computer books. As businesses of all sizes and types embrace connectivity, we are witnessing history in the making. This revolution of conducting business electronically will profoundly change businesses, markets, and society for decades. With years of service to the information systems discipline, this edition builds on the traditions and strengths of past successes, while keeping an eye on the needs of future managers and decision makers.

- *More Real-World Examples.* Our adopters and reviewers told us they wanted more real-world examples. The opening vignettes, supplemental interest boxes, and examples in the text have been updated to include more real-world examples that are current and interesting. We have also updated the cases at the end of each chapter.

- *New Supplemental Interest Boxes.* All supplemental interest boxes have been upgraded to include current issues and events. In addition, the "Making a Difference," "FYI," and "E-Commerce" boxes are new types for the fourth edition.

- *Current.* Every effort was made to make this edition the most up-to-date text on the market. New hardware and software concepts, the Internet and telecommunications, and many other current developments can be found throughout the text. This edition, however, goes beyond headline-making technologies to focus on how new technology can be used to solve business problems and further business goals. The newest, freshest, and most relevant examples, boxes, cases, and references possible have been included. Many were obtained just prior to the publication of this text.

- *Revised End-of-Chapter Material.* The material at the end of each chapter has been thoroughly updated. Information systems principles sections, summaries keyed to the learning objectives, key terms, review questions, discussion questions, problem-solving activities, team activities, and cases have been replaced and revised to reflect the themes of the fourth edition and to give students the opportunity to explore the latest technology in a business setting. Chapter references, found at the end of the book, are new and explore the latest developments in information systems. New to this edition, Web exercises further reinforce the themes of this text and help students explore organizations' use of this expanding technology.

- *Increased Emphasis on Performance-Based Management.* In many companies, we are seeing a trend toward performance-based management. There are at least three major stages in the business use of IS. The first stage started in the 1960s and was oriented toward cost reduction and efficiency. This stage generally ignored the revenue side. The second stage started in the 1980s and was oriented toward strategic advantage. In many cases, companies spent large amounts on IS and ignored the costs. The third stage is performance based, which carefully considers both strategic advantage and total IS costs.

- *Internet, Intranet, and Extranet Coverage.* In addition to the Internet chapter, Chapter 7, most chapters fully explore the use of Internet technology. The chapters on transaction processing, management information, and decision support systems have been revised to show the vast potential of this technology. Students see how real companies use the Internet, intranets, and extranets to help them satisfy customer needs and achieve organizational goals.

- *New Emphasis on Electronic Commerce.* The importance of electronic commerce is stressed throughout the book. The latest developments in conducting business electronically are covered, including electronic markets, product identification, product selection, and electronic product distribution. Current examples are included throughout the book, and the use of the Internet for electronic commerce is fully explored. In addition, a new box on electronic commerce has been included in every chapter.

- *New Coverage of Enterprise Resource Planning.* Chapter 8 covers the importance of enterprise resource planning (ERP). Platforms developed by SAP and others are explored. Corporations are investing millions of dollars in ERP to link their systems and provide crucial information to employees and managers. The costs and potential benefits are covered, as well as the disadvantages.

- *Greater Coverage of Ethical Issues.* In each chapter, important ethical issues and concerns are discussed in relation to the specific topics covered in the chapter. This increased coverage can be seen in the "Ethical and Societal Issues" boxes included in each chapter and woven into text discussion and end-of-chapter materials. Ethical issues are also covered in detail in Chapter 14.

PEDAGOGICAL FEATURES

In addition to the text, we revamped the pedagogy and the teaching resources for the fourth edition. Our emphasis throughout is on applying text concepts to the most up-to-date, real-world examples possible.

Chapter-Opening Material

The chapter-opening material introduces students to what is contained in each chapter. From the opening quotation to the vignette, we have developed this material to be interesting and motivational.

Opening Quotations. Each chapter starts with an opening quotation to stimulate interest in the material and set the stage for the chapter.

Chapter Outlines. Each chapter includes a chapter outline to show students and instructors the content of the chapter at a glance.

Learning Objectives. Carefully crafted learning objectives are included with every chapter. The learning objectives reflect what students should be able to accomplish after completing a chapter. The objectives are also integrated into the chapter summaries—each item in the summary starts with a learning objective from the beginning of the chapter.

Opening Vignettes. A brief opening vignette follows the learning objectives for each chapter. This vignette describes a real company or current business situation and is related to the concepts discussed in the chapter. Each vignette ends with a few questions for students to consider as they read the chapter.

Supplemental Interest Boxes

New supplemental interest boxes are interspersed throughout the text. Each chapter includes the following boxes: "E-Commerce," "Making a Difference," "Ethical and Societal Issues," "Technology Advantage," and "FYI." Like the previous editions, all boxes are designed to support the goals and themes of the fourth edition and the content of the specific chapter. Each box includes two discussion questions. Some questions tie the material to the text by asking students to relate chapter concepts to the topics in the supplemental boxes. Other questions challenge students to think "outside the box" to apply concepts to new situations in critical and creative ways.

E-Commerce. These new boxes contain a case study of a firm using the Internet, electronic data interchange, or other means of electronically conducting business. These boxes capture the many uses, benefits, and potential problems associated with the Internet and the World Wide Web. Each box reinforces the themes of connectivity and commerce.

Making a Difference. These new boxes show the role of business information systems to reduce costs, achieve a strategic advantage, or to improve overall performance. Each is current and reveals how a real company benefited from the use of information systems. Some boxes contain brief profiles of successful IS professionals to motivate students to consider IS as a major field of study.

Ethical and Societal Issues. Each "Ethical and Societal Issues" box presents a timely look at ethical challenges and the societal impact of information systems. Topics include classic ethical scenarios such as software piracy, data privacy, and other issues being raised by new technology. The dramatic impact of technology on business and society is also explored.

Technology Advantage. These boxes cover how technology is able to give a company an advantage in a competitive marketplace. New technologies and their use in a business setting are stressed. In addition, these boxes explore international aspects of information systems by highlighting companies that do business around the world.

FYI. These new boxes contain interesting and important developments in information systems to keep students informed. How these developments have affected individuals and corporations are stressed.

End-of-Chapter Material

To help students retain IS principles and to expand their understanding of important IS concepts and relationships, information systems principles, summaries, key terms, review questions, discussion questions, problem-solving exercises, team activities, Web exercises, and cases are included at the end of every chapter.

Information Systems Principles. "Information Systems Principles" summarize key concepts that every student should know. This important feature is a convenient summary of key ideas.

Summary. Every chapter includes a detailed summary. Each summary is tied to a learning objective to make sure students have mastered the material in the chapter.

Key Terms. A list of key terms with page numbers indicating where the term is defined follows the summary for each chapter. Each key term is also placed in bold in the text and defined in the margin. All key terms are also included in the glossary at the end of the book.

Review Questions. Directly linked to the text, these questions reinforce the key concepts and ideas within each chapter.

Discussion Questions. Picking up where the review questions leave off, discussion questions help instructors generate class discussion to move students beyond the concepts to explore the numerous aspects and principles of information systems.

Problem-Solving Exercises. Each chapter of the text contains a set of problem-solving exercises. These exercises enable students to continue to apply skills learned in application software courses to problems related to chapter material. Icons are provided to show students and instructors which applications are being tested. The primary purpose of these activities is to keep software skills fresh. The exercises also help prepare students for more rigorous applications in advanced business courses.

Team Activities. Because employees are increasingly involved in teamwork, these activities require students to work in small groups on a shared assignment. These activities foster teamwork, communication, and mutual accountability. Students work to create a joint product, such as a report, a database, or a group presentation. Some activities involve semistructured activities like visiting a local business, while others demand creative thinking, such as designing the perfect computer system with a given set of system constraints and price parameters.

Web Exercises. New to this edition, Web exercises reinforce the theme of connectivity and stress the use of the Internet, intranets, and extranets in a business setting. These exciting new exercises show students how businesses are reaching out to suppliers and customers to increase quality, service, and profitability.

Cases. We listened to our past adopters and reviewers and responded with four new cases at the end of each chapter. This expands the scenarios that instructors may use. These cases further reinforce important IS concepts and principles and show how real companies have applied information systems to achieve their goals. Questions focus students on the key issues of the cases and ask them to consider and apply the material.

TEACHING RESOURCE PACKAGE

The teaching tools that accompany this text offer many options for enhancing your course. In the fourth edition, we emphasize the importance of distance learning. And, as always, we are committed to providing one of the best teaching resource packages available in this market. Here are your options.

CyberClass

Course Technology is pleased to bring you *CyberClass* from HyperGraphics Corporation. *CyberClass* is a totally new Web-based tool for distance and on-campus settings. It is available with *Principles of Information Systems,* fourth edition, in three levels:

Level 1. This level is free when you use *Principles of Information Systems,* fourth edition. It has two parts: (1) access to the items below as read-only for users and (2) a chance for visiting professors to demo the capabilities of Level 2. Level 1 features:

- **Electronic Flash Cards.** A self-study tool for students to test their understanding of key computer concepts and terminology for each chapter.
- **Practice Tests.** Short tests that quiz students on key concepts from *Principles of Information Systems,* fourth edition. With each test offering 20 randomly generated questions per chapter, students can take these practice tests repeatedly to check their understanding of each chapter's content.
- **Link to Course Technology's Web Site for *Principles of Information Systems,*** fourth edition. Links to additional materials, reprint corrections, and updates to keep the book as timely as possible.

Level 2. Students may purchase a Level 2 diskette containing *CyberClass for Principles of Information Systems,* fourth edition. This diskette contains a password for the course that lasts for the duration of the term and the necessary software to access Level 2 features, which are:

- **Level 1 Features Plus Web Site.** All of Level 1 features, plus a customizable and secure Web site for instructors to use for administering their course.
- **Syllabus Posting.** A template where an instructor can type in a syllabus or copy and paste a syllabus from a word processing document. It is also possible to paste HTML into this template.
- **Hot Links.** Links that the instructor can post for students.
- **Assignment Posting.** An area where assignments can be posted for students to turn in.
- **Test Posting.** Along with practice tests from the book, an instructor can create tests online using Course Test Manager and then post them for students to take. The test is immediately graded and sent to the instructor of the class via e-mail.
- **Assignments.** A template that allows students to submit assignments to the instructor efficiently via e-mail.
- **Student Bulletin Board.** Threaded conferencing for class members, with the topics created by the instructor.
- **Messages.** Electronic messaging that allows sending and viewing messages among class members and instructor.
- **CyberChallenge.** An interactive learning game filled with multiple-choice questions where students compete to answer questions correctly in the shortest amount of time.
- **Instructor-supervised Text Chat.** Useful for such things as online real-time office hours, mini-lectures, group work, discussion groups, and so on.
- **Administration Utilities.** Accessible by instructors only, utilities allow them to view and edit the roster, edit user information, and so on.
- **Student's Administration Utilities.** Enables students to add or edit their own user information.

Level 3. Students may purchase a Level 3 diskette containing *CyberClass for Principles of Information Systems,* fourth edition. This diskette contains a password for the course that lasts for the duration of the term, and the necessary software to access Level 3 features, which are:

- **All Level 1 and 2 Features.**
- **Audio Classroom.** Runs off of instructors' Windows 95 Pentium computer (up to 30 students), the school's network (up to 200 students), or HyperGraphics's servers (upon sign-up with HyperGraphics).
- **Instructor Controlled and Monitored Synchronous Assessment.** Course Technology's *Course Test Manager* provides the technology backbone and bank of questions for real-time assessment during an audio session with real-time feedback to the instructor.

Instructor's Manual with Solutions

The *Instructor's Manual* is available in both electronic and printed formats. This all-new updated *Instructor's Manual* provides valuable chapter overviews; highlights key principles and critical concepts; offers sample syllabi, learning objectives, and discussion topics; and features possible essay topics, further readings or cases, and solutions to all of the end-of-chapter questions and problems as well as suggestions for conducting team activities. Additional end-of-chapter questions are also included, as well as the rationale, methodology, and solutions for each.

Course Test Manager and Testbank

This cutting-edge Windows-based testing software helps design and administer pretests, practice tests, and actual examinations. With *Course Test Manager*, students can randomly generate practice tests that provide immediate on-screen feedback and enable them to create detailed study guides for questions incorrectly answered. On-screen pretests help assess students' skills and plan instruction. *Course Test Manager* can also produce printed tests. In addition, students can take tests at the computer that can be automatically graded and can generate statistical information on students' individual and group performance.

Course Presenter

A CD-ROM–based presentation tool developed in Microsoft PowerPoint, *Course Presenter* offers a wealth of resources for use in the classroom. Instead of using traditional overhead transparencies, *Course Presenter* puts together impressive computer-generated screen shows including graphics and videos. All of the graphics from the book (not including photos) have been included.

Web Site

A dynamic site helps keep materials current. Visit us at http://www.course.com/sites/stair for additional and updated cases, and information about what is changing in the IS field. We provide information about the book, chapter-by-chapter updates, additional case studies, useful resources for instructors and students, and links to real-world companies showcased in the case studies throughout the book. Adopters will find it easy to remain current.

Supplemental Video Program

A number of high-quality videos are available to qualifying adopters of this text. These videos cover topics ranging from computer applications to the ways computers have changed how we live and run businesses. Videos range in length from 4 to 58 minutes. For more details, contact your sales representative.

ACKNOWLEDGMENTS

A book of this size and undertaking is always a team effort. We would like to thank every one of our fellow teammates at Course Technology for their dedication and hard work. Many thanks to our associate publisher, Kristen Duerr. We would also like to thank Cheryl Ouellette for her help. There were a number of people behind the scenes who made this book a reality;

thanks to Patty Stephan and Elizabeth Martinez. For their hard work on the manuscript, we would like to acknowledge and thank the team at Elm Street Publishing Services. Karen Hill helped with all stages of this project. Martha Beyerlein, Melissa Morgan, Barb Lange, and Abby Westapher helped with production and the final stages of the book.

Many thanks to the sales force at Course Technology. You make this all possible. You helped to get important feedback from current and future adopters. As Course Technology product users, we know how important you are.

Ralph Stair would like to thank the Department of Information and Management Sciences, College of Business Administration, at Florida State University for their support and encouragement. He would also like to thank his family, Lila and Leslie, for their support. George Reynolds thanks his family, Ginnie, Tammy, Kim, Kelly, and Kristy, for their patience and support in this major project.

TO OUR PREVIOUS ADOPTERS AND POTENTIAL NEW USERS

We sincerely appreciate our loyal adopters through the previous editions and welcome new users of *Principles of Information Systems: A Managerial Approach*. As in the past, we truly value your needs and feedback. We can only hope the fourth edition continues to meet your high expectations. In particular, we would like to thank the reviewers of the fourth edition, focus group members, and reviewers of previous editions.

Reviewers for the Fourth Edition
Warren Boe, *University of Iowa*
Thomas Browdy, *Washington University*
Roy Dejoie, *USWeb Corporation*
Karen Anne Forcht, *James Madison University*
William Harrison, *Oregon State University*
Roger McHaney, *Kansas State University*
Vikram Sethi, *Southwest Missouri State University*
Duane Truex, *Georgia State University*
David Whitney, *San Francisco State University*

Reviewers for the First, Second, and Third Editions
Robert Aden, *Middle Tennessee State University*
A. K. Aggarwal, *University of Baltimore*
Sarah Alexander, *Western Illinois University*
Beverly Amer, *University of Florida*
Noushin Asharfi, *University of Massachusetts*
Yair Babad, *University of Illinois—Chicago*
Charles Bilbrey, *James Madison University*
Thomas Blaskovics, *West Virginia University*
John Bloom, *Miami University of Ohio*
Glen Boyer, *Brigham Young University*
Mary Brabston, *University of Tennessee*
Jerry Braun, *Xavier University*
Thomas A. Browdy, *Washington University*
Lisa Campbell, *Gulf Coast Community College*
Andy Chen, *Northeastern Illinois University*
David Cheslow, *University of Michigan—Flint*
Robert Chi, *California State University—Long Beach*

Carol Chrisman, *Illinois State University*
Miro Costa, *California State University—Chico*
Caroline Curtis, *Lorain County Community College*
Roy Dejoie, *University of Oklahoma*
Sasa Dekleva, *DePaul University*
Pi-Sheng Deng, *California State University—Stanislaus*
John Eatman, *University of North Carolina*
Juan Esteva, *Eastern Michigan University*
Badie Farah, *Eastern Michigan University*
Karen Forcht, *James Madison University*
Carroll Frenzel, *University of Colorado—Boulder*
John Gessford, *California State University—Long Beach*
Terry Beth Gordon, *University of Toledo*
Kevin Gorman, *University of North Carolina—Charlotte*
Costanza Hagmann, *Kansas State University*
Bill C. Hardgrave, *University of Arkansas*
Al Harris, *Appalachian State University*
William L. Harrison, *Oregon State University*
Dwight Haworth, *University of Nebraska—Omaha*
Jeff Hedrington, *University of Wisconsin—Eau Claire*
Donna Hilgenbrink, *Illinois State University*
Jack Hogue, *University of North Carolina*
Joan Hoopes, *Marist College*
Donald Huffman, *Lorain County Community College*
Patrick Jaska, *University of Texas at Arlington*
G. Vaughn Johnson, *University of Nebraska—Omaha*
Grover S. Kearns, *Morehead State University*
Robert Keim, *Arizona State University*
Karen Ketler, *Eastern Illinois University*
Mo Khan, *California State University—Long Beach*
Michael Lahey, *Kent State University*
Jan de Lassen, *Brigham Young University*
Robert E. Lee, *New Mexico State University—Carlstadt*
Joyce Little, *Towson State University*
Herbert Ludwig, *North Dakota State University*
Jane Mackay, *Texas Christian University*
James R. Marsden, *University of Connecticut*
Roger W. McHaney, *Kansas State University*
Lynn J. McKell, *Brigham Young University*
John Melrose, *University of Wisconsin—Eau Claire*
Michael Michaelson, *Palomar College*
Ellen Monk, *University of Delaware*
Bijayananda Naik, *University of South Dakota*
Leah R. Pietron, *University of Nebraska—Omaha*
John Powell, *University of South Dakota*
Maryann Pringle, *University of Houston*
John Quigley, *East Tennessee State University*
Mary Rasley, *Lehigh-Carbon Community College*
Earl Robinson, *St. Joseph's University*
Scott Rupple, *Marquette University*
Dave Scanlon, *California State University—Sacramento*
Werner Schenk, *University of Rochester*
Larry Scheuermann, *University of Southwest Louisiana*

James Scott, *Central Michigan University*
Laurette Simmons, *Loyola College*
Janice Sipior, *Villanova University*
Harold Smith, *Brigham Young University*
Alan Spira, *University of Arizona*
Tony Stylianou, *University of North Carolina*
Bruce Sun, *California State University—Long Beach*
Hung-Lian Tang, *Bowling Green State University*
William Tastle, *Ithaca College*
Gerald Tillman, *Appalachian State University*
Jean Upson, *Lorain County Community College*
Misty Vermaat, *Purdue University—Calumet*
David Wallace, *Illinois State University*
Michael E. Whitman, *University of Nevada—Las Vegas*
David C. Whitney, *San Francisco State University*
Goodwin Wong, *University of California—Berkeley*
Myung Yoon, *Northeastern Illinois University*

Focus Group Contributors for the Third Edition
Mary Brabston, *University of Tennessee*
Russell Ching, *California State University—Sacramento*
Virginia Gibson, *University of Maine*
Bill C. Hardgrave, *University of Arkansas*
Al Harris, *Appalachian State University*
Stephen Lunce, *Texas A & M International*
Merle Martin, *California State University—Sacramento*
Mark Serva, *Baylor University*
Paul van Vliet, *University of Nebraska—Omaha*

OUR COMMITMENT

We are sincerely committed to serving the needs of our adopters and readers. Like the field of IS itself, the writing and publishing process is an evolutionary and participatory one. We encourage participation in our endeavor to provide the freshest, most relevant information possible. We pride ourselves on listening to instructors and developing creative solutions to problems and needs. Numerous individuals in the IS discipline have given us their time and insight during the process of this revision. They have offered valuable feedback on outstanding features of the previous editions and potential improvements in the fourth edition. We have listened to their comments and thank them for their time.

   As always, we welcome input and feedback. If you have any questions or comments regarding *Principles of Information Systems: A Managerial Approach,* fourth edition, please contact us through Course Technology or your local representative, via e-mail at **mis@course.com**, via the Internet at **www.course.com**, or address your comments, criticisms, suggestions, and ideas to:

Ralph Stair
George Reynolds
Course Technology
One Main Street
Cambridge, MA 02142

Contents

PART I

AN OVERVIEW   1

CHAPTER 1

An Introduction to Information Systems   2

WHIRLPOOL CORPORATION: AUTOMATED PRICING SYSTEM STREAMLINES WORK   3

ETHICAL AND SOCIETAL ISSUES
Information Technology's Impact on Society   4

Information Concepts   5
  Data vs. Information   5
  The Characteristics of Valuable Information   6
  The Value of Information   7

MAKING A DIFFERENCE
Natural Resources Conservation Service Uses IT to Avoid Flood Damage   8

System and Modeling Concepts   8
  System Components and Concepts   9
  System Performance and Standards   11
  System Variables and Parameters   13
  Modeling a System   13

What Is an Information System?   15
  Input, Processing, Output, Feedback   15
  Manual and Computerized Information Systems   17
  Computer-Based Information Systems   17

FYI
Can Less Information Result in Less Value?   19

TECHNOLOGY ADVANTAGE
Corporate Intranets—Needed to Stay Competitive   20

Business Information Systems   21
  Transaction Processing Systems and E-commerce   21

E-COMMERCE
Competing Electronically   24

  Management Information Systems   25
  Decision Support Systems   26
  Artificial Intelligence and Expert Systems   27

Systems Development   29
  Systems Investigation and Analysis   29
  Systems Design, Implementation, and Maintenance and Review   29

Why Study Information Systems?   29
  Computer and Information Systems Literacy   30
  Information Systems in the Functional Areas of Business   30
  Information Systems in Industry   31

CASE 1   Haworth Improves Furniture and Its Order Process   36

CASE 2   Sales Force Automation—Potential without a Payoff?   37

CASE 3   Liz Claiborne Upgrades Its Information Systems   38

CASE 4   Ticketmaster Sells in Cyberspace   39

CHAPTER 2

Information Systems in Organizations   40

FEDERAL EXPRESS: REDEFINES ITS SERVICES TO MAINTAIN COMPETITIVE ADVANTAGE   41

Organizations and Information Systems   42
   Organizational Structure   45
   Organizational Culture and Change   48

FYI
New Approach to Marketing   50

   Reengineering   51
   Continuous Improvement   53
   Total Quality Management   54
   Outsourcing and Downsizing   55
Competitive Advantage   57
   Factors That Lead Firms to Seek
      Competitive Advantage   57

E-COMMERCE
Barriers to E-commerce   58

   Strategic Planning for Competitive
      Advantage   58

TECHNOLOGY ADVANTAGE
Omaha Steaks   61

ETHICAL AND SOCIETAL ISSUES
Merger Mania: Consumer Dream or
Nightmare?   63

Performance-Based Information Systems   64
   Productivity   64
   Return on Investment and the Value of
      Information Systems   65
   Justifying Information Systems   66
Careers in Information Systems   67
   Roles, Functions, and Careers in the
      Information Systems Department   68

MAKING A DIFFERENCE
What Does It Take to be a Successful CIO?   72

   Other IS Careers   72

CASE 1   Textron Outsources Information
   Technology Services   77

CASE 2   Boscov Makes IS Investment
   Decision   78

CASE 3   Black & Veatch Uses Software Tool to
   Compete Globally   78

CASE 4   Continuous Replenishment: Revolution in
   Retailing   79

PART II

INFORMATION TECHNOLOGY CONCEPTS   81

CHAPTER 3
Hardware: Input, Processing, and Output Devices   82

DIGITAL EQUIPMENT CORPORATION AND
IBM: BREAKING THE GIGAHERTZ SPEED
BARRIER   83

**Computer Systems: Integrating the Power of
Technology**   84
   Hardware Components   85
   Hardware Components in Action   86
**Processing and Memory Devices: Power, Speed,
and Capacity**   87
   Processing Characteristics and Functions   87
   Memory Characteristics and Functions   92
   Multiprocessing   95
Secondary Storage   96
   Access Methods   96
   Devices   97
**Input and Output Devices: The Gateway to
Computer Systems**   103
   Characteristics and Functionality   103
   Input Devices   104
   Output Devices   109
   Special-Purpose Input and Output
      Devices   111
**Computer System Types, Standards, Selecting,
and Upgrading**   112
   Computer System Types   113

MAKING A DIFFERENCE
Michael Dell Sells Direct   114

FYI
And the Prices Just Keep Falling   116

E-COMMERCE
Eddie Bauer Adds Internet Site to In-Store and
Catalog Sales Channels   117

TECHNOLOGY ADVANTAGE
GTE Corp. Uses Mainframes and Servers to
Compete   119

ETHICAL AND SOCIETAL ISSUES
Information Technology Revolutionizes Biomedical
Science   120

Multimedia Computers   121

Standards   123

Selecting and Upgrading Computer
Systems   123

CASE 1   Chip Maker Reengineers Procurement
Process   131

CASE 2   Unisys Helps Bank Meet Customer
Needs   132

CASE 3   United Airlines Standardizes on Single
Workstation   133

CASE 4   Phillips Petroleum Aims for High
Reliability and Availability   134

CHAPTER 4

**Software: Systems and Application
Software   136**

DARIGOLD INC.: MEETING BUSINESS
CHALLENGES WITH SOFTWARE
PACKAGES   137

An Overview of Software   138

Systems Software   138

Application Software   138

Supporting Individual, Group, and
Organizational Goals   139

Software Issues and Trends   140

Systems Software   141

Operating Systems   141

Popular Operating Systems   147

FYI
Windows CE Applications   151

Utility Programs   154

Application Software   155

Types of Application Software   156

Personal Application Software   158

Object Linking and Embedding (OLE)   163

Workgroup Application Software   165

MAKING A DIFFERENCE
Illinois Power Uses Lotus Notes to Improve
Customer Service   166

Enterprise Application Software   167

ETHICAL AND SOCIETAL ISSUES
Trying to Champion ERP   168

Programming Languages   169

Standards and Characteristics   169

The Evolution of Programming
Languages   170

TECHNOLOGY ADVANTAGE
Java: Hot and Getting Hotter   175

E-COMMERCE
Hawaiian Greenhouse Combines New Technology
with Flower Power   177

Language Translators   178

CASE 1   Kellogg Implements Global Information
System   184

CASE 2   Gap Uses Object-Oriented
Programming   184

CASE 3   Tracking Software Licenses   185

CASE 4   Breathing Life into an Old System   186

CHAPTER 5

**Organizing Data and
Information   188**

WAL-MART: MINING DATA FOR
CUSTOMER GOLD   189

Data Management   190

The Hierarchy of Data   190

Data Entities, Attributes, and Keys   191

The Traditional Approach vs. the Database
Approach   192

E-COMMERCE
MasterCard International   193

ETHICAL AND SOCIETAL ISSUES
Privacy Standards in Credit Reporting   195

Data Modeling and Database Models   199

Data Modeling   199

Database Models   201

Database Management Systems (DBMS)   206

Providing a User View   206

Creating and Modifying the Database   207

Storing and Retrieving Data   209

Manipulating Data and Generating
Reports   211

Popular Database Management Systems for End Users   212

MAKING A DIFFERENCE
KeyCorp Develops Customer Relationship Databases   213

Selecting a Database Management System   215

TECHNOLOGY ADVANTAGE
Small but Powerful: Good Things Can Come in Small Databases   216

Database Developments   217
Distributed Databases   218
Data Warehouses, Data Marts, and Data Mining   219
On-Line Analytical Processing (OLAP)   223
Open Database Connectivity (ODBC)   225
Object-Relational Database Management System   226

FYI
Digital Images on Demand   229

CASE 1   Saab Cars USA   235

CASE 2   US West   236

CASE 3   MCI Communications Corporation   238

CASE 4   Sears   239

CHAPTER 6
Telecommunications and Networks   240

CITIBANK: UPGRADING GLOBAL NETWORKS TO MEET CUSTOMER NEEDS   241

An Overview of Communications Systems   242
Communications   242
Telecommunications   243
Networks   244

Telecommunications   244
Types of Media   245
Devices   248
Carriers and Services   251

ETHICAL AND SOCIETAL ISSUES
Snags Limit Increased Competition for Local Telephone Services   252

Networks and Distributed Processing   256
Basic Data Processing Strategies   256

TECHNOLOGY ADVANTAGE
Reach Out and Touch the World: The Promise of DSL   257

Network Concepts and Considerations   258
Network Types   258
Terminal-to-Host, File Server, and Client/Server Systems   262
Communications Software and Protocols   265
Bridges, Routers, Gateways, and Switches   268

Telecommunications Applications   269
Linking Personal Computers to Mainframes and Networks   269
Voice and Electronic Mail   270
Electronic Software and Document Distribution   270

FYI
Free E-Mail   271

E-COMMERCE
Comp-U-Card   272

Telecommuting   273
Videoconferencing   273
Electronic Data Interchange   274

MAKING A DIFFERENCE
Hospitals Use Telemedicine to Deliver Service to Rural Areas   275

Public Network Services   276
Specialized and Regional Information Services   277
Distance Learning   277

CASE 1   NTT Scales Up   283

CASE 2   Canada Privatizes Air Traffic Controller System   284

CASE 3   Hotel Vintage Park—A Haven for Telecommuting   285

CASE 4   Glendale Federal   287

CHAPTER 7
The Internet, Intranets, and Extranets   288

MICHELIN TIRE: CRUISING THE NET TO FINISH FIRST   289

Use and Functioning of the Internet   290
How the Internet Works   292

Accessing the Internet   293

Internet Service Providers   295

Internet Services   296

E-Mail   296

TECHNOLOGY ADVANTAGE
Pacific Bell Provides Customer Service via
Web   297

Telnet and FTP   298

Usenet and Newsgroups   298

Chat Rooms   300

Internet Phone and Videoconferencing
Services   300

Content Streaming   303

The World Wide Web   303

MAKING A DIFFERENCE
ENEN   304

FYI
Car Buyers Use the Internet for Bargain
Hunting   307

Web Browsers   308

Developing Web Content   309

Search Engines   310

Java   312

Push Technology   314

Business Uses of the Web   314

Intranets and Extranets   316

E-COMMERCE
Business on the Web Is Exploding   317

Net Issues   319

Management Issues   320

Service Bottlenecks   320

Privacy and Security   321

ETHICAL AND SOCIETAL ISSUES
U.S. Encryption Regulation   323

Firewalls   324

CASE 1 Dreamworks on the Web   330

CASE 2 Internet Travel Planning   330

CASE 3 US West Communications   331

CASE 4 Ford Uses Network to Gain Loyalty   332

PART III

BUSINESS INFORMATION SYSTEMS   333

CHAPTER 8

Transaction Processing, Electronic Commerce, and Enterprise Resource Planning Systems   334

BRITISH PETROLEUM: INTERNATIONAL
SYSTEMS PROGRAM HANDLES
COMMERCIAL TRANSACTIONS   335

**An Overview of Transaction Processing
Systems   337**

Traditional Transaction Processing Methods
and Objectives   337

Transaction Processing Activities   342

Control and Management Issues   345

**Traditional Transaction Processing
Applications   347**

Order Processing Systems   347

Purchasing Systems   355

MAKING A DIFFERENCE
Routing and Scheduling Software Cuts Distribution
Costs   356

FYI
Intelligent Agents Help Shoppers   359

Accounting Systems   360

Electronic Commerce   367

Electronic Markets and Commerce in
Perspective   368

Search and Identification   368

E-COMMERCE
E-Commerce System for the Foodservice
Industry   370

Selection and Negotiation   371

Purchasing Product and Services
Electronically   371

ETHICAL AND SOCIETAL ISSUES
The Electronic Check   371

Product and Service Delivery   372

After-Sales Service   373

Enterprise Resource Planning   373

An Overview of Enterprise Resource
Planning   373

Advantages and Disadvantages of ERP   374

Example of an ERP System   376

TECHNOLOGY ADVANTAGE
Core Business Activities for Enterprise Resource Planning   379

CASE 1   Orders for West Increase   384

CASE 2   Web-Based Purchasing   385

CASE 3   FedEx and SAP Team Up to Provide Integrated Logistics Solution   385

CASE 4   Florist Increases and Speeds Transactions   386

CHAPTER 9

Management Information Systems   388

GAF MATERIALS CORP.: FINANCIAL MIS   389

An Overview of Management Information Systems   390

Management Information Systems in Perspective   390

Inputs to a Management Information System   391

Outputs of a Management Information System   392

Characteristics of a Management Information System   395

Management Information Systems for Competitive Advantage   396

MIS and Web Technology   396

E-COMMERCE
New Holland North America Provides Web Access to MIS Data   397

Functional Aspects of the MIS   397

A Financial Management Information System   400

Inputs to the Financial MIS   401

Financial MIS Subsystems and Outputs   402

A Manufacturing Management Information System   404

Inputs to the Manufacturing MIS   405

Manufacturing MIS Subsystems and Outputs   407

A Marketing Management Information System   412

Inputs to the Marketing MIS   412

Marketing MIS Subsystems and Outputs   414

FYI
Rubric Revolutionizes Marketing MIS Software   418

A Human Resource Management Information System   418

Inputs to the Human Resource MIS   419

ETHICAL AND SOCIETAL ISSUES
Protecting Patient-Care Data   420

Human Resource MIS Subsystems and Outputs   421

Other Management Information Systems   424

Accounting MISs   424

MAKING A DIFFERENCE
Automating Recruiting at Ornda Health Corporation   425

Geographic Information Systems   425

TECHNOLOGY ADVANTAGE
California Water Districts Deploy Smallworld GIS   427

CASE 1   A Marketing MIS for the Greater Boston Convention Center   431

CASE 2   Toy Manufacturer Adopts Marketing MIS   432

CASE 3   Human Resources at the Bank of Montreal   433

CASE 4   Chrysler Implements Web-Based Manufacturing MIS   435

CHAPTER 10

Decision Support Systems   436

MILLER SQA: FACTORY-FLOOR DSS HELPS MANUFACTURER DELIVER THE GOODS   437

Decision Making and Problem Solving   438

Decision Making as a Component of Problem Solving   438

Programmed vs. Nonprogrammed Decisions   440

Optimization, Satisficing, and Heuristic
Approaches   441

■ **MAKING A DIFFERENCE**
Office Depot Optimizes Its Operations   442

Problem-Solving Factors   442

An Overview of Decision Support Systems   444

Characteristics of a Decision Support
System   445

■ **TECHNOLOGY ADVANTAGE**
First Chicago NBD Adopts DSS Software   446

Capabilities of a Decision Support
System   448

The Integration of TPS, MIS, and DSS   451

A Comparison of DSS and MIS   452

Web-Based Decision Support Systems   452

**Components of a Decision Support
System   453**

The Model Base   453

■ **E-COMMERCE**
Budgeting Decision Support System Helps Gulf
Canada   454

The Advantages and Disadvantages of
Modeling   456

■ **FYI**
DSS Helps Lighten United Airlines Load   457

The Dialogue Manager   458

The Group Decision Support System   458

Characteristics of a GDSS   458

Components of a GDSS and GDSS
Software   461

GDSS Alternatives   462

The Executive Support System   464

Executive Support Systems in Perspective   465

Capabilities of an Executive Support
System   466

■ **ETHICAL AND SOCIETAL ISSUES**
Use of Competitive Intelligence in ESS   467

CASE 1 Bank Uses Intranet to Support DSS   473

CASE 2 Decision Support for Individual
Investors   474

CASE 3 Getting Decision Support for Medical
Problems   475

CASE 4 Project Management Models at Bank of
America   476

CHAPTER 11
Artificial Intelligence and Expert Systems   478

EUROPEAN MEDIA BUYING:
COMPUTERS HELP SELECT
TV COMMERCIAL SLOTS   479

An Overview of Artificial Intelligence   480

Artificial Intelligence in Perspective   480

The Nature of Intelligence   481

The Difference between Natural and
Artificial Intelligence   483

The Major Branches of Artificial
Intelligence   483

■ **FYI**
Listening to Our Language   487

■ **ETHICAL AND SOCIETAL ISSUES**
Chicago Police Department Uses Neural Net to
Screen Officers   488

An Overview of Expert Systems   489

Characteristics of an Expert System   489

■ **TECHNOLOGY ADVANTAGE**
Software Vendor Uses Neural Network to
Enhance Products   490

Capabilities of Expert Systems   492

■ **E-COMMERCE**
Intelligent Software Agents Help Tame the
Internet   493

When to Use Expert Systems   494

Components of Expert Systems   495

The Knowledge Base   495

The Inference Engine   498

The Explanation Facility   499

The Knowledge Acquisition Facility   499

The User Interface   500

Expert Systems Development   500

The Development Process   500

■ **MAKING A DIFFERENCE**
Lotus Corporation Builds Knowledge Base to
Support Customers   501

Participants in Developing and Using Expert
Systems   502

Expert Systems Development Tools and
Techniques   503

Advantages of Expert System Shells and
Products   505

Expert Systems Development
Alternatives   505

Applications of Expert Systems and Artificial Intelligence   507

CASE 1 Using an Expert System to Improve Net Presence   513

CASE 2 Use of Fuzzy Logic to Predict Length of Patient Stay   514

CASE 3 Artificial Intelligence: An Intelligent Way to Schedule Jobs at Volvo   515

CASE 4 Immigration and Naturalization Service Applies High Technology   516

PART IV

SYSTEMS DEVELOPMENT   517

CHAPTER 12

Systems Investigation and Analysis   518

GERBER: DEVELOPS SYSTEM TO MANAGE CUSTOMERS' INVENTORY   519

An Overview of Systems Development   520

Participation in Systems Development   521

Initiating Systems Development   522

Information Systems Planning   523

MAKING A DIFFERENCE
Hyundai Motors Designs Extranet to Serve Dealers and Customers   526

Establishing Objectives for Systems Development   527

Systems Development and the Internet   528

Trends in Systems Development and Enterprise Resource Planning   529

Systems Development Life Cycles   530

TECHNOLOGY ADVANTAGE
After Learning Systems Development Techniques, Companies Become Consultants   531

The Traditional Systems Development Life Cycle   532

Prototyping   533

Rapid Application Development and Joint Application Development   534

The End-User Systems Development Life Cycle   536

Factors Affecting Systems Development Success   537

Degree of Change   537

Quality of Project Planning   538

ETHICAL AND SOCIETAL ISSUES
The IRS Modernization Project   539

Use of Project Management Tools   540

Use of Formal Quality Assurance Processes   542

Use of Computer-Aided Software Engineering (CASE) Tools   542

Systems Investigation   544

Initiating Systems Investigation   544

Participants in Systems Investigation   545

Feasibility Analysis   545

The Systems Investigation Report   547

Systems Analysis   547

General Considerations   547

Participants in Systems Analysis   548

Data Collection   548

Data Analysis   550

Requirements Analysis   553

E-COMMERCE
Defining Requirements for an On-line Service   555

The Systems Analysis Report   556

FYI
Utilities Perform Systems Analysis to Take Advantage of Deregulation   558

CASE 1 FAA Systems Development Project Slows   565

CASE 2 GATX Capital Corp.   565

CASE 3 CompUSA   566

CASE 4 Walgreens Pharmacies' Successful Completion of Strategic Project   567

CHAPTER 13

Systems Design, Implementation, Maintenance, and Review   568

EMPIRE DISTRICT ELECTRIC COMPANY: NEW CUSTOMER INFORMATION SYSTEM PRODUCES FLEXIBILITY IN CHANGING INDUSTRY   569

Systems Design 570

Logical and Physical Design 570

Special System Design Considerations 572

Emergency Alternate Procedures and Disaster Recovery 574

ETHICAL AND SOCIETAL ISSUES
Business Disruptions from Computer Problems 576

Systems Controls 578

The Importance of Vendor Support 580

Generating Systems Design Alternatives 580

E-COMMERCE
Geac Designs and Implements New Customer Support System 581

Evaluating and Selecting a System Design 582

Evaluation Techniques 584

Freezing Design Specifications 586

The Contract 586

The Design Report 587

Systems Implementation 588

Acquiring Hardware from an Information Systems Vendor 588

Acquiring Software: Make or Buy? 588

Externally Developed Software 588

In-House-Developed Software 590

Tools and Techniques for Software Development 592

FYI
The Cost of Software Defects 593

TECHNOLOGY ADVANTAGE
Travelers Designs Object-Oriented System to Improve Productivity 594

Acquiring Database and Telecommunications Systems 599

User Preparation 599

IS Personnel: Hiring and Training 600

Site Preparation 600

Data Preparation 600

Installation 601

Testing 601

Start-Up 602

User Acceptance 602

Systems Maintenance 602

Reasons for Maintenance 603

Types of Maintenance 604

The Request for Maintenance Form 604

Performing Maintenance 604

The Financial Implications of Maintenance 605

The Relationship between Maintenance and Design 606

MAKING A DIFFERENCE
Sanofi Reduces Maintenance with New Contract Management System 607

Systems Review 607

Types of Review Procedures 608

Factors to Consider during Systems Review 608

Systems Performance Measurement 609

CASE 1 Outsource It All 616

CASE 2 Measuring Return: A Systems Development Success Story 617

CASE 3 Slowing Systems Integration at Aetna 618

CASE 4 Company Frees IS Staff for Future Application Development 620

PART V

INFORMATION SYSTEMS IN BUSINESS AND SOCIETY 621

CHAPTER 14

Security, Privacy, and Ethical Issues in Information Systems and the Internet 622

CERT: POLICING THE INTERNET 623

Computer Waste and Mistakes 624

Computer Waste 624

Computer-Related Mistakes 625

Preventing Computer-Related Waste and Mistakes 626

Computer Crime 628

FYI
Y2K: This One Is Costing Billions and Taking Years to Fix 629

The Computer as a Tool to Commit Crime 630

The Computer as the Object of Crime 630

E-COMMERCE
The Dark Side of Electronic Commerce   636

Preventing Computer-Related Crime   637

MAKING A DIFFERENCE
New Center to Combat Terrorists Trading
Explosives for Computers   639

TECHNOLOGY ADVANTAGE
Becton Dickinson Implements Intranet   643

Privacy   643

Privacy Issues   644

ETHICAL AND SOCIETAL ISSUES
Personalization   646

Fairness in Information Use   647
Federal Privacy Laws and Regulations   648
State Privacy Laws and Regulations   650
Corporate Privacy Policies   650
Protecting Individual Privacy   650

The Work Environment   651

Health Concerns   651

Avoiding Health and Environmental
Problems   652

Ethical Issues in Information Systems   654

CASE 1 The GAO Finds Waste and Mistakes in
Federal Agencies   660

CASE 2 Computer Terrorism   661

CASE 3 AOL Works to Improve Security   662

CASE 4 The Politics of Technology   663

GLOSSARY   664
NOTES   680
INDEX   684

PART I

An Overview

1

An Introduction to Information Systems

"What many organizations don't realize is that if you don't manage the business part of a technology change, you can fail even if the technology part succeeds."

— Naomi Karten, president
Karten and Associates,
management consultants

Chapter Outline

Information Concepts
    Data vs. Information
    The Characteristics of Valuable Information
    The Value of Information

System and Modeling Concepts
    System Components and Concepts
    System Performance and Standards
    System Variables and Parameters
    Modeling a System

What Is an Information System?
    Input, Processing, Output, Feedback
    Manual and Computerized Information Systems
    Computer-Based Information Systems

Business Information Systems
    Transaction Processing Systems and E-commerce
    Management Information Systems
    Decision Support Systems
    Artificial Intelligence and Expert Systems

Systems Development
    Systems Investigation and Analysis
    Systems Design, Implementation, and
      Maintenance and Review

Why Study Information Systems?
    Computer and Information Systems Literacy
    Information Systems in the Functional Areas of Business
    Information Systems in Industry

Learning Objectives

After completing Chapter 1, you will be able to:

1. Distinguish data from information and describe the characteristics used to evaluate the quality of data.
2. Name the components of an information system and describe several system characteristics.
3. Identify the basic types of business information systems and discuss who uses them, how they are used, and what kinds of benefits they deliver.
4. Identify the major steps of the systems development process and state the goal of each.
5. Discuss why it is important to study and understand information systems.

Whirlpool Corporation
Automated Pricing System Streamlines Work

Whirlpool Corporation is the world's leading manufacturer and marketer of major home appliances, with headquarters in Benton Harbor, Michigan. Its 60,000 employees manufacture fine appliances in 13 countries and market them under 11 major brand names in approximately 140 countries around the world. Recent revenues exceeded $8.5 billion annually.

Responding to a competitive price change used to take a lot of effort for Whirlpool. When one of its competitors dropped prices, a flurry of faxes and overnight packages flew out of Whirlpool's headquarters to match them. Yet, it was often weeks before Whirlpool could adjust its prices. A price increase also created major problems. Customers, ranging from mega-retailers such as Sears Brand Central to little mom-and-pop stores, became understandably upset when they ordered a product and received an invoice with a higher price when the product arrived. Whirlpool had to issue a credit for the difference, but this created extra paperwork and, worse yet, dissatisfied customers.

Under this system, changing pricing on every product each quarter took over three months and was error prone. Performing the quarterly pricing required calculating new prices, reviewing them, printing them, reviewing them again, and feeding them into a mainframe-based computer system. After that, the new price lists were mailed, faxed, and sent by overnight delivery to trading partners and regional sales representatives.

To break free of the cumbersome pricing system, Whirlpool implemented a new system and streamlined work processes to make it respond to market changes more quickly or launch a special promotion of its own. Now sales agents access a centralized pricing database for quick reference when making their calls. The new system consolidates pricing and order entry systems for the entire company and halves the time it takes to reprice Whirlpool's entire product line of more than 2,000 models. The result is that it is easier to do business with Whirlpool.

Whirlpool's information technology overhaul is spreading to other systems

also. Not only is the company implementing a new corporate pricing system, it is also undergoing a massive reorganization to streamline all its business functions. Why is this necessary? To enable Whirlpool to compete with major companies such as General Electric, Maytag, Electrolux Corp., and Amana.

As you read this chapter, consider the following questions:

- How are companies using information technology to improve services, lower costs, and become more competitive?

- What are the fundamental components associated with a successful information system?

Sources: Adapted from Randy Weston, "Whirlpool to Try Pricing Systems," *Computerworld*, March 23, 1998, pp. 1, 14; Randy Weston, "No Room for Error," *Computerworld*, March 23, 1998, p. 14; and the Whirlpool Web site at http://www.whirlpoolcorp.com accessed on April 12, 1998.

An **information system (IS)** is a set of interrelated components that collect, manipulate, and disseminate data and information and provide a feedback mechanism to meet an objective. We all interact daily with information systems, both personally and professionally. We use automatic teller machines at banks, checkout clerks scan our purchases using bar codes and scanners, and we get information from kiosks with touchscreens. We saw how an information system improved communications and, as a result, improved the customer service and product pricing process at Whirlpool. Major Fortune 500 companies are spending in excess of $1 billion per year on information technology. In the future, we will depend on information systems even more. Knowing the potential of information systems and having the ability to put this knowledge to work can result in a successful personal career, organizations that reach their goals, and a society with a higher quality of life.

Information systems are everywhere. Automatic teller machines provide customers with up-to-date information about their bank accounts, including balance information and checks that have been paid.
(Source: Image copyright © 1998 PhotoDisc.)

information system (IS)

a set of interrelated elements or components that collect (input), manipulate (process), and disseminate (output) data and information and provide a feedback mechanism to meet an objective

Computers and information systems are constantly changing the way organizations conduct business. Today we live in an information economy. Information itself has value, and commerce often involves the exchange of information, rather than tangible goods. Systems based on computers are increasingly being used as a means to create, store, and transfer information. Investors are using information systems to make multimillion-dollar decisions, financial institutions are employing them to transfer billions of dollars around the world electronically, and manufacturers are using them to order supplies and distribute goods faster than ever before. Computers and information systems will continue to change our society, our businesses, and our lives (see the "Ethical and Societal Issues" box). In this chapter, we present a framework for understanding computers and information systems and discuss why it is important to study information systems. This understanding will help you unlock the potential of properly applied information systems concepts.

ETHICAL AND SOCIETAL ISSUES

Information Technology's Impact on Society

Information technology has had a profound impact on society—so much so that some people call this the Information Age. In his books *Megatrends* and *Megatrends 2000*, John Naisbitt points to the year 1956 as the beginning of the information society. That year, for the first time in American history, white-collar workers outnumbered blue-collar workers. Industrial society has given way to a new society, where most of us work with information rather than produce goods. These people who spend most of their working day creating, using, and distributing information are called *knowledge workers*.

Discussion Questions
1. List the different types of technology you have used in the past 24 hours. How many types did you use?
2. Would you consider yourself a knowledge worker? How about your parents? Your grandparents?

INFORMATION CONCEPTS

Information is a central concept throughout this book. The term is used in the title of the book, in this section, and in almost every chapter. To be an effective manager in any area of business, you need to understand that information is one of an organization's most valuable and important resources. This term, however, is often confused with the term *data*.

Data vs. Information

data

raw facts

information

a collection of facts organized in such a way that they have additional value beyond the value of the facts themselves

Data consists of raw facts, such as an employee's name and number of hours worked in a week, inventory part numbers, or sales orders. As shown in Table 1.1, several types of data can be used to represent these facts. When these facts are organized or arranged in a meaningful manner, they become information. **Information** is a collection of facts organized in such a way that they have additional value beyond the value of the facts themselves. For example, a particular manager might find the knowledge of total monthly sales to be more suited to his purpose (i.e., more valuable) than the number of sales for each individual sales representative.

Data represents real-world things. As we have stated, data—simply raw facts—has little value beyond its existence. For example, consider data as pieces of wood. In this state, the wood has little value beyond its inherent value as a single object. However, if some relationship is defined among the pieces of wood, they will gain value. By stacking the pieces of wood in a certain way, they can be used as a step stool (Figure 1.1a). Information is much the same. Rules and relationships can be set up to organize data into useful, valuable information.

The type of information created depends on the relationships defined among existing data. For example, the wood for our step stool could be stacked in a different way to be used as a box (Figure 1.1b). Adding new or different data means relationships can be redefined and new information can be created. For instance, adding nails to our wood can greatly increase

TABLE 1.1

Types of Data

Data	Represented by
Alphanumeric Data	Numbers, letters, and other characters
Image Data	Graphic images or pictures
Audio Data	Sound, noise, or tones
Video Data	Moving images or pictures

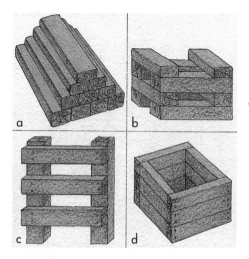

FIGURE 1.1

Defining and organizing relationships among data creates information. Defining different relationships results in different information. Here, wood can be organized differently to create two different structures—a step stool (a) and a box (b). Different data can be added to redefine the relationships and add value to the outcome. Adding nails (new data) to the wood results in a ladder (c) and a tighter-closing box (d), more valuable information.

process

a set of logically related tasks performed to achieve a defined outcome

knowledge

an awareness and understanding of a set of information and how that information can be made useful to support a specific task

knowledge base

the collection of data, rules, procedures, and relationships that must be followed to achieve value or the proper outcome

the value of the final product. We can now create a ladder (Figure 1.1c) that reaches higher than our step stool or a container (Figure 1.1d). Likewise, our manager could add specific product data to his sales data to create monthly sales information broken down by product line.

Turning data into information is a **process**, or a set of logically related tasks performed to achieve a defined outcome. The process of defining relationships among data requires knowledge. **Knowledge** is an awareness and understanding of a set of information and how that information can be made useful to support a specific task. Part of the knowledge needed for ladder building, for instance, is based on understanding that the rungs of a ladder must be placed horizontally and the legs vertically. The act of selecting or rejecting facts based on their relevancy to particular tasks is also based on a type of knowledge used in the process of converting data into information. Therefore, information can be considered data made more useful through the application of knowledge. The collection of data, rules, procedures, and relationships that must be followed to achieve value or the proper outcome is contained in the **knowledge base**.

In some cases, data is organized or processed mentally or manually. In other cases, a computer is used. In the earlier example, the manager could have manually calculated the sum of the sales of each representative, or a computer could calculate this sum. What is important is not so much where the data comes from or how it is processed but whether the results are useful and valuable. This transformation process is shown in Figure 1.2.

The Characteristics of Valuable Information

To be valuable to managers and decision makers, information should have the characteristics described in Table 1.2. These characteristics also make the information more valuable to the organization. If information is not accurate or complete, poor decisions can be made, costing the organization thousands, or even millions, of dollars. For example, if an inaccurate forecast of future demand indicates that sales will be very high when the opposite is true, an organization can invest millions of dollars in a new plant that is not needed. Furthermore, if information is not pertinent to the situation, not delivered to decision makers in a timely fashion, or too complex to understand, it may be of little value to the organization.

Useful information can vary widely in the value of each of these quality attributes. For example, with market-intelligence data, some inaccuracy and incompleteness is acceptable, but timeliness is essential. Market intelligence may alert us that our competitors are about to make a major price cut. The exact details and timing of the price cut may not be as important as being warned far enough in advance to plan how to react. On the other hand,

FIGURE 1.2

The Process of Transforming Data into Information

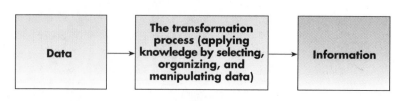

Characteristics	Definitions
Accurate	Accurate information is error free. In some cases, inaccurate information is generated because inaccurate data is fed into the transformation process (this is commonly called garbage in, garbage out [GIGO]).
Complete	Complete information contains all of the important facts. For example, an investment report that does not include all important costs is not complete.
Economical	Information should also be relatively economical to produce. Decision makers must always balance the value of information with the cost of producing it.
Flexible	Flexible information can be used for a variety of purposes. For example, information on how much inventory is on hand for a particular part can be used by a sales representative in closing a sale, by a production manager to determine whether more inventory is needed, and by a financial executive to determine the total value the company has invested in inventory.
Reliable	Reliable information can be depended on. In many cases, the reliability of the information depends on the reliability of the data collection method. In other instances, reliability depends on the source of the information. A rumor from an unknown source that oil prices might go up may not be reliable.
Relevant	Relevant information is important to the decision maker. Information that lumber prices might drop may not be relevant to a computer chip manufacturer.
Simple	Information should also be simple, not overly complex. Sophisticated and detailed information may not be needed. In fact, too much information can cause information overload, where a decision maker has too much information and is unable to determine what is really important.
Timely	Timely information is delivered when it is needed. Knowing last week's weather conditions will not help when trying to decide what coat to wear today.
Verifiable	Information should be verifiable. This means that you can check it to make sure it is correct, perhaps by checking many sources for the same information.
Accessible	Information should be easily accessible by authorized users to be obtained in the right format and at the right time to meet their needs.
Secure	Information should be secure from access by unauthorized users.

TABLE 1.2

The Characteristics of Valuable Information

accuracy, verifiability, and completeness are critical for data used in accounting for the use of company assets such as cash, inventory, and equipment.

The Value of Information

The value of information is directly linked to how it helps decision makers achieve their organization's goals. For example, the value of information might be measured in the time required to make a decision or in increased profits to the company. Consider a market forecast that predicts a high demand for a new product. If market forecast information is used to develop the new product and the company is able to make an additional profit of $10,000, the value of this information to the company is $10,000 minus the cost of the information. Valuable information can also help managers decide whether to invest in additional information systems and technology. A new computerized ordering system may cost $30,000, but it may generate an additional $50,000 in sales. The *value added* by the new system is the additional revenue from the increased sales of $20,000. Read the "Making a Difference" special interest box to learn about how valuable information can lessen the impact of natural disasters.

MAKING A DIFFERENCE

Natural Resources Conservation Service Uses IS to Avoid Flood Damage

Flooding costs U.S. citizens and businesses more than $5 billion a year, according to the Federal Emergency Management Agency. To avert flood disasters, the Natural Resources Conservation Service began conducting flood plain management studies using information systems including geographic information systems and satellites in the global positioning system (GPS). Geographic information systems enable users to pair existing maps or map outlines with tabular data to define a particular geographic region. The GPS satellites pinpoint the exact locations of landmarks and buildings on the map and ensure a high degree of accuracy.

The resulting information helps community planners and decision makers develop local flood plain management programs and determine whether proposed building sites are prone to flooding. Studies also help local and state governments enact or update flood plain zoning ordinances, secure flood insurance, correct flood plain maps, initiate improvements to bridges, and plan alternative routes for emergency vehicles during floods. Most recently, the agency studied the Thornapple River watershed in Michigan. Since the project began in 1970, 24 flood plain management studies have saved an estimated $12 million in flood losses.

(Source: Image copyright © 1998 PhotoDisc.)

DISCUSSION QUESTIONS:

1. What specific kinds of data are captured and what information is displayed using geographic information systems and the global positioning system?

2. What characteristics of valuable information must be present in the data to enable it to meet its purpose? Which characteristics of valuable information are not as important?

Sources: Adapted from "Watershed Surveys and Planning," http://www.nrcs.usda.gov/NRCSProg.html, accessed April 10, 1998; Allison Maxwell, "Online to Efficiency," *Government Executive Magazine*, December 1997; http:www/govexec.com/tech/articles/1297tech.htm.

SYSTEM AND MODELING CONCEPTS

system

a set of elements or components that interact to accomplish goals

Like information, another central concept of this book is that of a system. A **system** is a set of elements or components that interact to accomplish goals. The elements themselves and the relationships among them determine how the system works. Systems have inputs, processing mechanisms, outputs, and feedback. For example, consider the automatic car wash. Obviously, tangible *inputs* for the process are a dirty car, water, and the various cleaning ingredients used. Time, energy, skill, and knowledge are also needed as inputs to the system. Time and energy are needed to operate the system. Skill is the ability to successfully operate the liquid sprayer, foaming brush, and air dryer devices. Knowledge is used to define the steps in the car wash operation and the order in which those steps are executed.

The *processing mechanisms* consist of first selecting which of the cleaning options you want (wash only, wash with wax, wash with wax and hand dry, etc.) and communicating that to the operator of the car wash. Note that there is a *feedback mechanism* (your assessment of how clean the car is). Liquid

System	Elements			
	Inputs	**Processing mechanisms**	**Outputs**	**Goal**
Fast Food Restaurant 	Meat, Potatoes, Tomatoes, Lettuce, Bread, Drinks, Labor, Management	Frying, Broiling, Drink dispensing, Heating	Hamburgers, French Fries, Drinks, Desserts	Quickly prepared, inexpensive food
College 	Students Professors Administrators Textbooks Equipment	Teaching Research Service	Educated students Meaningful research Service to community, state, and nation	Acquisition of knowledge
Movie 	Actors, Director, Staff, Sets, Equipment	Filming, Editing, Special effects, Film distribution	Finished film delivered to movie theaters	Entertaining movie, Film awards, Profits

FIGURE 1.3

Examples of Systems and Their Goals and Elements
(Sources: Image copyright © 1998 PhotoDisc, Courtesy of 3M Visual Systems Division, Image copyright © 1998 PhotoDisc.)

sprayers shoot clear water, liquid soap, or car wax depending on where your car is in the process and which options you selected. The *output* is a clean car. It is important to note that independent elements or components of a system (the liquid sprayer, foaming brush, and air dryer) interact to create a clean car. Figure 1.3 shows a few systems with their elements and goals.

System Components and Concepts

Figure 1.4 shows a typical system diagram—a simple automatic car wash. The primary purpose of the car wash is to clean your automobile. The **system boundary** defines the system and distinguishes it from everything else (the environment).

system boundary

defines the system and distinguishes it from everything else

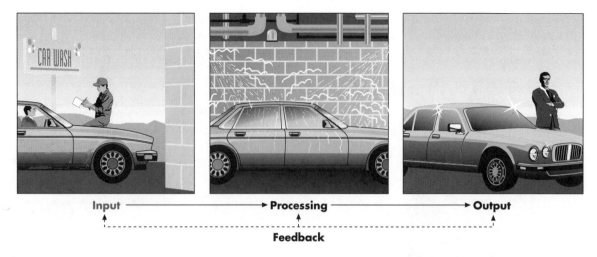

Input ──────────────► Processing ───────────────► Output

Feedback

FIGURE 1.4

Components of a System

A system's four components consist of input, processing, output, and feedback.

The way system elements are organized or arranged is called the configuration. Much like data, the relationships among elements in a system are defined through knowledge. In most cases, knowing the purpose or desired outcome of a system is the first step in defining the way system elements are configured. For example, the desired outcome of our system is a clean car. Based on past experience, we know that it would be illogical to have the liquid sprayer element precede the foaming brush element. The car would be rinsed and then soap would be applied, leaving your car a mess. As you can see from this example, knowledge is needed both to define relationships among the inputs to a system (your dirty car and instructions to the operator) and to organize the system elements used to process the inputs (the foaming brush must precede the liquid sprayer).

System types. Systems can be classified along numerous dimensions. They can be simple or complex, open or closed, stable or dynamic, adaptive or nonadaptive, permanent or temporary. Table 1.3 defines these characteristics.

TABLE 1.3

System Classifications and Their Primary Characteristics

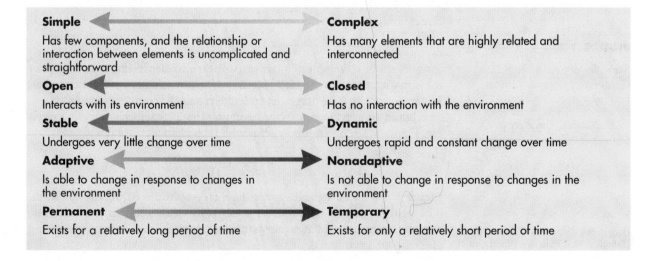

Simple	**Complex**
Has few components, and the relationship or interaction between elements is uncomplicated and straightforward	Has many elements that are highly related and interconnected
Open	**Closed**
Interacts with its environment	Has no interaction with the environment
Stable	**Dynamic**
Undergoes very little change over time	Undergoes rapid and constant change over time
Adaptive	**Nonadaptive**
Is able to change in response to changes in the environment	Is not able to change in response to changes in the environment
Permanent	**Temporary**
Exists for a relatively long period of time	Exists for only a relatively short period of time

Classifying organizations by system type. Most companies can be described using the classification scheme in Table 1.3. For example, a janitorial company that cleans offices after business hours most likely represents a simple, stable system because there is a constant and fairly steady need for its services. A successful computer manufacturing company, however, is typically complex and dynamic because it operates in a changing environment. If a company is nonadaptive, it may not survive very long. Many of the early computer companies, including Osborne Computer that manufactured one of the first portable computers and VisiCorp that developed the first spreadsheet program, did not change rapidly enough with the changing market for computers and software. As a result, these companies did not survive.

System Performance and Standards

System performance can be measured in various ways. **Efficiency** is a measure of what is produced divided by what is consumed. It can range from 0 to 100 percent. For example, the efficiency of a motor is the energy produced (in terms of work done) divided by the energy consumed (in terms of electricity or fuel). Some motors have an efficiency of 50 percent or less because of the energy lost to friction and heat generation.

Efficiency is a relative term used to compare systems. For example, a gasoline engine is more efficient than a steam engine because, for the equivalent amount of energy input (gas or coal), the gasoline engine produces more energy output. The energy efficiency ratio (energy input divided by energy output) is high for gasoline engines when compared with that of steam engines.

Effectiveness is a measure of the extent to which a system achieves its goals. It can be computed by dividing the goals actually achieved by the total of the stated goals. For example, a company may have a goal to reduce damaged parts by 100 units. A new control system may be installed to help achieve this goal. Actual reduction in damaged parts, however, is only 85 units. The effectiveness of the control system is 85 percent (85/100 = 85%). Effectiveness, like efficiency, is a relative term used to compare systems.

Efficiency and effectiveness are performance objectives set for an overall system. Meeting these objectives requires considering not only desired efficiency and effectiveness but also cost, complexity, and the level of control desired of the system. Cost includes the up-front expenses of a system, as well as any ongoing direct expenses. Complexity relates to how complicated the relationship among the system elements is. Control is the ability of a system to operate within predefined guidelines—such as policies, procedures, and budgets—and the managerial effort required to keep the system operating within those bounds. Meeting defined objectives for efficiency and effectiveness may involve trade-offs in terms of cost, control, and complexity.

Evaluating system performance also calls for the use of performance standards. A **system performance standard** is a specific objective of the system. For example, a system performance standard for a particular marketing campaign might be to have each sales representative sell $100,000 of a certain type of product each year (Figure 1.5a). A system performance standard for a certain manufacturing process might be to have no more than 1 percent defective parts (Figure 1.5b). Once standards are established,

efficiency
a measure of what is produced divided by what is consumed

effectiveness
a measure of the extent to which a system achieves its goals

system performance standard
a specific objective of the system

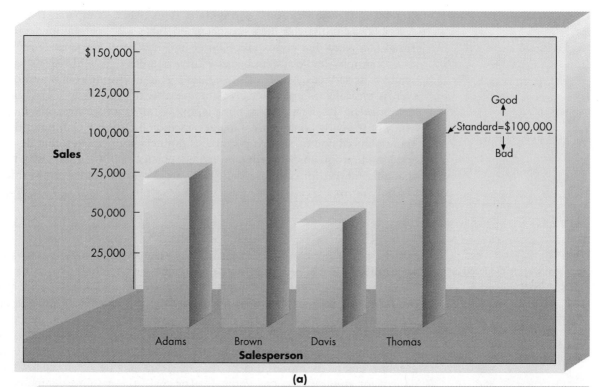

(a)

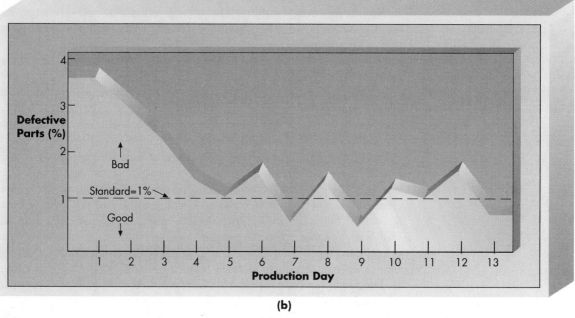

(b)

FIGURE 1.5

System Performance Standards

system performance is measured and compared with the standard. Variances from the standard are determinants of system performance. Achieving system performance standards may also require trade-offs in terms of cost, control, and complexity.

System Variables and Parameters

system variable

a quantity or item that can be controlled by the decision maker

system parameter

a value or quantity that cannot be controlled by the decision maker

Parts of a system are under direct management control, while others are not. A **system variable** is a quantity or item that can be controlled by the decision maker. The price a company charges for its product is a system variable because it can be controlled. A **system parameter** is a value or quantity that cannot be controlled, such as the cost of a raw material. The number of pounds of a chemical that must be added to produce a certain type of plastic is another example of a quantity or value that is not controlled by management (it is controlled by the laws of chemistry).

Modeling a System

model

an abstraction or an approximation that is used to represent reality

The real world is complex and dynamic. So when we want to test different relationships and their effects, we use models of systems, which are simplified, instead of real systems. A **model** is an abstraction or an approximation that is used to represent reality. Models enable us to explore and gain an improved understanding of real world situations.

Since the beginning of recorded history, people have used models. A written description of a battle, a physical mock-up of an ancient building, and the use of symbols to represent money, numbers, and mathematical relationships are all examples of models. Today, managers and decision makers use models to help them understand what is happening in their organizations and make better decisions.

There are various types of models. The major ones are narrative, physical, schematic, and mathematical, as shown in Figure 1.6.

A narrative model, as the name implies, is based on words. Both verbal and written descriptions of reality are considered narrative models. In an organization, reports, documents, and conversations concerning a system are all important narratives. Examples include the following: a salesperson verbally describing a product's competition to a sales manager, a written report describing the function of a new piece of manufacturing equipment, and a newspaper article explaining the economy or future sales of exports. Computers can be used to develop narrative models. Word processing programs, for example, can create written reports, and speech response software can store and play verbal messages like bank balances over the phone.

A physical model is a tangible representation of reality. Many physical models are computer designed or constructed. An engineer may develop a physical model of a chemical reactor to gain important information about how a large-scale reactor might perform. A builder may develop a scale model of a new shopping center to give a potential investor information about the overall appearance and approach of the development. In other examples, a marketing research department may develop a prototype of a new product, and a dentist may build a plastic tooth. These are all examples of physical models that can be used to provide information. Tupperware can produce a physical model (a plastic prototype) of a new product directly from a specialized computer system. After a product, such as a new plastic container, has been designed, the computer system controls equipment that produces the physical model, saving days of development time and reducing costs.

A schematic model is a graphic representation of reality. Graphs, charts, figures, diagrams, illustrations, and pictures are all types of schematic models.

Narrative

Physical

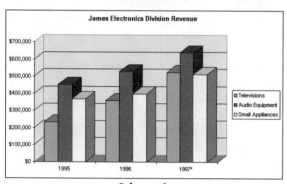

Schematic

Mathematical

FIGURE 1.6

Four Types of Models

Narrative (words, spoken or written), physical (tangible), schematic (graphic), and mathematical (arithmetic) models.
(Source: Image copyright © 1998 PhotoDisc, Image copyright © 1998 PhotoDisc.)

Schematic models are used to a great extent in developing computer programs and systems. *Program flowcharts* show how computer programs are to be developed. *Data-flow diagrams* are used to reveal how data flows through the organization. A blueprint for a new building, a graph that shows budget and financial projections, electrical wiring diagrams, and graphs that show when certain tasks or activities must be completed to stay on schedule are examples of schematic models used in business. Graphics programs can be used to develop simple or complex schematic models.

A mathematical model is an arithmetic representation of reality. Computers excel at solving mathematical models. Such models are used in all areas of business. For example, the following mathematical model might be developed to determine the total cost of a project:

$$TC = (V)(X) + FC$$

where:

TC = total cost
V = variable cost per unit
X = number of units produced
FC = fixed cost

In developing any model, it is important to make it as accurate as possible. An inaccurate model will usually lead to an inaccurate solution to the problem. In the mathematical model just presented, it is assumed that both the variable cost per unit and the fixed cost can be accurately measured. Most models contain many assumptions, and it is important that these be as realistic as possible. It is also important that potential users of the model be aware of the assumptions under which the model was developed.

WHAT IS AN INFORMATION SYSTEM?

An information system is a specialized type of system and can be defined in a number of different ways. As mentioned previously, an information system (IS) is a set of interrelated elements or components that collect (input), manipulate (process), and disseminate (output) data and information and provide a feedback mechanism to meet an objective. (See Figure 1.7.)

Input, Processing, Output, Feedback

input

the activity of gathering and capturing raw data

Input. In information systems, **input** is the activity of gathering and capturing raw data. In producing paychecks, for example, the number of hours worked for every employee must be collected before paychecks can be calculated or printed. In a university grading system, student grades must be obtained from instructors before a total summary of grades for the semester or quarter can be compiled and sent to the appropriate students.

Input can take many forms. In an information system designed to produce paychecks, for example, employee time cards might be the initial input. In a 911 emergency telephone system, an incoming call would be considered an input. Input to a marketing system might include customer survey responses. Notice that regardless of the system involved, the type of input is determined by the desired output of the system.

Input can be a manual process, or it may be automated. A scanner at a grocery store that reads bar codes and enters the grocery item and price into a computerized cash register is a type of automated input process. Regardless of the input method, accurate input is critical to achieve the desired output.

processing

converting or transforming data into useful outputs

Processing. In information systems, **processing** involves converting or transforming data into useful outputs. Processing can involve making calculations, making comparisons and taking alternative actions, and storing data for future use.

FIGURE 1.7

The Components of an Information System

Feedback is critical to the successful operation of a system.

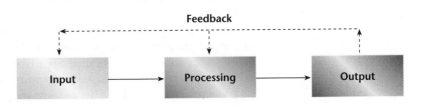

Processing can be done manually or with the assistance of computers. In the payroll application, the number of hours worked for each employee must be converted into net pay. The required processing can first involve multiplying the number of hours worked by the employee's hourly pay rate to get gross pay. If weekly hours worked are greater than 40 hours, overtime pay may also be determined. Then deductions are subtracted from gross pay to get net pay. For instance, federal and state taxes can be withheld or subtracted from gross pay; many employees have health and life insurance, savings plans, and other deductions that must also be subtracted from gross pay to arrive at net pay.

output
useful information, usually in the form of documents and/or reports

Output. In information systems, **output** involves producing useful information, usually in the form of documents and/or reports. Outputs can include paychecks for employees, reports for managers, and information supplied to stockholders, banks, government agencies, and other groups. In some cases, output from one system can become input for another. For example, output from a system that processes sales orders can be used as input to a customer billing system. Often output from one system can be used as input to control other systems or devices. For instance, the manufacture of office furniture is complicated with many variables. Thus, the salesperson, customer, and furniture designer go through several iterations of designing furniture to meet the customer's needs. Special computer software and hardware is used to create the original design and rapidly revise it. Once the last design mock-up is approved, the design workstation software creates a bill of materials that goes to manufacturing to produce the order.

Output can be produced in a variety of ways. For a computer, printers and display screens are common output devices. Output can also be a manual process involving handwritten reports and documents.

feedback
output that is used to make changes to input or processing activities

Feedback. In information systems, **feedback** is output that is used to make changes to input or processing activities. For example, errors or problems might make it necessary to correct input data or change a process. Consider a payroll example. Perhaps the number of hours worked by an employee was entered into a computer as 400 instead of 40 hours. Fortunately, most information systems check to make sure that data falls within certain predetermined ranges. For number of hours worked, the range might be from 0 to 100 hours. It is unlikely that an employee would work more than 100 hours for any given week. In this case, the information system would determine that 400 hours is out of range and provide feedback, such as an error report. The feedback is used to check and correct the input on the number of hours worked to 40. If undetected, this error would result in a very high net pay printed on the paycheck!

Feedback is also important for managers and decision makers. For example, output from an information system might indicate that inventory levels for a few items are getting low. A manager could use this feedback to decide to order more inventory. The new inventory orders then become input to the system. In this case, the feedback system reacts to an existing problem and alerts a manager that there are too few inventory items on hand. In addition to this reactive approach, a computer system can also be proactive by predicting future events to avoid problems. This concept, often called **forecasting**, can be used to estimate future sales and order more inventory before a shortage occurs.

forecasting
a proactive approach to feedback

Manual and Computerized Information Systems

As discussed earlier, an information system can be manual or computerized. For example, some investment analysts manually draw charts and trend lines to assist them in making investment decisions. Tracking data on stock prices (input) over the last few months or years, these analysts develop patterns on graph paper (processing) that help them determine what stock prices are likely to do in the next few days or weeks (output). Some investors have made millions of dollars using manual stock analysis information systems. Of course, there are many excellent computerized information systems as well. For example, many computer systems have been developed to follow stock indexes and markets and to suggest when large blocks of stocks should be purchased or sold (program trading) to take advantage of market discrepancies.

Many information systems begin as manual systems and become computerized. For example, consider the way the U.S. Postal Service sorts mail. At one time most letters were visually scanned by postal employees to determine the ZIP code and were then manually placed in an appropriate bin. Today the bar coded addresses on letters passing through the postal system are "read" electronically and automatically routed to the appropriate bin via conveyors. The computerized sorting system results in speedier processing time and provides management with information that helps control transportation planning. It is important to stress, however, that the simple computerization of a manual information system does not guarantee improved system performance. If the underlying information system is flawed, the act of computerizing it might only magnify the impact of these flaws.

Computer-Based Information Systems

technology infrastructure

a computer-based information system that consists of the shared IS resources that form the foundation of the information system

A computer-based information system (CBIS) is composed of hardware, software, databases, telecommunications, people, and procedures that are configured to collect, manipulate, store, and process data into information. The components are illustrated in Figure 1.8. A computer-based information system is also known as a business's **technology infrastructure** because it consists of the shared IS resources that form the foundation of the information systems. Read the "FYI" box to explore how a CBIS can add to or subtract from business processes.

Hardware. Hardware consists of computer equipment used to perform input, processing, and output activities. Input devices include keyboards, automatic scanning devices, equipment that can read magnetic ink characters, and many other devices. Processing devices include the central processing unit and main memory. There are many output devices, including secondary storage devices, printers, and computer screens.

Software. Software is the computer programs that govern the operation of the computer. These programs allow the computer to process payroll, to send bills to customers, and to provide managers with information to increase profits, to reduce costs, and to provide better customer service. There are two basic types of software: system software (which controls basic computer operations such as start-up and printing) and applications software (which allows specific tasks to be accomplished, such as word processing or tabulating numbers). A program that allows users to create a spreadsheet (for example, Excel, Lotus, QuattroPro, etc.) is an example of applications software.

Databases. A **database** is an organized collection of facts and information. An organization's database can contain facts and information on customers, employees, inventory, competitors' sales information, and much more. Most managers and executives believe a database is one of the most valuable and important parts of a computer-based information system.

Telecommunications, networks, and the Internet. Telecommunications is the electronic transmission of signals for communications and enables organizations to link computer systems into effective networks. **Networks** are used to connect computers and computer equipment in a building, around the country, or across the world to enable electronic communications.

Telecommunications and networks help people communicate using electronic mail (E-mail) and voice mail. These systems also help people work in groups. The **Internet** is the world's largest computer network, actually consisting of thousands of interconnected networks, all freely exchanging information. Research firms, colleges, universities, high schools, and businesses are just a few examples of organizations using the Internet. Anyone who can gain access to the Internet can communicate with anyone else on the Internet. The technology used to create the Internet is now being applied within companies and organizations to create an **intranet**, which allows people within an organization to exchange

database

an organized collection of facts and information

telecommunications

the electronic transmission of signals for communications and enables organizations to link computer systems into effective networks

networks

used to connect computers and computer equipment in a building, around the country, or across the world to enable electronic communications

Internet

the world's largest telecommunications network

intranet

a network that uses Internet technology within an organization

FIGURE 1.8

The Components of a Computer-Based Information System

A CBIS is also called a business's technology infrastructure because it is the foundation on which information systems are built.

FYI

Can Less Information Result in Less Value?

Computers have reduced costs, increased revenues, or both for most businesses. Few would disagree that computer-based information systems have been invaluable or even indispensable in today's competitive business environment. Yet, measuring the actual value of an information system has been difficult. Some experts believe that standard financial methods, such as payback period or rate of return, are the best approaches. Others argue that the value of an information system to a company is more difficult to measure and that its impact on effective decision making should be an important factor.

In healthcare, the computer plays a critical role. In addition to performing routine business functions, such as paying employees and suppliers, controlling inventory levels, and receiving payments for service, computer-based information systems have had a dramatic impact in the delivery of quality healthcare. Some doctors, for example, use a CBIS to help diagnose a patient's true problems, given a set of symptoms. In other cases, a CBIS can be used to suggest possible treatments and drugs. Most healthcare facilities use their CBIS to process increasingly complex state and federal reports.

Value, however, is a double-edged sword. A good CBIS can cut out competition and dramatically increase revenues and profits, thus providing at least a temporary competitive advantage. An inadequate CBIS can slice deeply into revenues and corporate profits. After spending millions of dollars on sophisticated computer-based information systems, a number of healthcare facilities have painfully discovered that their CBIS is full of errors and inadequate to handle the complex demands that are required. One healthcare facility was one of the fastest growing health maintenance organizations (HMOs) in the mid-1990s. Its stock price soared from about $4 a share

to almost $90 a share in a few short years. This staggering growth required a new CBIS to keep up with the huge processing demands. But the new CBIS was not up to the task. It did not produce the needed information. The system did not adequately keep up with millions of dollars it owed to doctors and other medical facilities. At the same time, the CBIS did not properly collect premiums from its members. With debts higher than expected and the inability to manage cash inflows, the HMO was in trouble. In October 1997, the problems were revealed and the stock price dramatically declined in what some have called one of the largest drops in shareholder wealth in a single day. The total loss in stock value was greater than $3 billion. At one point, over 80 percent of the stock value was lost.

Although it is not clear if the CBIS problems caused all of the $3 billion loss, the computer-related problems were certainly an important factor, perhaps the most important factor. Since the dramatic stock decline, the HMO has looked into the possibility of securing financing by selling stock or borrowing additional funds that it needs. The future profitability of the company is uncertain. An inadequate CBIS can not only hurt profits, it may even threaten the survival of the company.

DISCUSSION QUESTIONS

1. How would you measure the value of a CBIS?
2. Do you have any advice for companies experiencing dramatic growth?

Sources: Keith Hammonds, "Industry Outlook," *Business Week*, January 12, 1998, p. 114; Leslie Scism, "Oxford Health Discusses Sale of Stock," *The Wall Street Journal*, February 4, 1998, p. A3; Don Peppers, "Knocking on Healthcare's Door," *Healthcare Forum Journal*, January/February 1998, p. 29.

information and work on projects. For example, Arthur Andersen & Co. moved its entire body of knowledge to an intranet called KnowledgeSpace. It makes the equivalent of a 35,000 page three-ring binder available to its consultants around the world.[1] Read more about this intranet in the "Technology Advantage" box. The World Wide Web is a network of links to hypermedia (text, graphics, video, sound) documents. Information about the documents and access to them are controlled and provided by tens of thousands of Web servers. The Web is one of many services available over the Internet and provides access to literally millions of documents.

TECHNOLOGY ADVANTAGE
Corporate Intranets—Needed to Stay Competitive

Many professional service companies are global, multi-disciplinary organizations that provide clients, large and small, all over the world, the thing they need most to succeed: knowledge. Their work is to acquire knowledge and to share knowledge—knowledge of how to improve performance in management, business processes, operations, information technology, and finance so that their clients can grow and profit. This knowledge comes from three sources: experience, education, and research.

Leveraging corporate knowledge through an intranet has become a matter of staying competitive in the services industry. KPMG Peat Marwick, Coopers & Lybrand, Booz Allen & Hamilton, and many other professional services companies have implemented intranets to move corporate knowledge into databases on Web computers, enabling individual consultants to bring the company's collective knowledge to bear on their client's business problems. Development of a corporate intranet is a major undertaking. For example, it cost nearly $1 million and took eight months for Arthur Andersen & Co. to place the CD-ROM-based content from each of its major businesses—business practices, information technology, accounting, and vertical industries—onto its own intranet called KnowledgeSpace. Through KnowledgeSpace, Arthur Andersen delivers timely, relevant information to communities or to individuals, while connecting them to its own professionals, knowledge assets, and solutions.

These firms look upon their intranet as more of an investment than a cost. Without an intranet, consultants had to contact other consultants personally and days could go by before the necessary information was exchanged. Consultants had to read over the data and digest it into something meaningful to apply it to their situation. This whole process took several days. With a corporate intranet, consultants simply log onto the intranet and use software called a Web search engine to find relevant documents available on the intranet. Additional software called a Web browser is used to read and format the documents stored on the intranet so that tables and graphics can be included. Within minutes the consultant can find the company's collective knowledge on a particular subject or issue, saving time and tapping the best of the company's experience and research.

DISCUSSION QUESTIONS

1. One of the potential issues associated with the use of a corporate intranet is "data currency," or ensuring that the information is current and remains current. Do you think that this would be a major problem for a large professional service firm? How might this problem be addressed?
2. What other ways might a company use a corporate intranet?

Sources: Adapted from Justin Hibbard, "Spreading Knowledge," *Computerworld*, April 7, 1997, pp. 63–64; Arthur Andersen & Co. Web page at http://www.arthurandersen.com accessed April 13, 1998; and Web page for KnowledgeSpace at http://www.knowledgespace.com accessed April 7, 1998.

People. People are the most important element in most computer-based information systems. Information systems personnel include all the people who manage, run, program, and maintain the system. Users are any people who use information systems to get results. Users include financial executives, marketing representatives, manufacturing operators, and many others. Certain computer users are also IS personnel.

Procedures. Procedures include the strategies, policies, methods, and rules for using the CBIS. For example, some procedures describe when each program is to be run or executed. Others describe who can have access to facts in the database. Still other procedures describe what is to be done in case a disaster, such as a fire, an earthquake, or a hurricane, renders the CBIS unusable.

Now that we have looked at computer-based information systems in general, we briefly examine the most common types used in business today. These IS types are covered in more detail later in the book.

BUSINESS INFORMATION SYSTEMS

Workers at all levels, in all kinds of firms, and in all industries are using information systems to improve their own effectiveness. There are few workers who do not come into contact with a personal computer on at least a weekly, if not daily, basis to connect to a network, create presentations, prepare a memo, or develop a spreadsheet for analysis. At the corporate level, the most common types of information systems used in business organizations are transaction processing systems and electronic commerce (E-commerce), management information systems, decision support systems, and expert systems. Together, these systems help employees in organizations accomplish both routine and special tasks—from recording sales, to processing payrolls, to supporting decisions in various departments, to providing alternatives for large-scale projects and opportunities.

Transaction Processing Systems and E-commerce

transaction

any business-related exchange

transaction processing system (TPS)

an organized collection of people, procedures, software, databases, and devices used to record completed business transactions

---

Transaction processing systems are essential to grocery and retail stores. Items sold are automatically scanned into the system, allowing management to keep track of sales and inventory levels.
(Source: Chuck Keeler/Tony Stone Images.)

Transaction Processing Systems. Since the 1950s computers have been used to perform common business applications. The objective of many of these early systems was to reduce costs. This was done by automating many routine, labor-intensive business systems. A **transaction** is any business-related exchange such as payments to employees, sales to customers, or payments to suppliers. Thus, processing business transactions was the first application of computers for most organizations. A **transaction processing system (TPS)** is an organized collection of people, procedures, software, databases, and devices used to record completed business transactions. To understand a transaction processing system is to understand basic business operations and functions.

One of the first business systems to be computerized was the payroll system (Figure 1.9). The primary inputs for a payroll TPS are the numbers of employee hours worked during the week and pay rate. The primary output consists of paychecks. Early payroll systems were able to produce employee paychecks, along with important employee-related reports required by state and federal agencies, such as the Internal Revenue Service. Simultaneously, other routine processes, including customer billing and inventory control, were computerized as well. Because these early systems handled and processed daily business exchanges, or transactions, they were called transaction processing systems (TPSs). In improved forms, these transaction processing systems are still vital to most modern organizations: Consider what would happen if an organization had to function without its TPS for even one day. How many employees would be paid and paid the correct amount? How

FIGURE 1.9

A Payroll Transaction Processing System

The inputs (numbers of employee hours worked and pay rates) go through a transformation process to produce outputs (paychecks).

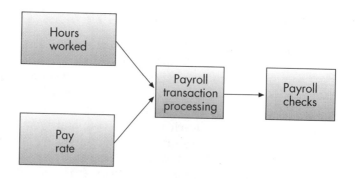

many sales would be recorded and processed? Transaction processing systems represent the application of information concepts and technology to routine, repetitive, and usually ordinary business transactions, but transactions that are critical to the daily functions of that business.

E-commerce

involves any business transaction executed electronically between parties such as companies (business-to-business), companies and consumers (business-to-consumer), business and the public sector, and consumers and the public sector

E-commerce. **E-commerce** involves any business transaction executed electronically between parties such as companies (business-to-business), companies and consumers (business-to-consumer), business and the public sector, and consumers and the public sector. People may assume that E-commerce is reserved for consumers visiting Web sites for on-line shopping. But Web shopping is only a small part of the E-commerce picture; the major volume of E-commerce is business-to-business transactions that make purchasing easier for big corporations. General Electric surpassed $1 billion worth of business with its suppliers in 1997, single-handedly greater than all of consumer spending in 1996.[2] E-commerce offers opportunities for small businesses too, by enabling them to market and sell at a low cost worldwide, thus offering small businesses an opportunity to enter the global market right from start-up.

E-commerce includes both business-to-business sales transactions and business-to-consumer sales transactions.

Results from a 1998 survey by Forrester Research estimates that the number of North American consumers investing and shopping on-line will increase from 5 to 10 percent by the end of 1998.[3] However, the real growth is expected to occur in business-to-business electronic commerce according to Terry Retter, director of strategic planning for Price Waterhouse. This growth is being stimulated by increased Internet access, user confidence, better payment systems, and rapidly improving Internet and Web security. By 2002, the value of goods and services traded via the Internet is estimated at an astounding $434 billion. The value of consumer purchases via the Internet is expected to increase to nearly $94 billion.[4]

Technically astute consumers who have tried on-line shopping appreciate the ease of E-commerce. They can avoid fighting the crowds in the malls, shop on-line at any time from the comfort of their home, and have the goods delivered to them directly. The typical E-commerce customer is a

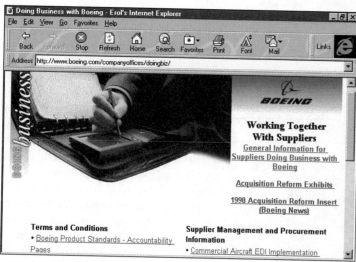

Prepare requisition

Obtain approval of requisition

Requisition

Purchasing department

In box

Purchase Order

please send us one each of the following:

100 boxes of 3/4" staples
24 sheets CDX pine plywood
10 lbs. Black drywall screws
47 Extra Heavy Duty hinges

ACME Inc.
7 That St.
Anywhere, USA
01800

Vendor

ACME

Traditional process for placing a purchase order

Electronic purchase order

Vendor

ACME

E-commerce process for placing a purchase order

FIGURE 1.10

E-commerce greatly simplifies the purchasing process.

30-something white male earning at least $78,000 per year, according to a survey of more than 1,000 on-line shoppers conducted by Binary Compass Enterprises. About 75 percent of E-commerce buyers are men.[5]

Here is a brief example of how E-commerce can simplify the purchasing process for, say buying new office furniture from an office supply company (Figure 1.10). Typically, a corporate office worker must get approval for a purchase that costs more than a certain amount. That request goes to the purchasing department, which generates a formal purchase order to procure the goods from the approved vendor. Business-to-business E-commerce automates that entire process. Employees go directly to the supplier's Web site, find the item in their catalog, and order what they need at a price prenegotiated by the employee's company. If approval is required, the approver is notified automatically. As the use of E-commerce systems grows, companies will reduce the use of more traditional transaction processing systems. There are many companies rushing to sell their products using E-commerce. These are discussed in the E-commerce special interest box.

E-COMMERCE
Competing Electronically

Companies are continually seeking better ways to achieve a competitive advantage. According to Michael Porter, the threat of new entrants into the market is an important factor in achieving a competitive advantage and an important part of Porter's five-force model.

Traditionally, companies used a sales force of men and women to alert customers to new and enhanced products and services. In addition to sales calls, brochures, newspaper ads, TV spots, and direct marketing all spread the word to the marketplace. Once a sale was made, factories, warehouses, and retail stores delivered goods and services to customers. Because it is expensive to maintain a sales force and an effective distribution channel for goods and services, the threat of new entrants—new companies entering the market—was low for many industries.

With the widespread use of the Internet, individuals and companies can market and sell new products quickly, without the high expenses of the more traditional approach to marketing and distributing products. Overnight, the Internet has forever changed how businesses operate, while dramatically increasing the threat of new companies entering the market. In the past year, about 7 million households purchased products over the Internet. This does not include businesses making Internet purchases. Although the number of purchases made over the Internet is still small, it is steadily climbing.

Well-established and new companies are effectively using the Internet to make sales and compete electronically. Computer manufacturers, such as Dell and Gateway, have been using the Internet for years to sell personal computers. Many investment firms, such as Schwab, also have active Internet sites to allow customers to buy and sell stocks, mutual funds, and other financial investments electronically. Sears has been very successful in marketing and selling its Craftsman line of tools and products over the Internet. J. Crew, a clothing chain, realized a 10-fold increase in their Internet sales in one year alone. And the phenomenal success of Amazon.com made Barnes and Noble and other book sellers take note of the new selling medium.

People tired of the corporate grind and politics are increasingly finding the Internet offers an alternate career path. In some cases, people have made more money and have had more time for their family and friends by selling products and services over the Internet. Some people, like Seth Godin, sell ideas and information over the Internet. Publishing and distributing books, such as "The Encyclopedia of Fictional People" and "E-Mail Addresses of the Rich & Famous," have made a nice profit for Godin. Godin's new Internet company, Yoyodyne, is looking to cash in on competing electronically.

There are at least two ways companies and individuals can sell products and services on the Internet. One approach is direct marketing. Customers contact the company or individual directly to purchase goods and services. Schwab, Dell, Gateway, and many other companies use this approach. Another alternative is to use an electronic market or an intermediary that connects buyers and sellers. BuyDirect.Com, for example, connects software buyers and software sellers. Most of the software purchased from BuyDirect can be downloaded instantly after the purchase. Filex.Com also acts as an intermediary by providing lists of free or inexpensive software, called shareware.

Regardless of how companies and individuals use the Internet, competing electronically is expected to soar in the years to come. Some people believe that more millionaires will be created through the Internet than any other market or technology.

DISCUSSION QUESTIONS

1. What are some of the companies that have been able to successfully compete electronically? Why do you think they have been successful?
2. If you decided to compete electronically, what products or services would you offer? What approach would you use?

Sources: Mary Kuntz, "Point, Click—And Here is the Pitch," *Business Week*, February 9, 1998, p. 8; Mark Millstein, "Internet's Role in Electronic Commerce," *Supermarket News*, January 26, 1998, p. 9; Sharon Nash, "Software Shopping Online," *PC Magazine*, February 10, 1998, p. 36.

FIGURE 1.11

Functional management information systems draw data from the organization's transaction processing system.

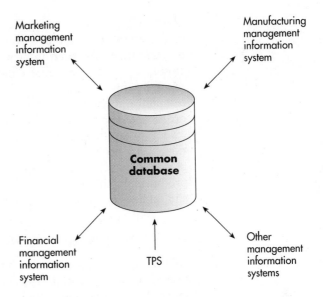

Management Information Systems

The benefits provided by an effective transaction processing system are tangible and can be used to justify their cost in computing equipment, computer programs, and specialized personnel and supplies. They speed the processing of business activities and reduce clerical costs. Although early accounting and financial transaction processing systems were valuable, it has become clear that the data stored in these systems can be used to help managers make better decisions in their respective business areas, whether human resources, marketing, or administration. Satisfying the needs of managers and decision makers continues to be a major factor in developing information systems.

A **management information system (MIS)** is an organized collection of people, procedures, software, databases, and devices used to provide routine information to managers and decision makers. The focus of an MIS is primarily on operational efficiency. Marketing, production, finance, and other functional areas are supported by management information systems and linked through a common database. Management information systems typically provide standard reports generated with data and information from the transaction processing system (Figure 1.11).

Management information systems began to be developed in the 1960s and are characterized by the use of information systems to produce managerial reports. In most cases, these early reports were produced periodically—daily, weekly, monthly, or yearly. Because they were printed on a regular basis, they were called *scheduled reports*. These scheduled reports helped managers perform their duties. For example, a summary report of total payroll costs might help an accounting manager control future payroll costs. As other managers learned the value of these reports, MIS began to proliferate throughout the management ranks. For instance, the total payroll summary report produced initially for the accounting manager might also be useful to a production manager to help monitor and control labor and job costs. Other scheduled reports could be used to help managers from a variety of departments control customer credit, payments to suppliers, the performance of sales representatives, inventory levels, and more.

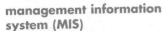

management information system (MIS)

an organized collection of people, procedures, software, databases, and devices used to provide routine information to managers and decision makers

Other types of reports were also developed during the early stages of management information systems. *Demand reports* were developed to give decision makers certain information upon request. For example, prior to closing a sale, a sales representative might seek a demand report on how much inventory exists for a particular item. This report would tell the representative if enough inventory of the item is on hand to fill the customer order. *Exception reports* describe unusual or critical situations, like low inventory levels. The exception report is produced only if a certain condition exists—in this case, inventory falling below a specified level. For example, in a bicycle manufacturing company, an exception report might be produced by the MIS if the number of bicycle seats is too low and more should be ordered.

Decision Support Systems

By the 1980s, dramatic improvements in technology resulted in information systems that were less expensive but more powerful than earlier systems. People at all levels of organizations began using personal computers to do a variety of tasks; they were no longer solely dependent on the information systems department for all their information needs. During this time, people recognized that computer systems could support additional decision-making activities. A **decision support system (DSS)** is an organized collection of people, procedures, software, databases, and devices used to support problem-specific decision making. The focus of a DSS is on decision-making effectiveness. Whereas an MIS helps an organization "do things right," a DSS helps a manager "do the right thing."

A DSS supports and assists all aspects of problem-specific decision making. A DSS goes beyond a traditional management information system. A DSS can provide immediate assistance in solving complex problems that were not supported by a traditional MIS. Many of these problems are unique and not straightforward. For instance, an auto manufacturer might try to determine the best location to build a new manufacturing facility, or an oil company might want to discover the best place to drill for oil. Traditional MIS systems are seldom used to solve these types of problems; a DSS can help by suggesting alternatives and assisting final decision making.

Decision support systems are used where the problem is complex and the information needed to make the best decision is difficult to obtain and use. A DSS also involves managerial judgment. In addition, managers often play an active role in the development and implementation of the DSS. A DSS operates from a managerial perspective, and it recognizes that different managerial styles and decision types require different systems. For example, two production managers in the same position trying to solve the same problem might require different information and support. The overall emphasis is to support rather than replace managerial decision making.

The essential elements of a DSS include a collection of models used to support a decision maker or user (model base), a collection of facts and information to assist in decision making (database), and systems and procedures (user interface) that help decision makers and other users interact with the DSS (Figure 1.12).

decision support system (DSS)

an organized collection of people, procedures, software, databases, and devices used to support problem-specific decision making

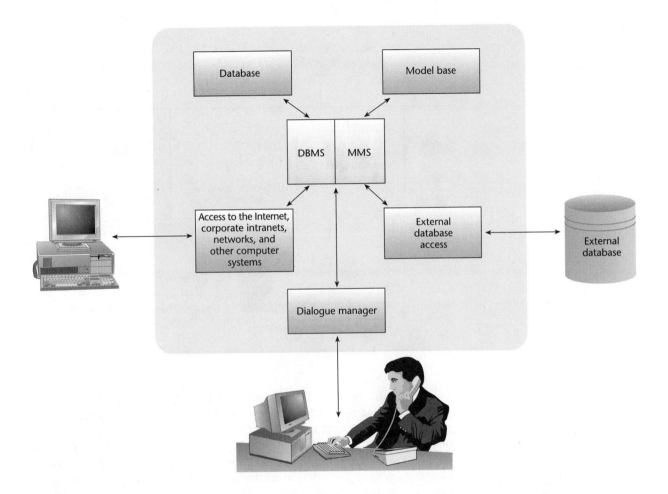

FIGURE 1.12

Essential DSS Elements

artificial intelligence (AI)

a field that involves computer systems taking on the characteristics of human intelligence

Artificial Intelligence and Expert Systems

In addition to TPS, MIS, and DSS, organizations often use systems based on the notion of **artificial intelligence (AI)**, where the computer system takes on the characteristics of human intelligence. The field of artificial intelligence includes several subfields (see Figure 1.13).

Robotics is an area of artificial intelligence where machines take over complex, routine, or boring tasks, such as welding car frames or assembling computer systems and components. Vision systems allow robots and other devices to have "sight" and to store and process visual images. Natural language processing involves the ability of computers to understand and act on verbal or written commands in English, Spanish, or other natural languages. Learning systems give computers the ability to learn from past mistakes or experiences, such as playing games or making business decisions, while neural networks is a branch of artificial intelligence that allows computers to recognize and act on patterns or trends. Some successful stock, options, and futures traders use neural networks to spot trends and make them more profitable with their investments. Finally, expert systems give the computer the ability to make suggestions and act like an expert in a particular field.

FIGURE 1.13

The Major Elements of Artificial
Intelligence

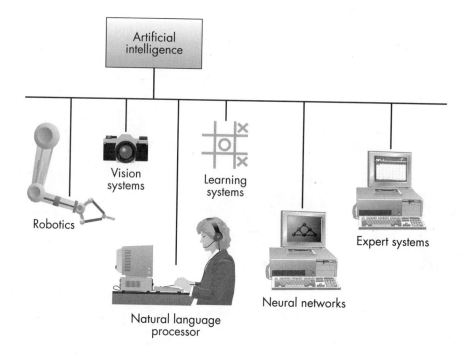

The unique value of expert systems is that they allow organizations to capture and use the wisdom of experts and specialists. Therefore, years of experience and specific skills are not completely lost when a human expert dies, retires, or leaves for another job. Expert systems can be applied to almost any field or discipline. Expert systems have been used to monitor complex systems like nuclear reactors, perform medical diagnoses, locate possible repair problems, design and configure information system components, perform credit evaluations, and develop marketing plans for a new product or new investment strategies.

The 1980s and 1990s have brought advances in both artificial intelligence and expert systems. More and more organizations are using these types of systems to solve complex problems and support difficult decisions. However, many issues remain to be resolved with these systems and more work is needed to refine their meaningful uses.

It is difficult to predict where information systems and technology will be in 10 to 20 years. It seems, however, that we are just beginning to discover the full range of its usefulness. Technology has been improving and expanding at an increasing rate; dramatic growth and change are expected for years to come. Without question, a knowledge of the effective use of information systems will be critical for managers both now and in the long term.

A medical expert system can help a physician determine the proper dosage for a patient's drug prescription.
(Source: Pete Saloutos/Tony Stone Images.)

SYSTEMS DEVELOPMENT

systems development

the activity of creating or modifying existing business systems

Systems development is the activity of creating or modifying existing business systems. Developing information systems to meet business needs is a highly complex and difficult undertaking. So much so that it is common for information systems projects to overrun budgets and exceed scheduled completion dates. Business managers would like the development process to be more manageable with predictable costs and timing. One strategy for improving the results of a systems development project is to divide it into several steps, each step with a well-defined goal and set of tasks to accomplish. These steps are summarized below.

systems investigation

gains a clear understanding of the problem to be solved or opportunity to be addressed

systems analysis

defines the problems and opportunities of the existing system

Systems Investigation and Analysis

The first two steps of systems development are systems investigation and analysis. The goal of the **systems investigation** is to gain a clear understanding of the problem to be solved or opportunity to be addressed. Once this is understood, the next question to be answered is "Is the problem worth solving?" Given that organizations have limited resources—people and money—this question deserves careful consideration. If the decision is to continue with the solution, the next step, **systems analysis**, defines the problems and opportunities of the existing system.

systems design

determines how the new system will work to meet the business needs defined during systems analysis

systems implementation

involves creating or acquiring the various system components (hardware, software, databases, etc.) defined in the design step, assembling them, and putting the new system into operation

systems maintenance and review

checks and modifies the system so that it continues to meet changing business needs

Systems Design, Implementation, and Maintenance and Review

Systems design determines how the new system will work to meet the business needs defined during systems analysis. **Systems implementation** involves creating or acquiring the various system components (hardware, software, databases, etc.) defined in the design step, assembling them, and putting the new system into operation. The purpose of **systems maintenance and review** is to check and modify the system so that it continues to meet changing business needs.

WHY STUDY INFORMATION SYSTEMS?

Studies have shown that the involvement of managers and decision makers in all aspects of information systems is a major factor for organizational success, including higher profits and lower costs. A knowledge of information systems will help you make a significant contribution on the job. It will also help you advance in your chosen career or field.

Information systems play a fundamental and ever-expanding role in all business organizations. If you are to have a solid understanding of how organizations operate, it is imperative that you understand the role of information systems within these organizations. Moreover, as we look to the next century, we see trends that business survival and prosperity will become more difficult. For example, increased mergers among former competitors to create global conglomerates, continued downsizing of corporations to focus on their core businesses and to improve efficiencies, efforts to reduce trade barriers, and the globalization of capital all point to the increased internationalization of business organizations and markets. In addition, business issues and decisions are becoming more complex. An understanding of information systems will help you cope, adapt, and prosper in this challenging environment.

Regardless of your chosen field or the organization for which you may work, it is likely that you will use information systems. Why study information systems? A knowledge of information systems will help you advance in your career, solve problems, realize opportunities, and meet your own personal goals.

Computer and Information Systems Literacy

computer literacy

a knowledge of computer systems and equipment and the ways they function

information systems literacy

a knowledge of how data and information are used by individuals, groups, and organizations

You must acquire both computer literacy and information systems literacy to be able to use information systems to meet personal and organizational goals. **Computer literacy** is a knowledge of computer systems and equipment and the ways they function. It stresses equipment and devices (hardware), programs and instructions (software), databases, and telecommunications.

Information systems literacy goes beyond a knowledge of the fundamentals of computer systems and equipment. **Information systems literacy** is a knowledge of how data and information are used by individuals, groups, and organizations. It includes not only a knowledge of computer technology but also aspects of the broader range of information technology. Most important, however, it encompasses *how* and *why* this technology is applied in business. Knowing about various types of hardware and software is an example of computer literacy. Knowing how to use hardware and software to increase profits, cut costs, improve productivity, and increase customer satisfaction is an example of information systems literacy. Information systems literacy can involve a knowledge of how and why people (managers, employees, stockholders, and other individuals) use information technology; a knowledge of organizations, decision-making approaches, management levels, and information needs; and a knowledge of how organizations can use computers and information systems to achieve their goals. Knowing how to deploy transaction processing, management information, decision support, and expert systems to help an organization achieve its goals is a key aspect of information systems literacy.

Information Systems in the Functional Areas of Business

Information systems are used in all functional areas and operating divisions of business. In *finance* and *accounting*, information systems are used to forecast revenues and business activity, determine the best sources and uses of funds, manage cash and other financial resources, analyze investments, and perform audits to make sure the organization is financially sound and that all financial reports and documents are accurate. In *sales* and *marketing*, information systems are used to develop new goods and services (product analysis), determine the best location for production and distribution facilities (place or site analysis), determine the best advertising and sales approaches (promotion analysis), and set product prices to get the highest total revenues (price analysis). As we saw at the beginning of the chapter, Whirlpool implemented an automated pricing system and streamlined work processes to enable it to respond more quickly to market changes or to launch a special promotion of its own.[6] In *manufacturing,* information systems are used to process customer orders, develop production schedules, control inventory levels, and monitor product quality. Procter & Gamble developed a continuous replenishment program to eliminate unnecessary warehousing and handling costs by keeping products moving through the supply pipeline instead of storing them, thus providing retailers with lower inventories, reduced warehouse space, and lower administration costs. In addition, information systems are used to design products (*computer-assisted design,* or *CAD*), manufacture items (*computer-assisted manufacturing,* or *CAM*), and integrate multiple machines or pieces of equipment (*computer-integrated manufacturing,* or *CIM*). Haworth, the second-largest office furniture maker in the world, has provided its sales force with three-dimensional software to allow a preview of what the customer's offices will look like, as well as an estimate of the total project's cost.[7] Information systems are also used in *human resource management* to screen applicants, administer performance tests to employees, monitor employee productivity, and more. *Legal information systems* are used to analyze product liability and warranties and to develop important legal documents and reports.

Information Systems in Industry

Information systems are used in almost every industry or field. The *airline industry* uses information systems to make seat reservations, to determine the best airfares and schedules, and even to determine which type of plane should fly which particular route. United Airlines, with 90,000 employees and thousands of destinations worldwide, uses decision support systems to improve customer service while reducing airline costs.[8] *Investment firms* use information systems to analyze stocks, bonds, options, the futures market, and other financial instruments, as well as to provide improved services to their customers. Charles Schwab & Co. formed an alliance with iVillage (a successful on-line community for adult women) to educate and empower people to become more informed investors.[9] *Banks and savings and loan companies* use information systems to help make sound loans and good investments. Citicorp and Traveler's Group announced their intention to merge and form a new global leader in financial services, generating substantial savings from cross-selling opportunities arising from improved information systems integration including customer databases.[10] The *transportation industry* uses information systems to schedule trucks and trains to deliver goods and services at the least cost. Federal Express, pioneer of Web package tracking capabilities, is linking its computers to those of its customers and taking over responsibility for their warehousing and distribution operations.[11] *Publishing companies* use information systems to analyze markets and to develop and publish newspapers, magazines, and books. *Healthcare organizations* use information systems to diagnose illnesses, plan medical treatment, and bill patients. HMOs use information systems to track the dollars owed physicians and

Information systems are used by Wall Street investment bankers to analyze stocks and bonds and provide improved service to investors.
(Source: The trading floor photo is a trademark of the New York Stock Exchange and is used with permission.)

medical facilities and to collect premiums from its members.[12] *Retail companies* use information systems to monitor customer needs and to produce the right products. *Power management* and *utility companies* use information systems to monitor and control power generation and usage. *Professional services* firms employ information systems to improve the speed and quality of services they provide to customers. Many of these firms have implemented intranets to enable individual consultants to bring the company's collective knowledge to bear on their client's business problems.[13] These industries will be discussed in more detail as we continue through the book.

Information Systems Principles

Knowing the potential impact of information systems and having the ability to put this knowledge to work can result in a successful personal career, organizations that reach their goals, and a society with a higher quality of life.

Computers and information systems are constantly making it possible for organizations to improve the way they conduct business.

The value of information is directly linked to how it helps decision makers achieve the organization's goals.

The goal of information systems is to enable the organization to achieve its business objectives.

■ SUMMARY

1. *Distinguish data from information and describe the characteristics used to evaluate the quality of data.*

Data consists of raw facts; information is data transformed into some meaningful form. The process of defining relationships between data requires knowledge. Knowledge is an awareness and understanding of a set of information and how that information can be made useful to support a specific task. To be valuable, information must have several characteristics: it should be accurate, complete, economical to produce, flexible, reliable, relevant, simple to understand, timely, verifiable, accessible, and secure. The value of information is directly linked to how it helps people achieve their organization's goals.

2. *Name the components of an information system and describe several system characteristics.*

A system is a set of elements that interact to accomplish a goal or set of objectives. The components of a system include inputs, processing mechanisms, and outputs. Systems also contain boundaries that separate them from the environment and each other. Feedback is used by the system to monitor and control its operation to make sure it continues to meet its goals and objectives. Systems may be classified in many ways. They may be considered simple or complex. A stable, nonadaptive system does not change over time, while a dynamic, adaptive system does. Open systems interact with their environments; closed systems do not. Some systems exist temporarily; others are considered permanent.

System performance is measured by its efficiency and effectiveness. Efficiency is a measure of what is produced divided by what is consumed; effectiveness is a measure of the extent to which a system achieves its goals. A systems performance standard is a specific objective.

Information systems (ISs) are sets of interrelated elements that collect (input), manipulate and store (process), and disseminate (output) data and information. Input is the activity of capturing and gathering new data; processing involves converting or transforming data into useful outputs; and output involves producing useful information. Feedback is the output that is used to make adjustments or changes to input or processing activities.

3. *Identify the basic types of business information systems and discuss who uses them, how they are used, and what kinds of benefits they deliver.*

Information systems play an important role in today's businesses and society. The key to understanding the existing variety of systems begins with learning their fundamentals. The types of systems used within organizations can be classified into four basic groups: (1) TPS, (2) MIS, (3) DSS, and (4) AI/ES. The most fundamental system is the transaction processing system (TPS). A transaction is any business-related exchange. The TPS handles the large volume of business transactions that occur daily within an organization. Another important system type is the management information system (MIS), which uses the information from a TPS to generate information useful for management decision making. Management information systems produce a variety of reports. Scheduled reports contain prespecified information and are generated on a regular basis. Demand reports are generated only at the request of the user. Exception reports contain listings of items that do not meet a predetermined set of conditions. A decision support system (DSS) is an organized collection of people, procedures, databases, and devices used to support problem-specific decision making. The DSS differs from an MIS in the support given to users, the decision emphasis, the development and approach, and system components, speed, and output. Still another type of system designed to support management

decision making, the expert system (ES), is designed to act as an expert consultant to a user who is seeking advice about a specific situation.

4. *Identify the major steps of the systems development process and state the goal of each.*

Systems development involves creating or modifying existing business systems. The major steps of this process and their goals include systems investigation (gain a clear understanding of what is the problem); systems analysis (define what the system must do in order to solve the problem); systems design (determine exactly how the system will work to meet the business needs); implementation (create or acquire the various systems components defined in the design step); and maintenance and review (maintain and then modify the system so that it continues to meet changing business needs).

5. *Discuss why it is important to study and understand information systems.*

Our society is becoming dependent on information technology. Computer and information systems literacy are prerequisites for numerous job opportunities, not only in the IS field. Computer literacy is a knowledge of computer systems and the way they function; information systems literacy is knowledge of how data, information, and information systems are used by individuals and organizations. Effective information systems can have a major impact on corporate strategy and organizational success. Businesses around the globe are enjoying better safety and service, greater efficiency and effectiveness, reduced expenses, and improved decision making and control because of information systems. Individuals who can help their businesses realize these benefits will be in demand well into the future.

■ KEY TERMS

artificial intelligence (AI) 27	E-commerce 22	information 5
computer literacy 30	effectiveness 11	information system (IS) 4
data 5	efficiency 11	information systems literacy 30
database 19	feedback 16	input 15
decision support system (DSS) 26	forecasting 16	Internet 19

intranet 19
knowledge 6
knowledge base 6
management information
  system (MIS) 25
model 13
networks 19
output 16
process 6

processing 15
system 8
system boundary 9
system parameter 13
system performance standard 11
system variable 13
systems analysis 29
systems design 29
systems development 29

systems implementation 29
systems investigation 29
systems maintenance and review 29
technology infrastructure 18
telecommunications 19
transaction 21
transaction processing
  system (TPS) 21

■ REVIEW QUESTIONS

1. What are some of the ways information systems can improve our lives?
2. Identify six characteristics of valuable information.
3. Define the term *system*. What is the difference between an open system and a closed system?
4. How have organizations changed as a result of the use of information systems?
5. How is system performance measured?
6. List and define four types of models. What is the purpose of using a model?
7. What is an information system?
8. Define *input, processing, output,* and *feedback* as they relate to information systems.
9. Describe the six components of a computer-based information system (CBIS).
10. What is the difference between a transaction processing system (TPS) and a management information system (MIS)?
11. What is a decision support system (DSS)? How is it different from an expert system?
12. What are computer literacy and information systems literacy? Why are they important?
13. What are some of the benefits organizations seek to achieve through using information systems?
14. What is the difference between a process and a procedure?
15. Identify the five steps in the systems development process and state the goal of each.

■ DISCUSSION QUESTIONS

1. Why is the study of information systems important to you? What do you hope to learn from this course to make it worthwhile?
2. What is the technology infrastructure of an organization? Give an example of one.
3. It is sometimes said that one person's information is another person's data. Explain and provide at least one example.
4. You are an architect assigned the task of designing a new high-rise office building. Why would you want to build a model of the building? What kinds of models would you create? Why might you create more than one type of model?
5. Describe the "ideal" automated class registration system for students at a university. Describe the input, processing, output, and feedback associated with this system.
6. Imagine that you are responsible for planning the marketing for one of the brands of a major consumer goods company. Describe some of the key decisions that you must make. Discuss the kinds of models and systems you would expect to have support you in your day-to-day work and planning activities.
7. What are the characteristics of valuable information? Give an example for each characteristic. Obtaining valuable information can involve trade-offs among these characteristics. Give an example of a potential trade-off.
8. Discuss how information systems are linked to the business objectives of an organization.

■ PROBLEM-SOLVING EXERCISES

1. Search through several business magazines (*Business Week, Computerworld, PC Week,* etc.) for a recent article that discusses the use of information technology to deliver significant business benefits to an organization. Now use other resources to find additional information about the same organization (Reader's Guide to Periodical Literature, on-line search capabilities available at your school's library, the company's pubic relations department, Web pages on the Internet, etc.). Use word processing software to prepare a one-page summary of the different resources you tried and their ease of use and effectiveness.

2. Prepare a data disk and a backup disk for the problem-solving exercises and other computer-based assignments you will complete in this class. Create one directory for each chapter in the textbook (you should have 14 directories). As you work through the problem-solving exercises and complete other work using the computer, save your assignments for each chapter in the appropriate directory. On the label of each disk be sure to include your name, course, and section. On one disk write "Working Copy," on the other write "Backup."

3. Create a simple spreadsheet to track your courses and grades throughout your college career.

a. Include in it the ability to automatically compute your grade-point average when given your letter grade for a course. If your school is on the +/– system, be sure to assign the correct point value for each grade. Don't forget to include the course number and the number of quarter or semester credit hours you received if these are necessary for the computation. Your spreadsheet might look something like the following:

Course	Hrs	Com-pleted	Grade	QP	QPA	Max QPA
Common						
Core (2)						
IS 100	3	3	A	12.0	4.0	12.0
IS 150	3	3	A-	11.1	3.7	11.1
IS Core (3)						
IS 200	3	3	B+	9.9	3.3	9.9
IS 250	3	3	B	9.0	3.0	9.0
IS 300						
Options (4)						
IS 620						
Project (2)						
IS 500						
IS Totals						

b. Enter sample data to test the formulas in your spreadsheet.

c. In this chapter, you learned about input, processing, and output. List the tasks you just completed to build the spreadsheet and categorize them as input, processing, or output.

4. Read the Help Wanted section of the Sunday newspaper from your hometown, the nearest metropolitan area, or an area of the country in which you might want to live. Using your word processor, prepare a report describing the number and types of information systems jobs available. In addition, note how many other occupations require some information systems technology and skills. Report your findings to the class.

■ TEAM ACTIVITY

1. Before you can do a team activity, you need a team! The class members may self-select their teams, or the instructor may assign members to groups. Once your group has been formed, meet and introduce yourselves to each other. You will need to find out the first name, hometown, major, and e-mail address and phone number of each member. Find out one interesting fact about each member of your team as well.

With the other members of your group, use word processing software to write a one-page report on how information technology has an impact on your daily lives. Send the report to your instructor via e-mail.

■ WEB EXERCISE

Throughout this book, you will see how the Internet provides a vast amount of information to individuals and organizations. We will stress the World Wide Web, or simply the Web, which is an important part of the Internet. Most large universities and organizations have an address on the Internet, called a Web site or home page. The address of the Web site for this publisher is http://www.course.com. You can gain access to the Internet through a browser, such as Internet Explorer or Netscape. Using an Internet browser, go to the Web site for this publisher. What did you find? Try to obtain information on this book. You may be asked to develop a report or send an e-mail message to your instructor about what you found.

■ CASES

1 Haworth Improves Furniture and Its Order Process

Haworth, Inc., is the second-largest office furniture maker in the world, with recent sales of $1.5 billion annually. Haworth operates internationally with offices, showrooms, dealers, and manufacturing facilities in virtually every global market: Europe, North America, Central America, South America, the Caribbean, Asia, and the Middle East. Based in Holland, Michigan, the international company employs 10,000 and has more than 60 showrooms worldwide. Haworth customers are both individuals and businesses. Corporations represent Haworth's core business, but it also provides products to institutions, theaters, hospitals, and schools.

Haworth credits its standing in the industry to its emphasis on customer needs. For example, Haworth's Next Generation Seating design team (a collection of designers, engineers, neurologists, and marketing experts) spent hundreds of hours researching how the body and mind affect one another while a person works. They uncovered some fascinating information: When people move while sitting they are considerably more alert and more productive. They also have fewer back problems associated with poor posture from lack of movement.

Designing office space for clients can be complicated. It requires analysis of many variables and understanding of customer needs. The baffling array of possible Haworth furniture combinations is so complex that many customers don't know exactly what they've bought until it's delivered. For example, an office chair alone could be assembled in 200 different ways with many possible color combinations. Until recently, the salesperson and computer-assisted design (CAD) operator typically had to go through a CAD mock-up of an order several times—with the sales representative returning to the customer each time to show the mock-up. Only after the last CAD mock-up was approved was the CAD workstation software used to create a bill of materials that went to Haworth's factory for manufacturing. What's more, having the sales representative relay information from the customer to the CAD operator sometimes allowed errors to creep into the mock-up.

Today, before the first piece of furniture even gets through a customer's door, Haworth's sales force uses three-dimensional software to provide a preview of what the customer's offices will look like, as well as an estimate of the project's total cost. On a typical $15,000 project, the system also provides pricing estimates accurate to within a couple hundred dollars.

Haworth has equipped sales representatives with computer-visualization software from the Trilogy Development Group. The software allows sales representatives with laptop computers to show a customer exactly what's being ordered and how it will look when assembled and installed at the customer's site. The Trilogy visualization software reduces the number of trips back and forth between the sales representatives, the customer, and the CAD operator and decreases the number of errors. Sales representatives who use the laptops equipped with Trilogy software can configure clusters of as many as 10 cubicles and provide the customer with a close cost estimate of what has been created on the laptop screen—all without returning to the dealership. The result is that the salesperson can do "what ifs" with the client all day long.

Haworth's executives hope that the Trilogy application, called the Sales Builder Engine, will shorten the company's sales cycle by eliminating some of the repetitive CAD work and repeated trips to the customers' sites by sales representatives. The software will also make the company's huge parts catalog more easily understood and increase order accuracy. Haworth spent more than $1 million on the Trilogy software.

The Trilogy application shouldn't be confused with traditional sales-force automation, which typically involves providing a salesperson a laptop computer and software tools for expense reporting, sales tracking, and e-mail. With the Trilogy software, Haworth is re-engineering its sales by shifting detailed parts-assembly information from the corporate level to the customer level. Haworth's project is one of several the furniture company is relying on to stay competitive with its number one rival, Steelcase, Inc. That company is also pushing cutting-edge technology for its sales force.

Haworth is an unusual customer for Trilogy because of its effort to get customers involved in the sales effort. Rather than wanting the salesperson to fill out a form and present the customer with a bill of materials to be ordered, Haworth wants its customers to be deeply involved in the whole process of purchasing office furniture.

1. What potential problems could be associated with the rollout of this new technology? What would you recommend to avoid these potential problems?
2. What additional features and capabilities might be added to this sales support system to meet customer needs even better?

Sources: Adapted from Kim Girard, "Want to See That Desk in 3-D?" *Computerworld*, April 6, 1998, pp. 55–56; Haworth, Inc. Web page at: http://www.haworth-furn.com accessed on April 12, 1998; and Trilogy Web page at http://www.trilogy.com accessed on April 12, 1998.

2   Sales Force Automation— Potential without a Payoff?

Software that helps salespeople keep track of customer information and sales calls has generally fallen short of its potential to boost sales, reduce costs, cut sales cycle time, and build better customer relationships. In short, it often does not improve business processes. Indeed, almost two-thirds of sales force automation projects end in failure. Why? The root cause of the problem seems to be a fundamental incompatibility between sales force culture and information systems. This results in a real mismatch between what the product is supposed to do and how the sales force chooses to use it. Users often complain that the software is difficult to use and requires entering too much data. So, they don't want to use the software at all. All too often, companies buy the leading software package with little thought to how well it fits with the way their company's salespeople do their job. Time spent entering customer data can reduce customer contact time. As a result, the software does not truly address sales issues and work styles, leading to frustrated users and plummeting profits. In some companies, moving from a manual to a computer-based system causes valuable salespeople to leave.

Imagine that you are a consultant working with a company about to embark on a sales force automation project. The company has asked your advice on the following topics:

1. Who should be involved in the project?
2. Using concepts presented in this chapter, how can the company avoid having its project end in failure?

Sources: Adapted from Tom Stein, "Software for the Hard Sell," *InformationWeek*, March 2, 1998, p. 139; Kim Girard, "Sales Force Automation, Users Clash," *Computerworld*, April 6, 1998, p. 6; and Kim Girard, "Stats Not Good for Sales Technology," *Computerworld*, April 6, 1998, p. 29.

3   Liz Claiborne Upgrades Its Information Systems

Liz Claiborne designs and markets an extensive range of women's fashion apparel and accessories, with versatile collections ranging from casual to dressy. The company also designs and markets men's apparel and furnishings, as well as fragrances for women and men. Net sales for a recent year were a record $2.4 billion.

It should come as no surprise to anyone who has ever tried to keep up with fashion trends that change is driving the apparel industry today. But change is occurring more than in the design of clothes. The structure and nature of retailing and manufacturing are also shifting. Geographical boundaries are disappearing. Limitations are dissolving. Above all, consumers today look for versatility and value, and they, not the retailers or manufacturers, define what constitutes those qualities. For example, the move toward casual dress is an attempt to simplify increasingly complex lives. These shifting priorities mean consumers are less loyal to brands or to stores, but more discerning and very time-constrained.

To keep pace with the rate of change, Liz Claiborne has put all business processes under the microscope. As a result, it is concentrating on streamlining the things it does best and teaming with others through licensing and outsourcing arrangements to perform activities in which it has less expertise, from marketing watches to producing footwear and home furnishings. Specific corporate goals include doubling revenue to more than $4 billion by the year 2000, cutting operations costs by $35 million per year, reducing time from product design to availability, and improving communications with customers.

To achieve these goals, Liz Claiborne is undergoing a major technology overhaul that will result in replacing over 80 percent of its business processes, business information systems, hardware, software, databases, and network capabilities. Even IS people are affected as they get training in the new technology and new roles are identified for current IS staff. Surprisingly, a key challenge to this transformation process is not installing or maintaining the new technology, but aligning technology and business needs, and teaching people how to cope with the change. "What many organizations don't realize is that if you don't manage the business part of a technology change, you can fail even if the technology part succeeds," states Naomi Karten, an adviser to the company.

Liz Claiborne has developed Web-based tools to improve communications with suppliers and retailers. A Web-based application allows retailers to track purchase orders and to check the status of transactions instantly—a process that used to be done over the phone. More than 60 percent of customer orders are now placed electronically. The company also invested heavily in software to track materials around the world and to help communicate better with service providers, manufacturing partners, and freight consolidators. The technology changes have also affected the design process. In the past, Claiborne relied on pen-based sketches from external organizations for designs. Now the company uses sophisticated software tools to help in the design process. While the company used to fly in retailers to view its new designs, it now sends the designs electronically via the Web, saving time and costs. This networking technology permits the global transfer of textile and design information and is dubbed LizCADalyst.

1. What do you think Karten means when she says: "What many organizations don't realize is that if you don't manage the business part of a technology change, you can fail even if the technology part succeeds"? How does this apply to Liz Claiborne?
2. Liz Claiborne has made a substantial investment in upgrading its information systems. If you had to justify this investment to the board of directors, what would you say?

Sources: Adapted from Jaikumar Vijayan, "IT Overhaul May Boost Fashion Profit," *Computerworld*, April 13, 1998, pp. 55–56; Tom Stein, "Going Global," *InformationWeek*, February 2, 1998, pp. 84–85; and Liz Claiborne Web page at http://www.lizclaiborne.com accessed on April 15, 1998.

4 Ticketmaster Sells in Cyberspace

Ticketmaster normally sells five million tickets a month through its phone and retail sales operation. In December 1996, its first month of Web sales, on-line visitors bought fewer than 18,000 tickets. However, Ticketmaster is patient and is taking a longer two-to-three year perspective for its on-line sales. Ticketmaster views the Web as a channel of commerce that will eventually complement its existing phone and retail sales outlets, not replace them. The goal is to combine databases and system operations on 17 similar, but separate, systems around the country. Unfortunately, this task is turning out to be more difficult and taking longer than expected—ten people working full time are needed to support the Web site.

In addition to ticket sales, the Web site hosts on-line "chats" with major artists such as LL Cool J, Counting Crows, and Tori Amos. It also publishes an electronic magazine called *Live Vibes* and provides a direct connection to the Internet Travel Network to help fans book flights to concerts. The Web site also enables customers to view theater seating charts, something not possible over the phone.

1. Visit this site at http://www.ticketmaster.com and write a brief report of what you like about it and how it might be improved.
2. If you were the senior vice-president of multimedia responsible for this "Web experiment," what success criteria would you set for deciding if the site should be continued?

Sources: Adapted from Rebecca Rohan, "Online at the Box Office," *Internet World*, May 1997, pp. 37–38; and the Ticketmaster Web page at http://www.ticketmaster.com.

CHAPTER 2

Information Systems in Organizations

A key to business success is to "build our products and organization according to the best interests of our customers."

—Lou Gerstner,
CEO of IBM

Chapter Outline

Organizations and Information Systems
  Organizational Structure
  Organizational Culture and Change
  Reengineering
  Continuous Improvement
  Total Quality Management
  Outsourcing and Downsizing
Competitive Advantage
  Factors That Lead Firms to Seek Competitive Advantage
  Strategic Planning for Competitive Advantage
Performance-Based Information Systems
  Productivity
  Return on Investment and the Value of Information System
  Justifying Information Systems
Careers in Information Systems
  Roles, Functions, and Careers in the Information Systems
    Department
  Other IS Careers

Learning Objectives

After completing Chapter 2, you will be able to:

1. Identify the seven value-added processes in the supply chain and describe the role of information systems within them.
2. Provide a clear definition of the terms *organizational structure*, *culture*, and *change* and discuss how they affect the implementation of information systems.
3. Identify some of the strategies employed to lower costs and improve service.
4. Define the term *strategic competitive advantage* and discuss how organizations are using information systems to gain such an advantage.
5. Discuss how organizations justify the need for information systems.
6. Define the types of roles, functions, and careers available in information systems.

Federal Express
Redefines Its Services to Maintain Competitive Advantage

Federal Express (FedEx), with headquarters in Memphis, has a fleet of 40,000 ground vehicles and 600 airplanes. It generates annual revenues in excess of $11 billion. The firm also has an impressive information system architecture that is driving FedEx's transformation from a package delivery company to a strategic provider of E-commerce, logistics, and other supply-chain services.

FedEx knows that building and leveraging its information systems and networks is key to its success in the twenty-first century. As a result, it spends about $1 billion a year on information technology. FedEx is not only reorganizing its internal operations around a more flexible technology infrastructure, but it is also attracting new customers and in many cases locking in existing customers with an unprecedented level of technology integration.

The Web package-tracking capabilities pioneered by FedEx have now become an industry norm rather than a competitive advantage. All major transportation and delivery companies, from United Parcel Service to Ryder System, are making major investments in information technology. Where FedEx is different is that it is using information technology to transform itself from a delivery service to a vital link in today's networked and increasingly electronic economy. FedEx seeks to become a fully integrated corporate partner that picks up, transports, warehouses, and delivers all of a company's finished goods from the factory floor to the customer's receiving dock—with status data available every step of the way.

Large companies such as National Semiconductor Corp. have hired FedEx to handle most of their warehousing and distribution operations. Today, virtually all of National Semiconductor's products, manufactured in Asia by three National Semiconductor factories and three subcontractors, are shipped directly to a FedEx distribution warehouse in Singapore.

National Semiconductor's order-processing application, running on an IBM mainframe in Santa Clara, California, sends a daily batch of orders directly to FedEx's inventory-management system running on a Tandem computer in Memphis. At this point, FedEx takes over; the orders are forwarded to the FedEx warehouse management application in Singapore, where they are fulfilled in a FedEx warehouse and shipped directly to customers via FedEx. Except for receiving a confirmation that the order was filled, National Semiconductor is done with the order transactions. National Semiconductor has gained significant benefits: the average customer delivery time has been reduced from four weeks to seven days, distribution costs have been cut from 2.9 to 1.2 percent of sales. In addition, seven regional warehouses in the United States, Asia, and Europe were closed, saving National Semiconductor costs for warehouse space and employees.

The tight information technology links between FedEx and National Semiconductor exemplify FedEx's strategy of technology integration with its corporate customers. FedEx stores the product, operates the warehouse, processes the order, and then hands it off to the carrier—which, of course, is FedEx.

FedEx is not alone in using information technology to move beyond package delivery. UPS, which has spent $9 billion on IT since 1986, has already formed five alliances to help disseminate its logistics software among E-commerce users with UPS providing order-entry, catalog, and inventory management. Ryder System has formed an alliance with IBM and Andersen Consulting to deliver logistics services to customers. IBM will lend technology expertise and Andersen its consulting personnel to Ryder projects worldwide.

As you read this chapter, consider the following questions:

- What is meant by strategic competitive advantage?

- How can information technology help a company to gain and maintain a strategic competitive advantage?

Sources: Adapted from Monua Janah and Clinton Wilder, "Special Delivery," *InformationWeek*, October 27, 1997, pp. 42–60; FedEx Web site at http://www.fedex.com, accessed April 8, 1998.

Technology's impact on business is growing steadily. Once used to automate manual processes, technology has now transformed the nature of work—and the shape of organizations themselves. During the late 1960s and early 1970s, many computerized information systems were developed to provide reports for business decision makers. The information in these reports helped managers monitor and control business processes and operations. For example, reports that listed the quantity of each inventory item in stock could be used to monitor inventory levels. Unfortunately, many of these early systems did not take the overall goals of the organization and managerial problem-solving styles into consideration. Some decision makers wanted detailed inventory reports of all the items, while others wanted a list of inventory items only when the number on hand was very low. Even more important, these early systems were not developed as part of the business process itself. As a result, many of the early systems failed or were not utilized to their potential. Today, businesses recognize that both important organizational concepts and processes must be considered and supported by effective information systems.

Business organizations use information systems for a number of purposes. Their use is strongly influenced by a business's organizational structure and the way particular businesses seek to achieve their goals. FedEx, for example, redefined the services it provides to reduce its customers' total costs. It linked its information systems with its customers' to keep products moving through the supply pipeline. FedEx's information systems did more than provide valuable shipment tracking information; its system functioned as part of the distribution process and actually changed the process itself.

To help provide an understanding of how information systems shape and are shaped by organizational structure, we now examine organizations and information systems.

ORGANIZATIONS AND INFORMATION SYSTEMS

organization

a formal collection of people and other resources established to accomplish a set of goals

An **organization** is a formal collection of people and other resources established to accomplish a set of goals. The primary goal of a for-profit organization is to maximize profits by increasing revenues or reducing costs. Nonprofit organizations include social groups, religious groups, universities, and other organizations that do not have profit as the primary goal.

An organization is a system. Money, people, materials, machines and equipment, data, information, and decisions are constantly in use in any organization. As shown in Figure 2.1, resources such as materials, people, money, and so forth are input to the organizational system from the environment, go through a transformation mechanism, and are output to the

FIGURE 2.1

A General Model of an Organization

Information system(s) support and work within all parts of an organizational process. Although not shown in this simple model, input to the process subsystem can come from internal and external sources. Just prior to entering the subsystem, data is external. Once it enters the subsystem, it becomes internal. Likewise, goods and services can be output to either internal or external systems.

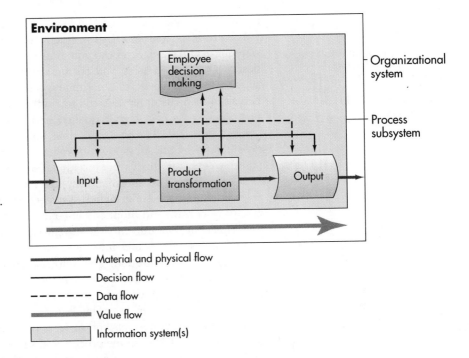

environment. The outputs from the transformation mechanism are usually goods or services. The goods or services produced by the organization are of higher relative value than the inputs alone. For example, the finished products delivered by FedEx (the output) are of greater relative value than the collection of inputs required to produce these goods. It is through this difference in value or worth that organizations attempt to achieve their goals.

How does this increase in value occur? Within the transformation mechanism, various subsystems contain processes that help turn specific inputs into goods or services of increasing value. These value-added processes increase the relative worth of the combined inputs on their way to becoming final outputs of the organization. Let us reconsider our simple car wash example from Chapter 1 (Figure 1.4). The first value-added process might be identified as washing the car. The output of this system—a clean, but wet, car—is worth more than the mere collection of ingredients (soap and water), as evidenced by the popularity of automatic car washes. Consumers are willing to pay more for the skill, knowledge, time, and energy required to wash their car. The second value-added process can be identified as drying—the transformation of the wet car into a dry one with no water spotting. Again, consumers are willing to pay more for the additional skill, knowledge, time, and energy required to accomplish this transformation.

In general, organizations establish these value-added processes to achieve their goals by exploiting opportunities and solving problems. At the beginning of the chapter, we saw how FedEx established a process to exploit an opportunity (provide better value to customers) and to solve a problem (the need to drive nonproductive costs out of the system).

All business organizations contain a number of value-added processes. Providing value to a stakeholder—customer, supplier, manager, or employee—is the primary goal of any organization. The value chain, first described by Michael Porter in a 1985 *Harvard Business Review* article, is a concept that

value chain

a series (chain) of activities that includes inbound logistics, warehouse and storage, production, finished product storage, outbound logistics, marketing and sales, and customer service

reveals how organizations can add value to their products and services. The **value chain** is a series (chain) of activities that includes inbound logistics, warehouse and storage, production, finished product storage, outbound logistics, marketing and sales, and customer service (Figure 2.2). Each of these activities is investigated to determine what can be done to increase the value perceived by a customer. Depending on the customer, value may mean lower price, better service, higher quality, or uniqueness of product. The value comes from the skill, knowledge, time, and energy invested by the company. By adding a significant amount of value to their products and services, companies will ensure further organizational success.

What role does an information system play in these value-added processes? A traditional view of information systems holds that they are used by organizations to control and monitor value-added processes to ensure effectiveness and efficiency. An information system can turn feedback from the value-added process subsystems into more meaningful information for employees' use within an organization. This information might summarize the performance of the systems and be used as the basis for changing the way the system operates. Such changes could involve using different raw materials (inputs), designing new assembly-line procedures (product transformation), or developing new products and services (outputs). In this view, the information system is external to the process and serves to monitor or control it.

A more contemporary view, however, holds that information systems are often so intimately intertwined with the underlying value-added process that they are best considered *part of* the process itself. From this perspective, the information system is internal to and plays an integral role in the process, whether by providing input, aiding product transformation, or producing output. Consider a phone directory business that creates phone books for international corporations. A corporate customer requests a phone directory listing all steel suppliers in Western Europe. Using its information system, the directory business can sort files to find the suppliers' names and phone numbers and organize them into an alphabetical list. The information system itself is an integral part of this process. It does not just monitor the process externally but works as part of the process to transform a product. In this example, the information system turns raw data input (names and phone numbers) into a salable output (a phone directory). The same system might also provide the input (data files) and output (printed pages for the directory).

The latter view brings with it a new perspective on how and why information systems can be used in business. Rather than searching to understand the value-added process independently of information systems, we consider the potential role of information systems within the process itself, often leading to the discovery of new and better ways to accomplish the process. Thus, the way an organization views the role of information systems will influence the ways it accomplishes its value-added processes.

3M Dental, a 1997 Malcolm Baldrige National Quality Award Winner, built an information system that tracks the purchasing decisions of dentists. The company's careful reading of customer requirements drives a finely tuned innovation process that delivers a steady stream of new or improved products.
(Source: Courtesy of 3M Dental.)

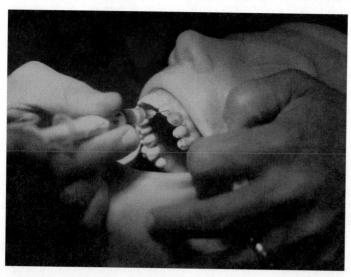

Organizational Structure

organizational structure

organizational subunits and the way they are related to the overall organization

Organizational structure refers to organizational subunits and the way they are related to the overall organization. Depending on the goals of the organization and its approach to management, a number of structures can be used. An organization's structure can have an impact on how information systems are viewed and what kind are used. Although there are many possibilities, organizational structure typically falls into one of these categories: traditional, project, team, or multidimensional.

traditional organizational structure

organizational structure in which major department heads report to a president or top-level manager

Traditional organizational structure. In the type of structure known as **traditional organizational structure**, major department heads report to a president or top-level manager. The major departments are usually divided according to function and can include marketing, production, information systems, finance and accounting, research and development, and so on (Figure 2.3). The positions or departments that are directly associated with the making, packing, or shipping of goods are called line positions. A production supervisor who reports to a vice president of production is an example of a line position. Other positions may not be directly involved with the formal chain of command but may assist a department or area. These are staff positions, such as a legal counsel reporting to the president.

The traditional organizational structure is often referred to as a *hierarchical structure*, since it can be viewed as a series of levels, with those at higher levels having more power and authority within the organization. Today, the number of management levels, or layers, in the traditional organizational structure have been reduced. A structure with a reduced number of management

FIGURE 2.2

The Value Chain of a Manufacturing Company

The management of raw materials, inbound logistics, and warehouse and storage facilities is called *upstream management*, and the management of finished product storage, outbound logistics, marketing and sales, and customer service is called *downstream management*.

Upstream management

Raw Materials → Inbound Logistics (Inbound Tracking Systems) → Warehouse and Storage (Raw Material Inventory Control Systems) → Production (Process Control Systems)

Customer Service (Customer Service Tracking and Control Systems) ← Marketing and Sales (Promotion Planning Systems) ← Outbound Logistics (Distribution Planning Systems) ← Finished Product Storage (Automated Storage and Retrieval Systems)

Downstream management

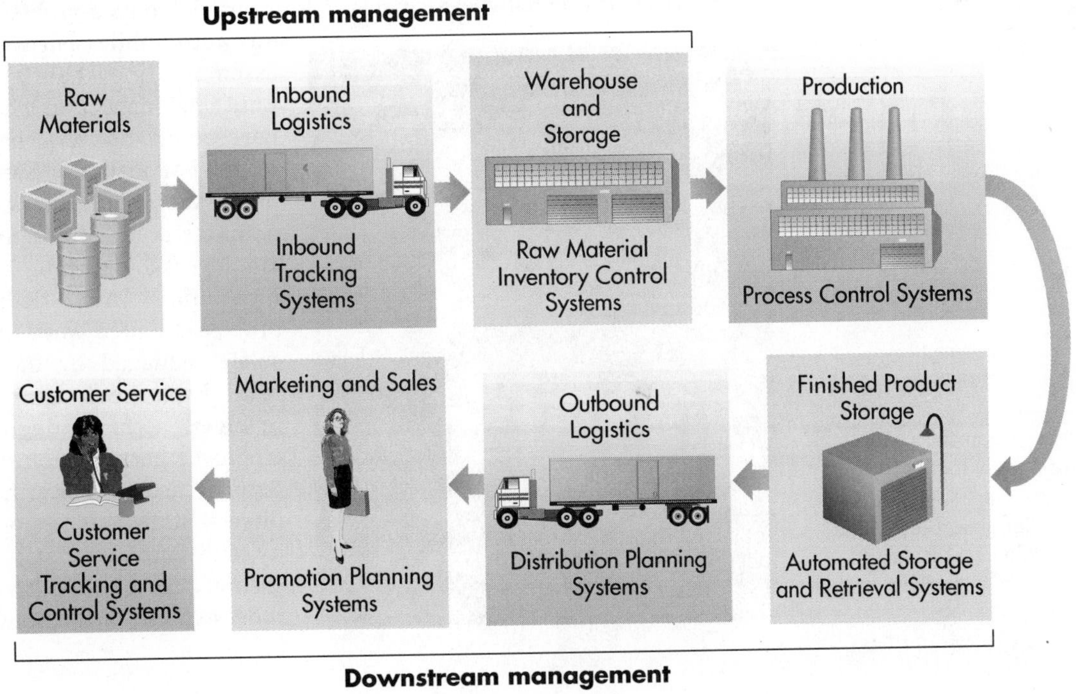

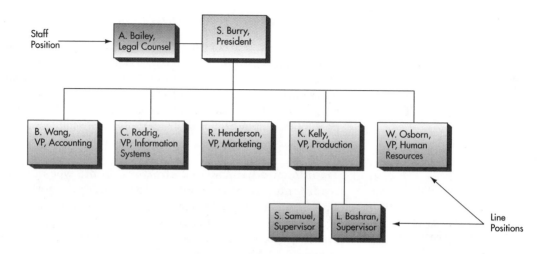

FIGURE 2.3

A Traditional Organizational Structure

flat organizational structure

organizational structure with a reduced number of management layers

empowerment

giving employees and their managers more power, responsibility, and authority to make decisions, take certain actions, and have more control over their jobs

layers, often called a **flat organizational structure**, is usually accomplished by empowering employees at lower levels to make decisions and solve problems without needing permission from midlevel managers. **Empowerment** involves giving employees and their managers more power, responsibility, and authority to make decisions, take certain actions, and have more control over their jobs. For example, an empowered salesclerk would be able to respond to certain customer requests or problems without needing permission from a supervisor. On the factory floor, empowerment can mean that an assembly-line worker has the ability to stop the production line to correct a problem or defect before the product is passed to the next station.

Empowerment usually results in faster action and quicker resolution of problems. It can also reduce costs and result in higher-quality products and services. It is usually less expensive, for example, to fix a problem with

an automobile part on an assembly line than to wait until final inspection when the part could be very hard to reach. For instance, Saturn, a highly successful American car manufacturer, empowered its employees by turning assembly lines into dedicated process-oriented work stations solely managed by the work team. Even the design process involves a high degree of employee participation. In the Saturn case, empowerment is directly linked to responsibility, and employees make suggestions for how to improve processes. In an empowered

Employees in Samsung Electronics' manufacturing facilities are empowered to stop the production line if they detect a problem or defect.
(Source: Courtesy of Samsung Electronics America, Inc.)

organization employees feel responsible beyond their own job; they feel the responsibility to make the whole organization work better.

Information systems can be a key element when companies empower their employees. Often, information systems make empowerment possible by providing information to employees at lower levels of the hierarchy. Corporate information systems may be used by a salesclerk to cancel an order and make adjustments to a customer's bill, by a factory floor operator to shut down the automated assembly line to fix a problem, or by midlevel managers to rapidly implement decisions. The employees may also be empowered to develop their own personal information systems, such as a simple forecasting model or spreadsheet. For example, Atrium Empowerment is a series of self-service applications that allow employees and managers to handle many functions formerly managed through the human resources department. The applications can be accessed by telephone, touchscreen kiosk, desktop, and web browser. All transactions are designed to allow for work flow and management approval. Atrium Empowerment is also integrated to human resource and payroll systems such as Ceridian Employer Services, PeopleSoft, MSA, Cyborg and many others, providing a complete solution to self-service.[1]

project organizational structure

structure centered on major products or services

Project organizational structure. A **project organizational structure** is centered on major products or services. For example, in a manufacturing firm that produces numerous types of baby food and products, each type is produced by a separate unit. Traditional functions like marketing, finance, and production are positioned within these major units (Figure 2.4). Many project teams are temporary. When the project is complete, the project members go on to new teams formed for the completion of some other project.

team organizational structure

structure centered on work teams or groups

Team organizational structure. The **team organizational structure** is centered on work teams or groups. In some cases, these teams are small; in other instances, they can be very large. Typically, each team has a team leader

FIGURE 2.4

A Project Organizational Structure

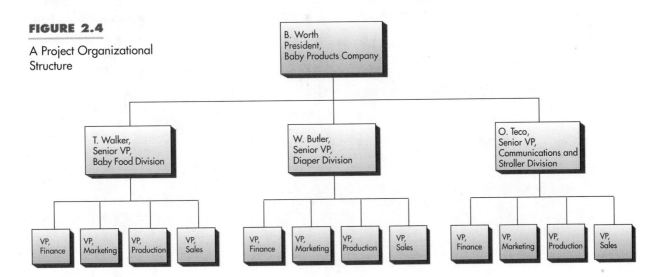

who reports to an upper-level manager in the organization. Depending on the tasks being performed, the team can be either temporary or permanent.

Multidimensional organizational structure. A **multidimensional organizational structure** may incorporate several structures at the same time. For example, an organization might have both traditional functional areas and major project units. When diagrammed, this structure forms a matrix, or grid (Figure 2.5).

One advantage of the multidimensional organizational structure is the ability to simultaneously stress both traditional corporate areas and important product lines. A potential disadvantage is multiple lines of authority. With this situation, an employee will have two bosses or supervisors: one functional boss and one project boss. As a result, conflicts may occur. One boss may want one thing, while the other boss may want something else. For example, the functional boss might want the employee to work on a new product in the next two days, while the project boss might want the employee to fly to a two-day meeting. Obviously, the employee cannot do both. One way to resolve this problem is to give one boss priority if there are problems or conflicts.

Organizational Culture and Change

Culture is a set of major understandings and assumptions shared by a group. **Organizational culture** consists of the major understandings and assumptions for a business, a corporation, or an organization. The understandings, which can include common beliefs, values, and approaches to decision making, are often not stated or documented in goal statements or formal policies. Employees, for example, might be expected to be clean-cut, wear conservative outfits, and be courteous in dealing with all customers. In some cases, organizational culture is formed over years. In a few cases, organizational culture is formed rapidly by top-level managers—for example, implementation of a "casual Friday" policy.

Like organizational structure, organizational culture can have a significant impact on the development and operation of information systems

multidimensional organizational structure

structure that may incorporate several structures at the same time

culture

a set of major understandings and assumptions shared by a group

organizational culture

the major understandings and assumptions for a business, a corporation, or an organization

FIGURE 2.5

A Multidimensional Organizational Structure

Employees in each group may have two bosses—a project boss and a functional boss.

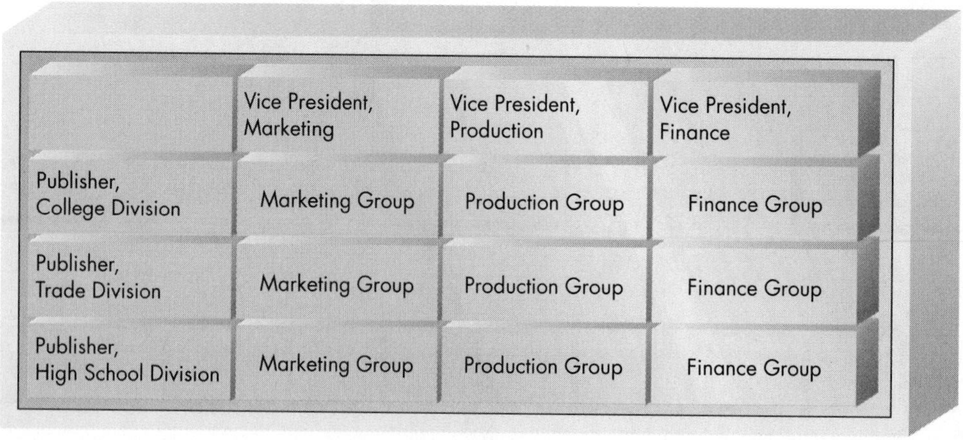

	Vice President, Marketing	Vice President, Production	Vice President, Finance
Publisher, College Division	Marketing Group	Production Group	Finance Group
Publisher, Trade Division	Marketing Group	Production Group	Finance Group
Publisher, High School Division	Marketing Group	Production Group	Finance Group

within the organization. A procedure associated with a newly designed information system, for example, might conflict with an informal procedural rule that is part of organizational culture. Organizational culture might also influence a decision maker's perception of the factors and priorities that must be considered in determining the objectives of a decision. For example, there might be an unwritten understanding that all inventory reports must be prepared before ten o'clock Friday morning. This understanding might cause the decision maker to reject an option to reduce costs by compiling the inventory report over the weekend.

organizational change

deals with how for-profit and nonprofit organizations plan for, implement, and handle change

Organizational change deals with how for-profit and nonprofit organizations plan for, implement, and handle change. Change can be caused by internal or external factors. Internal factors include activities initiated by all levels of employees. External factors include activities wrought by competitors, stockholders, federal and state laws, community regulations, natural occurrences (such as hurricanes), and general economic conditions. Introducing or modifying an information system will also cause change. Improving an organizational process through information systems requires changing the activities and tasks related to the process. Often, this means changing the way individuals, groups, and the entire enterprise work.

Overcoming the resistance to change can be the hardest part of bringing information systems into a business. Too many potential improvements have failed because managers and employees were not prepared for change. Occasionally, employees attempt to sabotage a new information system because they do not want to learn the required procedures and commands. In most of these instances, the employees were not involved in the decision to implement the change, nor were they fully informed about the reasons the change was occurring and the benefits that would accrue to the organization.

change model

a representation of change theories that identifies the phases of change and the best way to implement them

The dynamics of change can be viewed in terms of a change model. A **change model** is a representation of change theories that identifies the phases of change and the best way to implement them. Kurt Lewin and Edgar Schein propose a three-stage approach for change (Figure 2.6). *Unfreezing* is the process of ceasing old habits and creating a climate receptive to change. *Moving* is the process of learning new work methods, behaviors, and systems. *Refreezing* involves reinforcing changes to make the new process second nature, accepted, and part of the job.[2] When a company introduces a new information system, a few members of the organization must become agents of change. Understanding the dynamics of change can help them understand and overcome resistance, so that the new system can be used to maximum efficiency and effectiveness.

organizational learning

organizations adapt to new conditions or alter their practices over time

Organizational learning is closely related to organizational change. According to the concept of **organizational learning**, organizations adapt to new conditions or alter their practices over time. This means that assembly-line workers, secretaries, clerks, managers, and executives learn better ways of doing business and incorporate these approaches into their day-to-day activities. Collectively, these adjustments based on experience and ideas are called organizational learning. Read the "FYI" box for an example of two companies adapting their approach to marketing based on their understanding of how people are using the Web.

FYI

New Approach to Marketing

Web users' habits are changing profoundly, and the changes are transforming the Web from a publishing medium to a community medium. Users are surfing less and staying at sites where they find quality content. Increasingly, Web users themselves are creating content in the form of chat groups and user groups within on-line communities.

This shift in Web use is likely to make a huge impact on marketing and E-commerce, but how can marketers capitalize on the changes? Advertisers view chat rooms as a poor vehicle for selling. Chatters are usually engaged in their own conversations and are less likely to click on a banner ad. The solution for advertisers is to direct the discussions. Web experts encourage community building through moderated chats about goods and services. Charles Schwab and iVillage are pioneers in this approach to marketing.

iVillage is one of the most successful on-line communities. Established in 1995, its goal is to humanize cyberspace by providing a relevant and indispensable on-line experience for adult women. The company's flagship property, iVillage: The Women's Network at http://www.ivillage.com, is the world's largest destination for women, with more than 56 million pages viewed per month. Diverse people join the network to find others

with shared interests—to exchange information, advice, and support on the subjects in life that matter most, including parenting, work, and health. By actively participating in the network's well-known Parent Soup, ParentsPlace.com, Better Health, Armchair Millionaire, and other branded communities, members can learn from experts and from each other. They can also inspire other members to handle everyday challenges more effectively.

Charles Schwab & Co., the largest provider of on-line brokerage services, formed an alliance with iVillage to educate and empower people to become more informed investors. This Web site, dubbed The Investor Center, is located within iVillage's financial community, the Armchair Millionaire (http://www.armchairmillionaire.com). It includes a broad range of investing topics and tips and represents Schwab's first foray into developing and participating in an external on-line community that encompasses both novice and sophisticated investors. Both Schwab and the Armchair Millionaire community are committed to teaching people about the fundamentals and potential benefits of long-term investing. The sole intent is to show that the investing process does not have to be complicated. This site is truly an open forum for ideas, information, and peer-to-peer advice about investing culled from a variety of sources, including other members' experiences.

DISCUSSION QUESTIONS

1. Visit the Investor Center and write a paragraph summarizing your experience. Comment on the value of content, ease of use, and whether the site would encourage you to use the services of Charles Schwab.
2. Identify issues you think may have led to initial resistance to this marketing approach. How might these barriers have been overcome?

Sources: Adapted from "Charles Schwab & Co., Inc.," *PRNewswire*, March 20, 1998; Charles Schwab Web site at http://www.schwab.com, accessed April 16, 1998; iVillage Web site at http://www.iVillage.com, accessed April 16, 1998; "Chat Is Where It's At," *PC Magazine*, May 5, 1997.

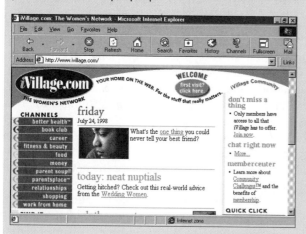

The change brought by information systems results in organizational learning. For instance, a company's decision-making process might become more data driven once employees know they can obtain accurate, complete, and relevant data for decision making. Organizational learning can also relate to work activities. A manager might find, for example, that a database is more efficient than a paper file for storing customer contacts.

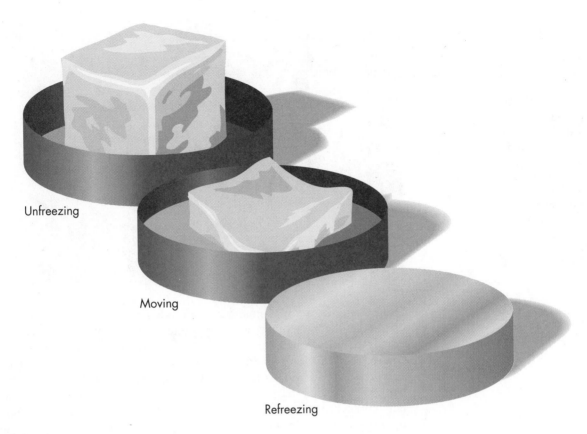

Unfreezing

Moving

Refreezing

FIGURE 2.6

A Change Model

reengineering (process redesign)

the radical redesign of business processes, organizational structures, information systems, and values of the organization to achieve a breakthrough in business results

Reengineering

To stay competitive, organizations must occasionally make fundamental changes in the way they do business. In other words, they must change the activities and tasks, or processes, that they use to achieve their goals. **Reengineering**, also called **process redesign**, involves the radical redesign of business processes, organizational structures, information systems, and values of the organization to achieve a breakthrough in business results (Figure 2.7). Reengineering can reduce delivery time, increase product and service quality, enhance customer satisfaction, and increase revenues and profitability.

A business process includes all activities, both internal (such as thinking) and external (such as taking action), that are performed to transform inputs into outputs. It defines the way work gets done. Despite the advent of the computer over 45 years ago, many organizations' sequential steps and assembly-line work processes have remained intact. In fact, many efforts to apply information systems have only further cemented the steps. For example, in some companies, a customer order is processed by several different people. The order moves from one step to the next, and people could make errors and create misunderstandings at any point. Other companies employ a customer service representative who oversees the entire process: taking the order, entering order processing data, and scheduling product delivery and setup. The customer service representative expedites and coordinates the process, and the customer has just one contact who always knows the status of the order.

Barclays Merchant Services completed its Darwin program, one of the largest new information system development projects ever undertaken at the bank. Handling close to 3 million card transactions from 130,000 merchants daily, Darwin involved reengineering support systems including customer services, collections, recruitment, settlement, charging, and statements. The Darwin system replaced a 25-year-old processing and servicing system. (Source: Courtesy of Barclays Merchant Services.)

This simple example illustrates the fundamental changes reengineering creates in the way things are done, often across multiple departments. Asking people to work differently often meets with stiff resistance, and change is difficult to maintain—the values of the organization and its employees must be changed also. In the previous example, the original work process may have evaluated employees on how many orders were entered each day. Under the reengineered process, they may be evaluated on different factors associated with customer service—percentage of orders delivered on time or accuracy of customer bills. Helping employees understand the benefits of the new system is a major hurdle.

In contrast to simply automating the existing work process, reengineering challenges the fundamental assumptions governing the design of work processes. It requires finding and vigorously challenging old rules blocking

FIGURE 2.7

Reengineering

Reengineering involves the radical redesign of business processes, organizational structure, information systems, and values of the organization to achieve a breakthrough in business results.

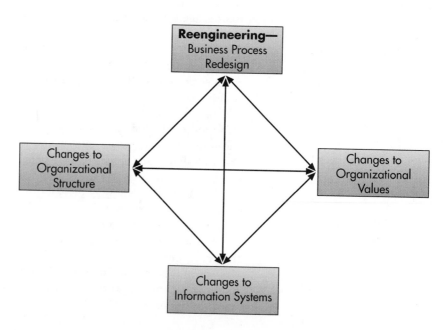

TABLE 2.1

Selected Business Rules That
Affect Business Processes

Rule	Original Rationale	Potential Problem
Small orders must be held until full-truckload shipments can be assembled.	Reduce delivery costs.	Customer delivery is slowed—lost sales.
No order can be accepted until customer credit is approved.	Reduce potential for bad debt.	Customer service is poor—lost sales.
All merchandising decisions are made at headquarters.	Reduce number of items carried in inventory.	Customers perceive organization has limited product selection—lost sales.

major business process changes. These rules are like anchors weighing a firm down and keeping it from competing effectively. Examples of such rules are given in Table 2.1. Today, many companies use reengineering to increase their competitive position in the market. Here are two examples.

Although many consumers are not yet ready to buy goods over the Internet, people who regularly use laptop batteries, cellular telephones, and other mobile devices are more receptive to E-commerce. So, although Internet sales are currently a relatively small percentage of many companies' sales, companies such as 1-800-Batteries expect more of their revenue to come from their Web site. As a result, companies are reengineering their sales processes to take advantage of this opportunity. Along the way, they are learning that the cost to process an order can be cut in half. Plus, orders entered by customers themselves tend to have fewer errors than those dictated over the phone. This type of reengineering enables firms to offer price discounts to E-commerce customers.[3]

Adverse drug reactions are a huge problem in terms of cost and quality of patient care. Nearly 25 percent of hospital patients suffer adverse drug reactions. Alameda Alliance for Health, an HMO, is reengineering its use of information systems to better support care providers. Using a software tool called MedIntelligence, Alameda's medical staff and pharmacists view a patient's records, including all prescription drugs the patient is taking. If it looks like the patient is taking conflicting medications or needs different medications, the system lets the medical professional know about it. At the same time, the software shows exceptions that allow doctors to prescribe drugs that Medicare might normally refuse. Doctors want to prescribe drugs without too many restrictions. The savings are passed on to local physicians as an inducement to take on more Medicare patients. By effectively using money that the state provides, Alameda can pay providers more for taking care of the sickest patients.[4]

Continuous Improvement

continuous improvement

constantly seeking ways to improve the business processes to add value to products and services

The idea of **continuous improvement** is to constantly seek ways to improve the business processes to add value to products and services. This in turn will increase customer satisfaction and loyalty and ensure long-term profitability. Manufacturing companies make continual product changes and

Business Process Reengineering	Continuous Improvement
Strong action taken to solve serious problem	Routine action taken to make minor improvements
Top-down-driven by senior executives	Worker-driven
Broad in scope; cuts across organizations	Narrow in scope; focus is on tasks in a given area
Goal is to achieve a major breakthrough	Goal is continuous, gradual improvements
Often led by outsiders	Usually led by workers close to the business
Information system integral to the solution	Information systems provide data to guide improvement team

TABLE 2.2

Comparing Business Process
Reengineering and Continuous
Improvement

improvements. Service organizations, as in the case of FedEx, regularly find ways to provide faster and more effective assistance to customers. By doing so, these companies increase customer loyalty, minimize the chance of customer dissatisfaction, and diminish the opportunity for competitive inroads.

Organizational commitment to goals such as continuous improvement can be supported by the strategic use of information systems. Continuous improvement involves constantly improving and modifying products and services to remain competitive and to keep a strong customer base. In doing so, companies can increase the quality of their products and services. Low-quality products can turn companies that once were the leaders in their industry into laggards that have lower profits and reduced market share; some have even gone out of business because of a perceived lack of quality products. Without question, quality will continue to be an important factor for profitability and survival. Table 2.2 compares reengineering and continuous improvement.

Total Quality Management

The definition of the term *quality* has evolved over the years. In the early years of quality control, firms were concerned with meeting design specifications—that is, conformance to standards. If a product performed as designed, it was considered a high-quality product. A product can perform its intended function, however, and still not satisfy customer needs. Today, **quality** means the ability of a product (including services) to meet or exceed customer expectations. For example, a computer that not only performs well but is easy to maintain and repair would be considered a high-quality product. This view of quality is customer oriented. A high-quality product satisfies customers by functioning correctly and reliably, meets needs and expectations, and is delivered on time with courtesy and respect.

To help them deliver high-quality goods and services, some companies have adopted continuous improvement strategies that require each major business process to follow a set of total quality management guidelines. **Total quality management (TQM)** consists of a collection of approaches, tools, and techniques that offers a commitment to quality throughout the organization. TQM involves developing a keen awareness of customer needs, adopting a strategic vision for quality, empowering employees, and rewarding employees and managers for producing high-quality products. As a result, processes may be redefined and restructured. General Electric Aircraft Engines adopted a variation of TQM and called it the Six Sigma effort. The goal is to achieve world-class quality of only 3.4 defects per million opportunities. Most companies, including GE, operate in the range of

quality

the ability of a product (including services) to meet or exceed customer expectations

total quality management (TQM)

a collection of approaches, tools, and techniques that offers a commitment to quality throughout the organization about three sigma, or about 60,000 defects per million opportunities. GE expects the program to save about $200 million per year.[5]

Information systems are fully integrated into business processes in organizations that adhere to continuous improvement or TQM strategies. Capturing and analyzing customer feedback and expectations and designing, manufacturing, and delivering quality products and services to customers around the world are only a few ways computers and information systems are helping companies pursue their goals of quality and continuous improvement.

Outsourcing and Downsizing

In an effort to control costs, organizations have looked at the number of people they have on the payroll. A significant portion of an organization's expenses go to hire, train, and compensate talented staff. So, organizations today are trying to determine the number of employees they need to maintain high-quality goods and services. With fierce competition in the marketplace, it is critical for organizations to use their resources wisely. Two strategies to contain costs are outsourcing and downsizing (sometimes called rightsizing).

outsourcing

contracting with outside professional services to meet specific business needs

Dupont & Company's global expansion of its energy operations and materials manufacturing prompted management to sign a contract with Computer Sciences and Andersen Consulting to support the necessary information systems to operate the overseas plants.

Outsourcing involves contracting with outside professional services to meet specific business needs. Often a specific business process is outsourced such as employee recruiting and hiring, development of advertising materials, product sales promotion, or global telecommunications network support. One reason organizations outsource a business process is to enable them to focus more closely on their core business—targeting limited resources to meet strategic goals. Other reasons for outsourcing are to obtain cost savings or to benefit from the expertise of the service provider.

Service providers can often add value to a process while cutting costs for a company. For example, pharmaceutical maker Eli Lilly & Company contracted management of its ambitious on-line healthcare network to EDS, conceding that a technology company is better suited for the effort. Thus a new joint effort, called Kinetra, is taking over Lilly's Integrated Medical Systems unit, which operates a private data network linking about 70,000 U.S. physicians to hospitals. Kinetra will expand the network to hundreds of thousands of physicians, pharmacies, hospitals, and healthcare players such as various state Blue Cross/Blue Shield organizations and Medicare systems. EDS already has existing relationships with the Blue Cross/Blue Shield organizations in 17 states, plus technical assets and expertise that Eli Lilly recognized it does not have.[6]

DuPont & Company has been a major user of outsourcing and recently signed a contract with Computer Sciences and Andersen Consulting valued at $4 billion. DuPont's goal was to provide greater speed and flexibility from its

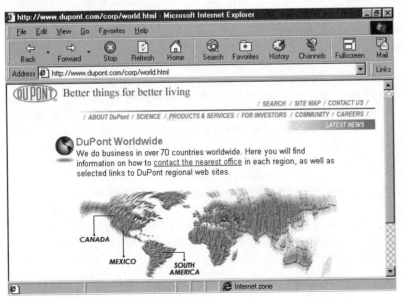

http://www.dupont.com/corp/world.html - Microsoft Internet Explorer

File Edit View Go Favorites Help

Back Forward Stop Refresh Home Search Favorites History Channels Fullscreen Mail

Address http://www.dupont.com/corp/world.html   Links

DUPONT Better things for better living

/ SEARCH / SITE MAP / CONTACT US /

/ ABOUT DuPont / SCIENCE / PRODUCTS & SERVICES / FOR INVESTORS / COMMUNITY / CAREERS /

LATEST NEWS

DuPont Worldwide
We do business in over 70 countries worldwide. Here you will find information on how to contact the nearest office in each region, as well as selected links to DuPont regional web sites.

CANADA

MEXICO

SOUTH AMERICA

Internet zone

IS organization than it could deliver alone. In addition, DuPont was embarking on a major overseas expansion, and management did not feel DuPont's IS group alone would be able to support the expansion.[7]

Companies considering outsourcing to cut the cost of their IS operations need to review this decision carefully. A growing number of organizations are finding that outsourcing does not necessarily lead to reduced costs. One of the primary reasons for cost increases is poorly-written contracts that result in additional charges from the outsourcing vendor for each additional task identified.

downsizing

reducing the number of employees to cut costs

Downsizing involves reducing the number of employees to cut costs. The term "rightsizing" is also used. Rather than pick a specific business process to be downsized, companies usually look to downsize across the entire company. Downsizing clearly drives down wages. However, there are often unwanted side effects. Employee morale hits rock bottom. Lines of communication within the company are weakened. Employee productivity drops. Often, high-priced consultants must be hired to help patch the business back together. The lost time, waning productivity, and devastated morale create hidden costs and can far outweigh the usual cost savings predicted from a layoff.

Employers need to be open to alternatives for reducing the number of employees, with layoffs viewed as the last resort. It is much easier and simpler to encourage people to leave voluntarily through early retirement or other incentives. Following this approach, the downsizing effort is accompanied with a "buyout package" offered to certain classes of employees (e.g., those over 50 years old). The buyout package offers employees certain benefits and cash incentives if they voluntarily retire from the company. Other options are job-sharing and transfers.

Under David Packard, Hewlett-Packard survived turbulent times by adopting a "fortnight" program: Every other Friday, it shut down almost all of its facilities and asked employees to take the day off without pay. Despite plenty opportunities to take on large defense contracts, Packard refused if it meant having to increase staff to handle the work and then lay off employees when it was complete. Other companies, such as Georgia-Pacific, avoid layoffs by using voluntary buyout programs. Employees have the option of accepting a certain amount of money for every year they have worked for the company; those who wish to retire can take the check and move on.[8]

The charge of age discrimination is frequently associated with downsizing, as older workers are most often affected by downsizing efforts. Companies describe their departure as "early retirement." As a result, job discrimination lawsuits have been filed against companies such as Pacific Telesis, AT&T, Digital Equipment, and IBM. Employers need a staff reduction policy that keeps laid-off employees from suing and winning in court. A key element is to develop and apply neutral, nondiscriminatory reduction criteria. Employees cannot be selected by a lottery nor can everyone over a certain age be downsized. Once an employer has established the criteria, they must be applied equally to all and with no exceptions. Otherwise, the employer runs the risk of a "disparate application" charge, which can lead to a judgment against the company.

COMPETITIVE ADVANTAGE

competitive advantage

a significant and (ideally) long-term benefit to a company over its competition

A **competitive advantage** is a significant and (ideally) long-term benefit to a company over its competition. Establishing and maintaining a competitive advantage is complex, but a company's survival and prosperity depend on its success in doing so.

Factors That Lead Firms to Seek Competitive Advantage

five-force model

a widely accepted model that identifies five key factors that can lead to attainment of competitive advantage including (1) rivalry among existing competitors, (2) the threat of new entrants, (3) the threat of substitute products and services, (4) the bargaining power of buyers, and (5) the bargaining power of suppliers

A number of factors can lead to the attainment of competitive advantage. Michael Porter, a prominent management theorist, suggested a now widely accepted **five-force model.** The five forces include (1) rivalry among existing competitors, (2) the threat of new entrants, (3) the threat of substitute products and services, (4) the bargaining power of buyers, and (5) the bargaining power of suppliers. The more these forces combine in any instance, the more likely firms will seek competitive advantage and the more dramatic the results of such an advantage will be.

Rivalry among existing competitors. The rivalry among existing competitors is an important factor leading firms to seek competitive advantage. Typically, highly competitive industries are characterized by high fixed costs of entering or leaving the industry, low degrees of product differentiation, and many competitors. Although all firms are rivals with their competitors, industries with stronger rivalries tend to have more firms seeking competitive advantage.

Threat of new entrants. The threat of new entrants is another important force leading an organization to seek competitive advantage. A threat exists when entry and exit costs to the industry are low and the technology needed to start and maintain the business is commonly available. For example, consider a small restaurant. The owner does not require millions of dollars to start the business, food costs do not go down substantially for large volumes, and food processing and preparation equipment is commonly available. When the threat of new market entrants is high, the desire to seek and maintain competitive advantage to dissuade new market entrants is usually high. The "E-commerce" box discusses some of the barriers to using E-commerce as a means for new entrants to sell their goods and services.

Threat of substitute products and services. The more consumers are able to obtain similar products and services that satisfy their needs, the more likely firms are to try to establish competitive advantage. Such an advantage often creates a "new playing field" in which "substitute" products are no longer considered as such by the consumer. Consider the personal computer industry and the introduction of low-cost computers. A number of consultants and computer manufacturers made much of the high cost of ownership associated with personal computers in the mid-1990s. They introduced low-cost network computers with minimal hard disk space, slower processing chips, and less main memory than some consumers

E-COMMERCE

Barriers to E-commerce

According to a survey conducted in March 1998 by CommerceNet, shoppers are not quite ready for E-commerce: they have trouble finding what they need, and there's no easy way to pay for things. Clearly, consumers are leery of E-commerce. As a result, many shoppers are not motivated to log on to the Internet, search for shopping sites, wait for the images to download, try to figure out the ordering process, and then worry about whether their credit card numbers may be stolen by a hacker. In addition, many Web shopping sites do not sell everything on-line that you can get from the company's printed catalog.

Business-to-business E-commerce is also encountering serious barriers. Businesses do not yet have good models for setting up their E-commerce sites, so they have trouble passing the orders and information collected on-line to their other transaction processing applications. Many companies continue to struggle with the idea of sharing proprietary business information with customers and suppliers, an important aspect of many business-to-business E-commerce systems.

DISCUSSION QUESTIONS

1. What do you think are the most serious issues with E-commerce—from a consumer-to-business standpoint and from a business-to-business standpoint?
2. What solutions do you propose for these issues?

Sources: Adapted from http://www.infowin.org/ACTS/, accessed April 8, 1998; Internet Commerce Explained section of IntCommSystems at http://www.int-comm.com, accessed September 21,1998.

---

desired, but at half the cost of a standard workstation. There was considerable interest in these new machines for awhile, but traditional personal computer manufacturers fought back. They developed a class of powerful workstations and implemented new pricing strategies to introduce these workstations for less than $1000. This eliminated the primary advantage of the stripped down network computers and regained lost customers.

Bargaining power of buyers and suppliers. Large buyers tend to exert significant influence on a firm. This influence can be diminished if the buyers are unable to use the threat of going elsewhere to influence the firm. Suppliers can help an organization obtain a competitive advantage. In some cases, suppliers have entered into strategic alliances with firms. When they do so, suppliers act like a part of the company. Using telecommunications, suppliers and companies can link their computers and personnel. This allows for fast reaction times and the ability to get the necessary parts or supplies when they are needed to satisfy customer needs.

Strategic Planning for Competitive Advantage

To be competitive, a company must be fast, nimble, flexible, innovative, productive, economical, and customer oriented. Given the five market forces just mentioned, Porter proposed three general strategies to attain competitive advantage: altering the industry structure, creating new products and services, and improving existing product lines and services. Subsequent research into the use of information systems to help an organization achieve a competitive advantage has confirmed and extended Porter's original work to include additional strategies—such as forming alliances with other companies, developing a niche market, maintaining competitive cost, and creating product differentiation.[9]

Altering the industry structure. Altering the industry structure is the process of changing the industry to become more favorable to the company or organization. This can be accomplished by gaining more power over suppliers and customers. Some automobile manufacturers, for example, insist that suppliers be located close to major plants and manufacturing facilities and that all business transactions be accomplished using electronic data interchange (EDI, direct computer-to-computer communications with minimal human effort required). This helps the automobile company control the cost, quality, and supply of parts and materials.

Creating barriers to new companies entering the industry can also be attempted. An established organization acquiring expensive new technology to provide better products and services can discourage new companies from getting into the marketplace. Creating strategic alliances may also have this effect. A **strategic alliance**, also called a **strategic partnership**, is an agreement between two or more companies that involves the joint production and distribution of goods and services. An example of a global strategic alliance is the merger of Chrysler and Daimler Benz to create a $130 billion automobile colossus known as Daimler-Chrysler AG. There is almost no overlap of products. Mercedes-Benz luxury cars compete in a market beyond Chrysler's more moderately priced offerings. Chrysler brings strength in minivans, profitable pickups, and sport-utility vehicles. Mercedes has hot sellers like the E-class sedan and SLK roadster. There is also little overlap in terms of geography—93 percent of Chrysler sales come from North America and 63 percent of Mercedes-Benz sales come from Europe.[10] With such alliances, existing rivals become partners, and each company benefits from the other's strengths. The power of the allied companies may discourage others in the industry from attempting to compete.

The pace of mergers and formation of strategic alliances is speeding up. Just five months into 1998, $679 billion worth of mergers and acquisitions had been announced in the United States, more than in all of 1996 and closing in on the total of $917 billion.[11] A number of mergers involving financial

**strategic alliance
(strategic partnership)**

an agreement between two or more companies that involves the joint production and distribution of goods and services

Chrysler Corporation's merger with Daimler-Benz will change the face of the automobile industry. Both companies will benefit through the exchange of components and technologies, combined purchasing power, and shared distribution logistics.
(Source: Courtesy of Chrysler Corporation.)

institutions were announced including Citibank and The Traveler's Group (creating the world's biggest financial institution with $700 billion in assets) and NationsBank and BankAmerica (creating the biggest bank in the United States). Several mergers were announced in the telecommunications industry including AT&T and TCI, SBC Communications and Ameritech, and Worldcom and MCI Communications.

Creating new goods and services. Creating new products and services is always an approach that can help a firm gain a competitive advantage. This is especially true of the computer industry and other high-tech businesses. If an organization does not introduce new products and services every few months, the company can quickly stagnate, lose market share, and decline. Companies that stay on top are constantly developing new products and services. For example, Compaq, Dell, IBM, Gateway, and other manufacturers are now delivering powerful personal computers that cost less than $1,000. New products are particularly important in the dynamic technology industry.

Improving existing product lines and services. Improving existing product lines and services is another approach to staying competitive. The improvements can be either real or perceived. Manufacturers of household products are always advertising new and improved products. In some cases, the improvements are more perceived than real refinements; usually, only minor changes are made to the existing product. Many food and beverage companies are introducing "Healthy" and "Light" product lines. A popular beverage company introduced "born on" dating for beer.

Using information systems for strategic purposes. The first IS applications attempted to reduce costs and to provide more efficient processing for accounting and financial applications, such as payroll and general ledger. These systems were seen almost as a necessary evil—something to be tolerated to reduce the time and effort required to complete previously manual tasks. As organizations matured in their use of information systems, enlightened managers began to see how IS could be used to improve organizational effectiveness and support the fundamental business strategy of the enterprise. Combining the improved understanding of the potential of information systems with the growth of new technology and applications has led organizations to use IS to gain a competitive advantage. In simplest terms, competitive advantage is usually embodied in either a product or service that has the most added value to consumers and that is unavailable from the competition or in an internal system that delivers benefits to a firm not enjoyed by its competition.

Although it can be difficult to develop information systems to provide a competitive advantage, some organizations have done so with success. A classic example is SABRE, a sophisticated computerized reservation system installed by American Airlines and one of the first CBISs recognized for providing competitive advantage. Travel

Sony will gain a distinct competitive advantage in the laptop computer market with its VAIO 505 notebook. Sony is one of the first manufacturers to offer a laptop that is less than one inch thick, weighs less than three pounds, and has the power and capabilities of a desktop computer.
(Source: Courtesy of Sony Electronics Inc.)

TECHNOLOGY ADVANTAGE
Omaha Steaks

The company began its mail order business in 1952 at the request of local meat lovers who wanted to send Omaha Steaks as gifts. It uses plenty of dry ice and a reusable, insulated cooler—a packaging combination designed to withstand any kind of weather. Omaha Steaks chose FedEx four years ago as its exclusive delivery service for mail-order steaks and other food because other rapid delivery carriers did not offer two-day air service. While others have since started such a service, Omaha Steaks has stayed with FedEx because of well-designed IT links with FedEx that allow Omaha Steaks' customer-service employees to easily track delivery status.

When customer orders—whether received via phone, mail, fax, the Web, or e-mail—are sent from Omaha Steaks' IBM AS/400 (midrange computer) to its warehouses, a copy of the order is sent to FedEx. Omaha Steaks generates a FedEx tracking label at the same time it prints its own shipping label. Next, Omaha Steaks delivers the warehouse-fulfilled orders by truck to the nearest FedEx regional hubs in Memphis, Indianapolis, or Columbus, Ohio. Once FedEx takes over, Omaha Steaks has full access to FedEx data on delivery status, planned routing, and planned delivery day.

Omaha Steaks, like many FedEx customers, also enables its customers to track their orders on its Web site by establishing a link to the FedEx Web-based tracking service.

This arrangement enables Omaha Steaks to deliver goods and services to its customers, and that's what they want. FedEx is very strong in the use of technology and gives Omaha Steaks very specific information that it can provide to its customers. Such information is a valuable tool in the marketplace.

DISCUSSION QUESTIONS

1. Visit the Omaha Steaks homepage at http://www.omaha-steaks.com. Find information about how customer packages are shipped and how you can track the status of your order.
2. Has Omaha Steaks achieved a permanent competitive advantage by outsourcing with FedEx? Do you think other shippers will take a similar approach? Why or why not?

Sources: Adapted from Monua Janah and Clinton Wilder, "Special Delivery," *Information Week*, October 27, 1997, pp. 42–60; Omaha Steaks Web page at http://www.omaha-steaks.com, accessed April 16, 1998.

agents used this system for rapid access to flight information, offering travelers reservations, seat assignments, and ticketing. The travel agents also achieved an efficiency benefit from the SABRE system. Because SABRE displayed American Airline flights whenever possible, it also gave the airline a long-term, significant competitive advantage.

A more recent example using information systems for strategic purposes is Kmart's use of a large central repository of data on its products to create a merchandising database. The database enables store managers to keep fast-selling items in stock, identify and purge slow-moving products, and increase individual store sales and profitability. Kmart lets suppliers access its data, treating them as extensions to its own purchasing department. The new system lets Kmart "micromanage" its local stores, while enhancing management's ability to analyze the big picture.[12]

Quite often, the competitive advantage a firm gains with a new information system is only temporary—competitors are quick to copy a good idea. So although the SABRE system was the first on-line reservation system, other carriers soon developed similar systems. Many retailers have also developed systems similar to Kmart's. However, SABRE has maintained a leadership position because it was the first system available, has been

Factors That Lead to Attainment of Competitive Advantage	Three General Strategies to Gain Competitive Advantage		
	Alter Industry Structure	Create New Products and Services	Improve Existing Product Lines and Services
Rivalry among existing competitors	Blockbuster Video changes nature of movie entertainment industry by allowing people to watch movies in their homes.	IBM, Oracle, and Sun Microsystems introduce new low-cost network computer.	Food and beverage companies offer "Healthy" and "Light" product lines.
Threat of new entrants	Chrysler and Daimler Benz form alliance to jointly manufacture and distribute autos.	Compaq enters the computer industry with one of the first portable computers.	Wendy's Old Fashioned Hamburgers competes in the fast-food market by offering improved hamburgers, fries, and other foods.
Threat of substitute products and services	Discount stock brokers change the structure of the investment industry by offering on-line trading at discounted prices.	Kmart uses large central repository of data on its products to create a merchandising database.	Laundry detergent manufacturers add bleach and/or fabric softeners to improve their base products.
Bargaining power of buyers	Auto manufacturers insist suppliers be located near plants and use EDI.	Investors and traders put pressure on the Chicago Board of Trade (CBOT) and other futures markets to offer new electronic trading services.	Distributors cause manufacturers to reduce order lead times, thus cutting inventory costs and improving service to customers.
Bargaining power of suppliers	American Airlines' SABRE system makes it possible for travel agents using this system to provide superior customer service.	Intel lowers chip prices and develops new multimedia computer chips. Intel supplies these chips to companies making personal computers.	Haworth, leading office furniture manufacturer, provides an automated tool that helps it design and visualize office systems, which ultimately improves customer products.

TABLE 2.3

Competitive Advantage Factors and Strategies

aggressively marketed, and has had continual upgrades and improvements over time. Maintaining a competitive advantage takes this sort of effort. Many companies have even instituted a new position—chief knowledge officer—to help them maintain a competitive advantage. Read the "Technology Advantage" box on page 61 to see how Omaha Steaks was able to gain a competitive advantage by outsourcing its delivery service and related information systems.

The extent to which companies are using computers and information technology for competitive advantage continues to grow. Forward-thinking companies must constantly update or acquire new systems to remain competitive in today's dynamic marketplace. FedEx, Charles Schwab, 1-800-Batteries, Alameda Alliance for Health, General Electric, Eli Lilly & Company, DuPont, American Airlines, and Kmart (all mentioned in this chapter) are examples of companies fighting to remain competitive. Table 2.3 lists several examples of how companies have attempted to gain a competitive advantage. The "Ethical and Societal Issues" box raises some interesting questions about two companies trying to gain a competitive advantage through sharing customer information.

ETHICAL AND SOCIETAL ISSUES

Merger Mania: Consumer Dream or Nightmare?

In April 1998, Citicorp and Travelers Group announced their intention to merge and form a new global leader in financial services. If approved, this combination joins two organizations whose core business is serving consumers, corporations, institutions, and governments globally, through a diverse array of sales and service channels. The merged company's principal operations will be traditional banking, consumer finance, credit cards, investment banking, securities brokerage and asset management, and property casualty and life insurance. The transaction is subject to a number of regulatory approvals, including the Federal Reserve Board, several state insurance commissions and various other bodies, as well as approvals by shareholders of both companies.

The combined company, to be named Citigroup, will serve over 100 million customers in 100 countries around the world. The company had recent year-end assets of almost $700 billion, net revenues of nearly $50 billion, operating income of approximately $7.5 billion, and equity of more than $44 billion, earning the number one ranking among the world's financial services companies.

The merger unites two leading financial services companies, creating a resource for customers like no other company in the world. Citigroup will be a diversified global consumer financial services company, a premier global bank, a leading global asset management company, a preeminent global investment banking and trading firm, and a broad-based insurance provider. The new company expects to generate substantial earnings from the cross-selling opportunities, as well as cost savings. For example, in the merged company, Traveler's Salomon Smith Barney financial could sell mutual funds and auto insurance to Citicorp banking customers. And Citicorp could offer home equity loans and credit-card services to Traveler's insurance clients.

Making the merged entity into a one-stop shopping experience requires tight systems integration to use each company's respective customer databases to their full capability. Citicorp and Travelers executives have yet to outline a plan for making cross-selling a reality. Furthermore, according to banking analysts, both companies have historically failed to persuade their business units to share proprietary customer data for that purpose. However, consolidation of customer information is inevitable.

Some consumer advocates fear that consolidation may hurt consumers. They claim that studies show larger banks mean higher fees, more fees, and higher balances required to avoid fees. They are also concerned that customers might come to a bank to get product A, but be offered products B, C, and D whether they want them or not. This will be irritating for some customers, while other customers may think that if they do not get multiple products, they may not be able to get them separately.

Discussion Questions

Imagine that you are the leader of a transition team set up to assess sharing of customer information.

1. Summarize the pros and cons of customer information sharing, adding points of your own not already mentioned above.

2. How can the concerns of consumer advocates be addressed without losing the advantages that the merged banks hope to gain?

Sources: Adapted from Thomas Hoffman and Kim S. Nash, "Titanic Tangle," *Computerworld*, April 13, 1998, pp. 1, 94; John Madden, "Megamerger Should Yield IT Economies," *PC Week*, April 13, 1998, p. 14; the Citicorp Web page at http://www.citicorp.com, accessed on April 6, 1998; Traveler's Group, Inc. Web page at http://www.travelers.com, accessed on April 15, 1998.

PERFORMANCE-BASED INFORMATION SYSTEMS

There have been at least three major stages in the business use of IS. The first stage started in the 1960s and was oriented towards cost reduction and productivity. This stage generally ignored the revenue side, not looking for opportunities to increase sales via the use of IS. The second stage started in the 1980s and was defined by Porter and others. It was oriented towards gaining a competitive advantage. In many cases, companies spent large amounts on IS and ignored the costs. We are seeing a shift from strategic management to performance-based management in many IS organizations today.[13] This third stage carefully considers both strategic advantage along with costs. This stage uses return on investment (ROI), net present value, and other measures of performance. Figure 2.8 illustrates these stages.

Productivity

Developing information systems that measure and control productivity is a key element for most organizations. **Productivity** is a measure of the output

productivity

a measure of the output achieved divided by the input required

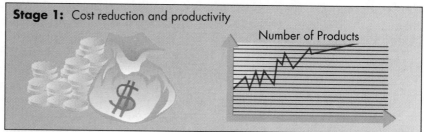

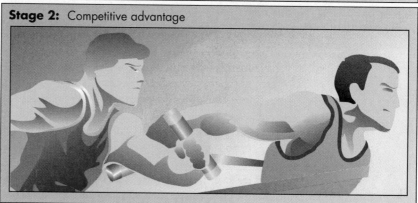

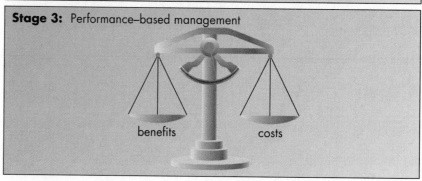

FIGURE 2.8

Three Major Stages in the Business Use of IS

achieved divided by the input required. A higher level of output for a given level of input means greater productivity; a lower level of output for a given level of input means lower productivity. Consider a tax preparation firm, for example, where productivity can be measured by the tax returns prepared divided by the total hours the employee worked. For example, in a 40-hour week, an employee may have prepared 30 tax returns. The productivity is thus equal to 30/40, or 75 percent. With administrative and other duties, a productivity level of 75 may be excellent. The numbers assigned to productivity levels are not always based on labor hours. Productivity may be based on factors like the amount of raw materials used, resulting quality, or time to produce the goods or service. In any case, what is important is not the value of the productivity number but how it compares with other time periods, settings, and organizations.

$$\text{Productivity} = (\text{Output/Input}) \times 100\%$$

Once a basic level of productivity is measured, an information system can monitor and compare it over time to see whether productivity is increasing. Then, corrective action can be taken if productivity drops below certain levels. In addition to measuring productivity, an information system can also be used within a process to significantly increase productivity. Thus, improved productivity can result in faster customer response, lower costs, and increased customer satisfaction.

Measuring productivity is important because improving productivity boosts a nation's standard of living. In an era of intense international competition, the need to improve productivity is critical to the well-being of any enterprise or country. If a company does not take advantage of technological and management innovation to improve productivity, its competitors will. The ability to apply information technology to improve productivity separates successful enterprises from failures.

It is important to understand that information technology is not productive by itself. It takes well-managed, superbly trained, and motivated people—with or without information technology—to deliver measurable gains in output. Many people think that real improvements in productivity come from a synergy of information technology and sweeping changes in management and organizational structure that redefine how work gets done. Largely produced in response to increasing global competition, these overhauls are known loosely as reengineering, which was previously discussed.

Once work has been redefined, information technology can be used to move information to the front lines—to give employees on the factory floor or in the customer service department the knowledge they need to act quickly. That is the formula for a productivity explosion. Such a productivity explosion leads to an increase in the world's standard of living.

Return on Investment and the Value of Information Systems

return on investment (ROI)

one measure of IS value that investigates the additional profits or benefits that are generated as a percentage of the investment in information systems technology

One measure of IS value is **return on investment (ROI).** This measure investigates the additional profits or benefits that are generated as a percentage of the investment in information systems technology. A small business that generates an additional profit of $20,000 for the year as a result of an investment of $100,000 for additional computer equipment and software would have a return on investment of 20 percent ($20,000/$100,000).

Earnings growth. Another measure of IS value is the increase in profit, or earnings growth, it brings. For instance, suppose a mail order company, after installing an order processing system, had a total earnings growth of 15 percent compared with the previous year. Sales growth before the new ordering system was only about 8 percent annually. Assuming that nothing else affected sales, the earnings growth brought by the system, then, was 7 percent.

Market share. Market share is the percentage of sales that one company's products or services have in relation to the total market. If installing a new on-line Internet catalog increases sales, it might help a company increase its market share by 20 percent.

Customer awareness and satisfaction. Although customer satisfaction can be difficult to quantify, about half of today's best global companies measure the performance of their information systems based on feedback from internal and external users. Some companies use surveys and questionnaires to determine whether the investment in information systems has increased customer awareness and satisfaction.

The preceding are only a few measures that companies use to plan for and maximize the value of their investments in information systems technology. In many cases, it is difficult to accurately measure ROI. For example, an increase in profits could be caused by an improved information system or by other factors, such as a new marketing campaign or a competitor that was late in delivering a new product to the market. Regardless of the difficulties, organizations must attempt to evaluate the contributions information systems make to be able to assess their progress and plan for the future. Information technology and personnel are too important to leave to chance.

Justifying Information Systems

Because information systems are so important to the work in organizations, businesses need to be sure that improvements or completely new systems are worthwhile. The process for reviewing IS changes involves justification that the change is necessary and will yield gains.

To avoid waste, each potential information systems project should be reviewed to ensure that the project meets an important business need, is consistent with corporate strategy, and leads to attainment of specific goals and objectives. A second check should be made to assess the degree of risk or uncertainty associated with each project. Risk can be assessed by answering questions such as:

1. How well are the requirements of the system understood?
2. To what degree does the project require pioneering effort in technology that is new to the firm?
3. Is there a risk of severe business repercussions if the project is poorly implemented?

The fundamental benefits for considering the project should be identified. Most IT projects fall into one of the following categories:

- *Tangible Savings.* Implementation of the project will result in hard dollar savings to the company that can be quantified (e.g., reduced staff, lowered operating costs, or increased sales).
- *Intangible Savings.* Implementation of the project will result in soft dollar savings to the company, the magnitude of which will be difficult to measure (e.g., help managers make better decisions or improve control over the operations of the business).
- *Legal Requirement.* Implementation of the project is required to meet a state or federal regulation (e.g., reporting information on the employment of handicapped or minorities).
- *Modernization.* Implementation of the project is needed to keep current with changing business requirements (e.g., systems changes needed due to a conversion from English to metric units of measure or conversion to the Eurodollar) or technology requirements (e.g., computer upgrade to improve work with new software).
- *Pilot Project.* Implementation of the project is required to gain experience in a new technology (e.g., the use of portable computers by salespeople to enhance customer presentations).

Most organizations today realize that they must look at both sides of the equation—benefits as well as costs—in evaluating potential information system investments. Furthermore, determining return on investment can help the IS organization prove its contribution to the organization and ensure that its efforts are aligned with the company's overall business objectives.[14]

CAREERS IN INFORMATION SYSTEMS

Work with information systems continues to be an exciting career choice. Business demand for IS professionals has continued to grow. Numerous schools have degree programs with such titles as information systems, computer information systems, and management information systems. These programs are typically in business schools and within computer science departments. A degree in information systems can provide a high starting salary for many students after graduation from college. For many schools and departments, information systems majors attain the highest starting salaries of all undergraduate business majors and show significant potential job growth. Information systems majors can earn salaries that are more than $40,000 per year.

In addition to having a relatively high starting salary, careers in information systems promise to expand even more than other business disciplines. Strong growth is projected for IS occupations as a result of the continuing spread of computer technology and information systems throughout government and business. Nearly 350,000 IS jobs are vacant, according to the Information Technology Association of America, a vendor trade group. In the future, there will be plenty of demand: 1.3 million IT jobs will be created over the next 10 years, predicts the Commerce Department.[15] Much of the growth has been in the so-called "packaged software" segment of the business, which includes companies such as Microsoft, IBM, Maxis, Intuit, and all the other producers of the software currently on sale at your local computer store. Systems analysts and computer scientists will be increasingly needed to meet the demand for computers in offices, factories, and research agencies and to support the rapid growth of telecommunications technology.

Roles, Functions, and Careers in the Information Systems Department

Information systems personnel typically work in an information systems department that employs computer programmers, systems analysts, computer operators, and a number of other information systems personnel. They may also work in other functional departments or areas in a support capacity. In addition to technical skills, information systems personnel also need skills in written and verbal communication, an understanding of organizations and the way they operate, and the ability to work with people (users). In general, information systems personnel are charged with maintaining the broadest perspective on organizational goals. For most medium- to large-sized organizations, information resources are typically managed through an IS department. In smaller businesses, one or more people may manage information resources, with support from outside services—outsourcing. As shown in Figure 2.9, the information systems organization has three primary responsibilities: operations, systems development, and support.

FIGURE 2.9

The Three Primary Responsibilities of Information Systems

Each of these elements—operations, systems development, and support—contains subelements critical to the efficient and effective operation of the organization.

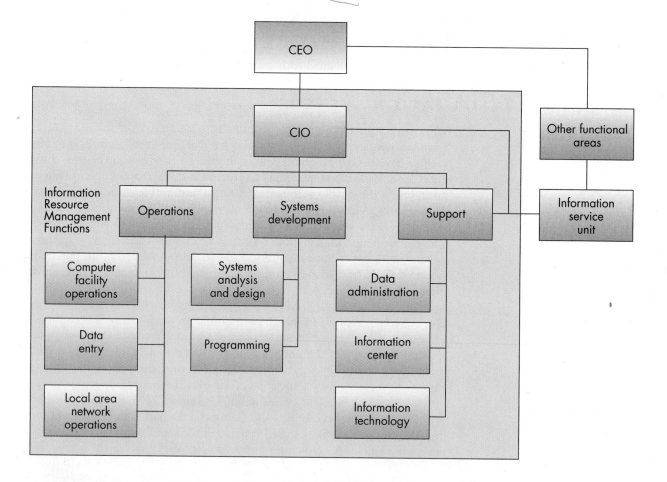

The system operator in this oceanographic data center maintains the institute's computer network and all computer hardware.
(Source: Image copyright © 1998 PhotoDisc.)

Operations. The operations component of a typical IS department focuses on the use of information systems in corporate or business unit computer facilities. It tends to focus more on the *efficiency* of information system functions rather than their effectiveness.

The primary function of a system operator is to run and maintain IS equipment. System operators are responsible for starting, stopping, and correctly operating mainframe systems, networks, tape drives, disk devices, printers, and so on. System operators are typically trained at technical schools or through on the job experience. Other operations include logging, scheduling, hardware maintenance, and preparation of input and output. Data-entry operators convert data into a form the computer system can use. They may use terminals or other devices to enter business transactions, such as sales orders and payroll data. Increasingly, data entry is being automated—captured at the source of the transaction rather than being entered later.

Systems development. The systems development component of a typical IS department focuses on specific development projects and ongoing maintenance and review. Systems analysts and programmers, for example, focus on these concerns.

The role of a systems analyst is multifaceted. Systems analysts help users determine what outputs they need from the system and construct the plans needed to develop the necessary programs that produce these outputs. Systems analysts then work with one or more programmers to make sure that the appropriate programs are purchased, modified from existing programs, or developed. The major responsibility of a computer programmer is to use the plans developed by the systems analyst to develop or adapt one or more computer programs that produce the desired outputs. The main focus of systems analysts and programmers is to achieve and maintain information system effectiveness.

Support. The support component of a typical IS department focuses on providing user assistance in the areas of hardware and software acquisition and use, data administration, and user training and assistance. Because information systems hardware and software is costly, especially if mistakes are made, standards for the acquisition of computer hardware and systems software is often managed by a specialized group within the support component. This group sets guidelines and standards for the rest of the organization to follow in making its purchases. Gaining and maintaining an understanding of available technology is an important part of the acquisition of information systems. Also, developing good relationships with vendors is important.

Firms may look to one outside source to supply all their information systems needs—a single-vendor solution. There are advantages to this approach, such as potential cost savings and built-in compatibility. There are also risks to this approach, including lack of flexibility, vendor complacency due to lack of competitive bidding, and the possibility of missing out on new products from other vendors.

Quite often, several vendors supply components of the information infrastructure to an organization. Coordination and communication are critical when multiple vendors are involved. Because many people and organizations are involved, problems are harder to pinpoint. If there is a compatibility problem, the hardware supplier may blame the software supplier, or vice versa. In any situation where an information system fails to perform as expected, vendor relationships can become strained. Thorough testing and review may alleviate and/or eliminate some of these problems. Having an in-house specialist who focuses on the acquisition of information systems may also be wise.

The database administrator focuses on planning, policies, and procedures regarding the use of corporate data and information. For example, it is the role of a database administrator to develop and disseminate information about the corporate databases for developers of information system applications. In addition, the database administrator is charged with monitoring and controlling database use.

The support component also typically operates the information center. An **information center** provides users with assistance, training, application development, documentation, equipment selection and setup, standards, technical assistance, and troubleshooting. Although many firms have attempted to phase out information centers, others have changed the focus of this function from technical training to helping users find ways to maximize the benefits of the information resource.

information center

provides users with assistance, training, application development, documentation, equipment selection and setup, standards, technical assistance, and troubleshooting

information service unit

a miniature IS department

Information service units. An **information service unit** is basically a miniature IS department attached and directly reporting to a functional area. Notice the information service unit shown in Figure 2.9. Even though this unit is usually staffed by IS professionals, the project assignments and the resources necessary to accomplish these projects are provided by the functional area to which it reports. Depending on the policies of the organization, the salaries of IS professionals staffing the information service unit may be budgeted by either the IS department or the functional area.

The growth of information service units may be directly attributed to the increased number of users doing their own computing. The increasing use of networks has put computers at nearly every desk. Communication between IS personnel and users is more effective the closer they work together. It is interesting that when such information service units are not part of the formal organizational structure, they tend to arise informally in organizations. That is, a particular functional manager might establish and maintain informal groups of employees who are more proficient with IS than other users. As more employees become computer users, such cooperation must be taken into consideration to manage resources properly. It is probably more productive to support and provide training to these informal groups than to attempt to interfere with or thwart their activities.

The organizational chart shown in Figure 2.9 is a simplified model of an IS department in a typical medium- or large-sized organization. Many organizations have even larger departments, with increasingly specialized positions such as librarian, quality assurance manager, and the like. Smaller firms often combine the roles depicted in Figure 2.9 into fewer formal positions.

The chief information officer. The overall role of the chief information officer (CIO) is to employ an IS department's equipment and personnel in a manner that will help the organization attain its goals. The CIO is usually a manager at the vice-president level concerned with the overall needs of the organization. He or she is responsible for corporatewide policy, planning, management, and acquisition of information systems. Some of the CIO's top concerns include integrating information systems operations with corporate strategies, keeping up with the rapid pace of technology, and defining and assessing the value of systems development projects in terms of performance, cost, control, and complexity. The high level of the CIO position is consistent with the idea that information is one of the organization's most important resources. This individual works with other high-level officers of the organization, including the chief financial officer (CFO) and the chief executive officer (CEO), in managing and controlling total corporate resources.

Depending on the size of the information systems department, there may be several people at senior IS managerial levels. Some of the job titles associated with information systems management are the CIO, vice president of information systems, and manager of information systems. A central role of all these individuals is to communicate with other areas of the organization to determine changing needs. Often these individuals are part of an advisory or steering committee that helps the CIO and other IS managers with their decisions about the use of information systems. Together they can best decide what information systems will support corporate goals. CIOs must work closely with advisory committees, stressing effectiveness and teamwork and viewing information systems as an integral part of the organization's business processes—not an adjunct to the organization. Read the "Making a Difference" box to learn more about the role of a CIO and how he or she can contribute to the success of an organization.

LAN administrator

person who sets up and manages the Local Area Network hardware, software, and security processes

webmaster

person who sets up and manages a company's Internet site

certification

process for testing skills and knowledge that results in a statement by the certifying authority that says an individual is capable of performing a particular kind of job

LAN Administrators and Webmasters. LAN administrators set up and manage the Local Area Network hardware, software, and security processes. They manage the addition of new users, software, and devices to the network. They isolate and fix operations problems. LAN administrators are currently in high demand. There is also an increasing need for trained personnel to set up and manage a company's Internet site, a role sometimes called **webmaster**.

Quite often the people filling these roles have completed some form of certification. **Certification** is a process for testing skills and knowledge that results in a statement by the certifying authority that says an individual is capable of performing a particular kind of job. It frequently involves specific, vendor-provided or vendor-endorsed course work. There are a number of popular certification programs including Novell Certified Network Engineer, Microsoft Certified Professional Systems Engineer, Certified Project Manager, and others.

One of the great fears of every IS manager is spending several thousand dollars to help someone get certified and then have them use their new certification as a springboard to get a higher-paying position with a new firm.

MAKING A DIFFERENCE
What Does It Take to be a Successful CIO?

Successful CIOs—

1. Think of the technological world for which they are responsible as a business. They run it as a business by making sure that their firm's operational needs are satisfied. They ensure the reliability of existing systems to protect the going business. This is where their credibility starts.
2. Look at the efficiency of business systems and reduce costs of information system. If they move too quickly, don't have appropriate controls, or lack a support structure, new technology can end up costing a lot more.
3. Do marketing. Not only do they market themselves and what they have accomplished, but they also explain what information system can do and how it can support the business. They are constantly selling ideas to senior management. They set a strategic direction for the use of information system that is aligned with the corporate strategy and structure.

4. Never underestimate the cultural changes of implementing new information system solutions. A common problem is that too little training is provided to the people affected, without comprehension of the psychological impact that it may have on the person adopting it. In addition to training, CIOs need to create a set of expectations, so that users actually take advantage of the system's full capabilities.

DISCUSSION QUESTIONS

1. Are the skills required to be a successful CIO much different than those required to be a successful manager in any other business discipline? Discuss.
2. Can a successful business manager become a successful CIO without advancing through the ranks of the IS organization? Discuss.

Sources: Adapted from "Success, Satisfaction to the Max," *Computerworld*, December 22, 1997, p. 60; Allan Alter, "The Word from the Top," *Computerworld*, December 15, 1997, pp. 74–75.

As a consequence, some organizations request a written commitment from the individual that he or she will stay on for a certain time after obtaining the certification. Needless to say, this can create some ill will with the employee. Other organizations provide salary increases based on additional acquired credentials.

Sometimes a certified individual may have depth in one or two areas but may not have the breadth required to function successfully in a complex IS environment. For example, an individual may be certified as a Novell Network Engineer but be supporting a network that ties into hubs, switches, and routers and have limited knowledge about those devices.

Other IS Careers

In addition to working for an information systems department in an organization, information systems personnel can work for one of the large information systems consulting firms, such as Andersen Consulting, EDS, and others. These jobs often entail a large amount of travel, because consultants are assigned to work on projects wherever the client is. Such roles require excellent people and project management skills in addition to IS technical skills.

Another IS career opportunity is with a hardware or software vendor developing or selling products. Such a role enables the individual to work on the cutting edge of technology and can be extremely challenging and exciting!

Information Systems Principles

The use of information systems is strongly influenced by the organizational structure and the way particular businesses seek to achieve their goals through problem solving or opportunity attainment.

Information systems are often so intimately intertwined with the underlying value-added process that they are best considered as part of the process itself.

Asking people to work differently often meets with stiff resistance; overcoming resistance to change can be the hardest part of bringing information systems into a business.

Because information systems are so important to the work in organizations, businesses need to be sure that improvements or completely new systems are worthwhile.

■ SUMMARY

1. *Identify the seven value-added processes in the supply chain and describe the role of information systems within them.*

Value-added processes increase the relative worth of the combined inputs on their way to becoming final outputs of the organization. The value chain is a series (chain) of activities that includes (1) inbound logistics, (2) warehouse and storage, (3) production, (4) finished product storage, (5) outbound logistics, (6) marketing and sales, and (7) customer service.

Organizations use information systems to support organizational goals. Before deciding on an information system for an organization, managers should identify the firm's critical success factors, which must be supported by the system. Because information systems typically are designed to improve productivity, methods for measuring the system's impact on productivity should be devised.

2. *Provide a clear definition of the terms organizational structure, culture, and change and discuss how they affect the implementation of information systems.*

An organization is a formal collection of people and various other resources established to accomplish a set of goals. The primary goal of a for-profit organization is to maximize profits by increasing revenues while reducing costs. Nonprofit organizations include social groups, religious groups, universities, and other organizations that do not have profit as the primary goal. Organizations are systems with inputs, transformation mechanisms, and outputs.

Organizational structure refers to organizational subunits and how they are related and tied to the overall organization. Several basic organizational structures exist: traditional, project, team, and multidimensional.

Organizational culture consists of the major understandings and assumptions for a business, corporation, or organization. Organizational change deals with how for-profit and nonprofit organizations plan for, implement, and handle change. Change can be caused by internal or external factors. The change model consists of these stages: unfreezing, moving, and refreezing. According to the concept of organizational learning, organizations adapt to new conditions or alter practices over time.

3. *Identify some of the strategies employed to lower costs and improve service.*

Business process reengineering involves the radical redesign of business processes, organizational structures, information systems, and values of the organization to achieve a breakthrough in business results. Continuous improvement involves constantly seeking ways to improve business processes to add value to products and services. Total quality management consists of a collection of approaches, tools, and techniques that offers a commitment to quality throughout the organization. Outsourcing involves contracting with outside professional services to meet specific business needs. This approach allows the company to focus more closely on its core business and to target its limited resources to meet strategic goals. Downsizing involves reducing the number of employees to reduce payroll related costs; however, it can lead to unwanted side effects.

4. *Define the term strategic competitive advantage and discuss how organizations are using information systems to gain such an advantage.*

Competitive advantage is usually embodied in either a product or service that has the most added value to consumers and that is unavailable from the competition or in an internal system that delivers benefits to a firm not enjoyed by its competition. A five-force model covers factors that lead firms to seek competitive advantage: rivalry among existing competitors, the threat of new market entrants, the threat of substitute products and services, the bargaining power of buyers, and the bargaining power of suppliers. Three strategies to address these factors and to attain competitive advantage include altering the industry structure, creating new products and services, and improving existing product lines and services.

The ability of the information system to provide or maintain competitive advantage should also be determined. Several strategies for achieving competitive advantage include enhancing existing products or services or developing new ones, as well as changing the existing industry or creating a new one.

5. *Discuss how organizations justify the need for information systems.*

The objectives of each potential information systems project are reviewed to ensure that the project meets an important business need, is consistent with corporate strategy, and leads to attainment of specific goals and objectives. A second check is made to assess the degree of risk or uncertainty associated with each project. The fundamental reason for considering the project should be identified. Developing information systems that measure and control productivity is a key element for most organizations. A useful measure of the value of an information system project is return on investment (ROI). This measure investigates the additional profits or benefits that are generated as a percentage of the investment in information systems technology. Most IS projects fall into one of the following categories: tangible savings, intangible savings, legal requirement, modernization, or pilot project.

6. *Define the types of roles, functions, and careers available in information systems.*

Information systems personnel typically work in an information systems department that employs a chief information officer (CIO), systems analysts, computer programmers, computer operators, and a number of other information systems personnel. The overall role of the chief information officer is to employ an IS department's equipment and personnel in a manner that will help the organization attain its goals. Systems analysts help users determine what outputs they need from the system and construct the plans needed to develop the necessary programs that produce these outputs. Systems analysts then work with one or more programmers to make sure that the appropriate programs are purchased, modified from existing programs, or developed. The major responsibility of a computer programmer is to use the plans developed by the systems analyst to develop or adapt one or more computer programs that produce the desired outputs. Computer operators are responsible for starting, stopping, and correctly operating mainframe systems, networks, tape drives, disk devices, printers, and so on. LAN administrators set up and manage the local area network hardware, software, and security processes. There is also an increasing need for trained personnel to set up and manage a company's Internet site, a role sometimes called webmaster. Information systems personnel may also work in other functional departments or areas in a support capacity. In addition to technical skills, information systems personnel also need skills in written and verbal communication, an understanding of organizations and the way they operate, and the ability to work with people (users). In general, information systems personnel are charged with maintaining the broadest enterprisewide perspective.

In addition to working for an information systems department in an organization, information systems personnel can work for one of the large information systems consulting firms, such as Andersen Consulting, EDS, and others. Another IS career opportunity is with a hardware or software vendor developing or selling products.

■ KEY TERMS

certification 71
change model 49
competitive advantage 57
continuous improvement 53
culture 48
downsizing 56
empowerment 46
five-force model 57
flat organizational structure 46
information center 70
information service unit 70

LAN administrator 71
multidimensional organizational structure 48
organization 42
organizational change 49
organizational culture 48
organizational learning 49
organizational structure 45
outsourcing 55
process redesign 51
productivity 64

project organizational structure 47
quality 54
reengineering 51
return on investment (ROI) 65
strategic alliance or strategic partnership 59
team organizational structure 47
total quality management (TQM) 54
traditional organizational structure 45
value chain 44
webmaster 71

■ REVIEW QUESTIONS

1. What is an organization?
2. What is a value-added process? Give several examples.
3. What role does an information system play in the value-added processes of an organization?
4. What is reengineering? What are the potential benefits of performing a process redesign? What is the difference between reengineering and continuous improvement?
5. What is quality? What is total quality management (TQM)?
6. What are organizational structure, organizational change, and organizational learning?
7. List and define the four basic organizational structures.
8. Sketch and briefly describe the three-stage organizational change model.

9. What is downsizing? How is it different from outsourcing?
10. What are some general strategies employed by organizations to achieve competitive advantage?
11. What are the five common justifications for implementation of an information system?
12. Define the term *productivity*. Why is it difficult to measure the impact that investments in information systems have on productivity?
13. Briefly describe six roles frequently found in the information systems organization.
14. What are the three primary responsibilities of the IS department?
15. What is an information systems unit?
16. How do you define quality?

■ DISCUSSION QUESTIONS

1. Providing high quality is a primary goal of all companies. Imagine that you work for a car dealership that is trying to implement its own TQM program. What are some of the measures of quality for the dealership? How would you identify which actions to initiate to improve quality?
2. What sort of information systems career would be most appealing to you—working as a member

of an IS organization, being a consultant, or working for a hardware or software vendor? Why?

3. As part of a TQM project initiated three months ago, you decided your company needed a new information system. The computer systems were brought in over the weekend. The first notice your employees received about the new information system was the computer located on each desk. How might the new system affect the culture of your organization? What types of behaviors might employees exhibit in response? As a manager, how should you have prepared the employees for the new system?

4. You have been asked to participate in preparing your company's strategic plan. Specifically, your task is to analyze the competitive marketplace using Porter's five-force model. Prepare your analysis, using your knowledge of a business you have worked for or have an interest in working for.

5. Based on the analysis you performed in Discussion Question 4, what possible strategies could your organization adopt to address these challenges? What role could information systems play in these strategies? Use Porter's strategies as a guide.

6. How would you recognize an IS organization that has an effective CIO?

7. Imagine that you are the CIO for a large, multinational company. Outline a few of your key responsibilities.

8. Discuss how the change model can be applied to breaking a bad habit—say, smoking or eating fatty foods. Some people have also related the stages in the change model to the changes one must go through to deal with a major life crisis, such as divorce or the loss of a loved one. Explain.

▪ PROBLEM-SOLVING EXERCISES

1. Using a presentation graphics package, design an organization chart for the school information systems department or an organization that you are familiar with from school, work, or recreation. Supermarkets, music stores, bookstores, video rental stores, student organizations, and retail stores are some examples of organizations that could be considered. This organization must have more than 15 employees. Use your word processor to create a description of the type of organization structure being used and the way decisions flow through the company.

2. A new IS project has been proposed that will produce not only cost savings but also an increase in revenue. The initial costs to establish the system are estimated at $500,000. The rest of the cash flow data is presented in the table below.

	Year 1	Year 2	Year 3	Year 4	Year 5
Increased Revenue	$0	$100	$150	$200	$250
Cost Savings	$0	$ 50	$ 50	$ 50	$ 50
Depreciation	$0	$ 75	$ 75	$ 75	$ 75
Initial Expense	$500				

(All amounts in thousands)

a. Using your spreadsheet program, calculate the simple return on investment (ROI) for this project.

b. How would the rate of return change if the project were able to deliver $50,000 in additional revenue and generate cost savings of $25,000 in the first year?

▪ TEAM ACTIVITY

With your team, interview one or more people or research articles in the library or on the Internet about companies that have undergone outsourcing or downsizing. Find out how the process was handled and what justification was given for taking this action. Try to get additional information from an objective source (financial reports, investment brokers, industry consultants) on how this action has affected the organization.

■ WEB EXERCISE

This book emphasizes the importance of information. You can get information from the Internet by going to specific addresses, such as http://www.ibm.com, http://www.whitehouse.gov, or http://www.fsu.edu. This will give you access to the home page of the IBM corporation, the White House, or Florida State University. Note that "com" is used for businesses or commercial operations, "gov" is used for governmental offices, and "edu" is used for educational institutions. Another approach is to use a search engine. Yahoo!, developed by two Tulane University students,

was one of the first search engines on the Internet. A search engine is a Web site that allows you to enter key words or phrases to find information. Lists or menus can also be used. The search engine will return other Web sites (hits) that correspond to a search request. Using Yahoo! at http://www.yahoo.com, search for information about a company or topic discussed in Chapter 1 or 2. You may be asked to develop a report or send an e-mail message to your instructor about what you found.

■ CASES

1 Textron Outsources Information Technology Services

Textron Inc. has its headquarters in Providence, Rhode Island, and ranks 130th on the Fortune 500 list of U.S. companies, with annual sales over $10 billion, assets of $18.6 billion, 64,000 employees, and a diverse, global customer base. Founded in 1923, Textron has expanded globally to provide customers worldwide with better value and improved service. Today, Textron's customers and shareholders benefit from the company's leadership positions in a diverse mix of business segments. Within that mix are aircraft, including Cessna Citation business jets and Bell commercial helicopters (29 percent of revenue); automotive, including instrument panels, plastic fuel tanks, and plastic interior and exterior trim (17 percent of revenue); industrial fasteners, golf carts, turf-care equipment, hand and machine tools (24 percent of revenue); and financial services, including global consumer and commercial (30 percent of revenue).

Textron's strategy for consistent growth is that it actively manages its mix of market-leading businesses to produce consistent earnings and revenue growth and to provide excellent returns to shareholders in constantly changing industry and economic environments. Textron plans to build on its core business segments and execute four strategies for growth:

1. Invest in new products—Nearly $1 billion per year is invested in research and development and capital expenditures.
2. Expand into international markets—In 1997, 39 percent of total revenues were generated outside the United States.
3. Make strategic acquisitions—12 market-leading businesses within its core segments have been added to Textron's family of companies since the beginning of 1997.
4. Drive operating excellence—Improve its operating margin by focusing on costs, quality, and operating efficiency.

Textron's chief information officer (CIO), Bill Gauld, has made sure his company's information technology systems and networks enable the company to launch new products ahead of its competition. However, he has had lots of help. Gauld has turned to a number of outsourcers to build advanced information systems for Textron. AT&T Solutions, Inc. handles Textron's corporate network. IBM takes care of network security and intranet applications. EDS handles its corporate office IT services, including software support and corporate services.

1. What principles might Bill Gauld follow in deciding whether a particular service should be performed internally or whether it should be outsourced?

2. Is outsourcing consistent with Textron's strategies for growth? Can you see any inconsistency?

Sources: Adapted from Esther Shein, "A New Flight Plan," *PC Week*, January 19, 1998, pp. 79–83; Textron, Inc. Web site at http://www.camcar.textron.com, accessed on April 14, 1998.

2   Boscov Makes IS Investment Decision

Boscov is a private, family-owned company that operates 29 department stores located primarily in malls throughout Pennsylvania, Delaware, New Jersey, New York, and Maryland. Annual sales are $850 million. There are 9,000 full- and part-time employees. Each store offers a full line of products including clothing, jewelry, cosmetics, shoes, toys, housewares, sporting goods, and appliances. Many of the stores offer special services such as travel agencies, hair salons, restaurants, and optical and hearing-aid centers. Its prices are in the moderate to high range, with lots of competition from many types of retailers, including national and regional department stores, specialty shops, and discount stores.

Boscov implemented a Web-based application that lets contracted delivery truckers check inventories and schedules. The system gives Boscov's line haulers, who deliver large appliances and furniture, direct access to data in a mainframe application developed by Boscov's IS organization. The project cost $16,500 and annual savings are projected at $33,000 from reductions in operations costs, equipment maintenance, and depreciation.

Boscov's original justification for this project was unusual: to retain its technical talent. The company's IS people wanted to learn and work on Web technology, and valuable employees might have left if not given this opportunity. With this application successfully completed, Boscov is looking to identify and implement more Web-related efforts.

1. What is the return on investment for this project?
2. Given the relatively modest cost and savings associated with this project, do you think there was sufficient project justification? Why or why not?

Sources: Adapted from the Boscov Web page at http://www.boscov.com, accessed on July 2, 1998; Clinton Wilder, "Retailer's Legacy App Gets Web Upgrade," *Information Week*, June 29, 1998, p. 92.

3   Black & Veatch Uses Software Tool to Compete Globally

Black & Veatch is one of Kansas City's largest engineering firms and was founded in 1913. The firm specializes in civil engineering projects, including waterworks, highways and bridges, utilities, and sewerage. The firm epitomizes the reach and diversity of a global engineering/construction firm. It is working on the largest ozone disinfection project in North America; developed the first modern coal-fired power plant in Central America; just completed the world's newest Planet Hollywood; and is wrapping up the reconstruction of a fifteenth-century building to a five-star, 72-room U Sixtu Hotel in the Czech Republic.

With revenue of $1.4 billion, Black & Veatch is a relatively small engineering consulting firm that competes with behemoths such as Raytheon ($11.7 billion) and Bechtel ($8.5 billion). One of the tools it uses to remain competitive is Powrtrak, a database including every component of a project, from the supplies used, to shipping times, to how long each part of the project takes, to partners involved in a project. It bridges all project management disciplines; including civil, chemical, electrical, mechanical, and structural engineering; project scheduling; cost estimating; procurement; and construction management planning.

Each component of a project is given an identification code in the Powrtrak database and is tied to the requirements for achieving a particular function. For example, if the user decides that a 10-inch copper pipe to transport water to a boiler is too small, Powrtrak generates reports to ensure that every item associated with the pipe is changed, along with the schedules for installation of the system. Without this application, designers could change the pipe to 12 inches, but keep the valves at 10 inches, resulting in a project in trouble. This ability to automatically generate altered reports to everyone affected slashes project costs. Powrtrak also enables Black & Veatch engineers to quickly determine the design, cost, and schedule impact of changes suggested by customers.

1. Does the Powrtrak software provide Black & Veatch with a strategic competitive advantage? Why or why not?

2. Is it possible that larger firms such as Raytheon or Bechtel have similar tools, but are not getting as much value from them as Black & Veatch? How could this be possible?

Sources: Adapted from Black & Veatch Web page at http://www.bv.com/, accessed on April 18, 1998; "The Art of Innovation," *InformationWeek*, December 1, 1997, pp. 36–64.

---

4 Continuous Replenishment: Revolution in Retailing

The U.S. grocery industry generates over $300 billion in annual sales, touches the lives of millions of people daily, and includes thousands of businesses—from the largest corporations to little neighborhood stores. Numerous industry studies indicate that closer cooperation among suppliers, distributors, and retailers can potentially save from 10 to 12 cents of every sales dollar, for an annual industry savings of $30 to $40 billion.

Some companies have developed approaches to both capture these savings and deliver lower cost and better quality products and services to consumers. A key strategy is an industrywide effort called efficient consumer response (ECR). This approach uses information systems and increased cooperation to reduce costs and drive inefficiencies from relationships among manufacturers, wholesalers, and retailers. Increased use of information systems is not a surprise, but increased cooperation is a major change from an adversarial climate where everyone concentrated on getting what they needed at the lowest cost. In the past, little consideration was given to delivering improved consumer value through more efficient movement of goods.

The new ECR strategy provides better value for consumers and offers new opportunities for any business willing to pioneer innovative approaches and systems. Indeed, companies following the ECR strategy have gained substantial competitive advantages through cost savings and increased influence with retailers.

As part of the ECR program, consumer packaged goods giant Procter & Gamble, with headquarters in Cincinnati, Ohio, developed a continuous replenishment program (CRP) that gained industrywide attention. CRP eliminates unnecessary costs by keeping products moving through the supply pipeline instead of storing them in warehouses or distribution centers. This approach replaced a multistep, paper-based ordering system with an electronic ordering system linking retailers' distribution centers with Procter & Gamble's Customer Service Centers. CRP runs on a timely, accurate, paperless flow of information. Shipment from a Procter & Gamble plant directly to a customer warehouse is automatic, based on actual store demand. Delivery by Procter & Gamble's carriers is on a "just in time" basis, minimizing the need to maintain large inventories.

CRP generates savings for retailers through lower inventories, reduced warehouse space, and lower administration costs. For example, in Michigan Procter & Gamble and Spartan Stores used CRP to reduce Spartan's inventories of Procter & Gamble products by about 30 percent, from $6.5 to $4.5 million. At the same time, Spartan's turnover—the cycle of purchase, sale, and replacement of merchandise—of Procter & Gamble products increased by 50 percent. Lower system costs ultimately make Procter & Gamble brands a better value for consumers.

1. Programs such as ECR and CRP require increased cooperation among manufacturers, wholesalers, and retailers in the grocery industry. What would be the pros and cons of manufacturers sharing their information technology resources—people, systems, and infrastructure—with wholesalers and retailers? Is this "going too far"? Discuss.

2. If Procter & Gamble's CRP program reduces the retailer's inventory, does this mean that P&G must carry the additional inventory? Explain.

Sources: Thomas Craig, "Logistics—Customer Says," Logistics Transportatioon Distribution at http://www.ltdmgmt.com, accessed September 21, 1998; "Vendor Managed Inventory," in the Electronic Commerce section of the Bussman Web page at http://www.bussman.com, accessed on September 21, 1998; Financial information in the Corporate Information section of the Procter & Gamble Web site at http://www.pg.com, accessed on September 21, 1998.

PART II

Information Technology Concepts

CHAPTER 3

Hardware: Input, Processing, and Output Devices

"The business we're in today is not providing a piece of software ... it's not providing a piece of hardware ... it's making our customers more successful.... The real key for anyone today in the service business is you've got to help your customer get to market faster, sooner, and cheaper than they ever did before."

— Larry Weinbach, Unisys chairman, president, and CEO

Chapter Outline

Computer Systems: Integrating the Power of Technology
   Hardware Components
   Hardware Components in Action

Processing and Memory Devices: Power, Speed, and Capacity
   Processing Characteristics and Functions
   Memory Characteristics and Functions
   Multiprocessing

Secondary Storage
   Access Methods
   Devices

Input and Output Devices: The Gateway to Computer Systems
   Characteristics and Functionality
   Input Devices
   Output Devices
   Special-Purpose Input and Output Devices

Computer System Types, Standards, Selecting, and Upgrading
   Computer System Types
   Multimedia Computers
   Standards
   Selecting and Upgrading Computer Systems

Learning Objectives

After completing Chapter 3, you will be able to:

1. Describe how to select and organize computer system components to support information system objectives.
2. Describe the power, speed, and capacity of central processing and memory devices.
3. Describe the access methods, capacity, and portability of secondary storage devices.
4. Discuss the speed, functionality, and importance of input and output devices.
5. Identify six classes of computer systems and discuss the role of each.
6. Define the term *multimedia computer* and discuss common applications of such a computer.

Digital Equipment Corporation and IBM
Breaking the Gigahertz Speed Barrier

In a race to produce the most powerful microprocessors, Digital Equipment Corporation (DEC) (acquired by Compaq) and IBM both claim to have figured out how to break the 1,000 MHz (gigahertz) barrier. That means processors, which control how fast a computer operates, will work at a billion cycles per second—at least three times faster than the fastest desktop chip currently available. Unfortunately, gigahertz chips won't be commercially available until early in the twenty-first century.

In general, the more transistors a chip has, the more commands it can handle at one time. Processor speed has boomed over the last few years as manufacturers have packed more and more transistors onto their chips. Intel Corp. launched its Pentium II processors with 1 million transistors in 1997, bringing the top speed for PCs to 400 MHz. Future processors will far exceed this number of transistors and speed.

DEC already has produced a 600 MHz processor for high-end scientific and commercial computer systems. Its next advance is an Alpha 21264 chip that will work at gigahertz speed. DEC's design approach is to work with its current technology while focusing on

making its chips increasingly slimmer. The thinner and faster the processor gets, the quicker the information on the chip can get where it needs to go. The first generation Alpha 21264 processors contain more than 15 million transistors, operate at a speed of 600 MHz, and have a mere .35 microns between components on the chip (a micron is one-millionth of a meter). This chip became commercially available in the summer of 1998. Later versions of Digital's Alpha will be produced at .25 microns (roughly 1/400 the width of a human hair) and finally at .18 microns, enabling chip efficiency to reach gigahertz status. One major disadvantage of DEC's approach is that the chip consumes more power at higher speeds and thus needs cooling fans to keep it from burning up.

IBM has taken a different design approach. IBM researchers revised each of the millions of pathways on the chip, making them more efficient and cutting the time between circuits. In addition, IBM designed the chip to allow each circuit to handle more functions. Those changes, combined with IBM's new copper chips, will significantly boost processing speed. An advantage of this approach is that

IBM's chip runs at room temperatures and thus doesn't require as much cooling as the DEC chip. Following its design approach, IBM has been able to produce a simplified 1,000 MHz chip using .25 micron technology.

Intel and Hewlett-Packard are jointly designing a chip called Merced, targeted for similar high-end applications. Intel has not stated how fast the Merced chip will run, but it will be produced at .18 microns by 1999.

So what will you be able to do with a gigahertz processor? Well, finding and updating data in large databases will certainly be much faster. More detail can be added to games and movie special effects while keeping computing time down to an acceptable level. Some applications that today are not quite feasible will become possible. For example, voice recognition software will be able to "hear" words spoken in one language, translate, and speak them in another language using voice synthesis. However, for many of us who perform basic tasks such as word processing and simple spreadsheet analysis, the difference in speed may be insignificant. For such applications, more powerful processors will simply save a few seconds each time a program does something.

Breaking the gigahertz barrier may equal the leap of jets cracking the sound barrier back in the early 1950s—an amazing technology breakthrough at the time, but only a preview of things to come!

As you read this chapter, consider the following questions:

- Is processor speed, measured in MHz, a good measure of the speed of a computer or should other factors be considered? If so, what are they?

- What features are important in selecting a computer to meet your business needs?

Sources: Adapted from Michael J. Martinez, "Gigahertz Barrier Broken," at http://www.ABCNEWS.com, February 6, 1998, and Lisa Dicarlo, "Digital Revs Alpha to 600 MHz," *PC Week*, February 8, 1998, p. 47.

Appropriate use of technology can reap huge benefits in business. Employing information technology and providing additional processing capabilities can increase employee productivity, expand business opportunities, and allow for more flexibility. As we already discussed, a computer-based information system (CBIS) is a combination of hardware, software, database(s), telecommunications, people, and procedures—all organized to input, process, and output data and information. In this chapter, we concentrate on the hardware component of a CBIS. **Hardware** consists of any machinery (most of which uses digital circuits) that assists in the input, processing, storage, and output activities of an information system. The overriding consideration in making hardware decisions in a business should be how hardware can be used to support the objectives of the information system and the goals of the organization.

hardware

any machinery (most of which uses digital circuits) that assists in the input, processing, storage, and output activities of an information system

COMPUTER SYSTEMS: INTEGRATING THE POWER OF TECHNOLOGY

A computer system is a special subsystem of an organization's overall information system. It is an integrated assembly of devices, centered on at least one processing mechanism utilizing digital electronics, that are used to input, process, store, and output data and information.

Putting together a complete computer system, however, is more involved than just connecting computer devices. In an effective and efficient system, components are selected and organized with an understanding of the inherent trade-offs between overall system performance and cost, control, and complexity. For instance, in building a car, manufacturers try to match the intended use of the vehicle to its components. Racing cars, for example, require special types of engines, transmissions, and tires. The selection of a transmission for a racing car, then, requires not only consideration of how much of the engine's power can be delivered to the wheels (efficiency and effectiveness) but also how expensive the transmission is (cost), how reliable it is (control), and how many gears it has (complexity). Similarly, organizations assemble computer systems so that they are effective, efficient, and well suited to the tasks that need to be performed.

Consider the FAA's need to replace its high-altitude air traffic control computers. The Federal Aviation Administration is getting new computers to move high-altitude air traffic around the nation. The decision to replace its computers before the year 2000 is being stepped up because of warnings that the old ones could cause unknown problems when the new millennium arrives. The problem is known as the Year 2000 problem, Y2K

One of the FAA's goals is to acquire new computer systems that are extremely fast and reliable.
(Source: Image copyright © 1998 PhotoDisc.)

problem, or millennium bug. Many older computers register the date only by the last two numbers, meaning that for some computers, the year 2000 would be a step back in time—"00" means 1900 to them. (The year 2002 would be stored as 02 and retrieved as 1902.) There have already been numerous instances of hardware failures with potentially disastrous results. Bringing the air traffic control system into the twenty-first century is a critical project for the FAA and one of the most visible issues that FAA administrator Jane Garvey has faced since taking office in 1997. Clearly, a goal of the FAA is to acquire new systems that have an extremely high level of reliability. It is simply not acceptable for the system to fail. However, reliability is not the only objective for the FAA's new computers. They must be extremely fast to support the hundreds of air traffic controllers and help them track thousands of flights daily.[1]

As shown in this example, assembling a computer subsystem requires an understanding of its relationship to the information system and the organization. While we generally refer to the computer subsystem as simply a computer system, we must remember that the computer system objectives are subordinate to, but supportive of, the information system and the organization.

The components of all information systems—such as hardware devices, people, procedures, and goals—are interdependent. Because the performance of one system affects the others, all of these systems should be measured according to the same standards of effectiveness and efficiency, given issues of cost, control, and complexity.

When selecting computer subsystem devices, you also must consider the current and future uses to which these systems will be put. Your choice of a particular computer system should always allow for later improvements in the overall information system. Reasoned forethought—a trait required for dealing with computer, information, and organizational systems of all sizes—is the hallmark of a true systems professional.

Hardware Components

Computer system hardware components include devices that perform the functions of input, processing, data storage, and output (Figure 3.1). To understand how these hardware devices work together, consider an analogy from a paper-based office environment. Imagine a one-room office occupied by a single individual. The human (the processor) is capable of organizing and manipulating data. The person's mind (register storage) and the desk occupied by the human (primary storage) are places to temporarily store data. Filing cabinets fill the need for a more permanent form of storage (secondary storage). In this analogy, the incoming and outgoing mail trays can be understood as sources of new data (input) or as a place to put the processed paperwork (output).

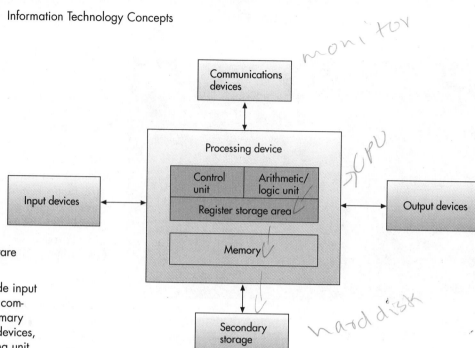

FIGURE 3.1

Computer System Hardware Components

These components include input devices, output devices, communications devices, primary and secondary storage devices, and the central processing unit (CPU). The control unit, the arithmetic/logic unit (ALU), and the register storage areas constitute the CPU.

central processing unit (CPU)

the part of the computer that consists of three associated elements: the arithmetic/logic unit, the control unit, and the register areas

arithmetic/logic unit (ALU)

the part of the CPU that performs mathematical calculations and makes logical comparisons

control unit

the part of the CPU that sequentially accesses program instructions, decodes them, and coordinates the flow of data in and out of the ALU, the registers, primary storage, and the secondary storage and various output devices

registers

high-speed storage areas used to temporarily hold small units of program instructions and data immediately before, during, and after execution by the CPU

primary storage

the part of the computer that holds program instructions and data, also called main memory or just memory

The ability to process (organize and manipulate) data is a critical aspect of a computer system. In a computer system, processing is accomplished by an interplay between one or more of the central processing units and primary storage. Each **central processing unit (CPU)** consists of three associated elements: the arithmetic/logic unit, the control unit, and the register areas. The **arithmetic/logic unit (ALU)** performs mathematical calculations and makes logical comparisons. The **control unit** sequentially accesses program instructions, decodes them, and coordinates the flow of data in and out of the ALU, the registers, primary storage, and the secondary storage and various output devices. **Registers** are high-speed storage areas used to temporarily hold small units of program instructions and data immediately before, during, and after execution by the CPU.

Primary storage, also called *main memory* or just *memory*, is closely associated with the CPU. Memory holds program instructions and data immediately before or immediately after the registers. To understand the function of processing and the interplay between the CPU and memory, let's examine the way a typical computer executes a program instruction.

Hardware Components in Action

The execution of any machine-level instruction involves two phases: the instruction phase and the execution phase. During the instruction phase, the following takes place:

- *Step 1: Fetch instruction.* The instruction to be executed is accessed from memory by the control unit.
- *Step 2: Decode instruction.* The instruction is decoded so the central processor can understand what is to be done, relevant data is moved from memory to the register storage area, and the location of the next instruction is identified.

FIGURE 3.2

Execution of an Instruction

In the instruction phase, the computer's control unit fetches the instruction to be executed from memory (1). Then the instruction is decoded so the central processor can understand what is to be done (2). In the execution phase, the ALU does what it is instructed to do, making either an arithmetic computation or a logical comparison (3). Then the results are stored in the registers or in memory (4). The instruction and execution phases together make up one machine cycle.

instruction time (I-time)

the time it takes to perform the instruction phase of the execution of an instruction

execution time (E-time)

the time it takes to complete the execution phase of the execution of an instruction

machine cycle

the instruction phase and the execution phase of an instruction

pipelining

a CPU operation in which multiple execution phases are performed in a single machine cycle

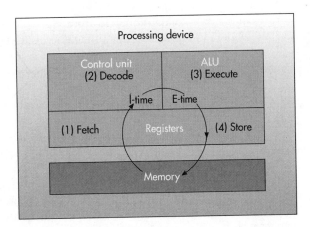

Steps 1 and 2 are called the instruction phase, and the time it takes to perform this phase is called the **instruction time (I-time)**.

The second phase is the execution phase. During the execution phase, the following steps are performed:

- *Step 3: Execute the instruction.* The ALU does what it is instructed to do. This could involve making either an arithmetic computation or a logical comparison.
- *Step 4: Store results.* The results are stored in registers or memory.

Steps 3 and 4 are called the execution phase. The time it takes to complete the execution phase is called the **execution time (E-time)**. After both phases have been completed for one instruction, they are again performed for the second instruction, and so on. The instruction phase followed by the execution phase is called a **machine cycle** (Figure 3.2). Some central processing units can speed up processing by using **pipelining**, which enables the processing unit to execute two instructions in a single machine cycle. The Pentium processor, for example, uses two execution unit pipelines.

PROCESSING AND MEMORY DEVICES: POWER, SPEED, AND CAPACITY

The components responsible for processing—the CPU and memory—are housed together in the same box or cabinet, called the system unit. All other computer system devices, such as the monitor and keyboard, are linked either directly or indirectly into the system unit housing. As discussed previously, achieving information system objectives and organizational goals should be the primary consideration in selecting processing and memory devices. In this section, we investigate the characteristics of these important devices.

Processing Characteristics and Functions

Because having efficient processing and timely output is important, organizations use a variety of measures to gauge processing speed. These include the time it takes to complete one machine cycle, clock speed, and others.

Machine cycle time. As we've seen, the execution of an instruction takes place during a machine cycle. The time in which a machine cycle occurs is measured in fractions of a second. Machine cycle times are measured in *microseconds* (one-millionth of one second) for slower computers to *nanoseconds* (one-billionth of one second) and *picoseconds* (one-trillionth of one second) for faster ones. Machine cycle time also can be measured in terms of how many instructions are executed in a second. This measure, called **MIPS**, stands for millions of instructions per second. MIPS is another measure of speed for computer systems of all sizes.

Clock speed. Each CPU produces a series of electronic pulses at a predetermined rate, called the **clock speed**, which affects machine cycle time. The control unit portion of the CPU controls the various stages of the machine cycle by following predetermined internal instructions, known as **microcode**. You can think of microcode as predefined, elementary circuits and logical operations that the processor performs when it executes an instruction. The control unit executes the microcode in accordance with the electronic cycle, or pulses of the CPU "clock." Each microcode instruction takes at least the same amount of time as the interval between pulses. The shorter the interval between pulses, the faster each microcode instruction can be executed (Figure 3.3).

Clock speed is often measured in megahertz. As seen in Figure 3.3, a **hertz** is one cycle or pulse per second. **Megahertz (MHz)** is the measurement of cycles in millions of cycles per second. The clock speed for personal computers can range from 200 MHz to 600 MHz or more. As we discussed in the opening vignette, Digital Equipment Corporation announced its new 21264 chip to support high-end motion video instructions that operates at 600 MHz initially.[2]

Because the number of microcode instructions needed to execute a single program instruction—such as performing a calculation or printing results—can vary, there is no direct relationship between clock speed measured in megahertz and processing speed measures such as MIPS and milliseconds.

MIPS

millions of instructions per second

clock speed

the predetermined rate at which the CPU produces a series of electronic pulses

microcode

predefined, elementary circuits and logical operations that the processor performs when it executes an instruction

hertz

one cycle or pulse per second

megahertz (MHz)

millions of cycles per second

FIGURE 3.3

Clock Speed and the Execution of Microcode Instructions

A faster clock speed means that more microcode instructions can be executed in a given time period.

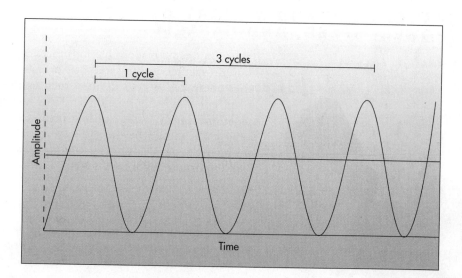

Wordlength and bus line width. Data is moved within a computer system not in a continuous stream but in groups of bits. A **bit** is a BInary digiT— 0 or 1. Therefore, another factor affecting overall system performance— particularly speed—is the number of bits the CPU can process at one time. This number of bits is called the **wordlength** of the CPU. A CPU with a wordlength of 32 (called a 32-bit CPU) will process 32 bits of data in one machine cycle.

Data is transferred from the CPU to other system components via **bus lines,** the physical wiring that connects the computer system components. The number of bits a bus line can transfer at one time is known as bus line width. For example, a bus line with a width of 32 will transfer 32 bits of data at a time. Common wordlength and bus line widths are 32 and 64. Bus line width should be matched with CPU wordlength for optimal system performance. It would be of little value, for example, to install a new 64-bit bus line if the system's CPU only had a wordlength of 32. Assuming compatible wordlengths and bus widths, the larger the wordlength, the more powerful the computer. Computers with larger wordlengths can transfer more data between devices in the same machine cycle. They can also use the larger number of bits to address more memory locations and hence are a requirement for systems with certain large memory requirements.

Because all these factors—machine cycle time, clock speed, wordlength, and bus line width—affect the processing speed of the CPU, comparing the speed of two different processors can be confusing. For this reason, Intel, the world's leading microprocessor chip maker, introduced the Intel Comparative Microprocessor Performance index, or iCOMP index as shown in Table 3.1, in 1992 in response to the widespread misperception among PC purchasers that a processor's megahertz rating is a direct measure of its performance. Although the megahertz rating has important consequences for the design of a PC system, and is therefore important to the PC design engineer, it is not necessarily a good measure of processor performance, especially when comparing one family of processors to the next.

bit

BInary digiT—0 or 1

wordlength

the number of bits the CPU can process at any one time

bus line

the physical wiring that connects the computer system components

TABLE 3.1

iCOMP Index 2.0 Ratings for a Representative Sample of Intel Microcomputer Chips (Source: "iCOMP Index 2.0 Performance Brief— Pentium II Processor Addendum," Intel Web site at http://www.intel.com/procs/perf/ icomp, accessed August 1, 1998.)

Processor	iCOMP Index 2.0
Pentium 150 MHz	114
Pentium 166 MHz	127
Pentium 200 MHz	142
Pentium with MMX Technology 166 MHz	150
Pentium with MMX Technology 200 MHz	182
Pentium with MMX Technology 233 MHz	203
Pentium II 233 MHz	267
Pentium II 266 MHz	303
Pentium II 300 MHz	332
Pentium II 333 MHz	366
Pentium II 350 MHz	386
Pentium II 400 MHz	440

The iCOMP index is a single number representing four industry-standard benchmarks weighted to reflect the mix of 16-bit and 32-bit software applications in use at that time. To reflect the trend in 1996 toward 32-bit software and the proliferation of multimedia applications, Intel updated the iCOMP index with four industry-standard, 32-bit benchmarks and a new multimedia benchmark that reflect today's computing environment. The updated index is called iCOMP index 2.0.

It is important to understand that the iCOMP index is a tool for making comparisons between different Intel processors, not systems. Thus, although two systems with a given processor will have exactly the same iCOMP index 2.0 rating, this does not mean that all systems with the same processor perform the same—differences in system design and configuration affect performance considerably. For instance, vendors sell systems with a wide variety of disk capabilities and speeds, system memory, system bus features and video and graphics capabilities, and all of these features influence how the processor and the system actually perform. However, given systems of comparable configuration and design, the one with the higher iCOMP index 2.0 rating will have more power and will run software faster. The iCOMP index 2.0 is most valuable for desktop and laptop systems. More complex systems typically require a measure of sophisticated I/O subsystems that the iCOMP index does not reflect.[3]

Physical characteristics of the CPU. CPU speed is also limited by physical constraints. Most CPUs are collections of digital circuits imprinted on silicon wafers, or chips, each no bigger than the tip of a pencil eraser. To turn a digital circuit within the CPU on or off, electrical current must flow through a medium (usually silicon) from point A to point B. The speed at which it travels between points can be increased by either reducing the distance between the points or reducing the resistance of the medium to the electrical current.

Reducing the distance between points has resulted in ever smaller chips, with the circuits packed closer together. In the 1960s, shortly after patenting the integrated circuit, Gordon Moore, former chairman of the board of Intel (the largest microprocessor chip maker), formulated what is now known as **Moore's Law**. This hypothesis states that transistor (the microscopic on/off switches or the microprocessor's brain cells) densities on a single chip will double every 18 months. Moore's Law has held up amazingly well over the years, as can be seen from Figure 3.4. To date, physicists see no reason why this trend cannot continue for several more years. Although the trend of getting more electronics at a cheaper cost will not be reversed, the rate of change in the technology may slow. By the time that slowdown comes, however, chips will be performing tasks we can only imagine today, and the world will be a far more interconnected place.[4]

To increase the speed of the CPU, the resistance of the medium to electrical current can be reduced. One approach is to substitute materials for silicon that allow the electrical current to pass through the medium more rapidly. For example, advances in gallium arsenide (GaAs) are giving scientists hope that the material will be a good replacement for silicon in the construction of computer chips. The advantages of GaAs chips are high speed (electrons zip through its silver-gray crystal at least five times faster than they move through silicon) and low power consumption compared

Moore's Law

a hypothesis that states transistor densities on a single chip will double every 18 months

FIGURE 3.4

Moore's Law

Transistor densities on a single chip will double every 18 months.
(Source: Reprinted by permission of Intel Corporation. Copyright Intel Corporation 1995.)

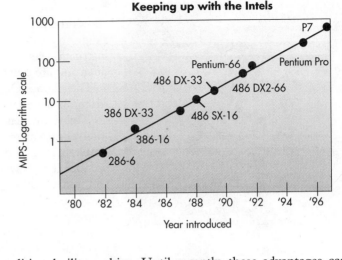

with traditional silicon chips. Until recently, these advantages came at a high price. The wafers on which GaAs chips are printed were so difficult to prepare that they once cost 10 times as much as silicon wafers. And in the chip factories, the wafers were difficult to process, causing further increases in costs. Recently, prices have been falling as both wafer suppliers and chip producers continuously improve their manufacturing techniques. Although GaAs chips still carry a premium cost, when new products are designed from the outset to exploit GaAs features, the final solution can be cheaper than a silicon-chip-based solution. Increasingly, GaAs chips are being used in cellular phones, cable TV equipment, and digital TV.

Another substitute material for silicon chips are superconductive metals. **Superconductivity** is a property of certain metals that allows current to flow with minimal electrical resistance. Traditional silicon chips create some electrical resistance that slows processing. Chips built from less resistant superconductive metals offer increases in processing speed.

Some companies are experimenting with chips called **optical processors** that use light waves instead of electrical current to represent bits. The primary advantage of optical processors is their speed. Optical processors have the potential of being 500 times faster than traditional electronic circuits. In addition, pathways for optical circuits do not have to be spaced as far apart as electronic circuits to avoid electrical interference. One company is experimenting with attaching a living cell membrane to a silicon chip. Doctors may be able to use the chip to analyze living tissue and blood samples.[5]

Complex and reduced instruction set computing. Most of the processors shown in Table 3.1 were designed based on **complex instruction set computing (CISC)**, which places as many microcode instructions into the central processor as possible. In the mid-1970s John Cocke of IBM recognized that most of the operations of a CPU involved only about 20 percent of the available microcode instructions. This led to an approach to chip design called **reduced instruction set computing (RISC)**. RISC involves reducing the number of microcode instructions built into a chip to this essential set of common microcode instructions. RISC chips are faster than CISC chips for processing activities that predominantly use this core set of instructions because each operation requires fewer microcode steps

superconductivity

a property of certain metals that allows current to flow with minimal electrical resistance

optical processors

computer chips that use light waves instead of electrical current to represent bits

complex instruction set computing (CISC)

a computer chip design that places as many microcode instructions into the central processor as possible

reduced instruction set computing (RISC)

a computer chip design based on reducing the number of microcode instructions built into a chip to an essential set of common microcode instructions

The PowerPC 604e microprocessor provides high levels of performance for desktop, workstation, and symmetric multiprocessing computer systems. (Source: © 1998 Motorola, Inc. Used by Permission. Motorola, and the Motorola logo are registered trademarks of Motorola, Inc. PowerPC, the PowerPC logo, and the PowerPC604e are trademarks of IBM Corp. and are used by Motorola under license therefrom.)

very long instruction word (VLIW)

a computer chip design based on further reductions in the number of instructions in a chip by lengthening each instruction

byte

eight bits together that represent a single character of data

random access memory (RAM)

a form of memory where instructions or data can be temporarily stored

prior to execution. Most RISC chips use pipelining, which allows the processor to execute multiple instructions in a single machine cycle. With less sophisticated microcode instruction sets, RISC chips are also less expensive to produce and are quite reliable.

The PowerPC chip is a RISC processor created by Motorola under agreement with IBM and Apple Computer. By almost any benchmark, RISC processors run faster than Intel's Pentium processor. And because RISC chips have a simpler design and require less silicon, they are cheaper to produce. The PowerPC chip is designed to provide portable and desktop personal computers the processing power normally associated with much more expensive computers. For example, the PowerPC has the ability to make functions such as voice recognition, dictation, pen input, and touch screens practical. Digital Equipment's Alpha chip and Sun Microsystems' Sparc chip are other examples of RISC processors.

Further reductions in the number of instructions in a chip can be made by lengthening each instruction. This approach—called **very long instruction word (VLIW)**—is being developed by Intel, Hewlett-Packard, and other companies. Chips built using VLIW are potentially even faster than RISC-based chips. In the future, RISC and VLIW will gradually replace CISC chips.

When selecting a CPU, organizations must balance the benefits of speed with the issues of cost, control, and complexity. CPUs with faster clock speeds and machine cycle times are usually more expensive than slower ones. This expense, however, is a necessary part of the overall computer system cost, for the CPU is typically the single largest determinant of the price of many computer systems. CPU speed can also be related to complexity. Having a less complex code, as in the case of RISC chips, can not only increase speed and reliability but also reduce chip manufacturing costs.

Memory Characteristics and Functions

Located physically close to the CPU (to decrease access time), memory provides the CPU with a working storage area for program instructions and data. The chief feature of memory is that it rapidly provides the data and instructions to the CPU.

Storage capacity. Like the CPU, memory devices contain thousands of circuits imprinted on a silicon chip. Each circuit is either conducting electrical current (on) or not (off). By representing data as a combination of on or off circuit states, the data is stored in memory. Usually eight bits are used to represent a character, such as the letter *A*. Eight bits together form a **byte**. Following is a list of storage capacity measurements. In most cases, storage capacity is measured in bytes, abbreviated with the letter *B*, with one byte usually equal to one character.

Types of memory. There are several forms of memory, as shown in Figure 3.5. Instructions or data can be temporarily stored in **random access memory (RAM)**. RAM is temporary and volatile. RAM chips lose

Name	Abbreviation	Number of Bytes	Approximate Number of Bytes
Byte	B	8 bits	One
Kilobyte	KB	1,024 bytes	One thousand
Megabyte	MB	1,024 x 1,024 bytes	One million
Gigabyte	GB	1,024 x 1,024 x 1,024 bytes	One billion
Terabyte	TB	1,024 x 1,024 x 1,024 x 1,024 bytes	One trillion

their contents if the current is turned off or disrupted (as in a power surge, brownout, or electrical noise generated by lightning or nearby machines). RAM chips are mounted directly on the computer's main circuit board or in chips mounted on peripheral cards that plug into the computer's main circuit board. These RAM chips consist of millions of switches that are sensitive to changes in electric current.

RAM comes in many different varieties. The mainstream type of RAM is Extended Data Out, or EDO RAM, which is faster than older types of RAM. Another kind of RAM memory is called SDRAM, or Synchronous Dynamic RAM, which exceeds the EDO RAM in performance. SDRAM also has the advantage of a faster transfer speed between the microprocessor and the memory. Dynamic RAM (DRAM) chips need high or low voltages applied at regular intervals—every two milliseconds (two one-thousandths of a second) or so—if they are not to lose their information.

Over the past decade, microprocessor speed has doubled every 18 months, but memory performance has not kept pace. In effect, memory has become the principal bottleneck to system performance. Thus, microprocessor manufacturers are working with memory vendors to keep up with the performance of faster processors and bus architectures. For example, Intel is working with Rambus Inc. to extend existing Rambus technology to improve total system performance. Developed in conjunction with Intel and in cooperation with other Rambus semiconductor partners, Direct Rambus (Direct RDRAM) technology is gaining broad industry support. The world's top ten DRAM makers have announced their intention to develop

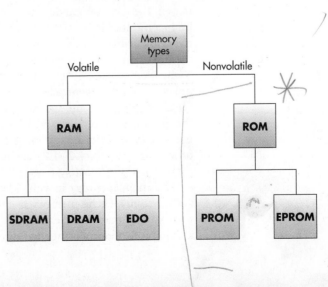

FIGURE 3.5

Basic Types of Memory Chips

Direct RDRAM products. Another two dozen companies representing the leaders in system-memory implementation products—including memory modules, connectors, clock chips, and test systems—announced their intention to support Direct Rambus technology. Planned applications include computer system memory, multimedia and graphics memory, communications system memory and consumer electronics memory.[6]

read-only memory (ROM)

a form of memory that provides permanent storage for data and instructions

Another type of memory, **ROM**, an acronym for **read-only memory**, is usually nonvolatile. In ROM, the combination of circuit states is fixed, and therefore its contents are not lost if the power is removed. ROM provides permanent storage for data and instructions that do not change, like programs and data from the computer manufacturer.

There are other types of nonvolatile memory as well. Programmable read-only memory (PROM) is a type in which the desired data and instructions—and hence the desired circuit state combination—must first be programmed into the memory chip. Thereafter, PROM behaves like ROM. PROM chips are used where the CPU's data and instructions do not change, but the application is so specialized or unique that custom manufacturing of a true ROM chip would be cost prohibitive. A common use of PROM chips is to store the instructions for popular video games such as those from Nintendo and Sega. Game instructions are programmed onto the PROM chips by the game manufacturer. Instructions and data can be programmed into a PROM chip only once.

Erasable programmable read-only memory (EPROM) is similar to PROM except, as the name implies, the memory chip can be erased and reprogrammed. EPROMs are used where the CPU's data and instructions change, but only infrequently. An automobile manufacturer, for example, might use an industrial robot to perform repetitive operations on a certain car model. When the robot is performing its operations, the nonvolatility and rapid accessibility to program instructions offered by EPROM is an advantage. Once the model year is over, however, the EPROM controlling the robot's operation will need to be erased and reprogrammed to accommodate a different car model.

~~cache memory~~

a type of high-speed memory that a processor can access more rapidly than main memory

Cache memory is a type of high-speed memory that a processor can access more rapidly than main memory (Figure 3.6). Cache memory functions somewhat like an address book used to record phone numbers. While a person's private address book may contain only 1 percent of all the numbers in the local phone directory, the chance that the person's next call will be to a number in his or her address book is high. Cache memory works on the same principle—frequently used data is stored in easily accessible cache memory instead of slower memory like RAM. Because there is less data in cache memory, the CPU can access the desired data and instructions more quickly than if it were selecting from the larger set in main memory. The CPU can thus execute instructions faster, and the overall performance of the computer system is raised. There are two types of cache memory present in the majority of systems shipped. The Level 1 (L1) cache is in the processor, the Level 2 (L2) cache memory is optional and found on the motherboard of most systems.

The main memory in your system that can move its information into your system's cache memory is called the cacheable memory. Memory in your system that is not cacheable performs as if your system is cacheless, moving information as needed directly to the processor without the ability

FIGURE 3.6

Cache Memory

Processors can access this type of high-speed memory faster than main memory. Located near the CPU, cache memory works in conjunction with main memory. A cache controller determines how often the data is used and transfers frequently used data to cache memory, then deletes the data when it goes out of use.

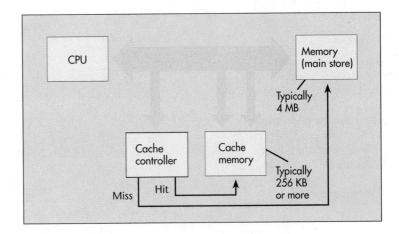

to use the cache memory as a fast retrieval storage bin. All systems have a main memory cacheable limit, typically 512 KB or greater.

Costs for memory capacity continue to decline. As we shall see, however, when considered on a megabyte-to-megabyte basis, memory is still considerably more expensive than most forms of secondary storage. Memory capacity can be important in the effective operation of a CBIS. The specific applications of a CBIS determine the amount of memory required for a computer system. For example, complex processing problems, such as computer-assisted product design, require more memory than simpler tasks like word processing. Also, because computer systems have different types of memory, other programs may be needed to control how memory is accessed and used. In other cases, the computer system can be configured to maximize memory usage. Before additional memory is purchased, all these considerations should be addressed.

Multiprocessing

multiprocessing

the simultaneous execution of two or more instructions at the same time

coprocessor

a part of the computer that speeds processing by executing specific types of instructions while the CPU works on another processing activity

parallel processing

a form of multiprocessing that speeds processing by linking several processors to operate at the same time or in parallel

There are numerous forms of **multiprocessing**, which is the simultaneous execution of two or more instructions at the same time. One form of multiprocessing involves **coprocessors**. A coprocessor speeds processing by executing specific types of instructions while the CPU works on another processing activity. Coprocessors can be internal or external to the CPU and may have different clock speeds than the CPU. Each type of coprocessor best performs a specific function. For example, a math coprocessor chip can be used to speed mathematical calculations, while a graphics coprocessor chip decreases the time it takes to manipulate graphics.

Parallel processing. Another form of multiprocessing, called **parallel processing**, speeds processing by linking several processors to operate at the same time or in parallel. The challenge is not connecting the processors but making them work effectively as a unified set. Accomplishing this difficult task requires software that can allocate, monitor, and control multiple processing jobs at the same time. With parallel processing, a business problem, such as designing a new product or piece of equipment, is divided into several parts. Each part is "solved" by a separate processor. The results from each processor are then assembled to get the final output (Figure 3.7).

FIGURE 3.7

Parallel Processing

Parallel processing involves breaking a problem into various subproblems or parts, then processing each of these parts independently. The most difficult aspect of parallel processing is not the simultaneous processing of the subproblems but the logical structuring of the problem into independent parts.

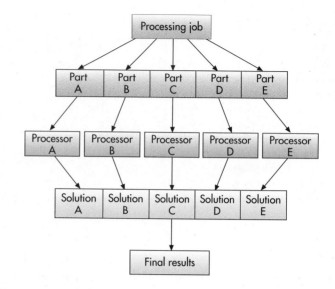

The most frequently used examples of parallel processing for businesses are modeling, simulation, and analysis of large amounts of data. In today's marketplace, the requirements and array of services that consumers demand have forced companies to find more creative and effective methods for gathering and reporting information about their customers. Collecting and organizing enormous amounts of data about consumers is no easy task. It is essential that the data be easily accessible. Parallel processing can provide the information necessary to build an effective marketing program based on existing consumer buying patterns and, as a result, can give a company a competitive advantage. Parallel processing systems can coordinate large amounts of data and access them with greater speed than was previously possible.

SECONDARY STORAGE

secondary storage

the part of the computer that stores large amounts of data, instructions, and information more permanently than main memory, also called permanent storage

As we have seen, memory is an important factor in determining overall computer system power. However, memory provides only a small amount of storage area for the data and instructions required by the CPU for processing. Computer systems also need to store larger amounts of data, instructions, and information more permanently than allowed with main memory. **Secondary storage**, also called permanent storage, serves this purpose.

Compared with memory, secondary storage offers the advantages of nonvolatility, greater capacity, and greater economy. As previously noted, on a megabyte-per-megabyte basis, most forms of secondary storage are considerably less expensive than memory (see Figure 3.8). Because of the electromechanical processes involved in using secondary storage, however, it is considerably slower than memory. The selection of secondary storage media and devices requires an understanding of their primary characteristics—access method, capacity, and portability.

Access Methods

As with other computer system components, the access methods, storage capacities, and portability required of secondary storage media are determined by the information system's objectives. An objective of a credit card company's

Device	DAT tape	Hard drive	External SCSI Jaz Drive	3.5" diskette	ZIP Plus Drive	RAM
Cost	$49.95	$349.95	$599.95	$.50	$199.95	$269.95
Storage	10,000MB	6,400MB	2,000MB	1.4MB	100MB	64MB
Cost per megabyte	$.005	$.05	$.30	$.35	$2.00	$4.21

FIGURE 3.8

Cost Comparison for Various Forms of Data Storage

All forms of secondary storage cost considerably less per megabyte of capacity than RAM, although they have slower access times. A diskette costs about 35 cents per megabyte, while RAM can cost around $4 per megabyte, 11 times more.
(Source: Data from CompUSA Direct Catalog, February 1998.)

sequential access

the process by which data must be retrieved in the order in which it is stored

direct access

the process by which data can be retrieved without the need to read or pass by other data in sequence

sequential access storage devices (SASDs)

devices used to sequentially access secondary storage data

direct access storage devices (DASDs)

devices used for direct access of secondary storage data

information system, for example, might be to rapidly retrieve stored customer data to approve customer purchases. In this case, a fast access method is critical. In other cases, such as sales force automation via laptop computers, portability and storage capacity might be major considerations in selecting and using secondary storage media and devices.

Storage media that provide faster access methods are generally more expensive than slower media. The cost of additional storage capacity and portability varies widely, but it is also a factor to consider. In addition to cost, organizations must address security issues to allow only authorized people access to sensitive data and critical programs. Because the data and programs kept in secondary storage devices are so critical to most organizations, all of these issues merit careful consideration.

Access methods: sequential and direct. Data and information access can be either sequential or direct. **Sequential access** means that data must be accessed in the order in which it is stored. For example, inventory data stored sequentially may be stored by part number, such as 100, 101, 102, and so on. If you want to retrieve information on part number 125, it is necessary to read and discard all the data relating to parts 001 through 124.

Direct access means that data can be retrieved directly without the need to pass by other data in sequence. With direct access, it is possible to go directly to and access the needed data—say, part number 125—without having to read through parts 001 through 124. For this reason, direct access is usually faster than sequential access. The devices used to sequentially access secondary storage data are simply called **sequential access storage devices (SASDs)**; those used for direct access are called **direct access storage devices (DASDs)**.

Devices

The most common forms of secondary storage include magnetic tapes, magnetic disks, and optical disks. Some of these media (magnetic tape) allow only sequential access, while others (magnetic and optical disks) provide

direct and sequential access. Figure 3.9 shows some different secondary storage media.

Magnetic tapes. One common secondary storage medium is **magnetic tape**. Similar to the kind of tape found in audio- and videocassettes, magnetic tape is a Mylar film coated with iron oxide. Portions of the tape are magnetized to represent bits. Magnetic tape is an example of a sequential access storage medium. If the computer is to read data from the middle of a reel of tape, all the tape before the desired piece of data must be passed over sequentially. This is one disadvantage of magnetic tape. When information is needed, it can take time for a tape operator to load the magnetic tape on a tape device and get the relevant data into the computer. Although access is slower, magnetic tape is usually less expensive than disk storage. In addition, magnetic tape is often used to back up disk drives and to store data off-site for recovery in case of disaster. Procom Technology's new Jetstream 1000 AIT Tape Array provides a total capacity of 200 GB, with a 20-MB-per-second data transfer rate, and 72-GB-per-hour backup capability, making it an ideal backup device.

Magnetic disks. Magnetic disks are also coated with iron oxide; they can be thin steel platters (hard disks; see Figure 3.10) or Mylar film (diskettes). As with magnetic tape, magnetic disks represent bits by small magnetized areas. When reading from or writing data onto a disk, the disk's read/write head can go directly to the desired piece of data. Thus, the disk is a direct access storage medium. While disk devices can be operated in a sequential mode, most disk devices use direct access. Because direct access allows fast data retrieval, this type of storage is ideal for companies that need to respond quickly to customer requests, such as airlines and credit card firms. For example, if a manager needs information on the credit history of a customer or the seat availability on a particular flight, the information can be obtained in a matter of seconds if the data is stored on a direct access storage device. If the data is stored on magnetic tape, it

magnetic tape

a common secondary storage medium that consists of Mylar film coated with iron oxide with magnetized portions to represent bits

magnetic disks

common secondary storage medium, with bits represented by magnetized areas

FIGURE 3.10

Hard Disk

Hard disks give direct access to stored data. The read/write head can move directly to the location of a desired piece of data, dramatically reducing access times, as compared with magnetic tape.
(Source: Courtesy of Western Digital Corporation.)

redundant array of independent/inexpensive disks (RAID)

a method of storing data that generates extra bits of data from existing data, allowing the system to create a "reconstruction map" so that if a hard drive fails, it can rebuild lost data

disk mirroring

a process of storing data that provides an exact copy that fully protects users in the event of data loss

optical disk

a rigid disk of plastic onto which data is recorded by special lasers that physically burn pits in the disk

compact disk read-only memory (CD-ROM)

a common form of optical disk on which data, once recorded, cannot be modified

could take from a few minutes to over half an hour to load the tape and get the information.

Magnetic disk storage varies widely in capacity and portability. Standard diskettes are portable but have a slower access time and lower storage capacity (1.44 megabytes) than fixed hard disks. Hard disk storage, while more costly and less portable, has greater storage capacity and quicker access time.

RAID. Companies' data storage needs are expanding rapidly. Today's storage configurations routinely entail many hundreds of gigabytes. However, putting the company's data on-line involves a serious business risk—the loss of critical business data can put a corporation out of business. The concern is that the most critical mechanical components inside a disk storage device—the disk drives, the fans, and other input/output devices—can break (like most things that move).

Organizations now require that their data storage devices be fault tolerant—the ability to continue with little or no loss of performance in the event of a failure of one or more key components. **Redundant array of independent/inexpensive disks (RAID)** is a method of storing data that generates extra bits of data from existing data, allowing the system to create a "reconstruction map" so that if a hard drive fails, it can rebuild lost data. With this approach, data is split and stored on different physical disk drives using a technique called stripping to evenly distribute the data. Since being developed at the University of Berkeley in 1987, RAID technology has been applied to storage systems to improve system performance and reliability.

RAID can be implemented in several ways. In the simplest form, RAID subsystems duplicate data on drives. This process, called **disk mirroring**, provides an exact copy that fully protects users in the event of data loss. However, if full copies are always to be kept current, organizations need to double the amount of storage capacity that is kept on-line. Thus, disk mirroring is expensive. Other RAID methods are less expensive because they only partly duplicate the data.[7] This allows storage managers to minimize the amount of extra disk space (or overhead) they must purchase to protect data.

Optical disks. Another type of secondary storage medium is the **optical disk**. Similar in concept to a ROM chip, an optical disk is simply a rigid disk of plastic onto which data is recorded by special lasers that physically burn pits in the disk. Data is directly accessed from the disk by an optical disk device, which operates much like a stereo's compact disk player. This optical disk device uses a low-power laser that measures the difference in reflected light caused by a pit (or lack thereof) on the disk.

Each pit represents the binary digit 1; each unpitted area (called a *land*) represents the binary digit 0. Thus, the presence or lack of a pit determines the bit. Once a master optical disk has been created, duplicates can be manufactured using techniques similar to those used to produce music CDs.

A common form of optical disk is called **compact disk read-only memory (CD-ROM)**. Once data has been recorded on a CD-ROM, it cannot be modified—the disk is "read only." CD-ROM disks and hardware have moved from being unique add-ons in the mid-1980s to become a standard

CD-rewritable (CD-RW)

a common form of optical disk that allows personal computer users to replace their diskettes with high-capacity CDs that can be written upon and edited over

write-once, read-many (WORM)

an optical disk that allows businesses to record customized data and information; once data and information has been recorded onto a WORM disk, it can be repeatedly accessed, but it cannot be altered

magneto-optical disk

a hybrid that combines magnetic disk technologies and CD-ROM technologies

feature of today's personal computers. **CD-rewritable (CD-RW)** technology allows personal computer users to replace their diskettes with high-capacity CDs that can be written upon and edited over. The CD-RW disk can hold 680 MB of data—roughly 500 times the capacity of a 1.44 MB diskette.

Write-once, read-many (WORM) optical disks allow businesses to record customized data and information onto an optical disk. Like a CD-ROM, data is written onto a WORM optical disk by a high-powered laser and read by a lower-powered laser that measures the reflected light. Once data and information have been recorded onto a WORM disk, it can be repeatedly accessed, but it cannot be altered—hence the name "write-once, read-many." WORM disks provide a relatively low-cost way for recording and storing data that needs to be kept but not altered.

Magneto-optical disk. The **magneto-optical disk** combines magnetic disk technologies with CD-ROM technologies. Like magnetic disks, MO disks can be read and written to. And like floppy disks, they are removable. However, their storage capacity can be more than 200 megabytes, much greater than magnetic floppies. In terms of data access speed, they are faster than floppies but not as fast as hard disk drives. This type of disk uses a laser beam to change the molecular configuration of a magnetic substrate on the disk, which in turn creates visual spots. In conjunction with a photodetector, another laser beam reflects light off the disk and measures the size of the spots; the presence or absence of a spot indicates a bit. The disk can be erased by demagnetizing the substrate, which in turn removes the spots, allowing the process to begin again. Some MO drives can store more than a gigabyte on a single, removable disk. PowerMO by Olympus and the DynaMo by Fujitsu are two examples of magneto-optical devices.

The primary advantage of optical disks is their huge storage capacities, compared with other secondary storage media. Because of this, optical disks can store large applications and programs that contain graphics and audio data. They also allow for storage of data on speculation; data not needed at a given moment can be easily stored for later possible use. Optical disks, however, suffer from some minor inconveniences, such as slow access time compared with diskettes and the lack of sufficient software to fully exploit the technology. In time, however, these inconveniences should diminish.

FIGURE 3.11

DVD Drive and Media

DVD is a storage format that offers the potential to be faster with greater storage capacity than today's CD-ROM.
(Source: Courtesy of Toshiba America Electronic Components, Inc.)

digital video disk (DVD)

storage format used to store digital video or computer data

Digital video disk. Another storage format that has exciting possibilities is the **digital video disk (DVD)**. The DVD brings together the hitherto separate worlds of home computing and home video. A DVD disk is a five-inch CD-ROM lookalike (Figure 3.11) with the ability to store about 135 minutes of digital video. Media gurus are predicting that the DVD will do for video what the compact disk did for sound. When used to store video, the picture quality far surpasses anything seen on tape, cable, or standard broadcast TV—sharp detail, true color, no flicker, no snow. The sound is recorded in digital Dolby, creating clear "surround" effects by completely separating all the audio channels in a home theater. The DVD costs less to duplicate and ship than videocassettes, takes less shelf space, and delivers higher quality.

DVD can double as a computer storage disk and provide up to 17-GB capacity. The physical disks resemble CD-ROMs, only they are thinner, so DVD players can also read current CD-ROMs, but current CD-ROM players cannot read the DVDs. Each DVD can hold at least 4.7 GB on a single side; some DVDs are double-layer disks, capable of holding 8.5 GB. Either type can be bonded back-to-back to create a two-sided disk with up to 17 GB of data. The access speed of a DVD drive is faster than the typical CD-ROM drive, with a data transfer rate clocked at 1.35 MB per second. DVD manufacturers include Sony, Philips, Toshiba, and others. These companies are also actively involved in making and improving standard CD-ROM drives. A number of companies are offering movie titles in the newer DVD format, including Time-Warner and Columbia. Eventually, DVD technology will provide write-once disks and rewrite versions capable of holding 2.6 GB per side.

flash memory

a silicon computer chip that, unlike RAM, is nonvolatile and keeps its memory when the power is shut off

Memory cards. A group of computer manufacturers formed the Personal Computer Memory Card International Association (PCMCIA) to create standards for a peripheral device known as a PC memory card. These PC memory cards are credit-card-size devices that can be installed in an adapter or slot in many personal computers. To the rest of the system, the PC memory card functions as though it were a fixed hard disk drive. Although the cost per megabyte of storage is greater than for a traditional hard disk storage, these cards are less failure prone than hard disks, are portable, and are relatively easy to use. Software manufacturers often store the instructions for their program on a memory card for use with laptop computers.

PC memory cards function like a fixed hard drive but are portable. They plug into an adapter or slot on a PC and provide up to 520 MB storage capacity.
(Source: Courtesy of Simple Technology.)

Flash memory. **Flash memory** is a silicon computer chip that, unlike RAM, is nonvolatile and keeps its memory when the power is shut off. Flash memory chips are small and can be easily modified and reprogrammed, which makes them popular in computers, cellular phones, and other products. Flash memory is also used in some handheld computers to store data and programs, in digital cameras to store photos, and in airplanes to store flight information in the cockpit. Compared with other types of secondary storage, flash memory can be accessed more quickly, consumes less power, and is smaller in size. The primary disadvantage is cost. Flash memory chips

FIGURE 3.12

Removable Storage

Removable storage drives allow one to add additional storage capacity by simply plugging in a removable disk or cartridge. Some people are also using these devices as a backup for their hard disks and as a way to archive important data. (Source: Courtesy of Iomega Corporation.)

removable storage devices

internal or external devices that provide additional storage capacity in the form of removable disks or cartridges

can cost almost three times more per megabyte than a traditional hard disk. Nonetheless, the market for flash memory has exploded in recent years.

Removable storage. Removable storage devices use removable disks or cartridges that offer unlimited capacity (Figure 3.12). When your storage needs increase, you can use more removable disks or cartridges. The storage capacity can range from under 100 MB to several gigabytes per cartridge. In recent years, the access speed of removable storage devices has increased. Some devices are about as fast as an internal disk drive.

Removable storage devices can be internal or external. A few personal computers are now including internal storage devices as standard equipment. Examples of removable storage drives include Zip and Jaz by Iomega, EZ135 by SyQuest, and Superdisk. The Superdisk storage device can use a standard 1.44 MB floppy disk or a new 120 MB disk. With so much critical data stored on your hard drive, it is wise to make frequent backups. However, use of the standard diskette would require over 200 disks and several hours to back up even a modest 300-MB hard drive. These removable storage devices are ideal for backups. They can hold at least 70 or more times as much data and operate five times faster than the existing 1.44-MB diskette drives. Although more expensive than fixed hard disks, removable disks or cartridges combine hard disk storage capacity and diskette portability. Some organizations prefer removable hard disk storage for the portability and control it provides. For example, a large amount of data can be taken to any location, or it can be secured so that access is controlled.

The overall trend in secondary storage is toward more direct-access methods, higher capacity, and increased portability. Organizations that select effective storage systems can greatly benefit from this trend. The specific type of storage should be chosen with consideration of the needs and resources of the organization. In general, the ability to store large amounts of data and information and access it quickly can increase organizational effectiveness and efficiency by allowing the information system to provide the desired information in a timely fashion. Table 3.2 lists the most common secondary storage devices and their capacities for easy reference.

TABLE 3.2

Comparison of Secondary
Storage Devices

Storage Device	Year First Introduced	Maximum Capacity
3.5-inch diskette	1987	1.44 MB
CD-ROM	1990	650 MB
Zip	1995	100 MB
DVD	1996	17 GB

INPUT AND OUTPUT DEVICES: THE GATEWAY TO COMPUTER SYSTEMS

A user's first experience with computers is usually through input and output devices. Through these devices—the gateways to the computer system—people provide data and instructions to the computer and receive results from it. Input and output devices are part of the overall user interface, which includes other hardware devices and software that allow humans to interact with a computer system.

As with other computer system components, the selection of input and output devices depends on organizational goals and information system objectives. For example, many restaurant chains use handheld input devices or computerized terminals that let waiters enter orders to ensure timely and accurate data input. These systems have cut costs by making inventory tracking more efficient and marketing to customers more effective.

Characteristics and Functionality

Rapidly getting data into a computer system and producing timely output can be very important for many organizations. The form of the output desired, the nature of the data required to generate this output, and the required speed and accuracy of the output and the input determine the appropriate output and input devices. Some organizations have very specific needs for output and input, requiring devices that perform specific functions. For example, American Express is testing and deploying a voice recognition system to take airline reservations so that customers will be able to check and book flights by talking to a computer on the phone.[8] The more specialized the application, the more specialized the associated system input and output devices.

The speed and functions performed by the input and output devices selected and used by the organization should be balanced with their cost, control, and complexity. More specialized devices might make it easier to enter data or output information, but they are generally more costly, less flexible, and more susceptible to malfunction.

The nature of data. Getting data into the computer—input—often requires transferring human-readable data, such as a sales order, into the computer system. Human-readable data is data that can be directly read and understood by humans. A sheet of paper containing adjustments to inventory is an example of human-readable data. By contrast, machine-readable data can be understood and read by computer devices (e.g., the universal bar code read by scanners at the grocery checkout) and is typically stored as bits or bytes. Data on inventory changes stored on a diskette is an example of machine-readable data.

Data can be both human readable and machine readable. For example, magnetic ink on bank checks can be read by humans and computer system input devices. Most input devices require some human interaction, because people most often begin the input process by organizing human-readable data and transforming it into machine-readable data. Every keystroke on a keyboard, for example, turns a letter symbol of a human language into a digital code that the machine can understand.

Data entry and input. Getting data into the computer system is a two-stage process. First, the human-readable data is converted into a machine-readable form through a process called **data entry**. The second stage involves transferring the machine-readable data into the system. This is **data input**.

Today, many companies are using on-line data entry and input—the immediate communication and transference of data to computer devices directly connected to the computer system. On-line data entry and input places data into the computer system in a matter of seconds. Organizations in many industries require the instantaneous update offered by this approach. For example, an airline clerk may need to enter a last-minute reservation. On-line data entry and input is used to record the reservation as soon as it is made. Reservation agents at other terminals can then access this data to make a seating check before they make another reservation.

Source data automation. Regardless of how data gets into the computer, it should be captured and edited at its source. **Source data automation** involves capturing and editing data where the data is originally created and in a form that can be directly input to a computer, thus ensuring accuracy and timeliness. For example, using source data automation, sales orders can be entered into the computer by the salesperson at the time and place the order is taken. Any errors can be detected and corrected immediately. If any item is temporarily out of stock, the salesperson can discuss options with the customer. Prior to source data automation, orders were written on a piece of paper to be entered later into the computer (often by someone other than the person who took the order). Often the hand-written information could not be read or, worse yet, got lost. If there were problems during data entry, it was necessary to contact the salesperson or the customer to "recapture" the data needed for order entry, thus leading to further delays and customer dissatisfaction.

Input Devices

[handwritten: Transfer data inside comp]

Literally hundreds of devices can be used for data entry and input. These range from special-purpose devices used to capture specific types of data to more general-purpose input devices. Some of the special-purpose data entry and input devices will be discussed later in this chapter. First, however, we will focus on devices used to enter and input more general types of data, including text, audio, images, and video for personal computers.

---

data entry

the process by which human-readable data is converted into a machine-readable form

data input

the process of transferring machine-readable data into the computer system

source data automation

the process of capturing and editing data where the data is originally created and in a form that can be directly input to a computer, thus ensuring accuracy and timeliness

An ergonomic keyboard is designed to be more comfortable to use.
(Source: Courtesy of Microsoft Corporation.)

voice-recognition devices

input devices that recognize human speech

digital computer cameras

input devices that record and store images and video in digital form

Personal computer input devices. A keyboard and a computer mouse are the most common devices used for entry and input of data such as characters, text, and basic commands. Some companies are developing newer keyboards that are more comfortable, adjustable, and faster to use. These keyboards, such as the split keyboard by Microsoft and others, are designed to avoid wrist and hand injuries caused by hours of keyboarding.

A computer mouse is used to "point to" and "click on" symbols, icons, menus, and commands on the screen. This causes the computer to take a number of actions, such as placing data into the computer system.

Voice-recognition devices. Another type of input device can recognize human speech. Called **voice-recognition devices**, these tools use microphones and special software to record and convert the sound of the human voice into digital signals. Speech recognition can be used on the factory floor to allow equipment operators to give basic commands to machines while they are using their hands to perform other operations. Voice recognition can also be used by security systems to allow only authorized personnel into restricted areas.

Voice-recognition devices analyze and classify speech patterns and convert them into digital codes. Some systems require "training" the computer to recognize a limited vocabulary of standard words for each user. Operators train the system to recognize their voices by repeating each word to be added to the vocabulary several times. Other systems are speaker independent and allow a computer to understand a voice it has never heard. In this case, the computer must be able to recognize more than one pronunciation of the same word. For example, recognizing the use of the phrase "Please?" spoken by someone from Cincinnati as meaning the same as "Huh?" spoken by someone from the Bronx or "I beg your pardon?" spoken by an English person.

Digital computer cameras. Some personal computers work with **digital computer cameras** that record and store images and video in digital form. These cameras look very similar to a regular camera. When you take pictures, the images are electronically stored in the camera. A cable is then used to connect the camera to the parallel port on the personal computer and the images can be downloaded. During the download process, the visual images are converted into digital codes by a computer board. Once downloaded and converted into digital format, the images can be modified as desired and included in other applications. For example, a photo of the company office recorded by a digital computer camera can be captured and then pasted into a word processing document used in a company brochure. You can even add sound and handwriting to the photo. Some personal computers, as shown in Figure 3.13, have a video camera that records full-motion video.

FIGURE 3.13

A PC Equipped with a
Computer Camera

Digital video cameras make it
possible for people at distant
locations to conduct videoconferences, thereby eliminating
the need for expensive travel to
attend physical meetings.
(Source: OnWAN desktop videoconferencing package; photo courtesy
of Zydacron.)

While the first generation of digital cameras could create photos with a resolution of 650 × 480 pixels, the current state-of-the-art cameras can deliver 2 megapixel resolution, enough resolution to deliver snapshot-sized photos that provide crisp, clear quality images. Dataquest estimates that by 2000, North American sales of digital cameras could exceed 7 million units. The number one advantage of digital cameras is saving time and money by eliminating the need to process film. Convenience, not photo quality, was a key deciding factor for Kodak digital-camera user Sentinel Real Estate Corp., which buys real-estate investment properties for clients and uses the cameras to photograph the properties.[9]

Terminals. Inexpensive and easy to use, terminals are input devices that perform data entry and data input at the same time. A terminal is connected to a complete computer system, including a processor, memory, and secondary storage. General commands, text, and other data are entered via a keyboard or a mouse, converted into machine-readable form, and transferred to the processing portion of the computer system. Terminals, normally connected directly to the computer system by telephone lines or cables, can be placed in offices, warehouses, and on a factory floor.

Scanning devices. Image and character data can be input using a scanning device. A page scanner is like a copy machine. The page to be scanned is typically inserted into the scanner or placed face down on the glass plate of the scanner, covered, and scanned. With a handheld scanner, the scanning device is moved or rolled manually over the image to be scanned. Both page and handheld scanners can convert monochrome or color pictures, forms, text, and other images into machine-readable digits. It has been estimated that U.S. enterprises generate more than one billion pieces of paper daily. To cut down on the high cost of using and processing paper, many companies are looking to scanning devices to help them manage their documents.

Optical data readers. A special scanning device called an optical data reader can also be used to scan documents. The two categories of optical

4756

19_____ 80-7068/3216

PAY TO THE
ORDER OF_____ $ [_____]

_____ DOLLARS

**First
National
Bank**

MEMO _____ _____

⑆221570687⑆ 002 710258 9⑈ 4756

Bank
identification
number

Account
number

Check
number

FIGURE 3.14

MICR Device

Magnetic ink character recognition is a process by which data is coded on the bottom of a check or other form using special magnetic ink, which is readable by both computers and humans. For examples, look at the bottom of a bank check or most utility bills.
(Source: Courtesy of NCR Corporation.)

data readers are for optical mark recognition (OMR) and optical character recognition (OCR). OMR readers are used for test scoring and other purposes when test takers use pencils to fill in boxes on OMR paper, which is also called a "mark sense form." In comparison, most OCR readers use reflected light to recognize various characters. With the use of special software, OCR readers can convert handwritten or typed documents into digital data. Once entered, this data can be shared, modified, and distributed over computer networks to hundreds or thousands of individuals.

Magnetic ink character recognition (MICR) devices. In the 1950s, the banking industry became swamped with paper in the form of checks, loan applications, bank statements, and so on. The result was the development of magnetic ink character recognition (MICR), a system for reading this data quickly. With MICR, data is placed on the bottom of a check or other form using a special magnetic ink. Data printed with this ink using a character set is readable by both people and computers (Figure 3.14).

point-of-sale devices (POS)

terminals used in retail operations to enter sales information into the computer system

Point-of-sale (POS) devices. Point-of-sale (POS) devices are terminals used in retail operations to enter sales information into the computer system. The POS device then computes the total charges, including tax. Many POS devices also use other types of input and output devices, like keyboards, bar code readers, printers, and screens. A large portion of the money businesses spend on computer technology involves POS devices.

Automatic teller machine (ATM) devices. The automatic teller machine (ATM) is a terminal used by most bank customers to perform withdrawals and other transactions concerning their bank accounts. The ATM, however, is no longer used only for cash and bank receipts. Companies use various ATM devices to support their specific business processes. Some are able to dispense tickets for airlines, concerts, and soccer games. Some colleges use them to output transcripts. For this reason, the input and output capabilities of ATMs are quite varied. Like POS devices, ATMs may combine other types of input and output devices. Unisys, for example, has developed an ATM kiosk where bank customers can not only make cash withdrawals and pay bills, but also receive advice on investments and retirement planning.[10]

Pen input devices. By touching the screen with a pen input device, it is possible to activate a command or cause the computer to perform a task; enter handwritten notes; and draw objects and figures. Pen input requires special software and hardware. Handwriting recognition software can convert handwriting on the screen into text.

Light pens. A light pen uses a light cell in the tip of a pen. The cell recognizes light from the screen and determines the location of the pen on the screen. Like pen input devices, light pens can be used to activate commands and place drawings on the screen.

Touch-sensitive screens. Advances in screen technology allow display screens to function as input as well as output devices. By touching certain parts of a touch-sensitive screen, you can execute a program or cause the computer to take an action. Touch-sensitive screens are popular input devices for some small computers because they preclude the necessity of keyboard input devices that consume space in storage or in use. They are frequently used at gas stations for customers to select grades of gas and request a receipt, at fast-food restaurants for order clerks to enter customer choices, at information centers in hotels to allow guests to request facts about local eating and drinking establishments, and at amusement parks to provide directions to patrons. They also are used in kiosks at airports and department stores.

Bar code scanners. A bar code scanner employs a laser scanner to read a barcoded label. This form of input is widely used in grocery store checkouts (fixed-position scanner) and in warehouse inventory control (hand-held scanner). Consumer goods manufacturers and

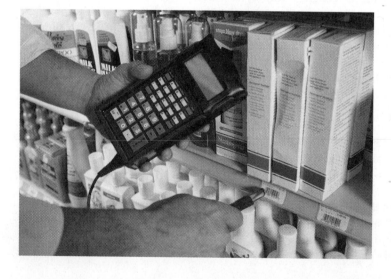

Computer readable bar codes on products provide retailers with data on the product shipment, size, and pricing.
(Source: © 1998 PhotoDisc.)

retailers employ bar codes as a means of source data automation to improve the accuracy and reduce the effort associated with inventory management and receiving. Labels with computer-readable bar codes are automatically applied to each container in which a product is shipped at the manufacturer. This bar code serves as a "license plate" and uniquely identifies that container. When the product is shipped from the manufacturer to the retailer, a computer-readable file is sent to the retailer with a list of the bar codes, quantities, and descriptions of all items on that shipment. Workers at the receiving dock of the retailer scan the label on each container with a laser gun, and the retailer's computer automatically matches that "license plate" to the data received electronically from the manufacturer, greatly speeding up shipments. Also, the process of reconciling what was shipped to what was received becomes simple.

Output Devices   *read*

Computer systems provide output to decision makers at all levels of an organization to solve a business problem or capitalize on a competitive opportunity. In addition, output from one computer system can be used as input into another computer system within the same information system. The desired form of this output might be visual, audio, and even digital. Whatever the output's content or form, output devices function to provide the right information to the right person in the right format at the right time.

Display monitors. The display monitor is a TV-screen-like device on which output from the computer is displayed. Because the monitor uses a cathode ray tube to display images, it is sometimes called a CRT. The monitor works in much the same way as a TV screen—one or more electron beams are generated from cathode ray tubes. As the beams strike a phosphorescent compound (phosphor) coated on the inside of the screen, a dot called a pixel lights up on the screen. The electron beam sweeps back and forth across the screen so that as the phosphor starts to fade, it is struck again and lights up again.

Many decision makers work directly off the monitor, or screen, of their computer systems. Even this fairly standard output device can provide organizational advantages. An upgrade from a 14-inch monitor to a 17-inch monitor, for example, can provide a productivity increase for managers working with certain applications. The larger display allows for easier, more precise reading of detailed reports. With today's wide selection of monitors, price and overall quality can vary tremendously.

Progress has been made in enabling one display device to display both TV and PC output. PC Theater from Compaq is a consumer entertainment device that merges computing and traditional forms of media and entertainment content. This system combines the best features of a TV with a 36-inch monitor and multimedia PC, delivering more entertainment options. The consumer can watch TV, use the PC, or do both at the same time.[11] The Gateway Destination Big Screen TV comes with a 31-inch monitor, which can be used with the personal computer and/or to watch TV. The suggested retail price for either system is about $5,000.

The quality of a screen is often measured by the number of horizontal and vertical pixels used to create it. A **pixel** is a point of light on a display

pixel

a point of light on a display screen

screen. It can be in one of two modes: on or off. A larger number of pixels per square inch means a higher resolution, or clarity and sharpness of the image. For example, a screen with a 1,024 × 768 resolution (786,432 pixels) has a higher sharpness than one with a resolution of 640 × 350 (224,000 pixels). The distance between one pixel on the screen and the next nearest pixel is known as dot pitch. The common range of dot pitch is from .25 to .31 mm. The smaller the number, the better the picture. A dot pitch of .28 mm or smaller is considered good. Greater pixel densities and smaller dot pitches yield sharper images of higher resolution. Display monitors can be either monochrome or color. Typically, characters in monochrome display screens appear in one of three colors: gray, green, or amber. Color monitors, also called RGB (red, green, and blue) monitors, have the ability to display the basic colors in a variety of shades. A monitor's ability to display color is a function of the quality of the monitor, the amount of RAM in the computer system, and the monitor's graphics adapter card. The color graphics adapter (CGA) was one of the first technologies to display color images on the screen. Today, super video graphics array (SVGA) displays are standard, providing vivid colors and superior resolution.

Liquid crystal displays (LCDs). Because CRT monitors use an electron gun, there must be a distance of one foot between the gun and screen, causing them to be large and bulky. Thus, a different technology, flat screen display, is used for portable personal computers and laptops. One common technology used for flat screen displays is the same liquid crystal display technology used for pocket calculators and digital watches. LCD monitors are flat displays that use liquid crystals—organic, oil-like material placed between two polarizers—to form characters and graphic images on a backlit screen.

The primary choices in LCD screens are passive-matrix and active-matrix LCD displays. In a passive-matrix display, the CPU sends its signals to transistors around the borders of the screen, which control all the pixels in a given row or column. In an active-matrix display, each pixel is controlled by its own transistor attached in a thin film to the glass behind the pixel. Passive-matrix displays are typically dimmer, slower, but less expensive than active-matrix ones. Active-matrix displays are bright, clear, and have wider viewing angles than passive-matrix displays. Active-matrix displays, however, are more expensive and can increase the weight of the screen.

LCD technology is also being used to create thin and extremely high-resolution monitors for desktop computers. Although the screen may measure just 13 inches from corner to corner, the display's extremely high resolution—1,280 × 1,280 pixels—lets it show as much information as a conventional

CRT monitors are large and bulky in comparison with LCD monitors (flat displays).
(Source: Courtesy of Sony Electronics, Inc.)

20-inch monitor. And while cramming more into a smaller area causes text and images to shrink, you can comfortably sit much closer to an LCD screen than the conventional CRT monitor. Unfortunately these monitors are expensive. A 14-inch flat-panel monitor costs around $2,000, while a top-of-the-line 21-inch CRT monitor costs around $1,000. However, prices are expected to continue dropping.

Printers and plotters. One of the most useful and popular forms of output is called hard copy, which is simply paper output from a device called a printer. Printers with different speeds, features, and capabilities are available. Some can be set up to accommodate different paper forms such as blank check forms, invoice forms, and so forth. Newer printers allow businesses to create customized printed output for each customer from standard paper and data input using full color.

The speed of the printer is typically measured by the number of pages printed per minute (ppm). Like a display screen, the quality, or resolution, of a printer's output depends on the number of dots printed per inch. A printer that prints 600 dots per inch (dpi) prints more clearly than one with 300 dpi. A recurring cost of using a printer is the inkjet or laser cartridge that must be replaced every few thousand pages of output. Figure 3.15 shows a laser printer and an example of its output.

Plotters are a type of hard-copy output device used for general design work. Businesses typically use these devices to generate paper or acetate blueprints, schematics, and drawings of buildings or new products onto paper or transparencies. Standard plot widths are 24 inches and 36 inches; the length can be whatever meets the need—from a few inches to feet.

Computer output microfilm (COM) devices. Companies that produce and store significant numbers of paper documents often use computer output microfilm (COM) devices to place data from the computer directly onto microfilm for future use. The traditional photographic phase of conversion to microfilm is eliminated. Once this is done, a standard microfilm reader can access the data. Newspapers and journals typically place their past publications on microfilm using COM, giving readers the ability to view past articles and news items.

Special-Purpose Input and Output Devices

Many additional input and output devices are used for specialized or unique applications. A **multifunction device** can combine a printer, fax machine, scanner, and copy machine into one device. Multifunction devices are less expensive than buying these devices separately, and they take less space on a desktop compared with separate devices. For example, the $240 Canon Inc. BJC-4304 inkjet printer can be converted to a color scanner for documents, photographs, and other images.[12] Special-purpose hearing devices can be used to detect manufacturing or equipment problems. The Georgia Institute of Technology has developed a

plotters

a type of hard-copy output device used for general design work

multifunction device

a device that combines a printer, fax machine, scanner, and copy machine into one unit

FIGURE 3.15

Laser printers, available in a wide variety of speeds and price ranges, have many features, including color capabilities. They are the most common solution for outputting hard copies of information. (Source: Courtesy of Hewlett-Packard Company.)

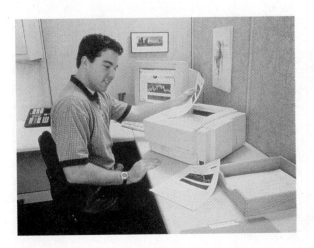

hardware device that can "listen to" equipment to detect worn or damaged parts. Voice-output devices, also called voice-response devices, allow the computer to send voice output in the form of synthesized speech over phone lines and other media. Some banks and financial institutions use voice recognition and response to give customers account information over the phone. The Smart Disk, now selling in Germany, allows you to place a smart card, which is similar to a credit card in size and function, into the Smart Disk device. The Smart Disk is then inserted into a standard diskette drive on a PC, which allows you to complete a variety of financial transactions using your smart card.

special-purpose computers

computers that are used for limited applications by military and scientific research groups

COMPUTER SYSTEM TYPES, STANDARDS, SELECTING, AND UPGRADING

In general, computers can be classified as either special purpose or general purpose. **Special-purpose computers** are used for limited applications by military and scientific research groups such as the CIA and NASA. Other

TABLE 3.3

Types of Computer Systems
(Source: Photos courtesy of IBM Corporation and Silicon Graphics, Inc.)

Characteristic	Network Computer	Personal Computer	Workstation	Midrange Computer	Mainframe Computer	Super-computer
Processor Speed	1–5 MIPs	5–20 MIPs	50–100 MIPs	25–100 MIPs	40–4,550 MIPs	60 billion–3 trillion instructions per second
Amount of RAM	4–16 MB	16–128 MB	32–256 MB	32–512 MB	256–1,024 MB	8,192 MB+
Approximate Cost	$500–$1,500	$1,200–$5,000	$4,000 to over $20,000	$20,000 to over $100,000	$250,000 to over $2 million	$2.5 million – $3.5 million
How Used	Supports "heads-down" data entry; connects to the Internet	Improves individual worker's productivity	Engineering; CAD; software development	Meets computing needs for a department or small company	Meets computing needs for a company	Scientific applications; marketing; customer support; product development
Example	Oracle Network computer	Compaq Pentium computer	Sun Microsystems computer	Hewlett-Packard HP-9000	IBM ES/9000	Cray C90

applications include specialized processors found in appliances, cars, and other products. Special-purpose computers are increasingly being used by businesses. For example, automobile repair shops connect special-purpose computers to your car's engine to identify specific performance problems.

General-purpose computers are used for a variety of applications and are the most common. The computers used to perform business applications discussed in this book are general-purpose computer systems. General-purpose computer systems combine processors, memory, secondary storage, input and output devices, a basic set of software, and other components. These systems can range from inexpensive personal computers to expensive supercomputers. These systems display a wide range of capabilities. Table 3.3 shows general ranges of capabilities for various types of computer systems.

general-purpose computers

computers that are used for a wide variety of applications

Computer System Types

Computer systems can range from desktop (or smaller) portable computers to massive supercomputers that require housing in large rooms. Let's examine the types of computer systems in more detail.

Personal computers. As previously noted, **personal computers (PCs)** are relatively small, inexpensive computer systems, sometimes called microcomputers. While personal computers are primarily designed for individual or single users, they are often tied into larger computer and information systems as well. Not only have PCs become ubiquitous in business, it is estimated that nearly 60 percent of all homes in the United States have a PC.[13]

personal computers (PCs)

relatively small, inexpensive computer systems, sometimes called microcomputers

There are several types of personal computers. Named for their size (small enough to fit on an office desk), *desktop computers* are the most common personal computer system configuration. Increasingly powerful desktop computers can provide sufficient memory and storage for most business computing tasks. Desktop PCs have become standard business tools; more than 30 million are in use in large corporations. The "Making a Difference" box profiles a businessman who made his career by selling PCs to businesses.

In addition to desktop personal computers, there are various smaller personal computers. A *laptop computer* is a small, lightweight PC about the size of a briefcase. Newer PCs include the even smaller and lighter *notebook* and *subnotebook* computers that provide similar computing power. Some notebook and subnotebook computers fit into docking stations of desktop computers to provide additional storage and processing capabilities. Small PCs continue to rise in popularity because of their portability and performance.

Handheld (palmtop) computers are PCs that provide increased portability because of their smaller size—some are as small as a credit card. These systems often include a wide variety of software and communications capabilities. Until 1996, these devices were fancy electronic organizers. Now Palm Computing® platform handhelds, such as the Palm III™ connected organizer, IBM® WorkPad® PC Companion, and similar handheld devices allow users to check electronic mail, browse the Internet and send faxes.[14]

Handheld (palmtop) computers include a wide variety of software and communications capabilities.
(Source: Courtesy of Palm Computing, Inc., a subsidiary of 3Com Corporation. 3Com and the 3Com logo are registered trademarks, and Palm III and the Palm III logo are trademarks of Palm Computer, Inc., 3Com Corporation, or its subsidiaries.)

MAKING A DIFFERENCE
Michael Dell Sells Direct

Michael Dell is the 32-year-old founder and chief executive of Dell Computer. His net worth of more than $4 billion makes him the richest man in Texas. Owning stock in his company over the past few years has been extremely profitable. From a split-adjusted low of $.39 a share in 1990, Dell Computer stock has soared to over $80. A $10,000 stock investment in Dell Computer at its 1988 initial offering is now worth more than $1 million. Dell Computer has become the driving force in the PC computer business.

The key reason for Dell's huge success is its approach of selling PCs directly to customers—90 percent of whom are businesses, schools, and government agencies. Dell builds PCs only after they have been ordered. Until recently, other PC manufacturers like Compaq, Hewlett-Packard, and IBM built PCs in advance of getting customer orders. Based on often inaccurate forecasts of customer demand, they build lots of PCs. Then they test, inspect, box, and store them before sending a batch to a reseller's warehouse, where the machines sit until customers place orders—on average six to eight weeks. Of course, each customer's needs are slightly different. So, once the reseller receives an order an individual computer is unpacked, its case opened, parts are installed or removed, the case is closed, software is loaded, and the whole thing is tested, inspected, packed, and shipped to the customer. (Each of these manufacturers, by the way, is in the process of trying to restructure their sales approach along the lines of the Dell model.)

This approach is obviously more costly because of the added steps in the process and the extra inventory that must be carried at both the manufacturer and reseller. Furthermore, there are additional hidden costs associated with the gradual obsolescence of the PCs. The rapid pace of innovation at Intel and other parts suppliers depreciates the value of newly built PCs—the same amount of money buys 2 percent more PC functionality every month. When PC manufacturers reduce their prices, they compensate resellers for the difference on any machine in stock. They also allow resellers to return unsold PCs for a full refund. Dell, on the other hand, has no U.S. resellers and virtually no inventory and thus does not have to pay compensation or refunds to resellers.

Dell's success has nothing to do with exotic software or cutting-edge chip technology. Instead, it's a matter of good strategy coupled with excellent execution. Sam Walton was another entrepreneur who was successful by selling products directly to customers by putting Wal-Mart megastores into rural areas. What made Walton so successful was an attention to detail and a vision of how the whole operation should work. Michael Dell, it turns out, is also a master at optimizing the details of his business.

DISCUSSION QUESTIONS

1. Do research (read trade journals, visit Web sites, talk to salespeople at computer stores) to learn how other PC manufacturers have adapted their sales approach to emulate Dell's. Briefly summarize the status of each of these company's efforts.

2. Visit the Dell Web site at http://www.dell.com and review the latest annual report, quarterly reports, and press releases. Can you find any evidence of a slowdown in the rate of earnings due to actions by competitors?

Sources: Adapted from Andrew E. Serwer, "Michael Dell Turns PC World Inside Out," *Fortune*, September 8, 1997, pp. 76–86; and David Kirkpatrick, "Now Everyone in PCs Wants to Be Like Mike," *Fortune*, September 8, 1997, p. 91.

Embedded computers are computers placed inside other products to add features and capabilities. In the case of automobiles, embedded computers can help with engine performance and braking. Household appliances, stereos, and some phone systems use embedded computers. Embedded computers and videos will be used in a "smart road" pilot project in Atlanta to send messages to oncoming vehicles to choose alternate routes if needed.

The **network computer** is a cheaper-to-buy and cheaper-to-run version of the personal computer that is used primarily for accessing networks and the Internet. These stripped-down versions of personal computers do not

network computer

a cheaper-to-buy and cheaper-to-run version of the personal computer that is used primarily for accessing networks and the Internet have the storage capacity or power of typical desktop computers, nor do they need it for the role they play. Unlike personal computers, network computers—or thin clients—download software across a network when needed. This can make it much easier and less expensive to manage the support, distribution, and updating of software applications. The initial target user is someone who performs what is called heads-down data entry—customer inquiry, phone order taking, and classic data entry. The network computer is designed to have no moving parts to avoid expensive equipment repairs. IBM, Oracle, and Sun Microsystems were the first companies to develop prototypes of such systems, with a purchase price in the $500–$1,500 range. Federal Express Corporation plans to replace tens of thousands of terminals with network computers.[15]

The Gartner Group (a computer industry consulting firm in Stamford, Connecticut) has estimated that the annual cost of personal computer ownership ranges from $9,000–$12,000.[16] These costs include hardware (15 percent), technical support (15 percent), administrative services (15 percent), and end user operations (55 percent), which include learning, application development, training, and data management (Figure 3.16). Advocates of the network computer argue that it has a much lower cost of ownership. Thus, it is not so much the low cost of purchasing a network computer that makes it so attractive but its low maintenance cost. However, the network computer's flexibility is extremely limited when compared with the personal computer.[17] In addition, PC companies have responded strongly to network computers with lower prices and more competitive products.[18] See the "FYI" box for a further discussion of PC costs.

Workstations are computers that fit between high-end microcomputers and low-end midrange computers in terms of cost and processing power. Workstation manufacturers use the RISC rather than the CISC computer chip to provide high computing power and reliability. They cost from $3,000–$40,000. For example, Hewlett-Packard's entry level workstation, the Kayak XA, is designed for budget-conscious business users running low-end Computer Aided Design (CAD) software, financial modeling, and software development. It comes with a 333-MHz processor, minimum of 64 MB of SDRAM, and a 6.4-GB hard drive for about $3,000.[19] Workstations are small enough to fit on an individual's desktop. A workstation may be

workstations

computers that fit between high-end microcomputers and low-end midrange computers in terms of cost and processing power

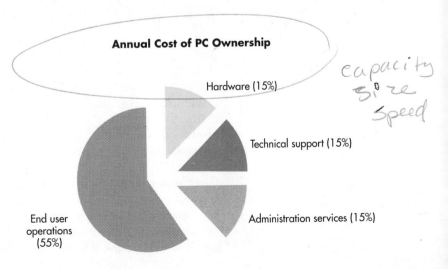

FIGURE 3.16

The Annual Cost of Owning a Personal Computer
(Source: Data from Bob Francis and Bruce Caldwell, "PC Ownership Cost Control," *InformationWeek,* January 27, 1997, p. 150.)

FYI

And the Prices Just Keep Falling

On a per megabyte or per megahertz basis, computer prices have dramatically fallen since the first computers were developed in the mid-1940s and early 1950s. Unlike any other industry, the computer industry has seen continuing price drops each year for decades. If the airline industry had the same price reductions, you would be able to fly around the world for about $3. Similar price reductions in the auto industry would mean that a great car would cost about $200. A four-bedroom house would cost about $2,500.

In the past, the per-megabyte or -megahertz price has fallen, but the cost of a typical computer remained about the same. You simply got more for your money. A desktop PC in the early 1980s with 512 KB of memory operating at a few megahertz cost about $3,000. Today, a desktop computer can also cost about $3,000, but you get 16 to 32 GB of memory with a processor operating at 300 megahertz or more.

Today, in addition to seeing a reduction in the per megabyte or megahertz cost, we are also seeing an absolute reduction in prices from thousands of dollars to hundreds of dollars. The early dreams of some PC pioneers of a PC in every business and home may eventually be reached or even exceeded. Like TVs, we might see multiple PCs in the typical home in the future. The number of PCs that cost less than $1,000 has dramatically increased, and the market is responding. In 1997, for example, the per unit shipments of these inexpensive computers jumped by more than 18 percent in a few short months. These sub-$1,000 computers are not expensive toys, but complete computers that can perform the same functions as more expensive models, but with less speed and capacity.

Falling PC prices can present a difficult dilemma for first time buyers or those considering an upgrade. When should a new PC be purchased? Do you wait for further price reductions or buy something now? Experts often suggest making a purchase now. Getting the tremendous benefits of a PC for first time users is usually worth a slightly higher price today. For those upgrading, the decision is more difficult. Unless you must have new features, additional capacity, or increased speed, older computers can often do everything you need.

DISCUSSION QUESTIONS

1. The PC industry is dynamic and constantly changing. What increased speed (in megahertz) and additional memory capacity (in gigabytes) will the typical PC have in a year?
2. If you don't currently have a PC, what price, features, or factors would cause you to make your first PC purchase? If you currently have a PC, under what conditions would you upgrade to a newer, faster machine?

Sources: Adapted from Jim Carlton, "Cheaper PCs Start to Attract New Customers," *The Wall Street Journal,* January 26, 1998, p. B1; Yardena Arar, "Oracle's $250 Computer Finally Arrives," *PC World,* February 1998, p. 76; Sean Fulton, "Net PCs Get Real," *Internet World,* February 1998, p. 35.

dedicated to support a single user or a small group of users. High-end personal computers are approaching the computing power of a workstation.

Workstations are used to support engineering and technical users who perform heavy mathematical computing, computer-aided design (CAD), and other applications requiring a high-end processor. Such users need very powerful CPUs, large amounts of main memory, and extremely high-resolution graphic displays to meet their needs. Engineers use CAD programs to create two- and three-dimensional engineering drawings and product designs. Although initially creating a design with CAD software may take as long as, if not longer than, creating a design in the traditional manner, CAD is much faster to revise. Instead of redrawing an entire plan, CAD allows the engineers to modify it with a few clicks of a mouse. They can also easily pan the design or zoom in to magnify a particular part. They can also rotate the view to examine it from different perspectives. Read the "E-commerce" box to see an example of how important it is to size the power of the computer to the task at hand.

E-COMMERCE

Eddie Bauer Adds Internet Site to In-Store and Catalog Sales Channels

Since 1920, Eddie Bauer has evolved from a single store in Seattle to an international company with nearly 500 stores and more than 120 million catalogs in the United States, Canada, Japan, Germany, and the United Kingdom with sales in excess of $1.6 billion. Today it is one of the country's premiere outdoor and casual clothing retailers. Recently, Eddie Bauer developed a thriving on-line business. At www.Eddiebauer.com, people can shop for clothing, outerwear, and shoes right from their PCs.

Eddie Bauer was an E-commerce pioneer when the company launched its first attempt at a Web site several years ago. The initial site was developed in house. But the company soon realized its own limitations. Despite the fact that the Web site was bringing in business, the platform had serious limitations in its flexibility. Users were complaining about how long it took to "visit the virtual store" because of slow Web site response time. As a result, sales over the Internet were not taking off as expected. So Eddie Bauer rebuilt the site, this time using software and hardware expertise from key vendors such as Microsoft.

The Eddie Bauer Web page provides customers with the ability to order items on-line, just as easily as flipping through one of their print catalogs. In addition, the company provides an on-line form that customers can use to order items from the print catalog.

The Eddie Bauer "Exclusive" section is completely customized to each registered user. As a member of EB Exclusive, you can customize your shopping experience. Size, name, and other fields on the site automatically default to your entered personal information. You can create a wish list of your favorite Eddie Bauer products while you shop. This service lets friends and family select gifts you want. You can also use the Eddie Bauer exclusive reminder service to help you remember special dates, such as an anniversary or birthday, when you need to purchase a gift for someone. Simply enter the dates you want to remember, leave yourself a message, and you will receive an e-mail reminder on the day, one week, or one month ahead of time.

Because the Web site software dynamically generates Web pages based on stored user preferences, Eddie Bauer can give customers only the information they want. The software also includes analysis tools that enable Eddie Bauer to gather and interpret important usage data. For instance, 50 percent of Eddie Bauer's on-line buyers are new customers.

DISCUSSION QUESTIONS

1. Imagine that you are responsible for selecting the hardware on which the Eddie Bauer Web site will run. Identify at least three key business requirements that the hardware must be able to meet.
2. Visit this Web site. What information are you asked to enter about yourself? How might this data be used to customize the way you use the site in the future?

Sources: Adapted from Microsoft Corporation Web site at http://www.microsoft.com/siteserver/commerce/showcase/bauer.asp, accessed August 1, 1998, and Company History at the Eddie Bauer Web site at http://www.eddiebauer.com, accessed August 1, 1998.

---

midrange computers

computer systems that are about the size of a small three-drawer file cabinet and can accommodate several users at one time

Midrange computers. Midrange computers (formerly called minicomputers) are systems about the size of a small three-drawer file cabinet and can accommodate several users at one time. These systems often have secondary storage devices with more capacity than workstation computers and can support a variety of transaction processing activities, including payroll, inventory control, and invoicing. Midrange computers often have excellent processing and decision-support capabilities. Many small to medium-size organizations—like manufacturers, real estate companies, and retail operations—use midrange computers.

mainframe computers

large, powerful computers that are often shared by hundreds of concurrent users connected to the machine via terminals

Mainframe computers. Mainframe computers are large, powerful computers often shared by hundreds of concurrent users connected to the machine via terminals. The mainframe computer must reside in an environment-controlled computer room or data center with special heating, venting, and air-conditioning (HVAC) equipment to control the temperature, humidity, and dust levels around the computer. In addition, most mainframes are kept in a secured data center with limited access to the room through some kind of security system. The construction and maintenance of such a controlled access room with HVAC can add hundreds of thousands of dollars to the cost of owning and operating a mainframe computer. Mainframe computers also require specially trained individuals (called system engineers and system programmers) to care for them. Mainframe computers can crunch numbers at a rate of over 300 million instructions per second and start at $200,000 for a fully configured system.

The traditional role of the mainframe computer was as the large, centrally located computer of a firm. Mainframes have been the cornerstone of computing in large corporations for many years. From the early 1950s until the mid-1970s, virtually all commercial computer processing was mainframe based. Mainframe computers were acquired by many companies to automate accounting and finance processes such as payroll, general ledger, accounts receivable, and accounts payable. Order processing, billing, and inventory control were other early computer applications.

Today the role of the mainframe is undergoing some remarkable changes as lower-cost midrange computers, workstations, and personal computers become increasingly powerful. (Read the "Technology Advantage" box to see how one company is using both mainframes and servers in its business.) Many computer jobs that used to be run on mainframe computers have migrated onto these smaller, less expensive computers. This information processing migration is called computer downsizing. The new role of the mainframe is a large information processing and data storage utility for the corporation—running jobs too large for other computers, storing files and databases too large to be stored elsewhere, and storing backups of files and databases created elsewhere (these large stores of data are sometimes called data warehouses). The mainframe is capable of handling the millions of daily transactions associated with airline, automobile, and hotel/motel reservation systems. It can process the tens of thousands of daily queries necessary to provide data to decision support systems. Its massive storage and input/output capabilities enable it to play the role of a video computer, providing full-motion video to users.

complementary metal oxide semiconductor (CMOS)

a semiconductor fabrication technology that uses special material to achieve low-power dissipation

Complementary metal oxide semiconductor (CMOS) is a semiconductor fabrication technology that uses special material to achieve low-power dissipation. IBM, Amdahl Corp., and Data General have all made a substantial commitment to this new technology and have announced CMOS-based mainframe products.[20] Over time, mainframes have been evolving into smaller, faster, less expensive systems as a result of CMOS processors and providing support for large packaged software products such as SAP, Web technologies, and communications protocols much like their smaller cousins, the midrange computers.[21]

TECHNOLOGY ADVANTAGE
GTE Corp. Uses Mainframes and Servers to Compete

With world headquarters in Palo Alto, California, Sun Microsystems has been described as "the last standing, fully integrated computing company adding its own value at the chip, operating system, and systems level." In a fiercely competitive industry, Sun has averaged 15 to 20 percent growth over the last several years and has a strong balance sheet with nearly $1 billion cash in the bank. Among the world leaders in workstation sales (with 35 percent in revenues and 39 percent in unit sales), Sun is attempting to transform itself into an enterprise computing firm focused on global network computing.

Sun's largest server, the Sun Enterprise 10000 (nicknamed Starfire), is one of the most powerful and scalable parallel processor servers in the industry—so powerful, in fact, that it rivals mainframe computers in features and functionality. Starfire supports from 16 to 64 UltraSPARC II processors (250 MHz processor), which enables customers to run the most demanding, multi-terabyte applications for data warehousing, decision support, on-line transaction processing and data analysis on a single scalable server. Pricing for an entry-level system starts around $870,000 for a 16 processor server with 2 GB of memory.

GTE is one of the largest publicly held telecommunications companies in the world, with annual sales exceeding $21 billion. GTE offers local and wireless service in 29 states and long-distance service and Internet access in all 50 states. GTE was the first among its peers to offer "one-stop shopping" for local, long-distance and Internet access services. Outside the United States, where GTE has operated for more than 40 years, the company serves over 6.5 million customers. GTE is also a leader in government defense communications systems and equipment, directories and telecommunications-based information services, and aircraft-passenger telecommunications.

Deregulation of the telecommunications industry has forced GTE to become a tougher competitor, needing better access to information for it to prosper in this new environment. As part of a $17 million deal, GTE purchased four Sun Ultra Enterprise 10000 Starfire Unix servers, several Sun Ultra 2 workstations, Enterprise 4000 servers, storage systems, and consulting from Sun's Professional Services organization. GTE plans to move data off its mainframe computers and onto these servers to provide a better platform for analyzing strategic financial and human resources information.

GTE says the Starfire platform handles its batch load requirements of more than 1 million transactions per hour. The Starfires will work alongside mainframe computers from IBM, Hitachi, and others, not replace them. Some data, including customer billing, will stay on the mainframes. But GTE officials say the daunting prospect of having to rewrite financial and human resources applications, in addition to challenges presented by the year 2000 problem, made the Starfire and Unix platform a better solution.

DISCUSSION QUESTIONS

1. Why does GTE think that the Starfire provides a better platform for analyzing strategic financial and human resources information?

2. Why isn't the Starfire replacing all GTE mainframe computers?

Sources: Adapted from the About GTE and Organizational Overview sections of the GTE Corp. Web site at http://www.gte.com, accessed August 1, 1998; "GTE Selects Sun's Scalable Starfire Servers," Sun Microsystem Web site at http://www.sun.com, accessed August 1, 1998; and Mary Hayes, "Sun Gets GTE Deal," *InformationWeek*, January 19, 1998, p. 131.

IBM has been the dominant world manufacturer of mainframe computers since the mid-1950s. IBM's speedy G5 mainframe can process 1.04 billion calculations per second. Unisys (a company formed by the merger of Burroughs Corporation and Sperry Corporation in 1986) is a distant second to IBM (see Table 3.4). Amdahl and Hitachi are examples of manufacturers that sell what are called plug-compatible systems to IBM. Their hardware can run the IBM software, including the operating system. The plug-compatible manufacturers (PCMs) attempt to deliver better price, performance, and service than does IBM. For PCMs to succeed, they must be able to react quickly to IBM's lead in new product development.

TABLE 3.4

Mainframe Computer
Manufacturers

Company	Product	MIPS (Millions of Instructions per Second)
IBM	Systems/390 Parallel	178–5,000
Unisys	A Series	1–250
Amdahl	Millennium	122–350
Hitachi Data Systems	Pilot	40–350

supercomputers

the most powerful computers, with
the fastest processing speeds

Supercomputers are the most powerful computer systems, with the fastest processing speeds.[22] Military and research organizations trying to solve complex problems use these very expensive machines. They are also used by universities and large corporations involved with research or high-technology businesses. Some large oil companies, for example, use supercomputers to perform sophisticated analysis of detailed data to help them explore for oil.

Originally, supercomputers were used primarily by government agencies to perform the high-speed number crunching needed in weather forecasting and military applications. With recent improvements in the cost and performance (lower cost and faster speeds) of these machines, they are being used more broadly for commercial purposes. For example, supercomputers are used to perform the enormous number of calculations required to draw and animate Disney films. To produce the special effects in such films required handling a gigabyte (one billion characters) of data for every second of film time. Obviously, this requires a very powerful computer.

Scientists often use computer models to simulate the problems that they are studying. These models typically omit certain details of the situation because they are not well understood or the effort required to include the details simply makes the model too difficult to create. Almost always, some degree of model completeness must be sacrificed to make the problem solvable in a reasonable amount of time. Lack of completeness leads to some loss of accuracy. For example, meteorologists studying how the oceans influence weather patterns know that ocean currents have a greater impact on weather patterns than the effect of sunlight reflected by the clouds. As a result, they may simplify their model to ignore the effect of the reflected sunlight to squeeze in more data on ocean currents. The degree of approximation is called the granularity of the computer model. Read the "Ethical and Societal Issues" box to learn how supercomputers are being used to model animals and humans.

ETHICAL AND SOCIETAL ISSUES

Information Technology Revolutionizes Biomedical Science

The San Diego Supercomputer Center (SDSC) at the University of California, San Diego, is developing a powerful new resource to allow scientists around the world to build better models of how animals and humans undertake the processes of life. The goal is to enable members of the biological modeling community to integrate their models and see them used more widely by experimental biologists, physiologists, physicians, and bioengineers.

This new resource, dubbed the BioNOME project, is an important step in computer-aided biology, medicine, and drug development. Just as engineers can now use computer models to design bridges and aircraft instead of building prototypes, biological models will be used to predict how humans or other organisms might respond to drugs, chemicals, and other physical factors.

Biological science is generating vast amounts of experimental data at an unprecedented rate. It is a tremendous challenge to organize and integrate this information into models that will drive the next round of biological advances. The BioNOME project organizes this data and provides researchers with access to integrated models. Through SDSC expertise, the project takes advantage of high-performance computers, data archives that can store hundreds of thousands of gigabytes, and the next generation of high-speed networks.

The center has established a globally accessible Web site, to which scientists can send mathematical models they have created. These models describe how various living systems are regulated and how they will interact—from molecules and cells to tissues and organs. Use of the Internet allows interactive links and accessibility, which encourages scientific collaboration. The project's initial efforts are focusing on models that describe how chemicals carry messages between and within cells and models of how the heart functions. These models might then be linked or combined to provide new knowledge on such issues as the causes of heart failure. Computer modeling also helps make possible the development of new drugs with more carefully targeted modes of action and fewer side effects, as well as better medical devices and improved medical diagnostic procedures.

Dr. Lynn F. Ten Eyck, a senior staff scientist at SDSC says, "Computer networking will be the key to the future of integrative biomedical science. It might take one scientist a lifetime to develop and validate an important biomedical model. But there are a lot of lifetimes out there. There are more scientists alive and practicing today than there have been in all of recorded human history. If we can find a way to bring this information together in a meaningful way, there is a potential to revolutionize biomedical science."

Discussion Questions

1. How do you feel about the potential of using biological models to predict how humans might respond to drugs, chemicals, and other physical factors? List some pros and cons.

2. What are the advantages of making these models broadly available through access of the BioNOME Web site? Are there any issues that might limit scientists from putting their biomedical models on the Web site?

Sources: Adapted from "SDSC Unveils the BioNOME Resource: Worldwide Systems for Modeling the Complexity of Life," October 15, 1997 joint news release from the San Diego Supercomputer Center and the University of California–San Diego Web site for the UCSD Supercomputer Center at http://www.sdsc.edu, accessed August 1, 1998; and the Introduction and Background portions of the Web site for the BioNOME project at http://bioNOME.sdsc.edu, accessed August 1, 1998.

Multimedia Computers

Multimedia merges sound, animation, and digitized video. The technology to bring these media to the desktop has existed for a long time—the early Apple Macintosh had impressive capabilities. The current technology emphasizes delivery of multimedia applications that are rich in content and features while taking up less than 600 MB of storage space. This is possible because of video, image, and audio compression technology, which

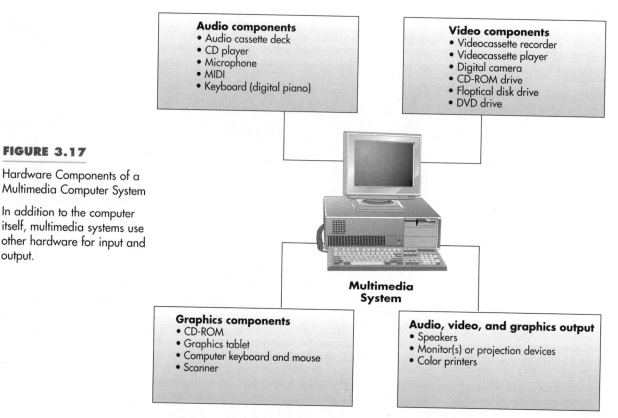

Audio components
- Audio cassette deck
- CD player
- Microphone
- MIDI
- Keyboard (digital piano)

Video components
- Videocassette recorder
- Videocassette player
- Digital camera
- CD-ROM drive
- Floptical disk drive
- DVD drive

Multimedia System

Graphics components
- CD-ROM
- Graphics tablet
- Computer keyboard and mouse
- Scanner

Audio, video, and graphics output
- Speakers
- Monitor(s) or projection devices
- Color printers

FIGURE 3.17

Hardware Components of a Multimedia Computer System

In addition to the computer itself, multimedia systems use other hardware for input and output.

plays a key role in the development and success of the multimedia industry. A popular delivery medium for multimedia is the CD-ROM because of its low production cost and large storage capacity. Figure 3.17 shows the typical components of a multimedia computer.

Multimedia is being used in a wide variety of ways. It is used to enhance presentations. Instead of a series of text and graphics slides, a multimedia business presentation includes video, sound, and animation. A multimedia presentation that allowed members of the International Olympic Committee to "walk through" proposed locations for sporting events was credited with helping Atlanta, Georgia, win its bid for the 1996 Summer Olympics.

The possibilities for multimedia are restricted only by our imagination. In fact, many industry leaders believe that multimedia systems may become as widespread as the television and VCR are today.

Audio. Audio involves converting sound to a digital recording for storage on a magnetic disk or CD-ROM and then converting the digital recording to sound when the multimedia program is executed. On the input, or sensing, side of the system are audio input devices that record or play analog sound and then translate it for digital storage and processing. Audio devices include CD-audio (just like your home CD players) and cassette players. The signal is sent from the audiocassette to a special audio board in the computer that digitizes the sound and stores it for further processing. Alternatively, you can use an audio board to create music in digital form, store it, and play it back using a special board in your computer.

Advanced sound systems use **digital signal processor (DSP)** chips to improve the analog-to-digital-to-analog conversion process. DSP chips take

digital signal processor (DSP)

a chip that improves the analog-to-digital-to-analog conversion process the signal conversion job over from the personal computer's CPU chip, thereby improving the entire sound process. The resulting sound may be equivalent to that produced by audio CDs and amplifiers.

Video. Video brings multimedia alive and makes it a complete education, presentation, and entertainment system. Video is also the most difficult element to display because a single uncompressed frame requires almost 1 MB of storage. At this rate, each second of a 30-frame-per-second full-screen video requires 27 MB, yielding less than 24 seconds of video per CD-ROM disk. This limitation can be avoided by reducing the number of bits required to represent a single video frame by using mathematical formulas (this process is called **video compression**), reducing the size of the video screen, reducing the rate at which the video is displayed, or a combination of all three methods. A common multimedia application displays compressed video in a quarter-screen window at 15 frames per second. Another approach is to add a video board to the personal computer like a sound board. This approach is more expensive but yields higher-quality video.

In the world of entertainment, movies can be converted to multimedia digital format that allows viewers to control the order in which sections of the movie are watched. These multimedia products often include interviews, scripts, and a capability to search the movie for particular scenes, actors, and songs.

Hardware companies like Intel have developed special-purpose chips and boards for compressing and processing full-motion video. The Intel standard is called digital video interactive (DVI). It compresses video at a 150:1 ratio using special chips on a DVI board enabling one hour of video to be stored on about 720 MB. Without compression, over 110 GB would be required. To play back DVI video, the board decompresses the frames and plays them back through the color graphics card for display on your screen.

video compression

a process that reduces the number of bits required to represent a single video frame by using mathematical formulas

Standards

The importance of hardware standards cannot be overemphasized. They diminish the cost of integration, help a developer determine which devices will be compatible with the rest of the system, provide increased options, and make the upgrading process less complex. Several common standards are summarized in Table 3.5. Note that in some cases there are competing standards.

In addition to industry standards, many large corporations also set their own internal standards by selecting specific computer configurations from a small set of manufacturers. The goal is to reduce hardware support costs and increase the organization's flexibility. Business units within the organization that adopt different hardware complicate future corporate information system projects that are based on a standard underlying hardware and software. For example, the installation of software is made much easier if installed on similar equipment from the same manufacturer rather than encountering something new and different at each installation site.

Selecting and Upgrading Computer Systems

computer system architecture

the structure, or configuration, of the hardware components of a computer system

The structure, or configuration, of the hardware components of a computer system is called the **computer system architecture**. This architecture can include a mixture of components, including processing, memory, storage, input, and output devices. As discussed in Chapter 2, organizations are

Standard	How Used
MultiMedia Extension (MMX)	Multimedia standard that enables software and hardware vendors to build products that will work well together.
Multimedia PC Council (MPC)	Multimedia standard that enables software and hardware vendors to build products that will work well together.
Ultimedia Solution	Multimedia standard that enables software and hardware vendors to build products that will work well together.
Musical Instrument Digital Interface (MIDI)	Standard system for connecting musical instruments and synthesizers to computers. Defines codes for musical events, including the start of a note and its pitch, length, volume, and other attributes.
Plug 'n' Play (PnP)	Standard system that consists of hardware and software components that card, personal computer, and operating system manufacturers incorporate into their products to eliminate the need for manual configuration so that hardware can be installed and used immediately.
Small Computer System Interface (SCSI)	System that ensures any storage, input, or output device that meets this standard can be quickly added to a system.
Fibre channel	An alternative to SCSI for connecting devices to a computer that allows a greater distance between devices and offers faster performance than SCSI.
Personal Computer Memory Card International Association (PCMCIA)	Standard that ensures compatibility between PC memory and communications cards. PCMCIA devices are also called PC cards.

TABLE 3.5

Industry Standards in Common Use

adaptive systems that must respond to changes in the business environment. A computer system that was once effective may need to be enhanced or upgraded to support new business activities. The ability to upgrade a system can be an important factor in selecting the best computer hardware.

Computer systems can be upgraded by installing additional memory, additional processors such as a math coprocessor, more hard disk storage, a memory card, or other devices. When upgrading or expanding an existing computer system, it is usually necessary to reconfigure the system.

Considerations When Selecting or Upgrading the Hard Drive. The optimal hard drive is a function of several overlapping features. Since its main role is for long-term data storage, capacity is a big plus. Most mobile PCs today come equipped with 2.5-inch removable hard drives. Look for at least 2 GB, depending on the type of data you will store on your hard drive. Other considerations are access speed (look for a minimum of 10 to 12 milliseconds), RAM, and hard drive cache size. It is also important to consider the type of data you will store. Today's business software applications and large video, audio, and graphics files require several megabytes of storage.

Considerations When Selecting or Upgrading Main Memory. Main memory stores software code, and the processor reads and executes the code. More RAM main memory means you can run more software programs at the same time. The minimum capacity needed to run most mainstream business software is from 16 MB to 32 MB. Systems with 32 MB are well-suited to take advantage of today's advanced personal productivity software (word processing, spreadsheet, graphics, and database) and multimedia programs.

As discussed earlier, your system's processor, main memory, and cache memory are heavily dependent on each other to achieve optimal system functionality. The original manufacturer of your computer takes this into consideration when designing and choosing the parts for the system. If you plan to upgrade your system's main memory above 64 MB, you should consult your PC supplier to understand your system's main memory limits on size of cache and the implication of exceeding those limits.

As mentioned throughout this chapter, a computer system's components and architecture should be chosen to support fundamental objectives, current business processes, and future needs of the organization and information system. Each computer system component—processing, memory, storage, input, and output devices—has a critical role in the successful operation of the computer system, the information system, and the organization. A thorough understanding of the broader system goals and the characteristics of the hardware as they relate to these goals will be an important guide for the future IS professional.

Information Systems Principles

Assembling an effective, efficient computer subsystem requires an understanding of its relationship to the information system and the organization. While we generally refer to the computer subsystem as simply a computer system, we must remember that the computer system objectives are subordinate to, but supportive of, the information system and the organization.

The components of all information systems—such as input devices, people, procedures, and goals—are all interdependent. Because the performance of one system affects the others, all these systems should be measured according to the same standards of effectiveness and efficiency, given issues of cost, control, and complexity.

When selecting computer subsystem devices, you also must consider the current and future needs of these systems. Your choice of a particular computer system should always allow for later improvements in the overall information system. Reasoned forethought—a trait required for dealing with computers, information, and organizational systems of all sizes—is the hallmark of a true systems professional.

Determine your hardware needs based on how the hardware will be used to support the objectives of the information systems and goals of the organization. For a personal computer user, this means knowing what software you want to run.

Do research to gain an understanding of the trade-offs between overall system performance and cost, control, and complexity.

■ SUMMARY

1. *Describe how to select and organize computer system components to support information system objectives.*

Hardware includes any machinery (often using digital circuitry) that assists with the input, processing, and output activities of a computer-based information system (CBIS). Hardware is a key component of a computer system, the heart of a CBIS. A computer system is an integrated assembly of physical devices with at least one central processing mechanism; it inputs, processes, stores, and outputs data and information. Computer system hardware performs many of these functions for a computer system.

Computer system hardware should be selected and organized to effectively and efficiently attain computer system objectives. These objectives should in turn support information system objectives and organizational goals. Balancing specific computer system objectives in terms of cost, control, and complexity will guide selection.

Hardware devices work together to perform input, processing, data storage, and output. Processing is performed by an interplay between the central processing unit (CPU) and memory. The CPU has three main components: the arithmetic/logic unit (ALU), the control unit, and register areas. The ALU performs calculations and logical comparisons. The control unit accesses and decodes instructions and coordinates data flow. Registers are temporary holding areas for instructions to be executed by the CPU.

2. *Describe the power, speed, and capacity of central processing and memory devices.*

Instructions are executed in a two-phase process. In the instruction phase, instructions are brought into the central processor and decoded. In the execution phase, the computer executes the instruction and stores the result. The completion of this two-phase process is a machine cycle. Processing speed is often measured by the time it takes to complete one machine cycle, which is measured in fractions of seconds.

Computer system processing speed is also affected by clock speed, which is measured in megahertz (MHz). Speed is further determined by a CPU's wordlength, the number of bits it can process at one time. (A bit is a binary digit, either 0

or 1.) A 32-bit CPU has a wordlength of 32 bits and will process 32 bits of data in one machine cycle. The iCOMP index, a measure of processing speed for Intel processors, averages the factors affecting processing speed into one number on a rating index.

Moore's Law is a hypothesis that states that the number of transistors on a single chip will double every 18 months. This hypothesis has held up amazingly well.

Processing speed is also limited by physical constraints, such as distance between circuitry points and circuitry materials. Most CPUs are collections of digital circuits on silicon chips. Advances in gallium arsenide (GaAs) and superconductive metals will result in faster CPUs. Most processors are based on complex instruction set computing (CISC) chips, which have many microcode instructions placed in them. With reduced instruction set computing (RISC) chips, only essential instructions are included, so processing is faster.

Primary storage, or memory, provides working storage for program instructions and data to be processed and provides them to the CPU. Storage capacity is measured in bytes. A common form of memory is random access memory (RAM). RAM is volatile. Loss of power to the computer will erase its contents. RAM comes in many different varieties. The mainstream type of RAM is Extended Data Out or EDO RAM, which is faster than older types of RAM. Two other variations of RAM include dynamic RAM (DRAM) and synchronous dynamic RAM (SDRAM). SDRAM also has the advantage of a faster transfer speed between the microprocessor and the memory. DRAM chips need high or low voltages applied at regular intervals—every two milliseconds (two one-thousandths of a second) or so—if they are not to lose their information.

Read-only memory (ROM) is nonvolatile and contains permanent program instructions for execution by the CPU. Other nonvolatile memory types include programmable read-only memory (PROM) and erasable programmable read-only memory (EPROM). Cache memory is a type of high-speed memory that CPUs can access more rapidly than RAM.

Together, a CPU and memory process data and execute instructions. Processing done using several processing units is called multiprocessing. One form of multiprocessing uses coprocessors; coprocessors execute one type of instruction while the CPU works on others. Parallel processing involves linking several processors to work together to solve complex problems.

3. *Describe the access methods, capacity, and portability of secondary storage devices.*

Computer systems can store larger amounts of data and instructions in secondary storage, which is less volatile and has greater capacity than memory. The primary characteristics of secondary storage media and devices include access method, capacity, and portability. Storage media can implement either sequential access or direct access. Sequential access requires data to be read or written in sequence. Direct access means data can be located and retrieved directly from any location on the media.

Common forms of secondary storage include magnetic tape, magnetic disk, and optical disk storage. Magnetic tape is an inexpensive sequential access storage medium. Magnetic disks are direct access media, including diskettes, fixed hard disks, and removable disk cartridges. Optical disks provide direct access storage and include compact disk read-only memory (CD-ROM), write-once, read-many (WORM) disks, and magneto-optical (MO) disks. Other storage alternatives are flash memory chips—silicon chips with nonvolatile memory—and PC memory cards, removable credit-card-size storage devices that function like fixed hard disk drives. RAID is a method of storing data that generates extra bits of data from existing data, allowing the system to more easily recover data in the event of hardware failure.

4. *Discuss the speed, functionality, and importance of input and output devices.*

Input and output devices allow users to provide data and instructions to the computer for processing and allow subsequent storage and output. These devices are part of a user interface through which humans interact with computer systems. Input and output devices vary widely, but they share common characteristics of speed and functionality.

Placing data into the computer system requires converting it from human-readable to machine-readable data. Data is thus placed in a computer system in a two-stage process: data entry converts human-readable data into machine-readable form; data input then transfers it to the computer. On-line data entry and input immediately converts and transfers data from devices to the computer system. Source data automation involves automating data entry and input so that data is captured close to its source and in a form that can be input directly to the computer.

Scanners are input devices that convert images and text into binary digits. Specialized scanners include magnetic ink character recognition (MICR) devices, optical mark recognition (OMR) devices, and optical character recognition (OCR) devices. Some input and output devices combine several into one. Point-of-sale (POS) devices are terminals with scanners that read and enter codes into computer systems. Automatic teller machines (ATMs) are terminals with keyboards used for transactions.

Output devices provide information in different forms, from hard copy to sound to digital format. Display monitors are standard output devices; monitor quality is determined by size, color(s), and resolution. Other output devices include printers, plotters, and computer output microfilm. Printers are a popular hard-copy output device whose quality is measured by speed and resolution. Plotters output hard copy for general design work. They produce charts 24 inches or 36 inches wide and whatever length is needed. Computer output microfilm (COM) devices place data from the computer directly onto microfilm.

5. *Identify six classes of computer systems and discuss the role of each.*

Computers may be classified as special-purpose or general-purpose. General-purpose computers are used for numerous applications and can be classified by processing speed, RAM capacity, and size. The six computer system types are personal computer, network computer, workstation, midrange computer, mainframe computer, and supercomputer. Personal computers (PCs) are small, inexpensive computer systems. Two major types of PCs are desktop and laptop computers. The network computer is a diskless, inexpensive computer used for accessing server-based applications and the Internet. Workstations are advanced PCs with greater memory, processing, and graphics abilities.

Filing cabinet-size minicomputers have greater secondary storage and support transaction processing. Even larger mainframes have higher processing capabilities, while supercomputers are extremely fast computers used to solve the most intensive computing problems.

The configuration of computer system hardware components is the computer system architecture. Computer systems can be upgraded by changing or adding memory, processors, and other devices. Standards are being created to lower the cost and complexity of upgrades. A hardware standard called plug and play (PnP) consists of hardware and software components that card, personal computer, and operating system manufacturers incorporate into their products to eliminate the need for manual configuration so that hardware can be installed and used immediately. When attaching CD-ROM players, scanners, and other devices, a standard called Small Computer System Interface (SCSI) ensures compatibility. MMX is standard to provide high performance for communications and multimedia applications.

6. *Define the term multimedia computer and discuss common applications of such a computer.*

Computer systems that input (and output) a combination of text, audio, and video data are called multimedia systems. These systems use various input devices, including keyboards and mouses, voice-recognition devices that recognize human speech, digital computer cameras that record and store images and video in digital form, pen input devices that enable a user to input via handwriting, light pens, touch-sensitive screens, and bar code scanners. Terminals, by contrast, are primarily used for on-line data entry and input of text and character data.

Multimedia involves the marriage of sound, animation, and digitized video. A multimedia computer can deliver multimedia applications that are rich in content and features. Multimedia is being used to enhance presentations, help customers find the products they are looking for, and alter the ways lawyers manage evidence in court cases. The possibilities for multimedia are restricted only by our imagination.

■ KEY TERMS

arithmetic/logic unit (ALU) 86
bit 89
bus line 89
byte 92
cache memory 94
CD-rewritable (CD-RW) 100
central processing unit (CPU) 86
clock speed 88
compact disk read-only memory
   (CD-ROM) 99
complementary metal oxide
   semiconductor (CMOS) 118
complex instruction set computing
   (CISC) 91
computer system architecture 123
control unit 86
coprocessor 95
data entry 104
data input 104
digital computer cameras 105
digital signal processor (DSP) 122
digital video disk (DVD) 101
direct access 97

direct access storage device
   (DASD) 97
disk mirroring 99
execution time (E-time) 87
flash memory 101
general-purpose computer 113
hardware 84
hertz 88
instruction time (I-time) 87
machine cycle 87
magnetic disks 98
magnetic tape 98
magneto-optical disk 100
mainframe computers 118
megahertz (MHz) 88
microcode 88
midrange computers 117
MIPS 88
Moore's Law 90
multifunction device 111
multiprocessing 95
network computer 114
optical disk 99

optical processors 91
parallel processing 95
personal computers (PCs) 113
pipelining 87
pixel 109
plotters 111
point-of-sale (POS) devices 107
primary storage 86
random access memory (RAM) 92
read-only memory (ROM) 94
reduced instruction set computing
   (RISC) 91
redundant array of
   independent/inexpensive disks
   (RAID) 99
registers 86
removable storage devices 102
secondary storage 96
sequential access 97
sequential access storage device
   (SASD) 97
source data automation 104
special-purpose computers 112

supercomputers 120
superconductivity 91
very long instruction word (VLIW) 92

video compression 123
voice-recognition devices 105
wordlength 89

workstations 115
write-once, read-many (WORM) 100

■ REVIEW QUESTIONS

1. What is a computer system and what is the role of hardware in the system?
2. What is the year 2000 problem?
3. Why is it said that the components of all information systems are interdependent?
4. Explain the two-phase process for executing instructions.
5. Identify the three components of the CPU and explain the role of each.
6. What is the iCOMP factor and how is it used?
7. What is Moore's Law?
8. What is the difference between CISC and RISC instruction sets?
9. What is the difference between wordlength and bus line width?
10. Describe the various types of memory.
11. Explain the difference between sequential and direct access.
12. Describe various types of secondary storage media in terms of access method, capacity, and portability.
13. Which secondary storage devices have the lowest cost per byte?
14. What is the difference between cache memory and main memory?
15. What is source data automation?
16. Discuss the speed and functionality of common input and output devices.
17. What are the computer system types? How do these types differ?
18. Discuss the methods of upgrading a computer system.
19. Discuss the role standards play in making it easier to use computer hardware.
20. List three common hardware standards and define how they are used.
21. What is microcode? How does complex instruction set vary from reduced instruction set computing?

■ DISCUSSION QUESTIONS

1. What are the implications of Moore's Law continuing the trend of increased computing power at lower costs? Use Moore's Law to forecast the computing power that could be available in three years. What sort of applications could benefit from that level of computer power?
2. What functions and capabilities is it reasonable to expect palm-top computers to perform? Are there functions that they should be able to perform that a desktop computer cannot perform?
3. What are the trade-offs between main memory and cache memory?
4. How many different classes of computer would you expect to find in a small office? How about in a medium-sized company with a few hundred employees? Describe the types of functions the various classes of computers might perform in the medium-sized company.
5. Imagine that you are the business manager for your university. What type of computer would you recommend for broad deployment in the university's computer labs—a standard desktop personal computer or a network computer? Why?
6. Identify at least four major concerns of a business traveler taking a laptop computer on a business trip. How might these concerns be dealt with?
7. If cost were not an issue, describe the characteristics of your ideal laptop computer.
8. What if you discovered that your favorite recording group composes, edits, and records all its music using multimedia computer technology? How would you feel? Does the use of computer technology to create original works of art or music diminish or enhance the accomplishment? Should such artists be considered as great as others who do not use computer technology?

■ PROBLEM-SOLVING EXERCISES

1. Do research (read various trade journals, search the Web pages of the chip manufacturers, and visit computer stores) to determine current chip speeds and the availability of even faster chips. Use your word processing program to write a short report summarizing your findings. Develop a simple spreadsheet with a row for each chip manufacturer (Compaq, IBM, and Intel) and two columns—one for the speed of their fastest available chip and one for the speed of their fastest announced chip.

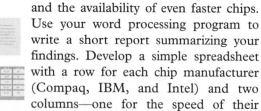

2. Over the upcoming year, your department is expected to add eight people to its staff. You will need to acquire eight personal computer systems and two additional printers for the new employees to share. Standard office computers have a Pentium (266 MHz) processor with 32 MB of RAM, SVGA color monitor, and a minimum of 2.4 GB hard disk drive. At least four of the new people will use their computers more than three hours per day. You would like to provide larger monitors and special ergonomic keyboards for these people—if it fits within your budget. You are not sure if you want to upgrade the machines to 64 MB of RAM and a 4.8-GB hard drive.

Your department budget will allow a maximum of $20,000 for computer hardware purchases this year, and you want to select only one vendor for all of the hardware. A price list from three vendors

appears in the table below, with prices for a single unit of each component. Use a spreadsheet to find the department's best solution; write a short memo explaining your rationale. Specify which vendor to choose and which items to be ordered as well as the total cost.

Component	Expert Solutions Ltd.	Business Processing Enterprises	Super Systems Inc.
266 MHz Pentium with: 32 MB RAM 2.4 GB HD	$1,245	$1,275	$1,200
Upgrade to 64 MB RAM	$ 250	$ 225	$ 245
Upgrade to 4.8 GB HD	$ 190	$ 215	$ 205
15-inch .28 dpi SVGA monitor	$ 350	$ 330	$ 340
17-inch .28 dpi SVGA monitor	$ 625	$ 600	$ 615
Ergonomic keyboard	$ 55	$ 50	$ 50
8 ppm color inkjet printer	$ 325	$ 325	$ 320
Surge protector/ power strip	$ 35	$ 32	$ 35
Three-year warranty (parts and labor)	$ 340	$ 300	$ 320

■ TEAM ACTIVITY

1. With two of your classmates, visit a major computer retail store (CompUSA, MicroCenter, etc.). Spend a couple of hours identifying the latest developments in processing, input, and output devices. Write a brief report summarizing your findings.
2. With two or three of your classmates, visit an office that uses a mainframe or midrange computer. Find out the manufacturer and model number as well as its specifications (speed of CPU, amount of main memory, disk drive capacity, etc.). How long has it been in use? How

much longer does the business plan to use it before replacing it with something different? What business processes have changed that spurred this alteration? Will the company upgrade the computer or buy a new one?
3. Identify a manager of a computer center and seek his or her permission to tour the facility. Make a list of various types of computers and input, output, and secondary storage devices that you see during your tour. Does the center have more than one class of computer? If so, why? What types of jobs are run on each class of computer?

■ WEB EXERCISE

Visit the Web sites of Intel, Sun Microsystems, Hewlett-Packard, IBM, and other computer manufacturers. Identify as many industry standards as possible that these manufacturers are following in the development of their products. Also, identify other standardization efforts in which they are involved.

■ CASES

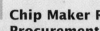

Chip Maker Reengineers Procurement Process

Advanced Micro Devices, Inc. (AMD) was founded in 1969 and today is a global supplier of integrated circuits for personal and network computers and communications markets. AMD manufactures processors for Microsoft Windows-compatible PCs, flash memories, products for communications and networking applications, and programmable logic devices. AMD has nearly 13,000 employees, with manufacturing facilities in the United States, Asia, and Japan, and sales offices throughout the world. Recent annual revenues were $2.4 billion.

AMD conducts microchip technology research and development at its Submicron Development Center (SDC) in Sunnyvale, California. This advanced wafer fabrication facility is also used for new product prototyping and production testing. There are three additional fabrication facilities in Austin, Texas, including the state-of-the-art Fab 25 facility, which cost $1.5 billion. These fabrication sites perform high-volume manufacturing of many kinds of logic integrated circuits, such as the AMD-K6 processor, programmable logic devices (PLDs), subscriber line interface circuits (SLICs), and the SLAC™ family of subscriber line audio-processing circuits.

The AMD-K6 processor is an advanced Windows-compatible processor that competes with the Pentium II processor. AMD also manufactures flash memory for control storage applications ranging from wireless communications to consumer products. Production capacity for flash memory is provided by the AMD-Fujitsu joint venture in Aizu-Wakamatsu, Japan. The AMD Dresden Microelectronics Center, including

AMD's first European manufacturing facility, Fab 30, is currently under construction in Dresden, Germany. The facility will include a design center and an 85,000-square-foot wafer fabrication area for AMD's next-generation processors and other advanced logic products. Production is expected to commence in 1999.

AMD designs, tests, analyzes and/or assembles products in Penang, Malaysia, Bangkok, Thailand, and Singapore. An additional test and assembly facility, now under construction in Suzhou, China, is planned to begin production in early 1999. Quality control organizations are located in Sunnyvale, California, and in Frimley, England.

In an effort to reduce costs, AMD began using the Internet to reengineer its entire purchasing function in May 1997. In a pilot project, AMD moved onto the Web its procurement of photo masks used in chip making. (A photo mask is a glass plate inserted into a machine that lets semiconductor engineers imprint an image of a circuit onto a silicon wafer). The new purchasing process is based on operating resource management (ORM) software from start-up software vendor Ariba Technologies. Under previous procurement procedures, engineers had to e-mail an AMD buyer and then the buyer wrote a purchase order that was sent to the supplier. Not only was the old approach time consuming, it resulted in lots of invoice discrepancies that took even more time to resolve. Using the Ariba software, authorized engineers are able to order photo masks directly from DuPont and Photronics. The streamlined purchasing process is intended to dramatically reduce processing and transaction costs, as well as free purchasing managers to allow them to negotiate better deals with suppliers.

The corporate director of supply management was not concerned about reengineering the procurement process based on software that was not yet proven in the marketplace. Ariba did not officially launch its product until two months after AMD began its pilot. As one of Ariba Technologies' first customers, AMD gets a generous discount from the $750,000 to $4 million price tag.

AMD believes that the Ariba solution is the procurement method in the twenty-first century. It allows AMD to focus more resources on supply management and strategic alignment with key suppliers. AMD expects to generate at least a 3 percent savings on several hundred million dollars in expenditures as a result of the Ariba implementation.

1. Can you identify potential issues that might arise as AMD expands use of the Internet procurement process from its corporate offices to worldwide? What actions would you take to avoid these potential problems?
2. Why has AMD built fab facilities and research centers worldwide? Wouldn't it be easier to manage the business if all facilities were located in the United States?

Sources: Adapted from Advanced Micro Devices, Inc. Web site at http://www.amd.com, accessed August 1, 1998; Ariba Web site at http://www.ariba.com, accessed August 1, 1998; and Clinton Wilder, "AMD to Move Procurement Onto the Net," *Information Week,* May 26, 1997, p. 70.

2   Unisys Helps Bank Meet Customer Needs

In 1986, Burroughs Corporation and Sperry Corporation combined to form Unisys. Today Unisys is one of the world's leading providers of professional information services and systems with current annual revenue of $6.6 billion and more than 32,000 employees. It does business in 100 countries with more than 60 percent of its revenue coming from countries outside the United States. One of its most popular hardware products is the Unisys ClearPath HMP NX4600 Series computers. This device meets midrange business needs while supporting transaction-intensive operations 24 hours a day, 7 days a week. The NX4600 Series has two processor subsystems to provide exceptional power and high-speed transaction processing plus a processor design that ensures high levels of reliability. The NX4600 allows performance to be expanded nearly seven times without a change in the cabinet. Such extensive scalability positions it as an entry- to medium-scale server. The ClearPath HMP NX4600 allows the use of redundant hardware components in two physically independent system environments, provides advanced levels of system availability, and offers the flexibility to use many types of computing resources.

Today most financial institutions follow a customer-oriented business approach. To achieve this successfully, however, they need support from their information systems suppliers. Unisys has developed its "Customerize" philosophy to enable its clients to become more responsive to customers' needs. Unisys works with its clients to assess how to re-organize their processes and systems to help them realize their strategic business objectives.

One strategy for achieving a customer-oriented business approach is the use of automated service channels to remove constraints (store hours, travel requirements, and scheduling). Automated channels appeal to customers who find the physical process restrictive and inconvenient. Offering these customers convenient, easy-to-use alternatives is important or they may defect to a competitor.

Just such a concern led Zurich-based Union Bank of Switzerland (UBS) to restructure its delivery channels. UBS launched the Marketing 2000 program to enable customers to conduct all basic banking transactions through kiosks located in public places and through home computers. UBS's full-service banking kiosks use Unisys ClearPath HMP NX4600 computers placed in shopping malls, parks, public plazas, and other central locations throughout the country. Customers can conduct transactions as simple as a quick withdrawal or as complicated as retirement planning and investment analysis. Like an ATM machine, the multimedia kiosks require a magnetic strip card and an identification number for access, but the kiosks support much more complex transactions than typical ATMs.

Unisys and UBS wanted to build from the ATM concept because people were already comfortable with it and knew how it works. By harnessing the power of the computer inside each kiosk, they were able to offer the convenience of the ATM and the depth of service of the bank teller in a single format. This essentially opens up the entire banking process so customers can conduct business on their own terms rather than on the bank's.

Bill payment is one of the most popular features of the kiosks. A customer who needs to pay a bill to a department store for any purchase can select the bill payment option, scan in the actual invoice using a built-in electronic imaging system, and identify which account to take the funds from. The kiosk does the rest including processing the debit, sending the payment to the store, and logging the transaction against the customer's account. If the store does not accept electronic payments, the system sends a message to the UBS central payment center where a physical check is prepared and mailed to the store—all by computer.

Initial use of the kiosks has been even higher than expected, and UBS is now rolling out new units to expand the program. The next step is to upgrade the systems to be able to provide additional information such as real estate listings, rental rates, stock exchange offers, and insurance products.

1. List some of the reasons UBS may have selected Unisys and its ClearPath HMP NX4600 computers to support the Marketing 2000 project.
2. Should Unisys continue to expand the level of services it provides to its customers or focus on improving and marketing its hardware products? Why?

Sources: Adapted from the Unisys Web site at http://www.unisys.com, accessed August 1, 1998; the About Union Bank of Switzerland and Investors' Fact Sheet portions of the Union Bank of Switzerland Web page at http://www.ubs.com, accessed August 1, 1998; and "Clear Path, NT, and Internet Hot Topics at Unite Conference," *Unisys World*, December 1997.

3   United Airlines Standardizes on Single Workstation

United Airlines offers more than 2,300 flights a day to 136 destinations in 30 countries and two U.S. territories. With 80,000 employees, United is the largest air carrier in the world and the largest majority employee-owned company. United is also an industry innovator with breakthroughs such as E-Ticket Service, United Connection, Self-Service Boarding Pass Machines, Shuttle by United, and the introduction of the technologically advanced Boeing 777. Net earnings recently exceeded $1 billion based on revenue of over $17 billion.

United is navigating the future using four major strategies under what it calls its Quality Flight Plan, a long-term business strategy formulated in 1995. It was designed to make United an attractive long-term investment by improving in four areas: (1) in the workplace by providing decision-making power and motivation to its workforce; (2) in onboard products and customer service by providing the right products and superior service; (3) in its fleet by retiring and replacing its older aircraft with newer models; and (4) in its balance sheet by strengthening the company's financial standing.

United Airlines' Operations Control Center is located at the company's headquarters in Chicago. The center is always open and handles flight planning, aircraft routing, crew scheduling and cancellations for United. Those who staff the center rely on a constant stream of data from multiple sources including weather reports from its own meteorological staff, flight and radar information from the Federal Aviation Administration and the Air Transport Association, and reports from United ground crews at various airports around the country. The data is used to set aircraft routes and crew schedules and also to warn pilots of expected delays or rough weather.

Data from all these sources is fed directly into the company's IBM mainframe computers. Until recently, the data had to be accessed by staff using three different terminals—each for different types of data. United has just implemented an integrated workstation platform that gives its staff at its Operations Control Center the ability to view and analyze all data using the same workstation. With information collected in one place, users are able to perform their jobs more efficiently because data is available quickly and in a format that is easy to understand. This also enables users to combine different data, something they could not do before. For example, overlay weather maps on radar images to provide flight dispatchers highly detailed views of weather patterns on an aircraft's route.

United officials would not say how much they spent to implement the new technology from Hewlett-Packard. But they said they hope to save at least $50 million per year in fuel and flight delay costs as a result of the integrated environment. In addition to these savings, United also will save money through improved flight planning and tracking capabilities offered by the new systems.

1. Is the move to a single workstation consistent with United's Quality Flight Plan strategy? Why or why not?
2. How is the single workstation environment able to save so much money? In addition to these tangible savings, can you identify additional, intangible benefits associated with moving to a single workstation environment?

Sources: Adapted from the Our Company section of the Web site for UAL Corp. and United Airlines at http://www.ual.com, accessed August 1, 1998 and Jaikumar Vijayan, "Airline Routes Arriving Data to One Terminal," *Computerworld*, September 8, 1997, p. 23.

4   Phillips Petroleum Aims for High Reliability and Availability

Phillips Petroleum Company is a major oil company based in the United States with worldwide operations. It has nearly $14 billion in assets and annual revenues of $15 billion generating over $1 billion in net income. Phillips employs around 17,200 employees. Worldwide crude oil production is 219,000 barrels per day, with U.S. crude oil production at 69,000 barrels per day. Worldwide natural gas production is 1.5 billion cubic feet per day with U.S. natural gas production at 1.1 billion cubic feet per day.

In implementing its new hardware and software, Phillips wanted to guarantee near-continuous systems and data availability. High information systems reliability and availability are required to meet corporate objectives of increased gas production, increased capacity of plastics and manufacturing, and improved refinery operations, which operate 24 hours a day, seven days per week. These objectives necessitated getting more current financial, sales, inventory, and production information to the business units in real time. In addition, high reliability and availability is critical to meet the needs of international users in different time zones. The requirement for continuous system operations was a key reason the company chose the IBM UNIX-based processing, storage, and backup solutions.

To help ensure the highest possible system availability, Phillips is using IBM's High Availability Cluster MultiProcessing (HACMP) software. This enables an application computer requesting data from a primary database computer to switch to a backup database computer if the primary computer fails. Although customers might experience a brief outage, the switchover occurs without operator intervention.

Phillips also implemented IBM's Serial Storage Architecture (SSA). SSA is a powerful high-speed serial interface designed to connect high-capacity data storage devices, subsystems, computers, and workstations. Each SSA link can send and receive data at a rate of 20 MB per second in each direction, with total throughput of 80 MB per second at each node. The combination of low power, fewer components, and fault isolation makes SSA highly reliable. Each node performs extensive error checking, and if errors occur, SSA provides immediate data recovery. In addition, the SSA disk subsystems—which Phillips mirrors for additional fault-tolerance—have dual connections to primary and backup database computers, so they can feed the backup computer during a failure.

Phillips also uses IBM's ADSTAR Distributed Storage Manager (ADSM) software to provide an enterprisewide solution integrating unattended network backup and archive with storage management and powerful disaster recovery planning functions. The ADSM software manages Magstart

3590 Tape Drives to store data. Using these high-capacity cartridges, Phillips needs fewer tapes and thus less storage space; plus less time is wasted on tape mounts and there are fewer tapes to ship off-site for disaster-recovery purposes.

1. Why is it so critical for Phillips to have virtually uninterrupted data processing capabilities?

2. Can you think of a way in which Phillips might experience an outage due to a disk failure? If so, how would you improve its current situation to avoid this potential problem?

Sources: Adapted from "Phillips Petroleum" Special Advertising section, *Profit*, February 1998; The Storage Products and Software Products sections of the IBM Web page at http://www.ibm.com, accessed August 1, 1998; and the Our Products and Services section of the Phillips Petroleum Web page at http://www.phillips66.com, accessed August 1, 1998.

CHAPTER  4

Software: Systems and Application Software

"We're not seeing any limit to where Java can go."

— Joan LeLuca, chief software strategist at Motorola's Semiconductor Products Sector, where Java will be used in pagers, cell phones, computer chips, and appliances such as toasters

Chapter Outline

An Overview of Software
Systems Software
Application Software
Supporting Individual, Group, and Organizational Goals
Software Issues and Trends

Systems Software
Operating Systems
Popular Operating Systems
Utility Programs

Application Software
Types of Application Software
Personal Application Software
Object Linking and Embedding (OLE)
Workgroup Application Software
Enterprise Application Software

Programming Languages
Standards and Characteristics
The Evolution of Programming Languages
Language Translators

Learning Objectives

After completing Chapter 4, you will be able to:

1. Identify and briefly describe the functions of the two basic kinds of software.
2. Outline the role of the operating system and identify the features of several popular operating systems.
3. Discuss how application software can support personal, workgroup, and enterprise business objectives.
4. Identify three basic approaches to developing application software and discuss the pros and cons of each.
5. Outline the overall evolution of programming languages and clearly differentiate between the five generations of programming languages.

Darigold Inc.
Meeting Business Challenges with Software Packages

Darigold Farms is a cooperative owned by approximately 900 dairy farm families in Washington, Oregon, Idaho, and northern California that exists to market the milk produced by its members. Darigold Inc. converts raw milk from the cooperative into marketable products for sale and distribution in national and international markets, with all earnings returned to its member owners. There are 1,300 employees operating a dozen processing plants and office facilities in the four states. Revenue for a recent year was $1 billion based on production of 5 million pounds of milk.

In planning for the future, Darigold developed a Business Systems Plan, a series of strategic initiatives designed to support the company's progress into the next century. A key step in that plan was to replace its homegrown information systems and standardize its business processes across the enterprise, from consolidation of financial statements to scheduling and production planning. Rather than replace its existing software with new software it developed, Darigold selected prebuilt software packages that are flexible enough to meet its business and information technology needs.

Oracle's Consumer Packaged Goods (CPG) software was selected based on its ability to save the company $25 million through more efficient, streamlined sales management, order fulfillment, purchasing, manufacturing, distribution, and asset management processes. These operational efficiencies help lower inventory costs and provide more flexible production control. The Oracle CPG software package also enables the company to increase efficiency through consolidation of customer orders, improved planning and decision support throughout the supply chain, more efficient labor planning, and improved vendor material management.

A second software package, System ESS from Industri-Matematik International Corp., was implemented to further improve efficiency by focusing on order processing and customer support activities. The company will streamline its order processing and display inventory levels at the time of order entry, enabling the company to reduce the average time from order entry to order validation from one to eight hours to five to ten minutes.

A third software package, Oracle Financials, was chosen to provide further savings from streamlining administration and accounting processes. Using this package, Darigold expects to reduce the time it takes to close the books each month from fifteen business days to five.

As you read this chapter, consider the following questions:

- What advantages does Darigold gain through implementation of existing software packages versus in-house development of software?

- Could Darigold have achieved the same level of improvement in business efficiency using its own software?

- Would Darigold achieve better results by selecting and implementing a complete set of software from a single vendor versus choosing the best software to perform a particular set of functions using multiple software vendors ("best of breed" approach)?

Sources: Adapted from Oracle's Web site at http://www.oracle.com and Darigold's Web site at http://www.darigold.com, accessed February 23, 1998.

FIGURE 4.1

The Importance of Software in Business

Since the 1950s, businesses have greatly increased their expenditures on software as compared with hardware.

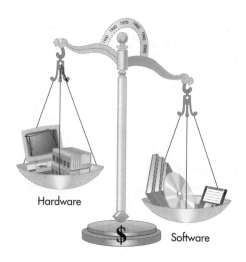

In the 1950s, when computer hardware was relatively rare and expensive, software costs were a comparatively small percentage of total information systems costs. Today, the situation has dramatically changed. Software can represent 75 percent or more of the total cost of a particular information system because of three major reasons: (1) advances in hardware technology have dramatically reduced hardware costs, (2) increasingly complex software requires more time to develop and so is more costly, and (3) salaries for software developers have increased because the demand for these workers far exceeds the supply.[1] In the future, as shown in Figure 4.1, software is expected to make up an even greater portion of the cost of the overall information system. The critical functions software serves, however, make it a worthwhile investment.

AN OVERVIEW OF SOFTWARE

computer programs

sequences of instructions for the computer

documentation

description of the program functions to help the user operate the computer system

systems software

the set of programs designed to coordinate the activities and functions of the hardware and various programs throughout the computer system

computer system platform

the combination of a particular hardware configuration and systems software package

application software

programs that help users solve particular computing problems

One of software's most critical functions is to direct the workings of the computer hardware. As we saw in Chapter 1, software consists of computer programs that control the workings of the computer hardware. **Computer programs** are sequences of instructions for the computer. **Documentation** describes the program functions to help the user operate the computer system. The program displays some documentation on screen, while other forms appear in external resources, such as printed manuals. There are two basic types of software: systems software and application software.

Systems Software

Systems software is the set of programs designed to coordinate the activities and functions of the hardware and various programs throughout the computer system. A particular systems software package is designed for a specific CPU design and class of hardware. The combination of a particular hardware configuration and systems software package is known as a **computer system platform**.

Application Software

Application software consists of programs that help users solve particular computing problems. Both systems and application software can be used to meet the needs of an individual, a group, or an enterprise. Application software can support individuals, groups, and organizations to help them realize business objectives. Application software has the greatest potential to affect the processes that add value to a business because it is designed for specific organizational activities and functions. The effective implementation and use of application software can provide significant internal efficiencies and support corporate goals. Before an individual, a group, or an enterprise decides on the best approach for acquiring application software, goals and needs should be analyzed carefully.

Supporting Individual, Group, and Organizational Goals

Every organization relies on the contributions of individuals, groups, and the entire enterprise to achieve business objectives. The organization also supports individuals, groups, and the entire enterprise with specific application software and information systems. As the power and reach of information systems expand, they promise to reshape every aspect of our lives: how we work and play, how we are educated, how we interact with others, how our businesses and governments conduct business, and how scientists perform research. Information systems will change virtually every method for capturing, storing, transmitting, and analyzing knowledge, including books, newspapers, magazines, movies, television, phone calls, musical recordings, and architectural drawings. One useful way of classifying the many potential uses of information systems is to identify the scope of the problems and opportunities addressed by a particular organization. This is called the **sphere of influence**. For most companies, the spheres of influence are personal, workgroup, and enterprise as shown in Table 4.1.

Information systems that operate within the **personal sphere of influence** serve the needs of an individual user. These information systems enable their users to improve their personal effectiveness, increasing the amount of work and its quality. Such software is often referred to as **personal productivity software**. There are many examples of such applications operating within the personal sphere of influence—a word processing application to enter, check spelling, edit, copy, print, distribute, and file text material; a spreadsheet application to manipulate numeric data in rows and columns for analysis and decision making; a graphics application to perform data analysis; and a database application to organize data for personal use.

A **workgroup** is two or more people who work together to achieve a common goal. A workgroup may be a large, formal, permanent organizational entity such as a section or department or a temporary group formed to complete a specific project. The human resources department of a large

sphere of influence

the scope of the problems and opportunities addressed by a particular organization

personal sphere of influence

information systems that serve the needs of an individual user

personal productivity software

information systems that enable their users to improve their personal effectiveness, increasing the amount of work and its quality

workgroup

two or more people who work together to achieve a common goal

TABLE 4.1

Classifying Software by Type and Sphere of Influence

Software	Personal	Workgroup	Enterprise
Systems software	Personal computer and workstation operating systems	Network operating systems	Midrange computer and mainframe operating systems
Application software	Word processing, spreadsheet, database, graphics	Electronic mail, group scheduling, shared work	General ledger, order entry, payroll, human resources

firm is an example of a formal workgroup. It consists of several people, is a formal and permanent organizational entity, and appears on a firm's organization chart. An information system that operates in the **workgroup sphere of influence** supports a workgroup in the attainment of a common goal. Users of such applications are operating in an environment where communication, interaction, and collaboration are critical to the success of the group. Applications include systems that support information sharing, group scheduling, group decision making, and conferencing. These applications enable members of the group to communicate, interact, and collaborate.

In the context of this book, an enterprise is a company organized for commercial purposes—a business firm. The word *enterprise* is a synonym for *firm* or *company*. Information systems that operate within the **enterprise sphere of influence** support the firm in its interaction with its environment. The surrounding environment includes customers, suppliers, shareholders, competitors, special interest groups, the financial community, and government agencies. Every enterprise has many applications that operate within the enterprise sphere of influence. The input to these systems is data about or generated by basic business transactions with someone outside the business enterprise. These transactions include customer orders, inventory receipts and withdrawals, purchase orders, freight bills, invoices, and checks. One of the results of processing transaction data is that the records of the company are instantly updated. For example, the processing of employee time cards updates individuals' payroll records used to generate their checks. The order entry, finished product inventory, and billing information systems are examples of applications that operate in the enterprise sphere of influence. These applications support interactions with customers and suppliers.

workgroup sphere of influence

information systems that support a workgroup in attainment of a common goal

enterprise sphere of influence

information systems that support the firm in its interaction with its environment

Lotus Notes enables a workgroup to schedule meetings and coordinate activities.
(Source: Courtesy of Lotus Development Corporation.)

Software Issues and Trends

Because software is such an important part of the computer system, issues such as licensing, upgrades, and global software support have received increased attention. We highlight three major software issues and trends here.

Software licensing. As software increases in importance, controlling its use is a concern. Vendors spend time and resources in software development. So, they must try to protect their software from being copied and distributed by individual users and other software companies. Today, companies can copyright software and programs, but this protection is limited. Another approach is to use existing patent laws, but this approach has not been fully accepted by legal experts and the courts.

Software upgrades. Software companies revise their programs and sell new versions occasionally. In some cases, the revised software can offer a number of new and valuable enhancements. In other cases, the software may use complex program code that offers little in terms of additional capabilities. Furthermore, revised software can contain bugs or errors. Deciding whether to purchase the newest software can be a problem for corporations and individuals with a large investment in software. Should the newest version be purchased when it is released? Some organizations and individuals do not always get the most current software upgrades or versions, unless there are significant improvements or capabilities. Instead, they may upgrade to newer software only when there are vital new features.

Global software support. Large, global companies have little trouble persuading vendors to sell them software licenses for even the most far-flung outposts of their company. But can those same vendors provide adequate support for their software customers in all locations? Taking into account the support requirements of local operations is one of the biggest challenges information systems teams face when putting together standardized, companywide systems. They must ensure that the vendor has sufficient support available for local operations. In slower technology growth markets, like Eastern Europe and Latin America, there may be no vendor presence at all. Instead, large vendors such as Sybase, IBM, and Hewlett-Packard contract out to local providers.

One approach that has been gaining acceptance in North America is to outsource global support to one or more third-party distributors. The software user company may still negotiate its software license with the software vendor directly but then hand over the global support contract to a third-party supplier. The supplier acts as a middleman between software vendor and user, often providing distribution, support, and invoicing. This is how American Home Products Corporation handles global support for both Novell NetWare and Microsoft Office applications in the 145 countries in which it operates. American Home Products, a pharmaceutical and agricultural products company, negotiated the agreements directly with the vendors for both purchasing and maintenance, but fulfillment of the agreement is handled exclusively by Philadelphia-based Softsmart, an international supplier of software and services.[2]

SYSTEMS SOFTWARE

Controlling the operations of computer hardware is one of the most critical functions of systems software. Systems software also supports the applications programs' problem-solving capabilities. Different types of systems software include operating systems and utility programs.

Operating Systems

operating system (OS)

a set of computer programs that control the computer hardware and act as an interface with applications programs

An **operating system (OS)** is a set of computer programs that controls the computer hardware and acts as an interface with application programs (Figure 4.2). The operating system, which plays a central role in the functioning of the complete computer system, is usually stored on disk. After a computer system is started, or "booted up," portions of the operating system are transferred to memory as they are needed. The collection of

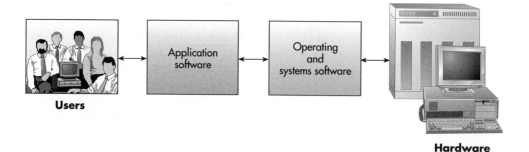

Users

Hardware

FIGURE 4.2

The role of the operating system and other systems software is as an interface or buffer between application software and hardware.

programs, collectively called the operating system, executes a variety of activities, including:

- Performing common computer hardware functions
- Providing a user interface
- Providing a degree of hardware independence
- Managing system memory
- Managing processing tasks
- Providing networking capability
- Controlling access to system resources
- Managing files

Common hardware functions. All application programs must perform certain tasks. For example:

- Getting input from the keyboard or some other input device
- Retrieving data from disks
- Storing data on disks
- Displaying information on a monitor or printer

Each of these basic functions may require detailed instructions to complete. The operating system converts a simple, basic instruction into a set of instructions that the hardware requires. For example, suppose an application program needs to read a piece of data:

- Retrieve Products from Drive C:

The operating system might translate this simple command to the hardware into a longer sequence, as follows:

- Check for Drive C: on the computer system
- If no such drive exists, inform the application program; otherwise, continue
- Start up Drive C:
- Find the block of data that represents Products
- Retrieve this data
- Send data to the billing application
- Stop Drive C:

The operating system simplifies the basic task of reading a block of data from a disk drive. The application program need only issue one basic instruction, rather than the more detailed sequence. In effect, the operating system acts as intermediary between the application program and the hardware. The typical OS performs hundreds of such functions, each of which is translated into one or more instructions for the hardware. The OS will notify the user if input/output devices need attention, if an error has occurred, or if anything abnormal occurs in the system.

User interface. One of the most important functions of any operating system is providing a **user interface**. A user interface allows individuals to access and command the computer system. The first user interfaces for mainframe and personal computer systems were command based. A **command-based user interface** requires that text commands be given to the computer to perform basic activities. For example, the command ERASE FILE1 would cause the computer to erase or delete a file called FILE1. RENAME and COPY are other examples of commands used to rename files and copy files from one location to another. Many mainframe computers use a command-based user interface. In some cases, a specific job control language (JCL) is used to control how jobs or tasks are to be run on the computer system.

A **graphical user interface (GUI)** uses pictures (called **icons**) and menus displayed on screen to send commands to the computer system. Many people find that GUIs are easier to use because users intuitively grasp the functions. Today, the most widely used graphical user interface is Windows by Microsoft. Alan Kay and others at Xerox PARC (Palo Alto Research Center, located in California) were pioneers in investigating the use of overlapping windows and icons as an interface. As the name suggests, Windows is based on the use of a window, or a portion of the display screen dedicated to a specific application. The screen can display several windows at once. For this reason, all GUI environments are sometimes referred to as "windows" environments, even though they are not Microsoft products. The popularity of Microsoft's Windows is due in part to the many advantages of using any graphical user interface, as listed in Table 4.2.

---

user interface

a part of the operating system that allows individuals to access and command the computer system

command-based user interface

a part of the operating system that requires text commands be given to the computer to perform basic activities

graphical user interface (GUI)

a part of the operating system that uses pictures (icons) and menus displayed on the screen to send commands to the computer system

icon

picture

TABLE 4.2

Advantages of Using a Graphical User Interface

- Performing tasks in a GUI environment is intuitive. To open a file, you click on a file icon or symbol. To delete a file, you drag it to a wastebasket icon.
- The applications are consistent. They have the same appearance and general operation for opening and closing files, editing, moving, printing, erasing, etc. Once you learn the basics for one application, such as a word processing program, the same basic commands and approaches work with other applications, such as spreadsheets and database programs.
- The applications are flexible. You can use either a mouse or the keyboard. In addition, you can save files in different ways, in different formats, and to different directories by using different save options.
- GUIs allow you to cut or copy something from one application and paste it into another application with ease.
- The applications can be easy to use. Detailed technical manuals describing complex commands are usually not needed. Some software companies, such as Microsoft, include assistance features, such as Wizard, that help with the creation of tables and forms.
- If you make a mistake, GUIs allow you to cancel or undo what you have done. Most menu boxes have a cancel option.
- GUIs often ask you to confirm important operations, like saving or deleting a file. Most have an OK or Yes option to check or click before an operation is done.

Applications written for command-based operating systems often can be run in a GUI environment, although applications written for GUIs often cannot run under a command-based environment. However, to maximize the capabilities of a GUI it is best to use applications written specifically for them.

Hardware independence. Looking back at Figure 4.2, notice that the application program communicates with the hardware through the operating system. Because of this, the OS serves as an interface between the application program and the hardware. This interface happens automatically so the user does not have to make it occur.

Suppose a computer manufacturer designs new hardware that can operate much faster than before. Let's further suppose that this new hardware functions differently from the old hardware, requiring different instructions to perform certain tasks. If operating systems did not exist, we would have to rewrite all our application programs to take advantage of the new, faster hardware. Fortunately, because many application programs usually share a single OS, we need only to redesign the OS layer so that it converts the commands from the application to the new group of instructions needed by the hardware. Having an OS layer allows us flexibility and reduces the dependency of applications on particular hardware configurations.

Memory management. The purpose of memory management is to control how memory is accessed and to maximize available memory and storage. The memory management feature of many operating systems allows the computer to effectively execute program instructions and speed processing.

Controlling how memory is accessed allows the computer system to efficiently and effectively store and retrieve data and instructions and to supply them to the CPU. Memory management programs that are part of operating systems are needed to convert a user's request for data or instructions (called a logical view of the data) to the physical location where the data or instructions are stored. A computer understands only the physical view of data, that is the specific location of the data in storage or memory and the techniques needed to access the data. This concept is described as logical versus physical access. For example, the current price of an item, say, a Texas Instruments BA-35 calculator with an item code of TIBA35, might always be found in the logical location "TIBA35$." If the CPU needed to fetch the price of "TIBA35" as part of a program instruction, the memory management feature of the operating system would translate the logical location "TIBA35$" into an actual physical location in memory or secondary storage (Figure 4.3).

Memory management is important because memory can be divided into different segments or areas. With some computer chips, memory is divided into conventional, upper, high, extended, and expanded memory. In addition, some computer chips provide "rings" of protection. An operating system can use one or more of these rings to make sure that application programs do not penetrate an area of memory and disrupt the functioning of the operating system, which could cause the entire computer system to crash. Memory management features of today's operating systems are needed to make sure that application programs can get the most from available memory without interfering with other important functions of the operating system or with other application programs.

FIGURE 4.3

An Example of the Operating System Controlling Physical Access to Data

The user prompts the application software for specific data. The operating system translates this prompt into instructions for the hardware, which finds the data the user requested. Having successfully completed this task, the operating system then relays the data back to the user via the application software.

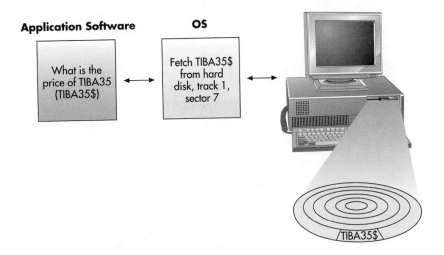

Application Software

What is the price of TIBA35 (TIBA35$)

OS

Fetch TIBA35$ from hard disk, track 1, sector 7

TIBA35$

virtual memory

memory that allocates space on the hard disk to supplement the immediate, functional memory capacity of RAM

paging

a function of virtual memory that allows the computer to store currently needed pages of a number of programs in RAM while the rest of these programs wait on the disk

Some operating systems support **virtual memory**, which allocates space on the hard disk to supplement the immediate, functional memory capacity of RAM. Virtual memory works by swapping programs or parts of programs between memory and one or more disk devices. The number of program segments held in RAM depends on the sizes of those segments and the computer's RAM capacity. Assume that the written instructions for a computer program would fill ten pages. If the computer stored only a segment of this program—perhaps the first few pages—in RAM, it might also be able to store another program segment from some other program there. Virtual memory allows the computer to store currently needed pages of a number of programs in RAM while the rest of these programs wait on the disk—a concept called **paging**.

As can be seen in Figure 4.4, six program parts have been stored in memory. The rest of these programs are stored on disk. What is the advantage of virtual memory? In general, more programs can be processed in the same amount of time needed by systems that do not use virtual memory. With virtual memory, only a few instructions are being executed for one program at any given moment. After these instructions are executed, a new subset of program instructions is transferred from storage to memory. In the meantime, however, the computer executes a few instructions from another program, and so on. Because the computer is executing only a few instructions at one time, the complete program is not needed. With more program segments, the CPU is less likely to have to wait for programs to be transferred from the disk to memory. This reduces CPU idle time and increases the number of jobs that can be run in a given time span. It is important to note that the physical size of memory remains the same. It only appears to be larger because more gets done in less time.

Processing tasks. Managing all processing activities is accomplished by the task management features of today's operating systems. Task management allocates computer resources to make the best use of each system's assets. Task management software can permit one user to run several programs or tasks at the same time (multitasking and multithreading) and allow several users to use the same computer at the same time (time-sharing).

FIGURE 4.4

Virtual Memory

Virtual memory uses a process by which partial programs are stored in memory and the remainder of the program is stored on disk. For this reason, virtual memory has also been called paging.

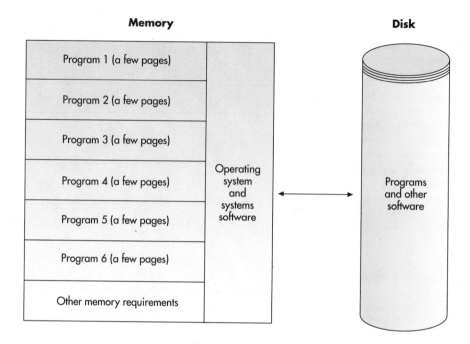

Memory   **Disk**

Program 1 (a few pages)

Program 2 (a few pages)

Program 3 (a few pages)

Program 4 (a few pages)

Operating system and systems software

Program 5 (a few pages)

Program 6 (a few pages)

Other memory requirements

Programs and other software

multitasking

a processing activity that allows a user to run more than one application at the same time

multithreading

a processing activity that is basically multitasking within a single application

time-sharing

a processing activity that allows more than one person to use a computer system at the same time

scalability

the ability of the computer to smoothly handle an increasing number of concurrent users

An operating system with **multitasking** capabilities allows a user to run more than one application at the same time. For example, a national sales manager might want to run an inventory control program, a spreadsheet program, and a word processing program at the same time. With multitasking, the manager can use data and results among all three of the programs. Spreadsheet results can be inserted into the inventory control program. Important tables and analysis from the inventory control program can then be inserted directly into the word processing program. **Multithreading** is basically multitasking within a single application—that is, several parts of the program can work at once.

Multitasking and multithreading can save users a considerable amount of time and effort. A number of programs can be open and running at the same time, sharing data and results. Changes are easy to make.

Time-sharing allows more than one person to use a computer system at the same time. For example, 15 customer service representatives may be entering sales data into a computer system for a mail-order company at the same time. In another case, thousands of people may be simultaneously using an on-line computer service to get stock quotes and valuable business news.

Time-sharing works by dividing time into CPU processing small time slices, which can be a few milliseconds or less in duration. During a time slice, some tasks for the first user are done. The computer then goes from that user to the next. During the next time slice, some tasks for the next user are completed. This process continues through each user and cycles back to the first user. Because the CPU processing time slices are small, it appears that all jobs or tasks for all users are being completed at the same time. In reality, each user is sharing the time of the computer with other users. The ability of the computer to smoothly handle an increasing number of concurrent users is called **scalability**.

Most operating systems for large computer systems, like mainframes, support time-sharing. Because personal computer operating systems usually are oriented toward single users, the management of multiple-user tasks often is not needed.

Networking capability. The operating system can provide features and capabilities that aid users in connecting to a computer network. For example, Apple computer users have built-in network access through the AppleShare feature, and the Microsoft Windows 98 operating system comes with the capability to link users to its own Internet service with no additional software required.

Access to system resources. Computers often handle sensitive data that can be accessed over networks. The operating system needs to provide a high level of security against unauthorized access to the users' data and programs. Typically, the operating system establishes a log-on procedure that requires users to enter an identification code and a matching password. If the identification code is invalid or if the password does not go with the identification code, the user cannot gain access to the computer. The operating system also requires that user passwords be changed frequently, say, every 20 to 40 days. If the user is successful in logging onto the system, the operating system records who is using the system and for how long. In some organizations, such records are also used to bill users for time spent using the system. The operating system also reports any attempted breaches of security.

File management. The operating system performs a file management function to ensure that files in secondary storage are available when needed and that they are protected from access by unauthorized users. Many computers support multiple users who store files on centrally located disk and/or tape drives. The operating system keeps track of where each file is stored and who may access it. The operating system must be able to resolve what to do if more than one user requests access to the same file at the same time. Even on stand-alone personal computers with only one user, file management is needed to keep track of where files are located, what size they are, when they were created, and who created them.

Popular Operating Systems

Early operating systems for computers were very basic. In the last several years, however, more advanced operating systems have been developed, incorporating some features previously available only with mainframe operating systems. This section reviews selected popular computer operating systems. These operating systems are classified in Table 4.3 by sphere of influence.

MS-DOS. The earliest IBM-compatible computers in the 1980s often used Microsoft Disk Operating System (MS-DOS) or a similar operating system. Command-based operating systems, such as MS-DOS, use commands like COPY, RENAME, and FORMAT to prepare a disk for use. DOS is a single-user, single-task operating system. Personal computer

TABLE 4.3

Popular Operating Systems

Operating systems are needed to support the computing needs of individuals, workgroups, and the enterprise.

Personal	Workgroup	Enterprise
MS-DOS	Novell Netware	IBM AIX, HP/UX, DEC Ultrix
DOS with Windows	Banyan Vines	IBM ESA/370
		IBM MVS/ESA
OS/2	Unix	HP MPE
Windows 95	Windows NT	DEC VMS
Windows 98		
Windows CE		
Mac OS 8.1		
Windows NT 5.0		
Unix		

users found DOS to be adequate as long as they worked with relatively slow computers and diskette systems. Now, with the heavy use of high-volume hard disks and faster processors, personal computer users need to be able to run multiple programs and work on multiple tasks concurrently. For example, they want to be able to do spell checking of a long word processing document while accessing a mainframe computer to retrieve electronic mail. The need for multitasking is one reason for interest in more advanced operating systems.

DOS with Windows. The original Windows was not technically an operating system but a shell that sat on top of the DOS operating system. The early versions of Windows, such as Windows 3.1, loaded on top of DOS, which provided many basic operating system services. On top of the DOS infrastructure, Windows adds a graphical operating environment and features like a simple cooperative multitasking scheme it uses to run multiple Windows and DOS applications. Windows was developed in response to the highly popular, easy-to-use, icon-driven operating system of the Apple Macintosh. Windows provides an intuitive GUI. It also adds some multitasking capabilities and makes it easier to move data from one application to another.

OS/2. In 1988, IBM announced an operating system for personal computers, called Operating System 2 (OS/2). This operating system, designed to run on more powerful personal computers, requires a minimum of 2 MB of memory and at least 5 MB of hard disk storage and a powerful CPU. OS/2 includes a shell program called the Workplace Shell that provides end users with a graphical user interface similar to that of Microsoft Windows. OS/2 is a multitasking operating system that can run multiple programs simultaneously, whether the programs were written for Windows, OS/2, or DOS. OS/2 comes in several versions: a standard edition, an extended edition, and a Warp Connect version. Warp is able to run Windows applications.

OS/2 can address up to 4 GB of main memory. It also provides protection between applications so that applications and data cannot write over other applications and data; error logging capability; trace utilities to isolate and report software problems; multimedia capabilities; enhanced graphics; and pen-computing compatibility. One of the disadvantages of IBM's OS/2 operating system is that not many application software packages have been written to run on it. IBM positioned Warp for consumers, not business users.

Windows 95. From a basic architecture viewpoint, Windows 95 is a true 32-bit, preemptively multitasking, multithreaded operating system. Windows 95 is designed to be able to run existing 16-bit Windows applications. This allows it to work with a far wider range of existing software. On the down side, though, the memory areas containing 16-bit applications remain unprotected, with all 16-bit Windows applications sharing a common address space. Thus, unfortunately, it is still relatively easy for a misbehaving 16-bit Windows application to crash the whole environment.

A Start button is used to launch applications or documents with their associated applications (Microsoft paid a large amount of money to use the Rolling Stones' hit song "Start Me Up" in early Windows 95 TV commercials). As you launch programs, they appear on a task bar normally positioned at the bottom of the screen. Clicking on any task bar button switches to that program. This approach is a very intuitive style of task switching.

Windows 95 uses a desktop metaphor—your files are shown as icons within folders (directories). It also includes Explorer, software that gives you a tree-based view of your computer and its connections. In addition, file names are no longer limited to eight characters plus a three-letter extension; you can use names up to 255 characters long. Windows 95 also comes bundled with Exchange, designed to be a universal mailbox. It works with Microsoft Mail for receiving electronic-mail messages and with Microsoft Fax for sending and receiving faxes. Windows 95 also comes with communications software to enable you to connect to the Internet. All these features can provide increased productivity and better control over the operation of the computer system.

Windows 98. Windows 98 is an improved version of Windows 95 with many end user productivity features, improved support for newer hardware devices, and additional enhancements.

Several improvements were made to Windows 98 to improve end user productivity. A new image-preparation tool lets an organization's IS department install and configure the operating system and all applications on one machine, back them up, and then copy the image to a destination machine. The combination of Intel's Application Launch Accelerator technology and key advances in the Microsoft Windows 98 operating system significantly shorten the time it takes for software applications to load from the hard disk drive. End users notice a dramatic reduction in loading time when they open applications on their PCs. Application Launch Accelerator uses new capabilities in Windows 98 to arrange data blocks used during application start-up in a sequential fashion, thereby reducing the amount of random disk access. On a hard disk drive, sequential disk accesses are many times faster than random disk accesses used under other approaches for application start-up.[3] System shutdown has also been speeded.

Windows 98 includes drivers and built-in support for new kinds of hardware that have shipped since Windows 95, such as the new Accelerated Graphics Port (AGP), DVD, FireWire (IEEE 1394) or USB devices, and ACPI-compliant notebook. Plug in a Kodak Digital Video Camera, and Windows 98 immediately recognizes the device and installs the driver for it.[4] Other features include some cosmetic improvements to the user interface, support for multiple monitors, a log-on that works for multiple family members, a better system information utility, an improved backup-and-restore utility, and a better way of handling updates and uninstalling.

One of the more exotic features is the inclusion of WavePhone Inc.'s innovative data broadcast and service. WavePhone's WaveTop technology enables PCs equipped with Windows 98 and TV tuner boards to receive selected multimedia Internet content via cable television or television broadcast signals. Users can receive information without having to pay the monthly cost of an Internet service provider or tie up their telephone line. You connect to a Web site, enter your ZIP code, and, if your cable company participates, view a listing of the shows playing on TV, watch videos in a window on your desktop, or view additional information about the video and even interact with the programming. WaveTop works by embedding data into an unseen portion of existing broadcast television signals, called the vertical blanking interval (VBI), and integrating them with the Windows 98 Broadcast Architecture to bring WaveTop content directly to a PC.[5] This feature requires a special tuner adapter in your PC, and your local cable operator, satellite service, or TV station must offer Web compatible TV services.[6]

Windows 98 support for portable computers includes Advanced Power Management (APM) 1.2, which allows for support of multiple batteries and the more forward-looking Advanced Configuration and Power Interface (ACPI).[7] ACPI is itself the basis for another Windows 98 innovation, OnNow. This collection of standards for systems and applications is designed to make computers behave more like consumer appliances—that is, like your stereo: When you turn on your PC, it's on, instead of taking several minutes to boot.

Microsoft Windows CE. The Windows CE (for Compact Edition) operating system is not designed to run on a desktop or laptop computer. Unlike the Windows 98 or Windows NT operating systems, you will not find Windows CE retailing as a software product in stores. Windows CE is preinstalled in read-only memory (ROM) on devices such as digital TV set-top devices, PCs in automobiles, and handheld PCs, which are available in computer and consumer electronics stores and from Microsoft Certified Solution Providers. The standards-based Windows CE platform makes possible new categories of non-PC business and consumer devices that can communicate with each other, share information with Windows-based PCs, and connect to the Internet. It provides a graphical user interface, incorporating many elements of the familiar Windows user interface, making it easy to use. Windows CE is able to synchronize, communicate, and exchange information with Windows-based PCs.[8]

A few of the interesting products to incorporate the Windows CE operating system are described in the "FYI" box.

Apple computer operating systems. While IBM system platforms traditionally use Intel processors and DOS or Windows, Apple computers typically use Motorola processors and a proprietary Apple operating system,

FYI

Windows CE Applications

Tele-Communications Inc. (TCI) and Microsoft Corporation signed an agreement in early 1998 under which TCI will license a version of the Microsoft Windows CE operating system for a minimum of 5 million digital set-top boxes. This agreement gives TCI an operating system that enables high-quality video and sound, as well as new interactive video services. This version of Windows CE has been developed for the television environment, with integrated support for Internet content and technology from WebTV Networks Inc. Microsoft and TCI expect that Windows CE will be available for the advanced digital set-top devices that TCI will begin to deploy in late 1998 or early 1999.

The Auto PC powered by Microsoft Windows CE 2.0 is a complete information and entertainment system for your automobile. The Auto PC features an intuitive speech interface that allows you to use simple voice commands to keep your hands and eyes focused on driving. You can use it to easily access phone numbers and addresses, which you can then use to control your cellular phone or prompt Auto PC for driving directions. You can also listen to e-mail and real-time traffic reports you receive through wireless support and control the high-end digital audio system—all this while you keep your eyes on the road. The Auto PC conveniently fits into the space normally occupied by your in-dash radio. The Auto PC faceplate and button design look similar to a standard radio. The advanced speech technology is speaker independent, recognizing voice commands from a variety of speakers, and works even in a noisy car. The speech interface can be tailored to meet your specific needs. A wide variety of manufacturers will offer Auto PCs powered by Microsoft Windows CE.

The Palm PC was designed specifically with the mobile professional in mind, so you can keep your most vital personal and business information up to date and close at hand. You can go anywhere and take your Calendar, Contacts, Tasks, Inbox, Internet, and intranet information with you. Palm PC carries it all in one convenient, palm-sized design with easy-to-use, one-handed operation. The Palm PC includes a compact disc with Windows CE Services and Microsoft ActiveSync technology—software you can use to effortlessly and automatically synchronize your important personal and business information. Installing Windows CE Services on your Windows-based computer at your home or office allows you to connect your Palm PC to your computer. ActiveSync continuously and automatically synchronizes your contacts, calendar, tasks, notes, electronic mail (e-mail) messages (with attachments), and even group-scheduling requests to and from your Palm PC.

DISCUSSION QUESTIONS

1. What additional Windows CE applications can you imagine or identify?
2. Imagine that you own a Palm PC. Describe how you would use this device to become more productive.

Sources: Adapted from "TCI Selects Microsoft Windows CE For 5 Million Digital Cable Set-Top Devices," *PC Computer,* January 10, 1998 and the Microsoft Corporation Web site at http://www.microsoft.com, accessed on January 21, 1998.

such as Apple OS or Mac OS. Although IBM and IBM-compatible computers hold the largest share of the business PC market, Apple computers are also quite popular, especially in the fields of publishing, education, and graphic arts. The Macintosh computer is the most popular Apple system for business applications.

Mac OS 8.1. Many people argue that the Mac OS has always been the easiest and most intuitive of all operating systems. The Mac OS 8.1 runs on all PowerPC and 68040 processor-based Macintosh computers with at least 16 megabytes of RAM. Perhaps the most noticeable improvement in

Mac OS 8.1 is the dramatic speed gains in applications launch and relaunch, as well as in routine tasks such as copying files over a network. By improving virtual memory caching and disk cache, Mac OS 8.1 performs up to 50 percent faster in launching and relaunching applications such as word processing and spreadsheet software. Mac OS 8.1 also includes an optimized version of MathLib that speeds the performance of mathematics-intensive applications.

New to Mac OS 8.1 is the Mac OS Extended format, an improved volume format that literally gives storage space back to users. Mac OS Extended format allows minimum file sizes more consistent with the actual amount of information being stored, providing much more efficient data storage. The Mac OS 8.1 also provides excellent support for PC compatibility via PC Exchange 2.2, which supports Windows 95 long file names and volume formats, including PC-formatted removable media such as Iomega Zip and Jaz cartridges. Mac OS 8.1 also supports DVD-ROMs (digital video disks), giving you access to an emerging media format used for DVDs and DVD-ROM interactive games.

The new Mac OS Finder lets you perform several activities simultaneously, so you can keep working while copying large files, emptying the Trash, or performing other tasks in the background. Whenever a window takes more than a few seconds to refresh, it happens in the background, and windows update continuously as you scroll. Thus you can launch applications while copying files in the background. The Finder also includes "spring-loaded" folders that automatically open and windows that pop up from the bottom of the screen, making it easy to find files and get organized. Spring-loaded folders are especially handy. Drag an item onto a folder and it springs open until you find the folder you're looking for, making it easier to find what you need and stay organized.

Mac OS 8.1 includes a complete suite of Internet services, so you can connect to, explore, and publish on the Internet and World Wide Web. The Mac OS 8.1 includes Netscape Navigator, Microsoft Internet Explorer, Claris Emailer Lite, and America Online. An Internet Setup Assistant speeds setup and configuration, making it easier for novice users to get on-line.

Contextual menus show only items appropriate to the job at hand to minimize distractions. Sticky menus open at a single click and stay open until you click again. Pop-up windows give you quick access to frequently used files, applications, and folders. Click on these windows and they pop open; click again, and they snap closed, displaying tabs at the bottom of your screen. Pop-up windows can help you get a handle on desktop clutter.

To support the novice through expert user, Mac OS 8.1 provides a variable user environment—a novice can simplify the interface by specifying short menus and a simplified Finder. Icons on the desktop and in windows can be turned into buttons that open or launch with just one click. The more experienced user can take advantage of such rich new features as contextual menus.[9]

Windows NT 5.0. The Windows NT 5.0 operating system is enormous—with over 27 million lines of code—making it larger than all previous versions of Windows combined. In recent speeches, Bill Gates and NT boss Jim Allchin have repeatedly stated, "We're betting the company on NT 5.0." To back up their bet, Microsoft invested more than $1 billion in research and development of the new OS in 1998.[10] Clearly Windows NT is strategic to Microsoft.

The NT operating system is designed to take advantage of the newer 32-bit processors and features multitasking and networking capabilities. NT is designed to run on multiple hardware platforms; thus, it is portable. For example, NT can run on DEC's Alpha and MIPS's R4000 RISC microprocessors. Using emulation software, NT can run programs written for other operating systems. NT supports symmetric multiprocessing, the ability to make simultaneous use of multiple processors. It also has built-in networking to support several communications protocols. NT also has a centralized security system to monitor various system resources. The many features and capabilities of NT make it very attractive for use on many computers.

With Windows NT 5.0, Plug and Play support is finally present, although only systems based on the new Advanced Configuration and Power Interface (ACPI) standard can truly exploit it. Windows NT 5.0 offers Windows 95 compatibility so that users can consider changing from a Windows 95 to Windows NT operating system rather than to Windows 98. Microsoft Management Console in Windows NT 5.0 replaces dozens of utility programs and dialog boxes with a single, easy-to-customize window. Tighter security controls help NT administrators lock intruders out of the network. A new version of NT's native file format, called NTFS file system, significantly enhances performance and reliability. NTFS enhancements let workstation users encrypt data. This version of Windows tightly integrates Internet Explorer into the Windows shell. These improvements make the computer faster and more reliable.

NT 5.0 is far more suited for road trips than any previous version, with improvements in storage and power management that will finally make it a credible operating system for users of portable PCs. Workstation users can use dialog boxes to create off-line copies of files and folders—a welcomed alternative to the unwieldy Briefcase utility in Windows 95 and NT 4.0. NT 5.0 does not support the current crop of notebooks that use Advanced Power Management; instead it must be installed on a machine built around the Advanced Configuration and Power Interface (ACPI).

Windows NT 5.0 also functions as a Web server and Web content authoring platform. Workstation users can tackle light publishing tasks with Personal Web Server (PWS) that integrates seamlessly with NT's security model. Additional features and capabilities included with the Personal Web Server make it powerful enough for use with a small workgroup over an intranet.

Interestingly, Microsoft has already set the year 2000 for replacing Windows 95/98 with a consumer version of Windows NT. At that time, the company will offer two major operating systems—NT and Windows CE with several variations of NT including consumer, desktop, server and multiuser versions.[11]

Unix. Unix is a powerful operating system originally developed by AT&T for minicomputers. Unix can be used on many computer system types and platforms, from personal computers to mainframe systems. Unix benefits companies using both small and large computer systems because it is compatible with different types of hardware and users only have to learn one operating system. This high degree of portability is one reason for Unix's popularity. A user can select a software application that runs under Unix and have a high degree of flexibility in choosing the hardware manufacturer. Unix also makes it much easier to move programs and data among computers or to connect mainframes and personal computers to share resources.

Unix is considered to have a complex user interface with strange and arcane commands. As a result, software developers have provided shells such as Motif from Open Systems Foundation and Open Look from Sun Microsystems. These shells do for Unix much of what older versions of Windows did for DOS—they provide a graphical user interface and shield the users from the complexity of the underlying operating system.

There are many variants of Unix—including HP/UX from Hewlett-Packard, AIX from IBM, Ultrix from DEC, UNIX SystemV from UNIX Systems Lab, Solaris from Microsystems, and SCO from Santa Cruz Operations—which have addressed many of the original Unix shortcomings. These variants have created a lack of common standards, although most organizations do not find this to be an insurmountable problem. Unix will likely lose market share to Windows NT in the low-end server and workstation arena, but it will continue to grow rapidly in the database and large application server arena.[12]

Multiple Virtual Storage/Enterprise Systems Architecture. Most mainframe computer manufacturers provide proprietary operating systems with their specific hardware. For example, Enterprise Systems Architecture/370 (ESA/370) and Multiple Virtual Storage/Enterprise Systems Architecture (MVS/ESA) are operating systems used on larger IBM mainframe computers. Hewlett-Packard (HP) and Digital Equipment Corporation (DEC) are also popular enterprise systems. These manufacturers have operating systems other than their Unix version.

Utility Programs

Utility programs are used to merge and sort sets of data, keep track of computer jobs being run, compress files of data before they are stored or transmitted over a network (thus saving space and/or time), and perform other important tasks. Utility programs often come installed on computer systems; a number of utility programs can also be purchased. Norton Utility, for example, is a set of utility programs that provides a number of capabilities, including checking your hard drive and diskettes for a virus, checking your hard drive for bad storage locations and removing them, and compressing data through a disk compression software package. Data or disk compression allows more data and programs to be stored on the same hard disk by eliminating repeated data and empty storage space. When programs and data are

Utility programs, such as those found in Norton Utilities, can provide detailed information about the user's entire system including hardware, peripherals, setup files, new Windows plug-n-play information, and system performance.
(Source: Courtesy of Symantec Corporation.)

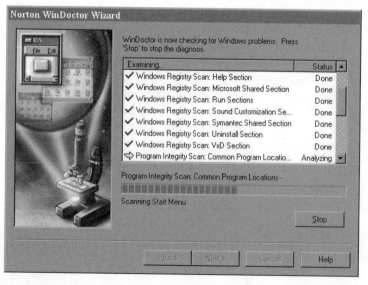

Personal	Workgroup	Enterprise
Software to compress data so that it takes less hard disk space	Software to provide detailed reports of work-group computer activity and status of user accounts	Software to archive contents of a database by copying data from disk to tape
Screen saver	Software that manages an uninterruptable power source to do a controlled shutdown of the work-group computer in the event of a loss of power	Software that compares the content of one file with another and identifies any differences
Virus detection software	Software that reports unsuccessful user log-on attempts	Software that reports the status of a particular computer job

TABLE 4.4

Examples of Utility Programs

needed, decompression routines make the programs and data available. In some cases, data compression software can almost double the amount of stored data, increasing capacity without high costs. Utility programs can meet the needs of an individual, a workgroup, or an enterprise, as shown in Table 4.4. They perform useful tasks—from tracking jobs to monitoring system integrity.

IBM has created systems management software that allows individual support people to monitor the growing number of desktop computers in a business. With this software, the support people can sit at their personal computers and check or diagnose problems, like hard disk failure, on individual computer systems on a network. The support people can even repair individual systems anywhere on the organization's network, often without having to leave their desks. The direct benefit comes to the system manager, but the business gains from having a smoothly functioning information system.

TestDrive is another example of a utility program. It allows you to try out a software package before you purchase it. Using TestDrive, you can download certain software you are thinking about purchasing from the Internet. TestDrive allows you to use the software for a set length of time. If you like the software, TestDrive allows you to purchase it using a credit card. If you do not, TestDrive will automatically delete it from your hard disk.

APPLICATION SOFTWARE

As discussed earlier in this chapter, the primary function of application software is to apply the power of the computer to give individuals, work-groups, and the entire enterprise the ability to solve problems and perform specific tasks. When you need the computer to do something, you use one or more application programs. The application programs then interact with systems software. Systems software then directs the computer hardware to perform the necessary tasks.

Suppose a manager is concerned that too many employees are getting overtime pay by working more than 40 hours each week, even though many others are working less than 40 hours per week. She would like to have those working below the 40-hour threshold replace those over the threshold, and hence avoid the time-and-a-half overtime pay rate. The manager can enlist a computer to print the names of all employees working significantly more or significantly less than 40 hours per week on average over the last three months.

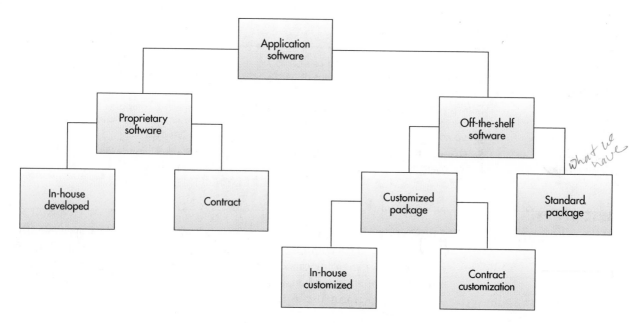

FIGURE 4.5

Sources of Software: Proprietary and Off-the-Shelf

Some off-the-shelf software may be modified to allow some customization.

Programs like this that complete sales orders, control inventory, pay bills, write paychecks to employees, and provide financial and marketing information to managers and executives are examples of application software. Most of the computerized business jobs and activities discussed in this book involve the use of application software.

Types of Application Software

The key to unlocking the potential of any computer system is application software. A company can either develop a one-of-a-kind program for a specific application (called **proprietary software**) or purchase and use an existing software program (sometimes called **off-the-shelf software**). It is also possible to modify some off-the-shelf programs, giving a blend of off-the-shelf and customized approaches. These different sources of software are shown in Figure 4.5. The relative advantages and disadvantages of proprietary software and off-the-shelf software are summarized in Table 4.5.

proprietary software

software designed to solve a unique and specific problem

off-the-shelf software

an existing software program

Proprietary application software. Software to solve a unique or specific problem is called *proprietary application software*. This type of software is usually built, but it can also be purchased from an outside company. If the organization has the time and IS talent, it may opt for **in-house development** for all aspects of the application programs. Alternatively, an organization may obtain customized software from external vendors. For example, a third-party software firm, often called a value-added software vendor, may develop or modify a software program to meet the needs of a particular industry or company. A specific software program developed for a particular company is called **contract software**.

in-house development

development of application software using the company's resources

contract software

software developed for a particular company

Off-the-shelf application software. Software can also be purchased, leased, or rented from a software company that develops programs and sells them to many computer users and organizations. Software programs developed for a

Proprietary Software		Off-the-Shelf Software	
Advantages	**Disadvantages**	**Advantages**	**Disadvantages**
You can get exactly what you need in terms of features, reports, and so on.	It can take a long time and significant resources to develop required features.	The initial cost is lower since the software firm is able to spread the development costs over a large number of customers.	An organization might need to pay for features that are not required and never used.
Being involved in the development offers a further level of control over the results.	In-house system development staff may become hard pressed to provide the required level of ongoing support and maintenance because of pressure to get on to other new projects.	There is a lower risk that the software will fail to meet the basic business needs—you can analyze existing features and the performance of the package.	The software may lack important features, thus requiring future modification or customization. This can be very expensive because users must adopt future releases of the software.
There is more flexibility in making modifications that may be required to counteract a new initiative by one of your competitors or to meet new supplier and/or customer requirements. A merger with another firm or an acquisition also will necessitate software changes to meet new business needs.	There is more risk concerning the features and performance of the software that has yet to be developed.	Package is likely to be of high quality since many customer firms have tested the software and helped identify many of its bugs.	Software may not match current work processes and data standards.

TABLE 4.5

A Comparison of Proprietary and Off-the-Shelf Software

general market are called off-the-shelf software packages because they can literally be purchased "off the shelf" in a store. Many companies use off-the-shelf software to support business processes.

Customized package. In some cases, companies use a blend of external and internal software development. That is, off-the-shelf software packages are modified or customized by in-house or external personnel. For example, a software developer may write a collection of programs to be used in an auto body shop that includes such features as generating estimates, ordering parts, and processing insurance. Body shops of all types have these needs. Designed properly—and with provisions for minor tailoring for each user—the same software package can be sold to many users. However, since each body shop has slightly different requirements, some modifications to the software may be needed. As a result, software vendors often provide a wide range of services, including installation of their standard software, modifications to the software required by the customer, installation of the software, training of the end users, and other consulting services.

Some software companies encourage their customers to make changes to their software. In some cases, the software company supplying the necessary software will make the necessary changes for a fee. Other software companies, however, will not allow their software to be modified or changed by those purchasing or leasing it.

Personal Application Software

There are literally hundreds of computer applications that can help individuals at school, home, and work. Personal application software includes general-purpose tools and programs that can support a number of individuals' needs. For example, a graphics program can be purchased to help a sales manager develop an attractive sales presentation to give to the sales force at its annual meeting. A spreadsheet program allows a financial executive to test possible investment outcomes. The primary programs are word processing, spreadsheet analysis, database, graphics, and on-line services. Advanced software tools—like project management, financial management, desktop publishing, and creativity software—are finding more and more use in business. The features of

TABLE 4.6

Examples of Personal Application Software

Type of Software	Explanation	Example	Vendor
Word processing	Create, edit, and print text documents	Word WordPerfect	Microsoft Corel
Spreadsheet	Provide a wide range of built-in functions for statistical, financial, logical, database, graphics, and data and time calculations	Excel Lotus 1-2-3 Quattro Pro	Microsoft Lotus/IBM Originally developed by Borland
Database	Store, manipulate, and retrieve data	Access Approach FoxPro dBASE	Microsoft Lotus/IBM Microsoft Borland
On-line information services	Obtain a broad range of information from commercial services	America Online CompuServe Prodigy	America Online CompuServe Prodigy
Graphics	Develop graphs, illustrations, and drawings	Illustrator FreeHand	Adobe Macromedia
Project management	Plan, schedule, allocate, and control people and resources (money, time, and technology) needed to complete a project according to schedule	Project for Windows On Target Project Schedule Time Line	Microsoft Symantec Scitor Symantec
Financial management	Provide income and expense tracking and reporting to monitor and plan budgets (some programs have investment portfolio management features)	Managing Your Money Quicken	Meca Software Intuit
Desktop publishing (DTP)	Works with personal computers and high-resolution printers to create high-quality printed output, including text and graphics; various styles of pages can be laid out; art and text files from other programs can also be integrated into "published" pages	QuarkXPress Publisher PageMaker Ventura Publisher	Quark Microsoft Adobe Corel
Creativity	Helps generate innovative and creative ideas and problem solutions. The software does not propose solutions, but provides a framework conducive to creative thought. The software takes users through a routine, first naming a problem, then organizing ideas and "wishes," and offering new information to suggest different ideas or solutions	Organizer Notes	Macromedia Lotus

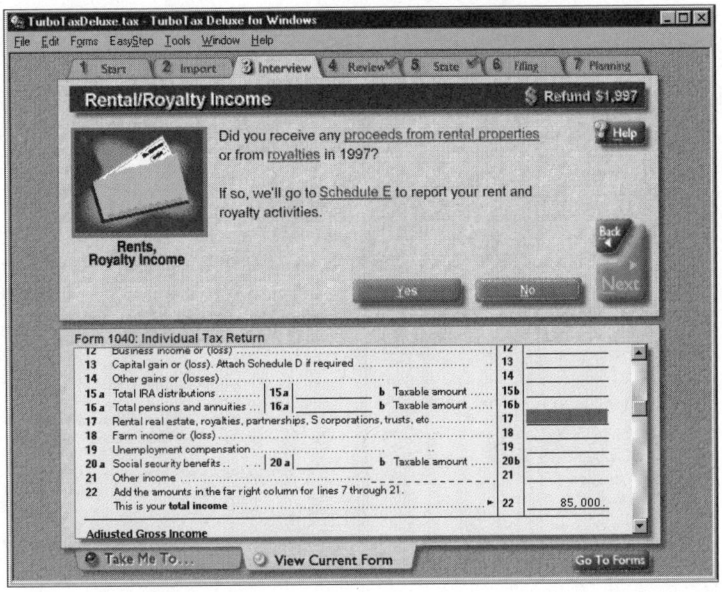

FIGURE 4.6

TurboTax

Tax preparation programs can save individuals or businesses hours of work. In addition, they are typically more accurate than doing a tax return by hand. Many tax programs input data from financial and check paying programs, check for potential problems, help you avoid audits caused by your return, and give you help and advice about what you may have forgotten to deduct. (Source: Courtesy of Intuit, Inc.)

FIGURE 4.7

Quicken

Most individuals and many businesses buy off-the-shelf programs to pay bills because it is less expensive, less risky, and of higher quality and requires fewer resources compared with developing or writing these types of programs or paying bills manually.

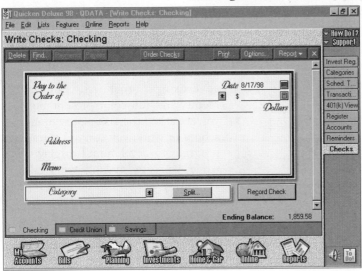

personal application software are summarized in Table 4.6. In addition, there are literally thousands of other personal computer applications to perform specialized tasks: to help you do your taxes, get in shape, lose weight, get medical advice, write wills and other legal documents, make repairs to your computer, fix your car, write music, and edit your pictures and videos (see Figures 4.6 and 4.7). This type of software, often called user software or personal productivity software, includes general-purpose tools and programs that can support a number of individuals' needs.

Word processing. If you write reports, letters, or term papers, word processing applications can be indispensable. The vast majority of personal computers in use today have word processing applications installed. Word processing applications can be used to create, edit, and print documents. Most come with a vast array of features, including those for checking spelling, creating tables, inserting formulas, creating graphics, and much more. This book (and most like it) was entered into a word processing application using a personal computer. (See Figure 4.8.)

Spreadsheet analysis. People use spreadsheets to prepare budgets, forecast profits, analyze insurance programs, summarize income tax data, and analyze investments. Whenever numbers and calculations are involved, spreadsheets should be considered. Features of spreadsheets include graphics, limited database capabilities, statistical analysis, built-in business functions, and much more. (See Figure 4.9)

Database applications. Database applications are ideal for storing, manipulating,

FIGURE 4.8

Word Processing Program

Word processing applications can be used to write letters, holiday greeting cards, work reports, and term papers.

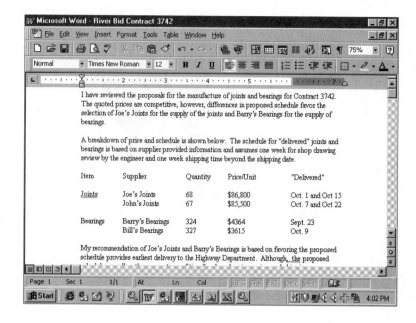

and retrieving data. These applications are particularly useful when you need to manipulate a large amount of data and produce reports and documents. Database manipulations include merging, editing, and sorting data. The uses of a database application are varied. You can keep track of a record or CD collection, the items in your apartment, tax records, and expenses using a database application. A student club can use a database to store names, addresses, phone numbers, and dues paid. In business, a database application can help process sales orders, control inventory, order new supplies, send letters to customers, and pay employees. A database can also be a front end to another application. For example, a database application can be used to enter and store income tax information. The stored results can then be exported to other applications, such as a spreadsheet or tax preparation application. (See Figure 4.10.)

FIGURE 4.9

Spreadsheet Program

Spreadsheet programs should be considered when calculations are required.

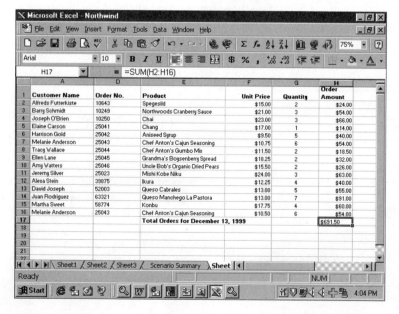

Graphics programs. It is often said that a picture is worth a thousand words. With today's graphics programs, it is easy to develop attractive graphs, illustrations, and drawings. Graphics programs can be used to develop advertising brochures, announcements, and full-color presentations. If you are asked to make a presentation at school or work, you can use a graphics program to develop and display slides while you are making your talk. A graphics program can be used to help you make a presentation, a drawing, or an illustration. (See Figure 4.11).

On-line information services. On-line services allow you to connect a

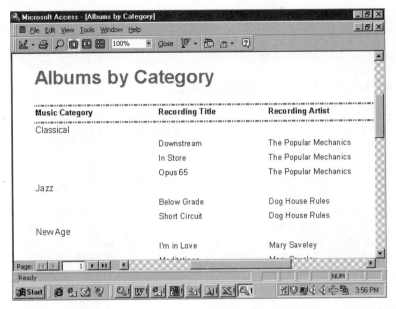

FIGURE 4.10

Database Program

Once entered into a database application, information can be manipulated and used to produce reports and documents.

software suite

a collection of single application software packages in a bundle

personal computer to the outside world through phone lines. Using an on-line service, you can get investment information, make travel plans, and check on news from around the world. You can also get prices and features for most consumer items, learn about companies, send electronic mail to friends and family, learn about degree programs offered by colleges and universities around the world, and search for job openings in your area. (See Figure 4.12.)

Software suites. A **software suite** is a collection of single application software packages in a bundle. Software suites can include word processors, spreadsheets, database management systems, graphics programs, communications tools, organizers, and more. There are a number of advantages to using a software suite. The software programs have been designed to work similarly, so that once you learn the basics for one application, the other applications are easier to learn and use. Buying software in a bundled suite is cost-effective: the programs usually sell for a fraction of what they would cost individually.

Microsoft Office, Corel Office, Novell Perfect Office, and Lotus SmartSuite are examples of popular general-purpose software suites for personal computer users. (See Figure 4.13.) Each of these software suites includes a spreadsheet program, word processor, database program, and graphics package with the ability to move documents, data, and diagrams among them. Thus, a user may

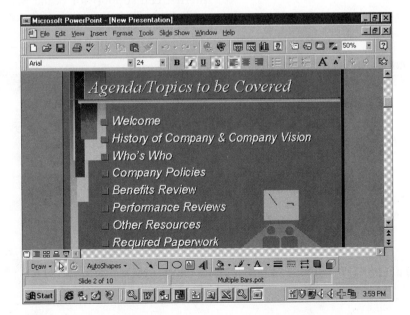

FIGURE 4.11

Graphics Program

Graphics programs can help you make a presentation at school or work. They can also be used to develop attractive brochures, illustrations, drawings, and maps.

FIGURE 4.12

On-Line Services

On-line services provide instant access to information. Prices of cars and trucks, travel discounts, information about companies, stock market data, and much more are available with a few key strokes using an on-line service.

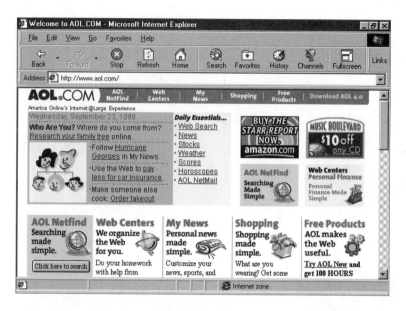

create a spreadsheet and then cut and paste that spreadsheet into a document created using the word processing application.

The Microsoft Office suite includes Excel for spreadsheet, Word for word processing, and PowerPoint for presentation preparation. It also includes Outlook, a personal information manager with e-Mail, a calendar, and a group scheduler. Office goes beyond its role as a mainstream package of ready-to-run applications with the extensive custom development facilities of Visual Basic for Applications and the additional tools provided in the Developer Edition of the suite.[13] Visual Basic for Applications (VBA) is the built-in scripting facility that is part of every Office application. VBA provides a means of enhancing off-the-shelf applications with powerful scripting facilities.[14] Office offers hyperlinks to and from any Office file; automatic conversion of Word, Excel, and Access files to a format usable on the Internet and support for Visual Basic for Applications across all products in the suite; Web FastFind to search and index across files and servers;

FIGURE 4.13

Software Suite

A software suite, such as Microsoft Office, offers a collection of powerful programs, including word processing, spreadsheet, database, graphics, and other programs. The programs in a software suite are designed to be used together. In addition, the commands, icons, and procedures are the same for all programs in the suite. (Source: Courtesy of Microsoft Corporation.)

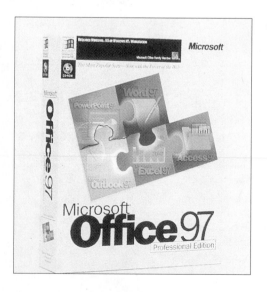

Office Web tool bar to navigate among linked documents; and access to ActiveX controls from Word, Excel, and PowerPoint.

Since one or more applications in a suite may not be as desirable as the others, some people still prefer to buy separate packages. Another issue with the use of software suites is the large amount of main memory required to run them effectively. For example, many users find that they must spend hundreds of dollars for additional internal memory to upgrade their personal computer to be able to run a software suite. Continual debates rate one vendor's spreadsheet superior to another vendor's, and yet a third vendor may have the best word processing package. Thus, some users prefer using individual software packages from different vendors rather than a software suite from a single vendor.[15]

Object Linking and Embedding (OLE)

To enable different application programs to work together, many operating systems support **object linking and embedding (OLE)**. With OLE, you can copy text from one document to another or embed graphics from one program into another program or document, such as a word processing report. You can also link many documents or files, such as a table developed in a spreadsheet program that can be linked to a word processing document. When you make a link, you are actually referencing the original document or program. Changes are always made to the source document or program and transferred to the linked documents, such as one or more word processing reports.

If the operating system and the application support OLE, data can be shared among applications in three ways: copying, linking, and embedding. In the following explanation, the data you transfer among applications is called an object. An object can be a picture, graph, text, spreadsheet, or other data. The **server application** is the application that supplies objects that you place into other applications. The **client application** is the application that accepts objects from other applications. An application can be a server, a client, or both. Figure 4.14 shows a graphic image, spreadsheet, and project schedule being incorporated into a word processing document.

Copying. You can copy data from one application and place it in another. Both applications will then display the data. The copy method is used when you do not want to change data shared between applications. To change data, you must first update the data in the client application and then repeat the copy procedure.

Linking. You **link** an object when you want any changes made to the server object to automatically appear in all linked client documents. You must first save an object in the server application before you can link it to any client documents. To link an object, use the Past Link command. Linking an object provides the client documents with a screen image of the object. The real object resides in the server document. Double clicking the object in any client document opens the server application that supplied the object so that you can then edit that object. Any changes you make to the object in the server application automatically appear in all linked client documents.

object linking and embedding (OLE)

a software feature that allows you to copy text from one document to another or embed graphics from one program into another program or document

server application

the application that supplies objects you place into other applications

client application

the application that accepts objects from other applications

link

procedure used when you want any changes made to the server object to automatically appear in all linked client objects

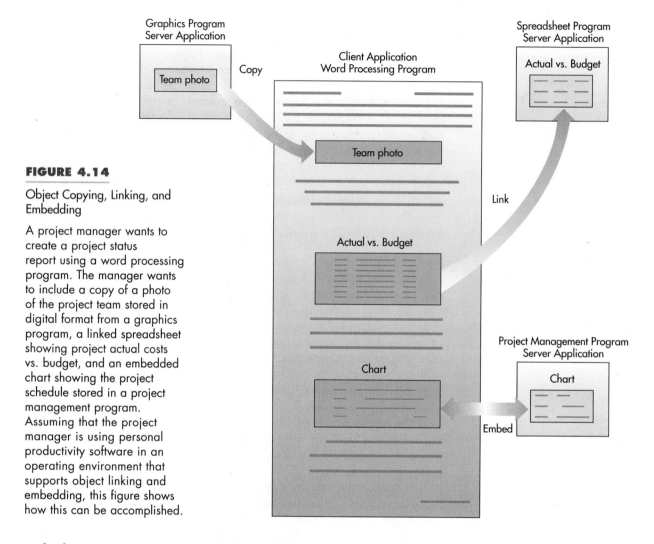

FIGURE 4.14

Object Copying, Linking, and Embedding

A project manager wants to create a project status report using a word processing program. The manager wants to include a copy of a photo of the project team stored in digital format from a graphics program, a linked spreadsheet showing project actual costs vs. budget, and an embedded chart showing the project schedule stored in a project management program. Assuming that the project manager is using personal productivity software in an operating environment that supports object linking and embedding, this figure shows how this can be accomplished.

embed

procedure used when you want an object to become part of the client document

Embedding. You **embed** an object when you want it to become part of the client document. To embed an object, use the Edit Paste command. (If the applications you are using support OLE, using the Paste command embeds the object. If the applications do not support OLE, using the Paste command copies the object.) You do not have to save an object in the server application before embedding it in a client document. This makes the document more portable since the server document is not required. Double clicking on the object opens the object in the client server application that supplied the object. Any changes you make to the object appear only in the client documents containing the object. However, to see those changes in any client application, you must use the File-Update command. When you embed an object, each client document receives a copy of the object. The object becomes part of the client document. Documents containing embedded objects require more memory and hard disk storage than documents linked to objects. After you embed an object in the client document, the server document is no longer required.

Workgroup Application Software

groupware

software that helps groups of people work together more efficiently and effectively

collaborative computing software

software that helps teams of people work together toward a common goal

The software class known as **groupware** cannot be concretely defined but is, in general, software that helps groups of people work together more efficiently and effectively. **Collaborative computing software**, an only slightly better term, at least conveys the sense that teams are working toward a common goal. Collaborative computing software can support a team of managers working on the same production problem, letting them share their ideas and work via connected computer systems. Examples of such software include group scheduling software, electronic mail, and other software that enables people to share ideas. New York–based Swiss Reinsurance America deployed a claims processing application to more than 300 workers. By streamlining its work processes and using workgroup software, it was able to reduce the number of steps needed to process a claim from 18 to 7 and reduce the time it takes to process a claim from three days to one.[16]

Lotus Notes. Lotus has defined knowledge management as the ability to provide individuals and groups of users with a method to find, access, and deliver valuable information in a coherent fashion. Its Lotus Notes product is an attempt to provide this ability. Lotus Notes gives companies the capability of using one software package, and one user interface, to integrate many business processes. For example, it can allow a global team to work together from a common or shared set of documents, have electronic discussions using common threads of discussion, and schedule team meetings. A key design feature is to make Notes 5 very easy to use.[17]

As Lotus Notes matured, Lotus added services to it and renamed it Domino, and now an entire third-party market has emerged around building collaborative software based on Domino. These products, which include Changepoint's Involv, remove the burden of Notes administration and broaden the application's scope to better support the Internet.[18] For example, Domino.Doc is a Domino-based document management application with built-in workflow and archiving capabilities. Its "life cycle" feature tracks a document through the review, approval, publishing, and archiving processes. Similarly, the workflow integration adds support for multiple roles, log tracking and distributed approval.[19] Read the "Making a Difference" box to see how one utility uses Lotus Notes to streamline work processes.

Group scheduling. Group scheduling is another form of groupware, but not all software schedulers approach their tasks in the same way. Some schedulers, known as personal information managers (PIMs), tend to focus on personal schedules and lists, as opposed to coordinating the schedules and meetings of a team or group. Schedulers do not suit everyone's needs, and if they are not truly required, they could impede efficiency. The "Three Cs" rule for successful implementation of groupware is summarized in Table 4.7.

MAKING A DIFFERENCE

Illinois Power Uses Lotus Notes to Improve Customer Service

Illinois Power, a subsidiary of Illinova, provides gas and electric power to 500,000 customers in southern Illinois. The utility business is changing rapidly due to deregulation and increased competition. To prosper under these changing circumstances, investor-owned Illinois Power must build flexible, high-quality processes for delivering the right information to the right people, in the right fashion, and at the right time to improve organizational performance. Such processes allow Illinois Power to adapt quickly and effectively to marketplace conditions. Illinois Power is using Lotus Notes to become more customer-focused and enhance the speed and quality of its customer service (to both customers and employees) but without increasing operating costs.

The customer call center is the primary point of contact for customers. This operation must work effectively and efficiently to make a good impression on customers. With this in mind, Illinois Power improved the response time of its call center using a Notes-based application to provide access to information. The customer service reps can perform full-text searches against this database to find answers to customers' questions. This application replaced the cumbersome set of 3-ring binders that operators used to organize reference materials. Customer service reps can now provide complete, accurate, and consistent answers more quickly. This makes customers happier and improves customer service productivity. Illinois Power has also been able to lower the costs to keep information current by distributing updates through straightforward changes to a single Notes database instead of replacing pages in more than 30 separate binders.

Illinois Power also uses Notes to improve the productivity and responsiveness of groups that provide service to employees. For example, the field computer support function uses a Notes-based application for its problem resolution process. Notes logs each incident and the responsible service technician and tracks progress and closure. In addition, customers can attach high-value information to the service request such as images taken directly from the malfunctioning system. This type of data exchange was not possible before Notes, and it assures a clearer, more accurate transmission of information. This accelerates resolution of the problem, so the department is more efficient and its customers are back on track faster.

Illinois Power uses Notes' discussion databases extensively to create short-term teams targeted on specific projects. Because Notes makes it so easy to set up applications that tie teams together, regardless of its members' location, the company can pool the expertise of its best people on projects where they can have the greatest impact, then have the flexibility to move them quickly to other areas as needed.

Illinois Power currently has 1,700 Notes users and is extending Notes to another 1,000. The company uses eight servers from different manufacturers and the Windows NT operating system as its platform for Notes. All applications are built and maintained in house, either by Illinois Power's programming staff or by the individual user departments.

DISCUSSION QUESTIONS

1. Illinois Power has done all Lotus Notes-based application development using in-house resources. Why haven't they sought application software packages to meet their needs?
2. What other potential Lotus Notes applications can you envision?

Source: Adapted from Lotus Development Corporation's Web site at http://www.lotus.com; and the Illinova home page at http://www.illinova.com, accessed February 1998.

TABLE 4.7

Ernst & Young's "Three Cs" Rule for Groupware

Convenient	If it's too hard to use, it doesn't get used; it should be as easy to use as the telephone.
Content	It must provide a constant stream of rich, relevant, and personalized content.
Coverage	If it isn't close to everything you need, it may never get used.

TABLE 4.8

Examples of Enterprise
Application Software

Accounts receivable	Sales ordering
Accounts payable	Order entry
Airline industry operations	Payroll
Automatic teller systems	Human resource management
Cash-flow analysis	Check processing
Credit and charge card administration	Tax planning and preparation
Manufacturing control	Receiving
Distribution control	Restaurant management
General ledger	Retail operations
Stock and bond management	Invoicing
Savings and time deposits	Shipping
Inventory control	Fixed asset accounting

Enterprise Application Software

Software that benefits the entire organization can also be developed or purchased. A fast-food chain, for example, might develop a materials ordering and distribution program to make sure that each fast-food franchise gets the necessary raw materials and supplies during the week. This materials ordering and distribution program can be developed internally using staff and resources in the IS department or purchased from an external software company. Table 4.8 lists a number of applications that can be addressed with enterprise software.

Many organizations are moving to integrated enterprise software that supports supply chain management (movement of raw materials from suppliers through shipment of finished goods to customers), as shown in Figure 4.15.

FIGURE 4.15

Use of Integrated Supply Chain
Management Software

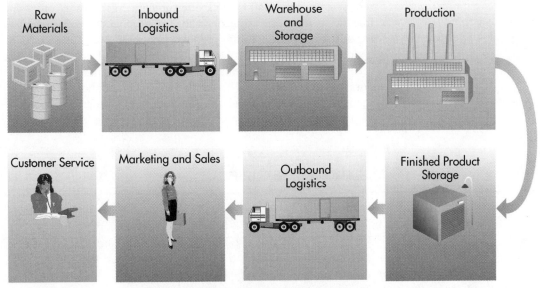

**Integrated Enterprise Software to Support
Supply Chain Management**

Vendors	
SAP	Baan
Oracle	SSA
PeopleSoft	Marcam
Dun & Bradstreet	QAD
JD Edwards	Ross Systems

TABLE 4.9

Selected Enterprise Resource Planning Software Vendors

enterprise resource planning (ERP)

a set of integrated programs that manage a company's vital business operations for an entire multisite, global organization

Organizations can no longer respond to market changes using nonintegrated information systems based on batch processing, conflicting data models, and obsolete technology. As a result, many corporations are turning to **enterprise resource planning (ERP)** software, a set of integrated programs that manage a company's vital business operations for an entire multisite, global organization. Thus, an ERP must be able to support multiple legal entities, multiple languages, and multiple currencies. While the scope of an ERP system may vary from vendor to vendor, most ERP systems provide integrated software to support manufacturing and finance. In addition to these core business processes, some ERP systems may be capable of supporting additional business functions such as human resources, sales, and distribution.

Successful linking of these business functions means that business requirements must come first and technology constraints second. Software will need to be rethought and reconfigured to reflect that reality. Unfortunately, the effort to implement ERP software is great and the path to success is full of hazards. Software vendors that provide integrated enterprise software are listed in Table 4.9.

Most ERP vendors specialize in software that addresses the needs of well defined markets such as automotive, semiconductor, petrochemical, and food/beverage manufacturers with solutions targeted to meet specific needs in those industries. ERP has become one of the most lucrative segments of the software market, as companies continue to switch from old information systems to more efficient client/server-based systems. In fact, ERP has become one of the hottest software trends in recent years, and industry sources project it will grow from a $4 billion industry in 1995 to an $11 billion industry by the year 2000.[20]

Increased global competition, new needs of executive management for control over the total cost and product flow through their enterprises, and more customer interactions are driving the demand for enterprisewide access to current business information. ERP offers integrated software from a single vendor that helps meet those needs. The primary benefits of implementing ERP include solving the software problem associated with the year 2000, eliminating inefficient systems, easing adoption of improved work processes, improving access to data for operational decision making, standardizing technology vendors and equipment, and enabling the implementation of supply chain management. The "Ethical and Societal Issues" box describes a scenario that can help you understand the problems of implementing ERP software.

ETHICAL AND SOCIETAL ISSUES

Trying to Champion ERP

As the information systems manager at a Global 1000 company, you have been told by the senior vice president of operations to implement an ERP system within the next two years. After several months of investigation, you have selected an ERP system from a large, reputable vendor who boasts of hundreds of successful implementations. The system would address the complete scope of your company's supply chain management activities.

For the past month, you have tried to get middle-level managers interested in the implementation of an ERP system; however, you have encountered unexpected, stiff resistance from these managers. Several of them have challenged you to quantify all costs as well as the business payoff. Some of them are able to cite examples of other companies that tried to implement ERP and were unsuccessful on their first attempt. Others feel there are more important business initiatives that need to be addressed and will not commit the necessary resources to ERP.

You realize that you will not be able to estimate costs and benefits for the project accurately. You are only too well aware that other companies have tried and failed to implement ERP. You realize the company has a number of opportunities for improvement that require both human and financial resources.

It is time to update the vice president with your findings and results.

Discussion Questions
1. What recommendations can you make to break the impasse over proceeding with this project?
2. What specific expectations do you need to set with the senior vice president to ensure the success of this project if it proceeds?

PROGRAMMING LANGUAGES

programming languages

coding schemes used to write both systems and application software

Both systems and application software are written in coding schemes called **programming languages**. The primary function of a programming language is to provide instructions to the computer system so that it can perform a processing activity. Specialized IS professionals work with programming languages, which are sets of symbols and rules used to write program code. Programming involves translating what a user wants to accomplish into a code that the computer can understand and execute. Like writing a report or a paper in English, writing a computer program in a programming language requires that the programmer follow a set of rules. Each programming language uses a set of symbols that have special meaning. Each language also has its own set of rules, called the **syntax** of the language. The language syntax dictates how the symbols should be combined into statements capable of conveying meaningful instructions to the CPU.

syntax

a set of rules associated with a programming language

Standards and Characteristics

Programming languages, like computers, have evolved over time. For the most part, their evolution has been driven by the desire to efficiently apply the power of information processing to as wide a variety of problem-solving activities as possible. This evolution has been somewhat regulated through standards. Earlier programming languages suffered from a lack of consistency in statements and procedures, resulting in one form of a programming language not working with another program written in a different form of the same language. To avoid this inefficiency, the American National Standards Institute (ANSI) developed standards for popular programming languages.

A program language standard is a set of rules that describe how programming statements and commands should be written. A rule that "variable names must start with a letter" is an example of a standard. (A variable is an item that can take on different values.) Program variable names such as

TABLE 4.10

Programming Language
Attributes

Extreme 1	Extreme 2
Supports programming of batch processing systems with data collected into a set and processed at one time.	Supports programming of real-time systems with each data transaction processed when it occurs.
Requires programmer to write procedure-oriented code, describing step by step each action the computer must take.	Enables a programmer to write nonprocedure-oriented code, describing the end result desired without having to specify how to accomplish it.
Supports business applications that require the ability to store, retrieve, and manipulate alphanumeric data and process large files.	Supports sophisticated scientific computations.
Programmers write code with a relatively high level of errors.	Programmers write code with a relatively low level of errors.
Programmers are less productive and able to create only a small amount of code per unit time.	Programmers are more productive and are able to create a large amount of code per unit time.

TABLE 4.10

Programming Language
Attributes

SALES, PAYRATE, and TOTAL follow the standard because they start with a letter, while variables such as %INTEREST, $TOTAL, and #POUNDS do not. By following program language standards, organizations can focus less on code writing and more on efforts to use programming languages to most effectively solve business problems.

Programming languages were developed to help solve a particular type of problem. Since they were each designed for different problems, they contain different attributes. Each of the attributes in Table 4.10 represents two extremes, with most languages falling somewhere between these extremes.

The Evolution of Programming Languages

The desire to use the power of information processing efficiently in problem solving has pushed development of newer programming languages. The evolution of programming languages is typically discussed in terms of generations of languages.

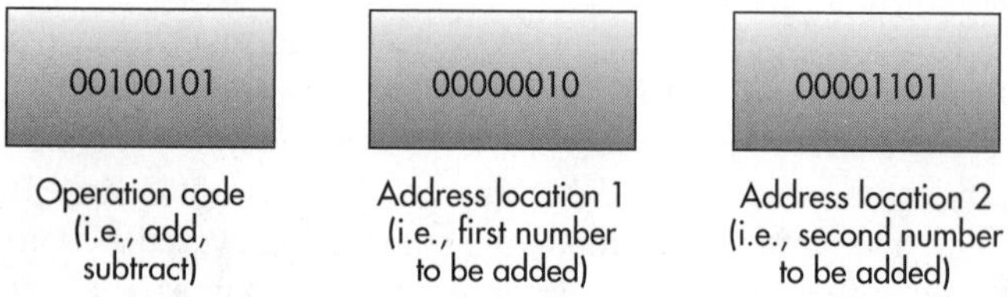

First-generation languages. The first generation of programming languages is **machine languages**. To give you an idea of the complexity of machine language, Figure 4.16 shows how even a simple instruction requires the use of many binary symbols. Machine language is considered a **low-level language** because there is no program coding scheme less sophisticated than that using the binary symbols 1 and 0. ASCII (American Standard Code for Information Interchange) uses all 0s and 1s to represent letters of the alphabet. As this is the language of the CPU, text files translated into ASCII binary sets can be read by almost every computer system platform.

Second-generation languages. Developers of programming languages attempted to overcome some of the difficulties inherent in

machine language

the first-generation programming language

low-level language

basic coding scheme using the binary symbols 1 and 0

FIGURE 4.16

A Simplified Machine Language Instruction

A machine language instruction consists of all 0s and 1s. Here, just a few elements of a single instruction are presented.

00100101	00000010	00001101
Operation code (i.e., add, subtract)	Address location 1 (i.e., first number to be added)	Address location 2 (i.e., second number to be added)

machine language by replacing the binary digits with symbols programmers could more easily understand. Assembly languages use codes like A for add, MVC for move, and so on. This second-generation language was termed **assembly language**, after the system programs used to translate it into machine code called assemblers. Systems software programs such as operating systems and utility programs are often written in an assembly language.

Procedure language

Third-generation languages. These languages continued the trend toward greater use of symbolic code and away from specifically instructing the computer how to complete an operation. BASIC, COBOL, C, C++, and FORTRAN are examples of third-generation languages that use English-like statements and commands. This type of language is easier to learn and use than machine and assembly languages because it more closely resembles everyday human communication and understanding. Third-generation languages use statements such as the following: PRINT TOTAL_SALES, READ HOURS_WORKED, and NORMAL_PAY=HOURS_WORKED*PAYRATE.

With third-generation programming languages each statement in the language translates into several instructions in machine language. In addition, third-generation languages take the programmer one step further away from directing the actual operation of the computer. While easier to program, third-generation languages are not as efficient in terms of operational speed and memory.

The various languages have certain characteristics that make them appropriate for certain types of problems or applications. For example, COBOL has excellent file- and database-handling capabilities for manipulating large volumes of business data, while FORTRAN is better suited for scientific applications. Although other languages are used to write new business applications, there are more lines of code of existing business applications written in COBOL than any other programming language.

Third-generation languages are relatively independent of computer hardware. This means that the same program can be used on a number of different computers by different computer manufacturers, with only small modifications or no modifications at all. This characteristic is gaining importance as companies connect different computers into distributed processing systems and as changing hardware technologies drive rapid upgrades.

While the use of low-level programming languages (machine and assembly) enables a programmer to write efficient code for programs that require the utmost speed, most companies use third-generation languages for the majority of their programs. This is primarily because it takes less effort to write a program using the more intuitive English-like third-generation languages. Note that all these programs are general-purpose and procedure-oriented languages.

Fourth-generation languages. These are programming languages that are less procedural and even more English-like than third-generation languages. They emphasize what output results are desired rather than how programming statements are to be written. As a result, many managers and executives with little or no training in computers and programming are using fourth-generation languages (4GLs). Some of the features of 4GLs

assembly language

second-generation language that replaced binary digits with symbols programmers could more easily understand

fourth-generation languages (4GLs)

programming languages that are less procedural and even more English-like than third-generation languages include query and database abilities, code-generation abilities, and graphics abilities. Prime examples include Visual C++, Visual Basic, PowerBuilder, Delphi, Forte, and many others. Visually-oriented application development environments will become more important than languages. You don't have to be a software engineer to become productive with these "abstract" tool sets.

With some 4GLs, a user can give simple commands or perform simple procedures to retrieve information from a database. These commands are stated in the form of simple questions or queries. With a billing database written in a 4GL, for example, you can click on "AMOUNT" and enter > $250 to get a listing of all bills that are greater than $250. These languages have been called **query languages** because they ask a computer questions in English-like sentences. Many of these languages also operate only on organized databases and are thus called database languages. Some examples of queries in a fourth-generation database or query language are shown here:

PRINT EMP_ NO IF GROSS PAY > 1000
PRINT CUS_NAME IF AMOUNT > 5000 AND IF DUE_DATE > 90
PRINT INV_NOSUB IF ON HAND < 50

Creating, manipulating, and using graphics and illustrations can be easier and simpler with a fourth-generation language than with a lower-level language. For example, fourth-generation languages can be used to develop trend lines and pie charts from data.

One popular fourth-generation language is a standardized language called **Structured Query Language (SQL)**, which is often used to perform database queries and manipulations. An example of a SQL statement follows:

SELECT OCCUPANT FROM ROOM_CHG WHERE RM_CHG > 1000

Many other fourth-generation languages—such as FOCUS, Powerhouse, and SAS—are used by end users and programmers alike to develop programs.

Object-oriented programming languages. A newer type of programming language—**object-oriented languages**—allows the interaction of programming objects. This approach to programming is called object-oriented programming. While most other programming languages separate data elements from the procedures or actions that will be performed on them, object-oriented programming languages tie them together into objects. Thus, an object consists of data and the actions that can be performed on the data. For example, an object could be data about an employee and all the operations (such as payroll calculations) that might be performed on the data.

In high-level, query, and database languages, programs consist of procedures to perform actions on each data element. However, in object-oriented systems, programs tell objects to perform actions on themselves. For example, a video display window does not need to be drawn on the screen by a series of instructions. Instead, a window object could be sent a message to open and the window will appear on the screen. That is because the window object contains the program code for opening itself.

As already stated, in object-oriented programming, data, instructions, and other programming procedures are grouped together into an item called an object. The process of grouping items into an object is called

query languages

used to ask the computer questions in English-like sentences

structured query language (SQL)

a standardized language often used to perform database queries and manipulations

object-oriented languages

languages that allow interaction of programming objects, including data elements and the actions that will be performed on them

encapsulation

the process of grouping items into an object

polymorphism

a process allowing the programmer to develop one routine or set of activities that will operate on multiple objects

inheritance

property used to describe objects in a group of objects taking on characteristics of other objects in the same group or class of objects

reusable code

the instruction code within an object that can be reused in different programs for a variety of applications

FIGURE 4.17

By combining existing program objects with new ones, programmers can easily and efficiently develop new object-oriented programs to accomplish organizational goals. Note that these objects can be either commercially available or designed internally.

encapsulation. **Encapsulation** means that functions or tasks are captured (encapsulated) into each object, which keeps them safe from changes because access is protected. Objects often have properties of polymorphism and inheritance. **Polymorphism** allows the programmer to develop one routine or set of activities that will operate or work with multiple (poly) objects. **Inheritance** means that objects in a group of objects can take on, or "inherit," characteristics of other objects in the same group or class of objects. This helps programmers select objects with certain characteristics for other programming tasks or projects.

Building programs and applications using object-oriented programming languages is like constructing a building using prefabricated modules or parts. The object containing the data, instructions, and procedures is a programming building block. Unlike constructing a building, however, the same objects (modules or parts) can be used repeatedly. An object can relate to data on a product, an input routine, or an order-processing routine. An object can even direct a computer to execute other programs or to retrieve and manipulate data. One of the primary advantages of an object is that it contains **reusable code**. In other words, the instruction code within that object can be reused in different programs for a variety of applications, just as the same basic type of prefabricated door can be used in two different houses. Thus, a sorting routine developed for a payroll application could be used in both a billing program and an inventory control program. By reusing program code, programmers are able to write programs for specific application problems more quickly (see Figure 4.17). Object-oriented programming is also an excellent programming approach for database applications.

Object-oriented programming languages offer the potential advantages of reusable code, lower costs, reduced testing, and faster implementation

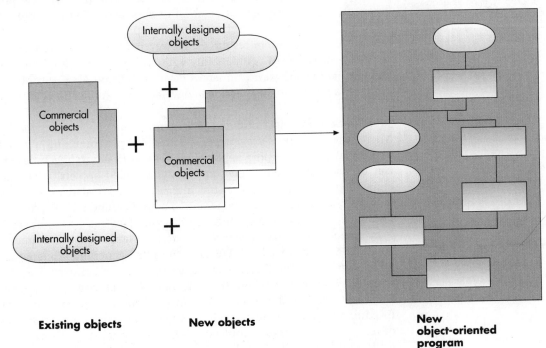

Existing objects   **New objects**   **New object-oriented program**

times. Instead of writing detailed lines of code, programmers can combine, modify, and integrate predeveloped modules into a unified program. Even with modifications, using the modular approach can be significantly faster than writing all the statements from scratch. Potential disadvantages of object-oriented programs include slower execution times and higher memory requirements. However, as hardware costs come down and programmer salaries increase, many managers are willing to trade off these disadvantages for much faster program development time.

There are several object-oriented programming languages; some of the most popular include Smalltalk, C++, and Java.

Smalltalk

a popular object-oriented programming language

Smalltalk. Smalltalk was developed by Alan Kay at Xerox at the Palo Alto Research Center in the 1970s. For more than a decade, the language was not used to a great extent in business settings. With the implementation of the language on desktop computers, this is changing.

USF&G retrained its COBOL programmers for Smalltalk development. All the programmers go through a nine-week Smalltalk training program prior to joining an object-oriented project. In 1994, the company standardized on Visual Smalltalk Enterprise tools from ParcPlace-Digitalk. Using these tools, USF&G developers produced an all-new, object-oriented policy-writing application known as Foundation. Developed in 12 months, the system contains 17,000 policies representing $31 million in insurance premiums. Application development productivity has increased from the mainframe-based rate of five function points per staff month to 18 function points under the object-oriented approach. **Function points** are standard measures used to gauge IS developers' productivity.[21]

function points

standard measures used to gauge IS developers' productivity

C++

a popular programming language that is an enhancement of the original C programming language

C++. C++ is a programming language that is an enhancement of the original C programming language. Like C, C++ is a real-time, general-purpose language, which has been used for business and scientific applications. Unlike C, C++ uses the object-oriented approach. C++ applications are efficient in using system resources. However, C++ uses a complex syntax to maintain backward compatibility with C. Because C++ is so inherently complex, some software companies have introduced C++ software development tools that make it easier to develop C++ applications.[22]

Java

an object-oriented programming language developed by Sun Microsystems

Java. Java is an object-oriented programming language developed by Sun Microsystems. It gives programmers a powerful programming environment and the ability to develop applications that work across the Internet. Java can be used to develop small applications, called applets, which can be embedded into Web pages on the Internet. The Java language includes a debugger, a documentation generator, a compiler, and a viewer for running Java applications without an Internet browser. Java has a structure and uses conventions similar to the C++ programming language. It is a much more complex and abstract language than the third-generation languages. Sun Microsystems hopes that the language will be the primary language for developing applications on the Internet. Several companies, including Microsoft, IBM, and Netscape, have incorporated Java into their products. In addition, other products such as ActiveX by Microsoft can be used to develop network-ready applets.[23] The "Technology Advantage" box explores the possibilities of Java.

TECHNOLOGY ADVANTAGE

Java: Hot and Getting Hotter

To many, Sun Microsystems' Java is much more than a programming language. It is also a software development tool and a complete platform used to implement sophisticated applications. Although only a few years old, the language is gaining acceptance and its use is exploding. This path, however, has not always been smooth. Some of the early users of Java encountered many bumps on the road, but today the language is moving beyond experimentation to becoming the language of choice for Internet and other related applications. The number of developers using Java has been estimated at 750,000 and rising rapidly.

One of the key advantages of Java is its cross-platform capabilities. This means that one Java application can be developed to run on a variety of computers and operating systems, including Unix, Windows NT, Windows 98, and Macintosh operating systems. The Java programming code does not have to be modified for these different systems. Another advantage of Java is its power and functionality. Some developers believe that the language is easier to use than other languages like C++. One disadvantage of Java is its slow speed. To overcome this speed problem, Sun Microsystems is expected to release HOTSpot Virtual Machine, which should speed Java performance to levels similar to C++ and other languages.

With its increased popularity, a number of companies are offering enhancements. For example, Developer 2000 Forms allows a developer to go from an application on some computer systems directly to a Java application without any additional programming. Developer 2000 Forms is an enhancement to a database development program from Oracle. Speed, cost, and reduced testing are all advantages of this approach. Converting an existing Oracle application into a Java application is faster than developing the Java application from scratch. Weeks or months could be saved by this approach. Developer 2000 Forms can dramatically cut programming costs. With rising salaries and a shortage of good programmers, this advantage is particularly attractive. Reducing the need to test the software and fix problems is another advantage to Developer 2000. One of the greatest benefits of this approach is being able to bring the full power of database and fourth-generation languages to Java and the development of Internet applications. Once implemented in a database language, these types of translation programs allow developers to place the application on the Internet using Java without any new programming. This also means that programmers don't have to learn Java to implement a powerful Internet application. Sun Microsystems is also developing enhancements to Java, including Java Studio and Java Workshop, to make building powerful Internet applications even easier and more efficient.

DISCUSSION QUESTIONS

1. Imagine that you are your company's representative to a Java User Group meeting at Sun Microsystems. How would you summarize the strengths and weaknesses of Java to your company?

2. What specific enhancements would you request of Sun Microsystems?

Sources: Sharon Gaudin, "Java Holds Its Own, Starts to Make Inroads," *Computerworld*, January 19, 1998, p. 32; Jeff Jurvis, "Sun Rolls Out Next-Generation Java Tools," *InformationWeek*, January 19, 1998; Lloyd Gray, "Center Gets Java Beans," *PC Week*, January 12, 1998, p. 60.

---

visual programming languages

languages that use a mouse, icons, or symbols on the screen and pull-down menus to develop programs

Visual programming languages. Many programming languages can be used in a visual or graphical environment. Often called **visual programming languages**, these languages use a mouse, icons, or symbols on the screen and pull-down menus. This visual environment can make programming easier and more intuitive. Visual Basic, PC COBOL, and Visual C++ are examples of visual programming languages. Despite their similar names, programming languages that use a visual or graphic environment are usually substantially different from the languages developed for a strictly character- and text-based environment.

Nabisco developed a Visual Basic 5.0 application tying together several Microsoft applications to track collaboration and workflow on development projects for its snack-food products. Other areas where Nabisco has enterprisewide applications based on Visual Basic include vehicle-purchase tracking and changing end-user passwords from the help desk.[24]

Unfortunately, these programming languages have never been considered appropriate for developing critical enterprise applications. For example, Visual Basic code runs slowly, its screen display performance is considered fair, and it takes major effort to connect to various databases. Also, it lacks the power of C++ and cannot be used for full Web development. Still, developers and novices flocked to Visual Basic because it was easy to use. And the industry that Visual Basic spawned is overtaking it. Visual Basic-style scripting is now integrated into a wide variety of applications. Visio—a drawing package—includes Visual Basic-style scripting. So does SQA TeamTest, a GUI testing tool that ironically is used to make robust and error-free Visual Basic applications. Likewise, all of the applications in Microsoft Office are based on Visual Basic for Applications, a derivative of Visual Basic. On the high end of the spectrum, Microsoft has provided Visual C++ with an integrated development environment that is more Visual Basic-like with each version. Eventually, the two environments will be combined, giving Visual C++ the ability to develop enterprise applications rapidly.[25]

Fifth-generation languages. Fifth-generation language tool kits appeared around mid-1998. They combine rules-based code generation, component management, visual programming techniques, reuse management and other advances. In general, software tool vendors call this approach **knowledge-based programming**, which means that you do not tell the computer how to do a job, but what you want it to do, and it figures out what you need. SunSoft is currently beta testing Java Studio, a connect-the-dots style of visual development toolset that uses components known as Java Beans. Java Studio is so simple to use that you can easily develop a working program without any previous programming experience.[26]

Rules-based programming may be the best way to develop intelligent, knowledge-based applications. You take the human decision-making process and define clear cut rules to be followed, thus eliminating the problem of inconsistency. Are artificial intelligence-based development tools and languages likely to become universally accepted? Given the rapid innovation in tools and languages, it appears so. In fact, it is hard to imagine a programmer learning one language and one development environment, and sticking with it for the next decade or two.[27]

Selecting a programming language. Selecting the best programming language to use for a particular program involves balancing the functional characteristics of the language with cost, control, and complexity issues. Read the "E-Commerce" box to see how one company helped solve its software selection and business problems by outsourcing its Web site development.

Machine and assembly languages provide the most direct control over computer hardware. For this reason, many vendors of popular application software programs take the time and effort to code portions of their leading programs in assembly language to maximize their speed. When a programmer requires a high degree of control over how various hardware components are used, these languages should be used. In selecting any programming language, the amount of direct control that is needed over the operation of the hardware can be an important factor to consider.

knowledge-based programming

an approach to development of computer programs in which you do not tell a computer how to do a job, but what you want it to do

E-COMMERCE

Hawaiian Greenhouse Combines New Technology with Flower Power

Hawaiian Greenhouse is a family-owned business located in Pahoa on the island of Hawaii. The flowers grown there offer the highest quality due to ideal weather conditions. Hawaiian Greenhouse has been growing anthuriums and other tropical flowers and foliage and shipping them worldwide since 1965. Although it takes a small-town, personal approach to the superior quality of its products and the special attention it gives customers, Hawaiian Greenhouse recognized that quality and service alone can not keep business thriving in increasingly competitive conditions. It needed a new way to boost the growth of its business.

Hawaiian Greenhouse's traditional approach to business involves a sales chain of wholesalers and retailers. As global competition intensified, Hawaiian Greenhouse found its margin of profitability decreasing year after year. Not only that, the company saw distance from its ultimate consumer increase, which made it difficult to expand its business by learning about customers' interests and patterns of purchase. While it didn't want to damage or eliminate the distributor relationships developed over 30 years in the industry, it saw an obvious advantage to reaching its customers directly—higher margins and the opportunity to build a loyal customer base.

In 1995, Hawaiian Greenhouse implemented the first phase of its strategy to reach its customers directly: a 1-800 mail-order service. Software consultant DataHouse provided the Lotus Notes technology to automate the mail-order service: routing orders, printing confirmations, generating work orders for flower pickers, and even printing mailing labels for order fulfillment.

After about two years, Hawaiian Greenhouse found that its mail-order business was achieving its objectives and it was ready to take the concept one step further. To cut costs even more and reach an even broader audience, Hawaiian Greenhouse began exploring the possibility of doing business over the Internet. It did not want to spend extravagantly on the experiment and it did not want a solution that would require advanced technical knowledge. DataHouse was able to get the Greenhouse site up and running in only a few weeks. Hawaiian Greenhouse's site includes features such as a catalog, buyer security, and complete order processing capabilities. DataHouse designed the solution to handle order fulfillment, customer information and contact databases, and a sales reporting system. Best of all, this solution requires absolutely no technical knowledge on the part of the Greenhouse staff, yet it can control the content of its site by simply logging on via a browser and adding or editing content.

With record-keeping done electronically and automation of many tasks, the Greenhouse staff is now free to concentrate on its real skill—growing flowers. With the new E-commerce site, Hawaiian Greenhouse expects to be able to widen its reach to customers considerably. It can now gain direct feedback on what customers want, allowing them to more closely tailor flower offerings to consumer desires and seasonal needs. Although it is still early to give exact figures, Hawaiian Greenhouse expects to generate significantly increased revenues thanks to its direct link to flower-lovers throughout the world.

DISCUSSION QUESTIONS

1. What role did the software consulting firm, DataHouse, play in the success of this project? What advantage did it provide to Hawaiian Greenhouse?
2. Visit the Hawaiian Greenhouse Web site. How might Hawaiian Greenhouse modify its current system to obtain more feedback from its customers and better understand their needs?

Source: Adapted from Lotus Development Corporation Web site at http://www.lotus.com, accessed August 4, 1998; the Hawaiian Greenhouse Web site at http://www.hawaiian-greenhouse.com, accessed August 4, 1998; and "Hawaiian Greenhouse" Internet Portfolio on the DataHouse Web site at http://www.datahouse.com, accessed August 4, 1998.

More recent programming languages are typically more complex than earlier programming languages. Although these newer languages appear to be simpler because they are more English-like, each command can drive complex routines and functions that operate behind the scenes. It takes less time to develop computer programs using higher-level languages than with lower-level languages. This means that the cost to develop computer programs can be substantially less with these more recent programming languages. While training programmers to use these higher-level programming languages may produce high up-front costs, using higher-level languages can reduce the total costs to develop computer programs in the long run.

C++ and Java both have advantages and disadvantages, but Java may be the future of programming. Java is far easier to learn and, as a result, people become productive much sooner. Programmers who learn C++ must spend a lot of time debugging rather than learning software engineering techniques. An increasing number of colleges in the United States are using Java as their first programming language. Java also provides built-in automatic garbage collection where allocated memory is released at run time—a common cause of bugs in C++ programs. Java is also more portable with the ability to run on more operating systems and hardware. However, C++ will not disappear anytime soon. There is a large base of C++ programs installed and a large user base specifically because Microsoft uses it for its programming. The ANSI and ISO standards committees have also been working on C and C++ since 1990, and it is apparent that people will continue to develop in C++ in or outside a Microsoft environment.[28]

Language Translators

Because machine language programming is extremely difficult, very few programs are actually written in machine language. However, machine language is the only language capable of directly instructing the CPU. Thus, every non-machine language program instruction must be translated into machine language prior to its execution. This is done by systems software called **language translators**. A language translator converts a programmer's source code into its equivalent in machine language. The high-level program code is referred to as the **source code**, whereas the machine language code is referred to as the **object code**. There are two types of language translators—interpreters and compilers.

An **interpreter** translates one program statement at a time, as the program is running. It will display on screen any errors it finds in the statement (Figure 4.18). This line-by-line translation makes interpreters ideal for those who are learning programming, but it does slow down the execution process.

---

language translator

systems software that converts a programmer's source code into its equivalent in machine language

source code

high-level program code written by the programmer

object code

another name for machine language code

interpreter

a language translator that translates one program statement at a time into machine code

FIGURE 4.18

How an Interpreter Works

An interpreter translates each program statement or instruction in sequence. The CPU then executes the statement, erases it from memory, and translates another statement. An interpreter does not produce a complete machine-readable version of a program.

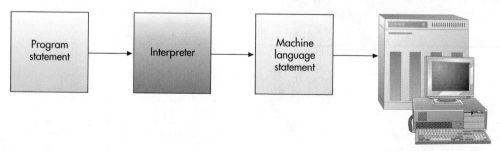

Statement execution

Stage 1:   **Convert program**

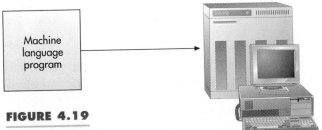

Stage 2:   **Execute program**

FIGURE 4.19

How a Compiler Works

A compiler translates a complete program into a complete set of binary data (Stage 1). Once this is done, the CPU can execute the converted program in its entirety (Stage 2).

Program execution

compiler

a language translator that converts a complete program into a machine language to produce a program that the computer can process in its entirety

A **compiler** is a language translator that converts a complete program into a machine language to produce a program that the computer can process in its entirety (Figure 4.19). Once the compiler has translated a complete program into machine language, the computer can run the machine language program as many times as needed. A compiler creates a two-stage process for program execution. First, it translates the program into a machine language; second, the CPU executes that program.

Most defects in computer software are unintentional mistakes in syntax or logic. During the course of developing computer programs, the programmer uses an interpreter or a compiler to detect syntax errors. Detecting logic errors usually requires programmers to run through test data and check the results of the program against the results of running the same data by hand or calculator. For example, if you made the logic error A = B + C instead of the correct A = B – C, then you tell the computer that B equals 42 and C equals 13 and ask it to find A, you'll quickly find the erroneous + sign when the answer comes out 55 instead of 29.

Unfortunately, few logic errors are so simple or so easily detected. Moreover, most computer programs involve thousands of lines of code. Thus, it can take years, even for teams of programmers, to debug programs such as those used to control emergency shutdown systems on nuclear reactors. Yet, even with the use of sophisticated debugging programs and years of study, experts estimate that one in every 5,000 lines of software code contains an error. Some programming languages are more prone to the introduction of program defects than others.

■ SUMMARY

1. *Identify and briefly describe the functions of the two basic kinds of software.*

Software consists of programs that control the workings of the computer hardware. There are two main categories of software: systems software and application software. Systems software is a collection of programs that interacts between hardware and application software. Systems software includes utility programs and operating systems. Application software enables people to solve problems and perform specific tasks. Application software may be proprietary or off-the-shelf. While there are literally hundreds of computer applications that can help individuals at school, home, and work, the primary ones are word processing, spreadsheet analysis, database, graphics, and on-line services.

All software programs are written in coding schemes called programming languages, which provide instructions to a computer to perform some processing activity. Programming languages provide instructions to the computer system so that it can perform a processing activity.

2. *Outline the role of the operating system and identify the features of several popular operating systems.*

An operating system (OS) is a set of computer programs that controls the computer hardware to support users' computing needs.

OS hardware functions involve converting an instruction from an application into a set of instructions needed by the hardware. The OS also serves as an intermediary between application programs and hardware, allowing hardware independence. Memory management involves controlling storage access and use by converting logical requests into physical locations and by placing data in the best storage space, perhaps expanded or virtual memory.

Task management allocates computer resources through multitasking, multithreading, and time-sharing. With multitasking, users can run more than one application at a time. Multithreading is basically multitasking within a single application—several parts of the program can work at once. Time-sharing allows more than one person to use a computer system at the same time.

An OS also provides a user interface, which allows users to access and command the computer. A command-based user interface requires text commands to send instructions; a graphical user interface (GUI), like Windows, uses icons and menus.

Over the years, several popular operating systems have been developed. These include several proprietary operating systems used primarily on mainframes. MS-DOS is an early OS for IBM-compatibles.

Older windows operating systems, such as Windows 3.1, are GUIs used with DOS. Newer versions, like Windows 95, Windows 98, and Windows NT, are fully functional operating systems that do not need DOS. Apple computers use proprietary operating systems like the Mac OS. Unix is the leading portable operating system, usable on many computer system types and platforms.

3. *Discuss how application software can support personal, workgroup, and enterprise business objectives.*

Application software applies the power of the computer to solve problems and perform specific tasks. Application software can support individuals, groups, and organizations. User software, or personal productivity software, include general-purpose programs that enable users to improve their personal effectiveness, increasing the amount of work and its quality. Software that helps groups work together is often referred to as groupware. Enterprise software that benefits the entire organization can also be developed or purchased.

4. *Identify three basic approaches to developing application software and discuss the pros and cons of each.*

Three approaches are: (1) build proprietary application software, (2) buy existing programs off-the-shelf, or (3) use a combination of customized and off-the-shelf application software.

Building proprietary software (in-house or contracting out) has the following advantages: the organization will get software that more closely matches it needs; by being involved with the development, the organization has further control over the results; and the organization has more flexibility in making changes. The disadvantages include the following: it is likely to take longer and cost more to develop; the in-house staff will be hard pressed to provide ongoing support and maintenance; and there is a greater risk that the software features will not work as expected or that other performance problems will occur.

Purchasing off-the-shelf software has its advantages: The initial cost is lower; there is a lower risk that the software will fail to work as expected; and the software is likely to be of higher quality. Some of the disadvantages are that the organization may pay for features in the software it does not need; the software may lack important features requiring expensive customization; and the system may work in such a way that work process reengineering is required.

Some organizations have taken a third approach—customizing software packages. This approach can combine all of the above advantages and disadvantages and must be carefully managed.

5. *Outline the overall evolution of programming languages and clearly differentiate between the five generations of programming languages.*

There are several classes of programming languages, including machine, assembly, high-level, query and database, natural and intelligent, object-oriented, and visual programming languages.

Programming languages have gone through changes since their initial development in the early 1950s. In the first generation, computers were programmed in machine language, or binary code, a series of statements written in 0s and 1s. The second generation of languages was termed assembly languages; these support the use of symbols and words rather than 0s and 1s. The third generation consists of many high-level programming languages that use English-like statements and commands. These also must be converted to machine language by systems software but are easier to write than assembly or machine language code. These languages include BASIC, COBOL, FORTRAN, and others. A fourth-generation language is less procedural and more English-like than third-generation languages. The fourth-generation languages include database and query languages like SQL. Fifth-generation programming languages combine rules-based code generation, component management, visual programming techniques, reuse management, and other advances. These languages offer the greatest ease of use yet.

Object-oriented programming languages, like C++, Smalltalk, and Java use groups of related data, instructions, and procedures called objects, which serve as reusable modules in various programs. These languages can reduce program development and testing time. Java can be used to develop applications on the Internet.

◢ KEY TERMS

application software 138
assembly language 171
C++ 174
client application 163
collaborative computing software 165
command-based user interface 143
compiler 179
computer programs 138
computer system platform 138
contract software 156
documentation 138
embed 164
encapsulation 173
enterprise resource planning (ERP) 168

enterprise sphere of influence 140
fourth-generation language (4GL) 171
function points 174
graphical user interface (GUI) 143
groupware 165
icon 143
inheritance 173
in-house development 156
interpreter 178
Java 174
knowledge-based programming 176
language translator 178
link 163
low-level language 170

machine language 170
multitasking 146
multithreading 146
object code 178
object linking and embedding (OLE) 163
object-oriented languages 172
off-the-shelf software 156
operating system (OS) 141
paging 145
personal productivity software 139
personal sphere of influence 139
polymorphism 173
programming languages 169
proprietary software 156

query languages 172
reusable code 173
scalability 146
server application 163
Smalltalk 174
software suite 161

source code 178
sphere of influence 139
Structured Query Language (SQL) 172
syntax 169
systems software 138
time-sharing 146

user interface 143
utility programs 154
virtual memory 145
visual programming languages 175
workgroup 139
workgroup sphere of influence 140

■ REVIEW QUESTIONS

1. Identify the two basic types of software and give a specific example of each one.
2. What is the difference between source code and object code? Which code do you need to run a computer program?
3. What is a sphere of influence? What are the three primary spheres of influence for most companies?
4. State the primary characteristics of each of the five different generations of programming languages, and identify an example programming language of each generation.
5. What are visual programming languages?
6. What is groupware? Give several examples of such software.
7. What are object-oriented programming languages? What are their advantages?
8. What is an operating system? What functions does it perform?
9. What are the trade-offs between using a command-based interface and a graphical user interface (GUI)?
10. Discuss the advantages and disadvantages of individual applications vs. a software suite.
11. List three factors to consider when selecting an operating system.
12. What is a computer system platform?
13. Describe the term *enterprise resource planning (ERP) system*. What functions does such a system perform?

■ DISCUSSION QUESTIONS

1. Assume that you must take a computer programming course next semester. What language do you think would be best for you to study? Why? Do you think that a professional programmer needs to know more than one programming language? Why or why not?
2. Your organization has decided that it will no longer do in-house development of software. Defend this position. Might there be any exceptions to this guideline?
3. Imagine that you are designing the ideal operating system. Describe this operating system in terms of its features.
4. Explain the difference between a language translator, an interpreter, and a compiler.
5. Some people believe that computers will eventually understand the human language well enough to follow verbal instructions. At that point, will computer programs and programmers no longer be needed? Why or why not?
6. What are some of the benefits associated with implementation of groupware? What are some of the issues that may arise that could keep a groupware project from being successful?
7. If planning for the CBIS begins with software, what do we start with: programming languages, systems software, or application software?
8. Many organizations have mandated that off-the-shelf application software be the first choice when implementing new software and that the software not be customized at all. What are the advantages and disadvantages of such an approach? What would you recommend for your firm if you were the manager in charge of application development? If you were a senior vice president of a business area?

■ PROBLEM-SOLVING EXERCISES

1. Choose a personal computer spreadsheet program to track the scores of your favorite sports team. The spreadsheet must contain information about the date, time, location, and opponent for each of the games. It must also include the final score as well as a brief comment about each game. Each row in the spreadsheet should contain all the information for a single game. Enter information for each game (make up scores and data if the game has not yet been played).

a. Calculate your team's average score and the opponent's average score for all games.

b. Use one of the built-in functions of the spreadsheet program to count the number of home games and the number of away games.

c. Calculate your team's average score differential for all away games and then for all home games.

d. Sort the data in order of highest to lowest points scored by your team.

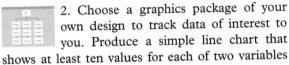

2. Choose a graphics package of your own design to track data of interest to you. Produce a simple line chart that shows at least ten values for each of two variables over a two-week period—high and low stock price of a particular company, high and low temperature at your favorite vacation location, etc. The graph must have a title, and the two line graphs must be labeled with the name of the variables. Use a word processing package to write a paragraph summarizing the data. Cut and paste the graph into the word processing document.

■ TEAM ACTIVITY

Form a group of three or four classmates. Find articles from business periodicals or interview managers from three firms implementing enterprise resource planning (ERP) systems. What is the primary reason the company is doing this? How long have they been at it and how much longer until the project is complete? Which specific vendor is each firm using and how was this vendor's software selected? Is the scope of the effort global or national? Do they have any concerns about the success of the project? Compile your results for an in-class presentation or a written report.

■ WEB EXERCISE

Go to the Web pages of Microsoft, Lotus, and Corel. Read about their latest software products. Choose the product you think will be the most successful in terms of software licenses sold. Write a two-page memo summarizing the products you considered, and explain why you chose the particular software product you think will have the most potential for success.

■ CASES

1 Kellogg Implements Global Information System

With products like Corn Flakes, Eggos, Pop-Tarts, and Rice Krispies, the Kellogg Co. controls more than 40 percent of the breakfast food industry. Revenue has doubled in the past ten years from $3.3 billion in 1987 to $6.8 billion in 1997. However, Kellogg is feeling pressure from customers such as Wal-Mart to continue to provide them the same service as they go global. Kellogg must be able to provide seamless integration so data can be entered in one location and flow to another.

The plan is to implement Oracle Corporation's Consumer Packaged Goods (CPG) software—a mix of financial and manufacturing management software applications from Oracle and other vendors including Manugistics, Industri-Matematik International Corp., Indus International, and Information Resources. Currently these multiple packages are not as tightly integrated as Kellogg would like. Version 3.0 of CPG is scheduled for release in 1999 and promises tighter integration.

Kellogg plans to roll out the package in North America, Latin America, Europe, and the Asia-Pacific region. Each region will implement the same software but be able to determine which modules will be implemented first based on local needs. Implementation is scheduled to be complete sometime in 2001, with key benefits being improved inventory management, absorption of acquisitions in a common information system, and support for multiple product launches. The goal is to be able to take an order anywhere, make it anywhere, stock it anywhere, and ship it from anywhere.

Jay Shreiner, chief information officer at Kellogg in Battle Creek, Michigan, believes that there is a big difference between global and central systems. Global systems means giving people in the various areas the tools to make the right decisions for Kellogg. Headquarters needs access to certain information, but does not believe in running the worldwide supply chain explicitly from Battle Creek.

1. What risks do you see with the course of action on which Kellogg has embarked? What could be done to reduce some of these risks?
2. Why does the CIO of Kellogg draw a distinction between global and central systems? Which do you think this new system will be? Why?

Sources: Adapted from Randy Weston, "Software to Tame Supply Chain Tiger," *Computerworld*, February 16, 1998, p. 53; Randy Weston, "Software Rattles the Supply Chain," *Computerworld*, February 16, 1998, p. 33; and the Kellogg Web site at http://www.kellogg.com, accessed February 9, 1998.

2 Gap Uses Object-Oriented Programming

The Gap started in 1969, when San Francisco real estate developer Donald G. Fisher tried to exchange a pair of blue jeans. He visited store after store, only to find jeans departments that were disorganized, poorly stocked, and difficult to shop. As a result of this experience, Fisher, and his wife Doris, created the antithesis: a jeans-only store that was neatly organized by size. The name they gave it was an allusion to the generation gap.

Gap expanded rapidly—starting first in California and then spreading quickly throughout the country. Gap also began selling more than just jeans, and in 1974 it began creating its own private-label clothing and accessories. Over the years, the percentage of private-label merchandise sold increased, until 1991 when Gap announced that by year-end everything it sold would be under the company's own label. In 1983 the company acquired Banana Republic, a two-store chain with a thriving catalog business. Gap, Inc. expanded on the chain's established line of travel and safari wear, creating product development and production teams to add innovative private-label fashions and accessories.

The first GapKids store opened in 1986, featuring fashionable clothes appealing to parents who wore Gap clothing themselves. In addition to the kid-sized versions of Gap basics, GapKids' own product development team began to design clothing in styles specifically for kids. In 1990, GapKids launched the BabyGap line of infant and toddler clothing, which is now available in virtually all GapKids stores and departments.

The first Gap store outside the United States opened in 1987 in the London metropolitan area. Today, many Gap and GapKids stores are located in the United Kingdom, Canada, France, Germany, and Japan.

Gap's newest division, Old Navy Clothing Co., started in 1994. Old Navy stores offer quality value-priced clothing for the entire family, with little extras like toys, accessories, and special displays. The division has plans for major expansion throughout the United States.

Gap is unique in that almost every aspect of its business, from product development to sales, is done by people within the corporation. Most items in each of Gap's divisions are designed, sourced, inspected, shipped, distributed, displayed, advertised, and sold by or under the supervision of Gap's employees. This vertically integrated structure allows the company to ensure quality, uniformity, and loyalty for each of the brands.

The Gap has nearly 2,000 stores around the world and earned $534 million on sales of $6.5 billion in 1997. Its fast growth (in excess of 20 percent per year) poses some serious issues for the international chain's 3,000 users and its information technology department. Part of the challenge of rapid growth is keeping everyone connected and giving them immediate access to the most up-to-date information available. And at the Gap, that information is spread across a wide array of computer hardware. Gap uses IBM mainframes with Sun Microsystems Solaris servers and OS/2 and Windows NT desktop computers.

The Gap uses object-oriented architecture to provide its sales force the information it needs to maintain the company's rapid growth. "Our real goal is to be a leading retailer," says Phil Wilkerson, director of technical architecture at The Gap in San Francisco. "We knew we needed to get away from mainframes and get to the Web. . . . That's the key for us."

The Gap has merchants and planners in offices around the world. They need access to purchase order information, such as styles and quantities. Because the Gap deals with a variety of suppliers in various locations, they need to be able to get information wherever they are and whenever they need it.

1. What are the Gap's primary business objectives? How will the use of object-oriented programming help the Gap achieve these business objectives?
2. Do you think that most of the Gap's application software is developed in-house? Why or why not?

Sources: "Frequently Asked Questions—About the Company in General," on the Gap Web site at http://www.gap.com, accessed August 4, 1998; and Sharon Gaudin, "Gap Styles Its Objects to Keep Up with Growth," *Computerworld*, June 9, 1997, p. 45.

3   Tracking Software Licenses

Since the mid-1980s, the Software Publisher's Association (SPA) has been trying to educate corporations about software copyright laws. Its work is beginning to pay off. Large corporations are now so aware of the legal requirement to comply with software licensing regulations that many react by purchasing too many licenses. As IS managers inventory their software assets they find that, in many cases, they have more software licenses for some popular programs than they have employees.

The SPA has taken legal action against well over 1,000 organizations since 1995, thus encouraging IS managers at large corporations everywhere to follow SPA guidelines, which suggest that when organizations cannot find proof of purchase for the software they are using, they get rid of it or buy new licenses. As the number of desktop computers keeps growing, businesses continue to decentralize and empower employees. The software-licensing landscape keeps changing, and it is hard to keep a handle on your existing software, let alone try to make an accurate prediction of what is needed in the future.

Being caught with illegal software can result in SPA fines of $100,000 or more. On the other hand, buying too many licenses at $200 to $500 each can also increase IS costs unnecessarily. As a result, major software resellers now specialize in tracking volume licenses like Microsoft's Select program. For a fee—sometimes an overt 1.5 percent of software costs, often hidden in the software price—software resellers such as ASAP Software Express, Software Spectrum, Stream International, and others will track your software licenses for you.

Although it seems it should be simple for a company to track the number of software licenses it purchases, it is not. Each major software vendor offers numerous volume discounts and maintenance plans, new versions of software packages come out on average every 16 months, and licensing programs change constantly. To add to the confusion, companies have decentralized purchasing functions, as in a large multinational firm with many subsidiaries. The resellers keep track of the number of licenses a company has consumed against its corporate contract. If a company commits to buying 5,000 NT licenses over the next two years, it needs to track its progress toward that goal. Resellers also ensure that every subsidiary, and every division, always buys at the low contract price.

Some people think that having a reseller track your software licenses is akin to having a fox patrol your henhouse. The easier resellers make it for you to do one-stop shopping, the more likely you are to consolidate all purchases through one reseller. When someone sells you software and its upgrades and is also policing your software, there is an inherent conflict of interest.

1. An IS manager has decided it is cheaper simply not to manage software licensing and instead pay any fines if caught by the SPA. Is this a sound strategy? Support your position.
2. If you decided to use a software reseller to help manage software licensing in your firm, how would you manage the inherent conflict of interest?

Sources: April Jacobs, "Scrutiny of Software Licenses Pays Off," *Computerworld*, March 16, 1998; Web site for the Software Publisher's Association at http://www.spa.org/piracy, accessed July 30, 1998; and Web site for GLOBEtrotter at http://www.globetrotter.com/gsico.htm, company background accessed July 30, 1998.

4   Breathing Life Into an Old System

Introduced in 1963 as the Semi-automated Business Research Environment, the SABRE reservation system runs on IBM's Transaction Processing Facility, a specialized mainframe operating environment. The system is huge, containing 4 terabytes of information, including data on 400 airlines, 50 car rental companies, and 35,000 hotels. At peak demand it handles over 5,200 messages per second. Last year SABRE processed 350 million reservations with an average transaction time of under two seconds.

When originally completed in the early 1960s, at a cost of over $250 million, the SABRE system gave American Airlines a competitive advantage. Prior to the SABRE system, reservation clerks had to rely on telex messages and phone calls to track seat availability—data that was several hours or even days old. Without current seating information, booking space for flights was pretty much a gamble.

With the SABRE system, each time a seat is sold a transaction is generated to update a database of seating data for all flights. Reservation clerks and travel agents access this database of up-to-the-minute data for booking seats. SABRE enabled American Airlines to reduce the number of staff required to handle reservations, increase the number of passengers on a flight without fear of overbooking, and guarantee passengers a seat.

Strategic competitive advantage is often short-lived. Competition must react or go out of business. So other airlines that could afford the investment soon developed similar systems. These airlines earned additional revenue by selling or renting their system to other airlines and travel agencies to handle their bookings.

The aging SABRE system, now nearing 40 years old, is inflexible and expensive to operate. It is also difficult to add new applications and features. To circumvent these barriers, the SABRE Group is developing a next-generation SABRE that involves surrounding the core transaction-processing engine with high-performance, special-purpose systems.

Travelocity is an example of an application that pages a subscriber when flights are delayed or sends an electronic message when airfares drop. This application runs on a Unix-based Silicon Graphics computer and an Oracle database. Travelocity communicates directly with the SABRE database to obtain airline reservation information. Users can access the service via the Internet. More than 1 million users have signed up for this service generating more than $1.5 million a week in revenue.

Another application is Business Travel Solutions introduced in October 1996. This application runs on a Windows NT–based Digital Equipment Corporation computer with Microsoft's SQL Server database. The application provides an electronic booking tool that taps directly into SABRE so users can purchase airline tickets directly.

1. What problems might arise because of the variety of hardware and software that is used in the SABRE system and related applications?
2. What additional applications can you identify as potential development efforts?

Sources: Adapted from "The SABRE Group and Genisys Reservation Systems Improve Surface Transportation Booking for Travel Professionals," SABRE Group Inc. press release, *PR Newswire*, March 16, 1998; and John Foley, "SABRE's Challenge," *InformationWeek*, August 18, 1997, p. 83.

CHAPTER 5

Organizing Data and Information

"One of the major barriers to effective management and execution is that, except in times of crisis, most organizations do not have an agreement on the state of the organization (the data) and what to do about it (the linkage between action and results). The act of creating a data warehouse generates the opportunity to get everyone on the same page, both in terms of definitions and how the business works."

—Doug Neal, CSC Index consultant

Chapter Outline

Data Management
  The Hierarchy of Data
  Data Entities, Attributes, and Keys
  The Traditional Approach vs. the Database Approach

Data Modeling and Database Models
  Data Modeling
  Database Models

Database Management Systems (DBMS)
  Providing a User View
  Creating and Modifying the Database
  Storing and Retrieving Data
  Manipulating Data and Generating Reports
  Popular Database Management Systems for End Users
  Selecting a Database Management System

Database Developments
  Distributed Databases
  Data Warehouses, Data Marts, and Data Mining
  On-Line Analytical Processing (OLAP)
  Open Database Connectivity (ODBC)
  Object-Relational Database Management Systems

Learning Objectives

After completing Chapter 5, you will be able to:

1. Define general data management concepts and terms, highlighting the advantages and disadvantages of the database approach to data management.
2. Name three database models and outline their basic features, advantages, and disadvantages.
3. Identify the common functions performed by all database management systems and discuss the key features of three popular end user database management systems.
4. Identify and briefly discuss recent database developments.

Wal-Mart
Mining Data for Customer Gold

Aside from being recognized as a phenomenal success in a very tough industry, Wal-Mart pioneered use of massive databases of its business transactions to revolutionize retailing. Since the 1980s, Wal-Mart has gathered mountains of cash register data each night to feed to its decision support systems. Wal-Mart managers depend on detailed data, closely analyzing every cost and every line item to execute its business strategy. One strategy is to cut $500 million in inventory in a year, excluding new stores. Using 65 weeks of historical data kept by item by store by day, Wal-Mart buyers can query and analyze customer buying trends to make better decisions on inventory replenishment and markdowns. Wal-Mart's huge database incorporates information nationwide; however, data can also be selected for individual stores to allow buyers to gain insight into local purchasing patterns.

Retrieving and analyzing information wasn't always so easy. Although it employed one of the most powerful computers available—a specialized machine from NCR Corporation—Wal-Mart was unable to use all the data. For example, a forecast for each item in each of its 2,900 stores meant calcu-

lating a staggering 700 million forecasts. To make the job manageable, Wal-Mart was forced to prepare macro-forecasts lumping stores into regions and products into categories. The company upgraded its data warehousing capability by contracting with NCR to increase the size and capabilities of its database. NCR provided a new 5100 Massively Parallel Processor (MPP) computer employing hundreds of coordinated processors and the Teradata relational database management system. Wal-Mart also upgraded its existing NCR 5100M computer. This additional capacity enabled the retailer to include new customer preference data and tripled the size of Wal-Mart's database to over 24 terabytes (24 TB, or 24 billion bytes of data). With the increased capacity, Wal-Mart has one of the largest commercial databases in the world. More than 30 applications run on the system, which handles as many as 50,000 queries in one week.

In addition, Wal-Mart has begun using a system from NeoVista Solutions to predict demand for individual items in specific stores. Wal-Mart analyzes customer shopping patterns to see combinations of items that consumers tend to buy during one visit. By discovering links

among the items purchased, the retailer can make these products easier to buy together and meet customer needs. For example, analysis of shopping data showed that buyers of heavily advertised children's videos often pick up more than one video per trip. The lesson: Don't set up a hot kiddie video all by itself on an end-of-aisle display. Make sure the rest of the video collection is nearby. The result? More and larger purchases and increased profits.

As you read this chapter, consider the following questions:

- How can databases be used to support critical business objectives?

- What are some of the issues in compiling and managing massive amounts of data?

Sources: Adapted from Wendy Zellner, "A Grand Reopening for Wal-Mart," *BusinessWeek*, February 9, 1998, pp. 86–88; "Mining the Largest Data Warehouse," from "Data Mining: Plumbing the Depths of Corporate Databases," Special Advertising Supplement White Paper, *Computerworld*, April 21, 1997; John W. Verity, "Coaxing Meaning Out of Raw Data," *BusinessWeek*, February 3, 1997, pp. 134–138; and Craig Stedman, "Wal-Mart Triples Data Warehouse," *Computerworld*, February 17, 1997, p. 8.

The bane of modern business is too much data, not enough information. Computers are everywhere, accumulating gigabytes galore. Yet it only seems to get harder to find the forest for the trees—that is, to extract significance from the blizzard of numbers, facts, and statistics.[1] Like other components of a computer-based information system, the overall objective of a database is to help an organization achieve its goals. A database can contribute to organizational success in a number of ways, including the ability to provide managers and decision makers with timely, accurate, and relevant information based on data. As we saw in the case of Wal-Mart, a database can help growing companies organize data to learn from this valuable resource. Databases also help companies generate information that can help reduce costs, increase profits, track past business activities, and open new market opportunities. Indeed, the ability of an organization to gather data, interpret it, and act on it quickly can distinguish winners from losers in a highly competitive marketplace. It is critical to the success of the organization that database capabilities be aligned with the company's goals. Because data is so critical to an organization's success, many firms develop databases to help them access data more efficiently and use it more effectively. In this chapter, we will investigate the development and use of different types of databases.

As we saw in Chapter 1, a database is a collection of data organized to meet users' needs. Throughout your career, you will be directly or indirectly accessing a variety of databases, ranging from a simple roster of departmental employees to a fully integrated corporatewide database. You will probably access these databases using software called a **database management system (DBMS)**. A DBMS consists of a group of programs that manipulate the database and provide an interface between the database and the user or the database and other application programs. A database, a DBMS, and the application programs that utilize the data in the database make up a database environment. Understanding basic database system concepts can enhance your ability to use the power of a computerized database system to support IS and organizational goals.

database management system (DBMS)

a group of programs that manipulate the database and provide an interface between the database and the user of the database or other application programs

DATA MANAGEMENT

Without data and the ability to process it, an organization would not be able to successfully complete most business activities. It would not be able to pay employees, send out bills, order new inventory, or produce information to assist managers in decision making. As you recall, data consists of raw facts, like employee numbers and sales figures. For data to be transformed into useful information, it must first be organized in a meaningful way.

The Hierarchy of Data

Data is generally organized in a hierarchy that begins with the smallest piece of data used by computers (a bit) and progresses through the hierarchy to a database. As discussed in Chapter 3, a bit (a binary digit) represents a circuit that is either on or off. Bits can be organized into units called bytes. A byte is typically eight bits. Each byte represents a **character**, which is the basic building block of information. A character may consist of uppercase letters (A, B, C, . . . Z), lowercase letters (a, b, c, . . . z), numeric digits (0, 1, 2, . . . 9), or special symbols (.![+][-]/ . . .).

character

the basic building block of information, represented by a byte

FIGURE 5.1

The Hierarchy of Data

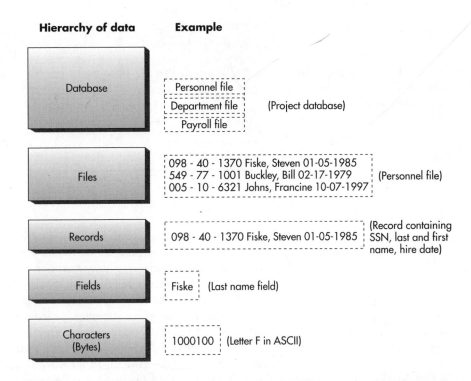

Hierarchy of data   **Example**

Database — Personnel file / Department file / Payroll file (Project database)

Files —
098 - 40 - 1370 Fiske, Steven 01-05-1985
549 - 77 - 1001 Buckley, Bill 02-17-1979
005 - 10 - 6321 Johns, Francine 10-07-1997
(Personnel file)

Records — 098 - 40 - 1370 Fiske, Steven 01-05-1985 (Record containing SSN, last and first name, hire date)

Fields — Fiske (Last name field)

Characters (Bytes) — 1000100 (Letter F in ASCII)

field

a group of characters

record

a collection of related fields

file

a collection of related records

database

a collection of integrated and related files

hierarchy of data

bit, characters, fields, records, files, and databases

entity

a generalized class of people, places, or things (objects) for which data is collected, stored, and maintained

attribute

a characteristic of an entity

Characters are put together to form a **field**. A field is typically a name, number, or combination of characters that describes an aspect of a business object (e.g., an employee, a location, a truck) or activity (e.g., a sale). A collection of related fields is a **record**. By combining descriptions of various aspects of an object or activity, a more complete description of the object or activity is obtained. For instance, an employee record is a collection of fields about one employee. One field would be the employee's name, another her address, and still others her phone number, pay rate, earnings made to date, and so forth. A collection of related records is a **file**—for example, an employee file is a collection of all company employee records. Likewise, an inventory file is a collection of all inventory records for a particular company or organization. PC database software often refers to files as tables.

At the highest level of this hierarchy is a **database**, a collection of integrated and related files. Together, bits, characters, fields, records, files, and databases form the **hierarchy of data** (Figure 5.1). Characters are combined to make a field, fields are combined to make a record, records are combined to make a file, and files are combined to make a database. A database houses not only all these levels of data but the relationships among them.

Data Entities, Attributes, and Keys

Entities, attributes, and keys are important database concepts. An **entity** is a generalized class of people, places, or things (objects) for which data is collected, stored, and maintained. Examples of entities include employees, inventory, and customers. Most organizations organize and store data as entities.

An **attribute** is a characteristic of an entity. For example, employee number, last name, first name, hire date, and department number are attributes for an employee (Figure 5.2). Inventory number, description,

FIGURE 5.2

Keys and Attributes

The key field is the employee number. The attributes include last name, first name, hire date, and department number.

Employee #	Last name	First name	Hire date	Dept. number
005-10-6321	Johns	Francine	10-07-1997	257
549-77-1001	Buckley	Bill	02-17-1979	632
098-40-1370	Fiske	Steven	01-05-1985	598

Entities (records)

Key field

Attributes (fields)

data item

the specific value of an attribute

key

a field or set of fields in a record that is used to identify the record

primary key

a field or set of fields that uniquely identifies the record

number of units on hand, and the location of the inventory item in the warehouse are examples of attributes for items in inventory. Customer number, name, address, phone number, credit rating, and contact person are examples of attributes for customers. Attributes are usually selected to capture the relevant characteristics of entities like employees or customers. The specific value of an attribute, called a **data item**, can be found in the fields of the record describing an entity.

As discussed, a collection of fields about a specific object is a record. A **key** is a field or set of fields in a record that is used to identify the record. A **primary key** is a field or set of fields that uniquely identifies the record. No other record can have the same primary key. The primary key is used to distinguish records so that they can be accessed, organized, and manipulated. For an employee record such as the one shown in Figure 5.2, the employee number is an example of a primary key.

Locating a particular record that meets a specific set of criteria may require the use of a combination of secondary keys. For example, a customer might call a mail-order company to place an order for clothes. If the customer does not know his primary key (such as a customer number), a secondary key (such as last name) can be used. In this case, the order clerk enters the last name, such as Adams. If there are several customers with a last name of Adams, the clerk can check other fields, such as address, first name, and so on, to find the correct customer record. Once the correct customer record is obtained, the order can be completed and the clothing items shipped to the customer. The "E-Commerce" box describes how MasterCard uses a database to provide credit services.

The Traditional Approach vs. the Database Approach

The traditional approach. Organizations are adaptive systems with constantly changing data and information needs. For any growing or changing business, managing data can become quite complicated. One of the most basic ways to manage data is via files. Because a file is a collection of related records, all records associated with a particular application (and therefore related by the application) could be collected and managed together in an

E-COMMERCE

MasterCard International

MasterCard has grown from a U.S.-based credit card system into a leader in the multitrillion dollar global payments industry. Today, MasterCard is a global payments company that provides consumer credit, debit, and other payment products together with 22,000 member financial institutions worldwide. Its role is to facilitate transactions among those who use its payment products, those who accept them, and the financial institutions that manage the relationships. MasterCard's family of brands, MasterCard, Maestro, and Cirrus, represent approximately 452 million cards in circulation and over 13 million acceptance locations, including 252,000 MasterCard/Cirrus ATMs worldwide, which handled 412 million transactions in a recent year.

MasterCard does not issue cards, set card fees, or establish annual percentage rates for cardholders, and it does not solicit individual merchants to accept the card or set discount rates. Instead, MasterCard licenses its brands to members and provides support services to maintain and enhance the value of its brand. Through focused brand development, advertising, and promotion, MasterCard develops brand preference and improves value to cardholders and merchants. MasterCard is the governing body that establishes and enforces policies and rules and evaluates and monitors the financial stability of members in the system.

In 1995, the company began a comprehensive effort to enhance its core systems to provide members with on-line, cost-effective processing support. As one of these initiatives, MasterCard International built one of the world's largest databases and has gradually expanded the system to its 22,000 bank, retailer, restaurant, and other partners. Called MasterCard On-line, the monster database started out in July 1995 with a whopping 1 terabyte (trillion bytes) of data and is growing. Up to 30,000 personal computers, workstations, and other devices have access to the database, which contains data on the company's 8.5 million daily credit- and debit-card transactions including authorization, clearing, and settlement. MasterCard has steadily added data on various subject areas and features such as the ability to set up alerts triggered by unusual cardholder activity.

MasterCard is challenged by stiff competition in the financial services industry and hopes that providing such unprecedented access to its customer information will keep its partners loyal. Prior to creation of the database, reports had to be requested through representatives and could be delivered only in hard-copy form. The goal of the database is to add value and increase service, thus improving existing business relationships with current members and building relationships with new members. Few companies have let outsiders, other than MasterCard, access their data warehouse whether they are close partners or customers or not.

The first application available to members was Market Advisor, a software program that enables members to query the database and analyze transactions and trends on-line. Market Advisor provides access to a 13-month historical database and report graphs. It can also trigger marketing alerts based on above- or below-average merchant or cardholder activity. In a typical use of Market Advisor, a marketing analyst can examine a trend in spending at aggregate levels for a particular store category, such as a hardware store, restaurant, car rental agency, or gas station. By using Market Advisor, an analyst can determine which states generated the bulk of sales revenue and identify which merchants within those states accounted for the greatest volume. The analyst could also discover which cardholder accounts were used at a particular store over time. With all this data, the analyst could then find common spending patterns among certain categories of cardholders, and tailor marketing promotions appropriately.

DISCUSSION QUESTIONS

1. What sort of problems could arise if the data in the database was used inappropriately?
2. How might the database be used to detect use of stolen credit/debit cards by criminals?
3. The database and Market Advisor tool are used by MasterCard customers worldwide. Can you think of any particular complications that might arise in interpreting or analyzing the data in the database given its global usage? What actions should be taken to avoid these problems?

Sources: "Security Dynamics Protects Mastercard Online," *Datamation*, April 1997, pp. 80–84 and "MasterCard Help Desk Vies for Service Award," *American Banker*, vol. 162, no. 95, May 19, 1997, p. 16.

application-specific file. At one time, most organizations had numerous application-specific data files; for example, customer records often were maintained in separate files, with each file relating to a specific process completed by the company, such as shipping or billing. This approach to data management, in which separate data files are created and stored for each application program, is called the **traditional approach**. For each particular application, one or more data files is created (Figure 5.3).

One of the flaws in this traditional file-oriented approach to data management is that much of the data, for example, customer name and address, is duplicated in two or more files. This duplication of data in separate files is known as **data redundancy**. The problem with data redundancy is that changes to the data (e.g., a new customer address) might be made in one file and not the other. The order-processing department might have updated its file to the new address, but the billing department is still sending bills to the old address. Data redundancy, therefore, conflicts with **data integrity**—the degree to which the data in any one file is accurate. Data integrity follows from the control or elimination of data redundancy. Keeping a customer's address in only one file decreases the possibility that the customer will have two different addresses stored in different locations. The efficient operation of a business requires a high degree of data integrity.

In many computerized database systems based on the traditional file approach, the data is organized for a particular application program (say, billing). These applications have **program-data dependence**—that is, programs and data developed and organized for one application are incompatible with programs and data organized differently for another application.

traditional approach

an approach to data management in which separate files are created and stored for each application program

data redundancy

the duplication of data in separate files

data integrity

the degree to which the data in any one file is accurate

program-data dependence

a situation in which programs and data organized for one application are incompatible with programs and data organized differently for another application

FIGURE 5.3

The Traditional Approach to Data Management

With the traditional approach, one or more data files is created and used for every application. For example, the inventory control program would have one or more files containing inventory data, such as the inventory item, number on hand, and item description. Likewise, the invoicing program can have files on customers, inventory items being shipped, and so on. With the traditional approach to data management, it is possible to have the same data, such as inventory items, in several different files used by different applications.

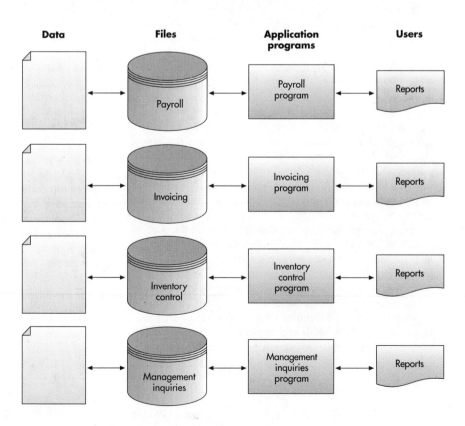

For example, one programmer might develop a billing program that stores ZIP code data in one format with five numbers, while another programmer might develop a separate order-processing program that stores ZIP code data in a nine-number format. In a computerized file-based environment, all the programs that access this ZIP code data would need to be changed. Bridging the gap between two files with different program-data dependencies is often difficult and expensive.

Despite the drawbacks of using the traditional file approach in computerized database systems, some organizations continue to use it. Many such organizations have identified and solved problems as they occurred with jointly developed application programs and data files. For these firms, the cost of converting to another other approach is too high.

The database approach. Because of the problems associated with the traditional approach to data management, many managers wanted a more efficient and effective means of organizing data. The result was the **database approach** to data management. In a database approach, a pool of related data is shared by multiple application programs. Rather than having separate data files, each application uses a collection of data that is either joined or related in the database.

The database approach offers significant advantages over the traditional file-based approach. For one, by controlling data redundancy, the database approach can use storage space more efficiently and increase data integrity. The database approach can also provide an organization with increased flexibility in the use of data. Because data once kept in two files is now located in the same database, it is easier to locate and request data for many types of processing. A database also offers the ability to share data and information resources. This can be a critical factor in coordinating organizationwide responses across diverse functional areas of a corporation. In sharing data, however, some consistency should exist among software programs. To examine some of the issues associated with gathering and sharing data about individuals, read the "Ethical and Societal Issues" box.

database approach

an approach to data management in which a pool of related data is shared by multiple application programs. Rather than having separate data files, each application uses a collection of data that is either joined or related in the database.

ETHICAL AND SOCIETAL ISSUES

Privacy Standards in Credit Reporting

The consumer credit reporting industry is a multibillion-dollar business. The three major credit service bureaus—experían, Equifax, and Trans Union—keep credit histories on 185 million Americans. They provide over 500 million consumer reports annually used primarily as a check on creditworthiness, but also to screen job applicants. The industry is governed by the 1970 Fair Credit Reporting Act. This act sets guidelines for the proper business uses to which credit bureaus can put their credit information. Critics of the industry maintain that the act is out of date. For example, it is not specific about some common industry practices such as prescreening for creditworthiness, reinvestigating inaccuracies based on consumer complaints, and providing consumer data to other firms for use in marketing campaigns.

Under pressure from Congress, the Individual Reference Services Group (IRSG)—whose members include the major credit service bureaus and other data gatherers such as LEXIS-NEXIS—has set privacy standards. These standards state that starting in 1999 its members won't sell social security numbers, mother's maiden names, birth dates, credit histories, unlisted phone numbers, or information about children to the public—only to qualified subscribers. Unfortunately, who is "qualified" is open to interpretation, while the effectiveness of safeguards against shifty operators is unclear. Moreover, the IRSG standard imposes no penalties for noncompliant companies.

The industry itself admits that this standard is a last-ditch defense against government regulation and that the standards are far from complete. However, the Federal Trade Commission has promised that the IRSG plan will forestall government regulation—at least for the time being.

The American Civil Liberties Union and other privacy groups are sharply critical of voluntary guidelines to limit the availability of personal information disseminated by on-line services. It would prefer a law to ensure against invasions of privacy and abuse of personally identifiable facts. "Incredibly, the right to information privacy remains largely unprotected by law," said Barry Steinhardt, associate director of the ACLU. "While the 1974 Privacy Act applies to the use of personal information by the government, it is poorly enforced and has too many exceptions," Steinhardt said. "Similarly, laws like the Fair Credit Reporting Act offer limited protection against abuses by the private sector."

Here are some examples of the problems resulting from lack of legal safeguards to protect privacy:

- Fraud: Ten current and former Social Security Administration employees had accepted bribes from a credit fraud ring in the business of selling mothers' maiden names to activate fraudulently obtained credit cards.
- Improper Prying: Congress passed a law in 1997 making improper prying, or so-called "browsing," a crime after an IRS employee targeted a state prosecutor he had a grudge against. The employee scrutinized the prosecutor's tax form, which included detailed information about the daycare center the prosecutor's children attended.
- Identity Theft: Because of the widespread availability of social security numbers, criminals are able to assume the identities of others and gain access to their victims' bank and charge accounts and steal their victims' government benefits.

The ACLU has encouraged Congress to enact legislation including the Personal Information Privacy Act of 1997, which would bar the commercial use of social security numbers, and the Social Security On-line Privacy Protection Act, which similarly covers use of social security numbers on the Internet.

Discussion Questions

1. What benefits do credit service bureaus provide to society?
2. Do you believe that credit service bureaus have the right to provide data about individuals to other companies that add the data to their mailing lists and telephone solicitation lists?
3. What sort of privacy legislation would you recommend? How would you identify and punish violators?

Sources: "If Your Privacy's Invaded—Tough Luck," *BusinessWeek*, February 16, 1998, p. 6. "ACLU, Others Sharply Critical of Government Privacy Report," December 24, 1997 from the Web site of the American Civil Liberties Union at http:// www.aclu.org.

To use the database approach to data management, additional software—a database management system (DBMS)—is required. As previously discussed, a DBMS consists of a group of programs that can be used as an interface between a database and the user or the database and application programs. Typically, this software acts as a buffer between the application programs and the database itself. Figure 5.4 illustrates the database approach.

The database approach to data management involves a combination of hardware and software. Tables 5.1 and 5.2 list some of the primary advantages and disadvantages of the database approach and explore some of these issues.

Because of the many advantages of the database approach, most businesses use databases to store data on customers, orders, inventory, employees, and suppliers. This data is used as the input to the various information systems throughout an organization. For example, the transaction processing system can use the data to support daily business processes like billing, inventory tracking, and ordering. This same data can be processed by a management information system to create reports or a decision support system to provide information to aid managerial decision making. Sears, Roebuck and Co. is building a large database to make it easier for merchandisers, accountants, and other users to measure performance, in terms of sales and other business metrics, against the retailer's plans. In the past, Sears' merchandisers looked only at merchandising information, while the retailer's accountants studied nothing but financial data. The change came about when the company recognized an increasing overlap between these two groups and their needs to look at each other's data.[2] Using the database approach continues to be important

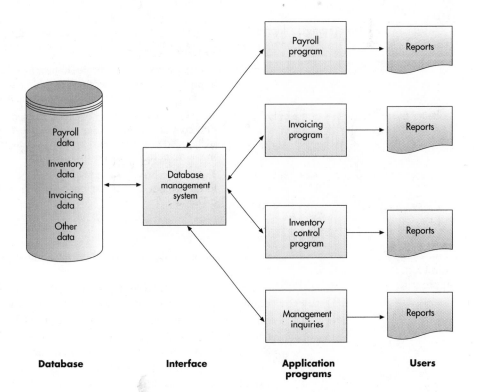

FIGURE 5.4

The Database Approach to
Data Management

Database   **Interface**   **Application**   **Users**
                                            programs

Advantages	Explanation
Improved strategic use of corporate data	Accurate, complete, up-to-date data can be made available to decision makers where, when, and in the form they need it.
Reduced data redundancy	The database approach can reduce or eliminate data redundancy. Data is organized by the DBMS and stored in only one location. This results in more efficient utilization of system storage space.
Improved data integrity	With the traditional approach, some changes to data were not reflected in all copies of the data kept in separate files. This is prevented with the database approach because there are no separate files that contain copies of the same piece of data.
Easier modification and updating	With the database approach, the DBMS coordinates updates and data modifications. Programmers and users do not have to know where the data is physically stored. Data is stored and modified once. Modification and updating is also easier because the data is stored at only one location in most cases.
Data and program independence	The DBMS organizes the data independently of the application program. With the database approach, the application program is not affected by the location or type of data. Introduction of new data types not relevant to a particular application does not require the rewriting of that application to maintain compatibility with the data file.
Better access to data and information	Most DBMSs have software that makes it easy to access and retrieve data from a database. In most cases, simple commands can be given to get important information. Relationships between records can be more easily investigated and exploited, and applications can be more easily combined.
Standardization of data access	A primary feature of the database approach is a standardized, uniform approach to database access. This means that the same overall procedures are used by all application programs to retrieve data and information.
A framework for program development	Standardized database access procedures can mean more standardization of program development. Because programs go through the DBMS to gain access to data in the database, standardized database access can provide a consistent framework for program development. In addition, each application program need only address the DBMS, not the actual data files, reducing application development time.
Better overall protection of the data	The use of and access to centrally located data are easier to monitor and control. Security codes and passwords can ensure that only authorized people have access to particular data and information in the database, thus ensuring privacy.
Shared data and information resources	The cost of hardware, software, and personnel can be spread over a large number of applications and users. This is a primary feature of a DBMS.

TABLE 5.1

Advantages of the Database Approach

as organizations rely more on data and information to achieve competitive advantage.

Many modern databases are enterprisewide, encompassing much of the data of the entire organization. Often, distinct yet related databases are linked to provide enterprisewide databases. Much planning and organization go into the development of enterprisewide databases. Allina Health System, a healthcare system that owns and manages 19 hospitals and 50 clinics and provides health plans to over 1 million customers, uses an enterprisewide database. Allina must try to reduce healthcare costs and hold the

Disadvantages	Explanation
Relatively high cost of purchasing and operating a DBMS in a mainframe operating environment	Some mainframe DBMSs can cost hundreds of thousands of dollars.
Increased cost of specialized staff	Additional specialized staff and operating personnel may be needed to implement and coordinate the use of the database. However, some organizations have been able to implement the database approach with no additional personnel.
Increased vulnerability	Even though databases offer better security because security measures can be concentrated on one system, they also make more data accessible to the trespasser if security is breached. In addition, if for some reason there is a failure in the DBMS, multiple application programs are affected.

TABLE 5.2

Disadvantages of the Database Approach

line on insurance premiums while providing its patients with the best treatment possible. It must also integrate the information systems and business practices of the organizations it has acquired through mergers. To address these issues, management called for an enterprise information systems strategy that included the development of enterprisewide databases. The goal was to enable Allina to pull information together and integrate it in ways never done before, for example, pulling cost information from the hospital and the health plans, comparing best practices in treatments, and matching cost to level of service.[3]

DATA MODELING AND DATABASE MODELS

Because there are so many elements in today's businesses, it is critical to keep data organized so that it can be effectively utilized. A database should be designed to store all data relevant to the business and provide quick access and easy modification. Moreover, it must reflect the business processes of the organization. When building a database, careful consideration must be given to these questions:

- Content: What data should be collected and at what cost?
- Access: What data should be provided to which users and when?
- Logical structure: How should data be arranged so that it makes sense to a given user?
- Physical organization: Where should data be physically located?

Data Modeling

Key considerations in organizing data in a database include determining what data is to be collected in the database, who will have access to it, and how might they wish to use the data. Based on these determinations, a database can then be created. Building a database requires two different types of designs: a logical design and a physical design. The logical design of a database shows an abstract model of how the data should be structured and arranged to meet an organization's information needs. The logical design of a database involves identifying relationships among the different

Lands' End, a mail-order clothing company, can provide customers with extensive information about its products by accessing a database that provides in-depth descriptions of merchandise and its availability.
(Source: Reprinted with permission from Lands' End, Inc.)

planned data redundancy

a way of organizing data in which the logical database design is altered so that certain data entities are combined, summary totals are carried in the data records rather than calculated from elemental data, and some data attributes are repeated in more than one data entity to improve database performance

data model

a map or diagram of entities and their relationships

enterprise data modeling

data modeling done at the level of the entire organization

entity-relationship (ER) diagrams

a data model that uses basic graphical symbols to show the organization of and relationships between data

data items and grouping them in an orderly fashion. Because databases provide both input and output for information systems throughout a business, users from all functional areas should assist in creating the logical design to ensure that their needs are identified and addressed. Physical database design starts from the logical database design and fine-tunes it for performance and cost considerations (e.g., improved response time, reduced storage space, lower operating cost). The person identified to fine-tune the physical design must have an in-depth knowledge of the DBMS to be used to implement the database. For example, the logical database design may need to be altered so that certain data entities are combined, summary totals are carried in the data records rather than calculated from elemental data, and some data attributes are repeated in more than one data entity. These are examples of **planned data redundancy**. It is done to improve the system performance so that user reports or queries can be created more quickly.

One of the tools database designers use to show the logical relationships among data is a data model. A **data model** is a map or diagram of entities and their relationships. Data modeling usually involves understanding a specific business problem and analyzing the data and information needed to deliver a solution. When done at the level of the entire organization, this is called **enterprise data modeling**. Enterprise data modeling is an approach that starts by investigating general data and information needs of the organization at the strategic level and then examining more specific data and information needs for the various functional areas and departments within the organization. Various models have been developed to help managers and database designers analyze data and information needs. An entity-relationship diagram is an example of such a data model.

Entity-relationship (ER) diagrams use basic graphical symbols to show the organization of and relationships between data. In most cases, boxes are used in ER diagrams to indicate data items or entities, and diamonds show relationships between data items and entities.

Figure 5.5 shows an ER diagram for a customer who places orders. ER diagrams can show a number of relationships. For example, one customer can place many orders. This is an example of a one-to-many relationship, as shown by the one-to-many symbol (1:N) used in Figure 5.5. Each order can include one or more line items where a line item specifies the product identification and quantity ordered. One-to-one, many-to-many, and other relationships can also be revealed using ER diagrams. ER diagrams help ensure that the relationships among the data entities in a database are logically structured so that application programs can be developed to best serve user needs. In addition, ER diagrams can be used as reference documents once a database is in use. If changes are to be made in the database, ER diagrams can help design them.

FIGURE 5.5

An Entity-Relationship (ER) Diagram for a Customer Ordering Database

Development of this type of diagram helps ensure the logical structuring of application programs that are able to serve users' needs and are consistent with the data relationships in the database.

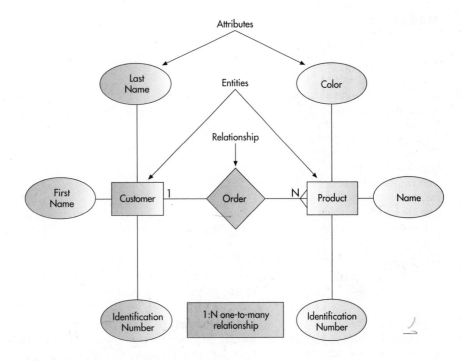

Database Models

The structure of the relationships in most databases follows one of three logical database models: hierarchical, network, and relational. Hierarchical and network models are still being used today, but relational models are the most popular. It is important to remember that the records represented in the models are actually linked or related logically to one another. These links dictate the way users can access data with application programs. Because the different models involve different links between data, each model has unique advantages and disadvantages.

Hierarchical (tree) models. In many situations, data follows a hierarchical, or treelike, structure. In a **hierarchical database model**, the data is organized in a top-down, or inverted tree, structure. For example, data about a project for a company can follow this type of model. Let's consider a typical project (Figure 5.6). Data about the project would form the beginning of the hierarchical database model. For a given project in an organization, a number of departments within the organization might be involved. In this case, Departments A, B, and C are involved with Project 1. In addition, certain employees from each department—in this case, Employees 1 through 6—will be involved in the project. The various departments and employees are represented as branches beneath the project. When this type of data is displayed in a logical fashion, it appears in a hierarchical model.

The hierarchical model is best suited to situations in which the logical relationships between data can be properly represented with the one parent-many children (one-to-many) approach. Unlike real-life parent-child relationships, however, the hierarchical model subordinate levels of data (children) can sufficiently define all relevant attributes of the superior data element (parent). Data is accessed logically by going through the appropriate "generations" of

hierarchical database model

a data model in which the data is organized in a top-down, or inverted tree, structure

A Hierarchical Data Model

Project 1 is the top, or root, element. Departments A, B, and C are under this element, with Employees 1 through 6 beneath them as follows: Employees 1 and 2 under Department A, Employees 3 and 4 under Department B, and Employees 5 and 6 under Department C. Thus, there is a one-to-many relationship among the elements of this model.

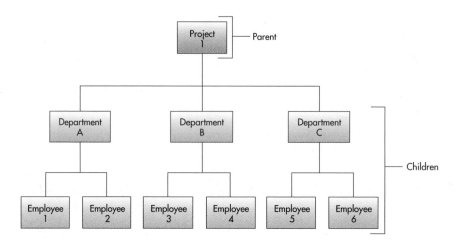

parents to get to the desired data element, and there is only one access path to any particular data element.

network model

an expansion of the hierarchical database model with an owner-member relationship in which a member may have many owners

Network models. A **network model** is an expansion of the hierarchical model. Instead of having only various levels of one-to-many relationships, however, the network model is an owner-member relationship in which a member may have many owners (Figure 5.7).

Consider two projects that require work from three departments. The projects (Projects 1 and 2) are placed at the top of the network. Below that, the various departments (Departments A, B, and C) required to do work on the projects would be listed. Then lines could be drawn that reveal which departments work on which projects. For instance, you can see that Department B performs work on both Project 1 and Project 2; hence, it is a member that is "owned" by Projects 1 and 2.

In a database structured as a network model, often a particular data element can be accessed through more than one path. In Figure 5.7, for example, a data element from Department B can be accessed through Project 1 or Project 2, a data element from Department C also can be accessed through Project 1 or Project 2. Both Departments B and C have two parent nodes—Project 1 and Project 2.

Databases structured according to either the hierarchical model or the network model suffer from the same deficiency: once the relationships are established between data elements, it is difficult to modify them or to create new relationships.

A Network Data Model

In this network model, two projects are at the top. Departments A, B, and C are under Project 1; Departments B and C are under Project 2. Thus, the elements of this model represent a many-to-many relationship.

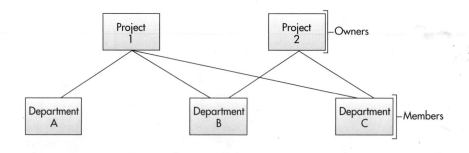

relational model

a data model in which all data elements are placed in two-dimensional tables, called relations, that are the logical equivalent of files

domain

the allowable values for an attribute

Relational models. Relational models have become the most popular database models, and use of these models will increase in the future. The overall purpose of the relational model is to describe data using a standard tabular format. In a database structured according to the relational model, all data elements are placed in two-dimensional tables, called relations, that are the logical equivalent of files. The tables in relational databases organize data in rows and columns, simplifying data access and manipulation. It is normally easier for managers to understand the relational model (Figure 5.8) than hierarchical and network models.

In the relational model, each row of a table represents a data entity with the columns of the table representing attributes. Each attribute can take on only certain values. The allowable values for these attributes are called the **domain**. The domain for a particular attribute indicates what values can be placed in each of the columns of the relational table. For instance, the domain for an attribute such as gender would be limited to male or female. A domain for pay rate would not include negative numbers. Defining a domain can increase data accuracy. For example, a pay rate of −$5.00 could not be entered into the database, because it is a negative number and not in the domain for pay rate.

Data table 1: Project table

Project number	Description	Dept. number
155	Payroll	257
498	Widgets	632
226	Sales Manual	598

Data table 2: Department table

Dept. number	Dept. name	Manager SSN
257	Accounting	005-10-6321
632	Manufacturing	549-77-1001
598	Marketing	098-40-1370

FIGURE 5.8

A Relational Data Model

In the relational model, all data elements are placed in two-dimensional tables, or relations. As long as they share at least one common element, these relations can be linked to output useful information.

Data table 3: Manager table

SSN	Last name	First name	Hire date	Dept. number
005-10-6321	Johns	Francine	10-07-1997	257
549-77-1001	Buckley	Bill	02-17-1979	632
098-40-1370	Fiske	Steven	01-05-1985	598

selecting

data manipulation that eliminates rows according to certain criteria

projecting

data manipulation that eliminates columns in a table

joining

data manipulation that combines two or more tables

linked

related tables in a relational database together

data cleanup

the process of looking for and fixing inconsistencies to ensure that data is accurate, complete, economical, flexible, reliable, relevant, simple, timely, verifiable, accessible, and secure

Once data has been placed into a relational database, users can make inquiries and analyze data. Basic data manipulations include selecting, projecting, and joining. **Selecting** involves eliminating rows according to certain criteria. Suppose a project table contains the project number, description, and department number for all projects being performed by a company. The president of the company might want to find the department number for Project 226, a sales manual project. Using selection, the president can eliminate all rows but number 226 and see that the department number for the department completing the sales manual project is 598.

Projecting involves eliminating columns in a table. For example, we might have a department table that contains the department number, department name, and the social security number (SSN) of the manager in charge of the project. The sales manager might want to create a new table with only the department number and the social security number of the manager in charge of the sales manual project. Projection can be used to eliminate the department name column and create a new table containing only department number and SSN. **Joining** involves combining two or more tables. For example, we can combine the project table and the department table to get a new table with the project number, project description, department number, department name, and the social security number for the manager in charge of the project.

As long as the tables share at least one common data attribute, the tables in a relational database can be **linked** to provide useful information and reports (see Figure 5.9). Being able to link tables to each other through common data attributes is one of the keys to the flexibility and power of relational databases. Suppose the president of a company wants to find out the name of the manager of the sales manual project and how long the manager has been with the company. The president would make the inquiry to the database, perhaps via a desktop personal computer. The DBMS would start with the project description and search the project table to find out the project's department number. It would then use the department number to search the department table for the department manager's social security number. The department number is also in the department table and is the common element that allows the project table and the department table to be linked. The DBMS then uses the manager's social security number to search the manager table for the manager's hire date. The manager's social security number is the common element between the department table and the manager table. The final result: the manager's name and hire date are presented to the president as a response to the inquiry. One of the primary advantages of a relational database is that it allows tables to be linked, as shown in Figure 5.9. This linkage is especially useful when information is needed from multiple tables, as in our example. The manager's social security number, for example, is maintained in the manager table. If the social security number is needed, it can be obtained by linking to the manager table.

Data cleanup. As discussed in Chapter 1, the characteristics of valuable data include that the data is accurate, complete, economical, flexible, reliable, relevant, simple, timely, verifiable, accessible, and secure. The purpose of **data cleanup** is to develop data with these characteristics. A 4 percent error rate may not sound like much. But a multibillion-dollar corporation

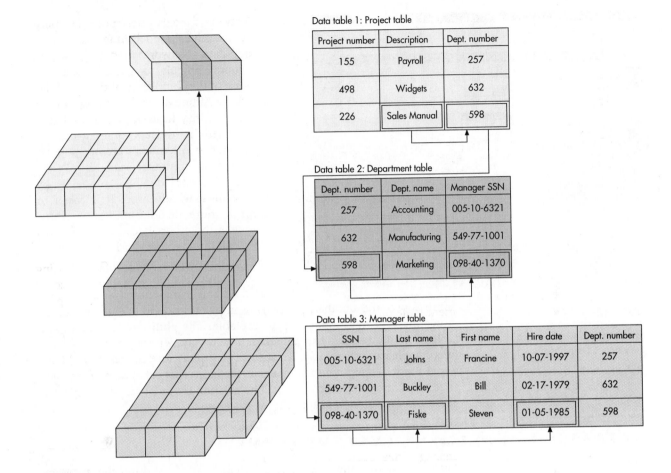

Data table 1: Project table

Project number	Description	Dept. number
155	Payroll	257
498	Widgets	632
226	Sales Manual	598

Data table 2: Department table

Dept. number	Dept. name	Manager SSN
257	Accounting	005-10-6321
632	Manufacturing	549-77-1001
598	Marketing	098-40-1370

Data table 3: Manager table

SSN	Last name	First name	Hire date	Dept. number
005-10-6321	Johns	Francine	10-07-1997	257
549-77-1001	Buckley	Bill	02-17-1979	632
098-40-1370	Fiske	Steven	01-05-1985	598

FIGURE 5.9

Linking Data Tables to Answer an Inquiry

In finding the name and hire date of the manager working on the sales manual project, the president needs three tables: project, department, and manager. The project description (Sales Manual) leads to the department number (598) in the project table, which leads to the manager's SSN (098-40-1370) in the department table, which leads to the manager's name (Fiske) and hire date (01-05-1985) in the manager table.

could lose millions of dollars if bad information caused the firm to bill all but 4 percent of its orders. When a database is created with data from multiple sources, those disparate sources may store different values for the same customer due to spelling errors, multiple account numbers, and address variations. The purpose of data cleanup is to look for and fix these and other inconsistencies that can result in duplicate or incorrect records ending up in the database.

A comparison of database models. Each of the database models we have discussed is used by a variety of organizations, and each has advantages and disadvantages. The primary advantage of the hierarchical model is processing efficiency. A hierarchical database system can take less time to manipulate data than other database models, because the data relationships are less complex (with each child having only one parent, so to speak). The hierarchical model is appropriate when the data forms a natural hierarchy, but this model generally is not flexible in terms of how the data is organized. Hierarchical models are also difficult to change, and the databases can be difficult to install. Even with these disadvantages, many large organizations use the hierarchical model because of processing efficiency or large investments in existing hierarchical database systems.

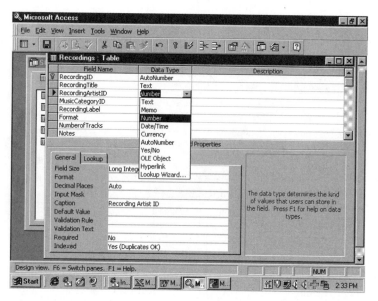

FIGURE 5.10

Building and Modifying a Relational Database

Relational databases provide many tools, tips, and tricks to simplify the process of creating and modifying a database.

Network models offer more flexibility than hierarchical models in terms of organizing data. Network models, however, are more difficult to develop and use because of the complexity of the data relationships. Although some network models are in use, they have not been used to a great extent in the corporate world. In addition, network models are not popular with personal computer users.

The relational database model is by far the most widely used. It is easier to control, more flexible, and more intuitive than the others because it organizes data in tables. As seen in Figure 5.10, a relational database, such as Access, provides a number of tips and tools for building and using database tables. This figure shows the database displaying information about data types and indicating that additional help is available. The ability to link relational tables also allows users to relate data in new ways without having to redefine complex relationships. Because of the many advantages of the relational model, many companies use it for large corporate databases, such as in marketing and accounting. The relational model can be used with personal computers and mainframe systems.

DATABASE MANAGEMENT SYSTEMS (DBMS)

Creating and implementing the right database system ensures that the database will support the business activities and goals. But how do we actually create, implement, use, and update a database? The answer is found in the database management system. As discussed, a database management system (DBMS) is a group of programs used as an interface between a database and application programs or a database and the user. DBMSs are classified by the type of database model they support. For example, a relational database management system follows the relational model. Access by Microsoft is a popular relational DBMS for personal computers. Popular mainframe relational DBMSs include DB2 by IBM, Oracle, Sybase, and Informix.[4] Regardless of the model they support, all DBMSs share some common functions, like providing a user view, physically storing and retrieving data in a database, allowing for database modification, manipulating data, and generating reports.

Providing a User View

schema

a description of the entire database

Because the DBMS is responsible for access to a database, one of the first steps in installing and using a database involves telling the DBMS the logical and physical structure of the data and relationships among the data in the database. This description is called a **schema** (as in schematic diagram). A schema can be part of the database or a separate schema file. The DBMS can reference a schema to find where to access the requested data in relation to another piece of data.

subschema

a file that contains a description of a subset of the database and identifies which users can perform modifications on the data items in that subset

A DBMS also acts as a user interface by providing a view of the database. A user view is the portion of the database a user can access. To create different user views, subschemas are developed. A **subschema** is a file that contains a description of a subset of the database and identifies which users can perform modifications on the data items in that subset. While a schema is a description of the entire database, a subschema shows only some of the records and their relationships in the database. Normally, programmers and managers need to view or access only a subset of the database. For example, a sales representative might need only data describing customers in her region, not the sales data for the entire nation. A subschema could be used to limit her view to data from her region. Subschemas mean that the underlying structure of the database can change, but the view the user sees might not change. For example, even if all the data on the southern region changed, the northeast region sales representative's view would not change if she accessed data on her region.

A number of subschemas can be developed for different managers or users and the various application programs. Typically, the database user or application will access the subschema, which then accesses the schema (Figure 5.11). Subschemas can also provide additional security because programmers, managers, and other users are typically allowed to view only certain parts of the database.

Creating and Modifying the Database

data definition language (DDL)

a collection of instructions and commands used to define and describe data and data relationships in a specific database

Schemas and subschemas are entered into the DBMS (usually by database personnel) via a data definition language. A **data definition language (DDL)** is a collection of instructions and commands used to define and describe data and data relationships in a specific database. A DDL allows the database's creator to describe the data and the data relationships that are to be contained in the schema and the many subschemas. In general, a DDL describes logical access paths and logical records in the database. Figure 5.12 shows a simplified example of a DDL used to develop a general schema. The Xs in Figure 5.12 reveal where specific information concerning the database

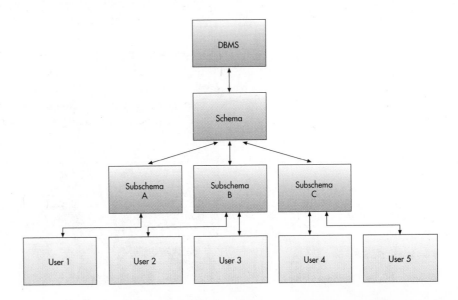

FIGURE 5.11

The Use of Schemas and Subschemas

FIGURE 5.12

Using a Data Definition
Language to Define a Schema

```
SCHEMA DESCRIPTION
SCHEMA NAME IS XXXX
AUTHOR         XXXX
DATE           XXXX
FILE DESCRIPTION
     FILE NAME IS XXXX
         ASSIGN XXXX
     FILE NAME IS XXXX
         ASSIGN XXXX
AREA DESCRIPTION
     AREA NAME IS XXXX
RECORD DESCRIPTION
     RECORD NAME IS XXXX
     RECORD ID IS XXXX
     LOCATION MODE IS XXXX
     WITHIN XXXX AREA FROM XXXX THRU XXXX
SET DESCRIPTION
     SET NAME IS XXXX
     ORDER IS XXXX
     MODE IS XXXX
     MEMBER IS XXXX
     .
     .
     .
```

is to be entered. File description, area description, record description, and set description are terms the DDL defines and uses in this example. Other terms and commands can be used, depending on the particular DBMS employed.

Another important step in creating a database is to establish a **data dictionary**, a detailed description of all data used in the database. The data dictionary contains the name of the data item, aliases or other names that may be used to describe the item, the range of values that can be used, the type of data (such as alphanumeric or numeric), the amount of storage needed for the item, a notation of the person responsible for updating it and the various users who can access it, and a list of reports that use the data item. Figure 5.13 shows a typical data dictionary entry.

data dictionary

a detailed description of all data used in the database

```
                 NORTHWESTERN MANUFACTURING

PREPARED BY:          D. BORDWELL
DATE:                 04 AUGUST 1998
APPROVED BY:          J. EDWARDS
DATE:                 13 OCTOBER 1998
VERSION:              3.1
PAGE:                 1 OF 1

DATA ELEMENT NAME:    PARTNO
DESCRIPTION:          INVENTORY PART NUMBER
OTHER NAMES:          PTNO
VALUE RANGE:          100 TO 5000
DATA TYPE:            NUMERIC
POSITIONS:            4 POSITIONS OR COLUMNS
```

FIGURE 5.13

A Typical Data Dictionary Entry

For example, the information in a data dictionary for the part number of an inventory item can include the name of the person who made the data dictionary entry (D. Bordwell), the date the entry was made (August 4, 1998), the name of the person who approved the entry (J. Edwards), the approval date (October 13, 1998), the version number (3.1), the number of pages used for the entry (1), the part name (PARTNO), other part names that may be used (PTNO), the range of values (part numbers can range from AAAAAAA to WWWWWWW), the type of data (alphabetic), and the storage required (four positions are required for the part number). Following are some of the typical uses of a data dictionary.

Provide a standard definition of terms and data elements. This can help in the programming process by providing consistent terms and variables to be used for all programs. Programmers know what data elements are already "captured" in the database and how they relate to other data elements.

Assist programmers in designing and writing programs. Programmers do not need to know which storage devices are used to store needed data. Using the data dictionary, programmers specify the required data elements. The DBMS locates the necessary data. More importantly, programmers can use the data dictionary to see which programs already use a piece of data and, if appropriate, copy the relevant section of the program code into their new program, thus eliminating duplicate programming efforts.

Simplify database modification. If for any reason a data element needs to be changed or deleted, the data dictionary would point to specific programs that utilize the data element that may need modification.

A data dictionary helps achieve the advantages of the database approach in these ways:

Reduced data redundancy. By providing standard definitions of all data, it is less likely that the same data item will be stored in different places under different names. For example, a data dictionary would reduce the likelihood that the same part number would be stored as two different items, such as PTNO and PARTNO.

Increased data reliability. A data dictionary and the database approach reduce the chance that data will be destroyed or lost. In addition, it is more difficult for unauthorized people to gain access to sensitive data and information.

Faster program development. With a data dictionary, programmers can develop programs faster. They don't have to develop names for data items because the data dictionary does that for them.

Easier modification of data and information. The data dictionary and the database approach make modifications to data easier because users do not need to know where the data is stored. The person making the change indicates the new value of the variable or item, such as part number, that is to be changed. The database system locates the data and makes the necessary change.

Storing and Retrieving Data

As just described, one function of a DBMS is to be an interface between an application program and the database. When an application program needs data, it goes to the DBMS. Suppose that to calculate the total price of a new car, an auto dealer pricing program needs price data on the engine option—six cylinders instead of the standard four cylinders. The application program thus seeks this data from the DBMS. In doing so, the application program follows a logical access path. Next, the DBMS, working in conjunction with various system software programs, accesses a storage device, such as disk or tape, where the data is stored. When the DBMS goes to this storage device to retrieve the data, it follows a path to the physical location (physical access path) where the price of this option is stored. In the pricing example, the DBMS might go to a disk drive to retrieve the price data for six-cylinder engines. This relationship is shown in Figure 5.14.

This same process is used if a manager (a user) wants to get information from the database. First, the manager requests the data from the DBMS. For example, a manager might give a command, such as LIST ALL OPTIONS FOR WHICH PRICE IS GREATER THAN 200 DOLLARS. This is the logical access path (LAP). Then the DBMS might go to the options price sector of a disk to get the information for the manager. This is the physical access path (PAP).

When two or more people or programs attempt to access the same record in the same database at the same time, there can be a problem. For example, an inventory control program might attempt to reduce the inventory level for a product by ten units because ten units were just shipped to a customer. At the same time, a purchasing program might attempt to

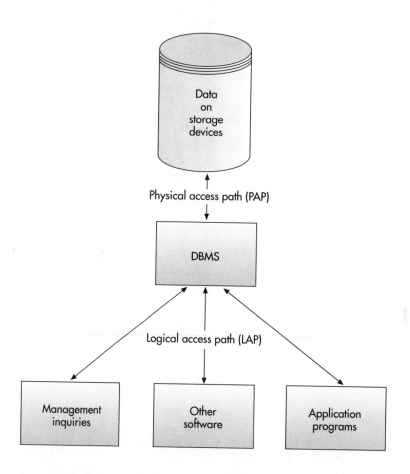

FIGURE 5.14

Logical and Physical
Access Paths

increase the inventory level for the same product by 20 units because more inventory was just received. Without proper database control, one of the inventory updates may not be correctly made, resulting in an inaccurate inventory level for the product. **Concurrency control** can be used to avoid this potential problem. One approach is to lock out all other application programs from access to a record if the record is being updated or used by another program.

Users increasingly need to be able to access and/or update databases via the Internet. Many database vendors are incorporating this capability into their products including Microsoft Corporation, Allaire Corporation, Inline Internet Systems Inc., Netscape Communications, EveryWhere Development Corporation, and StormCloud Development Corporation.[5] Such databases allow companies to create an Internet accessible catalog, which is nothing more than a database of items, descriptions, and prices.

Manipulating Data and Generating Reports

Once a DBMS has been installed, the system can be used by all levels of employees via specific commands in various programming languages. For example, COBOL commands can be used in simple programs that will access or manipulate certain pieces of data in the database. Here's another example of a DBMS query: SELECT * FROM EMPLOYEE WHERE JOB_CLASSIFICATION = "C2". The * tells the program to include all columns from the EMPLOYEE table. In general, the commands that are used to manipulate the database are part of the **data manipulation language (DML)**. This specific language, provided with the DBMS, allows managers and other database users to access, modify, and make queries about data contained in the database to generate reports. Again, the application programs go through subschemas, schemas, and the DBMS before actually getting to the physically stored data on a device such as a disk.

In the 1970s, D. D. Chamberlain and others at the IBM Research Laboratory in San Jose, California, developed a standardized data manipulation language, called **Structured Query Language (SQL)**, pronounced like the word *sequel*. The EMPLOYEE query shown earlier is written in SQL. In 1986, the American National Standards Institute (ANSI) adopted SQL as the standard query language for relational databases. Since ANSI's acceptance of SQL, interest in making SQL an integral part of relational databases on both mainframe and personal computers has increased.

As discussed in Chapter 4, SQL lets programmers learn one powerful query language and use it on systems ranging from PCs to the largest mainframe computers (Figure 5.15). Programmers and database users also find SQL valuable because SQL statements can be embedded into many programming languages, such as

---

concurrency control

a method of dealing with a situation in which two or more people need to access the same record in a database at the same time

data manipulation language (DML)

the commands that are used to manipulate the data in a database

structured query language (SQL)

a standardized data manipulation language

FIGURE 5.15

Structured Query Language

SQL has become an integral part of most relational database packages, as shown by this screen from Microsoft Access.

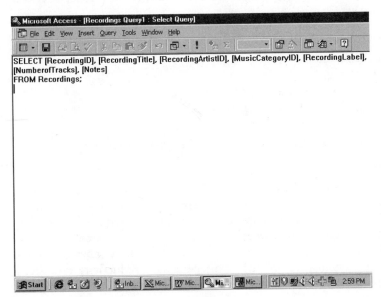

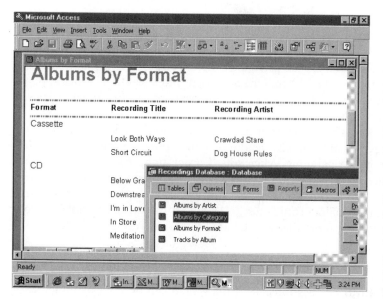

FIGURE 5.16

Database Output

A database application offers sophisticated formatting and organization options to produce the right information in the right format.

the widely used COBOL. Because SQL uses standardized and simplified procedures for retrieving, storing, and manipulating data in a database system, the popular database query language can be easy to understand and use.

Once a database has been set up and loaded with data, it can produce desired reports, documents, and other outputs (see Figure 5.16). These outputs usually appear in screen displays or hard-copy printouts. The output-control features of a database program allow you to select the records and fields to appear in reports. You can also complete calculations specifically for the report by manipulating database fields. Formatting controls and organization options (like report headings) help you to customize reports and create flexible, convenient, and powerful information-handling tools.

A database program can produce a wide variety of documents, reports, and other outputs that can help organizations achieve their goals. The most common reports select and organize data to present summary information about some aspect of company operations. For example, accounting reports often summarize financial data such as current and past-due accounts. Many companies base their routine operating decisions on regular status reports that show the progress of specific orders toward completion and delivery. Increasingly, companies are using databases to provide improved customer services, as the "Making a Difference" box demonstrates.

Exception, scheduled, and demand reports, first discussed in Chapter 1, highlight events that require urgent management attention. Database programs can produce literally hundreds of documents and reports. A few examples include:

- Form letters with address labels
- Payroll checks and reports
- Invoices
- Orders for materials and supplies
- A variety of financial performance reports

Popular Database Management Systems for End Users

The latest generation of database management systems makes it possible for end users to build their own database applications. End users are using these tools to address everyday problems like how to manage a mounting pile of information on employees, customers, inventory, or sales and fun stuff like wine lists, CD collections, and video libraries. These database management systems are an important personal productivity tool along with word processing, spreadsheet, and graphics software.

MAKING A DIFFERENCE
KeyCorp Develops Customer Relationship Databases

Increasingly, marketing companies employ what is called a customer relationship database to handle customers on a more personal and individual basis. Banks, brokerage firms, and other financial institutions are especially interested in such databases.

KeyCorp is a Cleveland-based banking company that reported earnings of $919 million for a recent year—an increase of $136 million over its previous year's earnings. In discussing KeyCorp's earnings improvement, Robert W. Gillespie, chairman and chief executive officer said: "With record earnings for the third consecutive quarter, we are realizing the payoff from significant re-engineering and investments at KeyCorp during the last few years. We are benefiting from an aggressive sales force supported by technology and marketing. Our re-engineering of commercial lending functions, begun early in 1996, was a major factor in our sales staff's ability to generate strong growth in middle market commercial lending throughout 1997."

Max is the name of KeyCorp's database for managing its customer relationships and is credited with improving profitability. It records all customer interactions and makes targeted marketing suggestions. The database assists the bank in handling each of its eight million accounts on a personal basis. Max has three "triggered immediacy marketing" (TIM) components for individual customers:

1. DirectTIM sends letters to customers who do any of 20 things, such as open an account, apply for a loan, use an automated teller machine for the first time, or inquire about an investment product for the first time. These letters reassure the customers that they did the right thing and relieve them of post-purchase anxiety.
2. Each customer service representative uses PCTIM to manage about 200 customer relationships. PCTIM tells agents who they should contact and which products to pitch on a weekly and daily basis.
3. TeleTIM greets customers by name and offers them personalized products when they call. The program will be expanded nationwide next year.

DISCUSSION QUESTIONS

1. Do you do business with a bank or other company that interacts with you as if it might be using a customer relationship database? Why do you think so?
2. Is there a possibility that use of customer relationship database may be viewed by some customers as being too intrusive and "all knowing?" What could a company that uses such a database do to guard against such a perception?

Sources: "KeyCorp Reports Record Earnings," *PRNewswire*, January 15, 1998; Tami Luhby, "A Database Named Max Helps KeyCorp Personalize Service and Marketing," *American Banker*, vol. 162, no. 199, October 15, 1997, p. 13.

A key to making DBMSs more useable for some databases is the incorporation of Wizards that walk you through how to build customized databases, modify ready-to-run applications, use existing record templates, and quickly locate the data you want. These applications also include powerful new features such as help systems and Web-publishing capabilities. Thus, users can create a complete inventory system and then instantly post it to the Web, where it does double-duty as an electronic catalog.

Some of the more popular database management systems for end users include Microsoft Access 98, Lotus Approach 98, and Inprise's (formerly Borland) dBASE 7. These DBMSs are summarized here. Other database management systems used by professional system developers include IBM's DB2 and Oracle.

Microsoft Access 98. Microsoft Access is available as a stand-alone product or bundled in the Office 98 suite. It provides new usability features for novice users as well as advanced capabilities for programmers. Microsoft rewrote much of the code for Access 98, providing the processing power for professionally developed multiuser applications. The end user language is greatly improved because Access 98 now shares Microsoft Visual Basic for Applications (VBA) with the rest of Office 98. VBA is much richer than Access' previous language for end user developed programs. Furthermore, tests show VBA end user programs are much faster than their Access 95 counterparts. New wizards simplify many tedious programming tasks as well. Unfortunately, Access has a steep learning curve and may be overwhelming at first. It also cannot publish live to the Web directly from within the program.[6]

Access includes predefined templates that can be edited to create specific kinds of databases. The new templates in Access 98 are wizards for creating databases geared toward both businesses and individuals—inventory control, expenses, household budgets, wine/CD/video collections and so on. The wizards give you a good deal of interactive control over what fields to include in the databases and what views to present, so you can define exactly the database you want.

You can convert older databases into Access 98 or interpret them, although you cannot make changes to underlying data objects without converting databases to Access 98 format. Fortunately, converting older databases is a one-step, risk-free process: Just choose the Convert option from the Tools menu, point at the database in question, type a new filename, and you're done.

The power of Access 98 makes it ideal for applications in workgroups. It's great when you want something that can handle multiple users easily, but you don't need to process an extremely high number of transactions for dozens of concurrent users. It costs around $125 stand-alone and is also sold as part of the Office Suite of software products.

Lotus Approach Millennium Edition is a relational database that provides an easy way to query, report on, and analyze data. Approach lets several members of a business team track, manage, and analyze business information.
(Source: Courtesy of Lotus Development Corporation.)

Lotus Approach 98. Lotus Approach has most of the features and capabilities of Access 98. The main differences between the two are in ease of learning and the types of applications they can support. Approach 98 is considered by many to be easier to learn, making it better suited for individuals without a strong programming background. But Access 98 can support more transactions and concurrent users. Thus, it is more appropriate for building multi-user and workgroup-level applications.[7] The Lotus scripting language is called LotusScript and is comparable to Microsoft Visual Basic for Applications.

Approach has a highly intuitive, notebook-style interface that includes an Action Bar with options such as Design, Find, and Report located right under the standard toolbar. It's easy to get started in Approach, and hard to get lost.

The New Record button makes entering data simple, and the ever-present Find Assistant makes searching for data effortless. Importing data is a little more difficult, but Approach offers precise options for doing so. Users can create tables easily and productively. Approach's reporting features are simple to find and use.

Approach 98 is more limited in its ability to link databases to the Web. It allows users to save reports in Web documents. Approach 98 integrates well with Lotus Notes. Approach users can open files in Notes or save Approach files directly to Notes. Approach's support for the Notes PowerKey function lets users create worksheets, reports, mailings, and even charts from live Notes data.

Inprise dBASE. Inprise's Visual dBASE 7 provides significant improvements over earlier versions, and includes features such as a graphical query builder, a powerful file format, report writer, and code editor. dBASE 7 provides a small selection of wizards relative to Access and Approach.

dBASE 7 is designed to run under 32-bit Windows operating systems. However, dBASE 7's support for 32-bit Windows operating systems sacrifices its Windows 3.1 applications. dBASE 7 is Win32-only and so are any programs it generates. Furthermore, while dBASE 7 works fine with earlier dBASE file formats, its new database format cannot be read by older dBASE versions.[8]

dBASE 7 can access dBASE, FoxPro, Paradox and Access databases. It also supports connections to SQL databases including Oracle, Sybase's SQL Server, Microsoft's SQL Server, Informix and DB2, as well as Borland's own InterBase. dBASE 7 also has a new, easy-to-use graphical query builder. Setting up SQL statements requires just dragging and dropping related fields; defining groups and adding calculated expressions are simple procedures.

dBASE 7 offers no automatic formatting styles for reports, one of the areas where automatic formatting is most useful, and the selection of report wizards is also very limited. Its tools for developing Web-based interfaces are limited. Access and Approach provide far more seamless and full-featured Web integration.

Selecting a Database Management System

Selecting the best database management system begins by analyzing database needs and characteristics. The information needs of the organization affect what type of data is collected and what type of database management system is used. Important characteristics of databases include the size of the database, number of concurrent users, performance, the ability of the DBMS to integrate with other systems, the features of the DBMS, vendor considerations, and the cost of the database management system.

Database size. The database size depends on the number of records or files in the database. The size determines the overall storage requirement for the database. Many database management systems can handle relatively small databases of less than 100 million bytes, fewer can manage terabyte size databases. The "Technology Advantage" box discusses the trend toward smaller databases.

TECHNOLOGY ADVANTAGE
Small but Powerful: Good Things Can Come in Small Databases

In the early days of database design, many believed that bigger was better. One huge database could process all business transactions and provide managers with the information they needed. These large databases, however, proved to be hard to develop, difficult to maintain, and nearly impossible to use by some managers. In addition, the need to process transactions quickly for routine operations often clashed with the managers' needs to manipulate and summarize data to help them make informed decisions. Large databases can also be very expensive to develop, use, and maintain.

To avoid some of these problems, more companies are implementing smaller databases. Lockheed Martin Corporation, for example, installed a small database for a collection of applications for a facility in Idaho. The remote facility was ideal for a small database. Using a product by Sybase, Lockheed Martin was able to provide critical information about hazardous materials to seven federal facilities outside Idaho Falls.

Because of the advantages of these smaller databases, a number of database companies are starting to market smaller versions of their larger databases. Oracle, a database company, markets Oracle Lite. Like other small databases, Oracle Lite is ideal for remote database needs. A distant field office, for example, can use a product like Oracle Lite to satisfy its processing and information needs. The small database does not require a mainframe computer or a large staff of database professionals. With this approach, each facility or center can have and use its own database.

Small databases, however, have caused some problems. Sometimes they are too small to handle processing or information requests. In other cases, the hardware these smaller databases use is not capable of providing enough processing power or speed to satisfy business needs. Union Gas Ltd. discovered another limitation. Using a small database by Sybase, Union was unable to effectively exchange data with its large database. To solve this problem, Union attempted to write its own programs to transfer needed data between the small database and the large one. Trying to integrate a small database package from one database company with a large database from another company can be extremely difficult. To avoid integration problems, companies often use the same vendor to supply their small and large databases. Union, for example, eventually decided to switch its small database from Sybase to Oracle to standardize its database vendor and reduce integration and compatibility problems.

Small databases offer many advantages over large ones for many applications, but the early users of these small databases struggled with speed, capacity, and integration problems. Many experts, however, believe that these problems will be eliminated as database companies continue to refine and improve their small database packages.

DISCUSSION QUESTIONS

1. What are the advantages and disadvantages of using a small database?
2. Describe a situation in which a small database would be ideal. Under what situations should a small database be avoided?

Sources: Craig Stedman, "Light Databases Get Some Muscle," *Computerworld*, January 26, 1998, p. 69; Bill Inmon, "Wherefore Warehouse," *Byte*, January 1998, p. 88; Richard Campbell, "Data Warehouses vs. Data Marts," *Database Web Advisor*, January 1998, p. 32.

Number of concurrent users. The number of simultaneous users that can access the contents of the database is also an important factor. Clearly a database that is meant to be used by anyone in a large workgroup must be able to support a number of concurrent users; if it cannot, then the efficiency of the members of the workgroup will be negatively impacted. The term scalability is sometimes used to describe how well a database performs as the size of the database and/or the number of concurrent users is increased. A highly scalable database management system is desirable to provide flexibility.

Performance. How fast the database is able to update records can be the most important performance criteria for some organizations. Credit card and airline companies, for example, must have database systems that can update customer records and check credit or make a plane reservation in seconds, not minutes. Other applications, such as payroll, can be done once a week or less frequently and do not require immediate processing. If an application demands immediacy, it also demands rapid recovery facilities in the event the computer system shuts down temporarily. Other performance considerations include the number of concurrent users that can be supported and how much main memory is required to execute the database management program.

Integration. A key aspect of any database management system is its ability to integrate with other applications and databases. Some companies use several databases for different applications at different locations. A manufacturing company with four plants in three different states might have a separate database at each location. The ability of a database program to import data from and export data to other databases and applications can be a critical consideration.

Features. The features of the database management system can also make a big difference. Most database programs come with security procedures, privacy protection, and a variety of tools. Other features can include how easy the database package is to use and the availability of manuals and documentation that can help the organization get the most from the database package. Additional features such as Wizards and ready-to-use templates help improve the product ease of use and are very important.

The vendor. The size, reputation, and financial stability of the vendor should also be considered in making any database purchase decision. Some vendors are well respected in the information systems industry and have a large staff of support personnel to give assistance, if necessary. A well-established and financially secure database company is more likely to remain in business.

Cost. Database packages for personal computers can cost a few hundred dollars, while large database systems for mainframe computers can cost hundreds of thousands of dollars. In addition to the initial cost of the database package, monthly operating costs should be considered. Some database companies rent or lease their database software. Monthly rental or lease costs, maintenance costs, additional hardware and software costs, and personnel costs can be substantial.

DATABASE DEVELOPMENTS

The types of data and information managers need change as business processes change. A number of developments in the use of databases and database management systems can help managers meet their needs, among them allowing organizations to place data at different locations, setting up data warehouses and marts, using the object-oriented approach in database development, and searching for and using unstructured data such as graphics and video.

Distributed Databases

Distributed processing involves placing processing units at different locations and linking them via telecommunications equipment. A **distributed database**—a database in which the actual data may be spread across several smaller databases connected via telecommunications devices—works on much the same principle. A user in the Milwaukee branch of a shoe manufacturer, for example, might make a request for data that is physically located at corporate headquarters in Milan. The user does not have to know where the data is physically stored. He or she makes a request for data, and the DBMS determines where the data is physically located and retrieves it (Figure 5.17).

Organizations often find that distributed databases provide some of the same advantages as distributed processing. Distributed databases give corporations more flexibility in how databases are organized and used. Local offices can create, manage, and use their own databases, and people at other offices can access and share the data in the local databases. Giving local sites a more direct way to access highly used data can provide significant organizational effectiveness and efficiency.

Despite its advantages, distributed processing creates additional challenges in maintaining data security, accuracy, timeliness, and conformance

FIGURE 5.17

The Use of a Distributed Database

For a clothing manufacturer, computers may be located at corporate headquarters, in the research and development center, the warehouse, and in a company-owned retail store. Telecommunications systems link the computers so that users at all locations can access the same distributed database no matter where the data is actually stored.

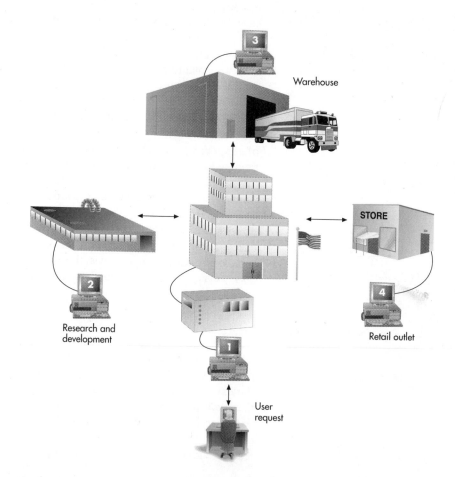

Warehouse

Research and development

STORE

Retail outlet

User request

to standards. Distributed databases allow more users direct access at different sites; thus, controlling who accesses and changes data is sometimes difficult. Also, because distributed databases rely on telecommunications lines to transport data, access to data can be slower. To reduce the demand on telecommunications lines, some organizations will build a replicated database. A **replicated database** holds a duplicate set of frequently used data. At the beginning of the day, the company will send a copy of important data to each distributed processing location. At the end of the day, the different sites send the changed data back to be stored in the main database.

replicated database

a database that holds a duplicate set of frequently used data

Another challenge created by distributed databases involves integrating the various databases. For example, some organizations use many database management systems. A hierarchical database management system may be used at the headquarters of a large manufacturing company, while different relational DBMSs may be used by various regional offices. Businesses must develop a solution that enables them to access data in these different DBMSs.

Data Warehouses, Data Marts, and Data Mining

The raw data necessary to make sound business decisions is stored in a variety of locations and formats—hierarchical databases, network databases, flat files, and spreadsheets, to name a few. This data is initially captured, stored, and managed by transaction processing systems that are designed to support the day-to-day operations of the organization. For decades, organizations have collected operational, sales, and financial data with their on-line transaction processing (OLTP) systems.

Traditional OLTP systems are designed to put data into databases very quickly, reliably, and efficiently. These systems are not good at supporting meaningful analysis of the data. Indeed, tuning a system to provide excellent performance for OLTP often renders rapid data retrieval for data analysis nearly impossible. Furthermore, data stored in OLTP databases is inconsistent and constantly changing. The database contains the current transactions required to operate the business, including errors, duplicate entries, and reverse transactions, which get in the way of a business analyst, who needs stable data. Historical data is missing from the OLTP database, which makes trend analysis impossible. Thus, because of the application orientation of the data, the variety of nonintegrated data sources, and the lack of historical data, companies were limited in their ability to access and use the data for other purposes. So, although the data collected by OLTP systems doubles every two years, it does not meet the needs of the business decision maker—they are data rich but information poor.

Data warehouses. Data warehousing is the current evolution of decision support systems (DSSs) that has emerged from improvements in database and network technologies. A **data warehouse** is a relational database management system designed specifically to support management decision making, not to meet the needs of transaction processing systems.[9] The data warehouse provides a specialized decision support database that manages the flow of information from existing corporate databases and external sources to end-user decision support applications. A data warehouse stores historical data that has been extracted from operational systems and external data sources (Figure 5.18). This operational and external data is

data warehouse

a relational database management system designed specifically to support management decision making

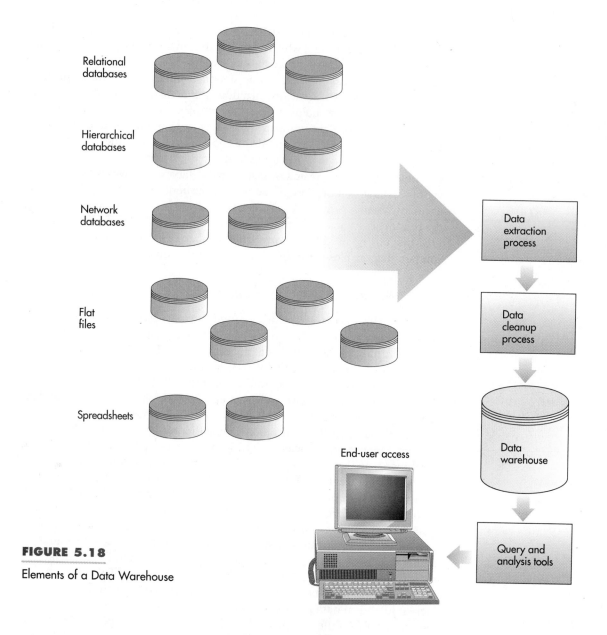

FIGURE 5.18

Elements of a Data Warehouse

"cleaned up" to remove inconsistencies and integrated to create a new information database that is more suitable for business analysis.

Data warehouses typically start out as very large databases (VLDBs), containing millions and even hundreds of millions of data records. As this data is collected from the various production systems, a historical base is built that the business analyst can use. To remain fresh and accurate, the data warehouse receives regular updates. Old data that is no longer needed is purged from the data warehouse. Updating the data warehouse must be fast, efficient, and automated or the ultimate value of the data warehouse is sacrificed. It is common for a data warehouse to contain 5 to 10 years of current and historical data. Data cleaning tools can merge data from many sources into one database, automate data collection and verification, delete unwanted data, and maintain data in a database management system.

The primary advantage of data warehousing is the ability to relate data in new, innovative ways. However, a data warehouse can be extremely difficult to establish. Table 5.3 provides some advice on how to create a data warehouse.

data mart

a subset of a data warehouse for small and medium-size businesses or departments within larger companies

Data Marts. A **data mart** is a subset of a data warehouse. Data marts bring the data warehouse concept—on-line analysis of sales, inventory, and other vital business data that has been gathered from transaction processing systems—to small and medium-size businesses and to departments within larger companies.[10] Rather than store all enterprise data in one monolithic database, data marts contain a subset of the data for a single aspect of a company's business—for example, finance, inventory, or personnel. In fact, there may even be more detailed data for a more specific area in a data mart than what a data warehouse would provide.

Data marts are most useful for smaller groups who want to access detailed data. A warehouse is used for summary data that can be used by the rest of the company. Because data marts typically contain tens of gigabytes of data, as opposed to the hundreds of gigabytes in data warehouses, they can be deployed on less-powerful hardware with smaller disks, delivering significant savings to an organization. Although any database software

TABLE 5.3

How to Design a Customer Data Warehouse

Sharply define your goals and objectives before you build the warehouse.
- Are you looking to increase your customer base?
- Do you want to double sales of a product line?
- Would you like to encourage repeat business?

Choose the software that best fits your goals.
- If you need a system that is tightly tied to sales rep activity in the field, choose a contact-management program with database capability.
- If you do a lot of telemarketing, you may need a telemarketing program.

Determine who should be in the database.
- First, figure out what types of customers have the most potential.
- Then construct the database, using everything from salespeople's contacts to lists bought from outside suppliers.

Develop a plan.
- Only after your objectives are laid out and your database is constructed is it time to devise a marketing program.
- Generally, it will fall into one of three categories: direct marketing, rewards for repeat purchases, and relationship-building promotions for long-time customers that generate profits.

Measure results.
- Generate periodic status reports in which you determine items like cost per contact and cost per sale.
- If you want to be really careful, do not launch your program at full tilt. Start with a small prototype; if you like what you see, expand to include the entire database.

Source: Adapted from Anne Field, "Precision Marketing," *Inc.*, vol. 18, no. 9, June 18, 1996, p. 54.

can be used to set up a data mart, some vendors deliver specialized software designed and priced specifically for data marts. Already, companies such as Sybase, Software AG, Microsoft, and others have announced products and services that make it easier and cheaper to deploy these scaled-down data warehouses. The selling point: data marts put targeted business information into the hands of more decision makers.[11]

Owens & Minor in Glen Allen, Virginia, a $3 billion distributor of medical and surgical supplies, is combining a series of data marts within a single Oracle database. Owens & Minor couldn't afford to take the time needed to plan and design a full data warehouse all at once, but it recognized that building separate data marts would make it harder for users to run queries that go across all of them. Using a combination approach, data marts, covering functions such as sales, inventory, and accounts receivable, are stored in their own tables within the Oracle database. End users are able to see different views of the information through their desktop query tools. Everything is stored in one database, but to the user, it looks as though they were separate data marts.[12]

data mining

the automated discovery of patterns and relationships in a data warehouse

Data mining. Another new information analysis tool is data mining. **Data mining** is the automated discovery of patterns and relationships in a data warehouse. Data mining represents the next step in the evolution of decision support systems. It makes use of advanced statistical techniques and machine learning to discover facts in a large database, including databases on the Internet.[13] Unlike query tools, which require users to formulate and test a specific hypothesis, data mining uses built-in analysis tools to automatically generate a hypothesis about the patterns and anomalies found in the data and then uses them to predict future behavior.

Data mining's objective is to extract patterns, trends, and rules from data warehouses to evaluate (i.e., predict or score) proposed business strategies, which in turn will improve competitiveness, improve profits, and transform business processes. It is used extensively in marketing to improve customer retention; cross-selling opportunities; campaign management; market, channel, and pricing analysis; and customer segmentation analysis (especially one-to-one marketing). In short, data mining tools help end users find answers to questions they never even thought to ask.

There are thousands of data mining applications. Credit card issuers and insurers mine their data warehouses for subtle patterns within thousands of customer transactions to identify fraud, often just as it happens. One U.S. cellular phone company is using Silicon Graphics MineSet software to dig through mountains of call data and pinpoint illegally cloned cell phone ID numbers. Manufacturers mine data collected from factory floor sensors to identify where an intermittent assembly line error is causing a defect that will show up only months after an appliance goes into use. And once shopping on the Web takes off, reams of data about the customers' behavior, tastes, and interests will be available for merchandisers to mine and react to nearly instantly.[14] Table 5.4 summarizes a few of the most frequent applications for data mining.

Traditional DBMS vendors are well aware of the great potential of data mining. Thus, companies like Oracle, Informix Software, Sybase, Tandem and Red Brick Systems are all incorporating data mining functionality into their products.

Application	Description
Market segmentation	Identifies the common characteristics of customers who buy the same products from your company.
Customer churn	Predicts which customers are likely to leave your company and go to a competitor.
Fraud detection	Identifies which transactions are most likely to be fraudulent.
Direct marketing	Identifies which prospects should be included in a mailing list to obtain the highest response rate.
Market basket analysis	Understands what products or services are commonly purchased together (e.g. beer and diapers).
Trend analysis	Reveals the difference between a typical customer this month versus last month.

Source: Vance McCarthy, "Strike It Rich," *Datamation*, February 1997, pp. 44–50.

TABLE 5.4

Common Data Mining Applications

on-line analytical processing (OLAP)

programs used to store and deliver data warehouse information

Hyperion Essbase 5 allows users to link rich multimedia content such as text, audio and video to any cell in an OLAP application.
(Source: Hyperion and Essbase are registered trademarks of Hyperion Solutions Corporation.)

On-Line Analytical Processing (OLAP)

Most industry surveys today show that the majority of data warehouse users rely on spreadsheets, reporting and analysis tools, or their own custom applications to retrieve data from warehouses and format it into business reports and charts. In general, these approaches work fine for questions that can be answered when the amount of data involved is relatively modest and can be accessed with a simple table lookup.

For nearly two decades, multidimensional databases and their analytical information display systems have provided flashy sales presentations and trade show demonstrations. All you have to do is ask where a certain product is selling well, for example, and a colorful table showing sales performance by region, product type, and time frame automatically pops up on the screen. Called **on-line analytical processing (OLAP)**, these programs are now being used to store and deliver data warehouse information. OLAP allows users to explore corporate data from a number of different perspectives.

OLAP servers and desktop tools support high-speed analysis of data involving complex relationships, such as combinations of a company's products, regions, channels of distribution, reporting units, and time periods. Speed is essential in a booming economy, as businesses grow and accumulate more and more data in their operational systems and data warehouses. Long popular with financial planners, OLAP is now being put in the hands of other professionals. The leading OLAP software vendors include Cognos, Comshare, Hyperion Solutions, Oracle, MineShare, WhiteLight, and Microsoft.[15]

Access to data in multidimensional databases can be very quick because they store the data in structures optimized for speed, and they avoid SQL and index processing. But multidimensional databases can take a great deal of time to update; in very large databases, update times can be so great that they force updates to be made only on weekends. Despite this flaw, multidimensional databases have continued to prosper because of their great retrieval speed. Some software providers are attempting to counteract this flaw through the use of partitioning and calculations-on-the-fly capabilities.

Consumer goods companies use OLAP to analyze the millions of consumer purchase records captured by scanners at the checkout stand. This data is used to spot trends in purchases and to relate sales volume to promotions and store conditions, such as displays, and even the weather. OLAP tools let managers analyze business data using multiple dimensions, such as product, geography, time, and salesperson. The data in these dimensions, called measures, is generally aggregated—for example, total or average sales in dollars or units, or budget dollars or sales forecast numbers. Rarely is the data studied in its raw, unaggregated form. Each dimension also can contain some hierarchy. For example, in the time dimension users may examine data by year, by quarter, by month, by week and even by day. A geographic dimension may compile data from city, to state, to region, to country, and even to hemisphere.

Sears, Roebuck and Co., which uses Hyperion Essbase OLAP Server to analyze data for 830 stores, is currently extending OLAP capability to another 2,500 stores. Sears began deploying Hyperion Essbase five years ago as a way of augmenting the company's monolithic Executive Information Systems (EIS). By extending OLAP to its retail outlets, Sears' local managers will be able to analyze everything from time sheets to product sales. The Hyperion Essbase servers will be linked to Sears' PeopleSoft applications, allowing store managers to track and compare sales performance and trends among locations.[16]

The value of data ultimately lies in the decisions it enables. Powerful information analysis tools in areas such as OLAP and data mining, when incorporated into a data warehousing architecture, bring market conditions into sharper focus and help organizations deliver greater competitive value. Table 5.5 compares OLTP, data warehousing, and OLAP. Data warehousing and OLAP provide top-down, query-driven data analysis; data mining provides bottom-up, discovery-driven analysis. Data warehousing requires repetitive testing of user-originated theories, while data mining requires no assumptions and instead identifies facts and conclusions based on patterns discovered. Data mining, however, is only as effective as the data it accesses. The data warehouse eliminates obstacles to effective data mining.

OLAP, or multidimensional analysis, requires a great deal of human ingenuity and interaction with the database to find information in the database. OLAP software tells users what happened in their business. Data mining is different. Data mining tells them why. As a user of a data mining tool, you do not need to figure out what questions to ask; instead, your approach is, "Here's the data, tell me what interesting patterns emerge." For example, a data mining tool in a credit card company's customer database can construct a profile of fraudulent activity from historical information. Then, this profile can be applied to all incoming transaction data to identify and stop fraudulent behavior, which may otherwise go undetected.

Characteristic	OLTP	Data Warehousing	OLAP
Purpose	Supports transaction processing	Supports information requests	Supports data analysis
Source of Data	Business transactions	Multiple files, databases—internal and external to firm	Multiple files, databases—internal and external to firm
Data Access Allowed Users	Read and write	Read only	Read only
Primary Data Access Mode	Simple database update and queries	Simple and complex queries with increasing use of data mining to recognize patterns in the data	Drill-down analysis and viewing of data from many perspectives
Primary Database Model Employed	Hierarchical and relational	Relational	Often proprietary database structures optimized for speed, moving to relational model
Level of Detail	Detailed transactions	Often summarized	Often summarized
Historical Data	Current data only	Multiple years of data	Multiple years of data
Update Process	On-line, ongoing process as transactions are captured	Periodic process, once per week or once per month	Periodic process, once per week or once per month
Ease of Update	Routine and easy	Complex, must combine data from many sources—both internal and external	Extremely time consuming for proprietary model
Data Integrity Issues	Each individual transaction must be closely edited	Major effort to "clean" and integrate data from multiple sources	Major effort to aggregate data to appropriate level

TABLE 5.5

A Comparison of Three Types of Database Usage

open database connectivity (ODBC)

a set of standards that ensures software written to comply with these standards can be used with any ODBC-compliant database

Open Database Connectivity (ODBC)

To help with database integration, many companies rely on **open database connectivity (ODBC)** standards. ODBC standards help ensure that software written to comply with these standards can be used with any ODBC-compliant database. This makes it easier to transfer and access data among different databases. For example, a manager might want to take several tables from one database and incorporate these tables into another database that uses a different database management system. In another case, a manager might want to transfer one or more database tables into a spreadsheet program. If all this software meets ODBC standards, this data can be imported, exported, or linked to other applications (see Figure 5.19). For example, a table in an Access database can be exported to a Paradox database or a spreadsheet. Tables and data can also be imported using ODBC. For example, a table in a dBASE database or an Excel spreadsheet can be imported into an Access database. Linking allows an application to use data or an object stored in another application without actually importing the data or object into the application. The Access database, for example, can link to a table in the Lotus 1-2-3 spreadsheet or the FoxPro database. This allows a database to have access to other data stored in different applications. Applications that follow the ODBC standard can use these powerful ODBC features to share data between different applications stored in different formats.

FIGURE 5.19

ODBC can be used to export, import, or link tables between different applications.

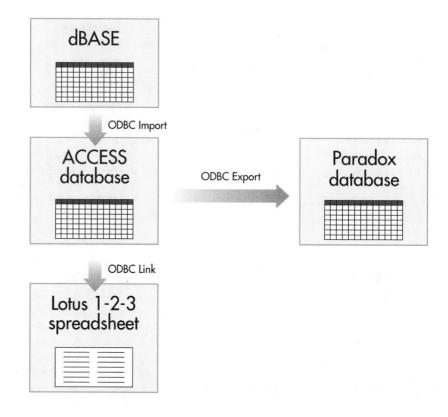

ODBC-compliant products suffer from their all-purpose nature. Their overall performance is usually less efficient than that of products designed for use with a specific database. Yet, more and more vendors are building ODBC-compliant products as businesses increasingly use distributed databases.[17] Many organizations are using such tools to allow their workers and managers easier access to a variety of databases and data sources. ODBC standards also make it easier for growing companies to integrate existing databases, to connect more users into the same database, and to move application programs from PC-oriented databases to larger, workstation-based databases, and vice versa.

Object-Relational Database Management Systems

Many of today's newer application programs require the ability to manipulate audio, video, and graphic data.[18] Conventional database management systems are not well suited for this, because these types of data cannot easily be stored in rows or tables. Manipulation of such data requires extensive programming so that the DBMS can translate data relationships. An **object-relational database management system (ORDBMS)** provides a complete set of relational database capabilities plus the ability for third parties to add new data types and operations to the database. These new data types can be audio, images, unstructured text, spatial, or time series data that require new indexing, optimization, and retrieval features.

In such a database, these types of data are stored as objects, which contain both the data and the processing instructions needed to complete the

object-relational database management system (ORDBMS)

a DBMS capable of manipulating audio, video, and graphical data database transaction. The objects can be retrieved and related by an ORDBMS. Businesses can then mix and match these elements in their daily search for clues and information. For example, by clicking on a picture of a red Corvette, a market analyst at General Motors might be able to call down a profile of red Corvette buyers. If he wants to break that up by geographic region, he might circle and click on a map. All in the same motion, he might view a GM sales-training film to see whether the sales pitch is appropriate, given the most recent market trends. Chances are the analyst will do all this using an Internet site tied into a database. As another example, MasterCard is interested in object-oriented technology to combine transactional data with cardholder fingerprints to prevent fraud.

Each of the vendors offering ORDBMS facilities provides a set of application programming interfaces to allow users to attach external data definitions and methods associated with those definitions into the database system. They are essentially offering a standard socket into which users can plug special instructions. DataBlades, Cartridges, and Extenders are the names applied by Informix, Oracle, and IBM to describe the plug-ins to their respective products. Other plug-ins serve as interfaces to Web servers.

Web-based applications increasingly require complex object support to link graphical and other media components back to the database. These systems make sense for developers of systems that are highly dependent on complex data types, particularly Web and multimedia applications. Because it supports so many applications, an ORDBMS is also called a universal database server.[19] Workers at *Time* magazine use an object-relational database connected to the Internet to provide a personalized news service on the company's Pathfinder Internet site.

An increasing amount of data organizations use is in the form of images, which can be stored in object-relational databases. Credit card companies, for example, input pictures of charge slips into an image database using a scanner. The images can be stored in the database and later sorted by customer, printed, and sent to customers along with their monthly statements. Image databases are also used by physicians to store X rays and transmit them to clinics away from the main hospital. Financial services, insurance companies, and government branches are also using image databases to store vital records and replace paper documents.

Image data has some disadvantages, one of which is the increased secondary storage requirements. Many businesses find that optical disk storage helps alleviate this problem. It is also sometimes difficult to locate the desired data in the database. Retrieval of image data from databases can be made easier by using the object-oriented approach. Other ways to index and cross-reference data are being developed.

Hypertext. The object-relational database provides greater flexibility in defining relationships between data. The previously discussed relational, hierarchical, and network models were designed for data that can be organized into fixed-length data fields and records with structured relationships. With hypertext, users can search and manipulate alphanumeric data in an unstructured way. In an object-relational database, text data is placed in chunks called nodes. The user then establishes links between the nodes. The relationships between the data can be created to user specifications, instead of following one of the more structured database models. Suppose a doctor is treating a new patient with symptoms similar to those of two other patients. The notes on each patient are placed in nodes, and the three nodes are linked. The doctor can use hypertext to retrieve and cross-reference these three cases when treating future symptoms of the new patient.

Hypermedia. Hypermedia is an extension of hypertext. Hypermedia allows businesses to search and manipulate multimedia forms of data—graphics, sound, video, and alphanumeric data. A marketing manager, for example, might store notes about the competition and new trends in the marketplace. These notes could include written material on new markets, images of products and advertising brochures, and TV commercials used by competitors. Using a hypermedia database management system, the marketing manager could organize this data into nodes and define the relationships among them. For example, she could link all TV commercials and written brochures about new products from several competitors. With the hypermedia database approach, many types of data can be organized into a web of nodes connected by links established by the user.

Spatial Data Technology. Spatial data technology involves the use of an object-relational database to store and access data according to the locations it describes and to permit spatial queries and analysis. NASA has set up a massive environmental monitoring program that calls for satellites and earth stations to deliver more than a terabyte of data every day to databases. The cumulative data will eventually multiply to a petabyte, which is 1,000 times a terabyte. Once the data is edited, it will be loaded into other databases for analysis. The primary database will offer up satellite photo images and measurements. If environmentalists wanted to compare carbon dioxide levels within a certain latitude, longitude, and altitude with another polygon of space, they could easily do so. Implications of spatial data technology are potentially significant. For example, the data derived from the NASA project and others like it could help find oil and mineral reserves or identify sources of pollutants. Builders and insurance companies can use spatial data to make decisions related to natural hazards. Spatial data can even be used to improve risk management with information stored by derivative type, currency type, interest rates, and time. The "FYI" box describes the use of an image database to store digital copies of the world's art treasures.

The painstaking process of ensuring accuracy, completeness, and currency of the data that's been extracted from operational databases and formatted for analysis by a variety of departments is key to the success of any data warehouse. Careleton, Evolutionary Technologies International, Platinum Technology, Prism Solutions, and other companies sell products for extracting and cleaning data. Also, a new class of tools is emerging for managing metadata, the data about the data in a warehouse. But these products address only part the problem—the harder part is having procedures in place to manage data as it is extracted from operational systems or imported from data providers and transformed into

Spatial data technology is used by NASA to store data from satellites and earth stations. Location-specific information can be accessed and compared. (Source: Courtesy of NASA.)

FYI

Digital Images on Demand

Corbis Corp. has an ambitious mission: "to capture the entire human experience throughout history." Corbis was founded by Microsoft's Bill Gates. Corbis negotiates licensing rights for images from a variety of sources, including photographers, museums, and private collectors. Creative professionals can license images for use in traditional print media, such as books, newspapers, magazines, and other electronic communications. For instance, while logged on to the Internet, a photo editor can call a Corbis researcher and describe the type of image he or she is looking for. The Corbis researcher then posts low-resolution images to a private page on the Internet for the editor to review. The editor makes his or her choices, and Corbis presses the images onto a CD-ROM or sends them directly over the Internet. Corbis provides professional image buyers with on-line access to its digital image collection through the Corbis View Service.

Corbis (http://www.images.corbis.com/members) already has a huge archive of 20 million images from 600 sources—the largest acquisition was The Bettmann Collection of 16 million images. Corbis technicians digitize the images in the archive at a rate of 40,000 per month. In addition, Corbis commissions photographers to capture specifically assigned images, such as the reinterment of the Romanovs in St. Petersburg or certain European battlefields. Using search-and-retrieval technologies based on text- and image-based links, Corbis can access digital images in seconds. The company has produced six award-winning multimedia CD-ROM titles that demonstrate the depth and breadth of the Corbis archive.

Corbis isn't the only firm with a digital image database, but it is the largest and probably has the highest profile because of Gates's involvement. Picture Network International (http://www.picturequest.com) has stock photos, illustrations, clip art, sound effects, and more. PhotoDisc (http://www.photodisc.com) offers 15,000 images, for which customers can pay using cybercash. Liaison International (http://www.liaisonintl.com) has 4 million photos, 2,400 of which are available on-line. Indeed, although Corbis seems to have cornered a huge chunk of the market, Josh Bernoff, senior analyst with Forrester Research, observes, "Regardless of how much Corbis acquires, they can't compete with the whole rest of the world." But maybe Corbis can, if it captures the entire human experience throughout history.

DISCUSSION QUESTIONS

1. How might businesses other than publishers or advertisers use digitized image databases such as those offered by Corbis?
2. How might you use this type of graphic?
3. Access one of the Web sites mentioned—either Corbis's or a competitors—and see what information you can find about the company's offerings.

Source: "The Corbis Story," http://www.corbis.com, March 17, 1997 and Bill Gates, "Corbis Putting Online Images in Focus," from Microsoft web site at www.microsoft.com/billgates, last updated May 13, 1998, accessed August 8, 1998.

meaningful information. Often data problems can be traced back to a transaction processing error that occurred in the process of updating an operational database. Warehouse implementers should expect to spend four to five times more time and effort on data cleaning activities than they would at first estimate.[20]

Information Systems Principles

Without data and the ability to process it, an organization would not be able to successfully complete most business activities. A database can help companies organize increased amounts of data to maximize this valuable resource.

Often, distinct, yet related, databases must be linked to provide enterprisewide databases to meet the needs of the modern organization.

Determining what data is to be collected in the database and who will have access to this data are two important considerations in organizing data in a database.

Distributed databases allow more users to gain direct access to data stored in different user sites. Thus, controlling who accesses and changes data is sometimes difficult.

Organizing data to support on-line transaction processing is significantly different than organizing it to support effective management decision making. As a result, special forms of databases and associated tools have been developed for each need.

▪ SUMMARY

1. *Define general data management concepts and terms, highlighting the advantages and disadvantages of the database approach to data management.*

Data is one of the most valuable resources a firm possesses. Data is organized into a hierarchy that builds from the smallest element to the largest. The smallest element is the bit, a binary digit. A byte (a character such as a letter or numeric digit) is made up of eight bits. A group of characters, such as a name or number, is called a field (an object). A collection of related fields is a record; a collection of related records is called a file. The database, at the top of the hierarchy, is an integrated collection of records and files.

An entity is a generalized class of objects for which data is collected, stored, and maintained. An attribute is a characteristic of an entity. Specific values of attributes—called data items—can be found in the fields of the record describing an entity. A data key is a field within a record that is used to identify the record. A primary key uniquely identifies a record, while a secondary key is a field in a record that does not uniquely identify the record.

The traditional approach to data management has been from a file perspective. Separate files are created for each application. This approach can create problems over time: as more files are created for new applications, data that is common to the individual files becomes redundant. Also, if data is changed in one file, those changes might not be made to other files, reducing data integrity.

Traditional file oriented applications are often characterized by program-data dependence, meaning that they have data organized in a manner that cannot be read by other programs.

To address problems of traditional file-based data management, the database approach was developed. Benefits of this approach include reduced data redundancy, improved data consistency and integrity, easier modification and updating, data and program independence, standardization of data access, and more efficient program development.

Potential disadvantages of the database approach include the relatively high cost of purchasing and operating a DBMS in a mainframe operating environment; specialized staff required to implement and coordinate the use of the database; and increased vulnerability if security is breached and there is a failure in the DBMS.

2. *Name three database models and outline their basic features, advantages, and disadvantages.*

When building a database, careful consideration must be given to content and access, logical structure, and physical organization. One of the tools database designers use to show the relationships among data is a data model. A data model is a map or diagram of entities and their relationships. Enterprise data modeling involves analyzing the data and information needs of the entire organization. Entity-relationship (ER) diagrams can be employed to show the relationships between entities in the organization. Entities may have a one-to-one (1:1), one-to-many (1:N), or many-to-one relationships (N:1).

Databases typically use one of three common models: hierarchical (tree), network, and relational. The hierarchical model has one main record type at the top, called the parent, with subordinate records, called children, below. Each parent may have several children, but each child may have only one parent. The network model, an expansion of the hierarchical structure, involves an owner-member relationship in which each member may have more than one owner. The newest, most flexible structure is the relational model. Instead of a hierarchy of predefined relationships, data is set up in two-dimensional tables. Tables can be linked by common data elements, which are used to access data when the database is queried. Each row represents a record. Columns of the tables are called attributes, and allowable values for these attributes are called the domain. Basic data manipulations include selecting, projecting, and joining.

The hierarchical model has the advantage of processing efficiency because the data relationships are less complex. However, it is not flexible in terms of how the data is organized, it is difficult to change and it can be difficult to install. The network model is more flexible than the hierarchical model in terms of organizing data, but it is difficult to develop and use because of the complexity of the data relationships and it is not popular with personal computer users. The relational model, the most widely used database model, is easier to control, more flexible, and more intuitive than the other models because it organizes data in tables.

3. *Identify the common functions performed by all database management systems and discuss the key features of three popular end user database management systems.*

To make certain types of changes to a database requires a database management system (DBMS). A DBMS is a group of programs used as an interface between a database and application programs. When an application program requests data from the database, it follows a logical access path (LAP). The actual retrieval of the data follows a physical access path (PAP). Records can be considered in the same way: a logical record is what the record contains; a physical record is where the record is stored on storage devices. Schemas are used to describe the entire database, its record types, and their relationships to the DBMS.

A database management system provides four basic functions: creating and modifying the database, providing user views, storing and retrieving data, and manipulating data and generating reports.

Subschemas are used to define a user view, the portion of the database a user can access and/or manipulate. Schemas and subschemas are entered into the computer via a data definition language (DDL), which describes the data and relationships in a specific database. Another tool used in database management is the data dictionary, which contains detailed descriptions of all data in the database.

Once a DBMS has been installed, the database may be accessed, modified, and queried via a data manipulation language (DML). A more specialized DML is the query language, the most common being Structured Query Language (SQL). SQL is used in several popular database packages today and can be installed on PCs and mainframes.

Popular end user DBMSs include Microsoft Access, Lotus Approach, and Inprise's dBASE 7. Lotus Approach and Microsoft Access provide Wizards to help end users create a new database and load data and to perform many other functions. dBASE 7 provides fewer Wizards and has less capability to publish a database that is accessible via the Web.

4. *Identify and briefly discuss recent database developments.*

With the increased use of telecommunications and networks, distributed databases, which allow multiple users and different sites access to data that may be stored in different physical locations, are gaining in popularity. To reduce the demand on telecommunications lines, some organizations will build a replicated database, which holds a duplicate set of frequently used data.

Organizations are also building data warehouses, which are relational database management systems specifically designed to support management decision making.

Multidimensional databases and on-line analytical processing (OLAP) programs are being used to store data and allow users to explore the data from a number of different perspectives.

Data mining, which is the automated discovery of patterns and relationships in a data warehouse, is emerging as a practical approach to generate a hypothesis about the patterns and anomalies in the data that can be used to predict future behavior.

An object-relational database management system (ORDBMS) provides a complete set of relational database capabilities plus the ability for third parties to add new data types and operations to the database. These new data types can be audio, images, unstructured text, spatial, or time series data that require new indexing, optimization, and retrieval features.

▪ KEY TERMS

attribute 191
character 190
concurrency control 211
data cleanup 204
data definition language (DDL) 207
data dictionary 208
data integrity 194
data item 192
data manipulation language (DML) 211
data mart 221
data mining 222
data model 200
data redundancy 194
data warehouse 219
database 191
database approach 195

database management system (DBMS) 190
distributed database 218
domain 203
enterprise data modeling 200
entity 191
entity-relationship (ER) diagrams 200
field 191
file 191
hierarchical database model 201
hierarchy of data 191
joining 204
key 192
linked 204
network model 202
object-relational database management system (ORDBMS) 226

on-line analytical processing (OLAP) 223
open database connectivity (ODBC) 225
planned data redundancy 200
primary key 192
program-data dependence 194
projecting 204
record 191
relational model 203
replicated database 219
schema 206
selecting 204
Structured Query Language (SQL) 211
subschema 207
traditional approach 194

▪ REVIEW QUESTIONS

1. Describe the hierarchy of data.
2. What are entities and attributes? What is a primary key?
3. What are the disadvantages of data redundancy?
4. What are the advantages of the database approach to data management, as opposed to the traditional file-based approach?
5. Describe the following three types of database models: hierarchical model, network model, and relational model.
6. What is a database management system?
7. Identify important characteristics in selecting a database management system.

8. Identify three popular database management systems for end users and summarize their strengths and weaknesses.
9. What is the purpose of a data definition language (DDL)? A data dictionary?
10. What is a distributed database system?
11. What advantages does the open database connectivity (ODBC) standard offer?
12. What is a data warehouse, and how is it different from a traditional database used to support OLTP?
13. What is OLAP? Does OLAP imply the use of a relational database?
14. What is data mining? How is it different from OLAP?
15. What is an ORDBMS? What kind of data can it handle?

■ DISCUSSION QUESTIONS

1. Why is a database a necessary component of the CBIS? Why is the selection of DBMS software so important to business organizations?
2. What database management systems have you used or are familiar with? Which ones would you classify as end user DBMSs? Which would you not? What is the basis for your distinction?
3. What is a data model and what is data modeling? Why is data modeling an important part of strategic planning?
4. You are going to design a database for your wine tasting club to keep track of its inventory of wines. Identify the database characteristics most important to you in choosing a DBMS to implement this system. Which of the database management systems described in this chapter would you choose? Why?
5. Identify at least three recent situations in which you needed to make a decision and were overwhelmed with data but lacking the real information you needed. How did you deal with this situation?
6. Make a list of the databases in which data about you exists. How is the data in each database captured? Who updates each database and how often? Is it possible for you to request a printout of the contents of your data record from each database?
7. Develop a list of conditions under which you would use the OLAP approach to analyze data in a data warehouse. Develop a list of conditions that favor the data mining approach.
8. You are the vice president of information technology for a large bank. You are to make a presentation to the board of directors recommending the investment of $5 million on the development of a data warehouse and associated tools for data analysis. What are your key points in favor of this investment? What counterpoints can you anticipate others might argue?

■ PROBLEM-SOLVING EXERCISES

1. You are a senior systems analyst in the information systems department of a large Fortune 500 company. Your organization is currently using a relational database management system (RDBMS) in business applications. Given the trends in database management systems, use your word processing program to write a letter to your manager(s) outlining the need to migrate toward object-relational database management systems (ORDBMSs). Using a word processing program, record the advantages of ORDBMSs over RDBMSs and discuss the disadvantages in migrating to the former.

2. You are a member of a research team developing an improved process for evaluating college applicants. The goal of the research is to predict which students will be most successful in their college career. Those that score well on the profile will be accepted, those that score exceptionally well will be considered

for scholarships. What data do you need on each college applicant? What sort of data might you need that is not typically requested on the college application form? How might you get this data?

Take a first cut at designing a database for this application. Using the chapter material on designing a database, show the logical structure of the relational tables for this proposed database. In your design, include those data attributes that you believe are necessary for this database, and show the primary keys in your tables. Keep the size of the fields and tables as small as possible to minimize required disk drive storage space. Fill in the database tables with the sample data for demonstration purposes (ten records). Once your design is complete, implement your design for a university database using a relational DBMS.

3. A video movie rental store is using a relational database to store the following information on movie rentals to answer customer questions. Movie types are comedy, drama, horror, science fiction, and western. MPPA ratings are G, PG, PG-13, R, X, and NR (not rated).

To improve service to their customers, the salespeople at the video rental store have proposed a list of changes being considered for this database. From this list, choose the database modifications you think would be necessary for each of the proposed changes. Explain your choices.

Movie ID No.	Movie Title	Year Made	Movie Type
MPPA Rating	Number of Copies on Hand	Quantity	

(primary key) (secondary key)

Proposed changes:
a. Adding the date that the movie was first available to help locate the newest releases
b. Adding the director's name
c. Adding the names of three actors in the movie
d. Adding a critic rating of one, two, three, or four stars
e. Adding the number of Academy Award nominations

Database modifications:
a. Adding an additional field
b. Adding an additional key to a field
c. Adding a new table
d. Building a new query

4. Based on the database design and the salespeople's proposed changes from the previous exercise, design a data-entry screen that could be used to enter information into this database. Also include some examples of typical queries the salespeople would use to respond to customers' requests.

■ TEAM ACTIVITY

In a group of three or four classmates, interview business managers from three different businesses that use databases to help them in their work. What data entities and data attributes are contained in each database? How do they access the database to perform analysis? Have they received training in any query or reporting tools? What do they like about their database and what could be improved? Do any of them use data mining or OLAP techniques? Weighing the information obtained, select one of these databases as being most strategic for the firm, and briefly present your selection and the rationale for the selection to the class.

■ WEB EXERCISE

Use a Web search engine to find information on one of the following topics: data warehouse, data mining, or OLAP. Find a definition of the term, an example of a company using the technology, and three companies that provide such software. Cut from the Web pages and paste into a word processing document to create a two-page report on your selected topic. At the home page of each software company, request further information from the company about its products.

■ CASES

1 SAAB Cars USA

Since GM paid $600 million for half of Saab in 1990 and invested another $700 million, the company has lost a cumulative $1.7 billion. GM expects Saab to be in the black by the 1999 model year. That means selling 150,000 vehicles annually worldwide. Much of the onus falls on the new 9-5 sedan, the successor to the 9000, and a wagon targeted for the 1999 model year. Those two models should account for 40 to 45 percent of the 150,000 vehicles sold by 2000, says Saab CEO Bob Hendry. Between 10,000 and 15,000 will be sold in the United States each year, he predicts.

The Saab 9-5 sedan is priced at $32,000 to $40,000 and is expected to help the struggling company compete with other luxury sedans, such as the Volvo S70, Audi A6, Mercedes E-Class, Lexus and BMW 5-series. The model debuted in the United States in April 1998 as a 1999 model. Saab executives say the new model was built with more concern for quality. It features a 2.3-liter, turbocharged engine.

Saab Cars USA rolled out an intranet to its dealers just in time to support the 1999 models. The intranet application allows Saab's 250 U.S. dealers to track every car they sell from the assembly line to the junkyard. The goal of the system, dubbed IRIS (Intranet Retail Information System), is to improve customer satisfaction by providing everyone in a car dealership with information about customers and their Saab automobiles. The business payoff comes from being able to answer the customer's questions on the spot and having access to timely, accurate information. Dealers use IRIS to view all the information about a car, including ownership, service history, and warranty.

Prior to IRIS, records about service, ownership, warranties, and parts were distributed across three separate systems: an AS/400 mid-range computer at Saab's U.S. headquarters in Norcross, Georgia, an IBM System 390 mainframe that held historical data including data about parts distributors, and dealer-management systems at each dealership. It was not possible to pull all this data together and integrate it in a way that was useful. Now all data is centralized in an IBM DB/400 relational database that resides on the AS/400. The database contains information from Saab's information systems and from the parts distributors' mainframe. Work is in progress, with the vendors that supply its dealers with dealer management software, to develop a system to forward warranty information from the dealers' servers to the IBM DB/400 database. When this is complete, IRIS will represent the repository for all customer information.

The effort required a high degree of collaboration among representatives of Saab Cars USA, several Saab dealerships, and technical resources from IBM Global Services, EDS, and CST Inc. IBM Global Services was the primary vendor on the project. EDS maintains Saab's information systems and worked with IBM to modify Saab's existing applications for the intranet. CST Inc. provided special-purpose software to pull data from the IBM DB/400 database and deliver it to the dealer via the intranet.

1. Would you assess IRIS as being strategic to the success of Saab Cars USA? Why or why not?
2. What issues may keep Saab from using IRIS worldwide?
3. What sort of questions would IRIS be used to answer? Identify at least eight data entities for which data should be collected to support this application. Specify at least three data attributes for each data entity.
4. What sort of issues might arise in trying to integrate data from a number of disparate dealer management systems?

Sources: Adapted from Justin Hibbard, "Saab's Driving Force," *InformationWeek*, January 26, 1998, p. 91 and Marjorie Sorge, "Yah shur, it's a Saab," *Automotive Industries*, vol. 177, no. 8, August 1997, p. 59.

2   US West

US West (USW) is a Denver-based provider of telecommunications services to 25 million customers in 14 Western and Midwestern states with 14.5 million access lines in service. It employs just over 51,000 people and had a 1996 net income of $1.1 billion on revenues of $10 billion. Back in 1994, USW came to a troubling conclusion—the company could not keep pace with the rapid changes in the industry. Employees were shackled by obsolete information systems and technology. As USW's IS organization began to analyze the situation, it realized that employees needed more efficient ways of tracking product revenues and expenses. They also needed better ways of determining which products customers were buying and why.

By late 1994, USW committed to create a data warehouse in which computer hardware, operating system, and database management system were specifically engineered to work as one system to provide data for decision support systems and processes. In building its data warehouse, the USW IS organization knew it must work closely with end users in the business units to consider how data should be organized and stored so that employees could obtain quick and accurate responses to their questions. To that end, consultants were brought in to conduct classes on information modeling for end users. These classes prepared end users in the business units not only to contribute to the design of the warehouse but to be more intelligent users once they began to use it. In these classes, users discussed how the data must be organized so that they could go through the higher, summary levels to the detailed data to answer questions. By the end of 1996, USW had implemented data marts where employees could take subsets of data from the warehouse and copy them to servers and desktop computers located in various business units and departments throughout USW.

USW's data warehouse and marts have had a major impact on marketing campaigns, particularly those targeted at large, multistate customers. Like other phone companies, USW is enjoying strong demand for second and third residential phone lines—customers want them for their home offices, teenagers, fax machines, and personal computers. But the carrier is reluctant to invest money into new network switches and trunk lines in a particular area unless it can be assured that the orders for extra lines will really materialize. Furthermore, USW wants to identify existing customers who will not only respond to introductory offers but will also keep their second or third lines long enough for USW to make a profit.

To achieve this business goal, USW uses its data warehouse to combine and reassemble data from its billing operations, the company's line provision units, and even outside databases from R.R. Donnelley and other sources. Gloria Farley, executive director of USW's market intelligence program, worked closely with USW's senior director of the Shared Capabilities Department, Brenda Moncla, to design a customer-centric warehouse containing a terabyte of data. "We needed to move from a corporate data warehouse that was merely storage to an actual customer-centered database," remembers Farley.

A program called Phoenix Additional Line Modeling System (PALMS) is used to analyze the customers' "household" data in the warehouse and enable US West to derive customer trends and needs based on descriptive details of household characteristics. PALMS was developed in conjunction with AT&T's NCR computer unit and SABRE Decision Technologies, a unit of AMR, which owns American Airlines. Running on a powerful NCR parallel-processing computer, PALMS first spends hours sifting through a sample of a few thousand customer records from the Phoenix area. Each record contains as many as 250 items about a household: family size, age of family members, income bracket, monthly phone bill, number of repair calls in the past year, and its history of trying and keeping such services as call-waiting, for instance. The result is a statistical model of the ideal prospect.

Next the PALMS computer program uses that model to search through millions more customer records taking up nearly 1 trillion bytes of data in its household data warehouse. By correlating data about the location of each home, the location of USW's trunk lines, and the capacity of local switches, the program identifies clusters of prospects—households that fit the model of the ideal prospect and that US West could provide service to without significant additional expense.

The use of the PALMS computer program helped USW discover that customers want more than lower phone rates. By offering nonprice or noncost related things, they were able to avoid a loss of 45 percent of customers who would otherwise have abandoned them. That fundamental finding helped USW rethink its business strategy. Says Farley, "When I came to US West, it was very clear that the service quality measurement was the kind of thing you'd expect from a company that had no competition. Basically it was asking customers, 'Are you satisfied with what you've got?'"

The design of the data warehouse allows users to access the data much more easily. If a user wants to match details associated with products and services with revenues, she might start with a particular state, then analyze a neighborhood wire center that serves several customers, and view data about an office or residence of a particular customer. This way, she can look at all the products or any one product used at a particular location and see all the associated costs and revenue to determine whether USW is making or losing money on specific products, services, or a specific customer. It used to take anywhere from a day to a month to perform this analysis; now it can be accomplished in an afternoon.

Farley believes that the long-term value of data mining for USW is that it enables people to think of data as the key to profitability that unlocks a series of relationships between the customer and US West products and services. The trick is no longer to focus on making products acceptable to customers, but to focus on customers, and what kinds of needs they have. This allows USW to identify its customers' current and future value in the new competitive marketplace. USW can treat each customer as an individual. The data warehouse will help USW answer questions about its customers and its industry that it cannot answer today. What trends will emerge to change the business environment of customers? How will these trends change the products and services offered and the way USW must position itself for the future? This is where the data warehouse will undoubtedly have its greatest impact.

1. Identify 20 specific pieces of data that the US West data warehouse would contain for each household.
2. What kind of data would US West obtain from R.R. Donnelley and other outside database sources?
3. What do you think Gloria Farley means when she says that the trick is no longer to focus on making products acceptable to customers but to focus on customers and what kinds of needs they have?
4. Imagine that an additional $1 million is needed to upgrade the data warehouse system. How would you justify this expenditure to senior management?

Sources: Adapted from Vance McCarthy, "Strike It Rich," *Datamation*, February 1997, pp. 44–50; John W. Verity, "Coaxing Meaning Out of Raw Data," *Business Week*, February 3, 1997, pp. 134–138; "Ameritech, BellSouth, Pacific Bell and US West Agree to Develop Single Yellow Pages and Information Search Software," *Communications Daily*, vol. 17, no. 124, June 27, 1997, p. 12; "US West, GTE and Bell Canada Sign Interoperability Agreement to Adopt Single Standard for Using Smart Cards," *Communications Daily*, vol. 17, no. 171, September 4, 1997, p. 6; and "Time Warner and US West Looking to Longer Term Partnership," *Communications Daily*, vol. 17, no. 138, July 18, 1997, p 6.

3   MCI Communications Corporation

Data mining can deliver huge business payoffs, as MCI Communications Corp. is learning. MCI is using data mining to analyze customer loss and predict those customers who might be considering jumping to a rival. If MCI can do that, the carrier can try to keep the customer with offers of special rates and services.

MCI has developed a data warehouse of 140 million households that includes many attributes such as income, age of family members, lifestyle, and details about past calling habits. But, which set of those attributes are the best indicators of a customer's loyalty and what are acceptable ranges for each variable? A rapidly declining monthly bill would seem to be a dead giveaway, but is there a subtler pattern in international calling to be looking for, too? Or in the number of calls made to MCI's customer service lines?

To find out, MCI performs data mining using its IBM SP/2 supercomputer, which houses its data warehouse, to identify the most telling variables to keep an eye on. So far, data mining has yielded a set of 22 detailed, and highly secret, statistical profiles based on repeated crunching of historical facts. None of these could have been developed without data mining programs.

1. Make a list of data entities and associated data attributes that are likely to be present in the MCI data warehouse. Draw a simple entity relationship diagram that represents a model of the data warehouse.
2. Do you think it is ethical for MCI to offer special programs and rates to customers it thinks are likely to move to a rival? Why or why not?

Source: John W. Verity, "Coaxing Meaning Out of Raw Data," *Business Week,* February 3, 1997, pp. 134–138; MCI 1997 Annual Report; and "MCI Measures its Marketing," DBMS Programming and Design 1997 Real Ware Awards, publisher David M. Kalman, found at http://www.dbpd.com, accessed August 8, 1998.

4   Sears

Sears Roebuck & Co. is joining the short but rapidly growing list of corporations with a terabyte-size customer data warehouse. The $37 billion retailer has implemented a multiterabyte data warehouse with advanced analytical capabilities to identify and capitalize on richly rewarding opportunities in its credit card business. Relationship marketing, risk analysis, and new product development are some of the applications that exploit the extensive customer and transaction information uniquely available to this industry leader. Joseph Smialowski, senior vice president and CIO with Sears, believes that use of data warehouse technology will provide Sears with a further competitive advantage in the industry.

The data warehouse also supports a Sears strategy of attracting new customers through credit account development. Sears's credit card business is expected to grow at over 18 percent for several years. However, write-offs of customer defaults have remained somewhat fixed and could rise if personal bankruptcies increase.

The multiterabyte data warehouse for Sears Credit holds information on 90 million households, 31 million Sears card users, transaction records, credit status, and related data. Nearly 60 percent of purchases within Sears stores are made with the company's credit card, a percentage that is growing. The warehouse will be used for sales analysis and target marketing. Use of the data warehouse to support relationship marketing will enable Sears to sell more goods by tailoring offers and marketing programs that are closely aligned with the needs of individual households.

The database will serve as the company's central repository of customer information. Starting at two terabytes, it is expected to grow several times beyond that over the next few years.

1. Why do you think Joseph Smialowski believes that the data warehouse will provide Sears with a competitive advantage within its industry?
2. If the database holds 2 terabytes of data and has information on about 31 million Sears card holders, how much data, on average, is there for each cardholder? Can you identify key pieces of data likely to be carried about each cardholder?

Sources: "Sears Roebuck to Open 110 Stores in NY State," *New York Times*, February 14, 1997; "Sears Plans: Sees '97 Earnings Growth In Mid-Teens," *Dow Jones/News Retrieval*, Copyright 1997 Dow Jones & Company, February 12, 1997; and "Sears Sales: Total January Revenues Up 11.8%," *Dow Jones/News Retrieval*, February 6, 1997.

CHAPTER 6

Telecommunications and Networks

"It's becoming increasingly more important to understand the fundamentals of LANs (Local Area Networks) and WANs (Wide Area Networks). Users are no longer computing in the "stand alone" mode—they're working as members of teams and groups sharing information in a dynamic environment that's interconnected across buildings, cities, continents, and the globe."

—CompuMaster sales literature

Chapter Outline

An Overview of Communications Systems
Communications
Telecommunications
Networks

Telecommunications
Types of Media
Devices
Carriers and Services

Networks and Distributed Processing
Basic Data Processing Strategies
Network Concepts and Considerations
Network Types
Terminal-to-Host, File Server, and Client/Server Systems
Communications Software and Protocols
Bridges, Routers, Gateways, and Switches

Telecommunications Applications
Linking Personal Computers to Mainframes and Networks
Voice and Electronic Mail
Electronic Software and Document Distribution
Telecommuting
Videoconferencing
Electronic Data Interchange
Public Network Services
Specialized and Regional Information Services
Distance Learning

Learning Objectives

After completing Chapter 6, you will be able to:

1. Define the terms *communications* and *telecommunications* and describe the components of a telecommunications system.
2. Identify several types of communications media and discuss the basic characteristics of each.
3. Identify several types of telecommunications hardware devices and discuss the role that each plays.
4. Identify the benefits associated with a telecommunications network.
5. Name three distributed processing alternatives and discuss their basic features.
6. Define the term *network topology* and identify five alternatives.
7. Identify and briefly discuss several telecommunications applications.

Citibank
Upgrading Global Networks to Meet Customer Needs

Citibank, a global provider of financial services, serves consumer, business, governmental, and institutional customers. Founded in 1812 in New York, the bank was a pioneer in establishing a worldwide network of offices to meet the growing needs of corporate customers in the early part of the twentieth century. Today it serves both major international corporations and local growth companies around the world. Citibank is a leader in applying advanced technology to bring individuals in 56 countries credit, payment and investment services through branches, electronic access, and charge and credit cards. All around the world, customers count on Citibank for global expertise and local insight, for quality and consistency, and for innovation and responsiveness. More than 90,000 employees are located in offices in 98 countries and territories. Earnings for 1997 exceeded $3.5 billion.

Citibank's global growth strategy depends on state-of-the-art technology. Its technology strategy is to integrate all systems into a standard global platform and to introduce new technology-based products and delivery systems in upcoming years. Citibank is standardizing its systems and processes, creating compatible infrastructures for everything from telecommunications to opening checking accounts. While Citibank's systems compete with the best available in any local market, it needs global integration of its systems and processes to serve its customers seamlessly across boundaries with consistent high quality. Once standardization is accomplished, Citibank can broaden the range of products it offers through its network and can speed distribution of "best practices" throughout the organization. Resulting efficiencies will enable Citibank to be the low-cost producer in many of its service areas. Response time to most customer inquiries will be within seconds.

To help implement this strategy, Citibank contracted with AT&T Solutions to manage a significant portion of its worldwide data network. This five-year agreement is valued at $750 million and is estimated to save Citibank approximately $250 million in operating costs. The savings come from transforming Citibank's multiple data networks into a single new, state-of-the-art networking platform. AT&T, under Citibank's direction, will merge multiple data networks to a common Global Data Network platform to be managed from AT&T Solutions' Global Client Support Center located in Durham, North Carolina. Once the network is implemented, AT&T Solutions will also manage Citibank's data networking requirements around the world.

Although the savings are substantial, the main goal is improved network performance, security, and reliability so that the bank can deliver new products to businesses and customers. The upgrade will make fund transfers to its branches worldwide more reliable and secure. Increased network capacity will let the bank open new commerce and home marketing services in nations where it has none. Much of Citibank's business in 2010 will come from customers it does not have today, and it will be in products and services that do not yet exist, delivered in totally new ways.

As you read this chapter, consider the following questions:

- What is a telecommunications network and what are its components?

- What benefits can telecommunications networks deliver to business organizations?

Sources: Adapted from Matt Hamblen, "Citibank Invests Big in Reliability," *Computerworld*, March 16, 1998, p. 4; AT&T Web site at http://www.att.com, accessed April 15, 1998; and Citibank Web site at http://www.citibank.com, accessed April 15, 1998.

No matter what the business, effective communication is critical to organizational success. Often, what separates good management from poor management is the ability to identify problems and solve them with available resources. Efficient communications is one of the most valuable of these resources, because it enables a company to keep in touch with its operating divisions, customers, suppliers, and stockholders. For example, Ford Motor Company is developing an integrated telecommunications system to reduce costs related to transporting auto parts and components to Ford's 20 North American plants and to streamline car production schedules. The system will integrate the plant's individual parts-ordering systems and connect them to suppliers for real-time information about component and part inventories, as well as real-time tracking of deliveries.[1] Memos, notices on bulletin boards, and presentations are all obvious examples of continual communication within a business organization. Other not so obvious examples include policy and procedure manuals and even salaries (they communicate the company's perception of the value of the contribution of the person being paid). Communication also exists in other forms—warning lights from a computer system that monitors manufacturing processes and signals from a building management system that monitors temperature, humidity, lighting, and security of a building are examples.

Communication is any process that permits information to pass from a sender to one or more receivers. Communications of all types form a major part of any business system. Therefore, managers must gain an appreciation of communication concepts, media, and devices—as well as an understanding of how these factors may best be employed to develop effective and efficient business systems.

AN OVERVIEW OF COMMUNICATIONS SYSTEMS

Communications

Communications is the transmission of a signal by way of a medium from a sender to a receiver (Figure 6.1). The signal contains a message composed of data and information. The signal goes through some communications medium, which is anything that carries a signal between a sender and receiver.

The components of communication can easily be recognized if you consider human communication (Figure 6.2). When we talk to one another face to face, we send messages to each other. A person may be the sender at one moment and the receiver a few seconds later. The same entity, a person in this case, can be a sender, a receiver, or both. This is typical of two-way communication. The signals we use to convey these messages are our spoken words—our language. For communication to be effective, both sender and receiver must understand the signals and agree on the way they are to be interpreted. For example, if the sender in Figure 6.2 is speaking in a language the receiver does not understand, or if the sender believes a particular word has one meaning and the receiver believes the word has some other meaning, effective communication will not occur.

FIGURE 6.1

Overview of Communications

The message (data and information) is communicated via the signal. The transmission medium "carries" the signal.

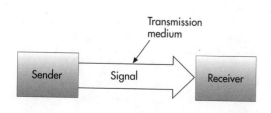

FIGURE 6.2

Communications and Telecommunications

In human speech, the sender transmits a signal through the transmission medium of the air. In telecommunication, the sender transmits a signal through the transmission medium of a cable.

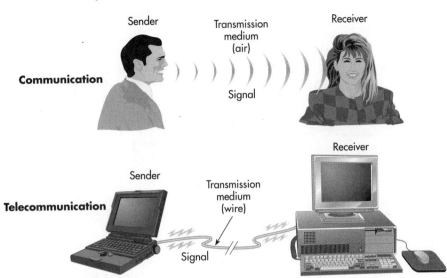

Telecommunications

telecommunications

the electronic transmission of signals for communications, including such means as telephone, radio, and television

data communications

a specialized subset of telecommunications that refers to the electronic collection, processing, and distribution of data—typically between computer system hardware devices

telecommunications medium

anything that carries an electronic signal and interfaces between a sending device and a receiving device

Telecommunications refers to the electronic transmission of signals for communications, including such means as telephone, radio, and television. Telecommunications has the potential to create profound changes in business because it lessens the barriers of time and distance. Telecommunications may change not only the way businesses operate, as with Citibank, but also may alter the nature of commerce itself. As networks are connected with one another and information is transmitted more freely, a competitive marketplace will make excellent quality and service imperative for success. **Data communications**, a specialized subset of telecommunications, refers to the electronic collection, processing, and distribution of data—typically between computer system hardware devices. Data communications is accomplished through the use of telecommunications technology.

Figure 6.3 shows a general model of telecommunications. The model starts with a sending unit (1), such as a person, a computer system, a terminal, or another device, that originates the message. The sending unit transmits a signal (2) to a telecommunications device (3). The telecommunications device performs a number of functions, which can include converting the signal into a different form or from one type to another. A telecommunications device is a hardware component that allows electronic communication to occur or to occur more efficiently. The telecommunications device then sends the signal through a medium (4). A **telecommunications medium** is anything that carries an electronic signal and interfaces between a sending device and a receiving device. The signal is received by another telecommunications device (5) that is connected to the receiving computer (6). The process can then be reversed and another message can go back from the receiving unit (6) to the original sending unit (1). In this chapter, we will explore the components of the telecommunications model shown in Figure 6.3. An important characteristic of telecommunications is the speed at which information is transmitted measured in bits per second or bps. Common speeds are in the range of thousands of bits per second (Kbps) to millions of bits per second (Mbps).

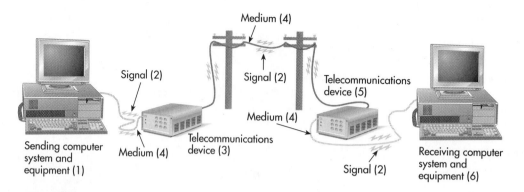

FIGURE 6.3

Elements of a
Telecommunications System

Telecommunications devices
relay signals between
computer systems and
transmission media.

computer network

the communications media,
devices, and software needed to
connect two or more computer
systems and/or devices

Telecommunications technology
enables businesspeople to communicate with co-workers and
clients from remote locations.
(Source: © 1998 PhotoDisc.)

Advances in telecommunications technology allow us to communicate rapidly with clients and co-workers almost anywhere in the world. Telecommunications also reduces the amount of time needed to transmit information that can drive and conclude business actions. A manufacturing sales representative, for example, can use telecommunications technology to get new product prices from the central sales office while working at a customer's location. This empowers the sales representative and often results in faster, higher-quality customer service. Telecommunications technology also helps businesses coordinate activities and integrate various departments to increase operational efficiency and support effective decision making. The far-reaching developments of telecommunications will have a profound effect on business information systems and on society in general.

Networks

A **computer network** consists of communications media, devices, and software needed to connect two or more computer systems and/or devices. Once connected, computers can share data, information, and processing jobs. More and more businesses are linking computers in networks to streamline work processes and allow employees to collaborate on projects.

The effective use of networks can turn a company into an agile, powerful, and creative organization, giving it a long-term competitive advantage. Networks can be used to share hardware, programs, and databases across the organization. They can transmit and receive information to improve organizational effectiveness and efficiency. They enable geographically separated workgroups to share documents and opinions, which fosters teamwork, innovative ideas, and new business strategies.

TELECOMMUNICATIONS

The use of telecommunications can help businesses solve problems and maximize opportunities. Using telecommunications effectively requires careful analysis of telecommunications media, devices, carriers, and services.

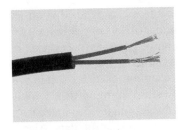

FIGURE 6.4

Twisted-Pair Wire Cable
(Source: Fred Bodin.)

Types of Media

Various types of communications media are available. Each type exhibits its own characteristics, including transmission capacity and speed. In developing a telecommunications system, the selection of media depends on the purpose of the overall information and organizational systems, the purpose of the telecommunications subsystems, and the characteristics of the media. As with other system components, the media should be chosen to support the goals of the information and organizational systems at the least cost and to allow for possible modification of system goals over time. The proper media will help a company link subsystems to maximize effectiveness and efficiency.

Twisted-pair wire cable. Twisted-pair wire cable is, as you might expect, a cable consisting of pairs of twisted wires (Figure 6.4). A typical cable contains two or more twisted pairs of wire, usually copper. Proper twisting of the wire keeps the signal from "bleeding" into the next pair and creating electrical interference. Because the twisted-pair wires are insulated, they can be placed close together and packaged in one group. Hundreds of wire pairs can be grouped into one large wire cable.

There are two kinds of twisted-pair wire cable: shielded and unshielded. Shielded twisted-pair wire cable has a special conducting layer within the normal insulation. This conducting layer makes the cable less prone to electrical interference, or "noise." Unshielded twisted-pair (UTP) wire cable lacks this special insulation shield. UTP cables have historically been used for telephone service and to connect computer systems and devices. Newer types of cable, however, have begun to replace UTP cable in both businesses and homes.

Coaxial cable. Figure 6.5 shows a typical coaxial cable, similar to that used in cable television installations. A coaxial cable consists of an inner conductor wire surrounded by insulation, called the dielectric. The dielectric is surrounded by a conductive shield (usually a layer of foil or metal braiding), which is in turn covered by a layer of nonconductive insulation, called the jacket. When used for data transmission, coaxial cable falls in the middle of the cabling spectrum in terms of cost and performance. It is more expensive than twisted-pair wire cable but less so than fiber-optic cable (discussed next). Coaxial cable offers cleaner and crisper data transmission (less noise) than twisted-pair wire cable. It also offers a higher data transmission rate.

FIGURE 6.5

Coaxial Cable
(Source: Fred Bodin.)

Fiber-optic cable. Fiber-optic cable, consisting of many extremely thin strands of glass or plastic bound together in a sheathing (a jacket), transmits signals with light beams (Figure 6.6). These high-intensity light beams are generated by lasers and are conducted along the transparent fibers. These fibers have a thin coating, called cladding, which effectively works like a mirror, preventing the light from leaking out of the fiber. Because it transmits via light rather than electricity, fiber-optic cable has some extraordinary abilities compared with other forms of cabling. Fiber-optic cable is capable of supporting tremendous data transfer rates—upward of 2.5 billion bits per second (bps), or 32,000 long-distance phone calls simultaneously. The

FIGURE 6.6

Fiber-Optic Cable
(Source: Greg Pease/Tony
Stone Images.)

much smaller diameter of fiber-optic cable makes it ideal in situations where there is not room for additional bulky copper wires—for example, in crowded conduits, which can be pipes or spaces used to house and carry the electrical and communications wires. In this case, the use of smaller fiber-optic telecommunications systems is very effective. Because fiber-optic cables are immune to electrical interference, signals can be transmitted over longer distances with fewer expensive repeaters to amplify or rebroadcast the data. A special plus for security-conscious applications is the difficulty of stealing information. With the right equipment installed, it is virtually impossible to tap into fiber-optic cable without being detected. Fiber-optic cable and devices are more expensive to purchase and install than their twisted-pair wire and coaxial counterparts, although the cost is coming down.

Microwave transmission. Microwave transmissions are sent through the atmosphere and space. Although these transmission media do not entail the expense of laying cable, the transmission devices needed to utilize this medium are quite expensive. Microwave is a high-frequency radio signal that is sent through the air (Figure 6.7). Microwave transmission is line-of-sight, which means that the straight line between the transmitter and receiver must be unobstructed. Typically, microwave stations are placed in a series—one station will receive a signal, amplify it, and retransmit it to the next microwave transmission tower. Such stations can be 30 to 70 miles apart (depending on the height of the towers) before the curvature of the earth makes it impossible for the towers to "see one another." Microwave signals can carry thousands of channels at the same time.

A communications satellite is basically a microwave station placed in outer space (Figure 6.8). The satellite receives the signal from the earth, amplifies the relatively weak signal, and then rebroadcasts it at a different frequency. The advantage of satellite communications is the ability to receive and broadcast over large geographic regions. Such problems as the

FIGURE 6.7

Microwave Communications

Because they are line-of-sight transmission devices, microwave dishes must be placed in relatively high locations such as atop mountains, towers, and tall buildings.

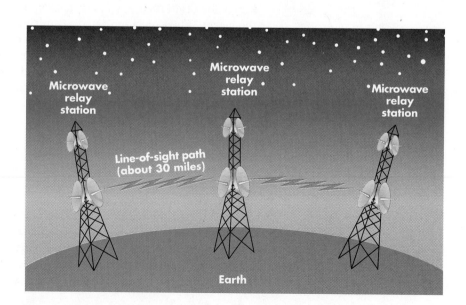

FIGURE 6.8

Satellite Transmission

Communications satellites are relay stations that receive signals from one earth station and rebroadcast them to another.

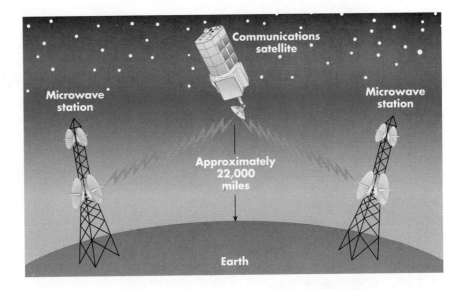

Large companies often use satellite transmissions to transmit data to remote offices.
(Source: Image copyright © 1998 PhotoDisc.)

curvature of the earth, mountains, and other structures that block the line-of-sight microwave transmission make satellites an attractive alternative.[2]

Most of today's communications satellites are owned by companies that rent or lease satellite communications capacity to other companies. However, several large companies are now using their own satellites for internal telecommunications. Some large retail chains, for example, use satellite transmission to connect their main offices to retail stores and warehouses throughout the country or the world. Holiday Inn, for one, has used satellites to improve customer service by sending the latest room and rate information to reservation desks throughout Europe and the United States. In addition to standard satellite stations, there are small mobile satellite systems that allow people and businesses to communicate. These portable systems have a dish that is a few feet in diameter and can operate on battery power anywhere in the world. This is important for news organizations that require the ability to transmit news stories from remote locations. Many people are investing in direct satellite dish technology to receive TV and to send and receive computer communications.[3]

Cellular transmission. With cellular transmission, a local area, such as a city, is divided into cells. As a car or vehicle with a cellular device, such as a mobile phone, moves from one cell to another, the cellular system passes the phone connection from one cell to another (Figure 6.9). The signals from the cells are transmitted to a receiver and integrated into the regular phone system. Cellular phone users can thus connect to anyone that has access to regular phone service, like a child at home or a business associate in London. They can also contact other cellular phone users. Because cellular transmission uses radio waves, it is possible for people with special receivers to listen to cellular phone conversations.

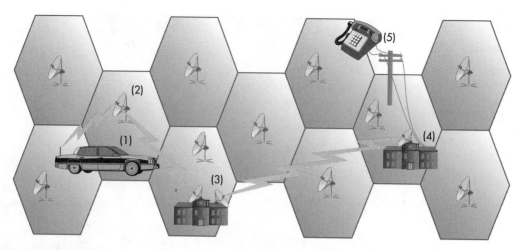

FIGURE 6.9

A Typical Cellular Transmission Scenario

Using a cellular car phone, the caller (1) dials a number. The signal is sent from the car's antenna to the low-powered cellular antenna located in that cell (2). The signal is sent to the regional cellular phone switching office, also called the mobile telephone subscriber office (MTSO) (3). The signal is switched to the local telephone company switching station located nearest the call destination (4). Now integrated into the regular phone system, the call is automatically switched to the number originally dialed (5), all without the need for operator assistance.

analog signal

a continuous, curving signal

digital signal

a signal represented by bits

modems

devices that translate data from digital to analog and analog to digital

Infrared transmission. Another mode of transmission, called infrared transmission, involves sending signals through the air via light waves. Infrared transmission requires line-of-sight transmission and short distances—under a few hundred yards. Infrared transmission can be used to connect various devices and computers. For example, infrared transmission has been used to allow handheld computers to transmit data and information to larger computers within the same room. Some special-purpose phones can also use infrared transmission. This means of transmission can be used to establish a wireless network with the advantage that devices can be moved, removed, and installed without expensive wiring and network connections.

Devices

A telecommunications device is one of various hardware devices that allow electronic communication to occur or to occur more efficiently. Almost every telecommunications system uses one or more of these devices to transmit or convert signals.

Modems. In data telecommunications, it is not uncommon to use transmission media of differing types and capacities at various stages of the communications process. If a typical telephone line is used to transfer data, it can only accommodate an **analog signal** (a continuous, curving signal). Because a computer generates a **digital signal** representing bits, a special device is required to convert the digital signal to an analog signal, and vice versa (Figure 6.10). Translating data from digital to analog is called modulation, and translating data from analog to digital is called demodulation. Thus, these devices are modulation/demodulation devices, or **modems**. Penril/Bay Networks, Hayes, Microcom, Motorola, and U.S. Robotics are examples of modem manufacturers.

Modems can automatically dial telephone numbers, originate message sending, and answer incoming calls and messages. Modems can also perform tests and checks on how they are operating. Some modems are able to vary their transmission rates, commonly measured in bits per second. Today, many modems transmit at 56,600 bps. Some of the more expensive modems, called smart modems, contain microprocessors that allow them

FIGURE 6.10

How a Modem Works

Digital signals are modulated into analog signals, which can be carried over existing phone lines. The analog signals are then demodulated back into digital signals by the receiving modem.

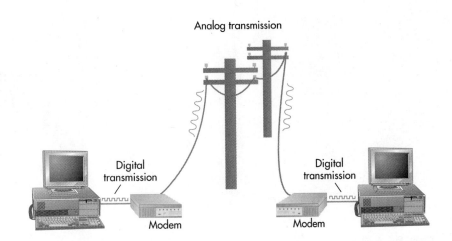

to operate and function under a variety of circumstances.[4] Many firms including Aware, Inc., Alcatel Alsthom, Texas Instruments, and Northern Telecom are competing to provide modems that will deliver fast Internet access (more than 1 Mbps) over ordinary copper phone lines with which most people's homes are wired.[5]

Fax modems. Facsimile devices, commonly called fax devices, allow businesses to transmit text, graphs, photographs, and other digital files via standard telephone lines. A fax modem is a very popular device that combines a fax with a modem, giving users a powerful communications tool.

Special-purpose modems. Various types of special-purpose modems are available. Cellular modems are placed in laptop personal computers to allow people on the go to communicate with other computer systems and devices.[6] With a cellular modem, you can connect to other computers while in your car, on a boat, or in any area that has cellular transmission service. Expansion slots used for PC memory cards can also be used for standardized credit-card-size PC modem cards, which work like standard modems. PC modems are becoming increasingly popular with notebook and portable computer users. Cable companies are promoting the cable modem, which has a low initial cost and transmission speeds up to 10 Mbps.

Multiplexers. Because media and channels are expensive, devices that allow several signals to be sent over one channel have been developed. A multiplexer is one of these devices. A **multiplexer** allows several telecommunications signals to be transmitted over a single communications medium at the same time (Figure 6.11).

Front-end processors. **Front-end processors** are special-purpose computers that manage communications to and from a computer system. Like a receptionist handling visitors at an office complex, communications processors direct the flow of incoming and outgoing jobs. They connect a midrange or mainframe computer to hundreds or thousands of communications lines. They poll terminals and other devices to see if they have any messages to send. They provide automatic answering and calling, as well as

multiplexer

a device that allows several telecommunications signals to be transmitted over a single communications medium at the same time

front-end processor

a special-purpose computer that manages communications to and from a computer system

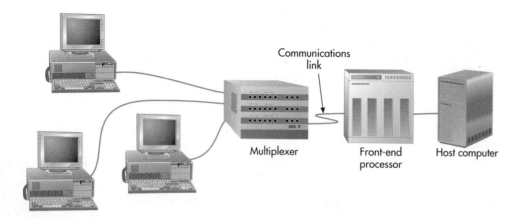

FIGURE 6.11

Use of Multiplexer to
Consolidate Data
Communications onto a Single
Communications Link

perform circuit checking and error detection. Front-end processors also
develop logs or reports of all communications traffic, edit data before it
enters the main processor, determine message priority, automatically choose
alternative and efficient communications paths over multiple data communications lines, and provide general data security for the main system CPU.
Because front-end processors perform all these tasks, the midrange or mainframe computer is able to process more work (Figure 6.12).

FIGURE 6.12

Front-End Processor

A front-end processor takes the
burden of communications
management away from the
main system processor.

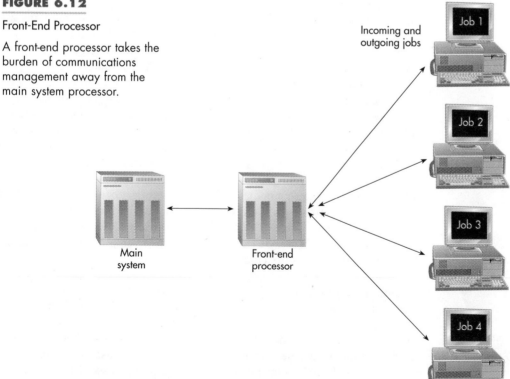

AT&T's Network Operations Center uses state-of-the-art computer software technology to continuously monitor the flow of voice, data, and image transmission over 2.3 billion circuit miles worldwide.
(Source: Property of AT&T Archives. Reprinted with permission of AT&T.)

Carriers and Services

Telecommunications carriers provide the telephone lines, satellites, modems, or other communications technology used to transmit data from one location to another. They also provide many types of services. Telecommunications carriers are classified as either common carriers or other special-purpose carriers. The **common carriers** are primarily the long-distance telephone companies. American Telephone & Telegraph (AT&T), one of the largest companies providing communications media and services, is a common carrier for long-distance service and a special-purpose carrier for other services. MCI, Sprint, and others make up a significant part of the telecommunications industry as well. **Value-added carriers** are companies that have developed private telecommunications systems and offer their services for a fee. Some value-added carriers that offer communications services include SprintNet and Telenet (developed by GTE) and Tymnet.

Switched and dedicated lines. Common carriers typically provide the use of standard telephone lines, called **switched lines**. These lines use switching equipment to allow one transmission device (e.g., your telephone) to be connected to other transmission devices (e.g., the telephones of your friends and relatives). A switched line is a special-purpose circuit that directs messages along specific paths in a telecommunications system. When you make a phone call, the local telephone service provider's switching equipment connects your phone to the phone of the person you're calling. Fees for a switched business line (versus residential line) can range from $25–$100 or more per month. A **dedicated line**, also called a leased line, provides a constant connection between two points. No switching or dialing is needed; the two devices are always connected. Many firms with high data transfer requirements between two points—say, an East Coast and a West Coast

common carriers

long-distance telephone companies

value-added carriers

companies that have developed private telecommunications systems and offer their services for a fee

switched lines

lines that use switching equipment to allow one transmission device (e.g., your telephone) to be connected to other transmission devices (e.g., the telephones of your friends and relatives)

dedicated line

a line that provides a constant connection between two points. No switching or dialing is needed; the two devices are always connected.

shared headquarters arrangement—utilize dedicated lines. The high initial cost of purchasing or leasing such a line is offset by eliminating long-distance charges incurred with a switched line. Monthly fees for a dedicated line can range from $100–$500 or more, but there is no additional charge for usage. Read the "Ethical and Societal Issues" box to learn more about options for obtaining local service from carriers.

ETHICAL AND SOCIETAL ISSUES

Snags Limit Increased Competition for Local Telephone Services

The Telecommunications Act of 1996 had as its goal to promote competition and eliminate monopolies for a wide range of communications services. Among the bill's provisions were deregulation of local phone service and opening the market to cable, utility, and long-distance phone companies. Options for choosing local-access telephone service included the current local-access telephone company, cable operators, and long-distance companies. The seven Regional Bell Operating Companies, created in 1984 following the breakup of AT&T, were permitted to offer long-distance services in other regions once they demonstrated they had fully opened their networks to local competition.

The promises of the Telecommunications Act of 1996—lower phone and faster data rates, new competitors offering innovative services, rapid implementation of new technologies such as high-speed Internet access—sounded good. But unfortunately, most telecommunications managers have not seen these benefits. The act was based on the flawed assumption that telecom giants like the Regional Bell Operating Companies, AT&T, and the cable companies would snap up the latest digital technologies and invade each other's markets. But so far, there's been more turf protecting than turf invading.

Those who want to compete in the local telephone service arena currently have two options: purchase/lease and then resell the local lines owned by the incumbents or construct their own bypass networks. In either case, the trouble and expense of building alternative local networks, combined with the difficulties of coordinating local facilities, have limited local phone competition primarily to the largest metropolitan markets and the most lucrative corporate accounts. In some cases, the local phone monopolies have gone to court to fight the FCC limits on rates they can charge resellers. Until the issue is resolved, potential competitors are holding off most plans for reselling.

Of course, hot new technologies were supposed to render such obstructions futile. But Congress and the industry underestimated the enormous hurdles that stood in the way. Cable companies, for instance, were supposed to upgrade their networks to start providing phone service. Local phone companies would make their own systems sophisticated enough to deliver video. While technically possible, the anticipated network upgrades remain dauntingly expensive. So both the local phone companies and the cable operators have all but scrapped their plans for immediate entry into the other's business.

Discussion Questions

1. Do research to find out the level of competition for local phone service in your area. Write three or four paragraphs summarizing your findings.

2. What suggestions can you make to improve the competitiveness of local telephone services?

Sources: Adapted from Charles Piller, Randy Ross, and Bill Snyder, "Telecom Reform, Where Are the Choices, Lower Prices?" *PC World*, February 1998, pp. 58–61; Mary E. Thyfault, "Local Connection," *InformationWeek*, January 19, 1998, pp. 48–56; John Rendleman, "Regional Disappointment," *PC Week*, December 15, 1997, pp. 1, 18; and Amy Barrett, "Regulators Should Discipline Telecom Brats," *BusinessWeek*, June 30, 1997, p. 40.

private branch exchange (PBX)

a communications system that can manage both voice and data transfer within a building and to outside lines

Private branch exchange. A **private branch exchange (PBX)** is a communications system that can manage both voice and data transfer within a building and to outside lines. In a PBX system, switching equipment routes phone calls and messages within the building. PBXs can be used to connect hundreds of internal phone lines to a few phone company lines. For example, an organization might have five phone lines coming in from the outside phone company. These five lines may be connected to 50 phones within the organization. Any of the 50 phones can use one of the five phone lines to make calls outside the organization. These same five lines may also be used for incoming calls. Furthermore, it is usually possible for any of the 50 phones to connect to another internal phone on an intercom system.

PBXs not only can store and transfer calls, but they can also serve as connections between different office devices. With PBX technology, a manager could connect his computer to the PBX via a modem and then send instructions to a copy machine through his PC. Another advantage of a PBX system is that it requires a business to have fewer phone lines coming in from the outside. The disadvantage is that the company has to purchase, rent, or lease the PBX equipment. Thus, there is a trade-off between the expense of the PBX equipment and the savings in the reduced number of incoming phone lines.

WATS service. Wide-area telecommunications service (WATS) is a billing method for heavy users of voice services. When you dial a company at a toll-free 800 or 888 number to place an order or make a query, you are using WATS. The company or organization you call via WATS pays a fee to the phone company, depending on the level of service and usage. The fee varies depending on the caller's geographic location within the United States and the number of incoming and outgoing calls. Companies that rely on phones for customer service typically use WATS services because customers can call the WATS number free of charge. For companies with a high volume of calls, WATS can also be substantially less expensive than a normal billing schedule. It is even possible for individuals to get a personal toll-free number.

Phone and dialing services. Common carriers are beginning to provide more and more phone and dialing services to home and business users. Automatic number identification (ANI), or caller ID, equipment can be installed on a phone system to identify and display the number of an incoming call. In a business setting, ANI can be used to identify the caller and link the caller with information stored in a computer. For example, when a customer calls FedEx, the customer service rep uses the ANI to identify the name and address of the customer, thus saving time when handling a request for a pickup. The ANI can be very useful in helping people screen calls before they are answered. Unwanted phone calls from other people and businesses can be identified before the phone is ever answered. Common carriers offer even more services to extend the capabilities of the typical phone system. Even with all the advances in computers and telecommunications, services offered by common carriers will remain important. Some of the services are listed below.

- The ability to integrate personal computers so that the telephone number of the caller is automatically captured and used to look up information in a database about this customer.
- Access codes to screen out junk calls, wrong numbers, and unwanted phone calls.
- Call priorities (e.g., only certain calls would be received during certain times of the day, such as after 10:00 P.M. to before 7:00 A.M.)
- The ability to use one number for a business phone, home phone, personal computer, fax, etc.
- Intelligent dialing (when a busy signal is received, the phone redials the number when your line and the line of the party you are trying to reach are both free).

Integrated Services Digital Network (ISDN)

technology that uses existing common-carrier lines to simultaneously transmit voice, video, and image data in digital form

ISDN. Many telephone companies now offer **Integrated Services Digital Network (ISDN)** service, a technology that uses existing common-carrier lines to simultaneously transmit voice, video, and image data in digital form. ISDN also offers high rates of transmission: the digital service has the capacity to send a 22-page document in about a second.[7] With ISDN, communications devices require a special ISDN board. These data communications systems use an ISDN network switch—a digital switch that allows different communications services to be connected to the system. For example, ISDN allows long-distance services, video and voice services, facsimile devices, telephones, and private branch exchanges to be integrated into one telecommunications system (Figure 6.13). ISDN digital networks are typically faster—from 64 thousand bits per second (64 kbps) up to 2 million bits per second (2 Mbps)—and can carry more signals than analog networks. They also allow for easier sharing of image, multimedia, and other complex forms of data across telephone lines.

Several organizations are attempting to take advantage of the benefits of ISDN. AT&T, for example, offers an ISDN service that will allow businesses to transmit full-motion video and computer files and continue voice communications at the same time. Using ISDN services, an advertising agent could send a consultant a draft of a full-motion TV commercial while talking with her on the phone about suggested changes. Modifications to the file could be returned using ISDN services. For companies that send a lot of video or multimedia data, ISDN can be a cost-effective and efficient way to connect to clients, customers, and business colleagues. Table 6.1 shows some of the costs, advantages, and disadvantages of different lines and services offered by communications carriers.

T1 carrier. The T1 channel was developed by AT&T to increase the number of voice calls that could be handled through the existing cables. For digital communications, T1 is the carrier used in North America. T1 is also

FIGURE 6.13

ISDN Network Switching

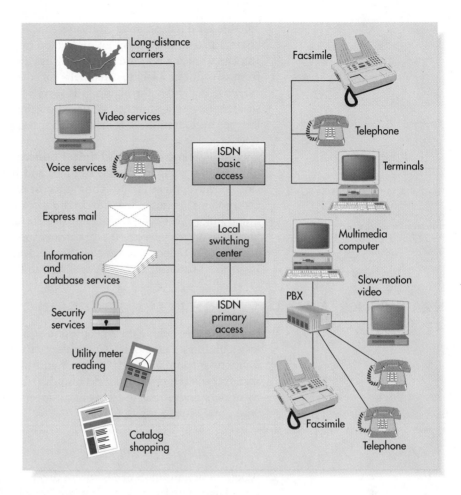

TABLE 6.1

Costs, Advantages, and Disadvantages of Several Line and Service Types

Line/Service	Speed	Cost per Month for Unlimited Connect Time	Advantages	Disadvantages
Plain old telephone service (POTS)	56 Kbps	$20–$60	Low cost. Broadly available.	Too slow for video.
ISDN	64–128 Kbps	$70–$120	Broad bandwidth for video, voice, and data. Good to very good quality. Wide product support.	High cost to install and connect costs. Not available everywhere.
DSL	500 Kbps 1.544 Mbps	$100–$300	Broad bandwidth for a variety of applications.	Requires expensive modem. Not broadly available.
Cable modem	Receive at up to 500 Kbps; send at up to 64 Kbps	Cost not yet set	Broad bandwidth for a variety of applications.	Just becoming available. Requires special modem.
T1	1.544 Mbps	$1,000	Broad bandwidth for digital, data, and image transmissions.	High cost; high installation fee, and subscribers pay monthly fee based on distance.

suitable for data and image transmissions. Large companies frequently purchase T1 lines to develop an integrated telecommunications network that can carry voice, data, and images. T1 has a speed of 1.544 Mbps developed from two dozen 64-Kbps channels, together with one 8-Kbps channel for carrying control information. T1 services are quite expensive with subscribers paying a monthly fee based on the distance (possible several dollars per mile) and there is also a high installation fee.

Digital Subscriber Line. A **Digital Subscriber Line (DSL)** uses existing phone wires going into today's homes and businesses to provide transmission speeds exceeding 500 Kbps at a cost of $100–$300 per month. A special modem costing a few hundred dollars is also required. DSL lines are not available everywhere.[8] Starting in February 1998, US West began offering Megabit Service in 40 cities, which upgrades a regular phone line by adding a DSL data channel that offers speeds from 256 Kbps to 7 Mbps.[9] Read the "Technology Advantage" box to learn more about DSL.

Digital Subscriber Line (DSL)

a line that uses existing phone wires going into today's homes and businesses to provide transmission speeds exceeding 500 Kbps at a cost of $100–$300 per month

NETWORKS AND DISTRIBUTED PROCESSING

Businesses are linking their people and equipment to enable people to work faster and more efficiently. Computer networks give organizations the flexibility to accomplish work wherever and whenever it is most beneficial. To take full advantage of networks and distributed processing, it is important to understand strategies, network concepts and considerations, network types, and related topics.

Basic Data Processing Strategies

When an organization needs to use two or more computer systems, one of three basic data processing strategies may be followed: centralized, decentralized, or distributed. With **centralized processing**, all processing occurs in a single location or facility. This approach offers the highest degree of control. With **decentralized processing**, processing devices are placed at various remote locations. The individual computer systems are isolated and do not communicate with each other. Decentralized systems are suitable for companies that have independent operating divisions. Some drug store chains, for example, operate each location as a completely separate entity; each store has its own computer system that works independently of the computers at other stores. With **distributed processing**, computers are placed at remote locations but connected to each other via telecommunications devices. Consider a manufacturing company with plants in Milwaukee, Chicago, and Atlanta with corporate headquarters in New York. Each location has its own computer system. By connecting all the computer systems into a distributed processing system, all the locations can share data and programs. Distributed processing also allows each plant to perform its own processing (say, for example, inventory) while the New York computer system coordinates and processes other applications, like payroll.

centralized processing

data processing that occurs in a single location or facility

decentralized processing

data processing that occurs when devices are placed at various remote locations

distributed processing

data processing that occurs when computers are placed at remote locations but are connected to each other via telecommunications devices

One benefit of distributed processing is that processing activity can be allocated to the location(s) where it can most efficiently occur. For example, the New York headquarters may have the largest computer system, but the Atlanta office might have hundreds of employees to input the data. The

TECHNOLOGY ADVANTAGE

Reach Out and Touch the World: The Promise of DSL

Connecting to other computers, networks, and the Internet offers the potential of utilizing the power of other machines and tapping into vast stores of information. The prospects are exciting. Like a football fan getting a free ticket to the Super Bowl, PC users are thrilled with the possibilities. But getting there is another story. Clogged telecommunications make the ramp to the information superhighway seem like a long single-lane dirt road that all ticket holders must take to get to the Super Bowl. In both cases, fun is quickly replaced with frustration and long delays. Telecommunications bottlenecks are making the promise of connectivity turn into disappointment. The long waits to get connected and get results have caused some to call the World Wide Web, an important part of the Internet, the World Wide *Wait*. Spectacular growth in the number of new people using networks and the Internet makes the problem even worse.

For networks and the Internet to be viable business tools, these connectivity problems must be solved. Realizing that someone's connectivity problem is another person's opportunity, a number of companies are now trying to solve the problem of slow speeds and bottlenecks by making the connection from PCs faster. One way to do this is to make modems that connect to standard phone lines operate at faster rates. But although modem speeds have increased, the increases are not enough to relieve bottlenecks. As a result, phone companies are now offering a number of alternatives to the standard phone line. While speeds can be increased over standard phone lines, ISDN lines can be expensive and are not available in all areas. In some areas, ISDN can cost hundreds of dollars per month. For businesses that need even more speed, a T1 line is available in some areas. While T1 lines are extremely fast, some people believe that the cost is also extreme—about $1,000 per month in some areas.

The Digital Subscriber Line (DSL) is a newer and potentially better alternative to standard phone lines, ISDN, and T1 lines. DSL service, which can use existing phone wires going into today's homes and businesses, offers speeds that are comparable to a T1 line with a cost that can be less than an ISDN line. Anchronous DSL (ADSL), for example, offers speeds up to 160 times faster than today's 56-Kbps modems. In addition, DSL allows phone calls and Internet connection at the same time on the same standard phone wire. With an initial cost that can range from $50 to a few hundred dollars per month, however, DSL will appeal primarily to businesses and individuals with a need for speed. In addition, DSL modems or connectors are required, which can cost a few hundred dollars. As prices come down, DSL will be more attractive to all network and Internet users. In addition, phone and computer companies are now developing standards for DSL that will allow equipment and line costs to come down. Even with the advantages of DSL, its future is not guaranteed. Agreement on standards may not be reached and it is possible that a new telecommunications breakthrough will make DSL obsolete before it is widely accepted.

DISCUSSION QUESTIONS

1. What are the advantages of DSL over other approaches to speeding communications?
2. Is speed or cost more important to you? How much more would you spend per month for local phone service to dramatically increase the speed of your network or Internet connection?
3. Would you invest in the equipment and line costs to use DSL if it becomes available in your area?

Sources: Adapted from Matt Hamblen, "Digital Subscriber Line Standard in the Works," *Computerworld*, January 26, 1998, p. 12; Dean Takahashi, "Bells Push a Modem Standard to Rival Cable's," *The Wall Street Journal*, January 21, 1998, p. B6; Andrew Kesslier, "Cancel That ISDN Order," *Forbes*, January 26, 1998, p. 88; Angela Hickman, "Get In the Fast Lane," *PC Magazine*, February 10, 1998, p. 28.

system's output may be most needed in Chicago, the location of the warehouse. With distributed processing, each of these offices can organize and manipulate the data to meet its specific needs, as well as share its work product with the rest of the organization. The distribution of the processing across the organizational system ensures that the right information is delivered to the right individuals, maximizing the capabilities of the overall information system by balancing the effectiveness and efficiency of each individual computer system.

Network Concepts and Considerations

Networks that link computers and computer devices provide for flexible processing. Building networks involves two types of design: logical and physical. A logical model shows how the network will be organized and arranged. A physical model describes how the hardware and software in the network will be physically and electronically linked.

Network topology. The number of possible ways to logically arrange the nodes, or computer systems and devices on a network, may seem limitless. Actually, there are only five major types of **network topologies**—logical models that describe how networks are structured or configured. These types are ring, bus, hierarchical, star, and hybrid (Figure 6.14).

The **ring network** contains computers and computer devices placed in a ring, or circle. With a ring network, there is no central coordinating computer. Messages are routed around the ring from one device or computer to another. A bus network is a cable or telecommunications line with devices attached to it. A **bus network** consists of computers and computer devices on a single line. Each device is connected directly to the bus and can communicate directly with all other devices on the network. The bus network is one of the most popular types of personal computer networks. The **hierarchical network** uses a treelike structure. Messages are passed along the branches of the hierarchy until they reach their destination. Like a ring network, a hierarchical network does not require a centralized computer to control communications. Hierarchical networks are easier to repair than other topologies because you can isolate and repair one branch without affecting the others. A **star network** has a central hub or computer system. Other computers or computer devices are located at the end of communications lines that originate from the central hub or computer system. The central computer of a star network controls and directs messages. If the central computer breaks down, it results in a breakdown of the entire network. Many organizations use a hybrid network, which is simply a combination of two or more of the four topologies just discussed. The exact configuration of the network depends on the needs, goals, and organizational structure of the company involved.

Network Types

Depending on the physical distance between nodes on a network and the communications and services provided by the network, networks can be classified as local area, wide area, or international. Local area networks tie together equipment in a building or local area; international networks are used to communicate between countries. Wide area networks operate over a broad geographic area.

Local area networks. A network that connects computer systems and devices within the same geographic area is a **local area network (LAN)**. A local area network can be a ring, bus, star, hierarchical, or hybrid network. Typically, local area networks are wired into office buildings and factories (Figure 6.15). They can be built around powerful personal computers, minicomputers, or mainframe computers. When a personal

network topology

a logical model that describes how networks are structured or configured

ring network

a type of topology that contains computers and computer devices placed in a ring, or circle. With a ring network, there is no central coordinating computer. Messages are routed around the ring from one device or computer to another.

bus network

a type of topology that consists of computers and computer devices on a single line. Each device is connected directly to the bus and can communicate directly with all other devices on the network. The bus network is one of the most popular types of personal computer networks.

hierarchical network

a type of topology that uses a treelike structure with messages passed along the branches of the hierarchy until they reach their destination

star network

a type of topology that has a central hub or computer system, and other computers or computer devices are located at the end of communications lines that originate from the central hub or computer

local area network (LAN)

a network that connects computer systems and devices within the same geographic area

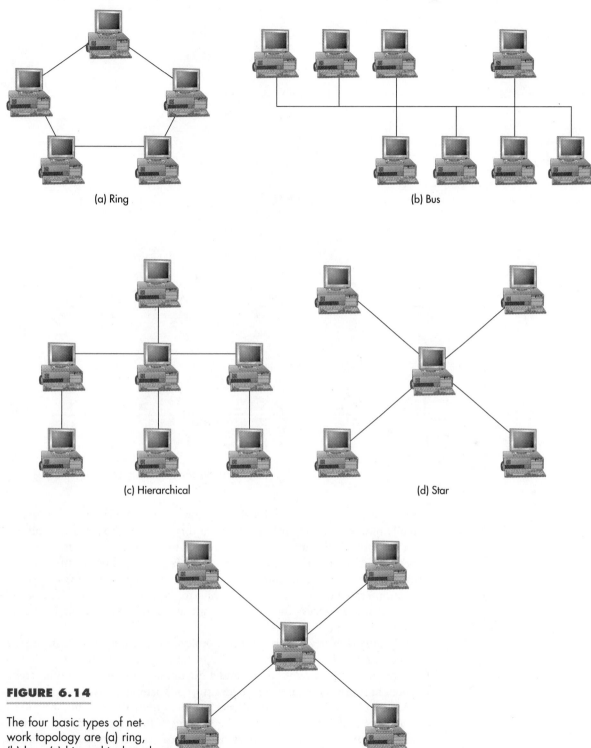

(a) Ring

(b) Bus

(c) Hierarchical

(d) Star

(e) Hybrid

FIGURE 6.14

The four basic types of network topology are (a) ring, (b) bus, (c) hierarchical, and (d) star. In addition, a hybrid configuration (e) can be formed from elements of any of these four topologies.

FIGURE 6.15

A Typical LAN in a Bus
Topology

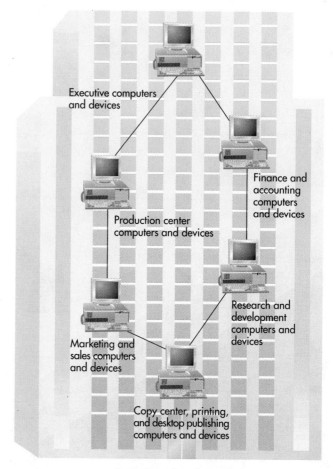

Executive computers
and devices

Finance and
accounting
computers
and devices

Production center
computers and devices

Research and
development
computers and
devices

Marketing and
sales computers
and devices

Copy center, printing,
and desktop publishing
computers and devices

Corporate headquarters

computer is connected to a local area network, a network interface card is
usually required. A network interface card is a card or board that is placed
in a computer's expansion slot to allow it to communicate with the network. A wire or connector from the network is plugged directly into the network interface card. For example, a salesperson whose notebook computer
has an interface card can establish a link to the corporate LAN. The salesperson can then access the network while at the office and download data
needed for the next sales call.

All network users within an office building can connect to each other's
devices for rapid communication. For instance, a user in the research and
development department could send a document from her computer to be
printed at a printer located in the desktop publishing center.

Another basic LAN is a simple peer-to-peer network that might be used
for a very small business to allow the sharing of files and hardware devices
such as printers. In a peer-to-peer network, each computer is set up as an
independent computer, except that other computers can access specific
files on its hard drive or share its printer. These types of networks have no
server. Instead, each computer uses a network interface card and cabling to
connect it to the next machine. Examples of peer-to-peer networks include
Windows for Workgroups, Windows 98, Windows NT, and AppleShare.

Many libraries are served by large networks such as the CARL System. More than 14,000 terminals are connected to CARL, either directly or through computer-to-computer network connections. (Source: Courtesy of CARL Corporation.)

Performance of the computers on a peer-to-peer network is usually slower because one computer is actually sharing the resources of another computer. These networks, however, are a good beginning network from which small businesses can grow—the software cost is minimal and network cards can be used if the company decides to enlarge the system.

It has been estimated that over 70 percent of business PCs in the United States are connected to a local area network. LANs provide excellent support for businesses whose main communications are internal or encompass only a small region. As these businesses expand their markets, they will often move to a network that allows communication across larger areas.

wide area network (WAN)

a network that ties together large geographic regions using microwave and satellite transmission or telephone lines

Wide area networks. A **wide area network (WAN)** ties together large geographic regions using microwave and satellite transmission or telephone lines. When you make a long-distance phone call, you are using a wide area network. AT&T, MCI, and others are examples of companies that offer WAN services to the public. (See Figure 6.16.)

Private WAN. Companies also design and implement WANs for private use. These WANs usually consist of computer equipment owned by the user, together with data communications equipment provided by a common carrier.

international network

a network that links systems between countries

International networks. Networks that link systems between countries are called **international networks**. As companies continue to globalize and use international networks, communications carriers are working to secure part of this $88 billion market. However, international telecommunications comes with special problems. In addition to requiring sophisticated equipment and software, global networks must meet specific national and international laws regulating the electronic flow of data across international boundaries, often called transborder data flow. Some countries have strict laws restricting the use of telecommunications and databases, making normal business transactions such as payroll costly, slow, or even impossible. Other countries have few laws

FIGURE 6.16

A Wide Area Network

Wide area networks are the basic long-distance networks used by organizations and individuals around the world. The actual connections between sites, or nodes (shown by dashed lines), may be any combination of satellites, microwave, or cabling. When you make a long-distance telephone call, you are using a WAN.

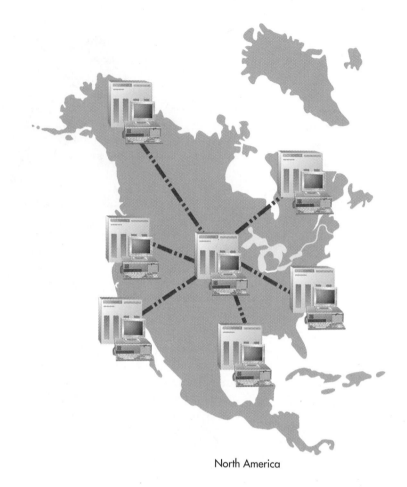

North America

restricting the use of telecommunications and databases. These countries, sometimes called data havens, allow other governments and companies to avoid their own country's laws by processing data within their boundaries.

Despite the obstacles, numerous private and public international networks exist. United Parcel Service, for example, has invested in an international network, called UPSnet. UPSnet allows drivers to use handheld computers to send real-time information about pickups and deliveries to central data centers. In addition to the 77,000 handheld computers for drivers, UPSnet is based on five mainframes, 60,000 personal computers, several satellite dishes, and enough fiber-optic cable to wrap around the earth 25 times. The huge network allows data to be retrieved by customers to track packages or to be used by the company for faster billing, better fleet planning, and improved customer service.[10] (The Internet, which we will discuss in Chapter 7, is the largest public international network.)

Terminal-to-Host, File Server, and Client/Server Systems

If an organization is selecting distributed information processing, it can connect computers in several ways, the most common of which are terminal-to-host, file server, and client/server architecture.

"Dumb" terminal

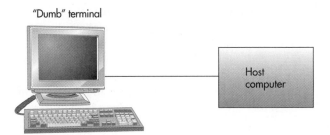

FIGURE 6.17

Terminal-to-Host Connection

terminal-to-host

an architecture in which the application and database reside on one host computer, and the user interacts with the application and data using a "dumb" terminal

file server

an architecture in which the application and database reside on one host computer, called the file server

client/server

an architecture in which multiple computer platforms are dedicated to special functions such as database management, printing, communications, and program execution

Terminal-to-host. With **terminal-to-host** architecture, the application and database reside on one host computer, and the user interacts with the application and data using a "dumb" terminal. (Even if you use a personal computer to access the application, you run terminal emulation software on the PC to make it act as if it were a dumb terminal with no processing capacity.) Since a dumb terminal has no data processing capability, all computations, data accessing and formatting, as well as data display, are done by an application that runs on the host computer (Figure 6.17).

File server. In **file server** architecture, the application and database reside on one host computer, called the file server. The database management system runs on the end user's personal computer or workstation. If the user needs even a small subset of the data that resides on the file server, the file server sends the user the entire file that contains the data requested, including a lot of data the user does not want or need. The downloaded data can then be analyzed, manipulated, formatted, and displayed by a program that runs on the user's personal computer (Figure 6.18).

Client/server. In **client/server** architecture, multiple computer platforms are dedicated to special functions such as database management, printing, communications, and program execution. These platforms are called servers. Each server is accessible by all computers on the network. Servers can be computers of all sizes; they store both application programs and data files and are equipped with operating system software to manage the activities of the network. The server distributes programs and data files to the other computers (clients) on the network as they request them. An application server holds the programs and data files for a particular application, such as an inventory database. Processing can be done at the client or server.

A client is any computer (often an end user's personal computer) that sends messages requesting services from the servers on the network. A client can converse with many servers concurrently. A user at a personal computer initiates a request to extract data that resides in a database somewhere on the network. A data request server intercepts the request and determines on which data server the data resides. The server then formats

FIGURE 6.18

File Server Connection

The file server sends the user the entire file that contains the data requested. The downloaded data can then be analyzed, manipulated, formatted, and displayed by a program that runs on the user's personal computer.

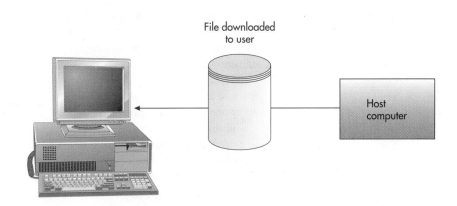

File downloaded
to user

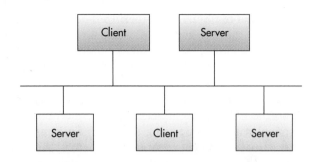

FIGURE 6.19

Client/Server Connection

Multiple computer platforms, called servers, are dedicated to special functions such as database management, data storage, printing, communications, network security, and program execution. Each server is accessible by all computers on the network. A server distributes programs and data files to the other computers (clients) on the network as they request them. The client requests services from the servers, provides a user interface, and presents results to the user. Once data is moved from a server to the client, the data may be processed on the client.

the user's request into a message that the database server will understand. Upon receipt of the message, the database server extracts and formats the requested data and sends the results to the client. Only the data needed to satisfy a specific query is sent—not the entire file (Figure 6.19). As with the file server approach, once the downloaded data is on the user's machine, it can then be analyzed, manipulated, formatted, and displayed by a program that runs on the user's personal computer.

There are several advantages of the client/server approach over both the terminal-to-host and file server approaches: reduced cost, improved performance, and increased security.

Reduced cost potential. The functionality achieved with client/server computing can exceed that provided by a traditional minicomputer or even a mainframe-based computer system at a lower cost. With client/server computing, a powerful workstation costing less than $25,000 may replace much of the function provided by a midrange computer costing over $100,000. In addition, vendor contracts for workstation software and hardware support are cheaper than for midrange and mainframe computers. Thus, many organizations view the migration of applications from mainframe computers and terminal-to-host architecture to client/server architecture as a significant cost savings opportunity. This downsizing (or, as some call it, "rightsizing") can yield significant savings in reduced hardware and software support costs.

Improved performance. The most important difference between the file server and client/server architecture is that the latter much more efficiently minimizes traffic on the network. With client/server computing, only the data needed to satisfy a user query is moved from the database to the client device, whereas the entire file is sent in file server computing. The smaller amount of data being sent over the network also greatly reduces the amount of time needed for the user to receive a response.

Increased security. Security mechanisms can be implemented directly on the database server through the use of stored procedures. These procedures execute faster than the password protection and data validation rules attached to individual applications on a file server. They can also be shared across multiple applications.

The type of application most appropriate for client/server architecture is one that uses large data files, requires fast response time, and needs strong security and recovery options. All these factors point to the kind of applications that are central to the operation and management of the business. On-line transaction processing and decision support applications are particularly good candidates for client/server computing.

While client/server systems have much to offer in terms of practical benefits, some problems are noticeable with such systems: increased cost, loss of control, and complexity of vendor environment.

Increased cost potential. If all costs associated with client/server computing are accounted for, expected savings may fail to materialize. Moving to client/server architecture is a major two- to five-year conversion process. Over that time period, considerable costs will be incurred for hardware, software, communications equipment and links, data conversion, and training. Costs will be even higher for multiple-site companies converting to client/server computing. These expenses are difficult for the IS organization to track because they are often paid by the end users directly. Thus, the move to client/server architecture may be much more expensive than the IS organization realizes.

Loss of control. Controlling the client/server environment to prevent unauthorized use, invasion of privacy, and viruses is also difficult. Despite these concerns, many companies expect to gain long-term efficiency and effectiveness by moving away from large mainframe systems. The overall use of mainframe computers may not decrease, however, because mainframes are often reconfigured to become primary servers for large-scale client/server systems.

Complex multivendor environment. Implementation of client/server architecture leads to operating in a multivendor environment with, in many cases, relatively new and immature products. Situations such as these make it likely that problems will arise. Often such problems are difficult to identify and isolate to the appropriate vendor.

Nevertheless, the dominance of single-vendor environments and terminal-to-host architecture is fading fast as corporations move into the much more complex client/server environment with multiple vendors for networks, hardware, and software. Open systems are essential to implementing a client/server architecture so that managers are free to choose clients and servers and be assured that their combinations will be able to communicate with one another.

Communications Software and Protocols

communications software

software that provides error checking, message formatting, communications logs, data security and privacy, and translation capabilities for networks

Communications software provides a number of important functions in a network. Most communications software packages provide error checking and message formatting. In some cases, when there is a problem, the software can indicate what is wrong and suggest possible solutions. Communications software can also maintain a log listing all jobs and communications that have taken place over a specified period of time. In addition, data security and privacy techniques are built into most packages.

In Chapter 4 you learned that all computers have operating systems that control many functions. When an application program requires data from a disk drive, it goes through the operating system. Now consider a situation in which a computer is attached to a network that connects large disk drives, printers, and other equipment and devices. How does an application program request data from a disk drive on the network? The answer is through the network operating system.

network operating system (NOS)

systems software that controls the computer systems and devices on a network and allows them to communicate with each other

A **network operating system (NOS)** is systems software that controls the computer systems and devices on a network and allows them to communicate with each other. A NOS performs the same types of functions for the network as operating system software does for a computer, such as memory and task management and coordination of hardware. When network equipment (such as printers, plotters, and disk drives) is required, the network operating system makes sure that these resources are correctly used. In most cases, companies that produce and sell networks provide the NOS. For example, NetWare is the NOS from Novell, a popular network environment for personal computer systems and equipment. Windows NT is another commonly used network operating system.

Software tools and utilities are available for managing networks. With **network management software**, a manager on a networked desktop can monitor the use of individual computers and shared hardware (like printers), scan for viruses, and ensure compliance with software licenses. Network management software also simplifies the process of updating files and programs on computers on the network—changes can be made through a communications server instead of being made on each individual computer. Network management software also protects software from being copied, modified, or downloaded illegally and performs error control to locate telecommunications errors and potential network problems. Some of the many benefits of network management software include fewer hours spent on routine tasks (like installing new software), faster response to problems, and greater overall network control.

Communications protocols. Communications protocols make communications possible. A number of communications **protocols** are used by companies and organizations of all sizes. Just as standards are important in building computer and database systems, established protocols help ensure communications among computers of different types and from different manufacturers.

Many protocols have layers of standards and procedures. The **Open Systems Interconnection (OSI) model** serves as a standard model for network architectures and is endorsed by the International Standards Committee. The OSI model divides data communications functions into seven distinct layers to promote the development of modular networks that simplify the development, operation, and maintenance of complex telecommunications networks. These layers are described in Figure 6.20.

TCP/IP. In the 1970s, the U.S. government pioneered the development of the **Transmission Control Protocol/Internet Protocol (TCP/IP)** to link its defense research agencies. The government has adopted OSI standards to replace TCP/IP, but TCP/IP remains the major network protocol used by schools and businesses. It is the primary communications protocol of the Internet.

SNA. IBM has also developed a communications protocol, called Systems Network Architecture (SNA), which is a protocol used for IBM systems. Because of the popularity of IBM systems, many other computer manufacturers and communications companies have made their systems compatible with the SNA protocol.

Ethernet. Ethernet is a popular communications protocol often used with local area networks. The Ethernet standard is designed for LANs that use a bus topology; the standard helps ensure compatibility among devices so that many people can attach to a common cable to share network facilities and resources.

X.400 and X.500. The X.400 and X.500 protocols are also used in many organizations. Many international companies have adopted one of these protocols as their standard. With more international acceptance of these protocols, telecommunications among countries will become simpler. As

network management software

software that enables a manager on a networked desktop to monitor the use of individual computers and shared hardware (like printers), scan for viruses, and ensure compliance with software licenses

protocol

rules that ensure communications among computers of different types and from different manufacturers

Open Systems Interconnection (OSI) model

a standard model for network architectures that divides data communications functions into seven distinct layers to promote the development of modular networks that simplify the development, operation, and maintenance of complex telecommunications networks

Transmission Control Protocol/Internet Protocol (TCP/IP)

standard originally developed by the U.S. government to link defense research agencies; it is the primary communications protocol of the Internet

FIGURE 6.20

The Seven Layers of the OSI Model

This Open Systems Interconnection (OSI) model is designed to permit communication among different computers from different manufacturers using different operating systems—as long as each conforms to the OSI model. (Source: Reprinted with permission from *Information Systems for Managers 3/e*, pp. 134–135, by George Reynolds, Copyright 1995 by West Publishing.)

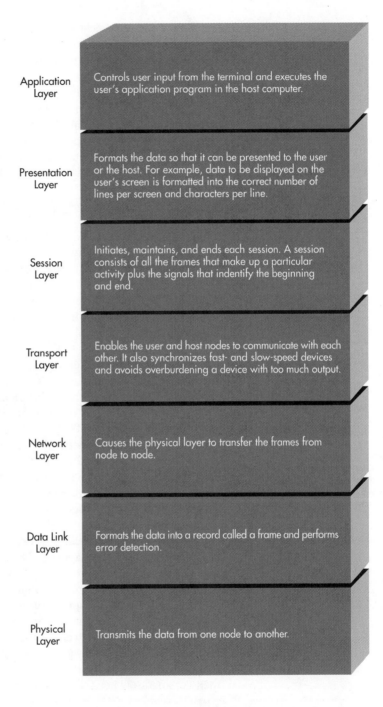

businesses continue to move toward global operations, the importance of adopting standard international protocols will increase.

X.400 is a set of messaging handling standards ranging from X.400 to X.440 that define a Message Handling System and a Message Transfer Service.

X.500 is a set of standards dictating the design of networking directories, which contain information on users—from names and e-mail addresses to job titles and resource-access privileges. Since X.500 is standards-based, it provides a technological foundation for connecting proprietary and non-proprietary directories, such as those used in Windows NT and Unix networks, as well as in Internet-based networks. For example, Unilever has built an X.500-based directory to streamline the maintenance of information system user information and virtually eliminate e-mail addressing errors.[11] The company manufactures and sells more than $50 billion a year worldwide in packaged goods, including Lipton tea, Popsicle ice-cream treats, Ragu tomato sauce, Aim toothpaste, and Calvin Klein cosmetics.

The complete list of standards in force in the X series can be found on the ITU Web site at http://www.itu.ch.

Bridges, Routers, Gateways and Switches

Many LANs have hardware and software devices that allow them to communicate with other networks that employ different transmission media and/or protocols. (See Figure 6.21.)

bridge

a device that connects two or more networks at the media access control portion of the data link layer; the two networks must use the same communications protocol

Bridge. A **bridge** connects two or more networks at the media access control portion of the data link layer. The two networks must use the same communications protocol.

router

a device that operates at the network level of the OSI model and features more sophisticated addressing software than bridges. Whereas bridges simply pass along everything that comes to them, routers can determine preferred paths to a final destination.

Router. A **router** operates at the network level of the OSI model and features more sophisticated addressing software than bridges. Whereas bridges simply pass along everything that comes to them, routers can determine preferred paths to a final destination. Routers also perform useful network management functions. They can break a network into subnets to create separate administrative network domains, thus helping to better distribute network management. Routers are also used as security firewalls between networks and the public Internet. The firewall keeps unwanted messages and/or users out of the organization's network. A specific router works with only one particular protocol.[12]

gateway

a device that operates at or above the OSI transport layer and links LANs or networks that employ different, higher-level protocols, thus allowing networks with very different architectures and using dissimilar protocols to communicate

Gateway. A **gateway** operates at or above the OSI transport layer and links LANs or networks that employ different, higher-level protocols. Data received by a gateway must be restructured into a format understandable by the destination network. Thus, a gateway allows networks with very different architectures and using dissimilar protocols to communicate. Corporations often have a gateway to other types of networks so that workers can access the programs and data contained on networks outside their geographic region. In most cases, a user can click an icon or give a few simple commands to access these other networks. The user's computer and the server on the local area network will then automatically perform the tasks needed to link into the other network.

switch

a device that routes or switches data to its destination

Switch. A data **switch** is a device that routs or switches data to its destination. A switch needs to have a way of establishing the desired connection. There are two main ways of doing this. In a matrix approach, each input channel has a predefined connection with each output channel. To pass something from an input channel to an output channel is merely a matter of following the connection. In a shared memory approach, the input controller writes material to a reserved area of memory and the specified output channel reads the material from this memory area.

FIGURE 6.21

Bridges, Routers, and Gateways

A switch can perform the role of either a router or a gateway. (Source: Reprinted with permission from *Information Systems for Managers* 3/e, pp. 134–135, by George Reynolds, Copyright 1995 by West Publishing.)

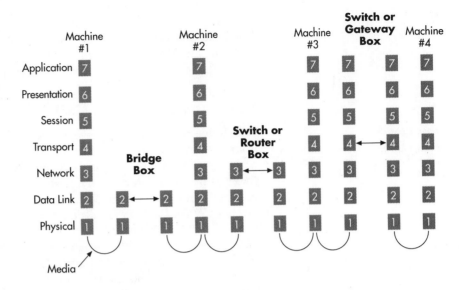

A switch may also need to translate the input before sending it to an output channel. In general, switches are replacing earlier, less flexible devices such as bridges and gateways. For example, a gateway may be able to connect two different architectures, but a switch may be able to connect several.[13]

TELECOMMUNICATIONS APPLICATIONS

It is easy to see how, perhaps more so than any other IS component, telecommunications and networks can be applied to support information systems and organizational goals. For example, suppose a business needs to develop an accurate monthly production forecast. This can require a manager to download data from databases of sales forecasts from its customers. Telecommunications can provide a network link so that the manager can access the data needed for the production forecast report, which in turn supports the company's objective of better financial planning.

The consumer goods giant Procter & Gamble uses local area networks in all its plants to link office and plant workers to common software and shared databases and to provide e-mail services. The result is faster, more cost-effective, higher-quality product manufacturing. Other organizations transfer millions of important and strategic messages from one location to another every day. Telecommunications has become a critical component of information systems. In some industries it is almost a requirement for doing business; most companies could not survive without it. This section will look at some significant business applications of networks.

Linking Personal Computers to Mainframes and Networks

One of the most basic ways telecommunications connect an individual to information systems is by connecting personal computers to mainframe computers so data can be downloaded or uploaded. For example, a data file or document file from a database can be downloaded to a personal computer for an individual to use. Some communication software programs will instruct the computer to connect to another computer on the network, download or send information, and then disconnect from the telecommunications line. These are called unattended systems because they perform these functions automatically, without user intervention.

Voice and Electronic Mail

voice mail

technology that enables users to leave, receive, and store verbal messages for and from other people around the world

With **voice mail**, users can leave, receive, and store verbal messages for and from other people around the world. Suppose Leslie May calls Paul Davis, who is out of his office. A digitized voice tells her to enter a code for Paul (perhaps his extension number). Leslie will then hear a voice message from Paul and can leave a message, just as she would if she reached an answering machine. When Paul Davis calls the voice mail system and enters his access code, he will hear his messages, including the one Leslie left. In some voice mail systems, a code can be assigned to a group of people instead of an individual. Suppose the code 100 stands for all 250 sales representatives in a company. If Leslie calls the voice mail system, enters the number 100, and leaves her message, all 250 sales representatives will receive Leslie's message.

e-mail

technology that enables a sender to connect his or her computer to a network, type in a message, and send it to another person on the network

People can also send messages to others via electronic mail, also called **e-mail**. With the right hardware and software, a sender can connect his or her computer to a network, type in a message, and send it to another person on the network. The receiving person is informed that there is a message waiting and can access it through communications software on his or her computer. For example, a plant manager in Milwaukee can use e-mail to send a message about a product change to a salesperson in Miami. The salesperson can read the e-mail and respond with an e-mail message asking specific questions about the new product. The plant manager can answer the questions using e-mail. All this can take place in less than an hour (provided both users are diligent in reading their messages!). It is also possible to attach files (spreadsheet, word processing document, etc.) to an e-mail message. Read the "FYI" box to learn about available free e-mail software.

Electronic Software and Document Distribution

electronic software distribution

a process that involves installing software on a file server for users to share by signing onto the network and requesting that the software be downloaded onto their computers over a network

Electronic software distribution involves installing software on a file server for users to share by signing onto the network and requesting that the software be downloaded onto the users' computers over a network. Electronic software distribution is faster and more convenient than traditional ways of acquiring software and significantly less costly and more efficient than having a network administrator constantly install upgrades. Wells Fargo Bank, for example, saves time and money by electronically sending software upgrades to the 10,000 users on its network. Companies are also developing services that allow buyers to order software and have it delivered and installed, without ever leaving their computer. One problem with electronic software distribution is the size of software programs—downloading large programs requires high-capacity telecommunications media and takes a lot of time. Controlling software piracy is another issue that must be addressed.

Networks also allow organizations to transmit documents without using paper at all. It is not known how many millions of dollars companies spend on printing, distributing, and storing documents of all types, but the amount

FYI

Free E-mail

"Free" e-mail services are readily available on the Internet. These services are simple to use and allow you to check your mail from any Web screen. Free e-mail services provide an e-mail address and a mailbox. As with any e-mail service, you can send messages back and forth to anyone on the Internet. Most services allow you to send and receive computer files as attachments, provide folders for storing related messages, and filters to screen mail. However, to use your free e-mail service, you need Web access, so more than likely you are paying some sort of fee whether it is to an on-line service provider like America Online or an Internet service provider like AT&T's WorldNet. As companies, schools, libraries and other institutions provide free Web access, this could change. For frequent travelers, free e-mail makes it possible to leave your laptop at home—simply get access to a Web browser anywhere and you can check your mail. This approach is particularly useful for international travelers, where differences in power outlets and Internet access instructions can make it extremely difficult to connect to the Internet.

One of the drawbacks is that free e-mail accounts typically show one or more ads on each new page that you call up—not just your mailbox window, but also on individual messages and the blank screens used to create new e-mail. (After all, the e-mail service provider has to make a profit somehow, so they charge for advertising.) Another drawback is that the free services may be overloaded at times and reading your mail can turn into a slow and aggravating wait, or worse yet, the e-mail service is down altogether. But then again, who has not experienced this with a service that charges for its e-mail?

To sign up for free e-mail, go to one of the sites listed below and fill out a brief registration form. Be sure to specify your preferences when you're asked if you wish to be excluded from marketing offers and the service's member directory.

Service	Address
JUNO	http://www.juno.com
EXCITE	http://mail.excite.com
YAHOO	http://mail.yahoo.com
HOTMAIL	http://www.hotmail.com
MAILCITY	http://www.mailcity.com
NETADDRESS	http://netaddress.com

DISCUSSION QUESTIONS

1. Visit one or more of these sites. What information are you required to provide to register? Read the service's privacy policy to see if the provider agrees to not sell your address or the information you provide them.
2. What other advantages or disadvantages are associated with free e-mail?

Sources: Adapted from Thomas E. Webber, "Free Mail May Be a Good Deal, Even if It Has a Hidden Price," *The Wall Street Journal*, January 22, 1998, p. B1; Juno Web site at http://www.juno.com, accessed May 31, 1998; and the Yahoo! Web page at: http://www.mail.yahoo.com, accessed May 31, 1998.

---

electronic document distribution

a process that involves transporting documents—such as sales reports, policy manuals, and advertising brochures—over communications lines and networks

is staggering. **Electronic document distribution** involves transporting documents—such as sales reports, policy manuals, and advertising brochures—over communications lines and networks. Electronic document distribution software allows word processing and graphics documents to be converted into binary code and sent over networks. Acrobat from Adobe, for example, is a software package that allows documents to be transmitted between different types of computer system platforms. For example, a color

E-COMMERCE
Comp-U-Card

Walter A. Forbes is the founder of Comp-U-Card and has built CUC International Inc. into two dozen mail-order shopping, travel, auto, entertainment, and financial-service clubs. All together these clubs have nearly 70 million members, and in 1996 revenues reached $2.4 million.

In June 1997, Forbes started an electronic superstore called netMarket, offering price discounts of 10 percent to 50 percent off manufacturers' list prices on some 250,000 brand name products—everything from light bulbs to new cars. Buyers must pay membership fees to get those discounts, just like CUC's regular shopping clubs. By the end of 1998, netMarket shoppers will be able to buy nearly 90 percent of what they need in their homes.

The question is: Will enough people make purchases at netMarket? CUC has offered shopper's clubs on the Web since 1995 at its CUC.com site. This site provides shoppers with on-line access to its traditional clubs—Shoppers Advantage, Traveler's Advantage, and AutoAdvantage. A $49.95 annual membership is charged for each club. But these Web sites are simply interactive versions of CUC's traditional 800-number approach that claim some 350,000 interactive members—a decent following, but not a record setter.

Forbes plans to turn netMarket into a mall site that links all these clubs—and more—into one Web site. In real malls, clustering stores makes shopping more convenient, but on the Web, zipping from one merchant to another is already just a mouse click away. Today, netMarket offers deals on new and used cars, travel services, a flea market, an auction site, books, games, and flowers. In the near future, Forbes plans to add videos, financial services, and classified advertising—all for a single fee of $69.95 a year. Furthermore, all the CUC Web sites will be linked to increase shopper traffic. A Web surfer who visits the travel section of CUC's bookseller, Book Stacks Unlimited, for example, might see a link to a vacation trip to Maui and find details just a click away in netMarket.

Unfortunately, malls have not done well on the Internet so far. In June 1997, IBM gave up on its cybermall, World Ave. And iMall, which has some 1,600 on-line storefronts, lost $1 million on revenues of $9.2 million in the first six months of 1997. "On-line malls are dead," says analyst Kate Dellagen of Forrester Research Inc.

Forbes does not agree. He predicts that in a few years, Web shopping habits will mimic those in the real world, where 80 percent of sales are concentrated in a handful of merchants. He states that netMarket is not just a cybermall but a discounter that promises to match the lowest prices of its rivals. Also, netMarket's source of revenue is different from most Web merchants. CUC earns profits on its membership fees, not on the products, which it delivers as the middleman. Based on an average $49.95 membership, CUC loses $58.95 the first year in marketing and overhead. However, in subsequent years, CUC earns $30 profit per member. On average 70 percent of all members renew, thus generating a steady cash flow. On-line members are even more lucrative, since they require even less telemarketing support.

In October 1997, CUC merged with competitor HFS, Inc., which owns brand names including Avis and Century 21 and makes its money on franchiser royalties. The combined company, Cendant Corp., has annual revenues of over $5 billion.

DISCUSSION QUESTIONS

1. Visit the CUC site at the CUC.com site. What are you able to do at this site without paying the membership fee?
2. Visit other mall sites on the Internet. How are they different from the CUC site?

Sources: Adapted from Susan Jackson, "Point, Click—and Spend," *BusinessWeek*, September 15, 1997, pp. 74–76 and the CUC Web site at http://www.cuc.com, accessed August 3, 1998.

advertising pamphlet can be created by an Apple Macintosh and sent electronically to an IBM personal computer. Electronic document distribution can save a substantial amount of time and money versus standard mailing and storage of hard-copy documents. Read the "E-Commerce" box to learn how one company is employing telecommunications to augment its mail-order operations.

Telecommuting

telecommuting

enables employees to work away from the office using personal computers and networks to communicate via electronic mail with other workers and to pick up and deliver results

More and more work is being done away from the traditional office setting. Many enterprises have adopted policies for **telecommuting** that enable employees to work away from the office using personal computers and networks to communicate via electronic mail with other workers and to pick up and deliver results. It is estimated that there are between six million and eight million U.S. workers who could be classified as telecommuters today.

There are several reasons why telecommuting is popular among workers. Single parents find that it helps in balancing family and work responsibilities. Telecommuting eliminates the daily commute. It enables qualified workers who may be unable to participate in the normal workforce (e.g., those who are physically challenged or who live in rural areas too far from the city office to commute on a regular basis) to become productive workers. Extensive use of telecommuting can lead to decreased need for office space, potentially saving a large company millions of dollars. Corporations are also being encouraged by public policy to try telecommuting as a means of reducing traffic congestion and air pollution.

Correctly implemented, telecommuting can provide your company with substantial competitive advantages in workforce recruitment and retention, productivity, real estate and office overhead, customer service, and corporate image and goodwill. Pacific Bell estimates that telecommuting will save the company over $20 million in office leasing over a five-year period. IBM has documented reductions of 40 to 60 percent in real estate costs per site, over $30 million annually. In a telecommuting project recently completed for the city of Los Angeles, "sick time" was reduced by an average of five days per year. At American Express, telecommuters reportedly handle 26 percent more calls and produce 43 percent more business (measured by sales per employee) than their office-based counterparts.[14]

Some types of jobs are better suited for telecommuting than others. These include jobs held by salespeople, secretaries, real estate agents, computer programmers, and legal assistants, to name a few. It also takes a special personality type to be effective while telecommuting. Telecommuters need to be strongly self-motivated, organized, able to stay on track with minimal supervision, and have a low need for social interaction. Jobs not good for telecommuting include those that require frequent face-to-face interaction, need much supervision, and have lots of short-term deadlines. Employees who choose to work at home must be able to work independently, manage their time well, and balance work and home life.

Videoconferencing

videoconferencing

systems that combine video and phone call capabilities with data or document conferencing

Videoconferencing enables people to have a conference by combining voice, video, and audio transmission. Not only are travel expenses and time reduced, but managerial effectiveness is increased through faster response to problems, access to more people, and less duplication of effort by geographically dispersed sites. Almost all **videoconferencing** (Figure 6.22) systems combine video and phone call capabilities with data or document conferencing. You can see the other person's face, view the same documents, and swap notes and drawings. With some of the systems, callers can make changes to live documents in real time. Many businesses find that the

FIGURE 6.22

Videoconferencing

Videoconferencing allows participants to conduct long-distance meetings "face to face" while eliminating the need for costly travel.
(Source: Courtesy of Zydacron, Inc.)

document and application sharing feature of the videoconference enhances group productivity and efficiency. Meeting over phone lines also fosters teamwork and can save corporate travel time and expense. Group videoconferencing is used daily in a variety of businesses as an easy way to connect work teams. Members of a team go to a specially prepared videoconference room equipped with sound-sensitive cameras that automatically focus on the person speaking, large TV-like monitors for viewing the participants at the remote location, and high quality speakers and microphones. It costs around $60,000 to set up a typical group videoconferencing room.[15] There are additional expenses associated with use of the telecommunications network to relay voice, video, data, and images.

Many companies are using desktop videoconferencing systems. Combining audio, video, and data applications for two or more personal computers can save a lot of travel expense and time. Virtual Mortgage Network, in Newport Beach, California, and Flagstar Bank, in Bloomfield Hills, Michigan, have cut the time required to get mortgage approval from weeks to a few hours. Home buyers use a personal computer located in a real estate broker's office to contact a mortgage counselor or an underwriter on another personal computer and talk face to face as they decide on the type of mortgage they want. Officials at Owens Corning Fiberglas, in Toledo, Ohio, cut the time it took to bring to market an insulation product used in Whirlpool appliances by sharing drawings at semiweekly desktop videoconferences. Kimmel Cancer Center, at Thomas Jefferson University in Philadelphia, has helped doctors at three suburban hospitals discuss diagnosis and treatment of cancer. They can view and diagram patient images on-screen. For more on the importance of telecommunications in the medical field read the "Making a Difference" box.

Electronic Data Interchange

electronic data interchange (EDI)

an intercompany, application-to-application communication of data in standard format, permitting the recipient to perform the functions of a standard business transaction

Electronic Data Interchange (EDI) is an intercompany, application-to-application communication of data in standard format, permitting the recipient to perform the functions of a standard business transaction. The EDI purchase order is the most common information transmitted according to a recent survey with 76 percent of manufacturers and 57 percent of wholesalers and chain retailers using EDI. Manufacturer and broker survey respondents are receiving 52 percent of their orders via EDI, while wholesalers and chain retailers are transmitting 43 percent of their orders by EDI.

Connecting corporate computers among organizations is the idea behind electronic data interchange. EDI uses network systems and follows standards and procedures that allow output from one system to be processed directly as input to other systems, without human intervention.

MAKING A DIFFERENCE
Hospitals Use Telemedicine to Deliver Service to Rural Areas

Healthcare in the twenty-first century will be built on the basis of expanded access to information, including distant access to medical professionals and services.

Telemedicine is specifically defined as the delivery of medical care to individuals throughout the world by combining advancing telecommunications technology with medical expertise. Its goal is to provide high-quality healthcare at a relatively low cost to patients throughout the world. It provides patients who otherwise may receive no care at all with quick access to high-quality medical services at a reasonable price. Telemedicine can also be utilized to educate future medical professionals. Supporters of this new type of medical practice envision a huge medical network that spans the globe and links medical providers to patients through various telecommunications devices ranging from telephones to digital compressed video.

The first telemedicine applications appeared in 1959 when X-ray images were sent across telephone lines. Today, technology has made tremendous strides, and physicians can now consult patients through digital compressed video. Viewing patients through the use of video is the most interactive method of doctor-patient interaction in telemedicine and has been used to treat injured workers at distant work locations. By initiating treatment immediately, the injured patient no longer has to wait to be transported to a medical facility, thus saving valuable time that can make the difference between life and death.

Networking technology is crucial for telemedicine. Much of what is done is image management and distribution including applications such as moving X rays around on the network. The equipment for capturing images is extremely expensive. Also, the expertise required to read the X rays is very limited and typically found in large metropolitan hospitals. It's cheaper and much faster to move the images over the network than to move the patients or the specialists.

There is a shortage of specialty care in rural areas. With the use of telemedicine, urban hospitals can provide rural hospitals with specialists' expertise without having to move patients from their own local hospital. Emotionally, this benefits patients since they no longer have to wait through long, anxious times to receive a consultation from a specialist. It also reduces medical expenses since the patient or insurance company no longer must pay for travel costs.

Of course there are some drawbacks to telemedicine. A virtual consultation cannot take the place of the traditional doctor-patient relationship, which is very important in many types of illnesses. When this doctor-patient relationship becomes virtual, the privacy rights of the patients also come into play. Many believe that with the transfer of so much patient information between hospitals the patient's privacy can no longer be guaranteed. However, telemedicine advocates argue that so much information is already stored and transferred that there simply is no real change in the way patient records are handled.

The Arizona Telemedicine Program is a multidisciplinary clinical program of The University of Arizona Health Sciences Center. The program was created in 1996 by the Arizona legislature to establish pilot projects demonstrating the effectiveness of telemedicine in delivering better healthcare to Arizona's medically underserved rural areas. The eight-site program began in July 1996. Site selection, needs assessments, and equipment/networking evaluations occurred over the first six months. The sites identified by the state legislature for inclusion in this project were Cottonwood, Holbrook, Parker, Payson, Sierra Vista, Springerville, Tuba City, and a site to be designated by the Department of Corrections. All these sites are in small, remote towns.

The program has been able to enhance healthcare delivery to medically underserved populations throughout the state. It has increased access to medical specialty services while decreasing healthcare costs. The program has also encouraged physicians, nurses, and other healthcare professionals to establish practices in underserved rural areas.

DISCUSSION QUESTIONS

1. Imagine that you are the administrator for a small, rural hospital. How would you learn more about how telemedicine could enhance your hospital's services? What sort of technology and expertise would you need to take full advantage of this revolution in medicine?

2. Imagine that you are the administrator for a large, urban hospital. What sort of technology and expertise would you need to take full advantage of this revolution in medicine? What potential issues and concerns might you have as your hospital begins to expand its use of telemedicine?

Sources: Adapted from Monu Janah, "Health Care by Cisco," *InformationWeek*, February 23, 1998, pp. 116–117; "What Is Telemedicine," the Telemedicine History Web site at http://tie.telemed.org, accessed August 16, 1998; the Arizona Telemedicine Program Web site at http://www.ahsc.arizona.edu, accessed August, 16, 1998; "The Arizona Rural Telemedicine Network," the University of Arizona College of Medicine and Arizona Health Service Web site at http://www.ece.arizona.edu, accessed August 16, 1998.

With EDI, the computers of customers, manufacturers, and suppliers can be linked (Figure 6.23). This technology eliminates the need for paper documents and substantially cuts down on costly errors. Customer orders and inquiries are transmitted from the customer's computer to the manufacturer's computer. The manufacturer's computer can then determine when new supplies are needed and can automatically place orders by connecting with the supplier's computer.

For some industries, EDI is a necessity. For many large companies, including General Motors and Dow Chemical, it is not uncommon for most computer input to originate as output from another computer system. Some companies will do business only with suppliers and vendors using compatible EDI systems, regardless of the expense or effort involved. As more industries demand that businesses have this ability to stay competitive, EDI will cause massive changes in the work activities of companies. Companies will have to change the way they deal with processes as simple as billing and ordering, while new industries will emerge to help build the networks needed to support EDI.

Public Network Services

public network services

services that give personal computer users access to vast databases and other services, usually for an initial fee plus usage fees

Public network services give personal computer users access to vast databases and other services, usually for an initial fee plus usage fees.

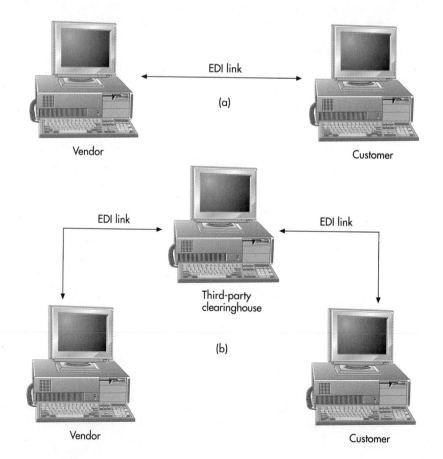

FIGURE 6.23

Two Approaches to Electronic Data Interchange

EDI is no longer just the wave of the future; many organizations now insist that their suppliers operate using EDI systems. Often, the EDI connection is made directly between vendor and customer (a); alternatively, the link may be provided by a third-party clearinghouse, which provides data conversion and other services for the participants (b).

Public network services allow customers to book airline reservations, check weather forecasts, get information on TV programs, analyze stock prices and investment information, communicate with others on the network, play games, and receive articles and government publications. Fees, based on the services used, can range from less than $15 to more than $500 per month.

The major providers of public network services are Dow Jones News/Retrieval, Microsoft, America Online, and Prodigy. Dow Jones News/Retrieval is operated by Dow Jones, the publisher of *The Wall Street Journal*. Primarily, it provides investment-related information to individuals and corporations. America Online (AOL) and Prodigy provide a vast array of services, including news, electronic mail, and investment information. The Microsoft Network, a newer service offered by Microsoft with its Windows 95 operating system, offers similar services. AOL is the number one provider of public network services in terms of size. With more than 11 million paying subscribers, it reaches as many homes as cable operators Time Warner or TeleComunications Inc. and is adding more than 10,000 users per day. It also acquired CompuServe in early 1998 with its 2.5 million members.[16] On weeknights during prime time, the number of people logged on to AOL peaks at around 650,000, putting it in a league with cable networks MTV and CNN.[17]

Specialized and Regional Information Services

With millions of personal computers in businesses across the country, interest in specialized and regional information services is increasing. Specialized services, which can be expensive, include legal, patent, and technical information. For example, investment companies can use systems, such as Quotron and Shark, to get up-to-the-minute information on stocks, bonds, and other investments. Regional services include local electronic bulletin boards and electronic mail facilities that offer information regarding local club, school, and government activities. An electronic bulletin board is a message center that displays messages in electronic form, much like a bulletin board that displays paper messages in schools and offices. An electronic bulletin board can be accessed by subscribers with a personal computer and network equipment and software. In addition to regional bulletin boards, national and international bulletin boards are available for people and groups with special interests or needs. These types of bulletin boards exist for many users, such as users of certain software packages or users with certain hobbies. Many public network services, including Prodigy and America Online, provide access to hundreds of different bulletin boards on a variety of topics and interest areas.

Distance Learning

Telecommunications can be used to extend the classroom. The University of Maine broadcasts introductory management classes on televisions and computers that are networked throughout the state of Maine. The National Technological University in Fort Collins, Colorado, uses a network to link thousands of engineering students with faculty from about 40 universities

distance learning

the use of telecommunications to extend the classroom

throughout the United States. Often called **distance learning** or cyber-class, these electronic classes are likely to thrive in the future.

Some vendors, such as HyperGraphics, are developing software to support distance learning. With this software, instructors can easily create course home pages on the Internet. Students can access the course syllabus and instructor notes on the Web page. E-mail mailing lists can be established so students and the instructor can easily mail one another as a means of turning in homework assignments or commenting and asking questions about material presented in the course. It is also possible to form chat groups so that students can work together as a "virtual team" that only meets electronically to complete a group project.

Information Systems Principles

Establishing a communications link between two hardware devices requires that they speak the same language; the characteristics of the medium are an important consideration.

An unmistakable trend of communications technology is that more people are able to send and receive all forms of information over greater distances at a faster rate.

Telecommunications reduces the amount of time needed to transmit information that can drive and conclude business actions.

The effective use of networks and their applications can turn a company into an agile, powerful, and creative organization, giving it a long-term competitive advantage.

Most communications protocols are described in terms of layers of standards or procedures.

Networks are key to delivering the right information to the right person in the right fashion.

■ SUMMARY

1. *Define the terms* communications *and* telecommunications *and describe the components of a telecommunications system.*

Communications is any process that permits information to pass from a sender to one or more receivers. Communications of all types form a major part of any business system. Telecommunications refers to the electronic transmission of signals for communications, including such means as telephone, radio, and television. Telecommunications

has the potential to create profound changes in business because it lessens the barriers of time and distance.

The telecommunications model starts with a sending unit, such as a person, a computer system, a terminal, or another device, that originates the message. The sending unit transmits a signal to a telecommunications device. The telecommunications device performs a number of functions, which can include converting the signal into a different form or from one type to another. A telecommunications device is a hardware component that allows electronic communication to occur or to occur more efficiently. The telecommunications device then sends the signal through a medium. A telecommunications medium is anything that carries an electronic signal and interfaces between a sending device and a receiving device. The signal is received by another telecommunications device that is connected to the receiving computer. The process can then be reversed and another message can go back from the receiving unit to the original sending unit.

2. *Identify several types of communications media and discuss the basic characteristics of each.*

The telecommunications media that physically connect data communications devices are twisted-pair wire cable, coaxial cable, and fiber-optic cable. Twisted-pair cable consists of pairs of twisted wires, either shielded or unshielded. A coaxial cable consists of an inner conductor wire surrounded by insulation (the dielectric) and a non-conductive insulating shield (the jacket.) Fiber-optic cable consists of thousands of extremely thin glass or plastic strands bound together in a sheathing (a jacket) for transmitting signals via light. Fiber-optic cables are faster and more reliable than other types, but they are expensive to install. Microwave transmission consists of a high-frequency radio signal sent through the air. Other transmission options that give organizations portability and flexibility in transmitting data are cellular and infrared transmission.

3. *Identify several types of telecommunications hardware devices and discuss the role that each plays.*

Telecommunications hardware devices include modems, fax modems, special-purpose modems, multiplexers, and front-end processors.
Modems convert signals from digital to analog for transmission, then back to digital. A fax modem allows businesses to transmit text, graphs, photographs, and other digital files via standard phone lines. Special-purpose modems include cellular modems placed in portable laptop computers to allow communication with other computer systems and devices and cable modems designed for use with coaxial cable media. A multiplexer allows several signals to be transmitted over a single

communications medium at the same time. A front-end processor is a special-purpose computer that connects a midrange or mainframe computer to a large number of communications lines and performs a number of tasks. Some of the tasks include polling, providing automatic answering and calling, performing circuit checking and error detection, developing logs or reports of all communications traffic, editing basic data entering the main processor, determining message priority, choosing alternative and efficient communications paths over multiple data communications lines, and providing general data security for the main system.

4. *Identify the benefits associated with a telecommunications network.*

The effective use of networks can turn a company into an agile, powerful, and creative organization, giving it a long-term competitive advantage. Networks can be used to share hardware, programs, and databases across the organization. They can transmit and receive information to improve organizational effectiveness and efficiency. They enable geographically separated workgroups to share documents and opinions, which fosters teamwork, innovative ideas, and new business strategies.

5. *Name three distributed processing alternatives and discuss their basic features.*

When an organization needs to use two or more computer systems, one of three basic data processing strategies may be followed: centralized, decentralized, or distributed. With centralized processing, all processing occurs in a single location or facility. This approach offers the highest degree of control. With decentralized processing, processing devices are placed at various remote locations. The individual computer systems are isolated and do not communicate with each other. With distributed processing, computers are placed at remote locations but are connected to each other via telecommunications devices. Three distributed processing alternatives include terminal-to-host, file server, and client/server.
With terminal-to-host architecture, the application and database reside on the same host computer, and the user interacts with the application and data using a "dumb" terminal. Since a dumb terminal has no data processing capability, all computations, data accessing and formatting, as

well as data display, are done by an application that runs on the host computer.

In the file server approach, the application and database reside on the same host computer, called the file server. The database management system runs on the end user's personal computer or workstation. If the user needs even a small subset of the data that resides on the file server, the file server sends the user the entire file that contains the data requested, including a lot of data the user does not want or need. The downloaded data can then be analyzed, manipulated, formatted, and displayed by a program that runs on the user's personal computer.

A client/server system is a network that connects a user's computer (a client) to one or more host computers (servers). A client is often a PC that requests services from the server, shares processing tasks with the server, and displays the results. Many companies have reduced their use of mainframe computers in favor of client/server systems using midrange or personal computers to achieve cost savings, provide more control over the desktop, increase flexibility, and become more responsive to business changes. The start-up costs of these systems can be high, and the systems are more complex than a centralized mainframe computer.

6. *Define the term network topology and identify five alternatives.*

Network topology is the manner in which devices on the network are physically arranged. Communications networks can be configured in numerous ways, but five designs are most prevalent: bus, hierarchical, star, ring, and hybrid (hybrid networks combine the basic designs of the four other topologies to suit the specific communication needs of an organization).

The physical distance between nodes on the network determines whether it is called a local area network (LAN) or wide area network (WAN), or an international network. The major components in a LAN are a network interface card, a file server, and a bridge and/or gateway. WANs tie together large geographic regions using microwave and satellite transmission or telephone lines. International networks involve communications between countries, linking systems together from around the world. These networks are also called global area networks. The electronic flow of data across international and global boundaries is often called transborder data flow.

7. *Identify and briefly describe several telecommunications applications.*

There are many applications of telecommunications, including the following: personal computer to mainframe links, voice and electronic mail, electronic software distribution, electronic document distribution, telecommuting, videoconferencing, electronic data interchange, public network services, specialized and regional information services, and distance learning. Personal computer to mainframe links enable people to upload and download data. Voice and electronic mail users can leave, receive, and store messages from other people around the world. Electronic software distribution involves installing software on a computer by sending programs over a network so they can be downloaded into individual computers. Electronic document distribution allows organizations to transmit documents without the use of paper, thus cutting costs and saving time. Telecommuting employs information technology to enable workers to work away from the office. Videoconferencing bring groups together in voice, video, and audio transmission. Electronic data interchange (EDI), another rapidly growing area, enables customers, suppliers, and manufacturers to electronically exchange data. EDI reduces the need for manual paper systems while speeding up the rate at which business can be transacted. Public network services give users access to vast databases and services, usually for an initial fee plus usage fees. Specialized services, which are more expensive, include legal, patent, and technical information. Regional services include local electronic bulletin boards that offer e-mail facilities and information regarding local activities. Distance learning is a way to support education of students who are unable to meet frequently with their instructor.

■ KEY TERMS

analog signal 248
bridge 268
bus network 258
centralized processing 256
client/server 263
common carriers 251
communications software 265
computer network 244
data communications 243
decentralized processing 256
dedicated line 251
digital signal 248
Digital Subscriber Line (DSL) 256
distance learning 278
distributed processing 256
electronic data interchange (EDI) 274
electronic document distribution 271
e-mail 270

electronic software distribution 270
file server 263
front-end processor 249
gateway 268
hierarchical network 258
Integrated Services Digital Network
    (ISDN) 254
international network 261
local area network (LAN) 258
modem 248
multiplexer 249
network management software 266
network operating system (NOS) 265
network topology 258
Open Systems Interconnection (OSI)
    model 266
private branch exchange (PBX) 253
protocol 266

public network services 276
ring network 258
router 268
star network 258
switch 268
switched line 251
telecommunications 243
telecommunications medium 243
telecommuting 273
terminal-to-host 263
Transmission Control
    Protocol/Internet Protocol
    (TCP/IP) 266
value-added carriers 251
videoconferencing 273
voice mail 270
wide area network (WAN) 261

■ REVIEW QUESTIONS

1. Define the term *communications*. How does telecommunications differ from data communications?
2. Describe the steps involved in the communications process.
3. What is a telecommunications medium? List three media in common use.
4. Discuss the function and use of a modem and a multiplexer.
5. What is the difference between a switched and a dedicated line?
6. Define the term computer network.
7. Identify three distributed data processing alternatives.

8. What advantages and disadvantages are associated with the use of client/server computing?
9. What is a T1 line? How might it be used?
10. What is a network operating system? What is network management software?
11. Identify five basic types of network topologies.
12. What role do the bridge, router, gateway and switch play in a network?
13. Describe a local area network and its various associated components.
14. What is a wide area network? What is a value-added network?
15. What is EDI? Why are companies using it?
16. What are the key elements of an effective distance learning course?

■ DISCUSSION QUESTIONS

1. Why is effective communications critical to organizational success?
2. What sort of issues would you expect to encounter in establishing an international network for a large, multinational company?

3. No large multinational company should employ centralized data processing. Do you agree or disagree? Why?
4. Client/server computing should always be employed versus other forms of distributed data processing. Do you agree or disagree? Why?

5. In practical terms, how would one distinguish between a LAN and a WAN?

6. If it were available, which would you rather have in your home—T1, ISDN or DSL? Why?

7. What factors are limiting the competitiveness for local telephone service?

8. Consider an industry that you are familiar with through work experience, coursework, or a study of industry performance. How could electronic data interchange be used in this industry? What limitations would EDI have in this industry?

9. What is telecommuting? What are the advantages and disadvantages of telecommuting? Do you anticipate that you will telecommute in your future career?

10. Discuss the pros and cons of conducting this course as a distance learning course.

■ PROBLEM-SOLVING EXERCISES

1. You have been hired as a telecommunications consultant for a small but growing consumer electronics manufacturer. The company wishes to establish a customer service operation for purchasers of its products. In addition to taking customer orders over the phone, reps will answer consumer questions about how to use the products, where to take the products for repairs, how to enhance performance, etc. The company plans to initially train about a dozen people to staff the customer service center. It has hired you to review its needs for the new system and to select a telecommunications solution. How would you proceed? What questions need to be asked? What telecommunications equipment is needed? What might be the annual cost for this operation? Use word processing software to prepare a list of at least ten questions that you need to answer to evaluate this project. Make some assumptions about the answers and write your opinion of this project. Embed a spreadsheet that details the approximate annual cost for the operation.

2. You work in the Los Angeles national headquarters of a large multinational company. After years of fighting the smog and traffic, you and your fellow workers have decided to do something about it. They have elected you as spokesperson to develop a recommendation for management to implement a telecommuting program. Use PowerPoint or similar software to make a convincing presentation to management to adopt such a program. Your presentation must address issues such as the benefits to the company, the costs of the hardware and software, selection process for individuals to participate, and benefits to the employees.

■ TEAM ACTIVITY

1. With a group of your classmates, visit one of the following: a cellular phone company, the college computing center, a phone or cable company, the police department, or another interesting business that uses telecommunications. Prepare a report on how the business is planning to use communications to enhance access to information—both yours and its. Find out what kind of telecommunications media and devices it currently uses and what changes the business might make to improve data and information access.

2. Have each member of your team contact a different provider to get information about local telecommunications service options. What services are available (T1, DSL, ADSL, etc.) to deliver high-speed data access into a home? What is the installation and ongoing operating cost of each of these options? Discuss your findings with the class. How would you evaluate the competitiveness of the local service providers in your area?

■ WEB EXERCISE

Digital Subscriber Lines (DSL) and Asynchronous Digital Subscriber Lines (ADSL) offer faster access to the Internet. Search the Internet to get additional information about this communications technology.

Describe a few companies that manufacture this type of equipment and their products. You may be asked to develop a report or send an e-mail message to your instructor about what you found.

■ CASES

1 NTT Scales Up

Nippon Telegraph and Telephone was established over a hundred years ago. Today it is the principal telecommunications company in Japan, with 58 million subscribers and over 500 branches throughout the country. Globally, over 3,200 multinational companies look to NTT for solutions to their telecommunications requirements. NTT America was founded in 1987 as a wholly-owned subsidiary of Nippon Telegraph and Telephone. Today, NTT America is a leading provider of state-of-the-art products and services, as well as a major player in communications research.

The current pace of technological innovation requires new information services that go far beyond conventional telephony. NTT has been transforming itself into a multimedia company to meet today's and tomorrow's needs. NTT America plays a significant role in that transformation.

NTT Mobile is by far the largest supplier of cellular services in the world. The Tokyo-based company had 10 million subscribers as of March 1997. That number is expected to grow to 17 million by March 1998, even with the economic slowdown in Japan. However, competition is increasing. For example, regional competitor TuKa Cellular is increasing the number of cellular subscribers by nearly 10 percent per month. Such competition is putting pressure on the one-time monopoly NTT Mobile to lower rates and improve customer service.

NTT Mobile's core mainframe systems were preventing the company from coping with rapid growth. Developed in the late 1980s, these systems were terminal-to-host applications that supported customer service representatives in 2,000 retail locations. A number of disparate, nonintegrated

systems were bundled together to support the reps in initiating new customer accounts, handling customer support calls, and performing administrative functions. The lack of integration across systems was a serious problem. One system handled new accounts processing, another tracked equipment inventory levels, and a third stored customer information. Not only were the systems not integrated, they relied on batch processing to update corporate data. Things were so bad that a new customer would buy an NTT service, begin to use it, then return to the store for repair before NTT could get the customer's information from the accounts processing system into the customer information system. Then, when the customer showed up for service, the NTT systems had no data about him or her. Obviously, this made it difficult to provide good customer service.

To improve customer service, NTT formed the ALADIN project team to provide an integrated, up-to-the-minute view of the company's distributed mobile telecommunications business. The company had to replace the stand-alone, mainframe based applications with new, fully integrated client/server applications running on a 40-processor, two-node cluster configuration of Sun Microsystem's large Enterprise 6000 SPARC-based symmetric multiprocessing server.

NTT Mobile completed a successful pilot of the ALADIN system using the new system in two branch offices serving 200,000 customers. However, performance slowed to a crawl when the company attempted to roll out the system to handle 1 million customers. The root cause of the performance problem was that the ALADIN applications on the two-node Enterprise 6000 cluster were bogging down as they contended for common blocks of data on disk (called data contention). This problem is not uncommon in large clustered environments, because nodes in systems such as an Enterprise 6000 cluster share disk storage and a database. The database system manages users' access to the data, keeping one user away from data while it is being used and updated by another.

The applications, operating systems, and the database had to be tuned to minimize contention. After struggling awhile, the team decided to move the ALADIN application logic to separate application servers—running on Sun SC 2000E hardware—thereby creating a three-tier client/server architecture. Additional tuning was done to the database management software and operating system to reduce data contention and boost the performance of the operating system. The tuning efforts took several months but eventually system performance improved dramatically and NTT Mobile completed the ALADIN rollout. Today the new system supports 13,000 users, generating almost 25,000 database updates every minute and managing some 40 terabytes of historical data. When NTT began the conversion, the number of users, updates, and data placed a greater workload on the Sun hardware than any other Sun customer.

Already the company has seen increased customer satisfaction and operational efficiency. Company agents are now able to process customer requests for service or support faster. The company has been able to substantially reduce the number of administrative workers needed to handle customer transactions, shifting them to other jobs. The system has been so successful, in fact, that NTT Mobile plans to extend ALADIN to its pager business, thus adding 5 million customers.

1. Would you describe NTT's new system as a two-tier or three-tier client/server system? What type of computer is at each tier?
2. NTT experienced a performance problem when the pilot system expanded. Would you expect to see a similar performance problem when the system is expanded to serve the 5 million pager customers? How might this be avoided?

Sources: Adapted from NTT America Web site at http:www.nttamerica.com, accessed May 26, 1998 and Jeff Moad, "Sun Rises to the Occasion," *PC Week*, March 2, 1998, pp. 77, 83.

2 Canada Privatizes Air Traffic Controller System

On November 1, 1996, Nav Canada became the first fully privatized corporation in the world to own and operate a national air navigation service. Privatizing the Canadian government's air navigation system allowed the new entity to raise $600 million for a three-year network upgrade that will make air traffic control operations safer, more efficient, and secure. And, it ultimately will enable Nav Canada to streamline operations and cut annual costs by $135 million.

At its birth, Nav Canada inherited a mature, functioning air navigation service (ANS) with one of the strongest safety records in the world. Transport Canada had built up a solid safety infrastructure during the nearly 60 years that it operated the ANS. The operational components that make up this safety infrastructure were transferred intact to Nav Canada and remain in place today. Since the transfer, however, Nav Canada

has implemented additional safeguards to further enhance the safety of the Canadian air navigation system.

Before privatization, the air navigation system was funded mainly by the Air Transportation Tax levied by the government. Furthermore, its tools were old—Digital Equipment Corp. VAX clusters that required air traffic controllers to hunt for information on six different systems. A major network upgrade was required, and increased taxes would have been insufficient to pay for necessary network upgrades. As a private, non-profit firm, Nav Canada will create a set of fees to charge its customers—the airline industry—for its services.

The upgrade, called the Canadian Air Traffic Control Systems (CATCS) will be implemented in the summer of 1998. Nav Canada will replace the VAX clusters with Hewlett-Packard Co. HP-9000 servers and HP C200 Unix-based workstations. The new systems will give controllers radar data, flight path information, computer-based conflict prediction, weather updates, and navigational aid data on a single system. CATCS also will automatically route flight plans to the appropriate people. Automating the task means there's less chance of human errors, such as losing critical information. The end result will be fewer flight delays and the ability to support routes that are more direct and fuel efficient, which in turn cuts costs. Ultimately, this will eliminate conflicts like several planes vying for the same runway at the same time.

Nav Canada intends to be recognized as a vital force in the international aviation community, noted for its use of world-class technology. When it comes on-line, CATCS will provide Nav Canada with enhanced data processing capability to provide both better information to its controllers and added value for its customers.

Nav Canada has built-in redundancy to achieve nearly 100 percent network up time. Air navigation systems cannot afford to be down even five minutes. Each of Nav Canada's 23 operational sites will be outfitted with three servers—a primary, a backup, and a third server for training purposes.

Ironically, CATCS is more open to security breaches than the old VAX systems. However, Nav Canada is not as concerned with hackers invading the network to steal data as it is with them launching a denial-of-service attack or virus that could crash a portion of the network. To thwart would-be hackers, Nav Canada is using Secure Frame Unit frame-relay encryptor from Sunnyvale, California-based Cylink Corp. and its Privacy Manager, a software key and device management device. The combination of the two will ensure that flight data is transmitted and received without modification or corruption.

The company expects that technology, process improvements, and sound management will allow it to operate with lower staffing levels over time. These reductions will be linked to specific organization change plans, or process improvements. Currently, there are about 1,300 licensed controllers in Nav Canada's Air Traffic Services. By the year 2001 only 800 licensed controllers will be required to support the more than 200,000 people who will fly in Canada each day. The challenge and excitement for Nav Canada is to make significant gains through advanced technology while maintaining control over costs.

1. Discuss what sort of network and telecommunications devices are required to link Nav Canada's 23 operational sites.
2. Imagine that you are an outside consultant called in to review the CATCS system just before it goes "live." Knowing that it is responsible for the safety of over 200,000 people each day, what specific aspects of the system would you want reviewed in detail, tested, or demonstrated?

Sources: Adapted from Laura DiDio, "$600M Net Upgrade Takes Flight," *Computerworld*, February 16, 1998, pp. 49–52; and Navigation Canada's Web site at http://www.navcanada.ca, accessed May 26, 1998.

3 Hotel Vintage Park—A Haven for Telecommuting

In celebration of the Washington State wine industry, each of Hotel Vintage Park's 126 guest rooms is named after local wineries and vineyards. The Seattle hotel's guest rooms are stylishly appointed in rich color schemes of hunter green, deep plum, cerise, and taupe. Custom cherry armoires, exquisite draperies, and tailored beds with columned headboards or unique canopies complete the regal decor. Each guest room features twice-daily maid service, hair dryer, iron, ironing board, complimentary European toiletries, and 100 percent cotton oversized bath towels. To assure a good night's sleep, guests are offered down and hypoallergenic pillows, soundproof windows, thick terry robes, and a stocked private bar for late-night snacking. With rates for a single room running from $185–$375 per night, the Hotel Vintage Park will never be confused with Motel 6.

Last year, 30 of the rooms at Vintage Park were outfitted with either wireless laptops or the IPort Communications System. IPort is a joint effort between Microsoft Corp. and ATCOM/Info that consists of a wall jack and a communications link providing a connection to a Windows NT network server. The IPort solution enables notebook users to plug into an ethernet cable in the wall jacks;

download the IPort software, which secures the connection to the server through encryption; connect to an ethernet hub; and venture out to the Internet over a high-speed communications link. For travelers without laptops, the hotel's wireless laptops are preprogrammed for Internet access, browsers, games, and Microsoft Office applications, including printing capabilities.

Executives at the Hotel Vintage Park initiated the in-room computer and server access due to customer demand. In past years, travelers wanted coffeemakers in their rooms and ironing boards. Then, more affluent travelers expected honor bars with shavers and vitamins. More recently, busy executives wanted two phone lines with a 15-foot phone cord, a fax, and a printer, as well as floppy diskettes and high-speed data communications lines. Despite the large investment in new technology, Vintage Park offers its computer services free in order to attract new business clientele and to keep them coming back.

Currently, staying at a "high-tech" hotel such as the Vintage Park is a novelty. But there is a movement afoot that will soon make public Internet access as ubiquitous as public telephones. Internet kiosks are popping up in hotel lobbies, convention center floors, airports, restaurants, convenience stores, and public transportation terminals. As a result, travelers will always have e-mail access or the Web's information resources at their fingertips.

1. How important is it for you when you travel to be able to stay connected with others through e-mail and be able to access the Web? If you had free access to these services, how would it change your stay at a hotel?
2. Call the national 800 number for several hotel chains and find out which of them provide rooms with personal computers and Internet access. Which hotel chains still do not provide this service or charge for this service?

Sources: Adapted from Stephanie Neil, "Business Checks In," *PC Week*, February 16, 1998, p. 73 and the Hotel Vintage Park Web site at http://www.hotelvintagepark.com, accessed May 27, 1998.

4   Glendale Federal

Glendale Federal is a medium-sized Southern California bank based in Glendale, a Los Angeles suburb. Through 182 branches in California, the bank serves more than 1 million customers and has just over $16 billion in assets. The bank provides a full range of traditional banking services, from checking and savings accounts to consumer and mortgage loans. It also has a discount brokerage subsidiary that provides full stock and mutual fund brokerage services to its existing banking customers.

Glendale Federal has embarked on a major overhaul of its information technology systems to provide high-quality customer service and an expanding array of financial products. The goal is to move key customer service and product information up from the back office to the front lines—putting technology at every point where employees interact with a customer. The bank provides telecommunications access to an on-line brochure that lists and explains the bank's services via a Web site. In the future, it wants telecommunications and the Web site to become important delivery channels to provide customer access to the full range of bank products, such as new accounts, loan applications, and brokerage transactions.

1. Visit the current Glendale Federal Web site at http://www.glenfed.com. and evaluate how well it seems to meet its bank's goal of providing access to the full range of financial products. Write several paragraphs on what you like and do not like about this Web site as well as ideas for how it might be improved.

2. Develop a list of issues that need to be addressed for customers to feel comfortable doing business with their bank via telecommunications over the Internet.

Sources: Adapted from Phillip J. Gill, "Banking on the Future," *Profit*, February 1998, pp. 74–82 and the Glendale Federal Web site at http://www.glenfed.com, accessed May 28, 1998.

CHAPTER 7

The Internet, Intranets, and Extranets

"This is a rocket that has been launched. There's no one who can stop it. The future of computing is defined by the Net."

—Eric E. Schmidt, chief technology officer at Sun Microsystems

Chapter Outline

Use and Functioning of the Internet
  How the Internet Works
  Accessing the Internet
  Internet Service Providers

Internet Services
  E-Mail
  Telnet and FTP
  Usenet and Newsgroups
  Chat Rooms
  Internet Phone and Videoconferencing Services
  Content Streaming

The World Wide Web
  Web Browsers
  Developing Web Content
  Search Engines
  Java
  Push Technology
  Business Uses of the Web

Intranets and Extranets

Net Issues
  Management Issues
  Service Bottlenecks
  Privacy and Security
  Firewalls

Learning Objectives

After completing Chapter 7, you will be able to:

1. Briefly describe how the Internet works, including alternatives for connecting to it and the role of Internet service providers.
2. Identify and briefly describe eight services associated with the Internet.
3. Describe the World Wide Web, including tools to view and search the Web.
4. Outline a process for creating Web content.
5. Describe Java and discuss its potential impact on the software world.
6. Identify who is using the Web to conduct business, and discuss some of the pros and cons of Web shopping.
7. Define the terms *intranet* and *extranet* and discuss how organizations are using them.
8. Identify several issues associated with the use of networks.

Michelin Tire
Cruising the Net to Finish First

You don't get to be the world's tire technology leader by chance: spotting the issues, visualizing the future, saying no to habit and refusing to conform—this is how Michelin has achieved a leadership position. One of Michelin's latest innovations is Bib Net, an intranet link to support its 1,700 North American independent tire dealers.

Bib Net lets the billion-dollar tire manufacturer's independent U.S. dealers easily order tires, check inventory and pricing, and monitor other business transactions. Michelin sees its approach not only as automating routine transactions with its dealers but also as forging stronger relationships in an industry that is marked by cut-throat competition. Development of Bib Net took nearly three years and involved visits to many of Michelin's 1,700 dealers nationwide to explain the concept and solicit ideas. Michelin expects about 300 dealers, who account for about 80 percent of its sales in the dealer channel, to be connected. Dealers can check inventory, order tires, and handle claims processing.

To support these transactions, Bib Net is linked to Michelin's inventory and other information contained in IBM DB2 databases running on an IBM RS/6000 midrange computer. Bib Net has security systems to protect it from unauthorized access, plus each dealer has a unique log-on ID and password and must follow strict authentication procedures to gain access.

Bib Net provides many benefits for dealers. They can check what's in stock and get immediate answers to questions (e.g., what tire fits a 1967 Camaro). Placing an order is simpler for the dealer. Order delivery is faster with orders arriving in just three days versus five days when orders are placed through a sales rep. In addition, dealers can conduct business on Bib Net without being dependent on Michelin's work schedule. For example, most tire dealers are open on Saturday, yet that's when tire manufacturers, such as Michelin, are closed.

The primary benefits for Michelin include reducing some order processing and customer service costs. There is also a savings from no longer printing as many copies of the product directory. However, the biggest benefit for Michelin is the way it is perceived by its dealers. This is a giant step for Michelin to share information with its dealers. This project shows that Michelin is dedicating itself to changing its reputation among dealers by becoming more dealer-friendly.

As you read this chapter, consider the following questions:

- What effect has Internet technology had on communications and business practices?

- How can organizations use links to the Internet to improve customer relations and increase revenue?

Sources: Adapted from Michelin Web site at http://www.michelin.fr, accessed April 23, 1998; and "Good Deal for Michelin Dealers," *InformationWeek*, October 20, 1997, pp. 99–100.

To speed communications and share information, businesses such as Michelin are linking employees, branch offices, and global operations via networks, whether they set up their own or use outside services. Increasingly they are turning to the Internet, the world's largest computer network. Actually, the **Internet** is a collection of interconnected networks, all freely exchanging information (see Figure 7.1). Research firms, colleges, and universities have long been part of the Internet, and now businesses, high schools, elementary schools, and other organizations are joining up as well. Nobody knows exactly how big the Internet is because it is a collection of separately run smaller computer networks with no single place where all the connections are registered. Current estimates are that there are more than 200,000 networks connected to form the Internet, with nearly 100 million computers worldwide connected to it.

Internet

a collection of interconnected networks, all freely exchanging information

USE AND FUNCTIONING OF THE INTERNET

The Internet is truly international in scope, with users on every continent—including Antarctica. However, the United States has the most usage, by far. The use of the Internet in Europe lags that in the United States by about two years. Canada is moving slower in its use of the Internet because of high communications costs and strict government control over its growth. In Africa, Internet connectivity is limited in every country except South Africa. But even there, its use is very limited, since even a slow 14.4-Kbps modem costs more than a month's salary for most people. In Russia, using the Internet's e-mail provides a timely mail service, whereas it may take weeks for an airmail letter to reach the United States. In China, the Internet has been available only since 1993, but plans are under way to connect all the country's universities, institutions, grade schools, and

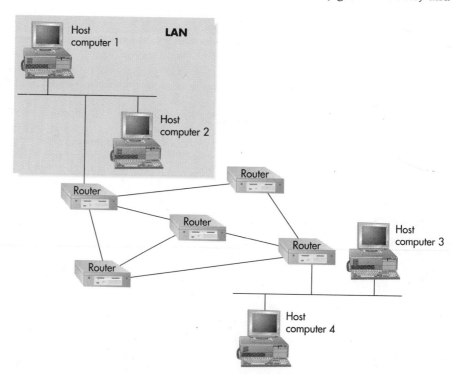

FIGURE 7.1

Routing Messages over the Internet

research organizations by the year 2000. Again, cost is a major issue in China, with the price of a computer and modem beyond the means of most of the country's academics and professionals. Thus, technically the Internet is global, but outside the United States it is seen as overwhelmingly U.S.-centric, inundated with English-language sites and U.S.-generated content.

ARPANET

a project started by the U.S. Department of Defense (DOD) in 1969 as both an experiment in reliable networking and a means to link DOD and military research contractors, including a large number of universities doing military-funded research

   The ancestor of the Internet was the **ARPANET**, a project started by the U.S. Department of Defense (DOD) in 1969. The ARPANET was both an experiment in reliable networking and a means to link DOD and military research contractors, including a large number of universities doing military-funded research. (*ARPA* stands for the Advanced Research Projects Agency, the branch of Defense in charge of awarding grant money. The agency is now known as DARPA—the added *D* is for *Defense*.)

   The ARPANET was highly successful, and every university in the country wanted to sign up. This wildfire growth made it difficult to manage the ARPANET, particularly the large and rapidly growing number of university sites on it. It was decided to break the ARPANET into two networks: MILNET, which included all military sites, and a new, smaller ARPANET, which included all the nonmilitary sites. The two networks remained connected, however, through use of the **Internet Protocol (IP)**, which enabled traffic to be routed from one network to another as needed. All the networks connected to the Internet speak IP, so they all can exchange messages.

Internet Protocol (IP)

conventions that enable traffic to be routed from one network to another as needed

   Unlike a corporate network with a backbone and a centralized infrastructure, the Internet is nothing more than an ad hoc linkage of many networks that adhere to basic standards. Since these networks are constantly changing and being improved, the Internet itself is in a perpetual state of evolution. However, since the Internet is such a loose collection of networks, there is nothing to prevent some participants from using outdated or slow equipment.

   Internet 2 (I2), launched in 1997, and Next Generation Internet (NGI), announced by President Clinton in 1996, are twin programs involving a high-speed network and application development for universities and government agencies. Both networks, which may eventually merge into one program, plan to tap the infrastructure run by the National Science Foundation and partially owned by MCI Communications Inc. That network runs at up to 622 Mbps, and there are plans to increase the speed to 2.4 Gbps in the next five years.[1]

   I2's primary goal is delivery of a fully operational multimedia network by early 1999, and its use will initially be restricted to government entities. There's no telling when, or if, IS shops and the public will get their nodes on the new networks. However, the companies involved and the government expect it to spawn major commercial advances in telecommunications. If history is any guide, these applications will then start cropping up in business networks.

   The Clinton administration has pledged $300 million over three years for NGI, while I2 is being funded privately and publicly through the NGI Initiative. I2 also has the backing of the University Corporation for Advanced Internet Development, a nonprofit group that partners with private enterprise with nearly 120 members and the backing of more than 100 public and private universities.[2] For more information about I2, visit the Web site at http://www.internet2.edu. NGI information can be found at http://www.ngi.gov.

How the Internet Works

The Internet transmits data from one computer (called a host) to another (see Figure 7.1). If the receiving computer is on a network to which the first computer is directly connected, it can send the message directly. If the receiving computer is not on a network to which the sending computer is connected, the sending computer sends the message to another computer that can forward it. The message may be sent through a router (see Chapter 6) to reach the forwarding computer. The forwarding host, which presumably is attached to at least one other network, in turn delivers the message directly if it can or passes it to yet another forwarding host. It is quite common for a message to pass through a dozen or more forwarders on its way from one part of the Internet to another.

The various networks that are linked to form the Internet work pretty much the same way—they pass data around in chunks called packets, each of which carries the addresses of its sender and its receiver. The set of conventions used to pass packets from one host to another is known as the Internet Protocol (IP), which operates at the network layer of the seven-layer OSI model discussed in Chapter 6. Many other protocols are used in connection with IP. The best known is the **Transport Control Protocol (TCP)**, which operates at the transport layer. TCP is so widely used as the transport layer protocol that many people refer to TCP/IP, the combination of TCP and IP used by most Internet applications. Adhering to the same technical standards allows the more than 100,000 individual computer networks owned by governments, universities, nonprofit groups, and companies to constitute the Internet. Once a network following these standards links to a **backbone**—one of the Internet's high-speed, long-distance communications links—it becomes part of the worldwide Internet community.

Each computer on the Internet has an assigned address called its **uniform resource locator**, or **URL**, to identify it from other hosts. The URL gives those who provide information available over the Internet a standard way to designate where Internet elements such as servers, documents, newsgroups, etc., can be found. Let's look at the URL for Course Technology, http://www.thomson.com

The "http" specifies the access method and tells your software to access this particular file using the HyperText Transport Protocol. This is the primary method for interacting with the Internet. Other access methods include ftp (File Transfer Protocol) for transferring files, telnet for logging onto a remote computer, news for bulletin boards or newsgroups, and gopher for accessing information via a Gopher menu tree.

The "www" part of the address signifies that the address is associated with the World Wide Web service discussed later.

The "thomson.com" part of the address is the domain name that identifies the Internet host site and must adhere to strict rules. It always has at least two parts separated by dots (periods). For all countries except the United States, the rightmost part of the domain name is the country code (au for Australia, ca for Canada, dk for Denmark, fr for France, jp for Japan, etc.). For example, the address for the Canadian Tourism Commission is http://info.ic.gc.ca. Within the United States, the country code is replaced with a code denoting affiliation categories (see Table 7.1). The leftmost part

Transport Control Protocol (TCP)

a protocol that operates at the transport layer and is used in combination with IP by most Internet applications

backbone

one of the Internet's high-speed, long-distance communications links

uniform resource locator (URL)

an assigned address on the Internet for each computer

TABLE 7.1

United States Top-Level
Domain Affiliations

Affiliation ID	Affiliation
arts	cultural and entertainment activities
com	business organizations
edu	educational sites
firm	businesses and firms
gov	government sites
info	information service providers
mil	military sites
nom	individuals
net	networking organizations
org	organizations
rec	recreational activities
store	businesses offering goods for purchase
web	entities related to World Wide Web activities

of the domain name identifies the host network or host provider, which might be the name of a university or business (thomson for International Thomson, parent company of Course Technology).

Herndon, Virginia–based Network Solutions Inc. (NSI) was the sole company in the world with the direct power to register addresses using .com, .net, or .org domain names. But this government contract ended in October 1998, as part of the U.S. government's move to turn management of the Web's address system over to the private sector. A process to set up a nonprofit corporation governing Web addresses is under way, but new top-level domains will likely stay on the drawing board until at least the summer of 1999, say those involved.[3]

Accessing the Internet

There are three ways to connect to the Internet (Figure 7.2). Which method is chosen is determined by the size and capability of the organization or individual.

Connect via LAN server. This approach requires the user to install on his or her PC a network adapter card and Open Datalink Interface (ODI) or Network Driver Interface Specification (NDIS) packet drivers. These drivers allow multiple transport protocols to run on one network card simultaneously. LAN servers are typically connected to the Internet at 56 Kbps or faster. Such speed makes for an exciting trip on the Internet but is also very expensive— $2,000 or so a month! However, the cost of this connection can be shared among several dozen LAN users to get to a reasonable cost per user. Additional costs associated with a LAN connection to the Internet include the cost of the software mentioned at the beginning of this section.

FIGURE 7.2

Three Ways to Access the Internet

There are three ways to access the Internet—using a LAN server, dialing into an Internet using SLIP or PPP, or using an on-line service with Internet access.

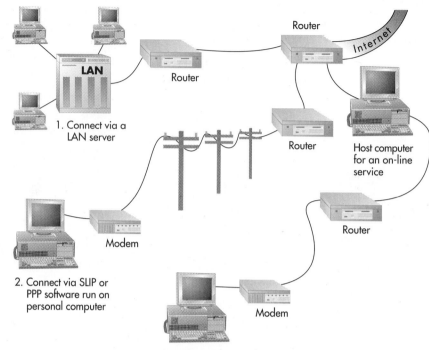

1. Connect via a LAN server

Router

Router

Router

Router

Modem

Modem

Host computer for an on-line service

2. Connect via SLIP or PPP software run on personal computer

3. Connect via an on-line service

Serial Line Internet Protocol (SLIP)

communications protocol software that transmits packets over telephone lines, allowing dial-up access to the Internet

Point to Point Protocol (PPP)

communications protocol software that transmits packets over telephone lines, allowing dial-up access to the Internet

On-line information services can provide access to the Internet, as well as financial information, news, shopping, and entertainment.
(Source: Courtesy of Prodigy, Inc.)

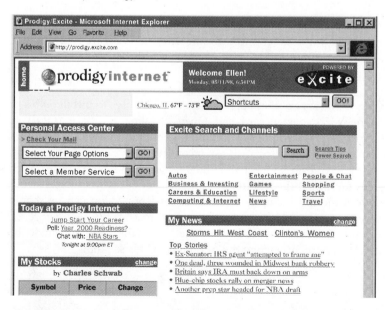

Connect via SLIP/PPP. This approach requires a modem and the TCP/IP protocol software plus **Serial Line Internet Protocol (SLIP)** or **Point to Point Protocol (PPP)** software. SLIP and PPP are two communications protocols that transmit packets over telephone lines, allowing dial-up access to the Internet. Users must also have an Internet service provider that lets them dial into a SLIP/PPP server. SLIP/PPP accounts can be purchased for $30 a month or less from regional providers. With all this in place, a modem is used to call into the SLIP/PPP server. Once the connection is made, you are on the Internet and can access any of its resources. The costs include the cost of the modem and software, plus the service provider's charges for access to the SLIP/PPP server. The speed of this Internet connection is limited to the slower of your computer's modem and the speed of the modem of the SLIP/PPP server to which you connect. Most Internet users agree that 14.4 Kbps is the minimum acceptable speed although faster rates are rapidly becoming the norm.

Connect via an on-line service. This approach requires nothing more than what is required to connect to any of the on-line information services—a modem, standard communications software, and an on-line information service account. The costs include the on-line service fee, a per-hour connect charge, and, where applicable, e-mail service charges. The on-line information services provide a wide range of services, including e-mail and the World Wide Web. America Online, Microsoft Network, and Prodigy are examples of such services.

Internet Service Providers

Internet service provider (ISP)

any company that provides individuals or organizations with access to the Internet

An **Internet service provider (ISP)** is any company that provides individuals or organizations with access to the Internet. ISPs do not offer the extended informational services offered by commercial on-line services such as America Online or Prodigy. There are literally thousands of Internet service providers, ranging from universities making unused communications line capacity available to students and faculty to major communications giants such as AT&T and MCI. To use this type of connection, you must have an account with the service provider and software that allows a direct link via TCP/IP.

In choosing an Internet service provider, the important criteria are cost, reliability, security, the availability of enhanced features, and the service provider's general reputation. Most providers score well on reliability, with 88 percent of users giving them favorable grades in that area.[4] Reliability is critical because if your connection to the ISP fails, it interrupts your communications with customers and suppliers. Among the value-added services ISPs provide are electronic commerce, networks to connect employees, networks to connect with business partners, host computers to establish a Web site, Web transaction processing, network security and administration, and integration services. Many corporate IS managers welcome the chance to turn to ISPs for this wide range of services because they do not have the in-house expertise and cannot afford the time to develop such services from scratch.[5] In addition, when organizations go with an ISP-hosted network, they can also tap the ISP's national infrastructure at minimum cost. This is important when a company has offices sprawled across the country. For example, Sport Rack went to AT&T's WorldNet IntraConnect Service unit to set up an Internet-based network that connects three locations with about 150 employees and offers remote access to the company's traveling sales force.

Despite the rapid growth of the Internet, analysts say there are more ISPs than the market can sustain. Considerable consolidation is expected. Today there are nearly 4,500 ISPs, but that number is expected to drop to fewer than 50 by 2001.[6] Table 7.2 identifies the major corporate Internet service providers.

An increasing number of ISPs are offering Internet connection via satellite. Hughes Network Systems Inc. began its DirecPC service in October 1996 to provide a high-speed connection to the Internet without the installation delays and cost of land lines. The cost of the service varies greatly depending on the type of service the customer chooses. For unlimited access at rates up to 200 Kbps from 6 PM to 6 AM on weekdays and unlimited access on weekends and holidays, the cost can be $19.95 per month.[7]

Startup Internet Satellite Systems (ISS) began domestic service in May 1998. It promises 400-Kbps download speeds for all customers for $25 per month, unlike DirecPC's service which offers various speeds and access times for various prices. ISS customers can rent or

Internet Service Provider	Web Address
AT&T's WorldNet Service	www.att.com
Digex, Inc.	www.digex.net
GTE Internetworking	www.gte.net
IBM Internet Connection	www.ibm.net
MCI Internet	www.mci2000.com
NetCom On-Line Communication Services, Inc.	www.netcom.com
PSINet, Inc.	www.psinet.com
Sprint Internet Services	www.sprint.net
UUnet Technologies, Inc.	www.us.uu.net

lease a satellite dish, a connector card, and other necessary equipment for less than $25 per month or buy equipment for less than $1,200. Loral Space and Communications Ltd. announced its Cyberstar service (http://www.loral.com) will provide electronic commerce, two-way communications, and Internet access at speeds of 200 Kbps to 400 Kbps.

An increasing number of businesses are using the Internet to provide better customer service (see the "Technology Advantage" box).

INTERNET SERVICES

The most commonly used Internet services include e-mail, Telnet, FTP, Usenet and newsgroups, chat rooms, Internet phone, and Internet video-conferencing and content streaming. Other commonly used services include connecting to other computers, finding and downloading files, and accessing different kinds of servers for information. These services are discussed next and summarized in Table 7.3.

E-Mail

Electronic mail, or e-mail, has been used internally in business networks for years, but with the spread of Internet use, it is now commonly used for national and international communications. E-mail is no longer limited to simple text messages. Depending on your hardware and software and the hardware and software of the recipient, you can embed sound and images in your message and attach files that contain text documents, spreadsheets, graphs, or executable programs. E-mail travels through the systems and networks that make up the Internet. Gateways can receive e-mail messages from the Internet and deliver them to users on other networks.

Many of these networks have agreements with the Internet and with each other to exchange e-mail, just as countries exchange regular mail across their borders. Similarly, an e-mail message may pass through a series of intermediate networks to reach the destination address. Since not all networks use

TECHNOLOGY ADVANTAGE
Pacific Bell Provides Customer Service via Web

Pacific Bell Internet Services provides Internet access to more than 100,000 customers. It charges its customers $19.95 per month and could not afford to spend the typical $25 to $50 (most of this is the cost of the employee's time) per call it costs to provide customer service through customer service representatives. So in July 1997, Pacific Bell launched Web-enabled customer service based on Silknet Software Inc.'s eService. The software supports a successful "call-avoidance strategy" that reduces the number of calls coming into the customer-service center while still providing excellent customer service. The main goal was to improve customers' experience by allowing them to help and educate themselves.

eService uses an innovative Web-based software designed to use multimedia and interactive capabilities of the Internet. It allows customers to serve themselves through the World Wide Web, thus eliminating waiting and relieving "live" service reps from performing many mundane duties. This allows the reps to focus on truly complex customer issues. To customers and companies, the benefits of eService are similar to those of other self-service models, such as the banking industry's automated teller machine: lower overhead, expanded hours of operation, and superior service.

If customers are unable to resolve a problem by themselves, they can use eService to submit information on the Web or initiate a telephone call with the customer support center. eService automatically identifies the best support agent to contact based on the description of the problem entered by the customer.

eService runs over the public Internet, and customer support agents use the same application as the customer. Using the telephone and the Web in tandem, the customer and support agent can work together to solve the problem. The traditional audio interaction over the telephone is enhanced by the multimedia capabilities of the Web and its ability to share visual information. In fact, eService allows each party to take remote control of the other's system so that they can more easily share the symptoms of—and solution to—a problem. This is particularly useful for software support operations.

Within the first two weeks of operation, the Pacific Bell Internet Services site was accessed more than 16,000 times. Only 200 customers needed to submit their problems for further review—that is about 1 percent.

DISCUSSION QUESTIONS

1. Is a Web-based customer service product like eService likely to replace the more traditional phone-based call center in all industries? Why or why not?
2. What technical problems can you identify with a Web-based customer service operation? What additional features and benefits can you identify besides those mentioned here?

Sources: Adapted from the Silknet Software Web site at http://www.silknet.com, accessed May 2, 1998; Candee Wilde, "Do-It-Yourself Service," *InformationWeek*, December 1, 1997, pp. 87–90; and "Silknet Software: Changing the Rules for Next-Generation Customer Support Applications," AberdeenGroup, Vol. 10, No. 9, July 8, 1997 at http://www.aberdeen.com.

the same e-mail format, a gateway translates the format of the e-mail message into one that the next network can understand. Each gateway reads the "To" line of the e-mail message and routes the message closer to the destination mailbox. Thus, you can send e-mail messages to literally anyone in the world if you know that person's e-mail address and if you have access to the Internet or another system that can send e-mail.

E-mail has changed the way people communicate. It improves the efficiency of communications by reducing interruptions from the telephone and unscheduled personal contacts. Furthermore, messages can be distributed or forwarded to multiple recipients easily and quickly without the inconvenience and delay of scheduling meetings. Because past messages can be saved, they can be reviewed, if necessary. And because messages are received

TABLE 7.3

Summary of Internet Services

Service	Description
E-mail	Enables you to send text, binary files, sound, and images to others
Telnet	Enables you to log on to another computer and access its public files
FTP	Enables you to copy a file from another computer to your computer
Usenet and newsgroups	Focuses on a particular topic in an on-line discussion group format
Chat rooms	Enables two or more people to carry on on-line text conversations in real time
Internet phone	Enables you to communicate with other Internet users around the world who have equipment and software compatible to yours
Internet videoconferencing	Supports simultaneous voice and visual communications
Content streaming	Enables you to transfer multimedia files over the Internet so that the data stream of voice and pictures plays more or less continuously

at a time convenient to the recipient, the recipient has time to respond more clearly and to the point. Opinions and feedback from remote experts and other interested people are easy to obtain, thus improving the quality of decisions and the probability of acceptance. For large organizations whose operations span a country or the world, e-mail allows people to work around the time zone changes. Some users of e-mail estimate that they eliminate two hours of verbal communications for every hour of e-mail use. But the person at the other end still must check their mailbox to receive messages.

Telnet and FTP

Telnet is a terminal emulation protocol that enables users to log on to other computers on the Internet to gain access to public files. Telnet is particularly useful for perusing library card files and large databases. It is also called remote logon.

File Transfer Protocol (FTP) is a protocol that describes a file transfer process between a host and a remote computer. FTP allows users to copy a file from one computer to another. FTP is often used to gain access to a wealth of free software on the Internet.

Usenet and Newsgroups

Usenet is a system closely allied with the Internet that uses e-mail to provide a centralized news service. It is actually a protocol that describes how groups of messages can be stored on and sent between computers. Following the Usenet protocol, e-mail messages are sent to a host computer that acts as a Usenet server. This server gathers information about a single topic into a central place for messages. A user sends e-mail to the server,

Telnet

a terminal emulation protocol that enables users to log on to other computers on the Internet to gain access to public files

File Transfer Protocol (FTP)

a protocol that describes a file transfer process between a host and a remote computer. FTP allows users to copy a file from one computer to another.

Usenet

a system closely allied with the Internet that uses e-mail to provide a centralized news service. It is actually a protocol that describes how groups of messages can be stored on and sent between computers.

This business import/export newsgroup has many participants who are offering to buy or sell various commodities.

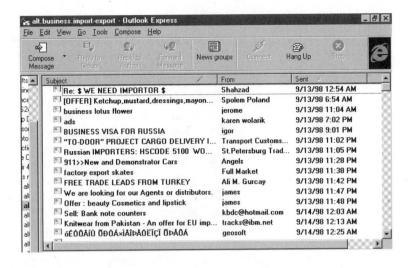

which stores the messages. The user can then log on to the server to read these messages or have software on the computer log on and automatically download the latest messages to be read at leisure. Thus, Usenet forms a virtual forum for the electronic community, and this forum is divided into newsgroups.

Newsgroups are what make up Usenet, a worldwide discussion system classified by subject. Articles or messages are posted to newsgroups using newsreader software and are then broadcast to other interconnected computer systems via a wide variety of networks. Estimates place the number of users with access to newsgroups between 10 and 30 million. A newsgroup is essentially an on-line discussion group that focuses on a particular topic. Newsgroups are organized into various hierarchies by general topic, and within each topic there can be many subtopics. On the Internet, there are tens of thousands of newsgroups, covering topics from astrology to zoology (see Table 7.4). Discussions take place via e-mail, which is sent to the newsgroup's address. A newsgroup may be moderated or unmoderated. If a newsgroup is moderated, e-mail is automatically routed to the moderator, a person who screens all incoming e-mail to make sure it is appropriate before posting it to the newsgroup.

Newsgroup servers around the world host newsgroups that share information and commentary on predefined topics. Each group takes the form

newsgroups

on-line discussion groups that focus on a specific topic

TABLE 7.4

Selected Usenet Newsgroups

alt.fan.addams.family	alt.life.itself
alt.pets	alt.fan.bevis-n-butthead
alt.autos.camaro	alt.fan.leonardo-dicaprio
alt.cloning	alt.history
alt.fan.u2	alt.music.blues
alt.sports.baseball.cinci-red	alt.music.zz-top
alt.sports.basketball.nba.la-lakers	alt.politics.socialism
alt.sports.college.sec	gov.us.fed.congress.record.house
alt.sports.soccer.european.uk	gov.us.fed.congress.record.senate

of a large bulletin board where members post and reply to messages, creating what is called a message thread. The open nature of newsgroups encourages participation, but the discussions often become rambling and unfocused. As a result of so much active participation, newsgroups can evolve into tight communities where certain members tend to dominate the discussions.

Here are some tips to consider when accessing newsgroups. When you join a newsgroup, first check its list of Frequently Asked Questions, or FAQs (pronounced "facks"), before submitting any questions to the newsgroup. The FAQ list will have answers to common questions the group receives. It is considered impolite to waste the group's time by asking common questions when FAQs are available. Most new users just read messages without responding at first. Many newsgroups include members from around the world, and in the interest of courtesy, you should pick up some sense of the audience and its culture before jumping in with questions and opinions. It is impolite to jump in the middle of a conversation (also called a thread). You may raise points and issues long since discussed and abandoned. Be concerned about what you say and the feelings of others. Remember, a person is receiving your messages. Do not use extreme words or repeat rumors (you could risk libel or defamation lawsuits). Do not post copyrighted material, and be careful how you use copyrighted material downloaded to your computer. Protect yourself by not offering personal information such as home address, employer, or phone number. Remember that this global on-line community has fragmented into thousands of different groups for a reason—to maintain the focus of each conference. Respect the specific subject matter of the group.

Chat Rooms

chat room

a facility that enables two or more people to engage in interactive "conversations" over the Internet

A **chat room** is a facility that enables two or more people to engage in interactive "conversations" over the Internet. Indeed, when you participate in a chat room there may be dozens of participants from around the world. Multiperson chats are usually organized around specific topics, and participants often adopt nicknames to maintain anonymity. One form of chat room, Internet Relay Chat (IRC), requires participants to type their conversation rather than speak. Voice chat is also an option, but requires that you have a microphone, sound card and speakers, a fast modem, and voice-chat software that is compatible with the other participants.

Internet Phone and Videoconferencing Services

Internet phone service enables you to communicate with other Internet users around the world who have equipment and software compatible to yours. This service is relatively inexpensive and can make sense for international calls. However, it may be a long time before conversing over the Internet replaces, or even begins to compete with, conventional telephone technology.

voice-over-IP (VOIP)

technology that enables network managers to route phone calls and fax transmissions over the same network they use for data

Using **voice-over-IP** technology (**VOIP**), network managers can route phone calls and fax transmissions over the same network they use for data—which means no more phone bills. Gateways installed at both ends of the communications link convert voice to IP packets and back. With the advent of widespread, low-cost Internet telephony services, traditional long-distance providers are being pushed to either respond in kind or trim their own long-distance rates.[8]

Here's how VOIP works (see Figure 7.3). Voice travels over the corporate intranet or Internet rather than the circuit-switched public network. Most corporate-class IP telephony applications use gateways that sit between the PBX and a router and convert calls into IP packets and shunts them onto the network. Using packets allows multiple parties to share digital lines so data transmission is much more efficient than traditional phone conversations, each of which requires a line. When the packets hit the destination gateway, the message is depacketized, converted back into voice and sent out via local phone lines. A number of major communications vendors such as Bay Networks Inc., Cisco Systems Inc., and 3Com Corp. all are integrating VOIP into their products.[9]

Net telephony first hit the Web in 1995—only to be quickly derided for poor voice quality and annoying delays. One big problem has been that data networks break speech into little packets so some packets may arrive out of order or too late to be included in the conversation. Another is the lag time inherent in the Net. Speech packets have to travel through a dozen or more routers, and each router takes a split second to do its job. However, improvements are

FIGURE 7.3

How Voice-Over-Internet Phone Technology (VOIP) Works

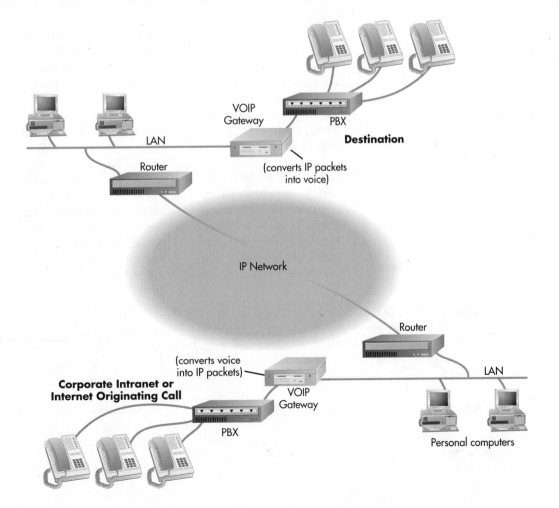

expected.[10] One major breakthrough is the development of a gateway server that connects data networks—such as the Internet and internal business networks—to the public telephone networks. This allows Internet calls from computer to phone or, with gateways on both ends, from phone to phone.

What is especially interesting about VOIP is the promise of new ways for merging voice with video and data communications over the Web or a company's data network. In the long run, it's not the cost savings that will boost the market, it's the multimedia capabilities it gives us and the smart call-management capabilities. Travel agents could use voice and video over the Internet to discuss travel plans; Web merchants could use it to show merchandise and take orders, customers could show suppliers problems with their products. Table 7.5 lists some current suppliers of Internet phone services.

Users who do not have access to a corporate PBX with the necessary VOIP equipment need a multimedia computer with a modem, sound card, speakers, a microphone, special software, and an Internet connection (you can download free Internet phone software from the Web at http:www.net2phone.com). The microphone serves as a mouthpiece and the speakers as a receiver. The sound card translates digital input into analog output, and vice versa.

With this form of Internet phone service, reaching another party is much more complicated than picking up a standard phone handset and dialing. Both parties need to run the same software. No standards currently exist whereby a WebPhone user can, say, talk to a VocalTel, TeleVox, or FreeTel user (as an AT&T user in the United States can call a British Telecom user in Great Britain). Moreover, both parties must be on-line and running the software when the call is made; otherwise, the phone will not ring. To connect to another party, you typically log on to a remote server, read through a list of current callers, and click on an individual of your choice.

Internet phone users must spend a fair amount of time and energy on each call to make their voices heard. If your system supports full-duplex conversation, both parties can talk at the same time. A half-duplex system limits the conversation to one direction at a time, like walkie-talkies. To talk in the full-duplex mode, both parties must have full-duplex sound boards with the appropriate drivers. However, half-duplex is not that much of an additional handicap, since Internet phone transmissions typically involve voice delays of a half-second to several seconds.

TABLE 7.5

Current IP Telephone Services

Company	Service	Availability	Per-Minute Rate
AT&T Corp.	AT&T WorldNet Voice	Three cities yet to be named, expanding to 16 cities by the end of 1998	$.075–$.09
Delta Three Inc.	PC-to_Phone	16 countries, including the United States	$.30–$.40 for international calls
IDT Corp.	Net2Phone Direct	Nationwide using "800" or local access	$.05–$.08 for domestic calls
Qwest Communications International Inc.	Q.talk	Nine U.S. cities	$.075

Intel® ProShare® Video System 500 is a desktop videoconferencing system that transmits audio and video over the Internet.
(Source: Courtesy of Intel Corporation.)

A codec (*compression-de*compression) device prepares the voice transmission by squeezing the recorded sound data and slicing it into packets for transfer over the Internet. On the receiving end, a codec reassembles and decompresses the data for playback. Different codecs are optimized for different uses and conditions, and the characteristics of a specific codec can affect voice quality.

Touch of Lace, a lace manufacturer located in New Jersey with 29 offices worldwide, shaved its long-distance phone bill by 30 percent using Internet phone services. The company does a lot of business in Asia, and employees in the United States were on the phone all the time. These users now pay $19.95 per month for Internet connections and can talk all they want with no additional costs.[11]

Internet videoconferencing, which supports both voice and visual communications, is another emerging service. Hardware and software are available to support a two-party conferencing system. The key here is a video codec to convert visual images into a stream of digital bits and translate them back again. The ideal video product will support multipoint conferencing in which multiple users appear simultaneously on the multiple screens.

Content Streaming

content streaming

a method for transferring multimedia files over the Internet so that the data stream of voice and pictures plays continuously, without a break, or very few of them. It also enables users to browse large files in real time.

Content streaming is a method for transferring multimedia files over the Internet so that the data stream of voice and pictures plays continuously, without a break, or very few of them. It also enables users to browse large files in real time. For example, rather than wait the half-hour it might take for an entire 5-Mbyte video clip to download before they can play it, users can begin viewing a streamed video as it is being received.

GE Information Services (GEIS) uses content streaming technology for communications and training applications that save money and improve the quality of training. GEIS holds quarterly meetings at which the CIO addresses all 2,000 employees, some in person, but many others scattered in 65 locations worldwide. The use of streaming technology has eliminated the logistical nightmare of trying to communicate with remote locations at the same time. GEIS now streams the meetings to participants who cannot attend in person. The company sends audio, video, and slide presentations across its intranet.[12] The "Making a Difference" box discusses one company that provides videoconference services.

THE WORLD WIDE WEB

World Wide Web

a collection of tens of thousands of independently-owned computers that work together as one in an Internet service

The World Wide Web was developed by Tim Berners-Lee at CERN, the European Center for Nuclear Research in Geneva. He originally conceived of it as an internal document-management system. This server can be located at http://www.cern.ch. From this modest beginning, the **World Wide Web** (the Web, WWW, or W3) has grown to a collection of tens of

MAKING A DIFFERENCE
ENEN

ENEN is the Education, News, and Entertainment Network, but you won't find it in your TV listings—it broadcasts on the Internet. The network provides business-to-business sales and education seminars. Rather than leave the office for an off-site meeting, employees simply log on to a Web site to attend the meeting.

ENEN is a subsidiary of Marshall Industries, a major distributor of industrial electronic components and production supplies in the United States. ENEN's founders were a team of Internet-savvy executives responsible for Marshall's pioneering usage of the Internet as a business tool. They recognized the vast potential of the Internet to deliver innovative and cost-saving services in 1995. Since then NetSeminar, which began as a value-added virtual broadcasting service for Marshall's 50,000 worldwide customers in 1998, has hosted nearly 1,000 seminars for its parent company as well as for outside customers. More than 10,000 people have attended these on-line seminars.

NetSeminar allows companies to broadcast live interactive multimedia presentations over the Internet. NetSeminar incorporates vocal and chat interactivity between speakers and participants. A key component to making this all work is the RealAudio streaming audio Web browser plug-in from RealNetworks. ENEN also includes an on-line chat session that lets attendees type in questions that other participants can see. The presenter then provides an audio response that goes out to all attendees. ENEN is able to convert videotapes into a form that can be shown via its Web presentation format.

Participants can attend the live broadcasts, or view an archived version any time from anywhere in the world. ENEN clients have used NetSeminar for public product announcements, sales training, engineering training, high-level business meetings, parenting workshops, and other private events.

Internet-based presentations offer participants the time savings of never having to leave town. Conference attendees simply initiate an Internet connection at the designated hour to become part of a live audience. ENEN can eliminate costs and save on the wear and tear of employees. Another advantage of this presentation approach is speeding time to market. Product launches, training, and other components of product introductions take weeks or months to plan. Hotels must be booked far in advance. With ENEN, time to market is reduced because presentations can be arranged and conducted within hours. Instead of waiting for the annual sales conference at U.S. headquarters in the fall, for example, the Latin American sales force can get product launch presentations and more as soon as the product is determined ready for release.

Internet presentations also enable the "delivery of experts" to an audience. Experts who might not have been available to attend a conference in Chicago may be more willing to participate if they can conduct a live broadcast from their own office. Because of the Internet's global reach, participants can log in live from anywhere in the world.

While productions can be taped at ENEN's Los Angeles studio, remote production at a site determined by the customer is also available. If the event is live, the presenter coordinates viewers to be ready and logged into the designated URL at a specified time. During presentations, live interaction is possible. Participants use standard off-the-shelf software. Otherwise, all that is required for participation is Internet access.

ENEN presentations vary greatly in length, reach, and complexity. Entry-level presentations can start as low as $4,000, while highly graphical, lengthy events geared for thousands of users can cost upwards of $50,000.

DISCUSSION QUESTIONS

1. What barriers to the use of Internet presentations such as NetSeminars can you identify? Can you identify a type of presentation that would not work well as an Internet presentation?
2. Identify three classes you have attended in the past year that could have been effectively conducted using Internet presentation technology.

Sources: Adapted from ENEN Web site at http://www.enen.com, accessed April 25, 1998 and Phillip J. Gill, "Seminars on the Net," *InformationWeek*, November 24, 1997, p. 82.

thousands of independently-owned computers that work together as one in an Internet service. These computers, called Web servers, are scattered all over the world and contain every imaginable type of data. Thanks to the high-speed Internet circuits connecting them and some clever cross-indexing software, users are able to jump from one Web computer to another effortlessly—creating the illusion of using one big computer. Because of its ability to handle multimedia objects, including linking multimedia objects distributed on Web servers around the world, the Web is emerging as the most popular means of information access on the Internet today. The number of Web servers is growing by about 6,000 per month.[13]

The Web is a menu-based system that uses the client/server model. It organizes Internet resources throughout the world into a series of menu pages, or screens, that appear on your computer. Each Web server maintains pointers, or links, to data on the Internet and can retrieve that data. However, unless you have the right hardware and telecommunications connections, the Web can be painfully slow. For example, graphics and photos take a long time to materialize on the screen. An ordinary phone line connection does not always provide sufficient speed to use the Web effectively. Serious Web users need to connect via the LAN server or SLIP/PPP approaches discussed earlier. An interesting development is the entry of cable TV companies into this field. They promise to offer on-line services at unheard-of data speeds. Some phone companies are starting new high-speed Internet access services based on new technology that allows traditional copper phone lines to connect to the Web.[14]

Data can exist on the Web as ASCII characters, word processing files, audio files, graphic images, or any other sort of data that can be stored in a computer file. A Web site is like a magazine, with a cover page called a **home page** that has graphics, titles, and black and blue text. All the blue type that is underlined is hypertext, which links the on-screen page to other documents or Web sites. **Hypermedia** connects the data on pages, allowing users to access topics in whatever order they wish. As opposed to a regular document that you read linearly, hypermedia documents are more flexible, letting you explore related documents at your own pace and navigate in any direction. For example, if a document mentions the Egyptian pharaohs, you can choose to see a picture of the pyramids, jump into a description of the building of the pyramids, and then jump back to the original document. Hypertext links are maintained using URLs. Table 7.6 lists some interesting Web sites. The client communicates with the server according to the hypertext transport protocol (http), which retrieves the document and presents it to the users. Web pages are loosely analogous to chapters in a book. Read the "FYI" box to learn about Web sites designed for potential car buyers.

Hypertext Markup Language (HTML) is the standard page description language for Web pages. One way to think about HTML is as a set of highlighter pens in different colors that you use to mark up plain text to make it a Web page—red for the headings, yellow for bold, and so on. The **HTML tags** let the browser know how to format the text: as a heading, as

home page

the cover page for a Web site that has graphics, titles, and black and blue text

hypermedia

tools that connect the data on Web pages, allowing users to access topics in whatever order they wish

Hypertext Markup Language (HTML)

the standard page description language for Web pages

HTML tags

codes that let the Web browser know how to format text: as a heading, as a list, or as body text and whether images, sound, and other elements should be inserted

Site	Description	URL (Uniform Resource Locator—Web Address)
Library of Congress	Provides access to five million photographs, rare books, maps, speeches, and historical documents	http://lcweb.loc.gov
PointCast	Delivers national, international, business, industry, and company news, plus stock quotes, sports scores, weather reports, and entertainment news from respected news sources such as Reuters, Business Wire, SportsTicker, and AccuWeather	http://www.pointcast.com
In-Box Direct	Allows users to create a personal newspaper from more than 20,000 news items from newspapers, news wires, trade journals, and newsletters by choosing topics, industries, and companies	http://www.netscape.com (and choose a link to In-Box Direct)
Online Career Center	Offers listings of jobs and profiles of companies; users can search the listing by key word and can place resumes on-line	http://www.occ.com
New York Times	The Web edition of the *New York Times*	http://www.nytimes.com
Project Gutenberg	Intends to revolutionize the way books are distributed; plans call for 10,000 books to be encoded into electronic form by the year 2001	http://www.gutenberg.org
Sportsline USA	Enables users to check the progress of favorite teams while games are in progress; displays results of games as they are being played	http://www.sportsline.com
White House	Users can take a virtual tour of the White House and gardens, as well as listen to the president's Saturday radio address	http://www.whitehouse.gov
MIT Lab for Computer Science	A highly technical site whose research categories include advanced network architecture, programming systems, and telemedia	http://www.lcs.mit.edu
The World Wide Web Virtual Library	A Web site maintained by the Indiana University School of Law that contains a wide range of information about law schools, law libraries, law firms, and law topics such as family law, environmental law, property law, torts, etc.	http://www.law.indiana.edu
Medscape	Contains an enormous amount of medical information and articles, news, and self-assessment tests; free to everyone but requires an ID and password	http://www.medscape.com
The Wall Street Journal	The on-line edition of *The Wall Street Journal*	http://www.wsj.com
Beverly Hills	The Beverly Hills travel shop where users can visit some of the prestigious locations in Beverly Hills, including Santa Monica Blvd., Beverly Drive, and Rodeo Drive	http://www.travelshop.com/trvshop3.htm
The Case	Provides the Cyberspace detective with the case, the clues, and the solution to a weekly mystery	http://www.thecase.com

TABLE 7.6

Several Interesting Web Sites

a list, or as body text. HTML also tells whether images, sound, and other elements should be inserted. Users mark up a page by placing HTML tags before and after a word or words. For example, to turn a sentence into a heading, you place the <H1> tag at the start of the sentence. At the end of the sentence, you place the closing tag </H1>. When you view this page, the sentence will be displayed as a heading. This means that a Web

FYI

Car Buyers Use the Internet for Bargain Hunting

A growing number of car buyers are searching the Internet for bargains. They use the Internet to look up information, get a dealer referral, actually make a purchase, and even price auto insurance. Market researcher J.D. Power & Associates, Inc. estimates that 16 percent of new-car buyers used the Web for shopping in 1997 and that by 2000 half of all new-car buyers will use the Internet to shop. About 2 percent of the 15 million cars sold in 1997 were a direct result of purchase requests funneled to dealers by such services as Auto-By-Tel, Autoweb.com, and CarPoint. These sales generated roughly $6 billion in business.

Here's how it works. Shoppers log on to the buying services or information sites like Kelley Blue Books to learn the invoice price of a car and its options and to figure out how much their trade-in is worth. Some stop here and simply use this data to negotiate a better deal when they visit the dealer's showroom. Some shoppers go the next step and submit purchase requests to on-line services such as CarPoint and Auto-By-Tel. These services sign up dealers who pay a fee to receive on-line referrals. In return, the dealers get the right to all the referrals from a group of Zip codes. Others, including CarPoint and Autoweb.com, are not as exclusive. They sign up competing dealers and let the shopper choose from them. The dealers agree to contact the shopper by phone with a best-price deal.

Auto-By-Tel invented low-cost auto buying on the Web and is the largest, most comprehensive Internet automotive purchasing program. Over one million customers have used the free, no obligation purchase request program to receive no-haggle, no-hassle, low-price quotes from 2,700 accredited dealers. Chrysler Corporation and General Motors Corporation are experimenting with their own Net sites that link on-line shoppers to dealers in a handful of regions. Chrysler is planning to go nationwide with a "Get A Quote" feature that allows buyers to specify the car they want, search out a dealer, and get a price. GM will offer something similar. Other car dealers, including Ford, Saab, and the Japanese manufacturers offer little more than electronic brochures with prices, specs, and addresses and phone numbers of local dealers on their Web sites.

On-line car-buying services are far from perfect. There is no guarantee the service will send the request to a dealer. Autoweb.com, for example, cannot deliver half the requests it gets simply because it does not have a dealer close enough to the buyer. Some dealers fail to respond to requests. These dealers bought in to secure the territory, but often fail to service it. As a result, Auto-By-Tel replaced 150 of its 2,600 dealers last year. On the other hand, many dealers support on-line buying. Inquiries get an immediate phone call or e-mail response. A special on-line sales team may even be formed that works on salary instead of commission.

This new approach can have a major impact on car dealerships and automakers because buyers can exercise more bargaining power than ever before. Dealers gross an average of $1,440, or 6.4 percent, on new car sales. Using the on-line services, savvy shoppers can determine the dealer's wholesale cost and use that as a starting point for bargaining rather than the sticker price. This strategy can save a few hundred bucks with each purchase, shaving the dealer's margin to a slim 5 percent. The result is a further consolidation among dealers that do not have enough volume to survive on lower margins.

With dealer related costs making up nearly a third of a car's costs, many manufacturers have been trying for years to make the system more cost-effective. Four or five large dealerships in a metropolitan area could cover the territory at a lower cost and higher profit than a dozen or more do now. The Internet is helping the trend toward consolidation. However, as much as manufacturers would like to use the Internet to steer sales to high-volume dealers and get rid of the weaker ones, they want to avoid dealer backlash caused by telling consumers where to get the best price.

DISCUSSION QUESTIONS

1. Access two or more of the Internet sites that provide on-line services to support the purchase of an automobile (see sources below). Write a brief paragraph describing your experience at each site. Which site do you think was most useful? Why?

2. Imagine that you are the president of the Auto Dealer's Association and you are making a presentation to dealers on the future of on-line shopping. What would you say?

Sources: Adapted from Larry Armstrong and Kathleen Kerwin, "Downloading Their Dream Cars," *BusinessWeek*, March 9, 1998, pp. 93–94; Auto-By-Tel Web site at http://www.autobytel.com, accessed April 25, 1998; Saab Web site at http://www.saab.com, accessed April 25, 1998; Autoweb.com Web site at http://www.autoweb.com, accessed April 25, 1998; Microsoft Carpoint Web site at http://www.carpoint.msn.com, accessed April 26, 1998; Chrysler Get a Quote Web site at http://www.chrysler.com, accessed April 26, 1998; and General Motors Buy Power Web site at http://www.gmbuypower.com, accessed April 26, 1996.

page is made up of two things: text and tags. The text is your message and the tags are codes that mark the way words will be displayed. All HTML tags are encased in a set of less than (<) and greater than (>) arrows, such as <H2>. The closing tag has a forward slash in it, such as </B> for closing bold. Figure 7.4 shows a simple document and its corresponding HTML tags. Software is available to help people interested in Web authoring.

Web Browsers

Web browser

software that creates a unique, hypermedia-based menu on your computer screen and provides a graphical interface to the Web

A **Web browser** creates a unique, hypermedia-based menu on your computer screen that provides a graphical interface to the Web. The menu consists of graphics, titles, and text with hypertext links. The hypermedia menu links you to Internet resources including text documents, graphics, sound files, and newsgroup servers. As you choose an item or resource, or move from one document to another, you may be jumping between computers on the Internet without knowing it, while the Web handles all the connections. The beauty of Web browsers and the Web is that they make surfing the Internet fun. Just clicking with a mouse on a highlighted word or a graphical button whisks you effortlessly to computers halfway around the world. Most browsers offer basic features such as support for backgrounds and tables, the ability to view a Web page's HTML source code, and a way to create hot lists of your favorite sites.

By 1996, Netscape Communications' Navigator had become the most widely used Web browser. It was setting the pace of browser development by embracing the HTML 3.0 standard and other hot new technologies. Navigator enables net surfers to view more complex graphics and 3-D models as well as audio and video material and runs small programs embedded in Web pages called **applets**. Many Web site developers boast that their sites use the latest Netscape innovations (such as frames that let

applet

a small program embedded in Web pages

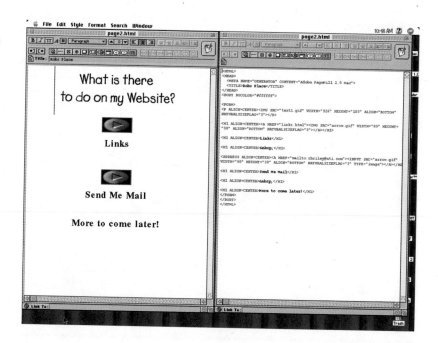

FIGURE 7.4

Sample Hypertext Markup Language

Shown at the left on the screen is a document, and at the right are the corresponding HTML tags.

the user split a screen into multiple, independent frames) and that their Web pages are best viewed with Navigator.

Starting with Version 3.0, Netscape has a new feature called CoolTalk that allows voice conversations, typed chat, and the sharing of "white-board" screens. These screens let people take turns making changes in a drawing or text while others watch. Netscape has also added a security feature called digital signatures that automatically identifies you to servers, a crucial step for business transactions on the Net. The latest version of Netscape can be downloaded free from the Internet.

Microsoft released Internet Explorer in the summer of 1995 to compete with Netscape. Explorer can be downloaded free and came bundled with Windows 95 computers. As of this writing, the Department of Justice and Microsoft are discussing whether it is appropriate for Explorer to come bundled with Windows 98. This type of giveaway is standard procedure—that's how Navigator became the number one browser. In addition to voice, chat, and white-board sharing, Microsoft's NetMeeting lets two or more people on the Internet work together in a Windows application such as Microsoft Excel. Microsoft developed and is marketing ActiveX as a method for creating network-ready applets. Considerable debate on the relative security of ActiveX and applets has been fueled by a demonstration on German TV exposing supposed security flaws in ActiveX—members of a computer club were allegedly able to transfer funds illegally from Intuit's Quicken accountancy software using an ActiveX applet to modify a user's Quicken transaction file.[15]

Developing Web Content

Web authors work with several standards to create their pages. Unfortunately, current Web technical limitations make it difficult to create a "singing and dancing" Web site with beautiful text layout; large, photographic-quality images; voice-over; music; and video clips. The two main problems—confusing HTML standards and slow communications speeds—can limit creativity.

The HTML standards are created by a committee of various people involved in the Web. Anyone can create tags, and others may adopt them, modify them, or reject them. Thus, the HTML standards are evolving. HTML 1.0 was the standard in 1994 and is now obsolete. HTML 2.0 introduced a forms feature for allowing users to enter data; it became the standard in 1995. HTML 3.0 allows banners, centering, right text alignment, tables, mathematical formulas, and image alignment. Netscape has added a number of new tags to this standard, such as <BLINK>, which causes blinking words that work only within Netscape. Microsoft wants to develop a new set of standards. Thus, not all browsers will work the same way when used to view the same Web page. For some browsers, a tag will work wonderfully. On others, that tag will not do anything at all, or, worse, it may cause problems. Some browsers are strictly HTML 2.0 and therefore ignore the Netscape extensions or newer HTML 3.0 features. Some browsers use the Netscape extensions and others are already using unapproved parts of the new HTML standards. Web authors need to keep these inconsistencies in mind when they develop pages. The art of Web design involves getting around the technical limitations of the Web and using a limited set of tools to make appealing designs. Following are tips for creating a Web page.

1. Your computer must be linked to a Web server, which can deliver Web pages to other browsers.
2. You will need a Web browser program to look at HTML pages you create.
3. The actual design activity can take one of the following approaches: (a) Write your copy with a word processor, then use an HTML converter to convert the page into HTML format complete with tags so the browser knows how it should format the page; (b) Use an HTML editor to write text and add HTML tags at the same time; (c) Edit an existing HTML template (with all the tags ready to use) to meet your needs; or (d) Use an ordinary text editor and type in the start and end tags for each item.
4. Open the page with the browser and see the result. You can correct mistakes by correcting the tags.
5. Add links to your home page to allow your readers to click on a word and be taken to a related home page. The new page may be either a part of your Web site or a home page on a different Web site.
6. To add pictures, you must first store them as a file on your hard drive. This can be done in one of several ways: draw them yourself using a graphics software package, copy pictures from other Web pages, buy a disk of clip art, scan photos, or use a digital camera.
7. You can add sound by using a microphone connected to your computer to record a sound file; adding links to the page will enable those who access your Web page to hear the sound file.
8. Upload the HTML file to your Web site using e-mail or FTP.
9. Review the Web page to make sure that all links are correctly established to other Web sites.
10. Advertise your Web page to others and encourage them to stop, take a look, and send feedback by e-mail.

A number of new Web standards are undergoing definition and early use. These include Extensible Markup Language (XML), cascading style sheets (CSS), and Dynamic HTML (DHTML). XML is a way to define and share document information over the Web, CSS improves Web page inheritance and presentation, and DHTML provides dynamic presentation of Web content.[16] These standards move more of the processing for animation and dynamic content to the browser, which reduces the need for plug-ins, and provide quicker access and displays. XML documents contain tags that pertain specifically to the information you requested. Thus, you are able to navigate to desired data more quickly. However, entire industries will need to agree on sets of tags to achieve this goal.

Search Engines

Looking for information on the Web is a little like browsing in a library—without the card catalog, it is extremely difficult to find information. Web search tools—called **search engines**—take the place of the card catalog.

search engine

a Web search tool

The Web is a big place, and even the largest search engine, HotBot, indexes only about a third of its pages. AltaVista, the second largest engine,

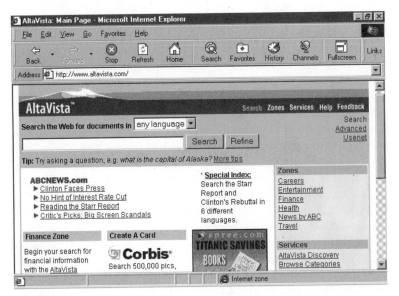

AltaVista is the second largest search engine on the Web, indexing about 28 percent of the Web's pages.

covers only about 28 percent, while some other well-known engines index only a single-digit percentage of the Web's pages. So, even if you do find a search site that suits you, your query might still miss the mark. When searching the Web, you may wish to try more than one search engine to expand the total number of potential Web sites of interest. Once you find a document that comes close to your goal, you can usually find related material by following the highlighted entries that take you to other Web pages when you click on them. And if you come across something you think you'll want to return to, you can add it to the "hot list" or "favorites" list on your Web browser to save time in the future.

Search engines that use keyword indexes produce an index of all the text on the sites they examine. Typically, the engine reads at least the first few hundred words on a page, including the title, the HTML "alt text" coded into Web-page images, and any keywords or descriptions that the author has built into the page structure. The engine throws out words such as "and," "the," "by," and "for." The engine assumes whatever words are left are valid page content; it then alphabetizes these words (with their associated sites) and places them in an index where they can be searched and retrieved.

This type of search engine usually does no content analysis per se, but will use word placement and frequency to determine how a page ranks among other pages containing the same or similar words. For example, when someone searches for the word alien, a page with "alien" in its title will appear higher in the search results than a site that does not mention "alien" in the title. Likewise, a page with 20 mentions of "alien" in the body text will rank higher than a page with one instance of the word.

Keyword indexes tend to be fast and broad; you'll typically get search results in seconds (faster than other kinds of engines). But, unless you are careful about how you construct your query, you are likely to be overwhelmed with data.

Subject directories operate like a card catalog: They assign sites to specific topic categories based on the site's content and human judgment. The advantage of this approach is that sites are pregrouped and easier to browse than those in a raw keyword index. A human-generated subject directory also allows more nuance and subtlety than machine-generated keyword indexes, and can offer advice not only on where the content is but how good or bad it is. However, human-generated directories can never be as comprehensive or up-to-date as machine-generated sites. In addition, human judgments are subjective. If you happen to think the same way as the site reviewers, you will find great value in these subject directories. But if you and the reviewers are on different wavelengths, the site's categorization might seem arbitrary and hard to understand and you might find that their top picks are not pertinent to your needs. There are a number of Web search tools as summarized in Table 7.7.

Java

Java

an object-oriented programming language from Sun Microsystems based on C++ that allows small programs—applets—to be embedded within an HTML document

Java is an object-oriented programming language from Sun Microsystems based on C++ that allows small programs—the applets mentioned earlier—to be embedded within an HTML document. When the user clicks on the appropriate part of the HTML page to retrieve it from a Web server, the applet is downloaded onto the client workstation, where it begins executing.

Java lets software writers create compact "just-in-time" programs that can be dispatched across a network such as the Internet. On arrival, the applet automatically loads itself on a personal computer and runs—reducing the need for computer owners to install huge programs anytime they need a new function. And unlike other programs, Java software can run on any type of computer. Thus far, Java is used mainly by programmers to make Web pages come alive, adding splashy graphics, animation, and real-time updates. Java-enabled Web pages are more interesting than plain Web pages.

The relationship among Java applets, a Java-enabled browser, and the Web is shown in Figure 7.5 and explained here. To develop a Java applet, the author writes the code for the client side and installs that on the Web server. The user accesses the Web page and pulls it down to his personal computer, which serves as a client. The Web page contains an additional HTML tag called APP, which refers to the Java applet. A rectangle on the page is occupied by the Java application. If the user clicks on the rectangle to execute the Java application, the client computer checks to see if a copy of the applet is already stored locally on the computer's hard drive. If it is not, the computer accesses the Web server and requests that the applet be downloaded. The applet can be located anywhere on the Web. If the user's Web browser is Java-enabled (e.g., Sun's HotJava browser or Netscape's Navigator product), then the applet is pulled down into the user's computer and is executed within the browser environment.

The Web server that delivers the Java applet to the Web client is not capable of determining what kind of hardware/software environment the client is running on, and the developer who creates the Java applet does not want

TABLE 7.7

Popular Search Engines

Search Engine	Web Address	Search Strategy
Altavista	http://www.altavista.digital.com	Keyword
Excite	http://www.excite.com	Keyword
Galaxy	http://www.einet.net	Subject
Hotbot	http://www.hotbot.com	Keyword
Infoseek	http://www.infoseek.com	Keyword
Lycos	http://www.lycos.com	Keyword
Webcrawler	http://webcrawler.com	Subject
Yahoo!	http://www.yahoo.com	Subject

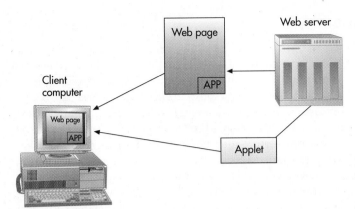

FIGURE 7.5

Downloading an Applet from a Web Server

The user accesses the Web page from a Web server. If the user clicks on the APP rectangle to execute the Java applications, the client's computer checks for a copy on its local hard drive. If the applet is not present, the client requests that the applet be downloaded.

to worry about whether it will work correctly on OS/2, Windows, UNIX, and MacOS. Java is thus often described as a "cross-platform" programming language.

The development of Java has had a major impact on the software industry. Sun Microsystems's strategy is to open up Java to any and all. Any software vendors and individual developers—from development tool vendors, language compiler developers, database management system vendors, and client/server application vendors to small businesses—could then use Java to create Internet-capable, run-anywhere applications and services (Figure 7.6). As a result, the Java community is becoming broader every day, encompassing some of the world's biggest independent software vendors, as well as users ranging from corporate CIOs, programmers, multimedia designers, and marketing professionals to educators, managers, film and video producers, and hobbyists.

Java could change the economics of paying for software. The software industry today is based on the concept of entire applications delivered to the marketplace in shrink-wrapped boxes for a fixed, one-time cost, for which the customer is given a license that allows him or her to use the software forever on a single computer platform. Java makes it possible to sell one-time usage of a piece of software. This usage could be defined for a single transaction, or for a single session, during which the user is connected to a Web server.

Security is still a big concern with the use of Java applets. Anytime you allow programs to be downloaded to your computer without verification of their source, there is concern. If one of Java's executable programs were infected with a virus, the virus could spread like wildfire through a corporate network. Sun Microsystems is working on ways to let Java applets be saved to a local computer disk when they arrive.

FIGURE 7.6

Web Page with a Java Applet

Push Technology

Push technology is used to automatically send information over the Internet rather than make users search for it with their browsers. Frequently the information, or "content," is customized to match an individual's needs or profile. The use of push technology is also frequently referred to as "Webcasting." Most push systems rely on HTTP (HyperText Transport Protocol) or Java technology to collect content from Web sites and deliver it to users' desktops. Before they can be "pushed," users must download and install software that acts like a TV antenna, capturing transmitted content. As with any new technology, the people paying for push have yet to venture beyond rudimentary applications. Most are focusing on improving communications with employees, customers, and business partners.

Content aggregators such as PointCast and InterMind, who provide Internet broadcasting or "Webcasting" services, collect information from Web sites of content creators and transmit continually. Based on their stated preferences, users can keep a constant watch on the latest press releases from their competitors, the prices of stocks in their portfolio, the weather at next week's vacation spot, or the latest European soccer score. Merchants like L.L. Bean can create their own push channels using software from Marimba, BackWeb Technologies, and others. Push technology also has potential applications for information transmission across corporate intranets, and could become an efficient means of distributing new or updated software.

Metropolitan Regional Information System (MRIS) is one of the nation's largest multiple listing services for the real estate industry and provides information to more than 30,000 member real estate agents. MRIS is using push technology to allow members to log on to MRIS and create profiles for each of their customers. The system then queries a database of prospective homes and "pushes" those that fit buyer profiles to the real estate agent. Push technology then provides HTML documents, photos, even video of homes that match the buyer's profile. This system costs hundreds of thousands of dollars, but makes MRIS members, who pay $35 per month for the on-line service, more efficient.[17]

There are drawbacks to the use of push technology. One issue, of course, is information overload. Another is the volume of data being broadcast is so great that push technology can clog up the Internet communications links with traffic. A recent study by Optimal Networks of Palo Alto, California, found that PointCast alone accounted for nearly 18 percent of Internet traffic, as measured in bytes.[18]

Business Uses of the Web

In 1991, the Commercial Internet Exchange (CIX) Association was established to allow businesses to fully connect to the Internet. This was the real beginning of the commercial use of the Internet. Since then, businesses have been using the Internet for a number of applications. Electronic mail is a major application for most companies. Some companies display products over the Internet, including catalogs and sample texts. Customers can place orders by keying in payment information and shipping addresses.

Athleta provides shoppers with access to an on-line catalog for ordering athletic clothing plus information for athletes including training tips, organizations, and events at its Web site at http://www.athleta.com. Holiday

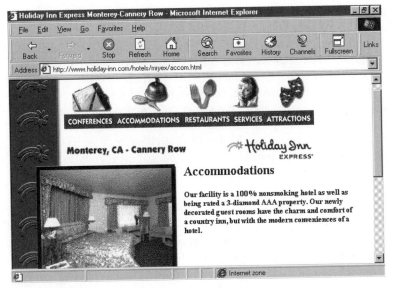

Holiday Inn's Web site provides information on each property and allows users to obtain rate and availability information.

Inn at http://www.holiday-inn.com allows you to view hotel rooms at various Holiday Inns around the country, obtain rate information, and make on-line reservations. UPS is the world's largest package distribution company with over 500 aircraft, 157,000 vehicles, and 2,400 facilities providing service in over 200 countries and territories. It delivers 18 million packages and documents per day to more than 1.6 million customers who receive daily pick-up service. These customers can access the UPS Web site at http://www.ups.com to request a pick-up, order supplies, and track the status of their package. The Microsoft Web site at http://www.microsoft.com allows visitors to download free software, obtain technical support, order product catalogs, and receive information about training.

Currently, the Web sites making the most money are those that sell advertising space directly on their sites. These Web sites, ranging from Microsoft's home page to The Dilbert Zone, sell the colorful pieces of on-screen real estate known as advertising banners. Five of the top ten Web advertising sites—Infoseek, Lycos, Yahoo!, Excite, and WebCrawler—are Web search engines or directory services. Web ad prices cost around $2–$10 per thousand "readers." One problem with banner ads is that their effectiveness is relatively low, with fewer than 5 percent of users clicking the banner to jump to the advertiser's site.

At the beginning, on-line shopping was slowly adopted. Lack of commerce infrastructure, fears about the security of on-line financial transactions, and poor execution severely hampered early on-line selling efforts. However, as software and procedures are perfected to eliminate these problems, use of the Web to support home shopping, pay-per-use entertainment, and other commercial ventures has grown rapidly. Table 7.8 lists products that have been successful and not so successful on the Web.

TABLE 7.8

On-line Shopping Choices

Base: 500 Internet Users; Multiple Responses Allowed

Products People are Likely to Buy on the Web	
Software	77%
Books	67%
CDs	64%
Computer hardware	63%
Airline tickets	61%
Magazine subscriptions	53%
Concert/theater tickets	48%
Flowers	45%

Source: Cybershoppers Research Report Survey, conducted by Greenfield Online, Inc. (www.greenfieldcentral.com) for the Better Business Bureau. Published in *Computerworld*, March 23, 1998, p. 48.

The great potential for the Web is that by linking buyers and sellers electronically, businesses will be able to establish a new and ongoing relationship with customers. Businesses can use the Web as a tool for marketing, sales, and customer support. The Web can also serve as a low-cost alternative to fax, express mail, and other communications channels. It also has the potential to eliminate paperwork and drive down costs per business transaction. Read the "E-Commerce" box to learn about the growth of electronic commerce.

The Internet's business potential has just begun to be tapped. As more and more people gain access to the World Wide Web, its functions will no doubt change drastically. A couple of these applications, corporate intranets and extranets, will be discussed next.

INTRANETS AND EXTRANETS

intranet

an internal corporate network built using Internet and World Wide Web standards and products that allows employees of an organization to gain access to corporate information

firewall

a device that sits between your internal network and the outside Internet and limits access into and out of your network based on your organization's access policy

---

An intranet is an internal corporate network used by employees to gain access to company information.

An **intranet** is an internal corporate network built using Internet and World Wide Web standards and products. It is used by the employees of the organization to gain access to corporate information. After getting their feet wet with public Web sites that promote company products and services, corporations are seizing the Web as a swift way to streamline—even transform—their organizations. These private networks use the infrastructure and standards of the Internet and the World Wide Web but are cordoned off from the public Internet through a device known as a **firewall** that sits between an internal network and the outside Internet. Its purpose is to limit access into and out of the network based on the organization's access policy. Employees can venture out onto the Internet, but unauthorized users cannot come in. A big advantage of this approach is that many people are already familiar with the Internet and Web, so they need little training to make effective use of their corporate intranet.

Most companies already have the foundation for an intranet—a network that uses the Internet's TCP/IP protocol. Computers using Web server software can store and manage documents built on the Web's HTML format. With a Web browser on your PC, you can call up any Web document—no matter what kind of computer it is on.

An intranet is an inexpensive yet powerful alternative to other forms of internal communications, including conventional computer setups. One of an intranet's most obvious virtues is its ability to slash the need for paper. Because Web browsers run on any type of computer, the same electronic information can be viewed by any employee. This means that all sorts of documents (such as internal phone books, procedure manuals, training manuals, and requisition forms) can be inexpensively converted to electronic form on the Web and be constantly

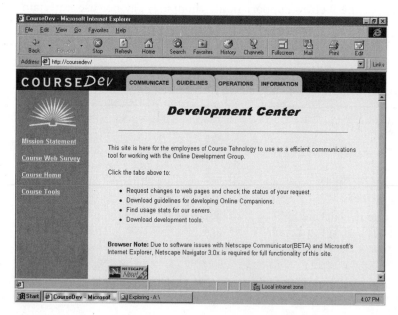

E-COMMERCE
Business on the Web Is Exploding

Once a communications channel for academics and governmental agencies, the Internet is making a dramatic impact on businesses of all sizes. Many of the early pioneers lost money or were disappointed with Internet generated sales. With a few short years of Internet experience, many companies are now finding the Internet to be indispensable.

Internet business has exploded in volume, and the growth is expected to continue into the foreseeable future. In 1997, sales of PCs and related hardware over the Internet has been estimated to be about $850 million. By 2000, sales of PCs and related equipment is expected to approach $3 billion annually. The travel industry is expected to see an even greater rate of increase in sales. In 1997, travel related business on the Internet was approximately $650 million. In three years, sales are projected to come close to $5 billion annually.

Other industries, including entertainment, gifts and greetings, and books and music, are also experiencing dramatic Internet growth rates. Entertainment sales should go from about $300 million to almost $2 billion. For gifts and greetings, the numbers are expected to jump from about $150 million to about $600 million. Books and music sales are expected to soar from under $300 million to almost $800 million in a few short years.

Products, services, and even promotional campaigns are jumping to the Internet. Office Depot is moving comic figure Dilbert from TV ads to the Internet. The Dilbert figure will be used to help Internet shoppers find the Office Depot products they want. Amazon.Com, one of the first computerized book stores, has seen Internet sales soar from about $16 million annually to over $130 million. Amazon offers about 2.5 million books for sale on-line.

One of the most successful Internet sites is netMarket, an on-line discount store. Members paying dues in one year spent more than $1 billion on this site alone. There are about 1 million members. The introductory fee is $1 for a three-month membership. With a growth rate of about 100,000 new members each month, the company is expecting millions of customers in the future.

Why is business on the Internet exploding? In a survey of *Forbes* readers, almost 60 percent of the respondents indicated that it is easier to shop on the Internet than through catalogs or retail stores. But, there are concerns and problems that could slow Internet growth. People are still afraid of Internet security. Will a credit card account number given over the Internet be secure? About 50 percent of the *Forbes* survey respondents indicated that security on the Internet is the biggest pitfall and problem with commerce on the Internet. Other problems include compatibility of technology and a lack of Internet shoppers. If the security concerns can be overcome, the problem of not enough Internet shoppers should disappear.

DISCUSSION QUESTIONS

1. Describe a business that uses the Internet to make sales. What safeguards do they have for security and privacy?
2. Would you purchase a product over the Internet using your credit card? What is your biggest concern in using the Internet to purchase products or services?

Sources: Mary Cronin, "Business Secrets of the Billion-Dollar Website," *Fortune*, February 2, 1998, p. 142; Katarzyna Moreno, et al., "E-Shopping," *Forbes*, February 9, 1998, p. 40. Heather Green, "The Virtual Mall Gets Real," *BusinessWeek*, January 26, 1998, p. 90.

updated. An intranet provides employees with an easy and intuitive approach to accessing information that was previously difficult to obtain. For example, it is an ideal way to provide information to a mobile sales force that needs access to rapidly changing information. Intranets can also do something far more important. By presenting information in the same way to every computer, they can do what computer and software makers have frequently promised but never actually delivered: pull all the computers, software, and databases that dot the corporate landscape into a single system that enables employees to find information wherever it resides.

Universal reach is what made the Internet grow so rapidly. But Internet enthusiasts tended to focus on how to link far-flung people and businesses.

When the Internet caught on, people did not consider it as a tool for running their business, but that is what it is quickly becoming. Just as the simple act of putting millions of computers around the world on speaking terms initiated the Internet revolution, connecting all the islands of information in a corporation is sparking unprecedented collaboration. Corporate intranets are breaking down the walls within corporations.

From AT&T to Levi Strauss to 3M, hundreds of companies are putting together intranets. At Compaq Computer, employees tap into a Web server to reallocate investments in their 401(k) plans. At Ford Motor, an intranet linking design centers in Asia, Europe, and the United States helped engineers craft the 1996 Taurus. At National Semiconductor, an engineer rigged a home page that allows his department to schedule meetings online. At Silicon Graphics, the company makes accessible more than two dozen corporate databases that employees can traverse by clicking on the hyperlinks. More sophisticated intranets are coming. They will let employees fill out electronic forms, query corporate databases, and hold virtual conferences over private Webs. Some companies are working with daily news services to provide intranet access for employees to publications ranging from *Aviation Week* to *Zoology Today*. This enables people to get information on current and prospective customers, suppliers, competition, the industry, and the economy in general.

Once a company sees the value of intranet access, it will want to move on to the next stage of intranet usage—interactive transaction-based applications. At this stage, employees can query corporate databases to see the status of a customer order, a raw material shipment, or a manufacturing production run of a finished product.

More advanced use of the corporate intranet supports what has come to be known as workgroup computing (see Chapter 6). Workgroup computing involves many aspects, but basically it is an approach to support people working together in teams. One of the key aspects is the ability to store and share information in any form—text, video, sound, graphics, handwritten memos, or hand-drawn figures. The idea is to be able to organize and retrieve all this data simply. Group calendaring and scheduling allows an employee to check others' schedules and set up meetings. Another aspect is support for real-time meetings with people linked over networks instead of people needing to be present in one place. Workgroup computing also supports work-flow processes, tracking the status of documents including who has them, who is behind, and who gets them next.

A rapidly growing number of companies have advanced beyond this stage to offer limited network access to selected customers and suppliers. Such networks are referred to as extranets, which connect people who are external to the company. An **extranet** is a network based on Web technologies that links selected resources of the intranet of a company with its customers, suppliers, or other business partners. For example, Compaq Computer and the Vanguard Group, a financial service group, collaborated to provide investment information. Compaq employees can use this extranet to gain access to Vanguard servers to obtain information about their accounts, as well as tips and educational materials on finance in general. A few companies allow select customers and suppliers to actually enter data into corporate transaction processing systems. Typically, these users are provided a form on the intranet to enter an order or request status information.

extranet

a network based on Web technologies that links selected resources of the intranet of a company with its customers, suppliers, or other business partners

TABLE 7.9

Summary of Internet, Intranet, and Extranet

	Users	Importance of Reliability and Performance	Is There a Need for User Authentication?
Internet	Anyone	Low	No
Intranet	Employees	Low	Yes
Extranet	Selected business partners	High	Yes

Security and performance concerns are different for an extranet than for a Web site or a network-based intranet. Authentication and privacy are critical on an extranet and are of no importance to a public Web site. Obviously, performance must be good to provide quick response to your customers and suppliers. Table 7.9 summarizes these differences between the Internet, intranets, and extranets.

Secured intranet and extranet access applications usually require the use of a virtual private network (VPN). A **virtual private network (VPN)** is a secure connection between two points across the Internet. VPNs transfer information by encapsulating traffic in IP packets and sending the packets over the Internet, a practice called **tunneling**. Most VPNs are built and run by Internet service providers. Companies that use a VPN from an Internet service provider have essentially outsourced their networks to save money on wide area network equipment and personnel. In using a VPN, a user sends data from his or her personal computer to the company's firewall, which also converts the data into a coded form that cannot be easily read by an interceptor. The coded data is then sent via an access line to the company's Internet service provider. From here, the data is transmitted through tunnels across the Internet to the recipient's Internet service provider and then over an access line to the receiving company's firewall where it is decoded and sent to the receiver's personal computer (Figure 7.7).

MacManus Group, a global media company, implemented an 18-site VPN to replace courier services, dial-up connections, and low-speed communications that had kept far-flung offices from working on large international accounts in a timely manner. Now employees can collaborate rapidly as a global company and have staff in 75 to 80 offices working on global accounts such as Procter & Gamble. In the past, all they could do was send e-mail a few times a day between international offices due to the cost. With the VPN, MacManus can now afford to send e-mail anytime it wants, as well as perform file transfers including the exchange of photos for print ads while working on the same account.

virtual private network (VPN)

a secure connection between two points across the Internet

tunneling

the process by which VPNs transfer information by encapsulating traffic in IP packets and sending the packets over the Internet

FIGURE 7.7

Virtual Private Network (VPN)

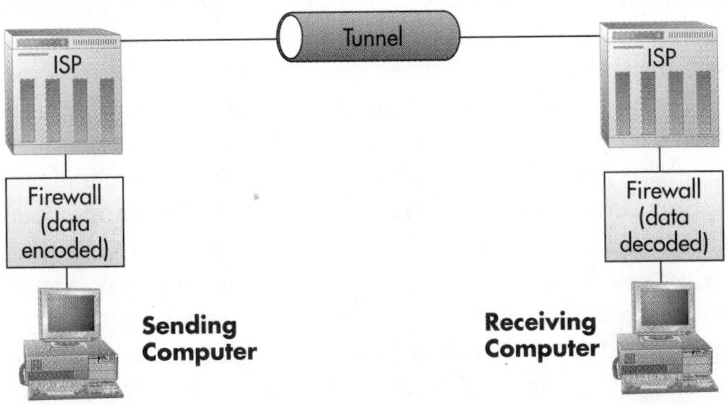

NET ISSUES

The topics raised in this chapter apply not only to the Internet and intranets but also to LANs, private WANS, and every type of network.

Control, access, hardware, and security issues affect all networks, so it is important to mention some of these management issues.

Management Issues

Although the Internet is a huge, global network, it is managed at the local level; no centralized governing body controls the Internet. While the U.S. federal government provided much of the early direction and funding for the Internet, the government does not own or manage the Internet. The Internet Society and the Internet Activities Board (IAB) are the closest things to centralized governing bodies. These societies were formed to foster the continued growth of the Internet. The IAB oversees a number of task forces and committees that deal with Internet issues. One of the main functions of the IAB is to manage the network protocols used by the Internet, including TCP/IP. Some universities and government agencies are investigating how the Internet can be controlled to prevent sensitive information and pornographic material from being placed on the Internet.

Service Bottlenecks

The primary cause of service bottlenecks is simply the phenomenal growth in traffic. The amount of traffic on the Internet is increasing at the astounding rate of 25 percent per month! It is estimated that by 2000, more than one million companies will be disseminating information and services on the Internet. According to Vinton Cerf, a senior vice president at MCI and codeveloper of the original Internet, every year the same amount of capacity is required as in all the years before. Traffic volume on company intranets is growing even faster than traffic on the Internet. Companies setting up an Internet or intranet Web site often underestimate the amount of computing power and communications capacity they need to service all the "hits" (requests for pages) they get from Web cruisers. Web server computers can be overwhelmed with thousands of hits per hour.

Slow modems and the copper-based telephone wire system that carries the signal into an office or home are the two current primary Internet bottlenecks. For most users, these two limit a user's maximum access speed to around 56 Kbps, which is still too slow for 16-bit stereo sound and smooth, full-screen video.

Connection agreements exist among the backbone companies to accept one another's traffic and provide a certain level of service. Some Internet providers do a good job and provide a high level of quality and service. Others do not do a good job, creating wide variations in the quality of the Internet. At some interconnect points where major Internet operators hand off to one another, one operator may not be able to accept incoming traffic fast enough because its lines are overloaded with traffic. For example, linking with Pacific Rim nations is especially difficult. Most providers lease their lines from phone companies, and the cost of even a medium-speed line to connect California with Australia is over $1.2 million per year—ten times the cost of a New York-to-San Francisco link, so some providers scrimp. This leads to inadequate capacity, which slows transmissions.

Routers, the specialized computers that send packets down the right network pathways, can also become bottlenecks. For each packet, every router along the way must scan a massive address book of about 40,000 area destinations (akin to Internet ZIP codes) to pick the right one. These routers can get overloaded and lose packets. The TCP/IP protocol compensates for this by detecting a missing packet and requesting the sending device to resend the packet. However, this leads to a vicious circle, as the network devices continually try to resend lost packets, further taxing the already overworked routers. This leads to long response times or loss of the connection to the network.

Several actions are being taken to open up the bottleneck. One solution involves the various backbone providers upgrading their backbone links. In some cases they are installing bigger, faster "pipes," and in other cases they are converting to newer transmission technology, such as asynchronous transfer mode (ATM), which can send a message down the right path more quickly than standard packet-switching technology. Each ATM transmission is preaddressed with its own route, so routing addresses do not have to be looked up and the packet can zip right through an ATM switch. Using this technology, MCI expanded its backbone capacity from 155 Mbps to 622 Mbps in 1996. A second solution to the bottleneck is provided by router manufacturers, who are working to develop improved models with increases in hardware capacity and more efficient software to provide quick access to addresses. Yet a third solution is to prioritize traffic. Today, all network traffic travels through the same big backbone pipes. There is no way to make sure that your urgent message is not stalled behind someone downloading a magazine page. With prioritized service, customers could pay more for guaranteed delivery speed, much like an overnight package costs more than second-day delivery. If implemented, this solution could also affect the cost of network services that generate a lot of traffic, such as Internet phone and videoconference services.

Privacy and Security

cryptography

the process of converting a message into a secret code and changing the encoded message back to regular text

encryption

the original conversion of a message into a secret code

Cryptography is the process of converting a message into a secret code and changing the encoded message back to regular text. The original conversion is called **encryption**. The unencoded message is called plaintext. The encoded message is called ciphertext. Decryption converts ciphertext back into plaintext (see Figure 7.8). For much of the cold war era, cryptography was the province of military and intelligence agencies; uncrackable codes were reserved for people with security clearance only.

Widespread deployment of cryptography requires additional hardware and software but is becoming increasingly necessary to support electronic commerce, copyright management, and electronic delivery of services. Without cryptography, people will not trust that electronic financial transactions, secret or private data, and valuable intellectual property will remain confidential across networks.

A cryptosystem is a software package that uses an algorithm, or mathematical formula, plus a key to encrypt and decrypt messages. The algorithm is calculated with the key and converts every character of the plaintext into other coded characters, thus creating the ciphertext. Only someone with the correct key should be able to decode the ciphertext. Good ciphertext appears to be nothing more than random characters. Encryption makes information useless to hackers and thieves.

FIGURE 7.8

Cryptography is the process of converting a message into a secret code and changing the encoded message back into regular text.

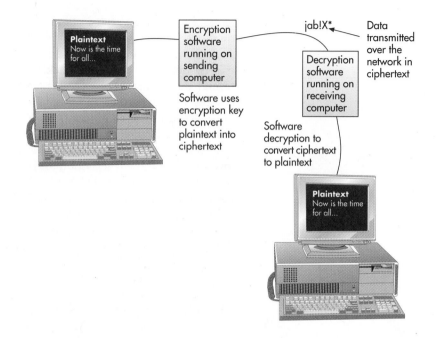

The Data Encryption Standard (DES), adopted as a federal standard in 1977 to protect classified communications and data, was designed by IBM and modified by the National Security Agency. It uses 56-bit keys, meaning a user must employ precisely the right combination of 56 1s and 0s to decode information correctly. Other technologies offer a range of key lengths of up to 2,048 bits in the case of RC5, for instance. The RSA protocol, meanwhile, has no limit on key length, but it can slow things down, since it uses separate keys for encryption and decryption. Many products mix technologies: They use a fast algorithm like DES for the actual encryption but send the DES key through a more secure method like RSA.

U.S. banks and brokerage houses use the federal government's Data Encryption Standard (DES) algorithm to protect the integrity and confidentiality of fund transfers totaling some $2.3 trillion a day worldwide. Organizations encrypt the words and videos of their teleconferencing sessions. Individuals encode their electronic mail. Researchers use encryption to hide information about new discoveries from prying eyes.

The Electronic Frontier Foundation (EFF) raised concerns by revealing that the Data Encryption Standard (DES) is insecure. To prove the insecurity of DES, EFF built the first unclassified hardware for cracking messages encoded with it. The EFF DES Cracker, which was built for less than $250,000, took less than 3 days to complete the challenge.[19]

Encryption is not just for keeping secrets. It can be used to verify who sent a message and to tell whether the message was tampered with en route. A **digital signature** is a technique used to meet these critical needs for the processing of on-line financial transactions. Digital signatures involve a complicated technique that combines the public-key encryption method with a "hashing" algorithm that prevents reconstructing the original message. The hashing algorithm provides further encoding by using rules to convert one set of characters to another set (e.g., the letter *s* is converted to

digital signature

an encryption technique used to meet the critical need for processing on-line financial transactions a *v*, 2 is converted to *7*, etc.). Thus, encryption also can prevent electronic fraud by authenticating senders' identities with digital signatures.

Microsoft and Visa expect to deliver secure-transaction software that works with the latter's Visanet payment system to authenticate buyers and sellers and protect clearing and settlement transactions. Credit card companies anticipate a new group of credit card holders on the Internet: corporate consumers. A significant volume of corporate procurement will use credit cards. Credit card companies foresee corporations putting out bids on the Internet and negotiating deals to buy everything from raw materials to computers. Read the "Ethical and Societal Issues" box to learn more about the issues associated with encryption.

ETHICAL AND SOCIETAL ISSUES

U.S. Encryption Regulation

The U.S. government issued a new export control policy in January 1997 that shifted the focus from maximum allowable encryption strength and encouraged vendors to implement key recovery—the ability of law enforcement officials to retrieve encryption keys from a central database and decode messages if necessary. Vendors that incorporate an approved key recovery mechanism in their products can export encryption of any length; those that commit to doing so within two years can export 56-bit encryption until then. Vendors that do not make the commitment are limited to exporting products with a maximum key length of just 40 bits, unless those products are intended for use by banks.

While most U.S. vendors have indicated they are willing to implement key recovery, few have provided solutions based on this approach. It is difficult until standards are set up. Work on standards cannot proceed until governments agree to a global encryption policy. The big sticking point is who should look after the central database of keys. The United States favors the use of trusted third parties or government-registered escrow agencies. The only problem is finding a third party that vendors and the government alike can put their faith in.

Other governments have chosen not to follow the lead of the United States. Worse, major vendors like Microsoft (Redmond, Washington) and Netscape Communications (Mountain View, California) have publicly scorned key recovery as impractical for the Internet. Users outside the United States are able to download higher-strength versions of browser software from Microsoft and Netscape. Customers with a top-level U.S. domain name like .com can download software boasting 128-bit RSA encryption while those with a non-U.S. Internet address automatically get only 40-bit RSA.

Senators John Ashcroft and Patrick J. Leahy introduced a new encryption bill, called the E-PRIVACY Act, in mid-1998. The bill authorizes the domestic use of strong encryption without "key recovery" back doors for government eavesdropping; eases export controls to allow U.S. companies to sell their encryption products overseas; strengthens protections from government access to decryption keys; and creates unprecedented new protections for data stored in networks. The act establishes a new research center to assist federal, state, and local police in dealing with encrypted data. The bill also makes it a crime to use encryption to obstruct justice. Overall, the E-PRIVACY Act presents a strong pro-privacy approach to the encryption issue. The bill makes more encryption more accessible to many more people. It also creates new privacy protections for data stored on networks—protections that will become increasingly important as more people go on-line.

Discussion Questions

1. Why does data encryption create a dilemma for our government?
2. Are you in favor of a data encryption approach that can be cracked by law enforcement agencies? Why or why not?

Sources: Adapted from Andrew Dornan, "Seeing Through Keyholes," *Data Communications,* July 1, 1998 and Alan Davidson, Press Release "Senators Introduce Pro-Privacy Encryption Bill, in Stark Contrast to Administration Position," May 12, 1998, the Center for Democracy and Technology Web site at http://www.cdt.org/press, accessed August 13, 1998.

Firewalls

When it comes to security on the Internet, it is essential to remember two things. First, there is no such thing as absolute security. Second, plenty of clever people consider it great sport to try to breach any security measures—the better your security, the greater the challenge to them. For example, officials at the Pentagon revealed that at least 11 U.S. military computers—all of them containing unclassified information—were violated early in February 1998. The break-ins, which took place via several different systems over the course of a week, were initially feared to be attempts by Iraqi supporters to disrupt the military's efforts to prepare for air strikes against Iraq. After more careful study, however, the attacks had "the quality of voyeurism or vandalism" and probably were perpetrated by U.S. hackers, the Pentagon officials said.[20]

The most popular method of preventing unauthorized access to corporate computer data is to construct a firewall between a company's computers and the Internet. According to International Data, over 70 percent of organizations using the Internet have firewalls. A firewall can be anything from a set of filtering rules set up on the router between a company and the Internet to an elaborate application gateway consisting of one or more specially configured computers that control access. The idea is to allow some services to pass but deny others. For example, you may be able to use the Telnet utility to log into systems on the Internet, but users on remote systems cannot use it to log into your local system because of the firewall.

A firewall can be set up to allow access only from specific hosts and networks or to prevent access from specific hosts. In addition, you can give different levels of access to various hosts; a preferred host may have full access, whereas a secondary host may have access to only certain portions of the host's directory structure.

For a higher level of security, you can have an assured pipeline, which uses more sophisticated methods to prevent access. A firewall looks at only the header of a packet, but an assured pipeline looks at the entire request for data and then determines whether the request is valid. Inappropriate requests can be routed away from the Internet, whereas files that meet specific criteria (such as those that contain the word *confidential*) can never be sent over the Internet.

The Federal Computer Incident Response Capability assists civilian government agencies with computer security incidents.

INTERSOLV, a Rockville, Maryland–based international software company, installed a firewall in its offices in Australia, Germany, the United Kingdom, Japan, and Belgium to secure international communication between sites. The firewall provides both application gateway security and the ability to use strong encryption. "Today's companies are no longer confined to one location, making secure communications instrumental to doing business and staying competitive," remarked Greg Gehring, INTERSOLV vice president of information services. "As one of the world's largest software companies, INTERSOLV has offices all over the globe, and we need to be able to assure the privacy and security of communications from every site."[21]

The Federal Computer Incident Response Capability, run by the National Institute of Standards and Technology, responds to virus attacks, network intrusions, and other threats. It also provides security training and consulting services to individual agencies. Vulnerability and threat information is shared with the public at this group's Web site (http://www.csrc.nist.gov).

Network management issues will take an increasing amount of time for IS personnel, but each user needs to be aware of the basics to function effectively in business. Communications, service, and daily work are all at stake.

Information Systems Principles

The Internet is like many other new technologies—it provides a wide range of services, some of which are effective and practical for use today, whereas others are still evolving, and still others will fade away from lack of use.

People need improved means to communicate and collaborate—the popularity of the Internet and other technology and services that make this happen.

As with every new technology, the Internet and corporate intranets and extranets are experiencing growing pains. Standards are needed, options must be sorted out, financial transactions must be made secure, privacy has to be guaranteed, and legal issues must be addressed.

The Java approach to delivering software to users is so radically different from current approaches that it will drive major changes in the software industry.

▪ SUMMARY

1. *Briefly describe how the Internet works, including alternatives for connecting to it and the role of Internet service providers.*

The Internet is the world's largest computer network. Actually, it is a collection of interconnected networks, all freely exchanging information. The Internet transmits data from one computer (called a host) to another. The set of conventions used to pass packets from one host to another is known as the Internet Protocol (IP). Many other protocols are used in connection with IP. The best known is the Transport Control Protocol (TCP). TCP is so widely used that many people refer to TCP/IP, the combination of TCP and IP used by most Internet applications. Each computer on the Internet has an assigned address called its uniform resource locator, or URL, to identify it from other hosts. There are three ways to connect to the Internet: via a LAN whose server is an Internet host, via SLIP or PPP, and via an on-line service that provides Internet access.

An Internet service provider is any company that provides individuals or organizations with access to the Internet. To use this type of connection, you must have an account with the service provider and software that allows a direct link via TCP/IP. Among the value-added services ISPs provide are: electronic commerce, intranets and extranets, Web-site hosting, Web transaction processing, network security and administration, and integration services.

2. *Identify and briefly describe eight services associated with the Internet.*

Internet services include:(1) e-mail, (2) Telnet, (3) FTP, (4) Usenet and newsgroups, (5) chat rooms, (6) Internet phone, (7) Internet videoconferencing, and (8) content streaming. E-mail is used to send messages. Telnet enables you to log on to remote computers. FTP is used to transfer a file from another computer to your computer. Usenet supports newsgroups, which are on-line discussion groups focused on a particular topic. Chat rooms let you talk to dozens of people at one time, who can be located all over the world. Internet phone service enables you to communicate with other Internet users around the world who have equipment and software compatible with yours. Internet videoconferencing enables people to conduct virtual meetings. Content streaming is a method of transferring multimedia files over the Internet so that the data stream of voice and pictures plays continuously.

3. *Describe the World Wide Web, including tools to view and search the Web.*

The Web is a collection of tens of thousands of independently-owned computers that work together as one in an Internet service. High-speed Internet circuits connect these computers, and cross-indexing software enables users to jump from one Web computer to another effortlessly. Because of its ability to handle multimedia objects and hypertext links between distributed objects, the Web is emerging as the most popular means of information access on the Internet today.

A Web site is like a magazine, with a cover page called a home page that has graphics, titles, and black and blue text. Hypertext links are maintained using URLs (uniform resource locator), a standard way of coding the locations of the HTML (Hypertext Markup Language) documents. Web pages are loosely analogous to chapters in a book.

The client communicates with the server according to a set of rules called HTTP (Hypertext Transfer Protocol), which retrieves the document and presents it to the users. HTML is the standard page description language for Web pages. The HTML tags let the browser know how to format the text: as a heading, as a list, or as body text. HTML also tells whether images, sound, and other elements should be inserted.

A Web browser reads HTML and creates a unique, hypermedia-based menu on your computer screen that provides a graphical interface to the Web. The browser uses data about links to accomplish this; the data is stored on the Web server. The hypermedia menu links you to other Internet resources, not just text documents, graphics, and sound files.

4. *Outline a process for creating Web content.*

Web authors work with several standards to create their pages. Unfortunately, current Web technical limitations make it difficult to create a "singing and dancing" Web site with beautiful text layout; large, photographic-quality images; voice-over; music; and video clips. The two main problems—confusing HTML standards and slow communications speeds—can limit creativity.

The steps to creating a Web page include getting space on a Web server; getting a Web browser program; writing your copy with a word processor, using an HTML editor, editing an existing HTML document, or using an ordinary text editor to create your page; opening the page using a browser, viewing the result, and correcting any tags; adding links to your home page to take viewers to another home page; adding pictures and sound; uploading the HTML file to your Web site; reviewing the Web page to make sure that all links are working correctly; and advertising your Web page.

5. *Describe Java and discuss its potential impact on the software world.*

Java is an object-oriented programming language from Sun Microsystems based on C++ that allows small programs—applets—to be embedded within an HTML document. When the user clicks on the appropriate part of the HTML page to retrieve it from a Web server, the applet is downloaded onto the client workstation environment, where it begins executing. The development of Java has had a major impact on the software industry and could change the economics of paying for software. Java makes it possible to sell one-time usage of a piece of software.

6. *Identify who is using the Web to conduct business, and discuss some of the pros and cons of Web shopping.*

A rapidly growing number of companies are doing business on the Web and enabling shoppers to search for and buy products on-line. The travel, entertainment, gift, greetings, book and music businesses are experiencing the fastest growth on the Web. For many people, it is easier to shop on the Web than to search through catalogs or trek to the shopping mall. However, other shoppers are concerned about the potential for credit card numbers to be stolen over the Internet.

7. *Define the terms* intranet *and* extranet *and discuss how organizations are using them.*

An intranet is an internal corporate network built using Internet and World Wide Web standards and products. It is used by the employees of the organization to gain access to corporate information. Computers using Web server software store and manage documents built on the Web's HTML format. With a Web browser on your PC, you can call up any Web document—no matter what kind of computer it is on. Because Web browsers run on any type of computer, the same electronic information can be viewed by any employee. This means that all sorts of documents can be converted to electronic form on the Web and be constantly updated at virtually no cost.

An extranet is a network that links selected resources of the intranet of a company with its customers, suppliers, or other business partners. It is based on Web technologies. Security and performance concerns are different for an extranet than for a Web site or a network-based intranet. Authentication and privacy are critical on an extranet and are of no importance to a public Web site. Obviously, performance must be good to provide quick response to your customers and suppliers.

8. *Identify several issues associated with the use of networks.*

Management issues, service bottlenecks, privacy and security, and firewalls are issues that affect all networks. No centralized governing body controls the Internet. Also, because the amount of Internet traffic is so large, service bottlenecks often occur. Cryptography techniques and firewalls are required to combat information thieves and provide as much security as possible.

▪ KEY TERMS

applet 308	HTML tags 305	search engine 310
ARPANET 291	hypermedia 305	Serial Line Internet Protocol
backbone 292	Hypertext Markup Language	(SLIP) 294
chat room 300	(HTML) 305	Telnet 298
content streaming 303	Internet 290	Transport Control Protocol (TCP) 292
cryptography 321	Internet service provider (ISP) 295	tunneling 319
digital signature 322	Internet Protocol (IP) 291	uniform resource locator (URL) 292
encryption 321	intranet 316	Usenet 298
extranet 318	Java 312	virtual private network (VPN) 319
File Transfer Protocol (FTP) 298	newsgroups 299	voice-over-IP (VOIP) 300
firewall 316	Point to Point Protocol (PPP) 294	Web browser 308
home page 305	push technology 314	World Wide Web 303

▪ REVIEW QUESTIONS

1. What is the Internet? Who uses it and why?
2. What is the TCP/IP protocol? How does it work?
3. Explain the naming conventions used to identify Internet host computers.
4. Briefly describe three different ways to connect to the Internet. What are the advantages and disadvantages of each approach?
5. What is an Internet service provider? What services do they provide?
6. What is a newsgroup? How would you use one?
7. What are Telnet and FTP used for?
8. How does an Internet phone work? What are some of its advantages and disadvantages?
9. How would you use Internet videoconferencing?
10. What is content streaming?
11. What is the Web? Is it another network like the Internet, or a service that runs on the Internet?
12. Which Internet services use the client/server model for communications?
13. What is hypermedia?
14. What is a URL and how is it used?
15. What is HTML and how is it used?
16. What is a Web browser? How is it different from a Web search engine?
17. Describe how one would initiate the use of a Java applet, and give several examples of applets.
18. What is push technology?
19. What is an intranet? Provide three examples of the use of an intranet.
20. What is an extranet? How is it different from an intranet?
21. What is a virtual private network? Why might an organization use one?
22. What is cryptography?
23. What are firewalls? How are they used?

▪ DISCUSSION QUESTIONS

1. The Internet is international in scope, with literally millions of users worldwide. What issues could arise when using a chat room?
2. Briefly describe how the Internet phone service operates. Discuss the potential for this service to impact traditional telephone services and carriers.
3. The U.S. federal government is against the export of strong cryptography software. Discuss why this may be so. What are some of the pros and cons of this policy?
4. Identify three companies with which you are familiar that use the Web to conduct business. Describe how each uses the Web.

5. Outline a process to create a Web page. What computer hardware and software do you need if you wish to create a Web home page containing both sound and pictures?

6. One of the key issues associated with the development of a Web site is getting people to visit it. If you were developing a Web site, how would you inform others about it and make it interesting enough so they would return and tell others about it?

7. How has the Java programming language changed the software industry?

8. Briefly summarize the differences in how the Internet, a company intranet, and an extranet are accessed and used.

9. Can you envision this course being delivered more effectively over the Internet than with the use of a textbook? Or, would this lessen the value of the course? Explain your position.

■ PROBLEM-SOLVING EXERCISES

1. Do research on the Web to find the most current status of U.S. legislature on data encryption. Write a brief paper summarizing your findings. Be sure to cite your sources of information.

2. Prepare a brief proposal for developing a business based on E-commerce. Describe what products and/or services you will offer. Describe how users will interact with the Web site. How will you get people to visit your site? Develop a simple spreadsheet to analyze income and expenses.

3. You are a manager in a Fortune 1000 company. You have been appointed to lead a project to develop an extranet linking your company to key suppliers. The extranet will provide suppliers with access to production planning and inventory data so that the suppliers can ship raw materials, packing materials, and supplies to your plant just-in-time. The goal is to reduce the level of inventory you must carry while not adversely affecting production. Develop a one-page project charter that defines the scope and purpose of the project, identifies the key members of the project team (by title), outlines the key technical and non-technical issues you will face, and defines the key project success criteria.

■ TEAM ACTIVITY

1. Identify a company that is making effective use of a company extranet. Find out all you can about its extranet. Try to speak with one or more of the customers or suppliers who use the extranet and ask what benefits it provides from their perspective.

2. Work with your team to draft an e-mail policy for your university.

■ WEB EXERCISE

This chapter covers a number of powerful Internet tools, including Internet phones, search engines, browsers, e-mail, newsgroups, Java, intranets, and much more. Pick one of these topics and get more information from the Internet. You may be asked to develop a report or send an e-mail message to your instructor about what you found.

▪ CASES

1 Dreamworks on the Web

Why would anyone want to have a Web site that explored something that happened over 100 years ago? If your company is Dreamworks, the answer is to promote the film *Amistad*. The movie, which tells the story of 53 slaves, traces historical developments in an action- and emotion-packed epic. The challenge is to develop a powerful Web page that introduces the characters of the movie, explores the plot, and generates interest in the film. The bottom line is to motivate more people to see the film.

To get a dynamic Web site, Dreamworks got the help of Media Revolution in Santa Monica, California. In addition to developing a compelling Web site, Media Revolution also wanted to use the latest technology. It started with a dynamic splash page. When first viewed, the splash page spelled Amistad in back and white. The letters then moved and transformed to form a quote from the film. The quote then faded into the background with the word Amistad again emerging. To accomplish this, Media Relations used Java. The program randomly retrieved a Flash 2.0 file that delivered the quote to the splash page.

In addition to the quotes displayed on the Web site, Media Relations also used plenty of graphics. To keep file sizes small and download times reasonable, Media Relations used vector-based graphics that draws lines between points. With this approach, the software defines the points on the graph or image and connects the points with lines. This takes far less storage space and is faster than using a bitmapped image that uses bits or pixels for each point of the graph. In addition to these visual images, Media Relations also used sound files to generate tribal music, which played in the background.

Media Relations also built a chat facility into the Web site. This allowed people to hear and see information from historians and experts on the actual event. People could also use the chat facility to share their ideas and feelings about the film and the historical event.

Developing a powerful Web site with text, moving graphics, sound, and chat facilities was not easy. Some people have Internet browsers that are newer and more powerful than others. To accommodate these differences, Media Relations had to build a feature into the site that could determine the specific Internet browser being used. If a newer and more powerful browser was being used, then the Web site was flexible enough to take full advantage of the capabilities of the newer browser. If an older browser was being used, the Web site would make sure to deliver only what that browser could handle. Dreamworks and Media Relations also faced another obstacle. Web sites for movies are usually temporary. In most cases, they only survive as long as the movie is playing and generating revenues. The Web site had to be good, but not too expensive.

1. What features were used in the *Amistad* Web site to generate interest in the movie and the actual historical event?
2. Visit another movie Web site on the Internet. What are the strengths and weaknesses of the site?

Sources: Adapted from Nicole Schotland, "Site of Epic Proportions," *Internet World*, January 1998, p. 94; Staff, "Hungry for e-D?" *PC Magazine*, January 21, 1998, p. 9; Gary Trudeau, "Amistad is Important," *Time*, January 5, 1998, p. 170.

2 Internet Travel Planning

SABRE Business Travel Solutions (SABRE BTS) is a comprehensive suite of software that enables corporations to manage their travel more effectively. It includes fully integrated applications that provide travel booking, travel policy management, expense reporting and decision-making tools. Travelers access a company-specific URL address using an industry-standard Web browser. Once at the SABRE BTS site, they can obtain the benefits of secure electronic commerce with preferred travel suppliers without having to install costly hardware and software on-site. Travelers can connect to

SABRE BTS from anywhere and book reservations with more than 420 airlines, 39,000 hotel properties, and 50 rental car companies.

Travelers access the SABRE Service Bureau in Tulsa, Oklahoma, via the Internet. This site is designed to support high-speed Internet access with more than 2,000 data lines. It also uses advanced Secure Socket Layer technology for data authentication and encryption from Web server to Web browser. SABRE BTS operates on a wide range of technology platforms so that users can access the SABRE BTS travel database through Windows, Lotus Notes, or the Internet.

The system works like this: Employees make reservations and order tickets on-line. They charge expenses using Diners Club cards, which sends them receipts via e-mail. Expense information is downloaded automatically into an electronic expense form. Employees simply type in items for which they paid cash. Once the report is complete, it can be e-mailed to managers for electronic approval. The company then electronically reimburses both the credit card company and employees. Summary reports are produced that help managers analyze companywide travel activity—corporations can track every travel dollar from planning and purchasing to expense reporting and reimbursement. The system has the potential to reduce total travel management costs by as much as 30 percent.

The Yankee Group, a Boston-based research firm, predicts that spending on business travel booked over the Internet or corporate intranets will soar from an estimated $236 million in 1997 to $945 million in 1998, $2.7 billion in 1999, and almost $10.4 billion in 2000. American Express's Interactive (AIX) and Internet Travel Network's Global Manager are other Internet-based travel planning packages.

1. Identify issues that must be overcome for travelers to adopt the use of this system. What new features could be incorporated into the system that would make it even more worthwhile?
2. Imagine that you are the manager in charge of corporate travel for your company. What steps would you take to implement such a system?

Sources: Adapted from Carol Silva, "Schwab Saves with 'Net Travel Planning," *Computerworld*, April 13, 1998, pp. 41–42; and Sabre Business Travels Solutions Web site at http://wwwsabrebts.com, press release, "The Sabre Group Adds Internet Access to Business Travel Suite," accessed April 30, 1998.

3 US West Communications

US West is a Denver-based provider of telecommunications services to 25 million customers in 14 Western and Midwestern states. It employs over 51,000 people and has over 14 million access lines in service.

US West's intranet began as a test to post some company information, foster a community of interest, and convince people that US West's network could essentially become an internal intranet. The test was successful beyond all expectations. Today US West's Global Village intranet consists of more than 300,000 Web pages and receives nearly 7 million hits per month from over 28,000 employees, with each department having its own Web page. In addition, there are more than 60 applications used for everything from posting internal job openings to programs that assist external business partners and customers.

There are 55 programmers who support the intranet with a relatively small annual budget of $5 million, considering the $900 million budget allotted to the 5,000-person IT staff.

Businesspeople and intranet programmers have been highly creative in finding valuable uses of the intranet. These applications often require less than three months to develop and cost under $100,000. Many of these applications link an easy-to-use Web interface to one or more corporate databases.

One of the most popular intranet sites is the Rumor Mill, a site that encourages employees to submit questions, anonymously if they wish. Another useful site is the Meet Me Bridge, which allows employees to schedule audio conference calls. But the most valuable application is one called Facility Check. This application gives US West employees accurate estimates of the date phone service will be installed, a key piece of information for customers when they call. The application saves an estimated $10 million by allowing the company to provide quicker service to customer sites. It also helps the company shave minutes off customer order calls, eliminate repeat calls, and reduce fines levied when US West can't meet pledged installation dates.

1. What other kinds of intranet applications would be of use to US West?

2. Do you think the initial proposal to launch an intranet test met much resistance when it was suggested back in 1995? Why or why not?

Sources: Adapted from Carol Silva, "Intranet Apps Applauded," *Computerworld*, March 23, 1998, pp. 47–48; Carol Silva, "How the Intranet Was Won: US West Pioneers Push Change," *Computerworld*, March 16, 1998, p. 14; US West Web site at http://www.uswest.com, accessed June 2, 1998; John W. Verity, "Coaxing Meaning out of Raw Data," *Business Week*, February 3, 1997, pp. 134–138.

4   Ford Uses Network to Gain Loyalty

Ford wants to increase customer satisfaction and owner loyalty by making people's experiences with Ford dealers more pleasant. One tactic to achieve this goal involved the implementation of a network to support the sales and servicing of automobiles, thus providing showroom-to-junkyard support for Ford customers. The network, called FocalPt, is used by more than 15,000 dealers worldwide.

Through FocalPt, salespeople use a PC to access promotional, inventory, and financing information to help close a deal. While talking to a customer, a salesperson can show pictures of various makes and models as well as explain options and associated prices. Salespeople can also use the network to locate the make, model, and configuration of the customer's choice anywhere in the world, and let the customer know when to expect delivery.

FocalPt also automates dealer service centers—so when a customer brings a car in for service, distributors can find repair information specific to that make and model as well as that car's repair history. Ford owners whose cars break down anywhere in the United States will find that the nearest Ford dealer has all the information on their car stored locally.

Ford selected a "push technology" offering from Wayfarer Communications as the software to send database updates from Ford headquarters to Microsoft SQL server databases in dealerships worldwide using the company's existing FordStar satellite network. With push technology, information is collected and automatically sent to the user without having to wait for a download request. Thus, price changes, car availability, financing, and other data are guaranteed to be current.

Ford has combined the power of the PC with network technology to provide new ways for interacting with customers. The current system enables different people to see the data they need in a timely fashion, without a lot of network overhead.

1. How might the FocalPt network be expanded to further improve the process of purchasing a Ford? How might it improve the repair experience for the customer?
2. How can other businesses use the Internet or their own networks to improve communications and service?

Sources: Adapted from Mitch Wagner, "Ford Extranet to Spark Loyalty," *Computerworld*, February 17, 1997, pp. 1, 16; Robert L. Scheier, "'Push' Software Has Yet to Come to Shove," *Computerworld*, February, 17, 1997, pp. 1, 66; Mitch Wagner, "Ford Pushes 'Net Data'," *Computerworld*, February 24, 1997, pp. 1, 73.

PART III

Business Information Systems

CHAPTER 8

Transaction Processing, Electronic Commerce, and Enterprise Resource Planning Systems

"A whole new world of buying and selling via the Internet is taking shape, and to venture capitalists, entrepreneurs, and purveyors of technology gear, it looks like the next great thing."

— Bernard Wysocki Jr.[1]

Chapter Outline

An Overview of Transaction Processing Systems
Traditional Transaction Processing Methods and Objectives
Transaction Processing Activities
Control and Management Issues

Traditional Transaction Processing Applications
Order Processing Systems
Purchasing Systems
Accounting Systems

Electronic Commerce
Electronic Markets and Commerce in Perspective
Search and Identification
Selection and Negotiation
Purchasing Products and Services Electronically
Product and Service Delivery
After-Sales Service

Enterprise Resource Planning
An Overview of Enterprise Resource Planning
Advantages and Disadvantages of ERP
Example of an ERP System

Learning Objectives

After completing Chapter 8, you will be able to:

1. Identify the basic activities and business objectives common to all transaction processing systems.
2. Describe the inputs, processing, and outputs for the transaction processing systems associated with order processing, purchasing, and accounting business processes.
3. Define the term *e-commerce* and discuss how an e-commerce system must support the many stages consumers experience in the sales life cycle.
4. Define the term *enterprise resource planning system* and discuss the advantages and disadvantages associated with the implementation of such a system.

British Petroleum
International Systems Program Handles Commercial Transactions

The British Petroleum Company (BP) is one of the world's largest petroleum and petrochemicals companies. It specializes in exploration for and production of crude oil and natural gas; refining, marketing, supply, and transportation; and manufacture and marketing of petrochemicals. BP has major operations in Europe, the United States, Australia, and parts of Africa, and is expanding into other areas, notably China, Southeast Asia, South America, and Eastern Europe. Founded in 1901, the London-based company had recent annual revenues of $57 billion.

Whenever you drive by a busy intersection with a gas station on each corner, you're reminded of how competitive the oil industry is. BP must control costs and offer exceptional service to compete in this intensely aggressive market. Neither task is easy, especially when you consider all the supply, distribution, and marketing activities that must occur before the customer even sees the product. If not carefully managed, these activities can involve high administrative costs. In addition, getting a handle on global operations was difficult because BP used comput-

ers from different manufacturers in different countries. Transmitting information among systems often required the development of expensive interfaces and/or duplicate data entry. So, BP decided to standardize and integrate its operations to reduce the high overhead stemming from administrative and computing costs.

Over a period of four years, BP developed and implemented a new application to streamline key business activities. Built on Oracle database technology and Sun and Sequent servers, the International Systems Program (ISP) handles all commercial transactions, including supply, distribution, sales and marketing, accounting, and taxation. Over 4,000 employees worldwide use their desktop and portable computers to access ISP across BP's enterprisewide network. They use ISP to take orders, schedule and record deliveries, track inventory, handle invoices, manage accounts receivable, produce monthly and quarterly expense reports, and perform compliance audits.

ISP runs worldwide on about 250 Oracle databases that range in size

from 10 to 90 GB. Information flows constantly among the various databases. After an order is taken in one location, that transaction triggers a message to another server where sales are handled. The delivery is scheduled, and a message is sent to another server to issue an import permit. Once the oil is delivered, details are sent to another server, where an invoice is generated. ISP supports business transactions in multiple currencies.

ISP has enabled BP to reduce the cost and complexity of business operations by standardizing tasks across countries. This makes it easier to transfer workers because they know exactly what they'll find when they access a server anywhere in the world. ISP also reduces overhead costs by eliminating redundant data entry and expensive interfaces. Once the data is entered correctly, it later reappears automatically to initiate and perform a task. ISP has enabled BP to standardize its computer technology globally. This has lowered hardware and software maintenance and support costs as well as provided volume discounts on new purchases. In the future, BP may link

ISP with systems that provide decision support. ISP can provide consistent views of management information.

As you read this chapter, consider the following questions:

• Why are business transaction pro-

cessing systems critical to the success of an organization?

• What are the advantages of having an integrated information system capable of linking business transactions to support multiple business processes?

Sources: Adapted from, press releases: "BP Targets Annual Profits of $6 Billion," the British Petroleum Company Web site at http://www.bp.com, accessed April 15, 1998; "Oracle at Work with British Petroleum," Oracle Web site at http://www.oracle.com/corporate, accessed April 15, 1998; "Oracle Announces Upgraded Energy Applications," November 11, 1997.

To operate efficiently, businesses have increasingly turned to information systems. As seen with British Petroleum, the many business activities associated with supply, distribution, sales, marketing, accounting, and taxation can be quickly performed while avoiding waste and mistakes. The goal of this computerization is ultimately to satisfy a business's customers and provide a competitive advantage by reducing costs and improving service.

Transaction processing was one of the first business processes to be computerized, and without information systems, recording and processing business transactions would consume huge amounts of an organization's resources. The transaction processing system (TPS) also provides employees involved in other business processes—the management information system (MIS), decision support system (DSS), and the artificial intelligence/expert systems (AI/ES)—with data to help them achieve their goals. A transaction processing system serves as the foundation for the other systems (Figure 8.1). Transaction processing systems perform routine operations such as sales ordering and billing, often performing the same operations daily or weekly. The amount of support for decision making that a TPS directly provides managers and workers is low.

These systems require a large amount of input data and produce a large amount of output without requiring sophisticated or complex processing. As we move from transaction processing to management information, decision support, and expert systems, we see less routine, more decision support, less input and output, and more sophisticated and complex processing and analysis. But the increase in sophistication and complexity in moving from transaction processing does not mean that it is less important to a business. In most cases, all these systems start as a result of one or more business transactions.

FIGURE 8.1

TPS, MIS, DSS, and AI/ES in Perspective

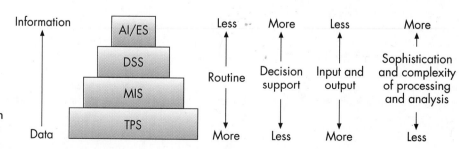

AN OVERVIEW OF TRANSACTION PROCESSING SYSTEMS

transaction processing systems (TPSs)

systems that process the detailed data necessary to update records about the fundamental business operations of the organization

transactions

the basic business operations such as customer orders, purchase orders, receipts, time cards, invoices, and payroll checks in an organization

British Petroleum Company's new application called the International Systems Program handles all commercial transactions, including supply, distribution, sales and marketing, accounting, and taxation.

Every organization has manual and automated **transaction processing systems (TPSs)**, which process the detailed data necessary to update records about the fundamental business operations of the organization. These systems include order entry, inventory control, payroll, accounts payable, accounts receivable, and general ledger, to name just a few. The input to these systems includes basic business **transactions** such as customer orders, purchase orders, receipts, time cards, invoices, and payroll checks. The result of processing business transactions is that the organization's records are updated to reflect the status of the operation at the time of the last processed transaction. Automated TPSs consist of all the components of a CBIS, including hardware, software, databases, telecommunications, people, and procedures used to process transactions. The processing includes data collection, data edit, data correction, data manipulation, data storage, and document production.

The opening vignette illustrates the importance of the processing activities of a TPS. British Petroleum developed and implemented an integrated set of applications to streamline key business activities. Their International Systems Program (ISP) handles all commercial transactions, including supply, distribution, sales and marketing, accounting, and taxation. Over 4,000 employees worldwide access ISP across BP's enterprisewide network to take orders, schedule and record deliveries, track inventory, handle invoices, manage accounts receivable, produce monthly and quarterly actual expense reports, and perform compliance audits.

For most organizations, TPSs support the routine, day-to-day activities that occur in the normal course of business that help a company add value to its products and services. Depending on the customer, value may mean lower price, better service, higher quality, or uniqueness of product. By adding a significant amount of value to their products and services, companies ensure further organizational success. Because the TPSs often perform activities related to customer contacts, such as order processing and invoicing, these information systems play a critical role in providing value to the customer. For example, by capturing and tracking the movement of each package, United Parcel Service (UPS) is able to provide timely and accurate data on the exact location of a package. Shippers and receivers can access an on-line database and, by providing the airbill number of a package, find the package's current location. Such a system provides the basis for added value through improved customer service.

British Petroleum Company Home Page - Microsoft Internet Explorer

World Energy - who's using it? who's producing it?

This website is about **Our Business** *and how we aim to make a positive contribution to* **The World**

Traditional Transaction Processing Methods and Objectives

When computerized transaction processing systems first evolved, only one

batch processing system

a system whereby business transactions are accumulated over a period of time and prepared for processing as a single unit or batch

on-line transaction processing (OLTP)

a system whereby each transaction is processed immediately, without the delay of accumulating transactions into a batch

method of processing was available. All transactions were collected in groups, called *batches,* and processed together. With **batch processing systems**, business transactions are accumulated over a period of time and prepared for processing as a single unit or batch (Figure 8.2a). The time period during which transactions are accumulated is whatever length of time is needed to meet the needs of the users of that system. For example, it may be important to process invoices and customer payments for the accounts receivable system daily. On the other hand, the payroll system may receive time cards and process them biweekly to create checks and update employee earnings records as well as to distribute labor costs. The essential characteristic of a batch processing system is that there is some delay between the occurrence of the event and the eventual processing of the related transaction to update the organization's records.

Today's computer technology allows another processing method, called *on-line, real-time processing,* or **on-line transaction processing (OLTP)**. With this form of data processing, each transaction is processed immediately, without the delay of accumulating transactions into a batch (Figure 8.2b). As soon as the input is available, a computer program performs the necessary processing and updates the records affected by that single transaction. Consequently, at any time, the data in an on-line system always reflects the current status. When you make an airline reservation, for instance, the transaction is processed and all databases, such as seat occupancy and accounts receivable, are updated immediately. This type of processing is

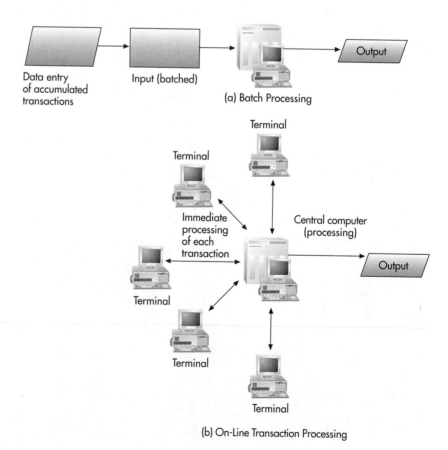

FIGURE 8.2

Batch vs. On-line Transaction Processing

Batch processing (a) inputs and processes data in groups. In on-line processing (b), transactions are completed as they occur.

When you order a product over the phone, the vendor may use on-line entry with delayed processing. Your order is entered into the computer at the time of the call but is not processed immediately.
(Source: Image Copyright © 1998 PhotoDisc.)

absolutely essential for businesses that require data quickly and update it often, such as airlines, ticket agencies, and stock investment firms. Many companies have found that OLTP helps them provide faster, more efficient service—one way to add value to their activities in the eyes of the customer.

A third type of transaction processing, called *on-line entry with delayed processing*, is a compromise between batch and on-line processing. With this type of system, orders or transactions are entered into the computer system when they occur, but they are not processed immediately. For example, when you call a toll-free number and order a product, your order is typically entered into the computer when you make the call. However, the order may not be processed until that evening after business hours.

Even though the technology exists to run TPS applications using on-line processing, it is not done for all applications. For many applications, batch processing is more appropriate and cost-effective. Payroll transactions and billing are typically done via batch processing. Specific goals of the organization define the method of transaction processing best suited for the various applications of the company. Figure 8.3 shows the total integration of a firm's major transaction processing systems.

Because of the importance of transaction processing, organizations expect their TPSs to accomplish a number of specific objectives, including the following:

Process data generated by and about transactions. The primary objective of any TPS is to capture, process, and store transactions and to produce a variety of documents related to routine business activities. These business activities can be directly or indirectly related to selling products and services to customers. Processing orders, purchasing materials, controlling inventory, billing customers, and paying suppliers and employees are all business activities that are a result of customer orders. These activities result in transactions that are processed by the TPS.

Maintain a high degree of accuracy. One objective of any TPS is error-free data input and processing. Even before the introduction of computer technology, employees visually inspected all documents and reports introduced into or produced by the TPS. Because humans are fallible, the transactions were often inaccurate, resulting in wasted time and effort and requiring resources to correct them.

Ensure data and information integrity and accuracy. Another objective of a TPS is to ensure that all data and information stored in computerized databases are accurate, current, and appropriate. As the volume of data being processed and stored increases, it becomes more difficult for individuals and machines to review all input data. Doing so is critical, however, because data and information generated by the TPS are often used by other information systems in an organization. So, a company must ensure both data integrity and accuracy. The processes of verification and editing are used to check whether data is accurate and up to date before it is stored.

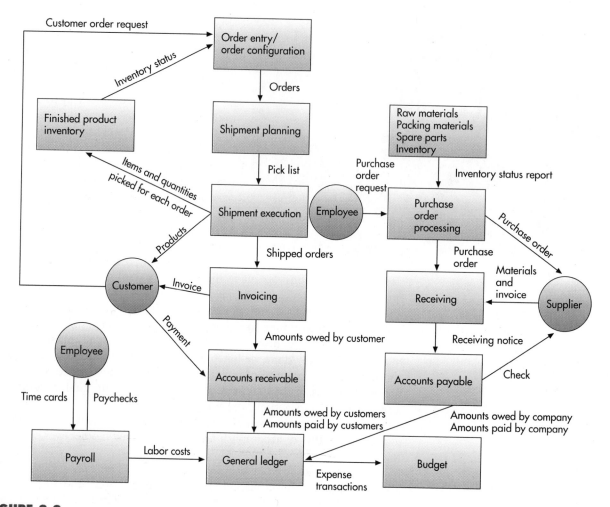

FIGURE 8.3

Integration of a Firm's TPSs

For example, an editing program has the ability to determine that an entry that should read "40 hours" is not entered as "400 hours" or "4000 hours" because of a data-entry error.

Produce timely documents and reports. Manual transaction processing systems can take days to produce routine documents. Fortunately, the use of computerized transaction processing systems significantly reduces this response time. Improvements in information technology, especially hardware and telecommunications links, allow transactions to be processed in a matter of seconds. The ability to conduct business transactions in a timely way can be very important for the profitable operation of the organization. For instance, if bills (invoices) are sent out to customers a few days earlier than usual, payment may be received earlier.

Timing is also crucial for related applications such as order processing, invoicing, accounts receivable, inventory control, and accounts payable. Because of electronic recording and transmission of sales information, transactions can be processed in seconds rather than overnight, thus improving companies' cash flow. Customers find credit card charges they made on the final day of the billing period on their current monthly bill.

Increase labor efficiency. Before computers existed, manual business processes often required rooms full of clerks and equipment to process the necessary business transactions. Today, transaction processing systems can substantially reduce clerical and other labor requirements. A small minicomputer linked to a company's cash registers can replace a room full of clerks, typewriters, and filing cabinets. Many TPSs are cost-justified by labor savings.

Help provide increased and enhanced service. Without question, we are quickly becoming a service-oriented economy. Even strong manufacturing companies, including household appliance makers and automobile manufacturers, realize the importance of providing superior customer service. One objective of any TPS is to assist the organization in providing this type of service. For example, some companies have EDI systems (see Chapter 6) that allow customers to place orders electronically, thus bypassing slower and more error-prone methods of written or oral communication.

Help build and maintain customer loyalty. A firm's transaction processing systems are often the means for customers to communicate. It is important that the customer interaction with these systems keeps customers satisfied and returning.

Achieve competitive advantage. A goal common to almost all organizations is to gain and maintain a competitive advantage. As discussed in Chapter 2, a competitive advantage provides a significant and long-term benefit for the organization. When a TPS is developed or modified, the personnel involved should carefully consider how the new or modified system might provide a significant and long-term benefit. Some of the ways that companies can use transaction processing systems to gain competitive advantage are summarized in Table 8.1.

Depending on the specific nature and goals of the organization, any of these objectives may be more important than another. By meeting these objectives, transaction processing systems can support corporate goals such as reducing costs; increasing productivity, quality, and customer satisfaction; and running more efficient and effective operations.

For example, overnight delivery companies such as UPS expect their transaction processing systems to increase customer service. These systems can locate a client's package at any time from initial pickup to final delivery. This improved customer information allows UPS to produce timely information and be more responsive to customer needs and queries.

Key success criteria for banking institutions include increasing customer service and labor efficiency, maintaining a high degree of accuracy, and

TABLE 8.1

Examples of Transaction Processing Systems for Competitive Advantage

Competitive Advantage	Example
Customer loyalty increased	Use of customer interaction system to monitor and track each customer interaction with the company
Superior service provided to customers	Use of tracking systems that are accessible by customers to determine shipment status
Superior information gathering performed	Use of order configuration system to ensure that products and services ordered will meet customer's objectives
Costs dramatically reduced	Use of warehouse management system employing scanners and bar-coded product to reduce labor hours and improve inventory accuracy

ensuring data and information integrity. Such institutions meet customer needs by placing automated bank terminals, called *automatic teller machines (ATMs)*, in grocery stores, on street corners, inside factories and warehouses, and at other convenient locations. To accommodate the use of ATMs, specialized computer programs have been developed to handle all sorts of transactions, such as withdrawals from and deposits to checking and savings accounts, credit card advances, transfers of funds from one account to another, inquiries into balances, and other related activities.

Brokerage firms employ specialized transaction processing systems that process orders, provide customer statements, and produce managerial reports. Increasingly, brokerage firms offer their clients software to monitor their own accounts on-line. These systems indicate all buy and sell orders and associated commissions, the securities held, the purchase price, the current market price, the dividends or interest per period, the dividends per year, and the anticipated yield on each security as well as the anticipated yield of each client's portfolio of securities. The objectives are to increase customer service and to provide timely reports. Obviously, maintaining a high degree of data accuracy and integrity is also important. Discount brokerage firms are even offering software that allows clients to enter their own buy and sell requests without first speaking with a broker. This increases labor efficiency but raises issues about the level of commissions that should be charged.

Transaction Processing Activities

Along with having common characteristics, all transaction processing systems perform a common set of basic data processing activities. TPSs capture and process data that describe fundamental business transactions. This data is used to update databases and to produce a variety of reports for use

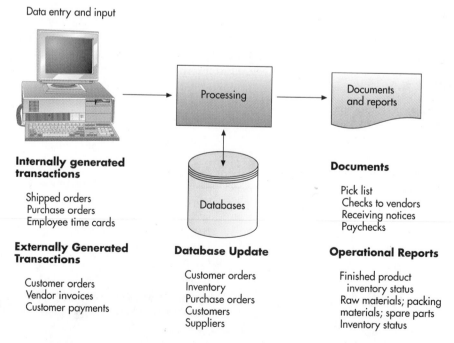

Data entry and input

Processing

Documents and reports

Internally generated transactions

Shipped orders
Purchase orders
Employee time cards

Externally Generated Transactions

Customer orders
Vendor invoices
Customer payments

Databases

Database Update

Customer orders
Inventory
Purchase orders
Customers
Suppliers

Documents

Pick list
Checks to vendors
Receiving notices
Paychecks

Operational Reports

Finished product
  inventory status
Raw materials; packing
  materials; spare parts
Inventory status

FIGURE 8.4

A Simplified Overview of a Transaction Processing System

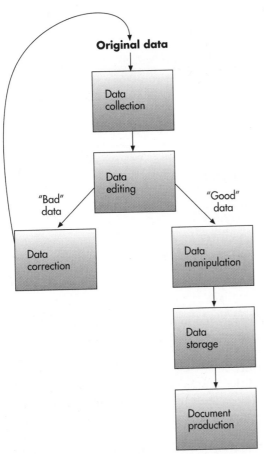

Original data

FIGURE 8.5

Data Processing Activities
Common to Transaction
Processing Systems

**transaction processing
cycle**

the process of data collection, data
editing, data correction, data
manipulation, data storage, and
document production

data collection

the process of capturing and
gathering all data necessary to
complete transactions

source data automation

the process of capturing data at its
source, recording it accurately, in
a timely fashion, with minimum
manual effort, and in a form that
can be directly entered to the
computer rather than keying the
data from some type of document

by people both within and outside the enterprise
(Figure 8.4). The business data goes through a
transaction processing cycle that includes data
collection, data editing, data correction, data manipulation, data storage, and document production
(Figure 8.5).

Data collection. The process of capturing and gathering all data necessary to complete transactions is
called **data collection**. In some cases this can be
done manually, such as by collecting handwritten
sales orders or changes to inventory. In other cases,
data collection is automated via special input devices
such as scanners, point-of-sale devices, and terminals.

Data collection begins with a transaction (e.g., taking a customer order) and results in the origination of
data that is input to the transaction processing system.
Data should be captured at its source, and it should be
recorded accurately, in a timely fashion, with minimal
manual effort, and in a form that can be directly
entered to the computer rather than keying the data
from some type of document. This approach is called
source data automation. An example of source data
automation is the use of scanning devices at the grocery
checkout to read the Universal Product Code (UPC)
automatically. Reading the UPC bar codes is quicker
and more accurate than having a cash register clerk
enter codes manually. The scanner reads the bar code
for each item and looks up its price in the item data-
base. The point-of-sale transaction processing system uses the price data to
determine the customer's bill. The number of units of this item purchased,
the date, the time, and the price are also used to update the store's inventory
database, as well as its database of detailed purchases. The inventory database
is used to generate an exception report notifying the store manager to reorder
items whose sales have reduced the stock below the reorder quantity. The
detailed purchases database can be used by the store (or sold to market
research firms or manufacturers) for detailed analysis of sales (Figure 8.6).

Many grocery stores combine point-of-sale scanners and coupon printers. The systems are programmed so that each time a specific product—say,
a box of cereal—crosses a checkout scanner, an appropriate coupon—perhaps a milk coupon—is printed. Other companies can also "rent time" on
the system, which is then reprogrammed to print those companies'
coupons if the customer buys a competitive brand. These TPSs help grocery stores to increase profits by improving their repeat sales and bringing
in revenue from other businesses.

As another example, industrial data collection devices allow employees
to scan their magnetized employee ID cards to enter data into the payroll
TPS when they start or end a job. Not only do these devices provide important employee payroll information, they also help an organization determine how many labor hours are spent on each job or project so that staffing
adjustments can be made or future projects can be accurately planned.

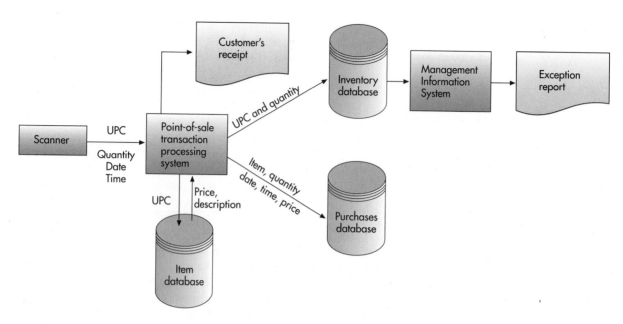

FIGURE 8.6

Point-of-Sale Transaction
Processing System

Scanning items at the checkout
stand results in updating a
store's inventory database and
its database of purchases.

data editing

the process of checking data for
validity and completeness

data correction

the process of reentering miskeyed
or misscanned data that was found
during the data editing

data manipulation

the process of performing
calculations and other data
transformations related to business
transactions

data storage

the process of updating one or
more databases with new
transactions

Data editing. An important step in processing transaction data is to perform **data editing** for validity and completeness to detect any problems with the data. For example, quantity and cost data must be numeric and names must be alphabetic; otherwise, the data is not valid. Often the codes associated with an individual transaction are edited against a database containing valid codes. If any code entered (or scanned) is not present in the database, the transaction is rejected.

Data correction. It is not enough to reject invalid data. The system should provide error messages that alert those responsible for the data edit function. These error messages must specify what problem is occurring so that corrections can be made. A **data correction** involves reentering miskeyed or misscanned data that was found during data editing. For example, a UPC that is scanned must be in a master table of valid UPCs. If the code is misread or does not exist in the table, the checkout clerk is given an instruction to rescan the item or key in the information manually.

Data manipulation. Another major activity of a TPS is **data manipulation**, the process of performing calculations and other data transformations related to business transactions. Data manipulation can include classifying data, sorting data into categories, performing calculations, summarizing results, and storing data in the organization's database for further processing. In a payroll TPS, for example, data manipulation includes multiplying an employee's hours worked by the hourly pay rate. Overtime calculations, federal and state tax withholdings, and deductions are also performed.

Data storage. Data storage involves updating one or more databases with new transactions. Once the update process is complete, this data can be further processed and manipulated by other systems so that it is available for management decision making. Thus, while transaction databases can be considered a "by-product" of transaction processing, they have a pronounced effect on almost all other information systems and decision-making processes in an organization.

document production

the process of generating output
records and reports

Document production. TPSs produce important business documents. **Document production** involves generating output records and reports. These documents may be hard-copy paper reports or displayed on computer screens (sometimes referred to as *soft copy*). Paychecks, for example, are hard-copy documents produced by a payroll TPS, while an outstanding balance report for invoices might be a soft-copy report displayed by an accounts receivable TPS. Often, results from one TPS are passed downstream as input to other systems (as shown in Figure 8.6), where the results of updating the inventory database are used to create the stock exception report of items whose inventory level is less than the reorder point.

In addition to major documents like checks and invoices, most transaction processing systems provide other useful information, such as reports that help employees perform various operational activities. A report showing current inventory is one example; another might be a document listing items ordered from a supplier to help a receiving clerk check the order for completeness when it arrives. A TPS can also produce reports required by local, state, and federal agencies, such as statements of tax withholding and quarterly income statements.

Throughout this chapter we will look at some specific examples of ways companies have employed TPSs to help them meet organizational goals.

Control and Management Issues

Transaction processing systems are the backbone of any organization's information systems. They capture facts about the fundamental business operations of the organization—facts without which orders cannot be shipped, customers cannot be invoiced, and employees and suppliers cannot be paid. In addition, the data captured by the transaction processing systems flow downstream to the other systems of the organization. Like any structure, an organization's information systems are only as good as the foundation on which they are built. Indeed, most organizations would grind to a screeching halt if their transaction processing systems failed.

business resumption planning

the process of anticipating and
providing for disasters

Business resumption planning. Business resumption planning is the process of anticipating and providing for disasters. A disaster can range from a flood, a fire, or an earthquake to labor unrest or erasure of an important file. Business resumption planning focuses primarily on two issues: maintaining the integrity of corporate information and keeping the information system running until normal operations can be resumed.

One of the first steps of business resumption planning is to identify potential threats or problems, such as natural disasters, employee misuse of personal computers, and poor internal control procedures. Business resumption planning also involves disaster preparedness. IS managers should occasionally hold an unannounced "test disaster"—similar to a fire drill—to ensure that the disaster plan is effective.

disaster recovery

the implementation of the business
resumption plan

Disaster recovery. Disaster recovery is the implementation of the business resumption plan. Although companies have known about the importance of disaster planning and recovery for decades, many do not adequately prepare. For example, in southern California, many businesses were flooded by the unusually heavy rains brought on by El Niño. In January 1998, prolonged

Companies like Iron Mountain provide a secure, off-site environment for records storage. In the event of a disaster, vital data can be recovered.

(Source: Photo courtesy of Iron Mountain.)

power failures forced many businesses to operate with skeleton staffs, shut down, move to disaster sites, or rely on backup generators.[2] The primary tools used in disaster planning and recovery are backups for hardware, software and databases, telecommunications, and personnel.

A common backup for hardware is a similar or compatible computer system owned by another company or a specialized backup system provided by an organization from which a written hardware backup agreement is obtained. Some companies achieve hardware backup with duplicate systems and equipment of their own. For example, a company might use two identical mainframes—one for systems development and the other for running business applications. The systems development processor (used to create new programs and to modify existing ones) can be used to run applications if the original applications processor fails. Two firms in different industries having compatible systems can also serve as hardware backups for one another.

Keeping a duplicate system that is operational or having immediate access to one through a specialized vendor is an example of a hot site. A hot site is often a compatible mainframe system that is operational and ready to use. If the primary mainframe has problems, the hot site can be immediately used as a backup.

Another approach is to use a cold site, also called a shell, which is a computer environment that includes rooms, electrical service, telecommunications links, data storage devices, and the like. If there is a problem with the primary mainframe, replacement computer hardware is brought into the cold site, and the complete system is made operational. For both hot and cold sites, telecommunications media and devices are used to provide for fast and efficient transfer of processing jobs to the disaster facility.

Software and databases can be backed up by making duplicate copies of all programs and data. At least two backup copies should be made. One backup copy can be kept in the information systems department in case of accidental destruction of the software. Another backup copy should be kept off-site in a safe, secure, fireproof, and temperature- and humidity-controlled environment in case of loss of the data processing facility. Service companies provide this type of backup environment.

Backup is also essential for the data and programs on users' desktop computers. The advent of more distributed systems, like client/server systems, means that many users now have important, and perhaps critical, data and applications on their desktop computers. Utility packages inexpensively provide backup features for desktop computers by copying data onto magnetic tape.

Some business recovery plans call for the backup of vital telephone communications. Complex plans might call for recovering whole networks. In other plans, the most critical nodes on the network are backed up by duplicate components. Using such fault-tolerant networks, which will not break down when one node or part of the network malfunctions, can be a more cost-effective approach to telecommunications backup.

There should also be a backup for information systems personnel. This can be accomplished in a number of ways. One of the best approaches is to provide cross-training for IS and other personnel so that each individual can perform an alternate job if required. For example, a company might train employees in accounting, finance, or other IS departments to operate the system if a disaster strikes. The company could also make an agreement with another information systems department or an outsourcing company to supply IS personnel if necessary.

transaction processing system audit

an examination of the TPS in an attempt to answer three basic questions: Does the system meet the business need for which it was implemented? What procedures and controls have been established? Are these procedures and controls being properly used?

Transaction processing system audit. A **transaction processing system audit** attempts to answer three basic questions:

- Does the system meet the business need for which it was implemented?
- What procedures and controls have been established?
- Are these procedures and controls being properly used?

In addition to these three basic auditing questions, other areas are typically investigated during an audit. These areas include the distribution of output documents and reports, the training and education associated with existing and new systems, and the time necessary to perform various tasks and to resolve problems and bottlenecks in the system. General areas of improvement are also investigated and reported during the audit. Two types of audits exist. An internal audit is conducted by employees of the organization; an external audit is performed by accounting firms or companies and individuals not associated with the organization. In either case, a number of steps are performed. The auditor inspects all programs, documentation, control techniques, the disaster plan, insurance protection, fire protection, and other systems management concerns such as efficiency and effectiveness of the disk or tape library. This is accomplished by interviewing IS personnel and performing a number of tests on the computer system.

audit trail

the trace to any output from the computer system back to the source documents

In establishing the integrity of the computer programs and software, an audit trail must be established. The **audit trail** allows the auditor to trace any output from the computer system back to the source documents. With many of the real-time and time-sharing systems available today, it is extremely difficult to follow an audit trail. In many cases, no record of inputs to the system exists; thus, the audit trail is destroyed. The auditor must investigate the actual processing in addition to the inputs and outputs of the various programs.

TRADITIONAL TRANSACTION PROCESSING APPLICATIONS

In this section we present an overview of several common transaction processing systems that support the order processing, purchasing, and accounting business processes (Table 8.2).

order processing systems

systems that process order entry, sales configuration, shipment planning, shipment execution, inventory control, invoicing, customer interaction, and routing and scheduling in an organization

Order Processing Systems

Order processing systems include order entry, sales configuration, shipment planning, shipment execution, inventory control, invoicing, customer interaction, and routing and scheduling. The business processes supported by

Order Processing	Purchasing	Accounting
• Order entry • Sales configuration • Shipment planning • Shipment execution • Inventory control (finished product) • Invoicing • Customer interaction • Routing and scheduling	• Inventory control (raw materials, packing materials, spare parts, supplies) • Purchase order processing • Receiving • Accounts payable	• Budget • Accounts receivable • Payroll • Asset management • General ledger

TABLE 8.2

Systems That Support Order Processing, Purchasing, and Accounting Business Functions

FIGURE 8.7

Order Processing Systems

these systems are so critical to the operation of the enterprise that the order processing systems are sometimes referred to as the "lifeblood of the organization." Figure 8.7 is a system-level flowchart that shows the various systems and the information that flows between them. A rectangle represents a system, a line represents the flow of information from one system to another, and a circle represents any entity outside the system—in this case, the customer.

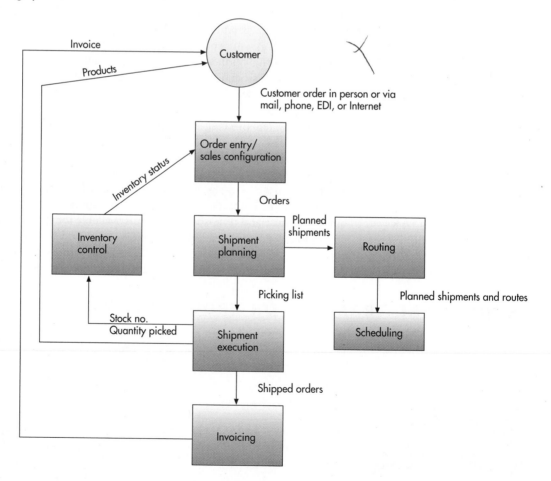

order entry system

a system that captures the basic data needed to process a customer order

FIGURE 8.8

Data-Flow Diagram of an Order Entry System

Orders are received by mail, phone, EDI, or the Internet from customers or sales reps and entered into the order processing system. This application affects accounting, inventory, warehousing, finance, and invoicing applications. Note that in an integrated order processing system, order entry personnel have access to back order, inventory, and customer information from separate data files or directly through the processing mechanism.
(Source: George W. Reynolds, *Information Systems for Managers*, 3rd ed., St. Paul, MN: West Publishing Co., 1995, p. 198. Reprinted with permission from Course Technology.)

Order entry. The **order entry system** captures the basic data needed to process a customer order. Orders may come through the mail or via a telephone ordering system, be gathered by a staff of sales representatives, arrive via EDI transactions directly from a customer's computer over a wide area network, or be entered directly over the Internet by the customer using a data entry form on the firm's Web site. Figure 8.8 is a data-flow diagram of a typical order entry system. The data-flow diagram is more detailed than the system-level flowchart. It shows the various business processes that are supported by a system and the flow of data between processes. A rectangle with rounded corners represents a business process.

With an on-line order processing system, such as one used by direct retailers, the inventory status of each inventory item (also called stock keeping unit, or SKU) on the order is checked to determine whether sufficient finished product is available. If an order item cannot be filled, a substitute item may be suggested or a back order is created—the order will be filled later when inventory is replenished. Order processing systems can also suggest related items for order takers to mention to promote "add-on" sales. Order takers also review customer payment history data from the accounts receivable system to determine whether credit can be extended.

Once an order is entered and accepted, it becomes an open order. Typically, a daily sales journal (which includes customer information, products ordered, quantity discounts, and prices) is generated.

Electronic data interchange (EDI) can be an important part of the order entry TPS. With EDI, a customer or client organization can place orders directly from its purchasing TPS into the order processing TPS of another organization. Or, the order processing TPS of the supplier companies and

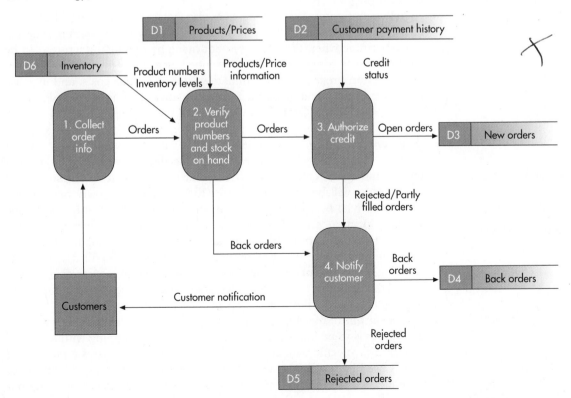

the purchasing TPS of the customers could be linked indirectly through a third-party clearinghouse. In any event, this computer-to-computer link allows efficient and effective processing of sales orders and enables an organization to lock in customers and lock out competitors through enhanced customer service. With EDI, orders can be placed any time of the day or night, and immediate notification of order receipt and processing can be made. Today, more and more companies are using electronic data interchange to make "paperless" business transactions a reality.

Sales configuration. Another important aspect of order processing is sales configuration. The **sales configuration system** ensures that the products and services ordered are sufficient to accomplish the customer's objectives and will work well together. For example, using a sales configuration program, a sales representative knows that a computer printer needs a certain cable and a LAN card so it can be LAN connected. Without a sales configuration program, a sales representative might sell a customer the wrong cable or forget the LAN card.

Sales configuration programs also suggest optional equipment. For example, if a customer orders a palmtop computer, the sales configuration program will suggest an AC adapter, back-up software and cables, and a modem to allow the palmtop computer the ability to connect to the Internet. If a company is buying a 747 aircraft from Boeing, a sales configuration program can help the sales representative work with the company to determine the number of seats that are needed, the most appropriate navigation systems to install, the type of landing gear that should be used, and hundreds of other available options that can be specified for the 747.

Sales configuration software can also solve customer problems and answer customer questions. For example, a sales configuration program can determine whether a factory robot made by one manufacturer can be controlled by a computer system developed by another manufacturer. Sales configuration programs can eliminate mistakes, reduce costs, and increase revenues. These advantages have led companies such as Hewlett-Packard, Boeing, and Silicon Graphics to implement these systems.

Shipment planning. New orders received and any other orders not yet shipped (open orders) are passed from the order entry system to the shipment planning system. The **shipment planning system** determines which open orders will be filled and from which location they will be shipped. This is a trivial task for a small company with lots of inventory, only one shipping location, and a few customers concentrated in a small geographic area. But, it is an extremely complicated task for a large global corporation with limited inventory (not all orders for all items can be filled), dozens of shipping locations (plants, warehouses, contract manufacturers, etc.), and tens of thousands of customers. The trick is to minimize shipping and warehousing costs and still meet customer delivery dates.

The output of the shipment planning system is a plan that shows where each order is to be filled and a precise schedule for shipping with a specific carrier on a specific date and time. The system also prepares a picking list that is used by warehouse operators to select the ordered goods from the warehouse. These outputs may be in paper form or they may be computer

---

sales configuration system

a system that ensures that the products and services ordered are sufficient to accomplish the customer's objectives and will work well together

shipment planning system

a system that determines which open orders will be filled and from which location they will be shipped

FIGURE 8.9

A Picking List

This document guides ware-house employees in locating items to fill an order. Note the second and third columns of the slip, which instruct the warehouse workers where to locate the items. Also note that in this instance, the third item ordered was back-ordered three cases. Once items are picked from inventory, this data is entered into the data trans-action processing system, and a packing slip and shipping notice are generated.

INDUSTRIAL FASTENING PRODUCTS, INC.										
3484 Seventh Street										
Calgary, USA	(182) 997-4567									

CUSTOMER NO.	012345678	ORDER NO.	C-654321	DATE	10/14/01	PAGE		1

SOLD TO:
DURA FURNITURE
COMPANY
PO BOX 491
478 ELM STREET
CINCINNATI, OHIO

SOLD TO:

CUSTOMER NO.	447918	SHIP VIA	WILSON FREIGHT ·ATTENTION· JIM JONSON

1	LOC	LINK	ITEM NUMBER	DESCRIPTION		ORDERED	SHIPPED	B/O	COMMENT
	8	105	10 L1L416028	FASTENING TOOL MODEL L	3	EACH	3		$
			20 S8276	STAPLE 3/4 INCH	15	CASE	15		$
			30 S8289	STAPLE 1 INCH	15	CASE	12	3	$
			40	SHIPPING CHARGE					$

*** END OF REPORT ***

records that are transmitted electronically. The picking list document, an example of which is shown in Figure 8.9, lists the customer name, number, order number, and all items that have been ordered. A description of all items along with the number to be shipped is also included.

Upgrading the order entry process and linking it with production scheduling has provided Anchor Glass with a competitive advantage. The company makes glass bottles for Coke, Pepsi, Smuckers, and other manufacturers who distribute their products in bottles. Before it implemented a new order processing system, the company was unable to provide a delivery date to customers who phoned in orders. But with the new system, order takers can check actual production schedules at the company's 15 factories, accept orders, and promise delivery—all during the initial phone call. Once the order is accepted, the system reserves production capacity for the order and the production schedule is updated, taking into account any changes in raw or packing material availability, plant capacity, and equipment downtime.

Anchor Glass improved its customer service by linking its order processing system with production scheduling. The company is now able to promise delivery dates when the order is placed.
(Source: Macduff Everton/© Corbis.)

shipment execution system

a system that coordinates the outflow of all products and goods from the organization, with the objective of delivering quality products on time to customers

Shipment execution. The **shipment execution system** coordinates the outflow of all products and goods from the organization, with the objective of delivering quality products on time to customers. The shipping department is usually given responsibility for physically packaging and delivering all products to customers and suppliers. This delivery can include mail services, trucking operations, and rail service. The system receives the picking list from the shipment planning system.

Sometimes orders cannot be filled exactly as specified. One reason is "out-of-stocks," meaning the warehouse does not have sufficient quantity of an item to fill a customer's order. This can result if a production run did not produce the expected quantity of an item because of manufacturing problems. The company policy may be not to ship any of the item, ship as many units of the item as are available and create a back order request for the remainder, or substitute another item. Thus, as items are picked and loaded for shipment, warehouse operators must enter data about the exact items and quantity of each that are loaded for each order. When the shipment execution system processing cycle is complete, it passes the "shipped orders" business transactions "downstream" to the invoicing system. These transactions specify exactly what items were shipped, the quantity of each, and to whom the order was shipped. This data is used to generate a customer invoice. The shipment execution system also produces packing documents, which are enclosed with the items being shipped, to tell customers what items are in the shipment, what is back-ordered, and the exact status of all items in the order.

inventory control system

a system that updates the computerized inventory records to reflect the exact quantity on hand of each stock keeping unit

Inventory control. For each item picked during the shipment execution process, a transaction providing the stock number and quantity picked is passed to the **inventory control system**. In this way, the computerized inventory records are updated to reflect the exact quantity on hand of each stock keeping unit. Thus, when order takers check the inventory level of a product, they receive current information.

FIGURE 8.10

An Inventory Status Report

This output from the inventory application summarizes all inventory items shipped over a specified time period.

Once products have been picked out of inventory, other documents and reports are initiated by the inventory control application. For example, the inventory status report (Figure 8.10) summarizes all inventory items shipped over a specified time period. It can include stock numbers, descriptions, number of units on hand, number of units ordered, back-ordered units, average costs, and related information. It is used to determine when to order more inventory and how much of each item to order, and it helps minimize "stockouts" and back orders. Data from this report is used as input to other information systems to help production and operations managers analyze the production process.

For almost all companies, inventory must be tightly controlled. One objective is to minimize the amount of cash tied up in inventory by placing just the right amount of inventory on the factory or warehouse floor.

To gain a competitive advantage, many manufacturing organizations are moving to real-time

PRD CLS	STOCK NUMBER	DESCRIPTION WH LOCN	UNITS ON HAND	UNITS ON ORDER	UNITS RESERVED	UNIT AVERAGE COST	VALUE AVERAGE COST	PHYSICAL QUANTITY
		DATE 5 30 01	JANUS TOOL COMPANY				PAGE 2	
		INVENTORY STATUS/STOCK TAKE REPORT						
10	1001 1 / 4" ELECTRIC DRILL . MO	1 0019	517.0000	0.0000	40.0000	29.50000	15,251.50	
40	1001 1 / 4" ELECTRIC DRILL . MO	1 0018	80.0000	30.0000	0.0000	29.00000	2,320.00	
10	1001 1 / 4" ELECTRIC DRILL . MO	1 0017	150.0000	25.0000	10.0000	28.75000	4,312.50	
20	1001 1 / 4" ELECTRIC DRILL . MO	1 0007	410.0000	25.0000	50.0000	10.25000	4,202.50	
20	1001 1 / 4" ELECTRIC DRILL . MO	1 0008	330.0000	0.0000	0.0000	27.45000	9,058.50	
25	1001 1 / 4" ELECTRIC DRILL . MO	1 0009	14,256.0000	0.0000	1,440.0000	0.45000	6,415.20	
32	1001 1 / 4" ELECTRIC DRILL . MO	1 0003	59.0000	40.0000	5.0000	41.50000	2,445.50	
100	1001 1 / 4" ELECTRIC DRILL . MO	1 0006	2,448.0000	0.0000	0.0000	4.10000	10,036.80	
25	1001 1 / 4" ELECTRIC DRILL . MO	1 0006	0.0000	0.0000	0.0000	1.75000	0.00	
30	1001 1 / 4" ELECTRIC DRILL . MO	1 0016	314.0000	100.0000	50.0000	5.85000	1,836.90	
15	1001 1 / 4" ELECTRIC DRILL . MO	1 0015	192.0000	240.0000	0.0000	2.05000	393.60	
20	1001 1 / 4" ELECTRIC DRILL . MO	1 0012	183.0000	50.0000	35.0000	4.75000	869.25	
XX	1001 1 / 4" ELECTRIC DRILL . MO	1 0011	105.0000	50.0000	0.0000	2.65000	278.25	
50	1001 1 / 4" ELECTRIC DRILL . MO	1 0004	109.0000	10.0000	0.0000	16.50000	1,798.50	
100	1001 1 / 4" ELECTRIC DRILL . MO	1 0001	960.0000	240.0000	0.0000	4.50000	4,320.00	
	GRAND TOTAL						62,754.80	

inventory control systems based on bar coding the finished product, scanners and radio display terminals mounted on forklifts, and wireless LAN communications to track each time an item is moved in the warehouse. One significant advantage is that the inventory data is more accurate and current for people performing order entry, production planning, and shipment planning. In addition, warehouse operations can be streamlined by providing directions to the forklift drivers.

In addition to being useful for physical goods such as automobiles and home appliances, inventory control is essential for industries in the service sector. Such organizations as hotels, airlines, rental car agencies, and universities, which primarily provide services, can use inventory applications to help them monitor use of rooms, airline seats, car rentals, and classroom capacity. Airlines face an especially difficult inventory problem. Empty airline seats (inventory) have absolutely no value after a plane takes off. Yet, overbooking can result in too many seats being sold and customer complaints. Sophisticated reservation systems allow airlines to quickly update and add seating assignments.

Invoicing. Customer invoices are generated based on records received from the shipment execution transaction processing system. This application encourages follow-up on existing sales activities, increases profitability, and improves customer service. Most invoicing programs automatically compute discounts, applicable taxes, and other miscellaneous charges (Figure 8.11). Because most computerized operations contain elaborate databases on customers and inventory, many invoicing applications require only information on the items ordered and the client identification number; the invoicing application does the rest. It looks up the full name and address of the customer, determines whether the customer has an adequate credit rating, automatically computes discounts, adds taxes and other charges, and prepares invoices and envelopes for mailing.

Invoicing in a service organization can be even more complicated than invoicing in manufacturing and retail firms. The trick is to match all services rendered with a specific customer and to include all appropriate rates and charges in calculating the bill. This is especially difficult if the data needed for billing has not been accurately and completely captured in a transaction processing system.

Customer interaction. Winning new customers and keeping existing ones happy are key to financial success. It is often said

FIGURE 8.11

An output of the invoicing system, a customer invoice reflects the value of the current invoice, as well as which products the customer purchased.

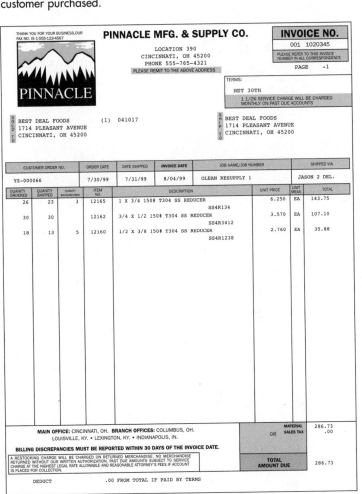

customer interaction system

a system that monitors and tracks each customer interaction with the company

that when customers have a pleasant experience with a firm, they may tell one or two people; however, when they have an unpleasant experience, they may tell 10 to 20 people! To keep current customers happy, some companies use a **customer interaction system** (Figure 8.12) to monitor and track each customer interaction with the company.[3] The goal of such systems is to build customer loyalty. These software systems capture data whenever a customer contacts the company. Often the initial contact is a request for a proposal or a request for product information from a potential customer. Valuable data about the potential customer can be gathered at this time. Obviously, additional data is captured at the time of each sale. After the sale, a customer may contact the firm with a request for customer service or may have an idea or request other information. At other times the customer may contact the firm with a complaint or a product improvement idea. The customer interaction system captures valuable data from each interaction and passes the data to others in the organization who can use it. The analysis of customer complaints provides ideas for market research and product development as well as useful measures for quality control. Sales and marketing employees gain a deeper understanding of what the customer wants through analysis of customer questions about the product and its use. The customer interaction data represents a gold mine of data

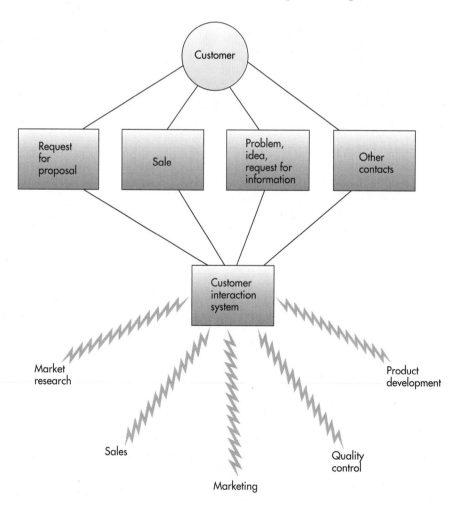

FIGURE 8.12

Customer Interaction System

that can be used to keep customers satisfied, generate new leads for future sales, and lead to new products or product improvements.

Consumer goods companies such as Procter & Gamble know the importance of building and maintaining customer loyalty. They recognize that the cost to acquire a new customer is tenfold the cost to keep an existing one. As a result, they have toll-free numbers on all their products for consumers to call for information on how to use the product, register a complaint, or find out where a new product is available. Consumer services representatives are supported by an on-line customer interaction system that captures data about the nature of the call and the consumer. The next time the consumer calls, the service rep knows that the customer is a repeat caller and can look up information about his or her previous interaction(s) with the company. Consumer comments about the product are provided to product development and manufacturing. Comments regarding commercials and promotions are forwarded to the advertising department.

Routing and scheduling. Many computer manufacturers and software firms have developed specialized transaction processing systems for companies in the distribution industry. Some distribution applications are for wholesale operations; others are for retail or specialized applications. Trucking firms, beverage distributors, electrical distributors, and oil and natural gas distribution companies are only a few examples.

Like airlines, distribution companies must also determine the best use of their resources. For example, a motor freight company might have 100 deliveries to make during the next week, including loads from Miami to Boston and Seattle to Salt Lake City. A **routing system** involves determining the best way to get goods and products from one location to another. Waste Management Systems, for example, now has terminals on board its garbage collection trucks to facilitate routing. Drivers can be notified of changes in the routes after they are out on the streets. For an oil and natural gas company, routing means finding the fastest and cheapest pipelines to carry the product from the source to a distant final destination.

The **scheduling system** determines the best time to deliver goods and services. For example, trucks can be scheduled to deliver automobile transmission systems from California to Michigan during the second week of September, when oil and gas prices are low. Other objectives are to carry a profitable load on the return trip and to minimize total distance traveled, which can result in lower fuel, driver, and truck maintenance costs. For these reasons, many distribution companies have designed TPSs to help determine which routes will allow for efficient service, while making cost-effective use of drivers and trucks. For firms such as these, scheduling and routing programs are connected to the organization's order and inventory transaction processing system. Read the "Making a Difference" box to learn more about routing and scheduling software.

Purchasing Systems

The **purchasing transaction processing systems** include inventory control, purchase order processing, receiving, and accounts payable (Figure 8.13).

---

routing system

a system that determines the best way to get goods and products from one location to another

scheduling system

a system that determines the best time to deliver goods and services

purchasing transaction processing systems

processing systems that include inventory control, purchase order processing, receiving, and accounts payable

MAKING A DIFFERENCE
Routing and Scheduling Software Cuts Distribution Costs

The Lightstone Group has its headquarters in Mineola, New York, and branch offices in Palo Alto, California; Glastonbury, Connecticut; and Westminster, Maryland. The company is dedicated to creating software for logistics and supply-chain decision support to businesses requiring route optimization and scheduling for their field personnel and delivery vehicles.

The firm was founded in 1989 by three people who initially developed the MachUP airline scheduling systems that was sold to the American Airlines SABRE Group. Today it is a leading routing and scheduling software developer, with Resources in Motion Management System (RiMMS) as its flagship product. A wide range of industries and customers, large and small, enjoy cost savings and improved customer service due to RiMMS.

For example, Tuscan-Leigh Dairies used to schedule its deliveries the old fashioned way, with an atlas, maps, and pins. That was until the company automated the tedious plotting process. With routing and scheduling software from tiny Lightstone Group, Inc., Tuscan expects to improve service to its 8,000 customers across six states. Many customers have narrow delivery time frames that must be met. Routing and scheduling software enables Tuscan-Leigh Dairies to meet its customer timing needs, an especially important task given that their products are perishable. The software eliminates overlapping routes and cuts mileage and driver time. Although the system costs more than $30,000, it will pay for itself within a year.

A typical system includes data from Navigation Technologies, which supplies navigable map databases. Users add their own data, such as routes, number of deliveries, and customer requirements. The software then determines which delivery person to send and which route is best. It also calculates the best delivery time and other details for every stop on the delivery path. It is also able to handle pickup and delivery time changes.

The use of RiMMS enhances routing efficiency, increases the number of stops made per day, and improves customer satisfaction. The system is designed for businesses with a need for high levels of customer service such as delivery, field service, pickup and sales force management. The system can account for multiple vehicle types, multiple depots, different starting and ending locations for the vehicles, time windows, multiple capacity constraints, driver skill levels, and many other variables.

In April 1998, Lightstone announced Easy Router, the first ever Internet-based routing service. Utilizing the RiMMS routing engine to generate optimal routes, the service enables subscribers to quickly receive optimized routes and schedules with localized maps and driving directions via the Internet. Subscribers send in specific delivery information via the Internet as frequently as needed, and Easy Router transmits optimized routes and schedules with localized maps and driving directions back to the user within an hour.

DISCUSSION QUESTIONS

1. If you were the logistics manager for a company, how would you decide which version of the Lightstone software to use—the basic RiMMS software or the Easy Rider service?
2. How might you be able to quantify the savings from the use of routing and scheduling software?

Sources: Adapted from the Lightstone Web site at http://www.lightstone.com, accessed May 6, 1998; Lightstone Press Release "Lightstone Group Introduces Easy Router, the First Internet-Based, Computerized Routing Service for Businesses," April 27, 1998; and Karen M. Carrillo, "Lightstone: On Schedule," *InformationWeek*, Nov. 24, 1997, pp. 90–93.

Inventory control. A manufacturing firm has several kinds of inventory, such as raw materials, packing materials, finished goods, and maintenance parts. We have already discussed the use of an inventory control system for finished product inventory. In addition, the firm needs to ensure that sufficient raw material, packing material, and maintenance parts are available. The same or a similar transaction processing system can be used to manage the inventory of these items.

Purchase order processing. An organization's purchasing department typically has a number of employees who are responsible for all purchasing

FIGURE 8.13

Purchasing Transaction
Processing System

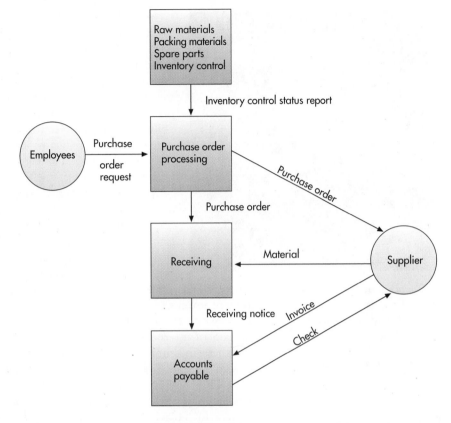

activities for the organization. Whenever materials or high-cost items are purchased, the purchasing department is involved. The **purchase order processing system** helps purchasing departments complete their transactions quickly and efficiently.

Every organization has its own policies, practices, and procedures for purchasing supplies and equipment. Most purchasing operations are flexible enough to handle requests by memo, phone, or formal purchasing documents. Many organizations allow inexpensive items to be purchased without a formal purchasing request or approval. For example, a need for low-cost office supplies may not require a formal purchase request. The cost of going through the formal process for a few inexpensive items can be greater than the cost of the items themselves. For larger purchases, once the request is received, formal authorization is sought, perhaps requiring signatures from managers at high levels in the organization. Once authorization is granted, the purchasing department submits bids for items and evaluates potential suppliers.

The purchasing department can facilitate the buying process by keeping data on suppliers' goods and services. The increased use of telecommunications has given many purchasing departments easier access to this information. For instance, technologies like the Internet and public networks allow purchasing managers to compare products and prices listed in Internet catalogs and large-scale consumer databases. Once the supplier is selected, the suppliers' computer systems might be directly linked to the systems of buyers. Orders can be sent via EDI, reducing purchasing costs and time spent and helping companies maintain low, yet adequate, inventory levels.

**purchase order
processing system**

a system that helps purchasing
departments complete their
transactions quickly and efficiently

Many companies use a sophisticated warehouse management system to speed order processing time and improve inventory accuracy. (Source: Roger Tully/Tony Stone Images.)

Instead of searching for the lowest prices from a list of suppliers, many companies form strategic partnerships with one or two major suppliers for important parts and materials. The partners are chosen based on prices and their ability to deliver quality products consistently on time. For example, the automotive companies request that their suppliers have plants or offices close to their operations in Michigan. The ability to conduct electronic commerce following EDI standards is also a key factor. Read the "FYI" box to learn about how intelligent agents help Internet shoppers find what they need.

Another approach to streamlining purchasing and order processing involves linking the supplier's and customer's transaction processing systems. For example, when the customer's computerized inventory control system determines that an inventory shortage exists, new products and materials can be automatically ordered from the supplier. By standardizing and integrating software and communications equipment, the customer's inventory system is first linked to its purchasing TPS, which in turn is linked to the supplier's order processing transaction system. This integration of internal and external systems allows the customer to operate with less inventory by reducing the delays from inventory use to order placement and inventory replenishment. Hence, manufacturers that have such TPSs and link them to their customers are able to provide improved customer service. For example, Wal-Mart uses a continuous replenishment system to replace a multistep, paper-based ordering system. The system links the retailer's distribution centers with suppliers' customer service centers for electronic transmission of orders based on actual store demand. Product shipment then occurs automatically from the appropriate supplier plant directly to a Wal-Mart distribution center or store.

Receiving. Like centralized purchasing, many organizations have a centralized receiving department responsible for taking possession of all incoming items, inspecting them, and routing them to the people or departments that ordered them. In addition, the receiving department notifies the purchasing department when items have been received. This may be done by a paper form called a receiving report or electronically through a business transaction created by entering data into the receiving transaction processing system.

An important function of many receiving departments is quality control by inspection. Inspection procedures and practices are set up to monitor the quality of incoming items. Any items that fail inspection are sent back to the supplier, or adjustments are made to compensate for faulty or defective products.

receiving system

a system that creates a record of expected and actual receipts

Many suppliers now send their customers advance shipment notices. This business transaction is input to the customer's **receiving system** to create a record of expected receipts. In addition, items are shipped with a bar code identification on the container. At the receiving dock, the worker scans the bar code on each container and a transaction is sent to the receiving system, where the bar code identification number is matched against the file of expected receipt records. This improves the accuracy of the

FYI

Intelligent Agents Help Shoppers

Intelligent agents are software programs that assist people and act on their behalf. An enormous amount of information is available from a wide variety of sources on the Internet. Intelligent agents can filter this flood of information, passing on to the user only those items in which the user is interested. In this way, intelligent agents reduce the information complexity of the networked world.

Over the past few years, intelligent agents have begun to transform the way many people shop for information and products on the Internet. These Internet shopping agents are constantly adapting to an individual's choices and styles and filtering and sorting through masses of information. Until recently, most agent technology was limited to helping consumers identify books, music, and movies they might be interested in. Shopping bots (short for robots) is another form of agent that scours Internet stores for the best price on a certain item.

In the past, shopping bots could be recognized by a Web site's software, and their access blocked, or else the information on their products could be displayed in a way difficult to compare with others. Netbot, an Internet company, has developed Jango, which uses the customer's own Web browser and behaves like a real customer. It quickly searches many Web shops and can deal with the complications of alternative product descriptions. Some industries, for example, travel or computer suppliers, have complex price structures, and it is still difficult to obtain clear direct comparisons between different companies. Other retailers are concerned that shopping bots and their users place too much emphasis on price and not enough on service.

IBM's Personal Shopping Assistant uses intelligent agent technology to help you find the items you want. Personal Shopping Assistant can customize both stores and merchandise. It learns your preferences, and whenever it enters a mall of stores or looks at specific merchandise, it can rearrange merchandise so that the items you like the most are the first ones you see. In addition, it can correlate buying patterns of whole groups of shoppers, finding and filtering where there are items that are often purchased together, for example. So, when you buy shoes, there will be socks for you to look at as well! Finally, Personal Shopping Assistant automates your shopping experience by reminding you of times you might want to go shopping, such as birthdays, anniversaries, or when items in which you are interested go on sale.

DISCUSSION QUESTIONS

1. What is the primary justification for using intelligent agents?
2. Make a list of items for which you would use the Netbot Jango shopping bot to help you find on the Internet. Make a separate list of items for which you would use the IBM Personal Shopping Assistant.

Sources: "Top Ten Information Agents," Agent Knowledge Base Associates, Inc., Web page at http://www.akainc.com accessed November 10, 1998; "Intelligent Shopping Agents," published June 14, 1997, accessed August 16, 1998; IBM Intelligent Agent Web site at http://www.networking.ibm.com, "Introduction to Intelligent Agents," accessed August 16, 1998.

receiving process, eliminates the need to perform manual data entry, and reduces the manual effort required.

Ford Motor reengineered its entire accounts payable/receiving operation by implementing invoiceless processing. When the purchasing department initiates an order, it enters the information into an on-line database. An electronic version of the order is sent to the supplier. When goods arrive at the receiving dock, the receiving clerk checks the database to see whether the items correspond to what was ordered. If they do, the order is accepted, a receiving transaction is entered, and the accounts payable system sets up everything to generate a check to the supplier by the payment due date. If the items do not match what was ordered, the order is returned. This process has completely eliminated the paper flow of purchase orders and invoices back and forth and the resulting clerical work to file, retrieve, and match various forms. It has also enforced a much higher level of order completeness among suppliers.

accounts payable system

a system that increases an organization's control over purchasing, improves cash flow, increases profitability, and provides more effective management of current liabilities

accounting systems

systems that include budget, accounts receivable, payroll, asset management, and general ledger

budget transaction processing system

a system that automates many of the tasks required to amass budget data, distribute it to users, and consolidate the prepared budgets

FIGURE 8.14

A Check Generated by an Accounts Payable Application

The check stub details items ordered, invoice dates, invoice numbers, cost of each item, discounts, and the total amount of the check.

Accounts payable. The **accounts payable system** attempts to increase an organization's control over purchasing, improve cash flow, increase profitability, and provide more effective management of current liabilities. Checks to suppliers for materials and services are the major outputs. Most accounts payable applications strive to manage cash flow and minimize manual data entry. Input from the purchase order processing system provides an electronic record to the accounts payable application that updates the accounts payable database to create a liability record. The liability record shows that the firm has made a commitment to purchase a specific good or service. Once the accounts payable department receives a bill from a supplier, the bill is verified and checked for accuracy. Upon receiving notice that the goods and/or services have been delivered in a satisfactory manner, the data is entered into the accounts payable application. A typical check from an accounts payable application is shown in Figure 8.14. In addition to containing standard information found on any check, most accounts payable checks include the items ordered, invoice date, invoice numbers, amount of each item, any discounts, and the total amount of the check. This allows the company to consolidate several invoices and bills into a single payment. In addition to checks, companies can also pay their suppliers electronically, using EDI or other electronic payment systems.

A common report produced by the accounts payable application is the purchases journal. As shown in Figure 8.15, this report summarizes all the organization's bill-paying activities for a particular period. Financial managers use this report to analyze bills that have been paid by the organization. This information is also used to help analyze current and future cash flow needs. Data is summarized for each supplier or parts manufacturer. Invoice number, description, amounts, discounts, and total checking activity are included. In addition, many purchasing reports include the total amount of checks generated by the accounts payable application on a daily, weekly, or monthly basis.

The accounts payable application ties into other information systems, including cash flow analysis, which helps an organization ensure that sufficient funds are available for the accounts payable application and shows the best sources of funds for payments that must be made via the accounts payable application.

Accounting Systems

The primary **accounting systems** include the budget, accounts receivable, payroll, asset management, and general ledger (Figure 8.16).

Budget. In an organization, a budget can be considered a financial plan that identifies items and dollar amounts that the organization estimates it will spend. In many organizations, budgeting can be an expensive and time-consuming process of manually distributing and consolidating information. The **budget transaction processing system** automates many of the tasks required to amass budget data, distribute it to users, and consolidate the prepared

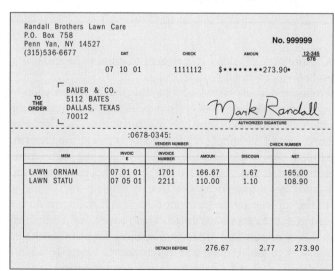

FIGURE 8.15

An Accounts Payable Purchases Journal

Generated by the accounts payable application, this report summarizes an organization's bill-paying activities for a particular period.

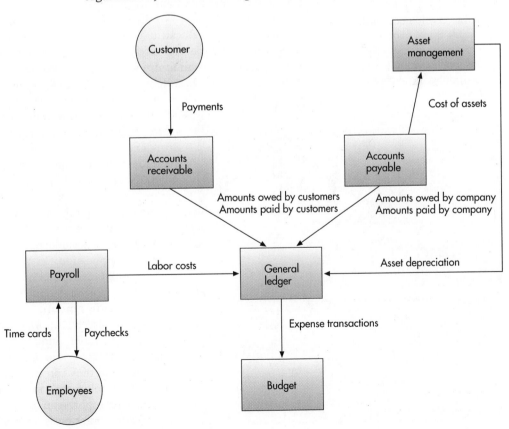

```
        DATE   07 10 01                  RANDALL BROTHERS LAWN CARE                                          PAGE

                                      ACCOUNTS  PAYABLE  PURCHASES  JOURNAL

  TRANS   PYMT  REV  TRANS   INVOICE   INVOICE  INVOICE     BNK REFERENCE           INVOICE    DISCOUNT VENDOR BALANCE
  CTRL NO  NO   CD   STATUS  NUMBER    DATE     DUE DATE  CD                        AMOUNT     ALLOWED  BEG.   */END **
                                                G/L NUMBER G/L DESCRIPTION   G/L AMT

  ***     1002  ALFRED PAINTS, INC.
           3          INVOICE  91105   07 03 01  08 03 01       PAINT                 1,000.00    20.00   1,000.00 *
           17         CR-ME    CM6725  07 10 01            DAMAGED MERCHANDISE 500.00   1.00
                      GL-DLS           1004200    INVENT.-TYPE 2         500.00
                                                                                                          500.00 **
                                                                                                             .00 *

  ***     1002  ALFRED PAINTS, INC.
           15         INVOICE  7216    07 10 01  08 10 01  1 LAWN MACH & FURN    1,700.00    54.00
                      GL-DIS    CM6725  1004100    INVENT.-TYPE 1       1,350.00
                      GL-DLS           1004200    INVENT.-TYPE 2         350.00
                                                                                                        1,700.00 **

  ***     1002  ALFRED PAINTS, INC.
           16         INV-MP   2455    07 09 01  07 20 01  1 FASTENERS             300.00     6.00
                      GL-DIS           5013000    BLDG, MAINT.           300.00
           16         MLT-PY   2445    07 09 01  07 29 01  1 FASTENERS             150.00     3.00
           16         MLT-PY   2445    07 09 01  07 29 01  1 FASTENERS             150.00     3.00
                                                                                                          300.00 **
                                                                                                             .00 *

  ***     1002  ALFRED PAINTS, INC.
           14         INVOICE  6211    07 10 01  08 10 01  1 LAWN MACH & FURN    1,050.00    54.00
                      GL-DIS           5018200    UTILITIES            1,050.00
                                                                                                        1,050.00 **
```

budgets. This allows financial analysts more time to manage the budget process to meet organizational goals by setting enterprisewide budgeting targets, ensuring a consistent budget model and assumptions across the organization, and monitoring the status of each department's spending.

FIGURE 8.16

Financial Systems

accounts receivable system

a system that manages the cash flow of the company by keeping track of the money owed the company on charges for goods sold and services performed

Accounts receivable. The **accounts receivable system** manages the cash flow of the company by keeping track of the money owed the company on charges for goods sold and services performed. When goods are shipped to a customer, the customer's accounts payable system receives a business transaction from the invoicing system, and the customer's account is updated in the accounts receivable system of the supplier. A statement reflecting the balance due is sent to active customers. Upon receipt of payment, the amount due from that customer is reduced by the amount of payment.

The major output of the accounts receivable application is monthly bills or statements sent to customers. As you can see in Figure 8.17, a bill sent to a customer can include the date items are purchased, descriptions, reference numbers, and amounts. In addition, bills can include amounts for various periods, totals, and allowances for discounts. The accounts receivable application should monitor sales activity, improve cash flow by reducing the time between a customer's receipt of items ordered and payment of bills for those items, and ensure that customers continue to contribute to profitability. Most systems can handle payment in a variety of ways, including standard bank checks, credit cards, money-wiring services, and electronic funds transfer via EDI.

The accounts receivable system is vital to managing the cash flow of the firm. This can be done through identification of overdue accounts. Reports are generated that "age" accounts to identify customers whose accounts are overdue by more than 30, 60, or 90 days. Special action may be initiated to collect funds or reduce the customer's credit limit with the firm, depending on the amount owed and degree of lateness.

An important function of the accounts receivable application is to identify bad credit risks. Because a sizable amount of an organization's assets can be tied up in accounts receivable, one objective of an accounts receivable application is to minimize losses due to bad debts through early identification of potential bad-debt customers. Thus, many companies routinely check a customer's payment history before accepting a new order. With advances in telecommunications, companies can search huge national databases for the names of firms and individuals who have been reported as delinquent on payments or as bad credit risks. When using external data like this in a TPS application, however, companies must be extremely cautious regarding the accuracy of the data.

The accounts receivable aging report, shown in Figure 8.18, is a valuable aspect of an accounts receivable application. In most cases, this report sorts all outstanding debts or bills by date. For unpaid bills that have remained outstanding for a predetermined amount of time, "reminder notices" can also be automatically generated. The accounts receivable aging report gives managers an immediate look into large bills that are long overdue so that they can be followed up to prevent more orders being sent to delinquent customers. This type of report can be produced on a customer-by-customer basis or in a summary format.

FIGURE 8.17

An Accounts Receivable Statement

Generated by the accounts receivable application, a bill is sent to a customer (usually monthly) and details items purchased, dates of purchase, and amounts due.

STATEMENT OF ACCOUNT

ABBEY WINDOW CLEANING CO.
163 BERKSHIRE AVE.
ANY CITY, ANYWHERE
12345

STATEMENT DATE	ACCOUNT NUMBER
05/27/01	1001001

Amount Paid $

PLEASE DETACH AT PERFORATION AND RETURN TOP PORTION WITH YOUR CHECK

STATEMENT DATE	ACCOUNT NO.	PREVIOUS BALANCE:
05-27-01	1001001	267.27

DATE	DESCRIPTION	REF.NO.	REFERENCE ONLY	DEBITS	CREDITS
05-01-01	INVOICE	1700		65.95	
05-15-01	INVOICE	3064		203.32	
05-01-01		3067		60.25	
05-15-01	INVOICE	3069		18.33	
05-01-01		3072		254.82	
05-15-01	INVOICE	3076		1,222.51	
05-15-01		3079		7.57	
			UNPAID BALANCE:	2,402.27	

SUM OF PERIODS ▪ CREDITS = UNPAID BALANCE

PERIOD 1	PERIOD 2	PERIOD 3	PERIOD 4	PERIOD 5
1,832.75	56.40	17.56	106.36	74.95

```
6588-DIAGE   TOOL DISTRIBUTORS INC.   0001 ACCOUNTS RECEIVABLE AGING ANALYSIS      SEP 25 2001      PAGE   1
                                            AS OF SEP 25 2001

TYPE OPEN-ITEM  ITM-DATE  REFERENCE INVOICE/PAYMENT CURRENT   31-60 DAYS  61-90 DAYS 91-120 DAYS OVER 120 DAYS
       NUMBER             NUMBER     AMOUNT


   CLASS 05    CUST 94367     NAME TOOLS OF AMERICA       3225 N PARKWAY     TEL 513-2864731   CONT B. BROWN

   INV  95361   03/17/01              23,058.37                                                   23,058.37
   DM   96853   03/20/01               5,589.42                                                    1,589.42
   INV  105395  07/04/01               2,923.45                2,923.45
   DM   116594  07/06/01                 198.32                 198.32
   INV  123984  07/15/01              23,087.28
   DSC  123984  08/17/01               1,204.36
   CA   968351  08/19/01              21,882.92
   INV  147296  07/23/01              19,709.57
   CA   83495   07/31/01              18,725.21
   CM   995473  08/19/01                 984.59
   INV  149384  08/29/01              23,831.37   21,831.59
   INV  158439  09/10/01              30,086.68   30,086.68
   INV  161236  09/23/01              25,520.37   25,520.37


   TOTL RECEIVABLE FOR CUST 94367    107,208.20   79,438.64    3,121.77       0.00      0.00     24,647.79
      CA                               3,121.77                3,121.77
      CA                              22,640.01   22,640.01
      CM                                 279.84      279.84


   UNAPPLIED CREDITS                  26,041.62   22,919.85    3,121.77       0.00      0.00          0.00


   NET RECEIVABLE FOR CUST 94367      81,166.58   56,518.79        0.00       0.00      0.00     24,647.79


   TOTAL FOR CLASS    05    42   PRINTED         TOTAL  RECEIVABLE  UNAPPLIED CREDITS  NET  RECEIVABLE

                                        CURRENT    161,506.12     67,832.57    93,673.55
                                        31-60 DAYS  31,494.14     18,984.68    12,509.46
                                        61-90 DAYS  11,569.32      4,267.91     7,301.59
                                        91-120 DAYS 27,764.18          0.60    27,746.18

                                        TOTAL      238,823.97     93,232.49   145,581.08
```

FIGURE 8.18

An Accounts Receivable Aging Report

The output from an accounts receivable application tells managers what bills are overdue, either customer by customer or in a summary format.

payroll journal

a report that contains employees' names, the area where employees worked during the week, hours worked, the pay rate, a premium factor for overtime pay, earnings, earnings type, various deductions, and net pay calculations

Payroll. The two primary outputs of the payroll system are the payroll check and stub, which are distributed to the employees, and the payroll register, which is a summary report of all payroll transactions. In addition, the payroll system prepares W-2 statements at the end of the year for tax purposes. In a manufacturing firm, hours worked and labor costs may be captured by job so this information can be passed on to the manufacturing costs system.

Responsibility for running this TPS application can be outsourced to a service company. Rather than write their own payroll application, many firms rely on a purchased software application for payroll processing. Some of these packages are tailored for one specific industry; others can accommodate a wide range of uses. In most cases, the number of hours worked by each employee is collected using a variety of data-entry devices, including time clocks, time cards, and industrial data collection devices. Once collected, payroll data is used to prepare weekly, biweekly, or monthly employee paychecks. Payroll systems can handle overtime, vacation pay, variable and multirate salary structures, incentive programs, and commissions. Most payroll applications automatically generate both federal and state tax forms related to payroll and process deductions, including tax-deferred annuities, savings plans, and U.S. government savings bonds. Often payroll applications have EDI arrangements with employees' banks to make direct deposits into employees' accounts.

Figure 8.19 shows the stub for the weekly paycheck produced by the payroll program. The stub includes the employee's total hours worked, regular pay, overtime pay, federal and state tax withholdings, and other deductions. In addition to paychecks, most payroll programs produce a **payroll journal**, shown in Figure 8.20. A typical payroll journal contains employees' names, the areas where employees worked during the week, hours worked, the pay rate, a premium factor for overtime pay, earnings, the earnings type, various deductions, and net pay calculations. Financial managers at the operational level use the payroll journal to monitor and control pay to individual employees. As you can see in Figure 8.20, the payroll journal also includes totals for hours worked, earnings, deductions, and net pay.

FIGURE 8.19

A typical paycheck stub details the employee's hours worked for the period, salary, vacation pay, federal and state taxes withheld, and other deductions.

```
TENDER CARE DAY CARE CENTER, INC.                                              4207
CINCINNATI, OH 45200
       Vacation Taken This Check       0.000     Sick Taken This Check          0.000
       Vacation Available              0.667     Sick Available                 0.667
                       Earnings                                  Deductions
       Description     Hours     Amount          Description                    Amount

       Regular Pay    36.320    199.76           FICA Withheld                  15.28
       Overtime        0.000      0.00           Fed. Tax W/H                   15.00
       SICK TIME       0.000      0.00           State Tax W/H                   1.67
       PERSONAL        0.000      0.00           Other W/H #1                    4.19
       VACATION        0.000      0.00           Other W/H #2                    0.00
       HOLIDAY PAY     0.000      0.00           Other W/H #3                    0.00
       Gross Pay#4     0.000      0.00           Other W/H #4                    0.00

                 Total          199.76                    Total                36.14
                 Net Pay      **163.62
       ---------------------------------YEAR TO DATE---------------------------------
                 Total Earnings 397.76           FICA                          30.43
                 Federal W/H     30.00           State W/H                      3.32
```

FIGURE 8.20

A Payroll Journal

Generated by the payroll
application, this report helps
managers monitor total payroll
costs for an organization and
the impact of those costs on
cash flow.

DATE 12 25 01   **RANDALL BROTHERS LAWN CARE**   PAGE   003

PAYROLL JOURNAL
PR020

EMPLOYEE NAME ERRORS WARNINGS	EMPLOYEE NUMBER	COST CENTER	HOURS	RATE	PREM FACTOR	EARNINGS	EARN TYPE	DEDUCTIONS	DED TYPE	NET PAY	PAYMENT LOCATION
DANIEL JACKSON											
	112	30	40.00	4.75		190.00	001			44.19	CIN
	112	30	10.00	4.75	1.5	71.25	030				
	112							5.00	220		
	112							15.81	801		
	112							44.19	802		
HOWARD SIMPSON											
	126	30	32.00	5.25		168.00	001			12.82	CIN
	126							5.00	220		
	126							2.00	223		
	126							10.16	801		
	126							12.82	802		

Payroll TPS applications also provide input into various weekly, quarterly, and yearly reports. Most of these reports are used by financial managers to help control payroll costs and cash flows. Payroll applications also provide necessary audit trails through the use of payroll master files and documents used by internal accountants and external auditors to make sure that the application is functioning as intended.

Like many other transaction processing applications, the payroll application interfaces with other applications. All payroll entries are entered to the general ledger system. Furthermore, there can be a direct link between payroll activities and production/inventory control operations. This is especially true for manufacturing operations or job-shop systems. Data collected on hours worked from the payroll application helps determine the total cost of completing various jobs. For example, if an employee who earns $15 per hour spends 20 hours completing a particular job, the labor costs for that job are $300. This type of information from the payroll application is useful in determining the cost to produce a product or render a service and thus in determining its profitability.

Asset management. Capital assets represent major investments for the organization whose value appears on the balance sheet under fixed assets. These assets have a useful life of several years, over which their value is depreciated, resulting in a tax reduction. The **asset management transaction processing system** controls investments in capital equipment and manages depreciation for maximum tax benefits. Key features of this application include efficient handling of a wide range of depreciation methods, country-specific tax reporting and depreciation structures for the various countries in which the firm does business, and work-flow-managed processes to easily add, transfer, and retire assets.

Increasingly, organizations are implementing their accounting applications as an integrated solution based on enterprisewide business processes. The accounting applications are designed to manage business processes and automatically route tasks and information among users, according to the work flow that is configured. For example, the accounts payable application recognizes an invoice for the purchase of a new asset. It then notifies

**asset management
transaction processing
system**

a system that controls investments
in capital equipment and manages
depreciation for maximum tax
benefits

a designated user of the asset management application of an incoming task. When the user opens the computerized file of "to-do" tasks, the work flow system opens the "add asset" activity—with information from the invoice already filled in—so that the asset management user can capitalize the new asset.

general ledger system

a system that produces a detailed list of business transactions designed to automate financial reporting and data entry

General ledger. Every monetary transaction that occurs within an organization must be properly recorded. Payment of a supplier's invoice, receipt of payment from a customer, and payment to an employee are examples of monetary transactions. A computerized **general ledger system** is designed to allow automated financial reporting and data entry. The general ledger application produces a detailed list of all business transactions and activities. Reports, including profit and loss (P&L) statements, balance sheets, and general ledger statements, can be generated (Figure 8.21). Furthermore, historical data can be kept and used to generate trend analyses and reports for various accounts and groups of accounts used in the general ledger package. Various income and expense accounts can be generated for the current period, year to date, and month to date as required. The reports generated by the general ledger application are used by accounting and financial managers to monitor the profitability of the organization and to control cash flows.

Financial reports that summarize sales by customer and inventory items can also be produced. These reports are used by marketing and financial managers to determine which customers are contributing to sales and inventory items that are selling as expected.

A key to the proper recording and reporting of financial transactions is the corporation's chart of accounts (Table 8.3). This chart provides codes for each type of expense or revenue. By entering transactions consistent with the chart of accounts, financial data can be reported in a simple and consistent fashion across all organizations of the enterprise, even if it is a multinational corporation.

Now that we have covered the major uses of transaction processing systems in organizations, we turn to new high-tech ways of accomplishing them. Organizations are applying the Internet to simplify and speed the results delivered by transaction processing systems.

TABLE 8.3

Sample Partial Chart of Accounts

Major Account Name	Type of Expense	Subaccount Code Used to Identify Transaction
Wages and Benefits	Management salaries and benefits	MSALSB
	Nonmanagement salaries and benefits	NMALSB
	Overtime	OVT
Travel and Training	Travel-related expenses	TRAVEL
	Tuition for training classes	TUITION
Professional Services	Fees paid to consultants, contractors, trainers, and other professionals	PROFSV
Maintenance Expense	Maintenance labor	MAINTL
	Maintenance parts	MAINTP
	Maintenance supplies	MAINTS

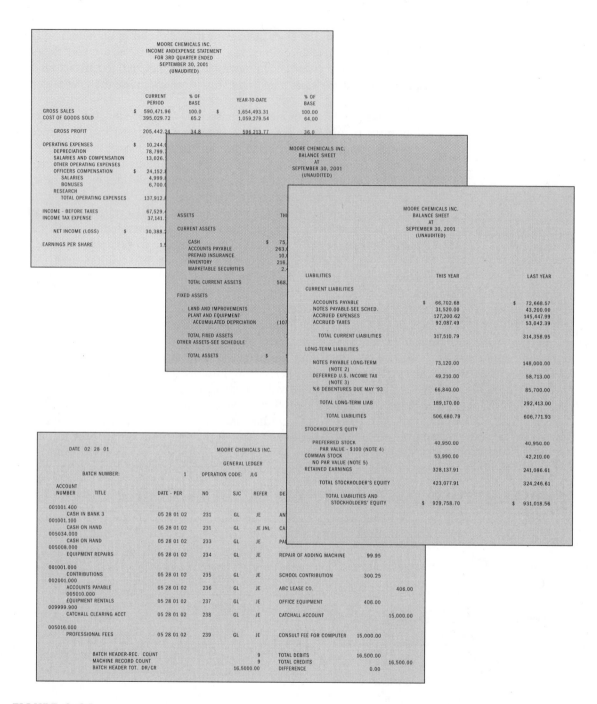

FIGURE 8.21

Outputs from a General Ledger System

An income statement (top) details sales, costs, and operating expenses to produce a statement of income for an organization. A balance sheet (center left and right) breaks down assets and liabilities so that managers can see at a glance whether income is covering expenses. A general ledger statement (bottom) tracks credits and debits.

ELECTRONIC COMMERCE

A successful E-commerce system must address the many stages consumers experience in the sales life cycle. At the heart of any E-commerce system is the ability of the user to search for and identify items for sale; select those items; negotiate prices, terms of payment, and delivery date; send an order to the vendor to purchase the items; pay for the product or service; obtain product delivery, and receive after-sale support. Figure 8.22 shows how E-commerce can support each of these stages. Product delivery may involve tangible goods delivered in a traditional form (e.g., clothing delivered via a package service) or goods and services delivered electronically (e.g., software downloaded over the Internet).

E-commerce for consumers is still in its early stages. Most shoppers are not yet convinced that it is worthwhile to connect to the Internet, search for shopping sites, wait for the images to download, try to figure out the ordering process, and then worry about whether their credit card numbers will be stolen by a hacker. But attitudes are changing.

As for business-to-business E-commerce, the issues are different but still serious. Many businesses do not yet have good models for setting up effective E-commerce sites. They also have trouble integrating the orders and information collected on-line with their old transaction processing systems. Companies continue to be concerned with the idea of sharing proprietary business information with customers and suppliers, an important component of many business-to-business E-commerce systems.

For most businesses, E-commerce is not about on-line catalogs, credit cards, or ordering sweaters. It is not really about selling at all, but about improving relationships among suppliers, distributors, and customers. For example, E-commerce could make it easier for a corporate customer to buy a new paper shredder from a stationery supply company. Typically, a corporate office worker must get approval for purchases that cost more than a certain amount. That request then goes to the purchasing department, which has to procure the goods from the sales representative of the approved vendor. Business-to-business E-commerce automates that entire process. Employees can go directly to their company's extranet, find the item at the stationery store's site, and get what they need at a price prenegotiated by their company. If approval is required, the employee's manager is notified automatically. Or, consider the case of Fruit of the Loom. The underwear manufacturer supplies plain white T-shirts to distributors who, in turn, sell the shirts to designers who add logos touting colleges and other organizations. Fruit of the Loom's E-commerce system automatically ships T-shirts to distributors at the negotiated price whenever stocks run low.

For business, E-commerce offers enormous opportunities. It allows consumers to buy at a low cost worldwide and it offers enterprises the chance to enter a global market right from start-up. Moreover, E-commerce offers great opportunities for developing countries. It can help them to enter the prosperous global marketplace, and hence serve to reduce the gap between rich and poor countries. However, the rapid development of E-commerce presents great challenges to society. Even though E-commerce will create new job opportunities, it could also cause a loss of employment in traditional job sectors. Many companies could fail to survive in the intense competitive environment of E-commerce and find themselves out of business before long. Therefore, it is vital that the opportunities and implications of E-commerce are communicated worldwide.[4]

For the consumer, it is easy to appreciate the importance of E-commerce. Why waste time fighting the very real crowds in supermarkets when from the comfort of home, one can shop on-line at any time in virtual Internet shopping malls and have the goods delivered home directly. Many goods and services are cheaper when purchased via the Web. It is cheaper to buy and sell stocks, to buy books, newspapers, and airline tickets. By providing more information about automobiles, cruises, and homes consumers have the data they need to cut better deals. More than a new way to place orders, the Internet is emerging as a paradise for comparison shoppers. Internet shoppers can unleash shopping bots or access sites such as www.jango.com to browse the Internet and obtain lists of items, prices, and merchants.[5]

Electronic Markets and Commerce in Perspective

Compared to the rest of the national economy, E-commerce is tiny. According to Forrester Research, consumers and businesses will funnel a total of $8 billion through E-commerce sites in 1998, and by the year 2000 more than $66 billion will be spent on-line. Everybody talks about retail sales moving on-line, but it is really the business-to-business market that is driving E-commerce (see Table 8.4). General Electric expects to do $1 billion worth of business-to-business E-commerce in 1998 by itself. By comparison, Forrester Research Inc. estimates that consumers will spend only about $3.3 billion via the Internet in 1998.[6]

As for the hottest areas of E-commerce, in terms of tangible goods sold via the Internet and other electronic means (such as interactive TV), Cowles/Simba says recently the biggest sellers are computer products ($196 million), consumer products ($186 million), books and magazines ($38 million), and music and entertainment products ($35 million). The agency reported total 1996 hard-goods sales at $993 million.

A five-stage model for purchasing over the Internet includes: search and identification, selection and negotiation, purchasing, product or service delivery, and after-sales support as shown in Figure 8.22.

Search and Identification

An employee ordering parts for a storeroom at a manufacturing plant would follow the steps shown in Figure 8.23. Such a storeroom stocks a wide range of office supplies, spare parts, and maintenance supplies. The

TABLE 8.4

Forecasted Volume of
E-Commerce

	1997	2000
Consumer E-commerce	$.5 billion	$7 billion
Business-to-business E-commerce	$8 billion	$66 billion

Source: "Survey: E-Commerce to Double by End of 1998," internet news.com at http://internetnews.com, accessed March 27, 1998 "Price Waterhouse Predicts Explosive E-Commerce Growth," internet news.com at http://internetnews.com, accessed March 26, 1998.

FIGURE 8.22

Five-Stage Model for
E-Commerce

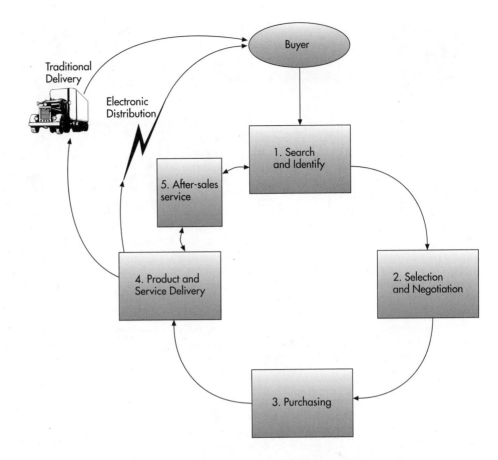

employee prepares a list of needed items, for example: fasteners, piping, and plastic tubing. Typically, for each item carried in the storeroom, a corporate buyer has already identified a preferred supplier based on the vendor's price competitiveness, level of service, quality of products, and speed of delivery. The employee then logs on to the Internet and goes to the Web site of the preferred supplier.

From the supplier's home page, the employee can access a product catalog and browse until finding the items that meet the storeroom's specifications. The employee fills out a request for a quotation form by entering the item codes and quantities of the items needed. When the employee completes the quotation form, the supplier's Web application prices out the

FIGURE 8.23

Buying Over the Internet

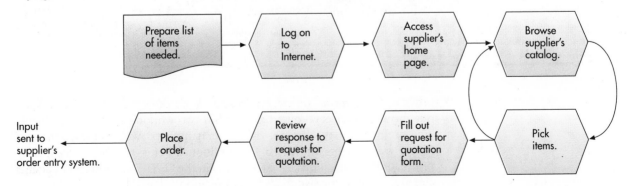

E-COMMERCE

E-Commerce System for the Foodservice Industry

Instill Corporation, with headquarters in Palo Alto, California, is a leading provider of electronic commerce and information services for the foodservice industry. Its mission is to develop easy-to-use services that cut costs and provide new information for all members of the food-service supply chain. Instill's services are used by over 1,000 foodservice operators, and leading customers include Marriott Distribution Services, Restaura, Inc., KinderCare, and Marriott Host.

In a prior career, Mack Tilling, one of the founders, was manager of a small chain of restaurants. He knew well the problems of ordering from multiple suppliers as well as the frustrations of constantly changing prices, illegible faxes, outdated catalogs, incomplete shipments, and rushed deadlines designed for the convenience of distributors, not operators. He incorporated his expertise and insights about how an electronic environment for the foodservice industry should work into E-store, a business-to-business on-line ser-vice for foodservice distributors and their customers.

With E-store, distributors post their catalogs on-line including customized pricing and incentives for individual cus-tomers. Customers, in turn, connect to their unique accounts to place orders and receive electronic billing for items shipped. In a bid to appeal to a wider network of distribu-tors and foodservice operators, Instill expanded the connec-tions to E-store to include an HTML interface and Web access. Instill customers can now use an industry-standard browser to log onto the Instill environment over the Internet.

From the customers' perspective, they are able to find the information they need for ordering food products from individual distributors catalogued and organized in one place. Using the system reduces the ordering time and effort. For one manager of a cafeteria the time it took to place twice-weekly food orders was reduced from two hours using a touch-tone phone system to ten minutes with E-store. Customer self-service is particularly important. Customers can enter their orders directly into the system, substantially reducing the keystroke errors of order clerks rekeying paper-based orders.

From the distributors' perspective, they can continue to

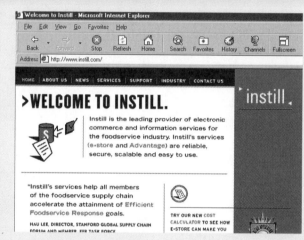

offer discounts, rebates, and other incentives to their best customers. Moreover, distributors can also begin to rein-force relationships with individual customers by improv-ing quality of service. Over time, they can begin to develop profiles of operators' buying preferences.

During 1997, its E-store system processed nearly 100,000 purchase orders with a total value of $180 mil-lion at a cost of about $2.50 each. Order volume is now growing so fast that Instill expects to transact more than $1 billion worth of orders per year.

DISCUSSION QUESTIONS

1. Summarize the key advantages of the E-store program for a large food distributor. What are some potential problems or issues associated with the program?

2. What are the advantages and disadvantages of the E-store program from the perspective of a foodservice operator such as Marriott or KinderCare?

Sources: Adapted from "Our Services," "About Us," and "Industry" pages found at the Instill Web site at http://www.instill.com, accessed May 7, 1998; Thomas Hoffman, "Food Service Duo Takes Orders Online," *Computerworld*, March 30, 1998, pp. 41–42; Geoffrey E. Bock, "Instill Corporation: Extending Electronic Commerce to the Foodservice Industry," Patricia Seybold Group Report, March 24, 1998.

order with the most current prices and shows the additional cost for vari-ous forms of delivery—overnight, within two working days, or next week. The employee may elect to visit other suppliers' Web home pages and repeat this process to search for additional items or obtain competing prices for the same items. As mentioned in the "FYI" box, intelligent agents can also be used for search and identification.

Read the "E-commerce" box to learn how a provider of electronic commerce and information services enables distributors to post their catalogs on-line, making it easier for shoppers to search for and identify sources to fill their needs.

Selection and Negotiation

The employee examines the price quotations from each supplier and checks off, on the request for quotation form, which of the items, if any, will be ordered from a given supplier. The employee also specifies the desired delivery date. This data is used to create input into the supplier's order processing TPS. In addition to price, quality, service, and speed of product delivery can be important in selection and negotiation.

Purchasing Products and Services Electronically

The employee completes the purchase order by sending a completed electronic form to the supplier. Complications may arise in paying for the products. Typically, in the case of a corporate buyer who makes several purchases from the supplier each year, the buyer has established credit with the firm in advance, and all purchases are billed to a corporate account. In the case of individuals who are making their first, and perhaps only, purchase from this supplier, additional safeguards and measures are required. Part of the purchase transaction can involve the customer providing a credit card number. However, alert computer criminals could capture this data and use the credit card information to make their own purchases. To avoid this potential problem, some companies have developed security programs and procedures. For example, Secure Electronic Transactions (SET) is endorsed by IBM, Microsoft, MasterCard, and others. Another approach to paying for goods and services purchased over the Internet is using electronic money, which can be exchanged for hard cash. CyberCash is an example. These and many other security procedures make buying products and services over the Internet easier and safer. Read the "Ethical and Societal Issues" box to learn about one approach to making electronic payments.

Smart cards are intelligent cards with an embedded computer. The card holds a great deal of information that identifies the bearer. Operating as an electronic purse, smart cards have many applications; they are commonly used in the telecommunications, banking, airline, transportation, and medical industries.
(Source: Image Copyright © 1998 PhotoDisc.)

ETHICAL AND SOCIETAL ISSUES

The Electronic Check

A group of inventors, under the auspices of BankBoston, Bellcore, and the Financial Services Technology Consortium (FSTC), received a patent for developing a highly secure, computer-based method of electronically transferring funds over an insecure public network. Their innovation has major implications for commerce on the Internet.

This patented technology is the foundation of an Internet-based payment mechanism called electronic check currently being tested in a pilot program by the U.S.

Department of the Treasury, BankBoston, the FSTC, and other FSTC members. During an 18-month market trial, the Treasury is offering electronic check payments to its suppliers who are also BankBoston customers. Small to medium-sized suppliers who currently receive paper checks for payment from the U.S. government will receive electronic checks for payments. These companies will be able to validate the authenticity of electronic checks, endorse them with a digital signature, and forward them through electronic mail to BankBoston for rapid deposit into their checking accounts.

The electronic check is an electronic version of a paper check. It is created on a computer, embedded in a secure file, digitally signed, and transmitted over a public network directly to the intended recipient. The electronic check is then endorsed by the payee and deposited electronically with his or her bank. The electronic check is then settled between the payee bank and payor bank using current settlement systems. This new electronic process effectively authenticates, verifies, and prohibits duplication of each individual transaction, and can be used in conjunction with existing payment systems.

The increased level of security and privacy are key to the potential wide-range use of the electronic check. The technology uses digital signatures and certificates, which are securely contained in a hardware token such as a smart card. Using the card, a recipient of an electronic check can verify that each check is authentic (that it was "signed" by the authorized person who sent it) and that it has not been tampered with. Digital signatures, unlike paper checks with handwritten or machine-generated signatures, are virtually impossible to forge.

The results of the market trial will help to identify what is required for electronic checks to mature into a ubiquitous, sustainable, and commercially viable payment mechanism. The results will also have broad implications for a more secure electronic funds transfer environment on the Internet.

Discussion Questions
1. How do you think the success of the market trial should be gauged? What distinguishes a successful test from an unsuccessful one?
2. Do you see any weak links in the electronic check process? If so, what are they?

Sources: Adapted from Laura Didio, "Beta Testers Endorse E-Checks," *Computerworld*, March 23, 1998, pp. 57–58 Bank of Boston Press Release, "BankBoston, FSTC and Bellcore Receive Patent for Electronic Commerce Security," BankBoston Web site at http://www.bankboston.com/today, accessed November 17, 1997.

Product and Service Delivery

The Internet can also be used to deliver products and services, primarily software and written material. Often called electronic distribution, getting software and written material from the Internet is faster and can be less expensive than with regular order processing. You can download software, reports on the stock market, information on individual companies, and a variety of other written reports and documents directly from the Internet. Electronic distribution can eliminate inventory problems for the manufacturer who does not have to stock hundreds or thousands of copies of the software, reports or documents when only one copy will do and can be downloaded to customers' computers. As more people start using the Internet, the electronic distribution of products and services could be a major revenue source for software and publishing companies. Most products cannot be delivered over the Internet. Such products can be delivered in a variety of ways: overnight carrier, regular mail service, truck, or rail. In some cases, the customer may elect to drive to the supplier and pick up the product.

After-Sales Service

In addition to capturing the information needed to complete the order, comprehensive customer information from the order is captured and stored in the supplier's customer database. This information can include customer name, address, telephone numbers, contact person, credit history, and more. If, for example, the employee later telephones the supplier to complain that not all items were received, or that some were received damaged, or just to ask information about how to use the product, all customer service representatives have this information available via a personal computer on their desks.

ENTERPRISE RESOURCE PLANNING

Flexibility and quick response are hallmarks of business competitiveness. Access to information at the earliest possible time can help businesses serve customers better, raise quality standards, and assess market conditions. Enterprise resource planning (ERP) is a key factor in instant access. While some think that ERP systems are only for the extremely large companies, this is not so. According to AMR Research Inc., 20,000 manufacturers, each with an annual worldwide revenue of $250 million or less, will license ERP from 1998 to 2002.[7] The leading vendors of ERP systems are listed in Table 8.5.

An Overview of Enterprise Resource Planning

The key to ERP is real-time monitoring of business functions. This permits real-time analysis of key issues such as quality, availability, customer satisfaction, performance, and profitability. Financial and planning systems receive "triggered" information from manufacturing and distribution. When something happens on the manufacturing line that affects a business

TABLE 8.5

Leading ERP Software Vendors

Software Vendor	Name of Software
Avalon Software	Avalon CIIM
qad.inc	MRG/PRO
Oracle	Oracle Manufacturing
SAP America	SAP R/3
Symix Computer Systems	Symix Advanced Manufacturing Inc.
Baan	Triton
PeopleSoft	PeopleSoft
J. D. Edwards	World

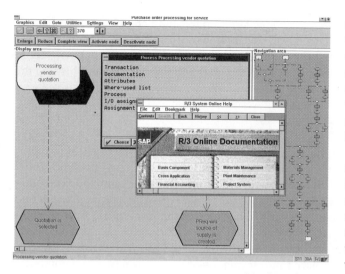

SAP R/3 ERP software provides comprehensive solutions for companies of all sizes and in all industry sectors.
(Source: Courtesy of SAP AG.)

situation—for example, packing material inventory drops to a certain level affecting the ability to deliver an order to a customer—it triggers a message for the appropriate person in purchasing. In addition to manufacturing and finance, ERP systems can also support human resources, sales, and distribution. This sort of integration is breaking through traditional corporate boundaries.

ERP systems accommodate the different ways each company runs its business by either providing vastly more functions than one business could ever need or including customization tools that allow shops to fine-tune what is hopefully an already close match. SAP America's R/3 is the undisputed king of the first approach. R/3, whose capabilities and number of data entities are roughly five times larger than those provided by its competitors, is easily the broadest and most feature-rich ERP system on the market. Thus, rather than compete on size, most competitors focus on customizability.[8] ERP systems have the ability to configure and reconfigure all aspects of the IS environment to support the way your company runs its business.

Advantages and Disadvantages of ERP

Increased global competition, new demands from executive management for control over total cost and product flow through their enterprises, and ever more numerous customer interactions are driving the demand for enterprisewide access to real-time information. ERP offers integrated software from a single vendor that helps meet those needs. The primary benefits of implementing ERP include eliminating inefficient systems, easing adoption of improved work processes, improving access to data for operational decision making, and technology standardization.[9] Surprisingly, most companies have found it difficult to justify implementation of an ERP system strictly based on cost savings.[10]

Eliminate costly, inflexible legacy systems. Adoption of an ERP system enables the organization to eliminate dozens or even hundreds of separate systems and replace them with a single integrated set of applications for the entire enterprise. In many cases, these systems are decades old, the original developers are long gone, and the systems are poorly documented. As a result, the systems are difficult to fix when they break and adapting them to meet new business needs takes too long. Such systems become an anchor around the organization that keeps it from moving ahead and remaining competitive. An ERP system helps match the capabilities of an organization's information systems to its business needs—even as these needs evolve.

Provide improved work processes. Competition requires companies to structure their business processes to be as effective and customer-oriented as possible. ERP vendors do considerable research to define the best business processes. They gather requirements of leading companies within the same industry and combine them with research findings from major research institutions and consultants. The individual application modules included in the ERP system are then designed to support these **best practices**, the most efficient and effective ways to complete a business process. Thus, implementation of an ERP system ensures good work processes based on best practices. For example, for managing customer payments, the ERP's finance module can be configured to reflect the most efficient practices of leading companies in an industry. This ensures that everyday business operations follow the optimal chain of activities, with all users supplied the information and tools they need to complete each step.

best practices

the most efficient and effective ways to complete a business process

Provide access to data for operational decision making. ERP systems operate via an integrated database and use essentially one set of data to support all business functions. Thus, decisions on optimal sourcing or cost accounting, for instance, can be run across the enterprise, rather than looking at separate operating units and then trying to coordinate that information manually or reconciling data with another application. The result is an organization that looks seamless, not only to the outside world, but also to the decision makers who are deploying resources within the organization.

The data is integrated to provide excellent support for operational decision making and allows companies to provide greater customer service and support, strengthen customer and supplier relationships, and generate new business opportunities. For example, once a salesperson makes a new sale, the business data captured during the sale is distributed to related transactions for the financial, sales and distribution, and manufacturing business functions in other departments.

Upgrade technology infrastructure. An ERP project provides an organization with the opportunity to upgrade and simplify the information technology it employs. In implementing an ERP, a company must determine which hardware, operating systems, and databases it wants to use. This enables the organization to eliminate the hodgepodge of multiple hardware platforms, operating systems, and databases it is using from a variety of vendors. Standardization on fewer technologies and vendors reduces ongoing maintenance and support costs as well as the training load for those who must support the infrastructure.

Disadvantages of ERP. Getting the full benefits of ERP is not simple or automatic. Although ERP offers many strategic advantages by streamlining a company's transaction processing system, ERP is time consuming, difficult, and expensive to implement. Some companies have spent years and tens of millions of dollars implementing ERP. Kodak, for example, will likely spend about $500 million installing ERP over two years. Chevron estimated that ERP software would cost $160 million to install and get running. ERP is so complex that many companies hire consultants at over $2,000 per day to get the system installed and running correctly.

In some cases, a company has to make radical changes in how it operates to conform with the work processes (best practices) supported by the ERP. These changes can be so drastic to long-time employees that they retire or quit rather than go through the change. This can leave the firm short of experienced workers.

The high cost to switch to another vendor's ERP system makes it highly unlikely that a firm will do so. Thus, the initial ERP vendor knows it has a "captive audience" and there is less of an incentive to listen to and respond to customer issues. The high cost to switch also creates a high level of risk associated with the ERP vendor allowing its product to become outdated or the vendor going out of business. The choice in picking an ERP system is not just to choose the best software product, but to choose the right long-term business partner.[11]

Even with the high cost, long installation times, and complexity, there is no indication that the enthusiasm for this powerful software is slowing.[12]

Example of an ERP System

SAP R/3 has been called one of the most complex packages ever written for use in corporations.[13] However, it is also the most widely used ERP solution in the world. The following sections provide a brief description of the fundamental design and architecture of the SAP R/3 ERP system.

The SAP ERP system was developed from the perspective of a corporation as a whole rather than any business department. All data is kept only once in the system and all SAP programs use the same database with little data redundancy. Each data item is clearly documented in a data dictionary. The software can be configured to meet the customer's business requirements. It is based on a three-level client/server architecture consisting of clients, application servers, and database servers (Figure 8.24). R/3 will run on a wide variety of hardware from a small Windows NT server up to massively parallel systems.

Clients in the SAP system. The R/3 system typically supports hundreds or even thousands of clients. Clients are usually Pentium desktop computers with at least 32 megabytes of RAM. Users of the clients request services from the application servers.

Application servers in the SAP system. There are many application servers in the typical R/3 system. The servers are powerful midrange or even mainframe computers. The job of the server is to reply to all requests made of it, including requests for data; to communicate messages; and to update master files. The request from the client travels along the network to the application server. A dispatcher program running on the application server manages the queues of user requests to determine which should be executed next. Application servers are grouped by the SAP R/3 system administrator into classes depending on the applications they run. One class might, for example, run the financial modules while another class might run the sales and distribution modules. The application server can contain either third-party or user-developed software, as long as it is written in ABAP/4, SAP's fourth-generation programming language.

Business application programming interfaces (BAPIs). Business application programming interfaces, or BAPIs, are public interfaces. These interfaces were developed with SAP customers, software development organizations, and standards organizations to enable SAP customers to develop their own applications to interface with SAP. SAP then has the

FIGURE 8.24

SAP Three-Tier Client/Server
Architecture

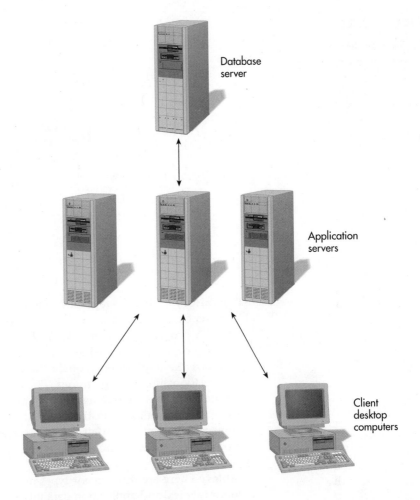

Database
server

Application
servers

Client
desktop
computers

flexibility to change the underlying software, as long as the interface itself
is not changed (Figure 8.25). Thus, new SAP software versions can be
introduced without invalidating existing running systems. An example of a
BAPI is a method of the SAP business object "customer order" to allow
checking the status of the customer order.

Database server in the SAP System. The database server in the R/3 system holds the data and is accessed and updated constantly. Depending on
the hardware selected, the database may be distributed among multiple
machines or reside on a single computer. The SAP update process is
designed to accommodate hundreds or even thousands of users on a single
database server and still provide satisfactory response times.

Objects in the SAP System. SAP has adopted objects as one of its key
implementation concepts. An SAP **object** is a collection of data and the
programs to process this data. "Purchase order" and "customer" are examples of SAP business objects used in business processes. Attributes contain
the data of an object, such as name, date of employment, and address of an
employee. An example of a method for the business object "applicant" is
"invite to an interview."

object

a collection of data and the
programs to process this data

FIGURE 8.25

Business Application
Programming Interface (BAPI)

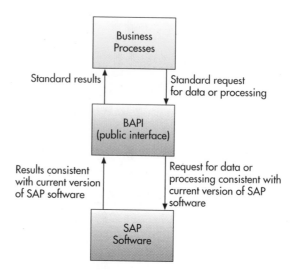

Repository. All development objects of the ABAP/4 Development Workbench are stored in the ABAP/4 repository. These objects include ABAP/4 programs, screens, documentation, etc. The SAP R/3 database makes constant use of the repository. The repository sits between the database and the application modules, and provides the logical mappings of data relationships and conditions. The repository serves primarily as a tool to enter, manage, and evaluate information about a company's data. It is an active data dictionary, so that new or changed data in the repository is immediately available to all system components. As a result, application programs and screens are always supplied with up-to-date information.

Tables. There are three major types of tables: system configuration tables, control tables, and application data tables. All are defined in the repository. System configuration tables are maintained primarily by SAP and define the structure of the system. Customers do not change these tables. To customize the system, the project team will use both control tables and application data tables. Control tables define functions that guide the user in his or her activities. For example, a control table might be set up to require that a customer service representative enter a line item to reference data about the product from the material master table before a purchase order is accepted. Application data tables are divided into two types—transaction and master data tables. Transaction tables are the largest, since they contain the daily operations data, such as orders, payments received, invoices, and shipments. Master data tables describe sets of basic business entities such as customers, vendors, products, materials and so on.

This section has focused on the key elements of the SAP R/3 ERP system; however, the systems from PeopleSoft, Oracle, and other vendors are based on similar basic design elements—the use of client/server architecture, standard business application interfaces, object-oriented programming, repository, and data tables. The "Technology Advantage" box summarizes the growth in use of ERP systems.

TECHNOLOGY ADVANTAGE
Core Business Activities for Enterprise Resource Planning

When powerful IBM and software giant Microsoft both use another company's software to run their basic transaction processing system, the software must be good. In both cases, the companies are using a new type of software called Enterprise Resource Planning (ERP) software.

ERP software runs all the major transaction processing activities for a business. The integrated nature of ERP allows a company to coordinate all core business activities using one powerful software package. From an order taken through the final delivery of the product or service, ERP software supports all critical activities among them: order processing, inventory control, production scheduling, supplier coordination, material flow during manufacturing, product delivery, and managerial decision making. Orders are recorded and processed using ERP. Inventory control is performed using ERP. Scheduling production is another function of ERP software. Coordinating suppliers can also be achieved. ERP excels at following the flow of items as they are manufactured and sent to customers. As a result, companies can give customers firm commitments for product delivery and service, while reducing costs. ERP also provides managers with a wealth of information about how transactions are being processed and basic business operations.

The growth in ERP systems has been phenomenal. In addition to IBM and Microsoft, many other large companies are installing ERP systems. Monsanto spent more than two years implementing an ERP system. Kodak is relying on ERP to help it compete in the tough photo and camera market. Procter & Gamble is also installing ERP worldwide. Chevron estimates that ERP could save it 25 percent annually.

SAP was one of the first companies to produce and market ERP software. Founded in 1972, by five former employees of IBM, SAP America is the market leader with its R/3 software. When R/3 was first released in 1993, 1,000 R/3 packages were sold. In a few short years, sales jumped to about 9,000 packages. Sales growth has averaged over 35 percent annually. SAP, however, is not alone in the ERP market. Other companies are starting to market and sell their own versions of this popular software. Oracle, for example, has started selling ERP software to work with its large and popular database applications. PeopleSoft is another ERP software vendor. Starting with a human resource planning package, PeopleSoft now sells ERP packages to large companies. Baan is yet another ERP provider.

DISCUSSION QUESTIONS

1. How does using an ERP system differ from using a conglomeration of transaction processing systems from a variety of vendors?

2. Pick a company and discuss how an ERP could process its transactions. What are the benefits and disadvantages of ERP?

Sources: Adapted from Michael Martin, "An Electronics Firm Will Save Big Money Using Enterprise Resource Planning," *Fortune*, February 2, 1998, p. 149; Randy Weston, "ERP Vendors Going with the Flow," *Computerworld*, January 26, 1998, p. 53; Michael Mecham, "SAP Targets Aerospace for R/3 Applications," *Aviation Week and Space Technology*, January 10, 1998, p. 61.

Information Systems Principles

An organization's TPSs must support the routine, day-to-day activities that occur in the normal course of business and help a company add value to its products and services.

Data should be captured at its source, and it should be recorded accurately, in a timely fashion, with minimal manual effort, and in a form that can be directly entered into the computer.

Business data goes through a transaction process that includes data collection, data editing, data correction, data manipulation, data storage, and document production.

Linking the transaction processing systems in one firm to those of another is an effective strategy to reduce costs and speed the flow of information.

As companies move into supporting E-commerce, they need to build systems that are capable of supporting all phases of the sales life cycle: search and identification, selection and negotiation, purchasing, delivery and after-sales support.

Implementation of an enterprise resource planning (ERP) system enables a company to achieve numerous business benefits by creating a highly integrated set of systems.

▪ SUMMARY

1. *Identify the basic activities and business objectives common to all transaction processing systems.*

Transaction processing systems (TPSs) are at the heart of most information systems in businesses today. TPSs consist of all the components of a CBIS, including hardware, software, databases, telecommunications, people, and procedures used to process transactions. All TPSs perform the following basic activities: data collection involves the capture of source data needed to complete a set of transactions; data edit checks for data validity and completeness; data correction involves providing feedback of a potential problem and enabling users to change the data; data manipulation is the performance of calculations, sorting, categorizing, summarizing, and storing for further processing; data storage involves placing transaction data into one or more databases; and document production involves outputting records and reports.

The methods of transaction processing systems include batch, on-line, and on-line with delayed processing. Batch processing involves the collection of transactions into batches, which are entered into the system at regular intervals as a group. On-line transaction processing (OLTP) allows transactions to be entered as they occur. Systems that use a compromise between batch and on-line processing are on-line with delayed-entry TPS. Transactions may be entered as they occur, but processing is not performed immediately.

Organizations expect TPSs to accomplish a number of specific objectives, including processing data generated by and about transactions; maintaining a high degree of accuracy; ensuring data and information integrity and accuracy; compiling timely reports and documents; increasing labor efficiency; helping provide increased and enhanced service; and building and maintaining customer loyalty.

2. *Describe the inputs, processing, and outputs for the transaction processing systems associated with order processing, purchasing, and accounting business processes.*

TPS applications are seen throughout an organization. The order processing systems include order entry, sales configuration, shipment planning, shipment execution, inventory control, invoicing, customer interaction, and routing and scheduling. Order entry captures the basic data needed to process a customer order. Once an order is entered and accepted, it becomes an open order. Sales configuration ensures that the products and services offered are sufficient. Shipment planning determines which open orders will be filled and from which location they will be shipped. The system prepares an order confirmation notice that is sent to the customer and a picking list used by warehouse operators to fill the order. The shipment execution system is used by the warehouse operators to enter data as to what was actually shipped to the customer. It passes shipped order transactions downstream to the invoicing system. For each item picked during the shipment execution process, a transaction providing stock number and quantity is passed to the finished product inventory control system. The invoicing system generates customer invoices based on the records received from the shipment execution system. The customer interaction system monitors and tracks each customer interaction. Routing and scheduling systems are used in distribution functions to determine the best use of a company's resources.

The purchasing information systems include inventory control, purchase order processing, accounts payable, and receiving. The inventory control system tracks the level of all packing materials and raw materials. It provides users with information about when to order additional materials. The purchase order processing system supports the policies, practices, and procedures of the purchasing department. The accounts payable system monitors and controls the outflow of funds to an organization's suppliers. The receiving system captures data about specific receipts of materials from suppliers so that approval for payment can be granted or refused.

The accounting systems include the budget, accounts receivable, payroll, asset management, and general ledger. The budget system automates many of the tasks required to amass budget data, distribute it to users, and consolidate the prepared budgets. The accounts receivable system manages the cash flow of the company by keeping track of the money owed the company. The payroll processing application processes employee paychecks and performs numerous calculations relating to time worked, deductions, commissions, and taxes. The outputs are used to help control payroll costs, cash flows, and develop reports to the federal government. The asset management system controls investments in capital equipment and manages depreciation for maximum tax benefits. The general ledger system records every monetary transaction and enables production of automated financial reporting.

3. *Define the term* E-commerce *and discuss how an E-commerce system must support the many stages consumers experience in the sales life cycle.*

E-commerce is a general concept covering any business transaction executed electronically between parties such as companies (business-to-business), companies and consumers (business-to-consumer), consumers and consumers, business and the public sector, and between consumers and the public sector. A successful E-commerce system must address the many stages consumers experience in the sales life cycle. At the heart of any E-commerce system is the ability of the user to search for and identify items for sale; select those items and negotiate prices, terms of payment, and delivery date; send an order to the vendor to purchase the items; pay for the product or service; and obtain product delivery.

4. *Define the term* enterprise resource planning system *and discuss the advantages and disadvantages associated with the implementation of such a system.*

Enterprise resource planning software (ERP) is a set of integrated programs that manage a company's vital business operations for an entire multisite, global organization. It must be able to support multiple legal entities, multiple languages, and multiple currencies. While the scope of an ERP system may vary from vendor to vendor, most ERP systems provide integrated software to support manufacturing and finance. In addition to these core business processes, some ERP systems are capable of supporting additional business functions such as human resources, sales, and distribution.

Implementation of an ERP system can provide many advantages including eliminating costly, inflexible legacy systems; providing improved work processes; providing improved access to data for operational decision making; and creating the opportunity to upgrade technology infrastructure. Some of the disadvantages associated with an ERP system are that it is time consuming, difficult, and expensive to implement.

■ KEY TERMS

accounting systems 360
accounts payable system 360
accounts receivable system 362
asset management transaction
  processing system 364
audit trail 347
batch processing system 338
best practices 375
budget transaction processing
  system 360
business resumption planning 345
customer interaction system 354
data collection 343
data correction 344

data editing 344
data manipulation 344
data storage 344
disaster recovery 345
document production 345
general ledger system 365
inventory control system 352
object 377
on-line transaction processing
  (OLTP) 338
order entry systems 349
order processing system 347
payroll journal 363
purchase order processing system 357

purchasing transaction processing
  systems 355
receiving system 358
routing system 355
sales configuration system 350
scheduling system 355
shipment execution system 352
shipment planning system 350
source data automation 343
transactions 337
transaction processing cycle 343
transaction processing systems
  (TPS) 337
transaction processing system audit 347

■ REVIEW QUESTIONS

1. Describe the basic activities common to all transaction processing systems (TPSs).
2. List several characteristics of transaction processing systems.
3. Define the term E-commerce. What component of E-commerce is growing the fastest?
4. What is an enterprise resource planning system?
5. List and briefly discuss the business objectives common to all TPSs.
6. What is the difference between batch processing, on-line processing, and on-line entry with delayed processing systems?
7. Identify some of the technology standards associated with E-commerce.
8. Describe an order entry system for taking orders over the Internet.

9. What systems are included in the order processing family of systems?
10. Identify the various stages consumers experience in the life cycle of a sale.
11. What systems are included in the purchasing family of systems?
12. Why is the general ledger application key to the generation of accounting information and reports?
13. What is the difference between a business resumption plan and a disaster recovery plan?
14. Give an example of how transaction processing systems can be used to gain competitive advantage.
15. What is the purpose of a transaction processing system audit?

■ DISCUSSION QUESTIONS

1. What advantages does the Internet-based approach to order processing offer the supplier? What advantages does this approach to order placement offer the customer? Does it appear that much of the burden for order processing has been moved from the supplier to the customer?
2. Your company is a medium-sized firm with sales of $500 million per year. It has been decided that

the organization will implement an ERP system to support all operations at headquarters, three plants, and four distribution centers. What are some of the key questions that must be answered to further define the scope of this effort?
3. Imagine that you are the new IS manager for a Fortune 1000 company. Your internal information systems audit has revealed that the firm's

systems are lacking adequate emergency alternate procedures, disaster recovery plans, and backup procedures. How would you justify to your manager spending a full year of people's time to implement these procedures?

4. Is each of the various stages experienced by a consumer in the life cycle of a sale equally important for all kinds of products? Why or why not?

5. What is the advantage of implementing ERP as an integrated solution to link multiple business processes? What are some of the issues and potential problems?

6. You are the key user of the firm's purchase order processing system and have been asked to perform an information system audit of this system. Outline the steps you would take to complete the audit.

7. You are building your firm's first-ever customer interaction system. Discuss the features you would design into the system. How might you include suggestions from your customers into your design? Should the system be built using Internet technology?

8. You are in charge of a complete overhaul of your firm's order processing systems. How would you define the requirements for this collection of systems? What features would you want to include?

9. How is the use of EDI to order and pay for items different than the use of an Internet-based system?

■ PROBLEM-SOLVING EXERCISES

1. Identify the differences between the use of an Internet Web site, intranet, and extranet to support the consumer in making a purchasing decision. Prepare a PowerPoint presentation that summarizes the key points.

2. The rental (order processing) application in your video store has three databases: video, title, and customer. The video database contains information about every tape available for rental. The title database contains information about each specific tape title; any one tape title (for example, *Casablanca*) may have multiple videos associated with it because more than one copy may be available for rental. The customer database contains each customer's ID number and address. Specific fields in each database are listed in the accompanying table. Key fields are indicated with an asterisk.

Title	Video	Customer
TapeNumber*	VideoNumber*	CustomerNumber*
Title	TapeNumber	CustomerName
Category	Status	Address
Rating	PhoneNumber	Rental Amount
Rental Rate		Y-T-D
Rental Amount Y-T-D		

Build a simple transaction processing system to support the video store operation. Enter the complete database definitions into your database management software and create a data-entry screen to allow store personnel to efficiently enter customer rentals and returns. This screen must include basic information such as tape name and number, customer name and number, date out, date returned, and charge for rental per day. The data from the screen updates all appropriate data in each database.

a. Enter several of your favorite movies to create at least ten entries for the title database.

b. For each title entry, create one to five entries in the video database.

c. Make up at least ten entries for the customer database.

d. Enter the data necessary to handle the checkout of at least six specific videos by different customers. Can your simple transaction processing system handle a situation where one customer checks out more than one video at a time? What happens if a customer wants to check out a video but there are no copies remaining?

e. Check to see whether the Y-T-D fields in the title and customer databases are updated correctly.

■ TEAM ACTIVITY

With two or three of your classmates, identify a local company that has established an Internet presence. Visit the site and become familiar with how it operates and learn about the company and its products and services. Conduct an interview with one or more managers from the organization and learn what it took to develop the site, what management's expectations were for the site, and whether the site is meeting these expectations. Are there plans for additional sites or improvements to the current one?

■ WEB EXERCISE

A number of companies, including SAP, sell enterprise resource planning (ERP) software to coordinate a company's transaction processing system. Search the Internet to get more information about ERP or one of the companies that makes and sells this powerful software. You may be asked to develop a report or send an e-mail message to your instructor about what you found.

■ CASES

1   Orders for West Increase

One of the most important transaction processing functions of any business is order processing. Without orders, no business can survive. Traditional order processing systems include phone and manual orders. The orders are processed with customer and inventory files to get the right product or service to the customer. Traditionally, the order processing function was viewed as a cost center. With electronic ordering and the Internet, however, this is changing.

To streamline order processing, the West Group decided to offer its products and services through the Internet. West publishes legal information and material. Some believe that West is the best source of legal information available. Used primarily by lawyers and legal professionals, West provides a vast array of services and products. The overall purpose of the Web site was to make order processing more efficient and more convenient for its customers. The results of its Web-based order processing system was not what West expected.

With a large surge in sales through its Web site, West Group was pleasantly surprised. The big surprise was the customer base. About half the people ordering products through the Internet site consisted of the usual customers. Half of the orders, however, represented new customers not in the legal profession. The new Web site made ordering more efficient and convenient. The big advantage of using the new order processing system is the huge increase in orders from the new customers. The new order processing system on the Web also allows West to sell its less expensive books and still make a profit. As a result of its new success, West is now exploring other ways to develop and deliver products and services to customers outside the legal profession. The new order processing system has created a new market and new opportunities. It has also caused West to rethink its basic business strategy.

What is next for the West Group? With its success in electronic marketing, West is now starting to distribute its products electronically, including books on CD ROM. The biggest problem facing West is not how to use new technology but how to get enough good IS people to implement its move into the electronic age.

1. Why did West decide to develop an order processing system on the Internet? What was the result?

2. What are the advantages and disadvantages of placing an order processing system on the Internet?

Sources: Sharon Machlis, "Legal Publisher Enters Online Court," *Computerworld*, February 16, 1998, p 47; Hope Viner Samborn, "The Costs of Growth," *ABA Journal*, January 1998; Noel Holston, "Good Legal Advice on the Net," *Minneapolis Star Tribune*, February 18, 1998, p. 1E.

2 Web-Based Purchasing

Most companies spend considerable time and money managing the paper and processes involved in buying office supplies, spare parts, and small equipment—all the things that keep the company running from day to day. Depending on the type and volume of items being bought, moving an order through the typical buying process (complete a requisition, obtain management approval, submit to purchasing, and buyer turns the requisition into a purchase order) can easily exceed the cost of the items being purchased. As a result, many companies are developing intranet-based approaches to streamline the purchasing process for many types of purchases.

One approach is to provide employees with direct access to an on-line catalog of products from vendors who have been preapproved by the company's buyers. Employees can go to the Web site, select an item, order it, and even track its delivery status. Potential benefits include cutting expenses by reducing the cost to place an order by 50 to 80 percent. Many companies believe that they can also cut costs by limiting what employees can buy and by driving tougher deals with vendors, who may agree to lower prices to win a spot on the Web-based corporate purchasing catalog. Employees will be able to order supplies 24 hours per day, seven days a week as needed.

Despite such potential benefits, most companies embracing on-line purchasing via Web-based catalogs are doing so cautiously and incrementally. Many still have concerns about security when it comes to executing transactions on the Web. Others are struggling with how to integrate new on-line purchasing applications with existing legacy and ERP applications such as SAP. And many are cautious about relying too heavily on relatively young vendors of on-line purchasing applications.

1. Imagine that you are the manager of purchasing for a major Fortune 1000 company. You are considering conducting a pilot project of Web-based purchasing for your employees. How would you design the pilot test to ensure that you are able to make a good assessment of the impact of this change on your firm?
2. What are the key success criteria upon which such a pilot test should be evaluated?

Sources: Adapted from Esther Shein, "Special Net Delivery," *PC Week*, March 16, 1998, pp. 70–71 and John Madden and Esther Shein, "Web Purchasing Attracts Some Pioneers," *PC Week*, March 16, 1998, p. 71.

3 FedEx and SAP Team Up to Provide Integrated Logistics Solution

Federal Express and SAP are working together to provide software that integrates FedEx's shipment tracking capability and logistics with SAP's R/3 ERP software. A module of the software runs on R/3 and automatically generates FedEx shipping documents and tracking numbers for products ordered using SAP's R/3 software. When the products are delivered by FedEx, transactions are sent from FedEx to update data in R/3. The goal is to make R/3 work like a standard business process and automatically trigger other processes, such as proof of delivery and payment. The applications are designed to give the 13,000 enterprises deployed on SAP R/3 real-time package life-cycle information.

The SAP/FedEx applications, dubbed The Product, dovetail with SAP's development of a suite of supply-chain products. "We think this is a turning point in the industry," said Paul Wahl, chief executive officer of SAP AG. Wahl says that applications for spotting bottlenecks, reacting to customer orders, and linking transportation with production schedules are among the five top priorities for SAP. Others include electronic commerce buying tools and systems for meshing Internet transactions with financial systems.

"Most of our customers are in the process of integrating an enterprise system," says Laurie Tucker, senior vice president of logistics, electronic commerce, and catalog, a $500 million FedEx unit. "They are looking for ways to improve return on investment and are interested in any business process that can be integrated, as opposed to being separately invoked. Our most innovative customers are pushing us and their other suppliers for that integration."

The functionality provided by the FedEx/SAP alliance is no small matter. The two companies are talking about systems that enable businesses to track components used in manufacturing processes from raw material to market. However, it is also in line with the rest of the industry. United Parcel Service of America says it is in pilot tests with a half-dozen corporate customers for similar functionality, and IBM Software recently rolled out a suite of Java applications for freight carrier J.B. Hunt. Fundamentally, SAP plans to extend its architecture to support multiple-carrier systems, Wahl said. He added that supply-chain planning is the largest single development effort at the billion-dollar software powerhouse, commanding the time of more than 250 engineers.

This type of software is an important addition to enterprise resource planning software. There is a real need to automate the logistics of delivering goods. Despite the advances in Internet connections, those companies dealing in tangible products still must rely on the same old business processes to physically deliver the products.

1. What sort of companies would be most interested in this integrated solution? Why?
2. Can you identify other important additions to enterprise resource planning software that would require similar partnering?

Sources: Adapted from John Evan Cook, "SAP AG Confirms Supply-Chain Partnership with FedEx," *InternetWeek,* April 8, 1998; Jeff Sweat, "FedEx and SAP Team on Tracking-Shipping APP," *InformationWeek,* April 13, 1998, p. 24.

4 Florist Increases and Speeds Transactions

Using a network of florists, 1-800-Flowers is able to deliver flowers around the country. Once a phone order is received from a customer, 1-800-Flowers contacts one of its florists to make and deliver the floral arrangement. Each florist pays 1-800-Flowers about $.50 for each order. In about 40 percent of the orders, the flowers are to be delivered on the same day the order is placed. Once the order is placed, customers often call one or more times to confirm that the order is correct and being delivered. The existing system used modems and a dial-up system to contact florists. Two key aspects of this transaction processing system that needed improvement were sales and the delivery of floral arrangements.

How does the largest on-line florist increase sales? For 1-800-Flowers, one answer was training. The training, however, was unusual. Instead of hiring traditional trainers, 1-800-Flowers hired a university drama instructor. In addition to normal sales tactics, such as increasing customer satisfaction, the drama instructor also coached sales personnel in voice control, concentration, articulation, breath control, and role-playing. The drama instructor also helped sales personnel handle a range of customer emotions from a humorous customer to a hysterical one. The bottom line was to treat each phone call like an on-stage performance.

The results of the drama coaching were positive. Many areas experienced an increase in sales. But a new on-line system may place the new training method in jeopardy. The new transaction processing system is based on the Internet. Florists now can use the Internet to contact 1-800-Flowers directly to receive orders. In addition, orders can be confirmed using the new Web site. Using this new system, florists can determine whether the orders can be modified or the delivery times can be changed. Florists can check the Web site several times during the day or remain connected to receive instantaneous information. With this new transaction processing system, florists can be more efficient with deliveries. Instead of sending three trucks for three different deliveries with the old system, florists can now coordinate their deliveries and use only one or two trucks. Florists are also paid $.10 for each order that the florist confirms using the new Web-based transaction processing system. There is also a chat facility on the Web site that allows florists to communicate with 1-800-Flowers or with other florists directly. Customers will also be given access to the chat facility to allow them to communicate with 1-800-Flowers and the many florists filling orders. This will greatly speed orders and allow customers to get their questions answered and their flowers delivered efficiently. The new Web-based transaction processing system will also be used for training. If the new on-line training program is successful, the drama instructor's training program may have a short run.

1. How was 1-800 Flowers able to develop a better transaction processing system?
2. What further improvements would you suggest?

Sources: Sharon Machlis, "Florists Use Web to Speed Deliveries," *Computerworld*, February 9, 1998, p. 45; Tricia Campbell, "The Drama of Selling: 1-800 Flower's Sales Personnel Training Program," *Sales and Marketing*, December 1997, p. 92; Dennis Cahill, "All Good Marketing is Interactive," *Marketing News*, January 19, 1998, p. 4.

CHAPTER 9

Management Information Systems

"Companies must be ready to examine and revamp practices before computerizing them. If you take a bad process and automate it, all you have is an automated bad process."

— Scott Nelson, analyst at Gartner Group in Stamford, Connecticut, discussing the use of marketing MIS software

Chapter Outline

An Overview of Management Information Systems
    Management Information Systems in Perspective
    Inputs to a Management Information System
    Outputs of a Management Information System
    Characteristics of a Management Information System
    Management Information Systems for Competitive Advantage
    MIS and Web Technology
    Functional Aspects of the MIS

A Financial Management Information System
    Inputs to the Financial MIS
    Financial MIS Subsystems and Outputs

A Manufacturing Management Information System
    Inputs to the Manufacturing MIS
    Manufacturing MIS Subsystems and Outputs

A Marketing Management Information System
    Inputs to the Marketing MIS
    Marketing MIS Subsystems and Outputs

A Human Resource Management Information System
    Inputs to the Human Resource MIS
    Human Resource MIS Subsystems and Outputs

Other Management Information Systems
    Accounting MISs
    Geographic Information Systems

Learning Objectives

After completing Chapter 9, you will be able to:

1. Define the term *MIS* and clearly distinguish the difference between a TPS and an MIS.
2. Discuss how organizations are enabling users to access MIS systems via Web technologies.
3. Describe the inputs, outputs, and subsystems associated with a financial MIS.
4. Describe the inputs, outputs, and subsystems associated with a manufacturing MIS.
5. Describe the inputs, outputs, and subsystems associated with a marketing MIS.
6. Describe the inputs, outputs, and subsystems associated with a human resource MIS.
7. Identify other functional management information systems.

GAF Materials Corp.
Financial MIS

GAF Materials Corp. (GAFMC) began modestly in 1886 as the Standard Paint Company, with 18 people working in a remodeled sawmill. Today it is one of the most fully-integrated roofing manufacturers in the United States with sales over $1 billion and 3,300 employees working in 26 plants across the country. GAFMC markets its diverse building materials products to contractors and distributors throughout the United States and in selected areas of the world. Its professional customer service representatives are trained to assist customers and coordinate a range of services. GAFMC has built its success on the diverse range of its building materials products along with key corporate acquisitions, research and development, and customer service. Its integrated operations and self-sufficiency are the mainstay of the company's future growth.

Until recently, GAFMC lacked a financial management information system. It was unable to obtain a consistent view of key financial measures such as production rates and labor utilization across its 26 plants. The problem was a lack of consistent standards for capturing and reporting data. Instead, GAFMC had 26 different ways of measuring production. As a result, plant managers and executives at the company's headquarters were unable to compare the manufacturing operations.

When a new chief financial officer came on board, he led an effort to standardize GAF's plant production data. The new financial system is based on the use of a database using Applix's TM1 software and was implemented at a cost of $2 million. The database ties the reporting systems of all the plants to one another and to the corporate offices in Wayne, New Jersey.

The financial database receives input from many sources. Cost data is captured from various accounting systems including payroll (labor hours and costs), accounts payable (cost of raw materials and services), and inventory control (stock counts and usage data). In addition, factory floor supervisors enter production statistics into workstations at the end of each shift. Each plant's data is kept on a Windows NT server running the TM1 software, and the data is copied weekly to a central NT midrange computer.

With the new consistent data and improved data-reporting capabilities, plant managers can view shift-by-shift manufacturing efficiency, costs, machine downtime, and other key indicators of plant operations. This enables them to see if another plant is doing something well or not. Production managers can view shift-by-shift production data broken down to the machine level. This enables them to determine which machines are responsible for production downtime and helps them take action to fix the problems. Buyers can review the cost of raw materials used in the shingles, insulation, vents, and other roofing products made by GAFMC. They can quickly determine whether materials cost less in different regions. The full cost to develop the system should be recovered within 18 months.

As you read this chapter, consider the following questions:

- What is the relationship between transaction processing systems and management information systems?

- How can management information systems be used to support the objectives of the business organization?

Sources: Adapted from Craig Stedman, "26 Plants Under One Roof," *Computerworld*, March 9, 1998, pp. 55, 60 and the "About GAFMC" section of the GAF Materials Corp. Web site at http://www.gaf.com, accessed May 22, 1998.

GAF Materials Corp. was able to improve its plant operations by standardizing the way its plants report their production statistics and by implementing a financial management information system. By capturing data about costs, machine downtimes, and shift-to-shift production data, the manufacturing transaction processing systems provide information on all plant operations for each of the company's 26 plants. This information allows the financial managers to develop new reports showing areas of opportunity for improvement. Thus, by using the financial MIS to analyze data provided by the firm's TPS, GAF Materials Corp.'s managers can meet their goals of reducing production costs while improving customer service.

AN OVERVIEW OF MANAGEMENT INFORMATION SYSTEMS

The system that delivered the data to GAF Materials Corp. managers in a form they could analyze effectively is a management information system (MIS). An important input into GAF Materials Corp.'s MIS, indeed, into the MIS of most organizations, is the transaction processing system. This link between the data generated by the transaction processing system and the reports generated by the management information system is vital to organizations and is growing stronger. For many organizations, as it was for GAF Materials Corp., a good MIS is the key to profitability.

Management Information Systems in Perspective

The primary purpose of an MIS is to help an organization achieve its goals by providing managers with insight into the regular operations of the organization so that they can control, organize, and plan more effectively and efficiently. In short, an MIS provides managers with information and support for effective decision making and provides feedback on daily operations. In doing so, an MIS adds value to processes within an organization. A manufacturing MIS, for example, is a set of integrated systems that can help managers monitor a manufacturing process to maximize the value of raw materials as they are assembled into finished products. For the most part, this monitoring is accomplished through various summary reports produced by the MIS. These reports can be obtained by filtering and analyzing the detailed data contained in transaction processing databases and presenting the results to managers in a meaningful way. These reports support managers by providing them with data and information for decision making in a form they can readily use. Figure 9.1 shows the role of MISs within the flow of an organization's information. Note that business transactions can enter the organization through traditional methods, via the Internet, or via an extranet connecting customers and suppliers to the firm's transaction processing systems.

As Figure 9.1 shows, the summary reports from the MIS are just one of many sources of information available to managers. As previously discussed, the use of management information systems spans all levels of management. That is, they provide support to and are used by employees throughout the organization.

Each MIS is an integrated collection of subsystems, which are typically organized along functional lines within an organization. Thus, a financial MIS includes subsystems that address financial reporting, profit and loss analysis, cost analysis, and the use and management of funds. Many functional subsystems share certain hardware resources, data, and often even

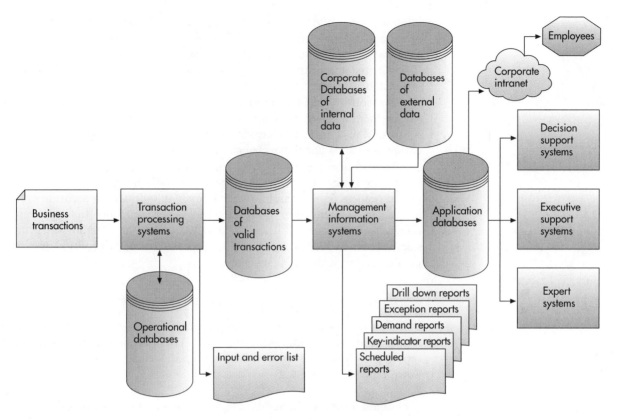

FIGURE 9.1

Sources of Managerial Information

The MIS is just one of many sources of managerial information. Decision support systems, executive support systems, and expert systems also assist in decision making.

personnel. Some subsystems, however, are self-contained within one functional area and are suited to a specialized purpose. One of the IS manager's roles is to increase the overall efficiency of the MIS by improving the integration of subsystems. For instance, similar data may be collected and maintained by two distinct functional departments (e.g., customer lists maintained by both sales and accounting). Or perhaps hardware resources are only partially used by one functional area and might be shared by another.

Even though increasing overall MIS efficiency is important, all managers (including IS managers) must appreciate that one important role of the MIS is to improve effectiveness by providing the right information to the right person in the right fashion at the right time. This aspect of MIS is too often forgotten. In their zest to improve system efficiencies through integration, some IS managers overlook the basic problem-solving needs of functional managers.

Inputs to a Management Information System

Data that enters an MIS originates from both internal and external sources (Figure 9.1). The most significant internal source of data for an MIS is the organization's various TPSs. One of the major activities of the TPS is to capture and store the data resulting from ongoing business transactions. With every business transaction, various TPS applications make changes to and update the organization's databases. For example, the billing application helps keep the accounts receivable database up to date so that managers know who owes the company money. These updated databases are a primary internal source of data for the management information system. In companies that have implemented an ERP system, the collection of databases associated with this system are an important source of internal data for the MIS. E-commerce applications can also be an important input to an MIS. Other internal data comes from specific functional areas throughout the firm. External sources of data can include customers, suppliers, competitors, and stockholders whose data is not already captured by the TPS, as well as other sources. Many companies have implemented extranets to link them to these entities and allow for the exchange of data and information.

The MIS uses the data obtained from these sources and processes it into information more usable to managers, primarily in the form of predetermined reports. For example, rather than simply obtaining a chronological list of sales activity over the past week, a national sales manager might obtain her organization's weekly sales data in a format that allows her to see sales activity by region, by local sales representative, by product, and even in comparison with last year's sales.

Outputs of a Management Information System

The output of most management information systems is a collection of reports that are distributed to managers. These reports include scheduled reports, key-indicator reports, demand reports, exception reports, and drill down reports (Figure 9.2).

scheduled reports

reports produced periodically, or on a schedule, such as daily, weekly, or monthly

Scheduled reports. Scheduled reports are produced periodically, or on a schedule, such as daily, weekly, or monthly. For example, a production manager could use a weekly summary report that lists total payroll costs to monitor and control labor and job costs. A manufacturing report produced once a day to monitor the production of a new product is another example of a scheduled report. Other scheduled reports can help managers control customer credit, the performance of sales representatives, inventory levels, and more.

key-indicator report

a report that summarizes the previous day's critical activities and is typically available at the beginning of each workday

Key-indicator report. A **key-indicator report** summarizes the previous day's critical activities and is typically available at the beginning of each workday. These reports can summarize inventory levels, production activity, sales volume, and the like. Key-indicator reports are used by managers and executives to take quick, corrective action on significant aspects of the business.

demand reports

reports developed to give certain information at a manager's request

Demand reports. Demand reports are developed to give certain information at a manager's request. In other words, these reports are produced on demand. For example, an executive may want to know the production of a particular item—a demand report can be generated to give the requested information. Other examples of demand reports include reports requested by executives to show the hours worked by a particular employee, total sales to date for a product, and so on.

exception reports

reports that are automatically produced when a situation is unusual or requires management action

Exception reports. Exception reports are reports that are automatically produced when a situation is unusual or requires management action. For example, a manager might set a parameter that generates a report of all inventory items with fewer than the equivalent of five days of sales on hand. This is an unusual situation that requires prompt action to avoid running out of stock on the item. The exception report generated by this parameter would

FIGURE 9.2

Reports Generated by an MIS

The five types of reports are (a) scheduled, (b) key-indicator, (c) demand, (d) exception, and (e-h) drill down.

(Source: George W. Reynolds, *Information Systems for Managers*, 3rd ed., St. Paul, MN: West Publishing Co.,1995. Reprinted with permission from Course Technology.)

(a) Scheduled Report

Daily Sales Detail Report

Prepared: 08/10/XX

Order #	Customer ID	Salesperson ID	Planned Ship Date	Quantity	Item #	Amount
P12453	C89321	CAR	08/12/96	144	P1234	$3,214
P12453	C89321	CAR	08/12/96	288	P3214	$5,660
P12454	C03214	GWA	08/13/96	12	P4902	$1,224
P12455	C52313	SAK	08/12/96	24	P4012	$2,448
P12456	C34123	JMW	08/13/96	144	P3214	$ 720
....			....	....		

(b) Key-Indicator Report

Daily Sales Key Indicator Report

	This Month	Last Month	Last Year
Total Orders Month to Date	$1,808	$1,694	$1,914
Forecasted Sales for the Month	$2,406	$2,224	$2,608

(c) Demand Report

Daily Sales by Salesperson Summary Report

Prepared: 08/10/XX

Salesperson ID	Amount
CAR	$42,345
GWA	$38,950
SAK	$22,100
JWN	$12,350
....	....
....	

(d) Exception Report

Daily Sales Exception Report—Orders Over $10,000

Prepared: 08/10/XX

Order #	Customer ID	Salesperson ID	Planned Ship Date	Quantity	Item #	Amount
P12345	C89321	GWA	08/12/96	576	P1234	$12,856
P22153	C00453	CAR	08/12/96	288	P2314	$28,800
P23023	C32832	JMN	08/11/96	144	P2323	$14,400
....	....	....	....	....	....	....
....	....	....	....	....	....	....

contain only items with fewer than five days of sales in inventory. As with key-indicator reports, exception reports are most often used to monitor aspects important to an organization's success. In general, when an exception report is produced, a manager or executive takes action. Parameters, or *trigger points*, for an exception report should be carefully set. Trigger points that are set too low may result in an abundance of exception reports; trigger points that are too high could mean that problems requiring action are overlooked. For example, if a manager wants a report that contains all projects over budget by $100 or more, he may find that almost every company project exceeds its budget by at least this amount. The $100 trigger point is probably too low. A trigger point of $10,000 might be more appropriate.

FIGURE 9.2 (continued)

Reports Generated by an MIS

(e) Manager sees actual earnings exceed forecast by 6.8 percent for 2nd quarter 1999. (f) Manager views sales and expenses for 2nd quarter 1999. (g) Manager views sales by division. (h) Manager views sales for the Health Care division.

(e) First-Level Drill Down Report

Earnings by Quarter (Millions)			
	Actual	Forecast	Variance
2nd Qtr. 1999	$12.6	$11.8	6.8%
1st Qtr. 1999	$10.8	$10.7	0.9%
4th Qtr. 1998	$14.3	$14.5	-1.4%
3rd Qtr. 1998	$12.8	$13.3	-3.8%

(f) Second-Level Drill Down Report

Sales and Expenses (Millions)			
Qtr: 2nd Qtr. 1999	Actual	Forecast	Variance
Gross Sales	$110.9	$108.3	2.4%
Expenses	$ 98.3	$ 96.5	1.9%
Profit	$ 12.6	$ 11.8	6.8%

(g) Third-Level Drill Down Report

Sales by Division (Millions)			
Qtr: 2nd Qtr. 1999	Actual	Forecast	Variance
Beauty Care	$ 34.5	$ 33.9	1.8%
Health Care	$ 30.0	$ 28.0	7.1%
Soap	$ 22.8	$ 23.0	-0.9%
Snacks	$ 12.1	$ 12.5	-3.2%
Electronics	$ 11.5	$ 10.9	5.5%
Total	$110.9	$108.3	2.4%

(h) Fourth-Level Drill Down Report

Sales by Product Category (Millions)			
Qtr: 2nd Qtr. 1999 Division: Health Care	Actual	Forecast	Variance
Toothpaste	$12.4	$10.5	18.1%
Mouthwash	$ 8.6	$ 8.8	-2.3%
Over-the-Counter Drugs	$ 5.8	$ 5.3	9.4%
Skin Care Products	$ 3.2	$ 3.4	-5.9%
Total	$30.0	$28.0	7.1%

Republic National Bank of New York uses exception reporting to manage its $15 million in annual legal bills. The bank receives invoices electronically from its law firms. Software checks the bills against Republic's fee guidelines and instantly generates an exception report for items that do not comply. Among the rules: photocopies must cost no more than 10 cents per page; travel, faxes, photocopying, messenger services, and other non-fee expenses should not exceed 10 percent of the total bill. The bank is adding a new rule that prohibits more than one attorney from attending a meeting, hearing, or trial unless approved in advance. Using such software became possible when the American Bar Association in 1995 standardized billing codes for all U.S. law firms, much like the standardized diagnosis codes that physicians use on insurance forms. For example, expenses connected with a lawsuit are now broken down by phase (case assessment, discovery), task (fact gathering, depositions), and activity (drafts/revisions, client communications). Each element is assigned a number. Drafting a document, for example, is coded A103.[1]

drill down reports

reports that provide detailed data about a situation

Drill down reports. Drill down reports provide increasingly detailed data about a situation. Nabisco Biscuit Company, with revenues of more than $8 billion, is known for its many top brands including Oreo and Chips Ahoy!

Guideline	Reason
Tailor each report to user needs.	This requires user involvement and input.
Spend time and effort producing only those reports that are used.	Once instituted, many reports continue to be generated even though no one uses them anymore.
Pay attention to report content and layout.	Prominently display the information that is most desired. Do not clutter the report with unnecessary data. Use commonly accepted words and phrases. Managers can work more efficiently if they can easily find desired information.
Use management by exception reporting.	Some reports should be produced only when there is a problem to be solved or an action that should be taken.
Set parameters carefully.	Low parameters may result in too many reports; high parameters mean valuable information could be overlooked.
Produce all reports in a timely fashion.	Outdated reports are of little or no value.

TABLE 9.1

Guidelines for Developing MIS Reports

cookies; Ritz and Premium crackers; and SnackWell's low-fat and fat-free products. Nabisco employs dozens of financial analysts and managers to identify and leverage every advantage to improve the profitability of each product. All costs associated with specific stock keeping units (SKUs) are tracked to determine profitability. Through the use of drill down reports, analysts are able to see data at a high level like total cookies, to a more detailed level like Oreo, and to a very detailed level like Oreo Double-Stuf cookies.[2]

Developing effective reports. Management information system reports can help managers develop better plans, make better decisions, and obtain greater control over the operations of the firm. It is important to recognize that various types of reports can overlap. For example, a manager can demand an exception report or set trigger points for items contained in a key-indicator report. Certain guidelines should be followed in designing and developing reports to yield the best results. Table 9.1 explains these guidelines.

Characteristics of a Management Information System

Scheduled, key-indicator, demand, exception, and drill down reports have all helped managers and executives make better, more timely decisions. When the guidelines for developing effective reports are followed, higher revenues and lower costs can be realized. In general, management information systems perform the following functions:

- *Provide reports with fixed and standard formats.* For example, scheduled reports for inventory control may contain the same types of information placed in the same locations on the reports. Different managers may use the same report for different purposes.
- *Produce hard-copy and soft-copy reports.* Some MIS reports are printed on paper and are considered hard-copy reports. Most output soft copy, using visual displays on computer screens. Soft-copy output is typically formatted in a report-like fashion. In other words, a manager might be able to call an MIS report up directly on the computer screen, but the report would still appear in the standard hard-copy format. Hard copy is still the most often used form of MIS report.

- *Use internal data stored in the computer system.* MIS reports use primarily internal sources of data that are contained in computerized databases. Some MISs use external sources of data about competitors, the marketplace, and so on. The Internet is a frequently used source for external data.
- *End users are able to develop their own custom reports.* While analysts and programmers may be involved in developing and implementing complex MIS reports that require data from many sources, end users are increasingly developing their own simple programs to query a database and produce basic reports.
- *Require formal requests from users.* When information systems personnel develop and implement MIS reports, a formal request to the information systems department for reports is usually required. End user developed reports require much less formality.

Management Information Systems for Competitive Advantage

An MIS provides support to managers as they work to achieve corporate goals. It provides them with feedback and insight into the regular operations of the organization. It enables them to compare results to established company goals and identify problem areas and opportunities for improvement. With this improved insight and understanding, they are able to control, organize, and plan more effectively and efficiently. Thus, an effective MIS can give an organization a competitive advantage and at least a temporary edge over another organization lacking such systems.

Developing a new MIS or modifying an existing one does not always result in a competitive advantage. In most cases, simply obtaining software or equipment that any competitor can acquire will not yield a long-term advantage. Putting the MIS to more effective use, however, may provide a firm with a competitive edge. In most cases, firms that know what data to obtain (and how to relate it properly) and when and in what form to present it to which managers achieve the most significant advantage through MISs. This advantage can have a significant impact on costs, profits, customer services, and new products.

Domino's Pizza implemented an integrated MIS including inventory management, order fulfillment, manufacturing, accounting, and human resources. Domino's likes a single, unified system so that everyone receives information from the same corporate database. The goal is for company managers to have complete, real-time access to important information for the entire business.[3]

MIS and Web Technology

Organizations are increasingly making data from their management information systems available to users via Web technology. Managers can access the data using their firm's intranet and Web browser software loaded on their personal computers. Data for use by different business functions can be accessed through a specially designed Web site for that data. From there, users can navigate through the data using the same technology used to surf the Web, including powerful search engines, HTML links, etc. Read the "E-commerce" box for an example of how one company is providing Web access to an MIS designed to support customer service agents and salespeople.

E-COMMERCE

New Holland North America Provides Web Access to MIS Data

New Holland NA is the agricultural-machinery unit of Italian automaker Fiat SpA. New Holland engineers, manufactures, markets, and distributes agricultural and construction equipment. It also provides wholesale dealer and retail financing. For the fiscal year ending December 31, 1997, total revenues rose 8 percent to $6 billion and net income increased 55 percent to $387.7 million. U.S. operations, headquartered in New Holland, Pennsylvania, manufactures primarily tractors, combines, and other farm equipment. These products are sold through 1,500 dealers nationwide.

Prior to implementing a new system, each customer service office handled a different function: dealer parts, locating inventory, tractor repairs, or dealer software system support. Each group also had its own way of logging dealer calls. The inventory locator group used a paper-based system and logged all dealer calls by hand. The other three used homegrown systems, and all were totally different. This made it impossible to share information among the groups or with others within New Holland. Salespeople had to go to the New Holland offices and manually comb through logged calls to get information. So they did without this potentially valuable data.

New Holland implemented a customer service MIS throughout its four service offices to improve service and

to enable users to see the same information regardless of where they are. Approximately 70 customer service agents in the four offices log dealer calls. MIS users have the capability of seeing various standard reports of the calls logged by any of the groups. As a result, the system gives the customer service agents a much better idea of what is going on. If a farm equipment dealer has a problem with a part for a New Holland tractor, customer service agents can respond quickly. Managers like it because they have improved knowledge of problems that are coming into New Holland and what dealers are experiencing in the field.

New Holland is also rolling out a sales force automation system to 500 regional salespeople that will provide the sales force with access to an expanded customer database including the data from the call logging system. Various reports provided by the system give the sales force full knowledge of any questions or problems a dealer may have before they visit the dealership. Salespeople will be able to access the system using a Web browser on their laptops and view customer reports to get a better idea of how to serve the dealers they support. They can see data about current issues for a specific dealer and review the status. Giving salespeople more information should help them avoid being blindsided by problems.

DISCUSSION QUESTIONS

1. What are the advantages of having salespeople access this data via the Web versus logging on to the system via other methods? Are there any disadvantages?
2. List some key points that make this system a success. What are some potential problems?

Sources: Adapted from Randy Weston, "Call Logging Improves Service," *Computerworld*, April 27, 1998, pp. 61–62; Company Information section of the Applix Web site at http://www.applix.com, accessed May 23, 1998; and "Fiat's New Holland Unit Reports 12% Advance In 1st-Quarter Profit," September 7, 1998, Dow Jones News Service.

Functional Aspects of the MIS

Most organizations are structured along functional lines or areas. This is usually apparent from an organization chart, which typically shows vice presidents under the president. Some of the traditional functional areas are accounting, finance, marketing, personnel, research and development (R&D), legal services, operations/production management, and information systems.

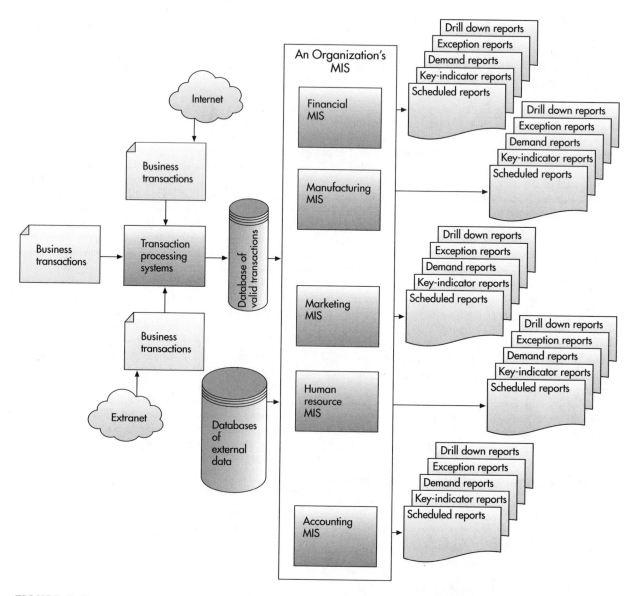

FIGURE 9.3

The MIS is an integrated collection of functional information systems, each supporting particular functional areas.

In addition, each of these functional areas within the organization contains the various levels of management (strategic, tactical, and operational). Thus, in addition to vertically categorizing management into the various functional areas, management is horizontally grouped into strategic, tactical, and operational levels. Each functional area utilizes its own set of function-specific subsystems, all of which interface with both the TPS and the MIS in some way. Each of these areas requires different information and support for decision making; they also share some common information needs. Using a functional approach, information is developed for the functional area managers, often spanning all levels of management within each functional area (Figure 9.3). Thus, a portion of the overall MIS is organized to support the financial functional area. Reports that are shared by all levels of accounting managers, as well as different reports for each level of accounting management, would be generated. This same strategy applies to top-, mid-, and lower-level managers

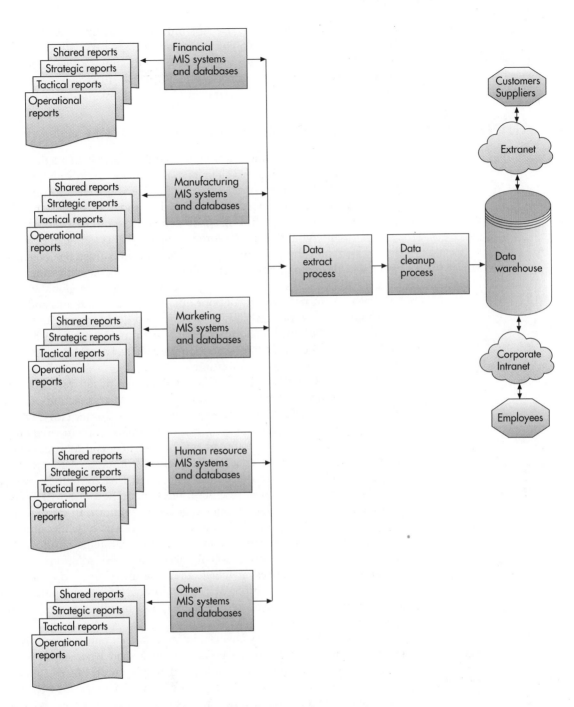

FIGURE 9.4

Functional area MISs are often integrated through a common database.

in marketing, and so on for each functional area. Each functional area receives reports supplying information focused on the particular needs of that group.

When a functional approach is taken, the various management information systems should be linked. See the large rectangle labeled "An Organization's MIS" that surrounds the five functional information subsystems in Figure 9.3. Otherwise, the organization might end up with a collection of disjointed and ineffective systems. One way to unify and integrate the various systems is through a shared database, as shown in Figure 9.4.

The use of a common database not only serves to integrate the various functional MISs but can also link the organization's TPS with various functional MISs. The integration of different information systems makes data and information sharing simpler—which can lead to reduced costs, more accurate reports, more secure data, and increased organizational efficiency.

A FINANCIAL MANAGEMENT INFORMATION SYSTEM

The world of corporate finance has changed from simply managing finance and administration operations such as monitoring cash flow and profitability to actually driving the organization's key decisions and helping define the strategies that make companies succeed. In today's complex and rapidly changing environment, filling this role demands a complete global finance system. The integrated financial software of an ERP system can play a decisive role in meeting this need. Such sophisticated financial systems are capable of providing financial managers and executives with timely information, which is critical to success in today's fast-moving global economy. History has shown the results of poor financial decisions. Banks and savings institutions have gone bankrupt because of bad decisions and unfavorable economic conditions. Corporations with too much debt have also gone bankrupt. On the other hand, good financial decisions have resulted in growing and prosperous organizations.

The financial MIS may be implemented as a set of software and several files that are updated by a large number of data transactions captured through various transaction processing systems (e.g., order entry, accounts payable, purchasing, etc.). In organizations that have implemented an ERP system, the financial MIS is based on the use of a common database of financial data shared by numerous functions. Multinational companies are adopting the financial modules of an integrated ERP system for two reasons. First, by enabling best practices, implementation of an ERP system helps the firm meet such corporate goals as minimizing administrative costs, reducing cycle times, and increasing productivity while simultaneously freeing the finance organization to focus on strategic business issues. Second, companies operating globally need enterprise solutions that span and support the countries in which they do and plan to do business—including operating in multiple languages and currencies, supporting local business practices and legal requirements, and handling business-critical operations across borders.

financial MIS

an MIS that provides financial information to all financial managers within an organization

A **financial MIS** provides financial information not only for executives but also for a broader set of people who need to make better decisions on a daily basis. Finding opportunities and quickly identifying problems can mean the difference between a business's success and failure. An integrated ERP system that encompasses the finance function maximizes decision-making capabilities by ensuring that the right business information reaches the right people, at the right time. Specifically, the financial MIS performs the following functions:

- Integrates financial and operational information from multiple sources including the Internet into a single MIS
- Provides easy access to data for both financial and nonfinancial users often through use of the corporate intranet to access corporate Web pages of financial data and information

- Makes financial data available on a timely basis to shorten analysis turn-around time
- Enables analysis of financial data along multiple dimensions—time, geography, product, plant, customer
- Analyzes historical and current financial activity
- Monitors and controls the use of funds over time

Figure 9.5 shows typical inputs, function-specific subsystems, and outputs of a financial MIS.

Inputs to the Financial MIS

Managerial actions supported by the financial MIS require diverse data and information. The input sources, both internal and external, are briefly discussed next.

Strategic plan or corporate policies. The strategic plan contains major financial objectives. Financial goals, debt and loan ratios, and expected returns are some of the measures that can be included in the strategic plan. The plan often projects the company's financial needs anywhere from one

FIGURE 9.5

Overview of a Financial MIS

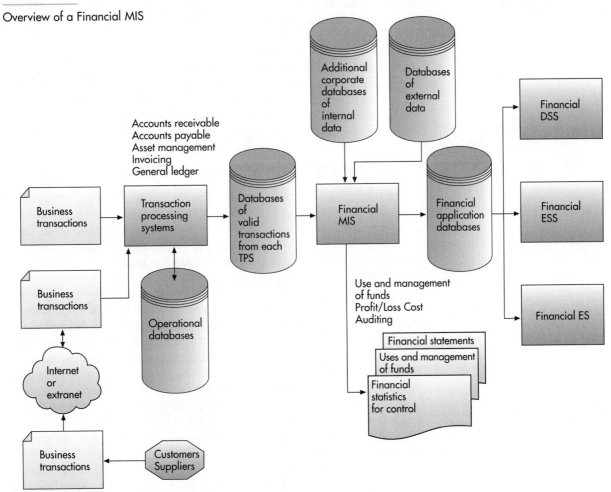

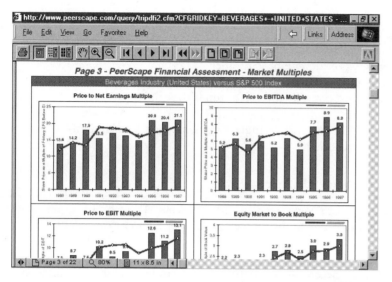

Deloitte & Touche PeerScape is an on-line financial MIS that provides investment bank analysis of over 7,000 companies.

to five years into the future. Many specific information needs, such as the return on investment for various projects, desired debt-to-equity ratios, and expected cash needs, come directly from the strategic plan.

The transaction processing system. Important financial information is collected from almost every transaction processing application—payroll, inventory control, order processing, accounts payable, accounts receivable, and general ledger. Total payroll costs, the investment in inventory, total sales over time, the amount of money paid to suppliers, the total amount owed to the company from customers, and detailed accounting data provide the basis for many financial reports.

External sources. Information from and about the competition can be critical to financial decision making. Annual reports and financial statements from competitors and general news items and reports can be incorporated into MIS reports to provide units of measure or as a basis of comparison. The Internet is used as a source for such information. Government agencies also provide important economic and financial information. Inflation, consumer price indexes, new housing starts, and leading economic indicators can help a company plan for future economic conditions. In addition, important tax laws and financial reporting requirements can also be reflected in the financial MIS.

Financial MIS Subsystems and Outputs

Depending on the organization and its needs, the financial MIS can include both internal and external systems that assist in acquiring, using, and controlling cash, funds, and other financial resources. These subsystems of the financial MIS have a unique role in adding value to a company's business processes. For example, a real estate development company might use a financial MIS subsystem to aid with the use and management of funds. Suppose the firm takes $10,000 deposits on condominiums in a new development. Until construction begins, the company will be able to invest these surplus funds. By using reports produced by the financial MIS, investment alternatives can be analyzed. The company might invest in new equipment or purchase global stocks and bonds. The profits generated from the investment can be passed along to customers in different ways. The company can pay stockholders dividends, buy higher-quality materials, or sell the condominiums at a lower cost.

Financial managers need data for better management reporting and decision making from their financial MIS. They also need access to additional internal reports and data to meet regulatory needs. All the data must be easily accessible.

Other important financial subsystems include profit/loss and cost accounting, and auditing. Each subsystem interacts with the TPS in a specialized way and has information outputs that assist financial managers in making better

Departments such as manufacturing or research and development are cost centers since they do not directly generate revenue.
(Source: Image copyright ©1998 PhotoDisc, Inc.)

decisions. These outputs include profit/loss and cost systems reports, internal and external auditing reports, and uses and management of funds reports.

Profit/loss and cost systems. Two specialized financial functional systems are profit/loss and cost systems, which organize revenue and cost data for the company. Revenue and expense data for various departments is captured by the TPS and becomes a primary internal source of financial information for the MIS.

Many departments within an organization are **profit centers**, which means they track total expenses and net profits. An investment division of a large insurance or credit card company is an example of a profit center. Other departments may be **revenue centers**, which are divisions within the company that primarily track sales or revenues, such as a marketing or sales department. Still other departments may be **cost centers**, which are divisions within a company that do not directly generate revenue, such as manufacturing or research and development. These units incur costs with little or no revenues. Data on profit, revenue, and cost centers is gathered (mostly through the TPS but sometimes through other channels as well), summarized, and reported by the profit/loss and cost subsystems of the financial MIS.

profit centers

departments within an organization that track total expenses and net profits

revenue centers

divisions within the company that track sales or revenues

cost centers

divisions within the company that do not directly generate revenue

Auditing. Managers and executives of the organization rely on financial statements produced by the financial MIS to determine whether the organization is reaching profit levels. Other people outside the organization also rely on financial statements developed by the financial MIS to evaluate the financial health of the organization. How do these people know whether the financial statements are correct and accurately reflect the financial conditions of the organization? They turn to auditing.

Auditing involves analyzing the financial condition of an organization and determining whether financial statements and reports produced by the financial MIS are accurate. Because financial statements, such as income statements and balance sheets, are used by so many people and organizations (investors, bankers, insurance companies, federal and state government agencies, competitors, and customers), sound auditing procedures are important.

auditing

the process that involves analyzing the financial condition of an organization and determining whether financial statements and reports produced by the financial MIS are accurate

internal auditing

auditing performed by individuals
within the organization

Internal auditing is performed by individuals within the organization. For example, the finance department of a corporation may use a team of employees to perform an audit. Typically an internal audit is conducted to see how well the organization is doing in terms of meeting established company goals and objectives, such as no more than five weeks inventory on hand, all travel reports completed within one week of returning from a trip, etc. **External auditing** is performed by an outside group, such as an accounting or consulting firm such as Arthur Andersen, Deloitte & Touche, or one of the other major, international accounting firms. The purpose of an external audit is to provide an unbiased picture of the financial condition of an organization. Auditing can also uncover fraud and other problems.

external auditing

auditing performed by an outside
group

Uses and management of funds. Another important function of the financial MIS is funds usage and management. Companies that do not manage and use funds effectively often have lower profits or face bankruptcy. Outputs from the funds usage and management subsystem, when combined with other subsystems of the financial MIS, can locate serious cash flow problems and help the organization increase profits.

Internal uses of funds include additional inventory, new or updated plants and equipment, additional labor, the acquisition of other companies, new computer systems, marketing and advertising, raw materials, land, investments in new products, and research and development. External uses of funds are typically investment related. On occasion, a company might have excess cash from sales that is placed into an external investment. External uses of funds often include bank accounts, stocks, bonds, bills, notes, futures, options, and foreign currency.

A MANUFACTURING MANAGEMENT INFORMATION SYSTEM

More than any functional area, manufacturing has been affected by great advances in technology. As a result, many manufacturing operations have been dramatically improved over the last decade. Furthermore, with the emphasis on greater quality and productivity, having an efficient and effective manufacturing process is becoming even more critical. The use of computerized systems is emphasized at all levels of manufacturing—from the shop floor to the executive suite. Figure 9.6 gives an overview of some of the manufacturing MIS inputs, subsystems, and outputs.

The manufacturing process includes a number of highly interdependent tasks with one small change affecting many tasks. The use of an integrated ERP that encompasses manufacturing provides the flexibility to make adjustments and improvements without risk, adapt to changes without delays, and customize processes to satisfy the most demanding customers. By integrating manufacturing functions with financial management and human resource considerations, an ERP provides improved control over every vital aspect of manufacturing operations. Administrative tasks are virtually eliminated, paperwork is transformed into on-line procedures, and communications tasks can be handled electronically through EDI and the Internet. Manufacturing ERP also uses the Internet and corporate intranet to connect domestic and international business units to facilitate decentralized operations and centralized control. In multiple languages, currencies, countries, or in a single manufacturing location, an integrated ERP

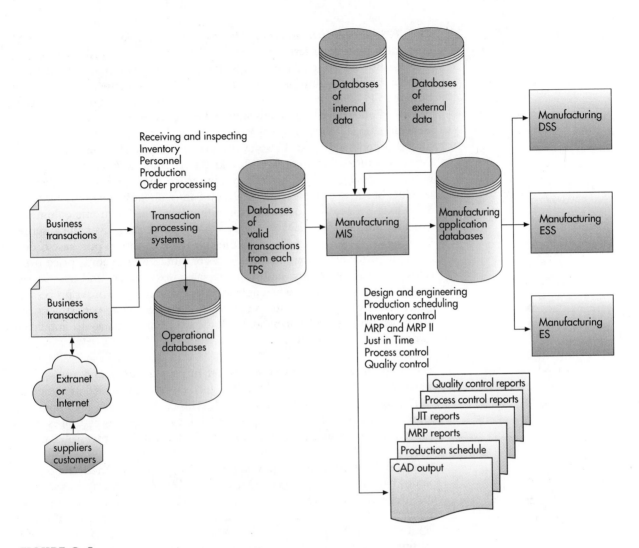

FIGURE 9.6

Overview of a Manufacturing MIS

offers countless opportunities to streamline operations, reduce overhead, and rapidly respond to competitive pressures.

Inputs to the Manufacturing MIS

Inputs to the manufacturing MIS are primarily from operations that deal with the flow and conversion of materials throughout the organization. Some sources of information are external, but most are internal. Information sources for the manufacturing MIS include the following:

Strategic plan or corporate policies. The manufacturing MIS gets overall direction from the strategic plan. This long-range planning document can specify quality, production, and service goals or constraints. Expansion into new facilities or the possibility of a plant shutdown are often reflected in this document. Increased manufacturing capacity, constraints on the allowable number of employees, altered inventory policy, and new quality-control programs and parameters are just a few of the inputs into the manufacturing MIS that can come from the strategic plan.

The TPS: Order processing. The manufacturing process begins with customers placing orders for goods and/or services. The order processing system, introduced in Chapter 8, is one of the first computerized applications in the manufacturing life cycle and an important subsystem spanning both the TPS and the MIS. Basically, this system is designed to handle order entry and all related aspects, including editing, verification of orders, database updating, canceling orders, and making adjustments to previously placed orders. American Hospital Supply, McKesson-Robbins, Wal-Mart, General Motors, Eastman Kodak, and Baxter International revolutionized order processing by linking their computers to those of their suppliers to conduct business electronically. U.S. companies already buy on the order of $500 billion worth of goods electronically each year. Electronic commerce using techniques like EDI (Chapter 6) and purchasing over the Internet (Chapter 7) can dramatically slash the cost of processing a purchase order.

Many order processing systems can compute commissions, price overrides, quantity discounts, adjustments, sales tax charges, and surcharges. Most can add comments when appropriate. Data from the order processing system can be put to a number of uses. The manufacturing department can use order processing information as it does inventory data—to anticipate raw material requirements.

The TPS: Inventory data. Information on the quantity and use of raw materials, work in process, and finished goods is typically stored in computer databases and updated continuously. Inventory reports are an important source of data for the manufacturing MIS—they can identify shortages, poor quality, and too much inventory before such elements become problems.

The TPS: Receiving and inspecting data. As you have learned, some companies have a receiving department responsible for accepting, inspecting, and delivering all incoming raw materials and supplies to the proper departments. Receiving and inspecting data is then input into the TPS. This input may also be used by the manufacturing MIS to report on quantity, quality, and arrival of all raw materials. Manufacturing executives use these reports to monitor and control suppliers and vendors that provide the company with raw materials. Reports that indicate whether raw materials were received on time can also be generated.

The TPS: Personnel data. Personnel data is captured using traditional payroll procedures, like time cards and industrial data collection devices. Once captured, labor and personnel data are used by the manufacturing MIS in a variety of reports and management actions. For example, determining actual labor costs for manufacturing jobs and defining the elapsed time to complete tasks can be accomplished with this type of data.

The TPS: The production process. The production process (involving assembly lines, equipment and machinery, inspection, and maintenance) produces a large amount of data for the manufacturing MIS, such as data on job costs, schedules, production problems, and the utilization of equipment. Data contained in these databases is captured during the production of goods and services. Production schedules, material flow reports, and job costing are typical types of data stored as a result of the production process.

External sources. External sources of data exist for the manufacturing MIS. New manufacturing processes and design techniques can be obtained from outside companies, journals and other publications, and seminars as well as the Internet. These outside sources can keep the company current with the latest manufacturing and fabrication processes and techniques. General economic data on labor supply and inflation can help predict labor and raw material costs. Informal sources, including friends, professional organizations, and business associates, can assist in getting information about the competition, enhanced production processes, and new customers and clients.

Manufacturing MIS Subsystems and Outputs

The subsystems and outputs of the manufacturing MIS monitor and control flow of materials, products, and services through the organization. The objective of the manufacturing MIS is to produce products that meet customer needs—from the raw materials provided by suppliers to finished goods and services delivered to customers—at the lowest possible cost. The activities of the manufacturing MIS subsystems support value-added business processes. As raw materials are converted to finished goods, the manufacturing MIS monitors the process at almost every stage. Take a car manufacturer that converts raw steel, plastic, and other materials into a finished automobile. In this case, the manufacturer has added thousands of dollars of value to the raw materials used in assembling the car. If the manufacturing MIS also lets the manufacturer provide customized paint colors on any of its models, it has further added value (although less tangible) by ensuring a direct customer fit. In doing so, the MIS helps provide the company the edge that can differentiate it from competitors. The success of an organization can depend on the manufacturing function. Some common information subsystems and outputs used in manufacturing are discussed next.

Design and engineering. During the early stages of product development, engineering departments are involved in many aspects of design. The size and shape of parts, the way electrical components are attached to equipment, the placement of controls on a product, and the order in which parts are assembled into the finished product are decisions made with the help of design and engineering departments. In some cases, computer-assisted design (CAD) assists this process. CAD can be used to determine how an airplane wing or fuselage will respond to various conditions and stresses while in use. The data from design and engineering can also be used to identify problems with existing products and help develop new products. For example, Boeing uses a CAD system to develop a complete digital blueprint of an aircraft before it ever begins its manufacturing process. As mockups are built and tested, the digital blueprint is constantly revised to reflect the most current design. Using such technology helps Boeing reduce its manufacturing costs and reduce the time required to design a new

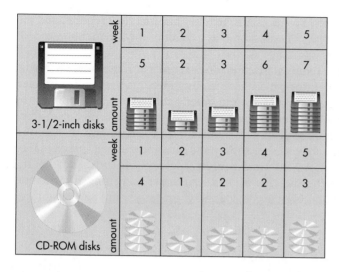

FIGURE 9.7

A master production schedule for computer disks and CDs indicates the quantity of each to be produced each week in thousands.

sensitivity analysis

analysis that allows a manager to determine how the production schedule would change with different assumptions concerning demand forecasts or costs figures

economic order quantity (EOQ)

a method used to determine the quantity of inventory that should be reordered to minimize total inventory costs

reorder point (ROP)

the critical inventory quantity level

aircraft. Bottle manufacturers use CAD to develop 3-dimensional models of their product so that they can be reviewed for aesthetics, stability, ease of handling, etc.

Master production scheduling. The overall objective of master production scheduling is to provide detailed plans for both short-term and long-range scheduling of manufacturing facilities (Figure 9.7). Master production scheduling software packages can include forecasting techniques that attempt to determine current and future demand for products and services. After current demand has been determined and future demand has been estimated, the master production scheduling package can determine the best way to engage the manufacturing facility and all its related equipment.

The result of the process is a detailed plan that reveals a schedule for every item that will be manufactured. Typically, the master production scheduling program determines total output for future periods by number of units or dollar equivalents. Most programs also perform **sensitivity analysis**, which allows a manager to determine how the production schedule would change with different assumptions concerning demand forecasts or cost figures.

The production schedule is critical to the entire manufacturing process. Information generated from this application is used with all aspects of production and manufacturing. Inventory control, labor force planning, product delivery, and maintenance programs depend on information generated from the master production schedule.

Inventory control. An important key to the manufacturing process is inventory control. Great strides have been made in developing cost-effective inventory control programs and software packages that allow automatic reordering, forecasting, generation of shop documents and reports, determination of manufacturing costs, analysis of budgeted costs versus actual costs, and the development of master manufacturing schedules, resource requirements, and plans. Many inventory control programs contain mathematical inventory control equations. Most of these formulas attempt to determine how much and when to order.

One method of determining how much inventory to order is called the **economic order quantity (EOQ)**. This quantity is determined in such a way as to minimize the total inventory costs. The "When to order?" question is based on inventory usage over time. Typically, the question is answered in terms of a **reorder point (ROP)**, which is a critical inventory quantity level. A reorder point is an excellent example of a parameter for an exception report. When the inventory level for a particular item falls to the reorder point, or critical level, a report might be output so that an order is immediately placed for the EOQ of the product. As companies improve the speed and accuracy of their inventory management systems, they also aim to cut the lead time required to replace items.

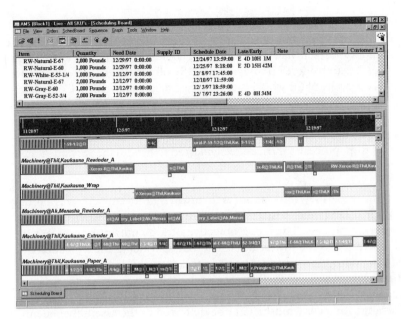

Production planners can interact in real time with the manufacturing schedule using the dynamic scheduling board in Manugistics' software solution for supply chain optimization.
(Source: Image copyright Manugistics.)

material requirements planning (MRP)

a set of inventory control techniques that help coordinate thousands of inventory items when the demand for one item is dependent on the demand for another

bill of materials

the description of the parts needed to make final products

manufacturing resource planning (MRPII)

an integrated, companywide system based on network scheduling that enables people to run their business with a high level of customer service and productivity, while lowering costs and inventories

Some inventory items are dependent on one another. For example, most automobiles come equipped with four tires and a spare, an engine, a transmission, and so on. The engine can be further broken down into the fuel injector, intake manifold, exhaust manifold, engine block, and so forth. Each of these inventory components can be further broken down into a specific number of nuts, bolts, and other pieces of hardware. As soon as a company knows that it wants to produce one automobile, it can directly compute all the subparts and subassemblies needed to produce that automobile. Today, sophisticated inventory control programs help coordinate thousands of inventory items when the demand for one item is dependent on the demand for another. These inventory techniques are called **material requirements planning (MRP)**. The basic goal of MRP is to determine when finished products are needed, then to work backward in determining deadlines and resources needed to complete the final product on schedule. This type of planning requires a knowledge of the **bill of materials**, which is a description of the parts needed to make final products. Once this is known, MRP analysis can determine the requirements for a subassembly or part. These models have saved manufacturing companies millions of dollars in inventory costs by reducing inventory levels while minimizing stockouts and delays due to shortages.

Most inventory control programs can be modified to allow managers to implement different inventory policies. Stocking policies, reorder points, and order quantities can be varied to provide good customer service while minimizing inventory costs.

MRPII. Manufacturing resource planning (MRPII) refers to an integrated, companywide system based on network scheduling that enables people to run their business with a high level of customer service and productivity, while lowering costs and inventories. MRPII is broader in scope than MRP; thus, the latter has been dubbed "little MRP." MRPII systems were started in the 1960s by individuals interested in researching why some companies were doing better than others. They found that the most successful companies had implemented very disciplined approaches to answering what are called the universal manufacturing questions. For finished products, these questions are: What does the customer need (demand forecasting)? What do we have (inventory control)? and What are we going to make (production planning)? For materials needed to make the product, these questions are: What does it take to make it (bill of materials)? What do we have (inventory control)? and What do we have to get (supply planning)? MRPII places a heavy emphasis on planning so that manufacturing processes ensure that the right product is in the right place at the right time.

In the mid-1990s, software vendors (SAP, PeopleSoft, Computer Associates, and Oracle) began to develop integrated MIS software and ERP systems to encompass the scope of the work processes associated with MRPII. Companies worldwide have begun to implement a manufacturing MIS based on the MRPII approach. However, the cost has been very high. Not only must companies pay upwards of several hundred thousand dollars for the software, the internal work processes must be reengineered to fit with the very integrated processes required with MRPII.

Just-in-time inventory and manufacturing. High inventory levels on the factory floor mean higher costs, the possibility of damage, and an inefficient manufacturing process. Thus, one objective of a manufacturing MIS is to control inventory to the lowest levels without sacrificing the availability of finished products. One way to do this is to adopt the **just-in-time (JIT) inventory approach**. With this approach, inventory and materials are delivered just before they are used in a product. A JIT inventory system would arrange for a car windshield to be delivered to the assembly line only a few moments before it is secured to the automobile, rather than having it sitting around the manufacturing facility while the car's other components are being assembled.

The JIT approach was originally implemented by many Japanese companies. One JIT approach is called *kanban*, Japanese for "card." In most cases, *kanban* involves a dual-card system. A conveyance card is used by workers and factory-floor employees; it is taken along with parts or inventory items to the work area. The production card initiates new production of inventory items. Workers and employees judge their own component consumption rates; when they start to run out of items, a production card is pulled, which causes component inventory to be produced or delivered to the factory floor. The automotive industry uses similar procedures to approach "zero inventory," in which parts that are needed arrive just in time to be added to the cars in various stages of production. For successful implementation of JIT, employees have to learn that it is not very smart to produce 100,000 parts when there is an immediate customer demand for only 10,000. Firms must make a large investment in training when making the shift to JIT manufacturing and integrated MIS systems.[4]

While JIT has many advantages, it also renders firms more vulnerable to process disruptions. If even one parts plant is shut down, an assembly plant might be left with no inventory to continue production. However, many organizations have had great success with JIT inventory control, and the overall concept has been expanded to include other aspects of the production process. Thus, materials (including raw materials and supplies) are delivered when they are needed rather than being stored for months or days in advance. The JIT manufacturing approach requires better coordination and cooperation between suppliers and manufacturing companies, substantially reducing inventory costs.

Wal-Mart is the leader in the area of inventory management, supplier relations, and working with vendors to make the whole supply chain more

just-in-time (JIT) inventory approach

an approach by which inventory and materials are delivered just before they are used in a product

A kanban card at Toyota signals that a delivery of parts is needed on the production line. (Source: Courtesy of Toyota Motor Manufacturing, Kentucky, Inc.)

computer-assisted manufacturing (CAM)

a system that directly controls manufacturing equipment

computer-integrated manufacturing (CIM)

systems that use computers to link the components of the production process into an effective system

flexible manufacturing system (FMS)

an approach that allows manufacturing facilities to rapidly and efficiently change from making one product to another

efficient. It completed a highly successful collaborative forecasting and replenishment pilot with Warner-Lambert and several technology firms. The goal was to improve forecasting by sharing customer and product information between suppliers and retailers over the Internet.[5]

Process control. Managers can use a number of technologies to control and streamline the manufacturing process. For example, the computer can be used to directly control manufacturing equipment using systems called **computer-assisted manufacturing (CAM)**. CAM systems have the ability to control drilling machines, assembly lines, and more. Some of them operate quietly, are easy to program, and have self-diagnostic routines to test for difficulties with the computer system or the manufacturing equipment.

Computer-integrated manufacturing (CIM) involves the use of computers to link the components of the production process into an effective system. CIM's goal is to tie together all aspects of production, including order processing, product design, manufacturing, inspection and quality control, and shipping. CIM systems also increase efficiency by coordinating the actions of various production units. In some areas, CIM is used for even broader functions. For example, it can be used to integrate all organizational subsystems, not just the production systems. In automobile manufacture, design engineers can have their ideas evaluated by financial managers before new components are built to see if they are economically viable. This saves not only time but also money.

Ford Motor implemented an ambitious computer-aided design project called C3P. It is based on a blend of CAD (computer-aided design) and CAM software from Structural Dynamics and Ford's own computer-aided engineering (CAE) applications. C3P enables Ford to develop prototypes of trucks and cars faster using computers than by building test models from the ground up. For example, it used to take two to three months to build, assemble, and test a prototype of a car's chassis—using the C3P technologies, Ford can do all that in less than two weeks. The new system detected a problem in the assembly of a European small car while it was still being designed. Had the error slipped out of design and into the factory, it could have cost Ford up to $60 million.[6]

A **flexible manufacturing system (FMS)** is an approach that allows manufacturing facilities to rapidly and efficiently change from making one product to making another. In the middle of a production run, for example, changes can be made to the production process to make a different product or change manufacturing materials. Often a computer is used to direct and implement the changes. By using an FMS, the time and cost to change manufacturing jobs can be substantially reduced and companies can react quickly to market needs and competition.

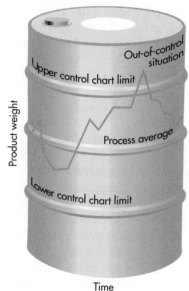

Product weight

Out-of-control situation

Upper control chart limit

Process average

Lower control chart limit

Time

FIGURE 9.8

Industrial Control Chart

This chart is used to monitor product quality in an industrial chemical application. Product variances exceeding certain tolerances cause those products to be rejected.

quality control

a process that ensures that the finished product meets the customers' needs

FMS is normally implemented using computer systems, robotics, and other automated manufacturing equipment. New product specifications are fed into the computer system, and the computer then makes the necessary changes. Although few companies have a fully implemented FMS system, recent years have seen increasing use of the overall FMS approach.

Quality control and testing. With increased pressure from consumers and a general concern for productivity and high quality, today's manufacturing organizations are placing more emphasis on **quality control**, a process that ensures that the finished product meets the customers' needs. For a continuous process, control charts are used to measure weight, volume, temperature, or similar attributes (Figure 9.8). Then, upper and lower control chart limits are established. If these limits are exceeded, the manufacturing equipment is inspected for possible defects or potential problems.

When the manufacturing operation is not continuous, sampling plans can be developed that allow the producer or consumer to accept or reject one or more products. Acceptance sampling can be used for material as simple as nuts and bolts or as complex as airplanes. The development of the control chart limits and the specific acceptance sampling plans can be fairly complex. Therefore, a number of quality-control software programs have been developed to generate the appropriate acceptance-sampling plans and control chart limits.

Whether the manufacturing operation is continuous or discrete, the results from quality control are analyzed closely to identify opportunities for improvements. Teams using the total quality management (TQM) or continuous improvement process (see Chapter 2) often analyze this data to increase the quality of the product or eliminate problems in the manufacturing process. The result can be a cost reduction or increase in sales.

Information generated from quality-control programs can help workers locate problems in manufacturing equipment. Quality-control reports can also be used to design better products. With the increased emphasis on quality, workers should continue to rely on the reports and outputs from this important application.

A MARKETING MANAGEMENT INFORMATION SYSTEM

marketing MIS

an MIS that supports managerial activities in product development, distribution, pricing decisions, and promotional effectiveness

A **marketing MIS** supports managerial activities in product development, distribution, pricing decisions, promotional effectiveness, and sales forecasting. Figure 9.9 shows the inputs, subsystems, and outputs of a typical marketing MIS.

Inputs to the Marketing MIS

More than the other functional areas, the marketing MIS relies on external sources of data. These sources include the Internet, competition, customers, journals and magazines, and other publications. Certain internal data sources are also important. This section gives an overview of these inputs.

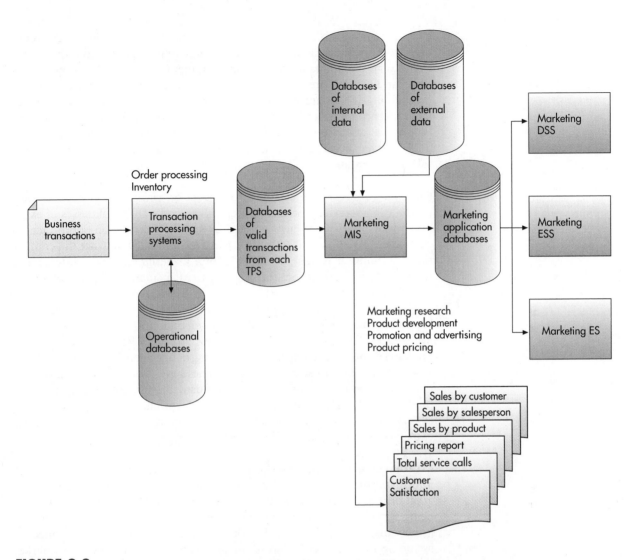

FIGURE 9.9

Overview of a Marketing MIS

Strategic plan or corporate policies. Marketing relies on the organization's strategic plan for sales targets and projections. For example, a strategic plan might indicate that sales are expected to increase by a steady 8 percent for the next five years. A marketing MIS report from such a firm might organize current sales performance in terms of this strategic goal. In addition to sales projections, the strategic plan can contain information about projected needs for the sales force, product and service pricing, distribution channels, promotion, and new product features. The strategic plan can provide a framework from which to analyze marketing information and to make marketing decisions.

The TPS. The TPS contains a large amount of sales and marketing data on products, customers, and the sales force. Data from an E-commerce system can be another important input to the marketing MIS. Sales data on products can reveal which products are selling well and which ones are slow sellers. Reports summarizing order data might include sales activity by customer, product, and area of the country. The marketing MIS might organize this information for use in formulating sales, developing promotional plans, or making product development decisions. Customer sales data can help determine which customers are contributing most to overall profitability. This data can also be analyzed to develop special customer incentive programs and to help the sales force reach and serve customers. Sales force effectiveness can also be monitored from data captured in the TPS, so that bonus and incentive programs can reward top-performing sales representatives.

External sources: The competition. Data on competitors—such as new products and services, pricing strategies, strengths and weaknesses of existing product lines, packaging, marketing, and the distribution of products to customers—is important to most organizations. Knowing what the competition is doing or is likely to do is helpful in designing new products and services. This information can also be used in marketing and selling an existing product line.

Competitive data can be obtained from many sources. Marketing materials, brochures, sales programs offered by the competition, and the Internet are typical sources. Customers and other business associates also provide data. The process of obtaining competitive marketing information is often called marketing intelligence. In some cases, companies have even used illegal means to obtain data on the competition—stealing data, illegally purchasing restricted products, and stealing industrial plans and blueprints are examples. This type of corporate espionage has been used by some companies and a few countries to gain competitive or technological advantages.

External sources: The market. An additional external source of useful information for the marketing MIS is the market for a firm's products. A great deal of useful data can be obtained from the TPS for markets already being served by the firm, but insights into buyer behaviors and preferences in new markets can be obtained only from sources outside the company, including the Internet. As we shall see, features of the functional market research system allow firms to obtain this data.

Marketing MIS Subsystems and Outputs

In the past two decades logistics, financial, and manufacturing business functions have benefited dramatically from process improvement and automation. Many factors, including intense competition and an increasingly sophisticated and time-constrained audience, are forcing companies to rethink their approach to marketing. As a result, traditional marketing techniques of "one-way" mass marketing are giving way to technology-based, interactive techniques aimed at market micro-segments. To improve overall marketing effectiveness, companies are applying the power of information technology to their marketing processes. As a result, many companies view the Web as holding significant potential. Hundreds or thousands of visitors may access a business Web site each day. Many of these individuals are potential customers, but only if marketing can cost-effectively capture, analyze, and respond to their needs. In this new environment, the challenge for the marketer is to be able to:

1. Quickly plan, create, and deliver the types of marketing programs different audiences require.
2. Capture market feedback on communication effectiveness and market preferences.
3. Fine-tune the program to better meet market needs.
4. Respond quickly to requests for information and action.

Subsystems for the marketing MIS include marketing research, product development, promotion and advertising, and product pricing. These subsystems and their outputs help marketing managers and executives increase sales, reduce marketing expenses, and develop plans for future products and services to meet the changing needs of customers.

As with other functional MISs, the marketing MIS adds value to business processes. From research and product development to the final distribution of goods and services, the marketing MIS helps provide superior goods, better service, and timely information to guide customers' purchasing decisions. For example, the marketing MIS can be used to analyze appropriate product prices. Suppose a bicycle manufacturing company wants to sell bikes at a low price while maintaining a certain level of profit. A marketing MIS can perform a price analysis and produce a report to guide the pricing decision. If the report shows that a lower price will increase sales and total profits, a manager might decide to reduce prices on the best-selling bikes. This decision, which was aided by information from the marketing MIS, translates into direct value to the customer. Marketing MISs also add value by helping companies analyze market data to match customer product and service requirements. For example, Kmart implemented a client/server marketing MIS to identify and distribute the right merchandise mix more effectively to its more than 2,100 stores.[7] As you can see from these examples, the marketing MIS is customer driven. An effective system supports the strategic sales and profit goals of a company by helping it win and maintain customer loyalty.

Marketing research. Surveys, questionnaires, pilot studies, and interviews are popular marketing research tools. The purpose of marketing research is to conduct a formal study of the market and customer preferences. Marketing research can identify prospects (potential future customers) as well as the features that current customers really want in a good or service. Such attributes as style, color, size, appearance, and general fit can be investigated through marketing research. Pricing, distribution channels, guarantees and warranties, and customer service can also be determined. Once entered into the marketing MIS, data collected from marketing research projects is manipulated to generate reports on key indicators like customer satisfaction and total service calls. Reports generated by the marketing MIS help marketing managers be better informed to help the organization meet its performance goals.

The Internet is changing the way many companies think about market research. Conventional methods of collecting data often cost millions of

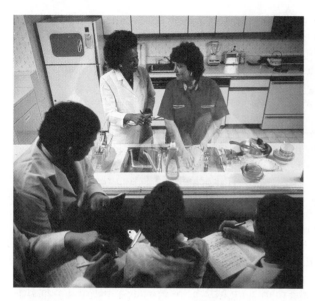

Marketing research data yields
valuable information for the
development and marketing of
new products.
(Source: Courtesy of Colgate-Palmolive
Company; photo: John S. Abbott.)

FIGURE 9.10

Typical Supply and Demand
Curve for Pricing Analysis

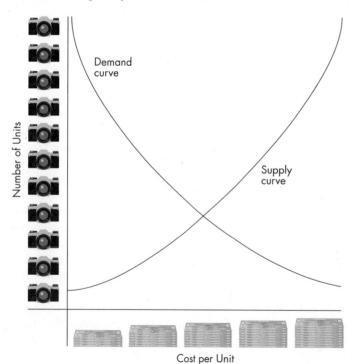

dollars. For a fraction of these costs, companies can put up Internet information servers and launch discussion lists on topics that their customers care about. These information sites must be well designed or they will not be visited; a frequently visited site can provide feedback worth a fortune. Companies that are viewed as credible, not just clever, will win enormous advantages. Presence and intelligent interaction, not just advertising, are the keys that will unlock the commercial opportunities on-line.[8]

Product development. Product development involves the conversion of raw materials into finished goods and services and focuses primarily on the physical attributes of the product. Many factors, including plant capacity, labor skills, engineering factors, and materials are important in product development decisions. In many cases, a computer program is used to analyze these various factors and to select the appropriate mix of labor, materials, plant and equipment, and engineering designs. Make-or-buy decisions can also be made with the assistance of computer programs.

Promotion and advertising. One of the most important functions of any marketing effort is promotion and advertising. Product success is a direct function of the types of advertising and sales promotion done. The size of the promotion budget and the allocation of this budget to various promotional campaigns are important factors in deciding on the type of campaigns that will be launched. Television coverage, newspaper ads, Internet ads, promotional brochures and literature, and training programs for salespeople are all components of these campaigns. Because of the time and scheduling savings they offer, computer programs are used to set up the original budget and to monitor expenditures and the overall effectiveness of various promotional campaigns.

Boise Cascade Office Products is a $2.6 billion distributor of office and computer supplies, office furniture, and paper products for Fortune 1000 companies and government entities. Boise conducts its customer and relationship management through database marketing, meaning the company looks for its best customers by collecting and analyzing data about buying habits. The company implemented a new marketing MIS to provide a consistent set of data companywide to deliver promotion information to its marketing analysts and managers, general managers, and sales managers. The MIS enables managers to act on fast-moving business trends and market

(a) Sales by Product

Product	August	September	October	November	December	Total
Product 1	34	32	32	21	33	152
Product 2	156	162	177	163	122	780
Product 3	202	145	122	98	66	633
Product 4	345	365	352	341	288	1,691

(b) Sales by Salesperson

Salesperson	August	September	October	November	December	Total
Jones	24	42	42	11	43	162
Kline	166	155	156	122	133	732
Lane	166	155	104	99	106	630
Miller	245	225	305	291	301	1,367

(c) Sales by Customer

Customer	August	September	October	November	December	Total
Ang	234	334	432	411	301	1,712
Braswell	56	62	77	61	21	277
Celec	1,202	1,445	1,322	998	667	5,634
Jung	45	65	55	34	88	287

FIGURE 9.11

Reports Generated to Help Marketing Managers Make Good Decisions

(a) This sales-by-product report lists all major products and their sales for the period August-December. (b) This sales-by-salesperson report lists total sales for each salesperson for the same time period. (c) This sales-by-customer report lists sales for each customer for the period. Like all MIS reports, totals are provided automatically by the system to show managers at a glance the information they need to make good decisions.

opportunities such as customer profitability and customer targeting for electronic ordering. It also gives them the basis for making precise assessments of customer and product contributions, location trends, customer buying patterns, and national accounts that lead to highly profitable operations and competitive position.[9]

Product pricing. Product pricing is another important and complex marketing function. Retail price, wholesale price, and price discounts must be determined. A major factor in determining pricing policy is an analysis of the demand curve, which attempts to determine the relationship between price and sales. Most companies try to develop pricing policies that will maximize total sales revenues. This is usually a function of price elasticity. If the product is very price sensitive, a reduction in price can generate a substantial increase in sales, which can result in higher revenues. A product that is relatively insensitive to price can have its price substantially increased without a large reduction in demand. Computer programs exist that help determine price elasticity and various pricing policies, such as supply and demand curves for pricing analysis (Figure 9.10). Typically, the marketing executive has the ability to make alterations in price on the computer system, which analyzes price changes and their impact on total revenues. The rapid feedback now obtainable through computer communications networks enables managers to determine the results of pricing decisions much more quickly than in the past. This facilitates more aggressive pricing strategies, which can be quickly adjusted to meet the market needs. Read the "FYI" box to learn about a new class of software that employs the Web and Internet technology designed to automate many of the marketing processes.

Sales analysis is also important to identify products, sales personnel, and customers that contribute to profits and those that do not. Several reports can be generated to help marketing managers make good sales decisions (Figure 9.11). The sales-by-product report lists all major products and their sales for a period of time, such as a month. This report shows which products are doing well and which ones need improvement or should be discarded altogether. The sales-by-salesperson report lists total sales for each salesperson for each week or month. This report can also be subdivided by product to show which products are being sold by each salesperson. The sales-by-customer report is a tool to use to identify high- and low-volume customers.

FYI

Rubric Revolutionizes Marketing MIS Software

Corporate marketing departments are starting to do what has already been done on the sales side: use MIS to support their activities and boost profits. Marketing MIS software can improve marketing efficiency and better track and qualify customer leads for the sales force. Other capabilities include responding to and tracking customer information requests, determining when an early lead is a qualified sales prospect, evaluating which marketing programs are most effective in generating sales, and tracking which customers respond to which kinds of events. The software helps companies to develop more comprehensive databases of customer preferences over time.

Rubric Inc. is one company offering a marketing MIS that helps organizations plan, implement, and measure their marketing campaigns. Its EMA (Enterprise Marketing Automation) product integrates marketing processes and data into a Web-based package. Written in Java, EMA streamlines the process by which organizations identify, cultivate, and retain customers through such features as time- and event-based triggers that communicate with sales prospects through mailings, faxes, or phone calls. The software also automates responses to e-mail inquiries by extracting information from a database of frequently asked questions. In addition, users can measure marketing effectiveness by tracking marketing expenses against revenue generated.

As such products come to market, some companies may find that integrating marketing MIS software into existing Web, sales, and other applications is not easy. Companies must be ready to examine and revamp their marketing practices before computerizing them. If you take a bad process and automate it, all you have is an automated bad process.

DISCUSSION QUESTIONS

1. Why might many companies be interested in this new approach to marketing?
2. Why do you think there may be difficulties in integrating such new marketing MIS software with existing Web, sales, and other applications? List several issues.

Sources: Adapted from Mark Hammond, "Startup Rubric Offers Marketing Automation Software," *PC Week Online,* April 10, 1998; Sharon Machlis, "Taking Automated Road to Market," *Computerworld,* April 6, 1998; and press release "Rubric Delivers First Web-Based Enterprise Marketing Automation Application," March 13, 1998 found at the Web site for Rubric, Inc. at http://www.rubricsoft.com, accessed September 3, 1998.

A HUMAN RESOURCE MANAGEMENT INFORMATION SYSTEM

human resource MIS

an MIS that is concerned with all of the activities related to employees and potential employees of the organization

A **human resource MIS,** also called the personnel MIS, is concerned with activities related to employees and potential employees of the organization. Because the personnel function relates to all other functional areas in the business, the human resource (HR) MIS plays a valuable role in ensuring organizational success. Some of the activities performed by this important MIS include workforce analysis and planning, hiring, training, job and task assignment, and many other personnel-related issues. An effective human resource MIS will allow a company to keep personnel costs at a minimum while servicing the required business processes needed to achieve corporate goals. Figure 9.12 shows some of the inputs, subsystems, and outputs of the human resource MIS.

The human resource software modules of an ERP provide the ability to maximize the potential of the workforce through effective recruitment, staffing, training, compensation, benefits, and planning. It provides a unified picture of the human resources within the enterprise. Everyone involved in managing the workforce benefits from easy access to the information they need to make timely and informed decisions.

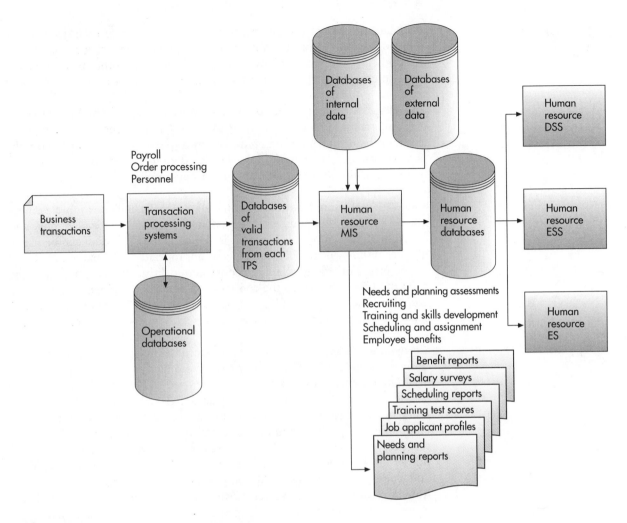

FIGURE 9.12

Overview of a Human Resource MIS

Inputs to the Human Resource MIS

Strategic plan or corporate policies. The strategic plan often contains critical human resource objectives and policies. Many companies are embarking on total quality programs that require additional training, employee empowerment (greater worker decision-making abilities and responsibilities), and the use of cross-functional teams (groups of workers from different areas or departments). Such a program initiated by top-level managers can have a profound impact on the human resource area. In addition, corporate restructuring, decentralization, mergers, and strategic alliances between organizations are likely to have a significant impact on human resource and personnel functions.

The TPS: Payroll data. The human resource area is often responsible for payroll costs, healthcare plans, and other fringe benefits. Many of these programs require data directly from the transaction processing system. Hours worked and pay rate for each employee are used to compute gross wages.

This data can be used to determine total payroll costs for each department and the organization as a whole. The amount that the organization invests in healthcare, retirement plans, savings programs, and other fringe benefits can also come directly from the transaction processing system.

The TPS: Order processing data. Sales ordering data is often used to provide important personnel planning data. Orders for products for the next few months or years can be converted into personnel needs. A large military contractor, for example, can use such data to predict its need for engineers, clerical staff, designers, and other personnel.

The TPS: Personnel data. Personnel data usually includes job classifications and skills, previous work experience, performance evaluations, and other important information stored in the database. This data can be useful in determining which employees may be appropriate for certain types of jobs or projects. For example, a large construction firm can use personnel data from the TPS to help plan the construction of a shopping mall or new office complex. The personnel TPS can be searched for the required number of qualified plumbers, electricians, bricklayers, designers, and so on.

External sources. Several important external data sources are used by the human resource MIS. The salaries paid to employees at other organizations and data on employment statistics can be used to help determine wages or salaries for employees in the organization. A research-and-development company, for example, can use the Internet to investigate what other R&D companies are paying research assistants to determine appropriate salaries for new employees. In addition, organizations must learn and follow numerous government regulations. Local, regional, and national unions and worker organizations are also important external sources of data for the human resource MIS. Read the "Ethical and Societal Issues" box to learn more about some of the issues in protecting sensitive personnel data.

ETHICAL AND SOCIETAL ISSUES

Protecting Patient-Care Data

Health information and medical records reveal some of the most intimate aspects of a person's life. In addition to diagnostic and testing information, a patient's medical records include the details of family history, genetic testing, history of diseases and treatments, history of drug use, sexual orientation and practices, and testing for sexually transmitted diseases. Subjective remarks about a patient's demeanor, character, and mental state are also part of the record.

An individual's medical records are the primary source for much of the healthcare information sought by parties outside the healthcare relationship. This data is important because healthcare information can influence decisions about an individual's access to credit, admission to educational institutions, and the person's ability to secure employment and obtain insurance. Inaccuracies in the information, or its improper disclosure, can deny an individual access to these basic necessities of life and can threaten an individual's personal and financial well-being.

At the same time, accurate and comprehensive information is critical to the quality of healthcare delivery and to the physician-patient relationship. Many believe that the efficacy of the healthcare relationship depends on the patient understanding that the information recorded by a physician will not be disclosed. Without these assurances, many patients might refuse to provide physicians with certain types of information needed to render appropriate care.

Privacy and confidentiality of patient medical information are receiving increased attention in the legislative arena and in the media. As the electronic storage and transmission of health information becomes more widespread, public concern over the use and security of patient-identifiable data has grown. This has triggered the promotion of new and expanded patient confidentiality legislation and regulation.

The Health Insurance Portability and Accountability Act of 1996 (HIPAA) was signed by President Clinton in August 1996. The act is designed to protect health insurance coverage for workers and their families when they change or lose their jobs. It also includes provisions that require the Secretary of Health and Human Services to adopt standards for protecting patient-care data. The Health Care Financing Administration has been charged with coordinating the plan and its implementation. This includes setting standards for unique identifiers for individuals, employers, health plans, and healthcare providers as well as security standards and safeguards for electronic information systems and electronic signatures.

The intent of the new regulations is to make it easier for healthcare providers to share electronic medical records with doctors, insurance companies, and patients. At the same time, these regulations are intended to improve the confidentiality of medical records by creating guidelines for keeping patient information secure. Once the regulations are set, firms will have two to three years to comply with them. Undoubtedly, these new rules will change the way patient information is stored and shared in hospital information systems nationwide. Indeed, for many companies, the new HIPAA regulations will necessitate a major overhaul of their healthcare information systems. Some firms may elect to modify their existing systems to make them HIPAA-compliant while others may elect to convert to an entirely new set of systems. At a minimum, healthcare firms will be required to convert their current patient identification system to one based on the new standard. Software and consulting companies are developing software products and consulting services designed to help companies meet the HIPAA standards.

Discussion Questions

1. What is the motivation for hospitals and patient-care facilities to adopt the emerging HIPAA standards?
2. List some suggestions for how a medical MIS system could protect patient privacy.

Source: Barb Cole-Gomolski, "Hospitals Face Info Overhauls," *Computerworld*, May 25, 1998, pp. 39, 42; The Health Insurance Portability and Accountability Act of 1996 found at the Health Care Financing Administration Web site at http://www.hcfa.gov/regs, last updated March 3, 1998, accessed May 30, 1998; and Louise Sargent, *Journal of Managed Care Pharmacy*, vol. 3, no. 4, July/August 1997.

Human Resource MIS Subsystems and Outputs

Human resource subsystems and outputs range from the determination of human resource needs and hiring through retirement and outplacement. Most medium and large organizations have computer systems to assist with human resource planning, hiring, training and skills inventory, and wage and salary administration. Outputs of the human resource MIS include reports such as human resource planning reports, job application review profiles, skills inventory reports, and salary surveys.

Value can be added to an organization's products and services through the human resource MIS. In the overnight mail industry, for example, proper scheduling during heavy delivery times, such as Christmas, is critical. Reports generated by the human resource MIS can help managers make effective and efficient use of drivers to ensure timely delivery to customers. Similar human resource MIS programs can be used to help trucking companies schedule drivers and help contractors schedule qualified plumbers, painters, and other construction workers. Value comes to customers in the form of on-time delivery or fast, quality construction work.

Human resource planning. One of the first aspects of any human resource MIS is determining personnel and human needs. The overall purpose of this MIS subsystem is to put the right number and kinds of employees in the right jobs when they are needed. Effective human resource planning requires defining the future number of employees needed and anticipating the future supply of people for these jobs. For companies involved with large projects, such as military contractors and large builders, human resource plans can be generated directly from data on current and future projects.

Suppose a construction company obtains a contract from a group of investors to build a 250-unit apartment complex. Forecasting programs and project management software packages can be used to develop reports that describe what people are needed and when they are needed during the entire construction project. A typical output would be a human resource needs and planning report, which might specify that ten employees will be needed in August to pour concrete slabs, and eight carpenters and four painters will be needed in October. Alternatively, a factory might use this report to list total number of employees needed, broken down according to skill level, such as highly technical, technical, semitechnical, and so on.

Personnel selection and recruiting. If the human resource plan reveals that additional personnel are required, the next logical step is recruiting and selection of personnel. This subsystem performs one of the most important and critical functions of any organization, especially in service organizations where employees can define the company's success. Companies seeking new employees often use computers to schedule recruiting efforts and trips and to test potential employees' skills. Some software companies, for example, use computerized testing to determine a person's programming skills and abilities. A human resource MIS can be used to help rank and select potential employees. For every applicant, the results of interviews, tests, and company visits can be analyzed by the system and printed. This report, called a job applicant review profile, can assist corporate recruiting teams in final selection. Some software programs can even analyze this data to help identify job applicants most likely to stay with the company and perform according to corporate standards.

Many companies now use the Internet to screen for job applicants. Applicants use a template to load their resume onto the Internet. HR managers can access these and identify the applicants they are interested in interviewing.

Training and skills inventory. Some jobs, such as programming, equipment repair, and tax preparation, require very specific training. Other jobs may require general training about the organizational culture, orientation, dress standards, and expectations of the organization. Today, many organizations conduct their own training, with the assistance of information systems and technology. Self-paced training can involve computerized tutorials, video programs, and CD-ROM books and materials.

At many companies, training managers are using the corporate intranet as a means to post information about classroom courses, videos, and CD-ROMs. Consolidated Edison, Dow Chemical Company, and The Boston Globe all have developed a corporate intranet site to support training that lists all the courses offered, the requirements for those courses, how to register, and student course evaluation forms.[10]

When training is complete, employees may be required to take computer-scored tests to reveal their mastery of skills and new material. The results of these tests are usually given to the employee's supervisor or boss in the form of training or skills inventory reports. In some cases, skills inventory reports are used for job placement. For instance, if a particular position in the company needs to be filled, managers might wish to hire internally before they recruit. The skills inventory report would help them evaluate current employees to determine their potential for the position. They can also be part of employee evaluations that can determine raises and/or bonuses.

Scheduling and job placement. Scheduling people and jobs can be relatively straightforward or extremely complex. For some small service companies, scheduling and job placements are based on which customers walk through the door. Determining the best schedule for flights and airline pilots, the placement of military recruits to jobs, and determining what truck drivers and equipment should be used to transport materials across the country normally require sophisticated computer programs. In most cases, various schedules and job placement reports are generated. Employee schedules are developed for each employee, showing their job assignments over the next week or month. Job placements are often determined based on skills inventory reports, which show which employee might be best suited to a particular job.

Performance Now! is an intelligent business management software package for writing employee performance evaluations.

(Source: Courtesy of KnowledgePoint.)

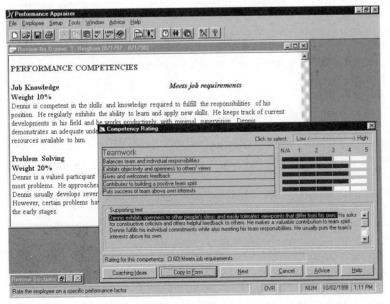

Wage and salary administration. The last of the major human resource MIS subsystems involves determining wages, salaries, and benefits, including medical payments, savings plans, and retirement accounts. Wage data, such as industry averages for positions, can be taken from the corporate database and manipulated by the human resource MIS to provide wage information and reports to higher levels of management. These reports, called salary surveys, can be used to compare salaries with budget plans, the cost of salaries versus sales, and the wages required for any one department or office. The reports also help show backup of key positions in the company. Wage and salary administration also entails designing retirement programs for employees. Some companies use computerized retirement programs to help employees gain the most from their retirement accounts and options.

Many organizations provide employee self-service processes designed specifically for access via the Web or the corporate intranet. These processes require little or no user training, knowledge of HR applications, or familiarity with the business transactions that the software automates. For users, which could span employees, line managers, or other third parties, the self-service interface is a seamless integration of the organization's procedures and the ease-of-use offered by access through the Web. For example, employees may use the self-service applications to change personal information, enroll in training courses, perform self-assessments, apply for a job, or update their resume anytime, anywhere. They may also query their own records to check on such information as year-to-date reports on training, vacation, or overtime.

Read the "Making a Difference" box to see how one company successfully implemented a human resource MIS system.

OTHER MANAGEMENT INFORMATION SYSTEMS

In addition to finance, manufacturing, marketing, and human resource MISs, some companies have other functional management information systems. For example, most successful companies have well-developed accounting functions and supporting accounting MISs. Also, many companies make use of geographic information systems for presenting data in a useful form.

Accounting MISs

accounting MIS

an MIS that provides aggregate information on accounts payable, accounts receivable, payroll, and many other applications

In some cases, accounting works closely with financial management. An **accounting MIS** performs a number of important activities, providing aggregate information on accounts payable, accounts receivable, payroll, and many other applications. The organization's TPS captures accounting data, which is also used by most other functional information systems.

Some smaller companies hire outside accounting firms to assist them with their accounting functions. These outside companies produce reports for the firm using raw accounting data. In addition, many excellent integrated accounting programs are available for personal computers in small companies. Depending on the needs of the small organization and its personnel's computer experience, using these computerized accounting systems can be a very cost-effective approach to managing information.

MAKING A DIFFERENCE
Automating Recruiting at Ornda Health Corporation

The human resources (HR) department often gets groans from employees and managers. Inefficiency and ineffectiveness are terms often used to describe this important but often neglected corporate function. Despite the complaints, the HR department remains a critical function. Today, some companies are transforming the HR department. Instead of concentrating on staffing and compensation, they are looking for HR to add a significant value to the organization in several ways. First, the HR department can bring value by helping the structure, organization, and execution of work. Second, the HR department can truly represent employees, making certain that employee issues and concerns are communicated to upper management. Finally, HR can form a partnership with senior management to help the company change and update work practices to stay competitive. This change and updating includes the globalization of work, the increasing use of technology, and the ability to tap into the intellectual capital of the workforce.

To capitalize on the potential of the HR function, corporations are increasingly investigating ways to automate HR activities. Traditionally, the IS department leads the way, but not in all cases. With huge backlogs of unfinished work, some IS departments resist initiating yet another project. This was the case with Ornda Health Corporation. In addition, Ornda was involved with a recent merger, which tends to increase the workload for the merged IS departments. For Ornda HealthCorp, the motivation to automate the HR function did not come from the IS department but from a recruiter. At the age of 39, Carolyn Schneider was hired by Ornda as the director of executive recruitment. With experience in automating the recruiting function at another company, Carolyn was confident that she could

also automate some aspects of the HR department at Ornda. She decided to use a human asset management software package that ran on a Unix operating system. Unfortunately, the IS department at Ornda did not support Unix and decided not to support the automation project. Getting support from top level management, Carolyn decided to develop a better HR system on her own. With theater as her major in college, Carolyn was inexperienced in executing an automation project. The object of the automation project was to match job openings at Ornda with resumes from incoming job applicants. After learning Unix, she was successful in installing the software package at a cost of $100,000. In the first year of operation, Ornda estimated that the new HR software package saved the company about $1 million. In the second year, another $250,000 was saved—not a bad return from a non-IS person with an interest in theater. In addition, her success showed other managers and executives at Ornda that they could also get projects successfully completed.

DISCUSSION QUESTIONS

1. How was the HR MIS automation project able to save money and improve recruiting for Ornda?
2. Would you undertake an automation project without help or support from the IS department? What are the advantages and disadvantages of this approach?

Sources: Thomas Hoffman, "Recruiter Drives Automation Project," *Computerworld*, January 19, 1998, p. 41; Dave Ulrich, "A New Mandate for Human Resources," *Harvard Business Review*, January/February 1998, p. 124; Staff, "1997 Hospital Mergers, Acquisitions, Joint Ventures, and Long Term Leases," *Modern Healthcare*, January 12, 1998, pp. 19.

Geographic Information Systems

geographic information system (GIS)

a computer system capable of assembling, storing, manipulating, and displaying geographically referenced information, i.e., data identified according to its location

Increasingly, managers want to see data presented in graphical form. A **geographic information system (GIS)** is a computer system capable of assembling, storing, manipulating, and displaying geographically referenced information, i.e., data identified according to its location. A GIS enables users to pair predrawn maps or map outlines with tabular data to describe aspects of a particular geographic region. For example, sales managers may want to plot total sales for each county in the states they service. Using a GIS, they can specify that each county be drawn with a degree of shading that indicates the relative amount of sales—no shading or light shading represents little or no sales, and deeper shading represents more sales. Because the GIS works with any data represented in tabular form,

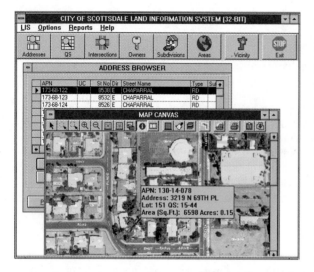

In Scottsdale, Arizona, many city functions such as zoning decisions, code enforcement, emergency response planning, work management, flood control, building permits, and inspections are supported by the GIS.
(Source: Courtesy of the City of Scottsdale.)

this capability is finding its way into spreadsheets. For example, Excel and Lotus include a mapping tool that lets you plot spreadsheet data as a demographic map. Such applications show up frequently in scientific investigations, resource management, and development planning. Retail, government, and utility organizations are frequent users of GIS. Retail chains, for example, need spatial analysis to determine where potential customers are located and where their competition is located.[11] (Read the "Technology Advantage" box to find out how GIS technology is being used by the water districts in San Diego County.)

Salt Lake City used a GIS to enable disaster planners to estimate emergency response times in the event of a natural disaster. The road network from a U.S. Geological Service digital line graph includes information on the types of roads, which range from rough trails to divided highways. The locations of fire stations were plotted on the road network, and a GIS function called network analysis was used to calculate the time required for emergency vehicles to travel from the fire stations to different areas of the city. The network analysis function considered the distance from the fire station and the speed of travel based on road type. The analysis showed that most of the area within the city can be served in less than eight minutes because of the distribution and density of fire stations and the continuous network of roads.[12]

Competitive pressures require that organizations operate with employees who make decisions with as much information as possible in a timely manner. Accuracy and speed are critical to organizational survival. Resources such as MISs help organizations formulate those decisions efficiently by tapping data and generating reports. Knowing what is happening as soon as possible can help managers monitor trends and fine-tune company functions, activities that ultimately support the organization's strategic goals.

Some organizations have special needs that go beyond the traditional MIS. An engineering company that constructs bridges might need a specialized MIS to get demand reports on the availability of building materials and their characteristics, including stress and strength data. A national charitable organization might require a specialized MIS to track current and potential donors and charitable programs that are under way. Although these specialized MISs perform unique and important functions, the overall approach to their development and use is the same as that for the functional MIS discussed in this chapter.

Both functional and specialized MISs provide various reports that can help managers make informed decisions about business problems and opportunities. The data used to produce these reports is generated primarily from the TPS. By providing such scheduled, demand, and exception reports, MISs can help managers monitor processes to make better decisions, find new ways to add value to products and services, and assist the organization in achieving its goals and gaining a competitive position in the marketplace.

TECHNOLOGY ADVANTAGE

California Water Districts Deploy Smallworld GIS

The Helix Water District provides water services to approximately 53,000 customers in San Diego County in a 50-square-mile area that includes the cities of La Mesa, El Cajon, Lemon Grove, and Spring Valley. Helix selected a geographic information system from Smallworld GIS to enable the water district to track 700 miles of pipe in its existing system of 224 quarter-section plats. In an unusual action, Helix and its sister utility, Otay Water District, collaborated in the selection of a GIS package. The water utilities used a joint competitive process with the intent of sharing resources and training costs.

At the core of any business application is a model of how the company's business processes operate and the information on which they depend. The GIS model is based on objects that can understand where they are and how they are connected to one another. The objects can represent anything from a customer location to a production work flow and pull together the information needed, whatever its source. Thus, the GIS software provides a platform for building applications that automate business processes where the spatial element is the key and integrating these with the other systems on which the company depends. The GIS software links and displays data stored in a database onto an interactive digital map. The software also provides powerful tools for querying, analyzing, and presenting information naturally in the form of maps, schematics, and plans.

DISCUSSION QUESTIONS

1. What sort of applications for the water districts require a spatial element as a key?
2. How might use of this system provide significant cost reductions or customer service improvements for the water districts?

Sources: Adapted from Mark Hammond, "Smallworld Taps Into Oracle Databases," *PC Week*, May 4, 1998, p. 72; Smallworld press releases: "GPU Selects Smallworld GIS," April 27, 1998; "Smallworld Doubles Customer Base," October 7, 1997, and "Helix Water District Selects Smallworld GIS," October 9, 1997 found at the Smallworld Web site at http://www.smallworld-us.com, accessed May 23, 1998.

Information Systems Principles

The MIS must provide the right information to the right person in the right fashion at the right time.

The most significant internal source of data for an organization's MIS is its TPS systems.

When designing and developing reports to yield the best results, consider the following guidelines: tailor each report to user needs, spend time and effort on producing only those reports that are used, pay attention to report content and layout, use management by exception in reporting, set parameters carefully, and produce all reports in a timely fashion.

In most cases, firms that best know what data to obtain (and how to relate it properly) and when and in what form to present it to which managers achieve the most significant advantage through MISs.

The integration of different information systems makes data and information sharing simpler, which can lead to reduced costs, more accurate reports, more secure data, and increased organizational efficiency.

▪ SUMMARY

1. *Define the term* MIS *and clearly distinguish the difference between a TPS and an MIS.*

A management information system is an integrated collection of people, procedures, databases, and devices that provide managers and decision makers with information to help achieve organizational goals. An MIS can help an organization achieve its goals by providing managers with insight into the regular operations of the organization so that they can control, organize, and plan more effectively and efficiently. The primary difference between the reports generated by the TPS and those generated by the MIS is that MIS reports support managerial decision making at the higher levels of management.

Data that enters the MIS originates from both internal and external sources. The output of most management information systems is a collection of reports that are distributed to managers. These reports include scheduled reports, key-indicator reports, demand reports, exception reports, and drill down. Scheduled reports are produced periodically, or on a schedule, such as daily, weekly, or monthly. A key-indicator report is a special type of scheduled report. Demand reports are developed to give certain information at a manager's request. Exception reports are automatically produced when a situation is unusual or requires management action. Drill down reports provide increasingly detailed data about situations.

Management information systems have a number of common characteristics, including producing scheduled, demand, exception, and drill down reports; producing reports with fixed and standard formats; producing hard-copy and soft-copy reports; using internal data stored in organizational computerized databases; and having reports developed and implemented by IS personnel or end users.

Most MISs are organized along the functional lines of an organization. Typical functional management information systems include finance, manufacturing, marketing, and human resources. Each system is composed of inputs, processing subsystems, and outputs. The primary sources of input to functional MISs include the corporate strategic plan, data from the TPS, information from other functional areas, and external sources including the Internet. The primary output of these functional MISs are summary reports that assist in managerial decision making.

2. *Discuss how organizations are enabling users to access MIS systems via Web technologies.*

Organizations are increasingly making data from their management information systems available to users via Web technology using the Internet and their firm's extranet, intranet, and Web browser software loaded on their personal computers. Customers, suppliers, and vendors provide and access data via the Internet or the corporate extranet. Employees can use the corporate intranet to obtain data and update some of their personal data. The corporate intranet enables rapid sharing of data among distributed units of an organization.

3. *Describe the inputs, outputs, and subsystems associated with a financial MIS.*

A financial management information system provides financial information to all financial managers within an organization, including the chief financial officer (CFO). Financial MIS inputs include the strategic plan, financial data from departments and the organization's TPS, competition, and state and federal agencies. Subsystems are profit/loss and cost systems, use and management of funds, and auditing.

4. *Describe the inputs, outputs, and subsystems associated with a manufacturing MIS.*

A manufacturing MIS accepts inputs from the strategic plan, the TPS, and external sources. The TPSs involved support the business processes associated with the receiving and inspecting of raw material and supplies; inventory tracking of raw materials, work in process, and finished goods; labor and personnel management; management of assembly lines, equipment and machinery, inspection, and maintenance; and order processing. The subsystems involved are design and engineering, master production scheduling, inventory control, just-in-time inventory and manufacturing, process control, and quality control and testing.

5. *Describe the inputs, outputs, and subsystems associated with a marketing MIS.*

A marketing MIS supports managerial activities in the areas of product development, distribution, pricing decisions, promotional effectiveness, and sales forecasting. The marketing MIS accepts inputs from sources such as the Internet, competition, market research, the strategic plan, and the organization's TPS. Subsystems include product development and reporting, promotion and advertising, product pricing, and marketing research.

6. *Describe the inputs, outputs, and subsystems associated with a human resource MIS.*

A human resource MIS is concerned with activities related to employees and potential employees of the organization. The inputs for human resource MIS include the strategic plan, the TPS (payroll, personnel, and order processing), and external sources. Subsystems include human resource planning,

personnel selection and recruiting, training and skills inventories, scheduling and job placement, and wage and salary administration.

7. *Identify other functional management information systems.*

An accounting MIS performs a number of important activities, providing aggregate information on accounts payable, accounts receivable, payroll, and many other applications. The organization's TPS captures accounting data, which is also used by most other functional information systems. Geographic information systems provide regional data in graphical form.

■ KEY TERMS

accounting MIS 424
auditing 403
bill of materials 409
computer-assisted manufacturing
 (CAM) 411
computer-integrated manufacturing
 (CIM) 411
cost centers 403
demand reports 392
drill down reports 394
economic order quantity (EOQ) 408
exception reports 392

external auditing 404
financial MIS 400
flexible manufacturing system
 (FMS) 411
geographic information system
 (GIS) 425
human resource MIS 418
internal auditing 404
just-in-time (JIT) inventory
 approach 410
key-indicator report 392

manufacturing resource planning
 (MRPII) 409
marketing MIS 412
material requirements planning
 (MRP) 409
profit centers 403
quality control 412
reorder point (ROP) 408
revenue centers 403
scheduled reports 392
sensitivity analysis 408

■ REVIEW QUESTIONS

1. Define the term *management information system (MIS)*.
2. What are the five basic kinds of reports produced by an MIS?
3. What is the primary source of input for an MIS?
4. What guidelines should be followed in developing reports for management information systems?
5. What is meant by the "functional approach" to management information systems?
6. Identify five functions performed by all MIS systems.
7. Why are multinational companies adopting the use of an integrated ERP system to meet their MIS needs?

8. What is the difference between a profit center and a revenue center?
9. Identify six typical inputs to a financial management information system.
10. Describe the functions of a manufacturing MIS.
11. What is MRPII? How is it different from MRP?
12. Identify six typical inputs to a marketing MIS. What are the subsystems of the marketing MIS?
13. What is a human resource MIS? What are its inputs and outputs?

▪ DISCUSSION QUESTIONS

1. What is the relationship between an organization's transaction processing systems and its management information systems? What is the primary role of management information systems?
2. How can management information systems be used to support the objectives of the business organization?
3. How are companies linking MIS systems to the Web? Why are they doing this?
4. Describe a financial MIS for a Fortune 1000 manufacturer of food products. What are the primary inputs and outputs? What are the subsystems? Does the company use an ERP system?

5. How can an effective MIS help an organization gain a competitive advantage?
6. What is a financial MIS? How can a strong financial MIS provide strategic benefits to the firm?
7. Why is there such keen interest in implementing an MRPII MIS system at many manufacturing firms?
8. What is the difference in roles played by an internal auditing group versus an external auditing group?
9. Imagine that you are the CFO for a services organization. You are concerned with the integrity of the firm's financial data. What steps might you take to ascertain the extent of problems?

▪ PROBLEM-SOLVING EXERCISES

1. You have been asked to develop your company's Web site to support employee training. Make a list of the kind of information that must be available at this site. Make a second list of the services that should be provided to the employees. Use your word processor or other software to design the home page for this Web site. In addition, design a Web page that could be used for employees to register for a course. Also, design a Web page for students to use to provide feedback on a course.

2. American Amalgamated is a 100-year-old, $50 million company specializing in the manufacture of bolts, screws, and other fasteners. The company's business transaction processing systems are relics originally created in the mid-1970s. Today these systems are patched and tattered from many program changes over the last 25 years. Management lacks timely data to make fundamental, day-to-day decisions. One bottleneck to providing managers with the data they need is the huge backlog of projects for the IS department. Any request for a special report is met with a chorus of groans from the programming staff who are already busy adding more patches to existing systems. Some managers have resorted to developing their own simple query language programs to access transaction files containing order, shipment, inventory, and customer data. Use your word processing software to draft a recommendation to solve American Amalgamated's management information problems.

▪ TEAM ACTIVITY

Divide the class into teams of three or four classmates and use the following role-playing scenario: Your consulting team is designing a management information system for a small manufacturing organization that produces ten different models of high-performance bicycles. These are sold to distributors and bicycle shops throughout North America. One of the major problems this company faces is poor inventory control. It always seems that there are too many bikes in stock and yet the "hottest selling" model is out-of-stock. The owner has suggested that your team look at the feasibility of implementing a manufacturing MIS to help deal with this problem.

Prepare a brief memo that describes the subsystems and outputs associated with the typical manufacturing MIS. Draw a systems level flowchart and outline how these subsystems need to be integrated, and discuss where the inventory control system fits. What are some of the issues that will need to be addressed to develop a single integrated MIS to meet the needs of this organization? Are there some prerequisites that must be completed before the inventory control system can be built? If so, what are they? What additional benefits, besides better inventory control, can be expected?

■ WEB EXERCISE

Companies typically have a number of functional MISs, such as manufacturing. Find the site of two manufacturing companies. Compare these sites. Which one do you prefer? How could these sites be improved? (Hint: If you are having trouble, try Yahoo! It should have a listing for Business and Economy on its home page. From here you can go to companies and then manufacturing. There will be several menu choices there.) You may be asked to develop a report or send an e-mail message to your instructor about what you found.

■ CASES

1 A Marketing MIS for the Greater Boston Convention Center

Boston is known for its history and tradition. The Boston Tea Party, Bunker Hill, and Paul Revere's ride are only a few things in its memorable past. These and the overall attraction of the area make Boston a popular place for tourists and visitors. In most years, about 11 million people visit this historically important and fun place. The greater Boston Convention center and Visitor's Bureau (BCVB) is the main group responsible for attracting visitors and coordinating visitor activities.

Traditionally, BCVB used an old marketing MIS to lure tourists and provide visitor information. The marketing MIS was based on a static Web site. Not only was the Web site not memorable, it was also extremely difficult to maintain. Tourists, convention organizers, and tour operators had to wade through the same data to try to find what they wanted. The information was not always up-to-date. With the BCVB, the right information was not being delivered to the right person at the right time. It was no wonder that the BCVB decided to invest the time and money necessary to develop a better marketing MIS.

The new marketing MIS strives to provide the right information to the right person at the right time. The new system recognizes that different people need different information. As a result, the BCVB has categorized its Web site into four areas: general visitors, press and media, tour operators, and meeting planners. Each group can get specific information to satisfy its needs on the Web. To provide the right information, BCVB has linked its Web site with an extensive database of information. The database not only provides the necessary information but allows easy updating of the Web site. This means that the BCVB Web site can be maintained by nontechnical people with little or no

Internet experience. To help pay for the Web site, BCVB is also placing ads on its Web site for a fee. A vendor, for example, can advertise on the BCVB Web site. The advertising is good for the vendor, and the additional revenue is good for the BCVB.

1. Describe the disadvantages of the old marketing MIS used by BCVB. What are the advantages of the new system?

2. How could BCVB further improve its marketing MIS? What additional features would you build into the Web site?

Sources: Lori Piquet, "New England's New Web Site," *ZD Internet Magazine*, February 1998, p. 86; Thomas Petzinger, "Harvard Webheads Aim to Save the City and Show a Profit," *The Wall Street Journal*, February 13, 1998, p. B1; Staff, "The Site Stuff," *Marketing Week*, March 5, 1998, p. 51.

2   Toy Manufacturer Adopts Marketing MIS

Most marketing software products fall into one of two camps—analytic or transactional. Analytic products can perform data-mining and sort through huge lists of addresses and contacts, looking for the demographics, buying history, and so forth, that will yield the best chance of a sale. Transactional applications include communication and action functions: for example, scheduling events; creating letters, forms, and Web pages; and tracking how these are used and what the results are. This area also includes telemarketing modules with interactive scripts, polling, market research, budgeting, and billing. Many products include or connect to sales and customer-service functions.

Connectivity is especially important for this type of software because marketing depends on getting information to and from other departments.

Community Playthings is a leading manufacturer of toys and furniture for early childhood daycare. Its equipment is an outgrowth of its experience in its own daycare centers and the input from many experts in the field of child play. While their products are very high quality, Community Playthings was having trouble with its marketing efforts. For example, most sales were from catalog orders. Community Playthings kept increasing the size of its mailing list; however, sending out more catalogs was increasing costs but not sales. Salespeople were tracking leads with a program that did not communicate with the customer service department's database.

And the product design department was off doing its own market research, unaware of customer comments and requests coming in from sales calls and customer service.

Community Playthings implemented a Marketing MIS from Onyx Software and was able to identify that three-fourths of its business came from repeat customers. This valuable information was used to refocus its sales target. As a result, Community Playthings reduced its catalog mailing list from 200,000 people to just 50,000, at a cost savings of $600,000 per year while sales kept climbing. In addition, the Marketing MIS brought together new sales-lead information and customer-service data in the same database so that the company could better use word-of-mouth promotion and direct feedback from its most influential customers to drive new product design.

1. Was the Marketing MIS software that Community Playthings implemented analytical or transactional in nature? Why?
2. What additional benefits might Community Playthings expect from the successful implementation of an effective Marketing MIS?

Sources: Adapted from Stannie Holt, "Marketing Gains Automation," *InfoWorld*, vol. 20, no. 29, July 20, 1998; the Community section of the Community Playthings Web site at http://www.bruderhof.com, accessed September 7, 1998; Corporate profile section of Onyx Software Web site at http://www.onyx.com, accessed September 7, 1998.

3 Human Resources at the Bank of Montreal

The human resource function for most firms involves the complete employment life cycle from hiring employees to their retirement (leaving the firm voluntarily or by dismissal). Interviewing potential employees, making employment decisions, scheduling jobs, managing employee benefits, overseeing compensation plans, and installing employee certification programs are just a few functions that the human resource MIS must perform. Information systems are often used to increase efficiencies and realize new opportunities. Like other firms, the Bank of Montreal had the option to develop the human resource information system internally or hire an outside company to develop the necessary software and systems. Internal development offers the option of more control and getting exactly what is needed, although this approach can be more expensive. In addition, there are a number of excellent software packages that can be purchased or leased. UltilPro, for example, is a human resource MIS that can run in a Windows environment. This type of software can cost from $50,000 to $1 million.

Another alternative for companies such as the Bank of Montreal is to outsource all or part of the human resource MIS. In addition to having human resource software developed by an outside company, it is possible to have the complete human resource function performed externally by an outsourcing company. With new technology and increasingly complex regulations, many companies are now starting to outsource most or all of their human resource activities.

In addition to the increasing complexity and the many options of acquiring an effective human resource MIS, the Bank of Montreal was also considering a major merger with the Royal Bank of Canada. The merger, if successful, would result in a company with more than $300 billion in managed assets. The merger would not only have a dramatic impact on the economy of Canada but would also result in changes in all functional areas, including human resources.

One of the biggest needs of the human resource MIS was to develop a compensation and certification system for information systems personnel. The special needs of the information systems department and a tight job market can often lead to conflicts between the human resource department and the information systems department, which is often given the responsibility of developing the human resource MIS. The typical IS department is changing rapidly. It has been estimated that 45 percent of all IS departments will restructure the human resource function and 40 percent will redefine salary and compensation plans in the near future. As a result, the Bank of Montreal decided to scrap its old human resource MIS and develop a new one. Mary Lou Hukezalie, director of human resources at the Bank of Montreal, realized the need for a system that could change with the changing needs of the IS department. The results are promising for the new compensation and certification system. In addition to improving efficiency and flexibility, the approach used by Hukezalie also improved the relationship between the human resources and information systems departments.

1. What are the options in acquiring a human resource MIS?
2. Describe the problems and complexities faced by the Bank of Montreal. If you were the director of human resources at the Bank of Montreal, what would be your highest priorities in developing a new system?

Sources: Tim Ouellette, "Human Resources, IT Develop Stronger Link," *Computerworld,* February 9, 1998, p. 20; Andrew Purvis, "The Royal and the Bank of Montreal Announce Megamerger," *Time,* February 2, 1998, p. 83; Peter Waal, "Size Over Substance," *Canadian Business,* February 13, 1998, p. 17.

4 Chrysler Implements Web-Based Manufacturing MIS

Chrysler Corporation is the number-three auto manufacturer in the United States, with recent annual revenue exceeding $60 billion. It implemented a manufacturing MIS that provides real-time access to manufacturing and production data. The system extends decision making to 2,000 end users including business analysts, line personnel, in fact, all employees—whether they are in Mexico, Canada, Argentina, or Detroit. All they need is an Internet browser and a connection to the corporate intranet to access the data. This system places Chrysler in a leadership role in making mainframe-based corporate data easily accessible from a company intranet.

The manufacturing MIS integrates data from two dozen different mainframe applications. Before the new MIS, users had to sign on to many different mainframe databases and request specific reports to get a comprehensive view of the carmaker's real-time operations. After that, the data often had to be downloaded or keyed into spreadsheets. With the Web-based system, users can easily view a wide range of manufacturing data including vehicle ID numbers, bill of materials, parts inventory, quality data, and warranty data. The data is used to make production, quality, scheduling, and even legal decisions. In the event of a shortage of bumpers, for example, the production scheduling manager can assess how the company's 21 assembly plants would be affected, then change the production schedule to minimize the impact.

Many factors go into making manufacturing related decisions. For example, a primary consideration for auto buyers is color. They often consider color as important as design, make, or model when shopping for a new vehicle. Thus, the manufacturing MIS must make data available as detailed as what color cars are in stock or scheduled to be produced. In addition, worldwide economic conditions must be considered. Recent economic problems have caused Chrysler's international sales to drop more than 10 percent.

1. Given that the data in the manufacturing MIS is used worldwide, what issues might arise over the standards of measure (U.S. dollars versus Eurodollars, English versus metric measure, etc.) in which data is expressed? How might these issues be addressed?

2. As of this writing, Chrylser and Daimler-Benz are planning to merge. What sort of impacts might this have on the use of the manufacturing MIS?

Sources: Adapted from Julia King, "Chrysler Soups up Data Access," *Computerworld*, July 13, 1998, pp. 1, 85; "Chrysler Announces Cool Colors For 1999; Classic Colors Will Be Popular In Paint Palette," *CNW*, September 8, 1998; "Chrysler International Reports August Sales," *PRNewswire*, September 3, 1998.

CHAPTER 10

Decision Support Systems

"Management is the task of staring reality straight in the eye and then having the courage to act."

— Jack Welch,
CEO, General Electric

Chapter Outline

Decision Making and Problem Solving
    Decision Making as a Component of Problem Solving
    Programmed vs. Nonprogrammed Decisions
    Optimization, Satisficing, and Heuristic Approaches
    Problem-Solving Factors

An Overview of Decision Support Systems
    Characteristics of a Decision Support System
    Capabilities of a Decision Support System
    The Integration of TPS, MIS, and DSS
    A Comparison of DSS and MIS
    Web-Based Decision Support Systems

Components of a Decision Support System
    The Model Base
    The Advantages and Disadvantages of Modeling
    The Dialogue Manager

The Group Decision Support System
    Characteristics of a GDSS
    Components of a GDSS and GDSS Software
    GDSS Alternatives

The Executive Support System
    Executive Support Systems in Perspective
    Capabilities of an Executive Support System

Learning Objectives

After completing Chapter 10, you will be able to:

1. Outline and briefly describe the stages of a problem-solving process.
2. List and discuss important characteristics of decision support systems (DSSs) that give them the potential to be effective management support tools.
3. Identify and describe the basic components of a DSS.
4. State the goal of a group decision support system (GDSS) and identify the characteristics that distinguish it from a DSS.
5. Identify fundamental uses of an ESS and list the characteristics of such a system.

Miller SQA
Factory-Floor DSS Helps Manufacturer Deliver the Goods

Miller SQA, a subsidiary of Herman Miller, Inc., is implementing a decision support system (DSS) at plants in Holland, Michigan and Rocklin, California. With DSS as part of its Quick Response Methods (QRM) program, Miller SQA hopes to achieve quicker and more reliable customer response and to reduce the lead time from receipt of a customer order to production of the finished product to just two days for a minimum of 25 percent of customer orders compared with 5 percent today. The company also expects to maintain its 100 percent on-time delivery performance, while achieving an expected 30 percent sales growth rate each year. Additionally, Miller SQA is looking to cut costs by significantly reducing its inventory levels and also making more effective use of floor space.

"Our customers want convenient buying, fast delivery, durability, just the right amount of options, and no hassles selecting office furniture," said Doug Bonzelaar, application development manager at Miller SQA. The Quick Response Method initiative is designed to support these customer goals. Miller SQA is implementing a DSS to provide decision-making support for production planners who work in a fast-paced

environment and must react quickly to customer changes, schedule changes, and process variation. The system also will improve allocation of raw materials by providing an accurate schedule and ensuring that the delivery of parts is coordinated with production schedules.

"The DSS will help us improve cycle times and reduce finished goods inventory through its detailed scheduling of sales order line items. We expect to eliminate our daily fire drills of resolving issues where we find a high-priority component is on schedule, only to learn a companion piece at another work center has a completely different schedule," said Bonzelaar.

With Microsoft Windows clients, the DSS software graphically displays the status of each job to Miller SQA personnel, supporting workers on the shop floor, as well as management, purchasing, and other departments. Shop floor workers also will use the DSS to determine what jobs to assemble, to input quantities produced, to view special instructions, to report status exceptions, and to post completed jobs. The software displays problem areas as they occur and notifies the responsible department.

At the heart of the DSS is graphical process modeling. This component fuels

Miller SQA's Bill of Process (BOP), a database that defines all the resources and activities necessary to produce goods and services. The BOP defines manufacturing operations, such as routings and resource needs, material requirements, and other support items such as specifications, approvals, tools, and programs. It also communicates to support centers regarding raw material procurement, engineering, and transportation activities.

The software translates plans or schedules into the details necessary for execution, including on-line dispatch lists with up-to-the-minute requirements for parts, people, machines, product specifications, and support activities. Because the system constantly runs and reevaluates actual status versus projected status, the software identifies variances—and their impact—as they occur. The software answers the questions, "Can I fulfill the customer's demand when he wants it?" and later, as the order is in process, "Do I need to make any schedule adjustments to meet the customer's need date?" Importantly, the system tells the end user what to do next, not what has already happened. It allows schedulers to perform their real jobs, plan for future activities, and not have to expedite.

As you read this chapter, consider the following questions:

- How do managers use a decision support system? How is it different from an MIS?

- How must the knowledge, experience, and insights of a manager be coupled with the outputs from a DSS to yield good decision making?

Sources: Adapted from "About Synquest," found at the SynQuest Web site at http://www.synquest.com, accessed May 27, 1998 and Craig Stedman, "Planning Systems Hit Shop Floor," *Computerworld*, April 27, 1998, pp. 61–62.

We saw in Chapters 8 and 9 that transaction processing systems (TPSs) and management information systems (MISs) provide useful summary reports to help solve structured and semistructured business problems. Decision support systems (DSSs), like the one used by Miller SQA, offer the potential to assist in solving both semistructured and unstructured problems. The emphasis of a DSS is on individual decision-making styles and techniques. This chapter covers the advantages, components, and use of decision support systems. We begin by investigating decision making and problem solving.

DECISION MAKING AND PROBLEM SOLVING

Every organization needs effective decision making to reach its objectives and goals. In most cases, strategic planning and the overall goals of the organization set the stage for value-added processes and the decision making required to make them work. Often, information systems assist with strategic planning and problem solving.

Decision Making as a Component of Problem Solving

decision-making phase

the phase of the problem-solving process that includes the intelligence, design, and choice stages

In business, one of the highest compliments you can get is to be recognized by your colleagues and peers as a "real problem solver." Problem solving is a critical activity for any business organization. Once the problem has been identified, the problem solving process begins with decision making. A well-known model developed by Herbert Simon divides the **decision-making phase** of the problem-solving process into three stages: intelligence, design, and choice. This model was later incorporated by George Huber into an expanded model of the entire problem-solving process (Figure 10.1).

Problem solving and decision making occur over a span of time and often include evaluation of several alternatives. Thus, one or more of the steps in the model shown in Figure 10.1 may be repeated. For example, if a problem solver uncovers new alternatives during the choice phase, he or she may have to go back to the intelligence and design phases to collect additional information on these alternatives. Feedback and continual adjustment is common during decision making.

intelligence stage

the first stage of the problem-solving process during which potential problems and/or opportunities are identified and defined

The first stage in the problem-solving process is the **intelligence stage**. During this stage, potential problems and/or opportunities are identified and defined. Information is gathered that relates to the cause and scope of the problem. During the intelligence stage, resource and environmental constraints are investigated. For example, exploring the possibilities of shipping tropical fruit from a farm in Hawaii to stores in Michigan would

FIGURE 10.1

How Decision Making Relates to Problem Solving

The three stages of decision making—intelligence, design, and choice—are augmented by implementation and monitoring to result in problem solving.

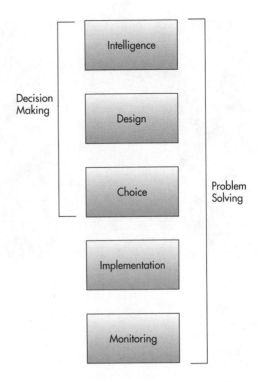

be done during the intelligence stage. The perishability of the fruit and the maximum price consumers in Michigan are willing to pay for the fruit are problem constraints. Aspects of the problem environment that must be considered in this case include federal and state regulations regarding the shipment of food products.

In the **design stage**, alternative solutions to the problem are developed. In addition, the feasibility of these alternatives are evaluated. In our tropical fruit example, the alternative methods of shipment, including the transportation times and costs associated with each, would be considered. During this stage the problem solver might determine that shipment by freighter to California and then by truck to Michigan is not feasible because the fruit would spoil.

The last stage of the decision-making phase, the **choice stage**, requires selecting a course of action. In our tropical fruit example, the Hawaiian farm might select the method of shipping by air to Michigan as its solution. The choice stage would then conclude with selection of the actual air carrier. As we will see later, various factors influence choice; the apparently easy act of choosing is not as simple as it might first appear.

Problem solving includes and goes beyond decision making. It also includes the **implementation stage**, when the solution is put into effect. For example, if the Hawaiian farmer's decision is to ship the tropical fruit to Michigan as air freight using a specific air freight company, implementation involves informing the farming staff of the new activity, getting the fruit to the airport, and actually shipping the product to Michigan.

The final stage of the problem-solving process is the **monitoring stage**. In this stage, decision makers evaluate the implementation to determine whether the anticipated results were achieved and to modify the process in

design stage

the second stage of the problem-solving process during which alternative solutions to the problem are developed

choice stage

the last stage of the decision-making phase during which a course of action is selected

problem solving

the process of combining intelligence, design, and choice stages along with implementation and monitoring to reach a solution

implementation stage

the stage where action is taken to put a solution into effect

monitoring stage

the final stage of the problem-solving process where decision makers evaluate the implementation

FIGURE 10.2

Ordering more inventory when inventory levels drop to specified levels is an example of a programmed decision.
(Source: Mitch Kezar/Tony Stone Images.)

light of new information. This can involve feedback and an adjustment process. For example, after the first shipment of fruit, the Hawaiian farmer might learn that the flight of the chosen air freight firm routinely makes a stopover in Phoenix, Arizona, where the plane sits exposed on the runway for a number of hours while loading additional cargo. If this unforeseen fluctuation in temperature and humidity adversely affects the fruit, the farmer might have to readjust his solution to include a new air freight firm that does not make such a stopover, or perhaps he would consider a change in fruit packaging.

Information systems can be an important part of all phases of decision making and problem solving. Computer-analyzed surveys and questionnaires can be used during the intelligence phase to determine overall problems and opportunities. Information systems can also help identify important new product areas. During the design phase, decision-making models can be used to explore and analyze alternatives. Finally, an information system can be used to assist in the final selection and to monitor the implementation of the decision. Selecting and installing information systems in business organizations also follows the same steps of the problem-solving process.

Programmed vs. Nonprogrammed Decisions

programmed decisions

decisions that are made using a rule, procedure, or quantitative method

In the choice stage, various factors influence the decision maker's selection of a solution. One such factor is whether the decision can be programmed. **Programmed decisions** are made using a rule, procedure, or quantitative method. For example, to say that inventory should be ordered when inventory levels drop to 100 units is to adhere to a rule (Figure 10.2). Programmed decisions are easy to computerize using traditional information systems. It is simple, for example, to program a computer to order more inventory when inventory levels for a certain item reach 100 units or fewer. Most of the processes automated through transaction processing systems share this characteristic: the relationships between system elements are fixed by way of rules, procedures, or numerical relationships. Management information systems are also used to solve programmed decisions by providing reports on problems that are routine and where the relationships are well defined (structured problems).

nonprogrammed decisions

decisions that deal with unusual or exceptional situations

Nonprogrammed decisions, however, deal with unusual or exceptional situations. In many cases, these decisions are difficult to quantify. Determining the appropriate training program for a new employee, deciding whether to start a new type of product line, and weighing the benefits and disadvantages of installing a new pollution control system are examples. Each of these decisions contains many unique characteristics for which the application of rules or procedures is not so obvious. Today, decision support systems are used to solve a variety of nonprogrammed decisions, where the problem is not routine and rules and relationships are not well defined (unstructured or ill-structured problems).

Optimization, Satisficing, and Heuristic Approaches

optimization model

an approach to decision support that involves finding the best solution, usually the one that will best help the organization meet its goals

In general, computerized decision support systems can either optimize or satisfice. An **optimization model** will find the best solution, usually the one that will best help the organization meet its goals. For example, an optimization model can find the appropriate number of products an organization should produce to meet a profit goal, given certain conditions and assumptions. Optimization models utilize problem constraints. A limit on the number of available work hours in a manufacturing facility is an example of a problem constraint. Some spreadsheet programs, such as Excel, have optimizing features including Solver and Goal Seek (Figure 10.3). Read the "Making a Difference" box for an example of a company applying an optimization model to increase profits and reduce inventory.

satisficing model

an approach to decision support that involves finding a good—but not necessarily the best—solution to a problem

A **satisficing model** is one that will find a good—but not necessarily the best—problem solution. Satisficing is usually used because modeling the problem properly to get an optimal decision would be too difficult, complex, or costly. Satisficing normally does not look at all possible solutions but only at those likely to give good results. Consider a decision to select a location for a new plant. To find the optimal (best) location, you would have to consider all cities in the United States or the world. A satisficing approach would be to consider only five or ten cities that might satisfy the company's requirements. This may not result in the best decision, but it will likely result in a good decision, without spending the time and effort to investigate all cities. Satisficing is a good alternative modeling method because it is sometimes too expensive to analyze every alternative to get the best solution.

heuristics

commonly accepted guidelines or procedures that usually find a good solution

Heuristics, often referred to as "rules of thumb"—commonly accepted guidelines or procedures that usually find a good solution—are very often used in decision making. An example of a heuristic rule in the military is to "get there first with the most firepower." A heuristic used by baseball team

FIGURE 10.3

Some spreadsheet programs, such as Excel, have optimizing routines. This figure shows Solver, which can find an optimal solution given certain constraints.

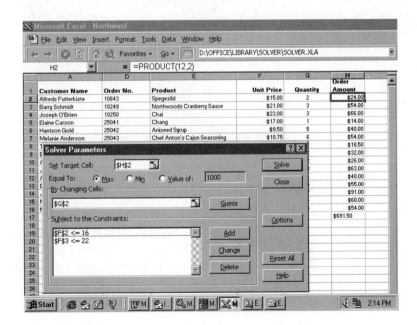

MAKING A DIFFERENCE
Office Depot Optimizes Its Operations

Office Depot is in a highly competitive business environment and has a rapid growth strategy. It is a leader in the growing office products retailing sector, and earnings for the quarter ending in March 1998 were $1,981 million compared to $1,772 million a year before. Office Depot generated a respectable 4 percent growth at its existing stores in the second half of last year. The company expects to open 80 to 100 new stores per year.

Office Depot uses decision support software for decision making. The company began using Essbase from Sunnyvale, California–based Arbor Software Corp. (now Hyperion Solutions Corporation) in early 1996. Use of these tools has allowed hundreds of merchandisers, salespeople, and executives to do their own data analysis and make data-based decisions using optimization models to move the business ahead.

For example, in the Fall of 1996, Office Depot's 100 merchandisers used Essbase to review the retailer's PC business by generating detailed analyses of profitability by store and by product type. They found they were carrying too much stock in the wrong stores. So the retailer used optimization models to narrow its assortment of PCs from 22 to 12 products. That helped the company eliminate unnecessary inventory and avoid costly markdowns on equipment that was not selling. As a result, Office Depot's return on assets for its computer business has improved substantially. The company's use of optimization models and decision support systems has translated into millions of dollars in profitability.

Office Depot invested less than $5 million for the Essbase software and Compaq 7000 servers. The company had already completed a five-year effort to replace a few IBM AS/400s with a 900-MIPS IBM CMOS mainframe computer running DB2. While a shift to mainframe processing may not sound forward thinking, Office Depot wound up with well-designed systems that were good at storing lots of detailed data, including 4 Tbytes of data stored in DB2 and another 250 Gbytes of Essbase information stored in databases.

Office Depot is considering providing suppliers with Web access to its databases. Office Depot already shares sales activity with two key suppliers through electronic data interchange transaction sets. A decision on which approach to take—EDI transaction sets or Web-based access to decision support data—is needed.

DISCUSSION QUESTIONS

1. Why would Office Depot consider providing suppliers with access to its databases?
2. What are the pros and cons of providing access to data via EDI transaction sets versus Web-based access?

Sources: Adapted from "Ratings of Office Depot Inc. Remain on Standard & Poor's Watch, Positive," *PRNewswire*, May 19, 1998; Thomas Hoffman, "Improved Analytics Drive Office Depot Sales," *Computerworld*, February 9, 1998, pp. 65, 68; and Press release "Now Shipping: Arbor Essbase Objects Version 1.1," found at the Arbor Software Web site at http://www.arborsoft.com, accessed August 22, 1998.

managers is to place batters most likely to get on base at the top of the lineup, followed by the power hitters who'll drive them in to score. An example of a heuristic used in business is to order four months' supply of inventory for a particular item when the inventory level drops to 20 units or fewer; even though this heuristic may not minimize total inventory costs, it may be a very good rule of thumb to avoid stockouts without too much excess inventory.

Problem-Solving Factors

Various factors are important in solving problems. An awareness of these factors will increase a manager's ability to properly analyze a problem and make good decisions. These factors include multiple decision objectives, increased alternatives, increased competition, the need for creativity, social and political actions, international aspects, technology, and time compression.

Multiple decision objectives. The goals of many organizations go beyond merely increasing profits or reducing costs. Some want to maintain certain production levels to keep a stable workforce, minimize worker accidents, contribute to the betterment of the community, or minimize the impact of their production processes on the environment. In addition, complex pollution regulations, legal conditions of clients and vendors, labor contracts that require intricate work rules, and suppliers' restrictions on products and materials must also often be considered. These multiple-objective, decision-making problems are typically more complex and can be extremely difficult to solve. Decision makers facing multiple-objective problems rely on experience, common sense, and a deep understanding of corporate goals.

Increased alternatives. One aspect of recent decision making is that there are more alternatives to consider than there were only a few years ago. Consider a financial manager of a firm seeking to obtain funds for expansion of the business. Not too long ago, such a manager faced a relatively simple choice between the issuance of new stock (equity) and the issuance of bonds (a form of debt usually secured by the assets of the firm). Today, myriad variations of each alternative exist.

Increased competition. Competition involves two or more organizations vying to reach similar goals through similar customer groups. The number and type of competitors in the marketplace have made it increasingly difficult for many organizations to meet defined goals. Increased competition from companies providing the same types of services and goods can be seen in most markets and businesses as a result of new technology and the Internet, improved transportation systems, and a mobile labor pool.

The need for creativity. The use of creativity and imagination in problem solving is one factor that can differentiate a company from its competition. Creativity involves the ability to originate or generate new ideas or approaches to add value to products and services. It is almost always a response to an opportunity to be exploited or a problem to be solved. In some cases, the new idea or approach is developed from scratch, such as the lightbulb, the zipper, and the telephone. In other cases, creativity can involve combining existing approaches or technology into a new form. The Nautilus exercise machine was the designer's creative response to his frustration with traditional weights that failed to increase his own slight build. The clipless bicycle pedal that is used by most bicycle racers today was created from the binding system used to attach a downhill ski to the ski boot.

Social and political actions. At all levels, social and political actions have a profound impact on problem solving. For example, a city council might pass an ordinance requiring local companies to meet certain pollution standards. Some firms might need to install new pollution control equipment to meet the standard; others might need to reconsider their choice of manufacturing processes; others may find the standard so onerous that they consider moving. If the organization decides to not relocate, it might need a sophisticated new information and monitoring system to comply with the regulations.

FIGURE 10.4

Many companies have global operations. Coca-Cola is manufactured and sold in more than 195 countries around the world. (Source: Keren Su/Corbis.)

International aspects. As businesses and markets shift from local concerns to national and international operations, international aspects have changed how businesses operate and compete (Figure 10.4). Concerns about sagging stock markets, creation of the euro-dollar, and the potential devaluation of currency relative to the U.S. dollar impact international trade and global communications and computing. Most of the world is aware of the international impact of major companies such as Unilever, General Electric, Toyota, and others in Japan, Germany, and the United States. Less obvious but just as important is the impact of the lesser-known nations, markets, and businesses. A clothing manufacturer in the United Kingdom, for example, might purchase raw materials from a developing country in Africa, have the products assembled in Asia, and sell them in France.

Technology. Reductions in the price of information technology and advances in its capabilities have provided increased decision alternatives to businesses and organizations of all sizes. For example, 10 to 15 years ago, a manufacturer might have had only a few local suppliers to consider. Today, technology and the Internet bring to the manufacturer's doorstep hundreds of suppliers throughout the world that could provide the necessary parts or materials. Or consider financial markets. A few decades ago, an investment banker could place excess funds in the stock market, bonds, or similar investments. Today, technology allows international markets to trade 24 hours a day in a variety of foreign currencies and investment areas.

time compression

a phenomenon whereby activities occur in a shorter time frame than was previously possible

Time compression. Today, major business events develop overnight or in a matter of hours. A rumor about a company going bankrupt can be reflected almost immediately in its stock market price. Fifty years ago, it could take days or weeks to transfer money from one location to another. Today, money and news travel in seconds, and the impact of events is seen just as fast. **Time compression** is a phenomenon whereby activities occur in a shorter time frame than was previously possible. Time compression increases the need for an information system that will supply the problem solver with the relevant data for decision making as quickly as possible. To remain competitive, organizations must learn to incorporate information systems into business processes to make them more efficient and effective.

AN OVERVIEW OF DECISION SUPPORT SYSTEMS

A DSS is an organized collection of people, procedures, software, databases, and devices used to support problem-specific decision making. The focus of a DSS is on decision-making effectiveness when faced with unstructured or semistructured business problems. Decision support systems offer the potential to generate higher profits, lower costs, and better products and services. For example, healthcare organizations use DSSs to clarify and reduce costs. As with a TPS and an MIS, a DSS should be designed, developed, and used to help the organization achieve its goals and objectives.

Decision support systems, although skewed somewhat toward the top levels of management, are used at all levels. This is true because, to some extent, managers at all levels are faced with somewhat less structured, nonroutine problems. The quantity and magnitude of these decisions increase as a manager rises higher in the organization. Many organizations face a bureaucracy of complex rules, procedures, and decisions. DSSs are used to bring more structure to these problems to aid the decision-making process. In addition, because of the inherent flexibility of decision support systems, managers at all levels are able to use DSSs to assist in some relatively routine, programmable decisions in lieu of more formalized management information systems.

Decision support systems are often also linked with managerial decision making regarding value-added business processes. The range of problems that can be supported is vast. For example, a DSS can be used to select an employee insurance healthcare plan. Another DSS can be used to predict how an increase in the price of paper might affect a newspaper's overall profits; the information generated by the decision support system might prompt the newspaper editor and marketing manager to increase advertising space in the paper, rather than raising the price to the customer.

These simple examples demonstrate only some aspects of the approach of a DSS. Overall, a decision support system should assist decision makers at all levels with all aspects of decision making, including those related to value-added business processes. Moreover, the DSS approach realizes that people, not machines, make decisions. DSS technology is used primarily to support making decisions that can solve problems and further corporate goals. Read the "Technology Advantage" box for an example of a bank using DSS technology to gain a competitive advantage in analysis of its credit portfolio.

Characteristics of a Decision Support System

Decision support systems have a number of characteristics that allow them the potential to be effective management support tools. Of course, not all DSSs work the same—some are small in scope and offer only some of these characteristics. In general, a decision support system can perform the following functions.

Handle large amounts of data from different sources. For instance, advanced database management systems and data warehouses have allowed decision makers to search databases for information when using a DSS, even when some data sources reside in different databases stored in different computer systems or networks. Other sources of data may be accessed via the Internet or over a corporate intranet.

Provide report and presentation flexibility. One of the reasons DSSs were developed was that TPSs and MISs were not flexible enough to meet the full variety of decision makers' problems and information needs. While other information systems output primarily fixed-format reports, DSSs have more widely varied formats. Managers can get the information they want, presented in a format that suits their needs. Furthermore, output can be presented on computer screens or produced on printers, depending on the needs and desires of the problem solvers.

TECHNOLOGY ADVANTAGE
First Chicago NBD Adopts DSS Software

With 34,000 employees and $114 billion in assets at year-end 1997, First Chicago NBD Corporation is the ninth largest U.S. bank holding company and the Midwest's number one provider of financial products and services to consumers, mid-sized companies, and large corporations. First Chicago operates banks with 700 offices in key markets domestically and internationally. Through its bank subsidiaries, First Chicago provides domestic retail banking, trust and investment management services, worldwide corporate and institutional banking, and credit card services.

First Chicago was an early adopter of the Pilot Commercial Credit Analysis software. "Delivering portfolio information to the desktop is a key component of our credit systems strategy," noted Paul Behrman, first vice president of First Chicago NBD. "Pilot's Commercial Credit Analysis Application integrates core portfolio reporting and analysis into a powerful, user friendly tool kit for our credit professionals. This tool kit provides the scalability and flexibility to respond to the changing needs of our credit analysts."

Pilot Software supplies customer and market analysis software. One of its most recent products is the Pilot Commercial Credit Analysis, a banking software decision support package that provides corporate lending executives the ability to track and analyze all activities of their commercial loan portfolio. The system is a decision support tool for banks, finance companies, and financing divisions of manufacturing companies. It provides a high-level view of the bank portfolio and how it is influenced by market trends, product distribution, and risk. This DSS allows bank management to gain insight into the risk, exposure, and profitability across the portfolio and identifies significant changes in the structure of the portfolio over time. The software's highly visual capabilities give managers the information they need to spot trends in the overall business and take proactive measures to increase performance.

According to John Diggins, Pilot's president and CEO, "Deregulation, globalization, and consolidation has led to highly competitive environments within the global financial industry. With Commercial Credit Analysis, banks can now gain competitive advantage with an easy-to-use software package that provides a high-level view and analysis of the portfolio. The potential savings from improving the management of a commercial credit portfolio are compelling."

DISCUSSION QUESTIONS

1. What key business benefits do you think First Chicago hopes to receive from use of this DSS?
2. As an early adopter of this new software, what sort of test or evaluation do you think First Chicago performed to ensure that the software will deliver the expected business benefits?

Sources: Adapted from the First Chicago NBD Corporation Web site at http://www.fcnbd.com, accessed May 23, 1998 and a Pilot Systems press release "Pilot Software Introduces Commercial Credit Analysis Application," November 24, 1997 found at the Pilot Systems Web site at http://www.pilotsw.com, accessed May 23, 1998.

Offer both textual and graphical orientation. A decision support system can provide whatever orientation a manager prefers, be it textual or graphical. Some decision makers prefer a straight text interface, while others want a decision support system that helps them make attractive, informative graphical presentations on computer screens and in printed documents. Today's decision support systems can produce text, tables, line drawings, pie charts, trend lines, and more. By using their preferred orientation, managers can use a DSS to get a better understanding of a true situation, if required, and to convey this understanding to others.

Support drill down analysis. A manager can get more levels of detail when needed by drilling down through data.[1] For example, he or she can view the overall project cost or drill down and see the cost for each project phase, activity, and task.

Perform complex, sophisticated analysis and comparisons using advanced software packages. Marketing research surveys, for example, can be analyzed in a variety of ways using analysis programs that are part of a DSS. Many of the analytical programs associated with a DSS are actually stand-alone programs. The DSS provides a means of bringing these together.

Support optimization, satisficing, and heuristic approaches. For smaller problems, decision support systems have the ability to find the best (optimal) solution. For more complex problems, satisficing or heuristic approaches are used. With satisficing and heuristics, the computer system can determine a very good—but not necessarily the best—solution. By supporting all types of decision-making approaches, a DSS gives the decision maker a great deal of flexibility in getting computer support for decision-making activities.

Perform "what-if," simulation, and goal-seeking analyses. **"What-if" analysis** is the process of making hypothetical changes to problem data and observing the impact on the results. Consider an inventory control application. Given the demand for products, such as automobiles, the computer can determine the necessary subparts and components, including engines, transmissions, windows, and so on. With "what-if" analysis, a manager can make changes to problem data (the number of automobiles needed for next month) and immediately see the impact on the requirements for subparts (engines, windows, and so on).

"what-if" analysis

the process of making hypothetical changes to problem data and observing the impact on the results

Decision Grid's multicriteria comparison table is used here by an interviewer to evaluate, rank, and select the best candidate.
(Source: Courtesy of Softkit, a division of CGI Inc.)

There are a number of software packages with these characteristics. For example, Decision Grid is a custom decision-support package originally developed for the engineers and managers at Hydro-Quebec, one of Canada's largest power companies. Thanks to its spreadsheet-like interface and unusually clear documentation, Decision Grid is relatively easy to use. Even new users can get to work immediately, laying out alternatives and criteria in Decision Grid's spreadsheet-like display. Users can either start from scratch by listing their alternatives (e.g., car models) in different columns and criteria (fuel economy, style, cost) in different rows; or they can use one of the 20 model templates that ship with the product. Creating a category of criteria (such as work experience) requires only that the user select a criterion in the table and choose the edit option from the main format menu. Users can set the relative importance of different criteria by typing in a number between 1 and 100. Once the users supply evaluations of how their alternatives measure up on all their criteria, they can compare them simply by clicking on the main tool menu's compare option. The program also supports group assessments. Up to 25 different people can separately assess the same set of alternatives. Users can then view and compare these different assessments

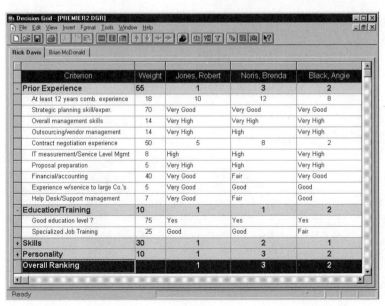

Criterion	Weight	Jones, Robert	Noris, Brenda	Black, Angie
- Prior Experience	**55**	**1**	**3**	**2**
At least 12 years comb. experience	18	10	12	8
Strategic planning skill/exper.	70	Very Good	Very Good	Very Good
Overall management skills	14	Very High	Very High	Very High
Outsourcing/vendor management	14	Very High	High	Very High
Contract negotiation experience	50	5	8	2
IT measurement/Service Level Mgmt	8	High	High	Very High
Proposal preparation	5	Very High	High	Very High
Financial/accounting	40	Very Good	Fair	Very Good
Experience w/service to large Co.'s	5	Very Good	Good	Good
Help Desk/Support management	7	Very Good	Fair	Good
- Education/Training	**10**	**1**	**1**	**2**
Good education level ?	75	Yes	Yes	Yes
Specialized Job Training	25	Good	Good	Fair
+ Skills	**30**	**1**	**2**	**1**
+ Personality	**10**	**1**	**3**	**2**
Overall Ranking		**1**	**3**	**2**

and combine them to form a final ranking of alternatives. Decision Grid also provides excellent tools for performing sensitivity analyses that determine how the alternatives compare with each other when users modify the relative importance of their criteria. (Find, for example, what happens if you increase the weight that you place on fuel economy and safety, and decrease the importance you place on style.)[2]

simulation

the ability of the DSS to act like or duplicate the features of a real system

Simulation is the ability of the DSS to act like or duplicate the features of a real system. In most cases, probability or uncertainty are involved. For example, the mean time between failure and the mean time to repair key components of a manufacturing line can be calculated to determine the impact on the number of products that can be produced each shift. Engineers can use this data to determine which components need to be reengineered to increase the mean time between failures and which components need to have an ample supply of spare parts to reduce the mean time to repair.

goal-seeking analysis

the process of determining the problem data required for a given result

Goal-seeking analysis is the process of determining the problem data required for a given result. For example, a financial manager is considering an investment with a certain monthly net income. Furthermore, the manager might have a goal to earn a return of 9 percent on the investment. Goal seeking allows the manager to determine what monthly net income (problem data) is needed to have a return of 9 percent (problem result). Some spreadsheets can be used to perform goal-seeking analysis (see Figure 10.5).

Capabilities of a Decision Support System

As the name implies, decision support systems support key decisions related to value-added business processes. They can be applied to most industries and business functions and can result in benefits to the organization. A DSS can be used by college administrators to effectively schedule classes into

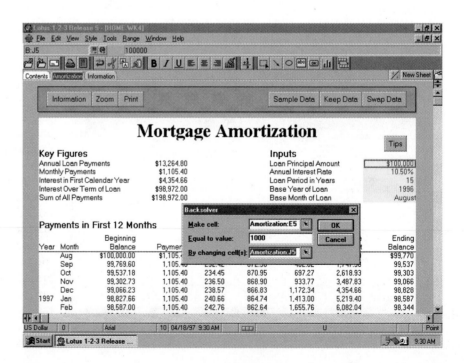

FIGURE 10.5

With a spreadsheet program, a manager can enter a goal, and the spreadsheet will determine the needed input to achieve the goal.
(Source: Courtesy of Lotus Development Corporation.)

- Cinergy Corporation, an electric utility company with headquarters in Cincinnati, Ohio, developed a DSS to significantly cut down the lead time and effort required to make decisions in the area of coal purchasing activity.
- RCA developed a DSS to deal with personnel problems and issues. The system, called Industrial Relation Information Systems (IRIS), can handle unanticipated or one-time only problems and can assist in difficult labor negotiations.
- Energy Plan (EPLAN) is a DSS being developed by the National Audubon Society to analyze the impact of U.S. energy policy on the environment.
- The U.S. Army developed an enlisted-force manpower DSS to help with recruitment, training, education, reclassification, and promotion decisions. The DSS uses optimization and simulation to model personnel needs and requirements. It includes "what-if" features and can interact with an on-line database and other statistical analysis programs.
- Hewlett-Packard developed Quality Decision Management to perform production and quality-control functions. It can help with raw materials inspection, product testing, and statistical analysis.
- The Transportation Evacuation Decision Support System (TEDSS) is a DSS used in nuclear plants in Virginia. This personal computer DSS analyzes and develops evacuation plans to help managers prepare for crisis management decisions. It helps employees make decisions regarding evacuation times and routes and the allocation of shelter resources.

TABLE 10.1

Selected DSS Applications

available classrooms. Sales forecasts, work schedules, and production flow data are fed into a production planning DSS to develop a detailed production schedule. In the investment arena, financial planners use a DSS to diversify a client's funds among a suitable set of investment alternatives to minimize risk and yet provide for an adequate rate of return on investment. Additional applications are listed in Table 10.1.

Developers of decision support systems strive to make them more flexible than management information systems and to give them the potential to assist decision makers in a variety of situations. DSSs can assist with all or most problem-solving phases, decision frequencies, and different degrees of problem structure. DSS approaches can also help at all levels of the decision-making process. In this section we investigate these DSS capabilities. An actual DSS may provide only a few of these capabilities, depending on the uses and scope of the DSS.

Support for problem-solving phases. The objective of most decision support systems is to assist decision makers with the phases of the problem-solving process. As previously discussed, these phases include intelligence, design, choice, implementation, and monitoring. A specific DSS might support only one or a few problem-solving phases.

Support for different decision frequencies. Decisions can range on a continuum from one-of-a-kind to repetitive decisions (Figure 10.6). One-of-a-kind decisions require flexible, efficient, and cost-effective computer support. For example, a company might consider merging with another firm. Many of today's decision support software packages can help with these types of decisions.

ad hoc DSS

a DSS concerned with situations or decisions that come up only a few times during the life of the organization

One-of-a-kind decisions are typically handled by an **ad hoc DSS**. An ad hoc DSS is concerned with situations or decisions that come up only a few times during the life of the organization; in small businesses, they may happen only once. For example, a company might be faced with a decision on whether to build a new manufacturing facility in another area of the country.

FIGURE 10.6

Decision Frequency

Business decisions fall somewhere on a continuum between one-of-a-kind decisions and repetitive, or routine, decisions.

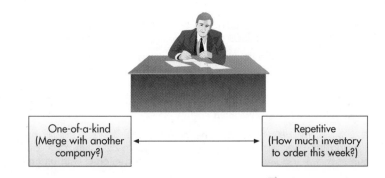

One-of-a-kind
(Merge with another
company?)

Repetitive
(How much inventory
to order this week?)

Repetitive decisions can be easier to support over time than one-of-a-kind decisions. These types of decisions are made daily, weekly, monthly, or yearly. For example, determining how much inventory to order once a week is a repetitive decision. Repetitive decisions are addressed by an institutional DSS. An institutional DSS handles situations or decisions that occur more than once, usually several times a year or more. An institutional DSS is used repeatedly and refined over the years. Examples of institutional DSSs include systems that support portfolio and investment decisions and production scheduling. These decisions may require decision support numerous times during the year. Between these two extremes are numerous decisions managers make several times, but not on a regular or routine basis.

Support for different problem structures. As discussed previously, decisions can range from highly structured and programmed to unstructured and nonprogrammed (Figure 10.7). **Highly structured problems** are straightforward, requiring known facts and relationships. Determining the best time-deposit account in which to place excess corporate funds for a few months based on a fixed interest rate is an example; the objective is to obtain the highest return, and the relationship between interest rate and return is known. The data required to make the decision is also known and readily available.

Semistructured or unstructured problems, on the other hand, are more complex. The relationships among the data are not always clear, the data may be in a variety of formats, and the data is often difficult to manipulate or obtain. In addition, the decision maker may not know the information requirements of the decision in advance. An example is a decision

highly structured problems

problems that are straightforward, requiring known facts and relationships

semistructured or unstructured problems

more complex problems wherein the relationships among the data are not always clear, the data may be in a variety of formats, and the data is often difficult to manipulate or obtain

FIGURE 10.7

Degree of Problem Structure

Business problems fall somewhere on a continuum between unstructured and highly structured.

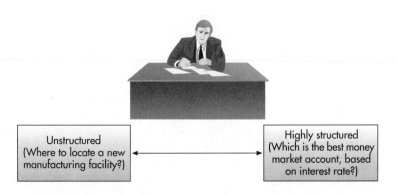

Unstructured
(Where to locate a new
manufacturing facility?)

Highly structured
(Which is the best money
market account, based
on interest rate?)

to locate a new manufacturing facility. This decision can involve numerous factors, including the available labor pool, current economic conditions, political climate, the types of schools appropriate for employees, the cultural programs offered by the community, and more.

Factoring all these concerns into a single measure, such as profits or costs, can be difficult. One of the primary functions of a DSS is to bring together a variety of data in varying formats in a single, usable set of relevant information. In doing so, decision support programs can help decision makers handle both semistructured and unstructured problems. These types of problems may have a greater impact on the organization than structured problems, another reason why a DSS is so important to an organization.

Support for various decision-making levels. Decision support systems can offer help for managers at different levels within the organization. Operational-level managers can be assisted with daily and routine decision making. Tactical-level decision makers can be supported with analysis tools that assist in proper planning and control. At the strategic level, DSSs can help managers by providing analysis for long-term decisions requiring both internal and external information (Figure 10.8).

The Integration of TPS, MIS, and DSS

Transaction processing, management information, and decision support systems are covered in different chapters of this text (Chapters 8, 9, and 10, respectively), but these systems overlap to provide complementary functions. Moreover, in many organizations they are integrated through the use of a common database. For example, a billing TPS that sends monthly bills to customers, a billing MIS that produces weekly reports to managers on overdue bills, and a billing DSS that performs "what-if" analysis to determine the impact of late bill paying on cash flows, revenues, and overall profit levels may all draw data from the same database. Using the same database for these different systems may require more powerful hardware and software; otherwise, extensive use of a decision support system could slow down the operation of the transaction processing system. In addition, ERP and E-commerce applications can help integrate an organization's TPS, MIS, and DSS. For many organizations a TPS has been expanded to provide management information, which in turn has evolved into a DSS. Thus, many applications discussed in this chapter, including inventory control and sales ordering, were introduced to you as TPS or MIS functions.

FIGURE 10.8

Decision-Making Level

Strategic-level managers are involved with long-term decisions, which are often made infrequently. Operational-level managers are involved with decisions that are made more frequently.

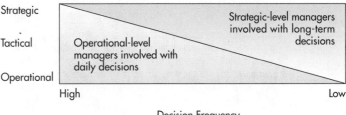

Decision Frequency

A Comparison of DSS and MIS

A DSS differs from an MIS in numerous ways, including the type of problems solved; the support given to users; the decision emphasis and approach; and the type, speed, output, and development of the system used. Table 10.2 lists brief descriptions of these differences.

Web-Based Decision Support Systems

Once the exclusive domain of sophisticated business analysts or senior IS staff, decision support has become a valuable tool for decision makers at all levels across the enterprise. Decision support system software provides business intelligence through Web browser clients that access databases either through an Internet or corporate intranet connection. This link enables employees to retrieve and analyze key operational and market data about customers, products, employees, and other data entities. This approach

TABLE 10.2

Comparison of DSSs and MISs

Factor	Comparison DSS	MIS
Problem Type	A DSS is good at handling unstructured problems that cannot be easily programmed.	An MIS is normally used only with more structured problems.
Users	A DSS supports individuals, small groups, and the entire organization. In the short run, users typically have more control over a DSS.	An MIS supports primarily the organization. In the short run, users have less control over an MIS.
Support	A DSS supports all aspects and phases of decision making; it does not replace the decision maker—people still make the decisions.	This is not true of all MIS systems—some make automatic decisions and replace the decision maker.
Emphasis	A DSS emphasizes actual decisions and decision-making styles.	An MIS usually emphasizes information only.
Approach	A DSS is a direct support system that provides interactive reports on computer screens.	An MIS is typically an indirect support system that uses regularly produced reports.
System	The computer equipment that provides decision support is usually on-line (directly connected to the computer system) and related to real time (providing immediate results). Computer terminals and display screens are examples—these devices can provide immediate information and answers to questions.	An MIS, using printed reports that may be delivered to managers once a week, may not provide immediate results.
Speed	Because a DSS is flexible and can be implemented by users, it usually takes less time to develop and is better able to respond to user requests.	An MIS's response time is usually longer.
Output	DSS reports are usually screen oriented, with the ability to generate reports on a printer.	An MIS, however, typically is oriented toward printed reports and documents.
Development	DSS users are usually more directly involved in its development. User involvement usually means better systems that provide superior support. For all systems, user involvement is the most important factor for the development of a successful system.	An MIS is frequently several years old and often was developed for people who are no longer performing the work supported by the MIS.

helps deliver the right information to the right people at the right time so that they can discover critical factors and trends that drive the business, then use that knowledge to optimize company performance. In decentralized enterprises, users can have access to the same business intelligence that they would if they were local and without dedicated network connections. For example, Pilot Software offers Pilot Internet Publisher (PIP), which delivers powerful decision support solutions on the World Wide Web to users at corporate or remote locations through popular Web browser interfaces.[3]

Many enterprises have a critical need to provide a growing population of decision makers with powerful access to and analysis of multidimensional data. For these enterprises, the combination of databases and Web technology delivers a complete Web-enabled decision support solution. This combination also provides users with application models and enables them to create sophisticated interactive decision support solutions themselves. Some decision support software vendors have even developed software that can inquire about the structure of any DSS data model and reflect that structure automatically in the Web interface. This approach provides extremely fast deployment of DSS tools to users. Indeed, users can be up and running as soon as the data model is loaded because no screen development time is required. The Kennedy Space Center in Cape Canaveral, Florida, has developed a Web system to integrate the databases involved in preparing and launching the space shuttle. The system uses Web-based decision support software to make data accessible to NASA engineers as well as contractors via a Web browser. Read the "E-Commerce" box for another example of a company using a Web-based decision support system.

COMPONENTS OF A DECISION SUPPORT SYSTEM

dialogue manager

a component of the DSS that allows decision makers to easily access and manipulate the DSS and to use common business terms and phrases

At the core of a DSS are a database and a model base. In addition, a typical DSS contains a **dialogue manager**, which allows decision makers to easily access and manipulate the DSS and to use common business terms and phrases. External database access allows the DSS to tap into vast stores of information contained in the corporate database, letting the DSS retrieve information on inventory, sales, personnel, production, finance, accounting, and other areas. Finally, access to the Internet, networks, and other computer-based systems permits the DSS to tie into other powerful systems, including the TPS or function-specific subsystems. Figure 10.9 shows a conceptual model of a DSS. Since database and database management system concepts were thoroughly covered in Chapter 5 and networks and the Internet were covered in Chapters 6 and 7, we begin with a discussion of the model base.

model base

a component of the DSS that provides decision makers with access to a variety of models and assists them in decision making

The Model Base

The purpose of the **model base** in a DSS is to give decision makers access to a variety of models and to assist them in the decision-making process. (You may want to refer to Chapter 1 for a discussion and examples of models.) The model base can include **model management software (MMS)** that coordinates the use of models in a DSS. Figure 10.10 on page 456 gives an overview of some of the models used in a DSS model base. Depending on the needs of the decision maker, one or more of these models can be used.

model management software (MMS)

software that coordinates the use of models in a DSS

E-COMMERCE

Budgeting Decision Support System Helps Gulf Canada

Gulf Canada, an oil and exploration corporation, has operations and interests around the globe. Units, such as Operator Ranger, are constantly seeking new oil and gas reserves and investment opportunities. This unit recently received licenses to explore for oil off Côte d'Ivoire in Africa. With subsidiaries, partnerships, and consortia operating in remote locations, Gulf Canada has to carefully allocate its financial resources to achieve the highest return on its exploration and extraction investments. One way to accomplish this is an effective budgeting system.

Whether a company is involved in the high-tech telecommunications industry or the capital-intensive oil and gas exploration business, budgeting is an effective way to distribute and monitor financial resources. For Gulf Canada, budgeting is critical. In addition to monitoring how financial resources are used in its operations, the budgeting process also helps determine what is needed to make acquisitions and enter into partnerships with other companies.

The old budgeting system for the $4.48 billion company was not adequate. Hundreds of managers did not receive timely budget information. To develop a better system, the company decided to acquire an effective budgeting decision support system (DSS). Gulf Canada also had a tight 90-day deadline to implement the new system before the next budgeting cycle started. Because of the tight deadline, Gulf Canada decided that developing its own intranet-based budgeting system would be faster and better than trying to have an outside consulting company develop one. Gulf Canada invested about $500,000 in an intranet budgeting system. A primary objective of the budgeting DSS was to make more accurate forecasts and help the company make acquisitions of natural resources faster and more effectively.

The intranet-based budgeting system starts with field foremen who compile budget data related to thousands of active oil and gas drilling sites. Using the intranet, the foremen are able to send budgeting data to corporate budgeting personnel. The old budgeting process involved several layers of management and communications delays. The foremen typically passed their budgeting data to their bosses. The data was reviewed and passed up the line to other bosses and managers until it reached the corporate level. With the budgeting DSS, the information can be received at the corporate offices immediately.

The budgeting DSS is based on an object-oriented software package by Cactus. The implementation of the DSS started with installing Internet browsers at the offices of the field foremen. Once this was accomplished, the Cactus software package was installed. The budgeting DSS runs on Sun Microsystems workstations. Each workstation is equipped with an Oracle database.

The new budgeting DSS has been worth the investment. With the new system, field foremen can continuously monitor the operation of thousands of drilling sites and other operations. The data can be immediately transferred to corporate level managers. This gives Gulf Canada managers access to accurate information and allows them to react to any budgeting irregularities, such as a site or operation that is spending too much for the revenues being generated. With the new system, managers are also able to compare forecasts to actual production numbers captured by the firm's transaction processing system. Not only can this help managers monitor current costs, but it can also help refine the forecasting process.

DISCUSSION QUESTIONS

1. Describe the old budgeting system and the new budgeting DSS. What hardware and software is being used in the new system?
2. What are the benefits of the new budgeting DSS? Are there any disadvantages or potential problems with the new system?

Sources: Thomas Hoffman, "Gulf Canada Refines Budgeting," *Computerworld*, January 12, 1998, p. 39; Alakun Hakhi, "Content by Committee," *InformationWeek*, January 12, 1998, p. 67; Robert Daly, "Budgeting the Bandwidth," *PC Magazine*, February 24, 1998, p. 42; "Ranger Group to Explore off Côte d'Ivoire," *Oil and Gas Journal*, January 26, 1998, pp. 42–44.

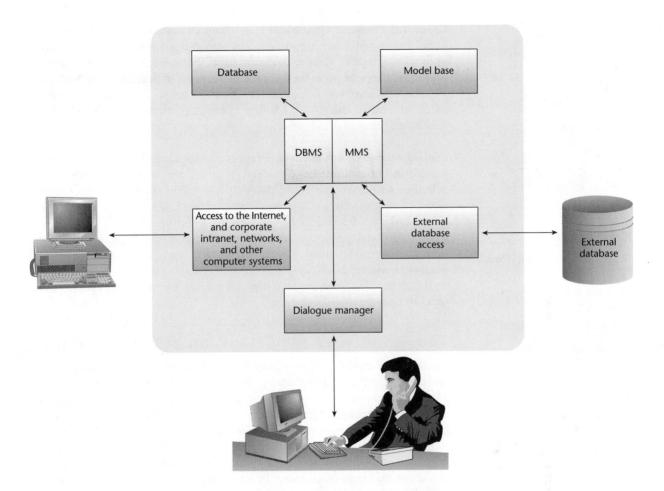

FIGURE 10.9

A Conceptual Model of a DSS

DSS components include a model base; database; external database access; access to the Internet and corporate intranet, networks, and other computer systems; and a dialogue manager.

financial models

DSS models that provide cash flow, internal rate of return, and other investment analysis

statistical analysis models

DSS models that provide such data as summary statistics, trend projections, and hypothesis testing

Financial models. Financial models provide cash flow, internal rate of return, and other investment analysis. Spreadsheet programs such as Excel are often used for this purpose. In addition, more sophisticated financial planning and modeling programs can be employed. Some organizations develop customized financial models to handle the unique situations and problems faced by the organization. However, as spreadsheet packages continue to increase in power, the need for sophisticated financial modeling packages may decrease.

Statistical analysis models. Statistical analysis models can provide summary statistics, trend projections, hypothesis testing, and more. These programs are available on both personal and mainframe systems. Many software packages, including SPSS and SAS, provide outstanding statistical analysis for organizations of all sizes. These statistical programs can compute averages, standard deviations, correlation coefficients, and regression analysis, do hypotheses testing, and use many more techniques. Some statistical analysis programs also have the ability to produce graphic displays that reveal the relationship between variables or quantities. Albertson's Inc. and other supermarket chains are making use of sophisticated statistical analysis models to better understand how effectively an advertising promotion worked—were the additional costs associated with the promotion offset by the increased revenue from increased sales?[4]

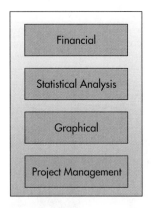

FIGURE 10.10

The Major Types of DSS Models

graphical modeling programs

software that assists decision makers in designing, developing, and using graphic displays of data and information

project management models

DSS models that are used to handle and coordinate large projects; they are also used to identify critical activities and tasks that could delay or jeopardize an entire project

Spreadsheet programs are often used for financial planning and modeling.

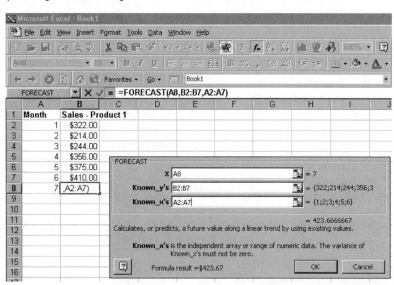

Graphical models. Graphical modeling programs are software packages that assist decision makers in designing, developing, and using graphic displays of data and information. Numerous personal computer programs that can perform this type of analysis, including PowerPoint and FreeLance Graphics, are available on the market. In addition, sophisticated graphical analysis, including computer-assisted design (CAD), is available (Figure 10.11).

Project management models. Project management models are used to handle and coordinate large projects; they are also used to identify critical activities and tasks that could delay or jeopardize an entire project if they are not completed in a timely and cost-effective fashion. Some of these programs can also determine the best way to speed up a project by effectively using additional resources, including cash, labor, and equipment. Project management models allow decision makers to keep tight control over projects of all sizes and types.

Read the "FYI" box to learn how airlines are using decision support systems to operate their businesses.

The Advantages and Disadvantages of Modeling

The use of a model base in a DSS has many advantages. Modeling can be less expensive than experimenting with custom approaches or real systems. Models can be more cheaply constructed and manipulated to determine the impact of various decisions. In a chemical processing system, for example, mathematical models can be developed to investigate new ways to refine chemicals or provide new products and materials, whereas developing small pilot plants and actually producing the materials can be both expensive and time consuming. In addition, modeling is usually faster than experimenting with real systems. In many cases, models can be developed, tested, and analyzed within a matter of weeks, while constructing real systems could take months or years. Modeling is less risky than experimenting on real systems yet still shows how a decision might affect the overall system. For example, a drug company experimenting with new procedures and approaches could develop models to test these new procedures before they are tried on animals or human beings. It is far better to have models show potential weaknesses and problems with new drugs than to have animals or humans risk their health or lives in actual testing.

Modeling can also provide managers with an excellent learning experience. Managers can experiment with the models and see the effects immediately. Various decision strategies and alternatives can be explored rapidly without causing

FYI

DSS Helps Lighten United Airlines Load

United Airlines, with 90,000 employees and thousands of destinations worldwide, uses decision support systems to improve customer service while reducing airline costs. United's Gate Assignment and Planning System and Customer Services Employee Management System are real-time DSSs designed to improve the efficiency of operations within major United hubs, including Chicago's O'Hare International Airport and Denver's International Airport. The systems display real-time information about flights around the world, including what equipment is used, where the delays are, and when arrivals and departures are scheduled.

The Gate Assignment and Planning System improves the management of gate conflicts, reduces traffic congestion, displays real-time information, and includes a planning model for future scheduling. The Customer Services Employee Management System determines the time and number of customer service representatives required for each arrival and departure, updates shift times and the status of each employee, accepts customizable rules by aircraft type and flight load, and displays flight status and capacity by class and service.

United is now in the second phase of developing its system, which runs on an IBM SP2 massively parallel processing computer. The goal of this system is to look at the origin and destination of passengers and apply that to future flights and fares to allocate the correct number of seats to maximize revenue.

DISCUSSION QUESTIONS

1. What are some real-life situations that may occur that might significantly throw off either the Gate Assignment and Planning System or the Customer Services Employee Management System?
2. Which characteristics of a DSS would be most important for a successful yield management system? Why? Which types of models are most likely to be used in the yield management system?

Sources: Adapted from the "Our Company" and "Products/Services" sections of the United Airlines Web site at http://www.ual.com, accessed May 23, 1998 and Stephanie Neil, "Guiding Airlines to Greater Efficiency," *PC Week*, January 19, 1998, pp. 79–80.

---

undue harm to the company, organization, clients, or the environment. This learning process can help companies avoid mistakes that could cost millions of dollars and years of wasted time. Models are also excellent at predicting future outcomes. With the modeling approach, complex relationships can be analyzed and projected into the future. With the globalization of communications and computer processing, it is becoming even more important to be able to predict future events. Finally, postmodeling support allows decision makers to test important assumptions of the model and to make sure that it is accurate and valid before it is used in decision making. Postmodeling support, for example, can remind a decision maker that the financial model used a 2.5 percent inflation factor and that this assumption may not be valid. Some models print warning screens or reports on key assumptions of the model, such as "Warning: The inflation factor of 2.5 percent may not be accurate. Check YES if you want to test other sales growth factors, such as 3 percent or 4 percent."

There are, however, a number of disadvantages to the modeling approach. A model by definition requires its builder to make simplifying assumptions. If the assumptions cause the model to deviate too much from reality, the results of using the model are highly suspect. With numerous choices of models, decision makers may spend a great deal of time deciding what model to use. In some cases, models do not accurately predict real systems, so results and conclusions may be false or misleading. Some models

FIGURE 10.11

Computer-assisted design (CAD) programs are powerful tools in developing products. (Source: Bob Krist/Corbis.)

group decision support system (GDSS)

a system that consists of most of the elements in a DSS, plus GDSS software needed to provide effective support in group decision-making settings

computerized collaborative work system

another term for a group decision support system

require a high degree of mathematical sophistication, making them extremely difficult to build and the results very hard to interpret. In addition, models can be expensive to develop if they are used only once.

The Dialogue Manager

The dialogue manager allows users to interact with the DSS to obtain information. It assists with all aspects of communications between the user and the hardware and software that constitute the DSS. In a practical sense, to most DSS users, the dialogue manager is the DSS. Upper-level decision makers are often less interested in where the information came from or how it was gathered than the fact that the information is both understandable and accessible.

THE GROUP DECISION SUPPORT SYSTEM

The DSS approach has resulted in better decision making for all levels of individual users. However, many DSS approaches and techniques are not suitable for a group decision-making environment. Although not all workers and managers are involved in committee meetings and group decision making sessions, some tactical and strategic-level managers can spend more than half their decision-making time in a group setting. Such managers need effective approaches to assist with group decision making. A **group decision support system (GDSS)**, also called a **computerized collaborative work system**, consists of most of the elements in a DSS, plus GDSS software needed to provide effective support in group decision-making settings (Figure 10.12).

The popularity of GDSSs comes from the trend for more teamwork.[5] Two-thirds of the nearly 2,000 companies surveyed by Hewitt Associates use formal teams to conduct work. Internal Data, a marketing research firm, estimates that the use of groupware will explode from about 30 million users in 1995 to over 250 million users by the end of this decade. Without question, GDSSs are becoming a vital decision-making tool for businesses of all sizes.

Characteristics of a GDSS

A GDSS has a number of unique characteristics that go beyond the traditional DSS. Developers of these systems try to build on the advantages of individual support systems while realizing that new and additional approaches are needed in a group decision-making environment. For example, some GDSSs have the ability to allow the exchange of information and expertise among people without meetings or direct face-to-face interaction. The following are some characteristics of a typical GDSS.

Special design. The GDSS approach acknowledges that special procedures, devices, and approaches are needed in group decision-making

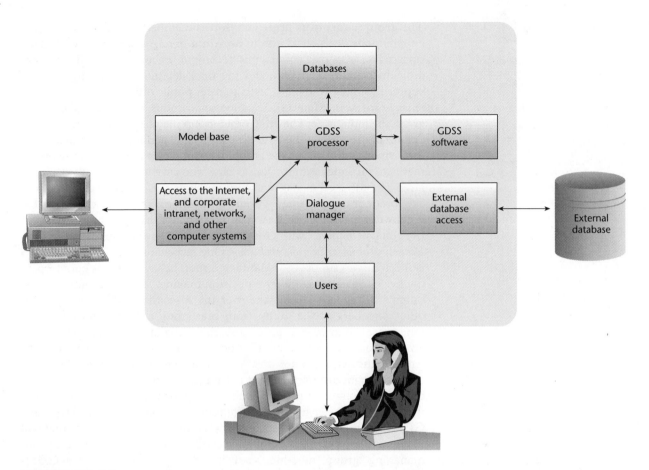

FIGURE 10.12

Configuration of a GDSS

A GDSS contains most of the elements found in a DSS, plus GDSS software to facilitate group member communications.

delphi approach

a decision-making approach in which group decision makers are geographically dispersed throughout the country or the world

brainstorming

a decision-making approach that fosters creativity and free thinking by allowing members of the group to offer ideas "off the top of their heads"

settings. These procedures must foster creative thinking, effective communications, and good group decision-making techniques.

Ease of use. Like an individual DSS, a GDSS must be easy to learn and use. Systems that are complex and hard to operate will seldom be used. Many groups have less tolerance than do individual decision makers for poorly developed systems.

Flexibility. Two or more decision makers working on the same problem may have different decision-making styles and preferences. Each manager makes decisions in a unique way, in part because of different experiences and cognitive styles. An effective GDSS not only has to support the different approaches that managers use to make decisions but also must find a means to integrate their different perspectives into a common view of the task at hand.

Decision-making support. A GDSS can support different decision-making approaches, including the **delphi approach**, in which group decision makers are geographically dispersed throughout the country or the world. This approach encourages diversity among group members and fosters creativity and original thinking in decision making. Another approach, called **brainstorming**, which often consists of members offering ideas "off the top of their heads," fosters creativity and free thinking.

group consensus approach

a decision-making approach that forces members in the group to reach a unanimous decision

nominal group technique

a decision-making approach that encourages feedback from individual group members, and the final decision is made by a voting approach similar to the one used to elect public officials

The **group consensus approach** forces members in the group to reach a unanimous decision. With the **nominal group technique**, each decision maker can participate; this technique encourages feedback from individual group members, and the final decision is made by a voting approach similar to the one used to elect public officials.

Anonymous input. Many GDSSs allow anonymous input, where the person giving the input is not known to other group members. For example, some organizations use a GDSS to help rank the performance of managers. Anonymous input allows the group decision makers to concentrate on the merits of the input without considering who gave it. In other words, input given by a top-level manager is given the same consideration as input from employees or other members of the group. Some studies have shown that groups using anonymous input can make better decisions and have superior results compared with groups that do not use anonymous input.

Reduction of negative group behavior. One key characteristic of any GDSS is the ability to suppress or eliminate group behavior that is counterproductive or harmful to effective decision making. In some group settings, dominant individuals can take over the discussion, which can prevent other members of the group from presenting creative alternatives. In other cases, one or two group members can sidetrack or subvert the group into areas that are nonproductive and do not help solve the problem at hand. Other times, members of a group may assume they have made the right decision without examining alternatives—a phenomenon called groupthink. If group sessions are poorly planned and executed, the result can be a tremendous amount of wasted time. Today, many GDSS designers are developing software and hardware systems that will reduce these types of problems. Procedures for effectively planning and managing group meetings can be incorporated into the GDSS approach. A trained meeting facilitator is often employed to help lead the group decision-making process and to avoid groupthink.

Parallel communication. With traditional group meetings, people must take turns addressing various issues. One person normally talks at a time. With a GDSS, it is possible for every group member to address issues or make comments at the same time by entering them into a PC or workstation. These comments and issues are displayed on every group member's PC or workstation immediately. Parallel communication can speed meeting times and result in better decisions.

Automated record keeping. Most GDSSs have the ability to keep detailed records of a meeting automatically. Each comment that is entered into a group member's PC or workstation can be anonymously recorded. In some cases, literally hundreds of comments can be stored for future review and analysis. In addition, most GDSS packages have automatic voting and prioritization features. After group members vote, the GDSS records each vote and makes the appropriate rankings.

Cost, control, and complexity factors. Before using a GDSS, the characteristics just mentioned should be compared with cost, control, and complexity issues. A GDSS can be expensive, requiring a large number of personal computers, sophisticated GDSS software, networks, personnel, and support. The complexity of the various GDSS components, including software, and the way the GDSS is to be controlled and used are also factors to consider before acquiring and implementing a GDSS.

Components of a GDSS and GDSS Software

As noted earlier, a typical GDSS configuration includes some of the components of a DSS—a database, a model base, and a dialogue manager. However, the components used in GDSSs and their interaction vary from single-user DSSs. Many of these systems allow multiple users simultaneous access to common files, databases, and the Internet, allowing group members to work on the same task while in group settings. Also, systems that provide group decision support are generally networked. Once connected, the GDSS can support group decisions among geographically separated individuals.

GDSS software is the heart of the GDSS. It offers many useful tools for group work (Figure 10.13). For example, **compound documents** that can include information from spreadsheet programs, database packages, word processors, and other applications can be created, used, and shared by the group. Compound documents can even include multimedia data, like audio and video clips. Compound documents are stored in a single file, while traditional applications require separate files for each different application (say, one for word processing, one for graphics, and so on). Some GDSS programs also allow compound documents to include applications from different software companies.

GDSS software, often called groupware or workgroup software, helps with joint workgroup scheduling, communication, and management. One popular package, Lotus Notes, can capture, store, manipulate, and distribute memos and communications that are developed during group projects.[6] (See Figure 10.14.) Microsoft's NetMeeting product supports application sharing in multiparty calls. It gains performance by sharing one task at a time. The user designated to be the "operator" must select one previously launched application, and each participant has to choose whether or not to "collaborate." Any collaborating participant can assume mouse control and work in the shared application while others watch.[7] Exchange from Microsoft is another example of groupware. This software allows users to set up electronic bulletin boards, schedule group meetings, and use e-mail in a group setting. Other GDSS software packages include Collabra Share, OpenMind, and TeamWare. All these tools can aid in group decision making.

A number of companies are using GDSS software to their advantage. Price Waterhouse uses groupware to coordinate group work and consulting. The company posts notes on over 1,000 bulletin boards on a variety of subjects. Any one of the company's 18,000 workers in 22 different countries can access the appropriate bulletin board through the groupware. CIGNA Corporation of Philadelphia holds workgroup meetings over a network to encourage brainstorming. Because group members remain unseen, workers can be freely creative without fearing rejection or worrying about which ideas were the president's. Domecq Importers, a liqueur importer in Connecticut, uses a similar approach to help generate ideas for promotions.

compound documents

components of GDSS software that include information from spreadsheet programs, database packages, word processors, and other applications

FIGURE 10.13

A GDSS can be a useful tool in the selection of a job applicant. This screen from Ventana Group Systems shows the group results matrix. Green cells indicate consensus, while red cells indicate lack of consensus. As participants complete their votes, the cells' colors change dynamically to show the latest consensus.
(Source: Courtesy of Ventana Corporation. Group Systems is a registered trademark of Ventana Corporation.)

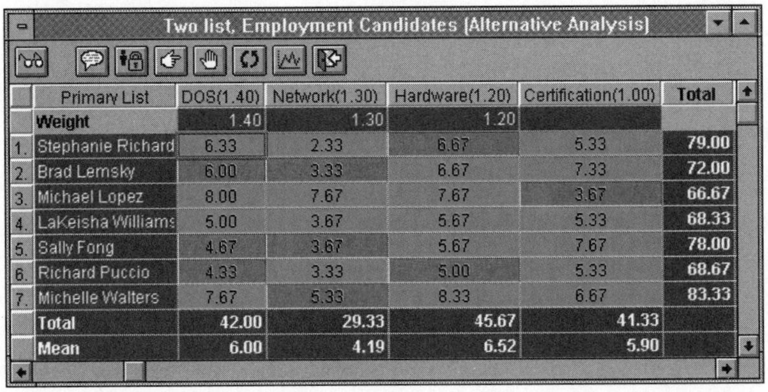

	Primary List	DOS(1.40)	Network(1.30)	Hardware(1.20)	Certification(1.00)	Total
	Weight	1.40	1.30	1.20		
1.	Stephanie Richard	6.33	2.33	6.67	5.33	79.00
2.	Brad Lemsky	6.00	3.33	6.67	7.33	72.00
3.	Michael Lopez	8.00	7.67	7.67	3.67	66.67
4.	LaKeisha Williams	5.00	3.67	5.67	5.33	68.33
5.	Sally Fong	4.67	3.67	5.67	7.67	78.00
6.	Richard Puccio	4.33	3.33	5.00	5.33	68.67
7.	Michelle Walters	7.67	5.33	8.33	6.67	83.33
	Total	42.00	29.33	45.67	41.33	
	Mean	6.00	4.19	6.52	5.90	

Two list, Employment Candidates (Alternative Analysis)

FIGURE 10.14

Workgroup software, such as
Lotus Notes, allows people
located around the world to
work on the same project,
documents, and files efficiently
and at the same time.
(Source: Courtesy of Lotus
Development Corporation.)

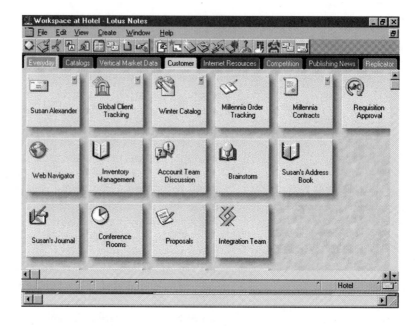

GDSS Alternatives

Group decision support systems can take on a number of alternative network configurations, depending on the needs of the group, the decision to be supported, and the geographic location of group members. The frequency of GDSS use and the location of the decision makers are two important factors (Figure 10.15).

The decision room. The **decision room** is ideal for situations in which decision makers are located in the same building or geographic area and the decision makers are occasional users of the GDSS approach. In these cases, one or more decision rooms or facilities can be set up to accommodate the GDSS approach. Groups, such as marketing research teams, production management groups, financial control teams, or quality-control committees, can use the decision rooms when needed. The decision room alternative combines face-to-face verbal interaction with technology-aided

decision room

a facility that supports decision
making when decision makers are
in the same building, combining
face-to-face verbal interaction with
technology-aided formalization to
make the meeting more effective
and efficient

FIGURE 10.15

GDSS Alternatives

This figure demonstrates that
the decision room may be
the best alternative for group
members who are located
physically close together and
who need to make infrequent
decisions as a group. By the
same token, group members
who are situated at distant
locations and who frequently
make decisions together may
require a wide area
decision network to
accomplish their goals.

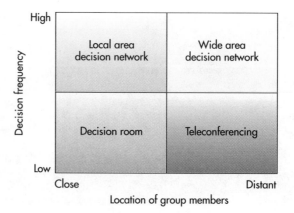

FIGURE 10.16

The GDSS Decision Room

For group members who are located in the same physical location, the decision room is an optimal GDSS alternative. By use of networked computers and computer devices, such as project screens and printers, the meeting leader can pose questions to the group, instantly collect their feedback, and, with the help of the governing software loaded on the control station, process this feedback into meaningful information to aid in the decision-making process.

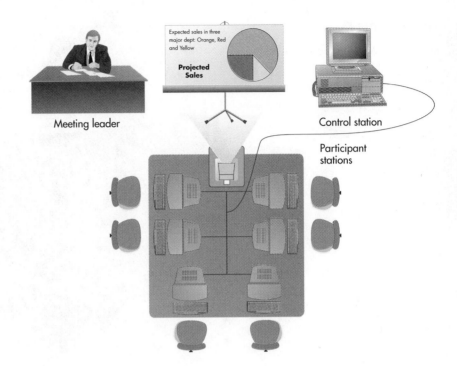

formalization to make the meeting more effective and efficient. A typical decision room is shown in Figure 10.16.

The local area decision network. The local area decision network can be used when group members are located in the same building or geographic area and under conditions in which group decision making is frequent. In these cases, the technology and equipment of the GDSS approach is placed directly into the offices of the group members. Usually this is accomplished via a local area network (LAN).

The teleconferencing alternative. The teleconferencing alternative is used for situations in which the decision frequency is low and the location of group members is distant. These distant and occasional group meetings can tie together multiple GDSS decision-making rooms across the country or around the world. Using long-distance communications technology, these decision rooms are electronically connected in teleconferences and videoconferences. This alternative can offer a high degree of flexibility. The GDSS decision rooms can be used locally in a group setting or globally when decision makers are located throughout the world.

The wide area decision network. The wide area decision network is used for situations in which the decision frequency is high and the location of group members is distant. In this case, the decision makers require frequent or constant use of the GDSS approach (Figure 10.17). This situation requires decision makers located throughout the country or the world to be linked electronically through a wide area network (WAN). In most cases, the GDSS components shown in Figure 10.17 are used. The group facilitator

FIGURE 10.17

A Wide Area Decision Network

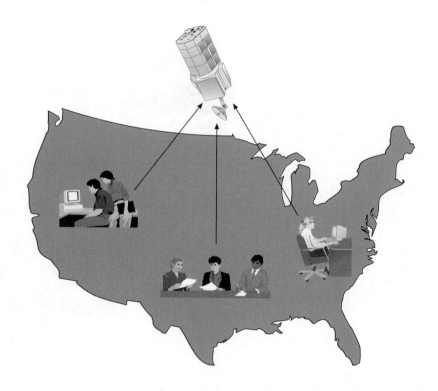

virtual workgroups

teams of people located around the world working on common problems

and all group members are at geographically dispersed locations. In some cases, the model base and database are also geographically dispersed. This GDSS alternative allows people to work in **virtual workgroups**, where teams of people located around the world can work on common problems.[8]

THE EXECUTIVE SUPPORT SYSTEM

executive support system (ESS) or executive information system (EIS)

a specialized DSS that includes all hardware, software, data, procedures, and people used to assist senior-level executives within the organization

Because top-level executives often require specialized support when making strategic decisions, many companies have developed systems to assist executive decision making.[9] This type of system, called an **executive support system (ESS)**, is a specialized DSS that includes all hardware, software, data, procedures, and people used to assist senior-level executives within the organization. In some cases, an ESS, also called an **executive information system (EIS)**, supports the actions of members of the board of directors, who are responsible to stockholders. These top-level decision-making strata are shown in Figure 10.18.

An ESS can also be used by individuals farther down in the organizational structure. Once targeted at the top-level executive decision makers, ESSs are now marketed to—and used by—employees at other levels in the organization. In the traditional view, ESSs give top executives a means of tracking critical success factors. Today, all levels of the organization share information from the same databases. However, for our discussion, we will assume ESSs remain in the upper management levels, where they indicate important corporate issues and new directions the company may take, and help executives monitor the company's progress.

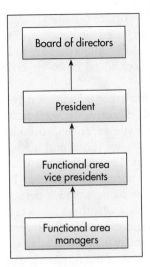

FIGURE 10.18

The Layers of Executive Decision Making

Executive Support Systems in Perspective

An ESS is a special type of DSS, and, like a DSS, an ESS is designed to support higher-level decision making in the organization. The two systems are, however, different in important ways. DSSs provide a variety of modeling and analysis tools to enable users to thoroughly analyze problems—that is, they allow users to *answer* questions. ESSs present structured information about aspects of the organization that executives consider important—in other words, they allow executives to *ask* the right questions.

The following are general characteristics of ESSs.

- *Tailored to individual executives.* ESSs are typically tailored to individual executives; DSSs are not tailored to particular users. As such, ESSs are truly representative of the overall objective of information systems: to deliver the right information to the right person at the right time in the right format. An ESS is an interactive, hands-on tool that allows an executive to focus, filter, and organize data and information.
- *Easy to use.* A top-level executive's most critical resource can be his or her time. Thus, an ESS must be easy to learn and use and not overly complex. Early ESSs were noted for their extreme ease of use, since they were targeted for decision makers who were often not technically oriented. The extensive use of color and graphics is common with an ESS.
- *Have drill down abilities.* An ESS allows executives to "drill down" into the company to determine how certain data was produced. Drill down allows an executive to get more detailed information if needed. For example, after seeing a report summarizing the progress and costs of a large project, an executive might want to drill down to see the status of a particular activity within the project.
- *Support the need for external data.* The data needed to make effective top-level decisions is often external—information from competitors, the federal government, trade associations and journals, consultants, and so on. This data is often "soft" as well. An effective ESS is able to extract data useful to the decision maker from a wide variety of sources including the Internet and other electronic publishing sources such as Lexis/Nexus. Often they may employ intelligent agents to search the Web for news about competitors, suppliers, customers, and key business trends. Typically, decisions made at top levels are also complex and unstructured. Traditional computer programs are normally ineffective for these types of decisions. As a result, more advanced ESSs are required to help support all types of strategic decisions.
- *Can help with situations that have a high degree of uncertainty.* There is a high degree of uncertainty with most executive decisions. What will happen if a new plant is started? What will be the consequences of a potential merger? How will union negotiators take a new proposal? The answers to these questions are not known with certainty. Handling these unknown situations using modeling and other ESS procedures helps top-level managers measure the amount of risk in a decision.

- *Have a futures orientation.* Executive decisions are future oriented, meaning that decisions will have a broad impact for years or decades. The information sources to support future-oriented decision making are usually informal—from golf partners to members of social clubs or civic organizations.
- *Are linked with value-added business processes.* Like other information systems, executive support systems are linked with executive decision making about value-added business processes. For instance, an ESS used by the Hertz car rental company allows executives to see rental and leasing trends for a particular area by touching a map. The system has been especially profitable in analyzing Hertz's corporate rate policy. By detecting which firms generate enough business to be worth a certain discount, executives can ask questions to determine why these firms create high revenue and how this business can be continued or bettered. They might also use these factors to identify similar firms with business potential. Strategic planning of this type is just one of the functions that can be aided by an ESS.

Capabilities of an Executive Support System

The responsibility given to top-level executives and decision makers brings unique problems and pressures to their jobs. The following is a discussion of some of the characteristics of executive decision making that are supported through the ESS approach. As you will note, most of these are related to an organization's overall profitability and direction. An effective ESS should have the capability to support executive decisions with many of these capabilities, such as strategic planning and organizing, crisis management, and more.

Support for defining an overall vision. One of the key roles of senior executives is to provide a broad vision for the entire organization. This vision includes the organization's major product lines and services, the types of businesses it supports today and in the future, and its overriding goals.

Support for strategic planning. ESSs also support strategic planning. **Strategic planning** involves determining long-term objectives by analyzing the strengths and weaknesses of the organization, predicting future trends, and projecting the development of new product lines. It also involves planning the acquisition of new equipment, analyzing merger possibilities, and making difficult decisions concerning downsizing and the sale of assets if required by unfavorable economic conditions.

Support for strategic organizing and staffing. Top-level executives are concerned with organization structure. For example, decisions concerning the creation of new departments or downsizing the labor force are made by top-level managers. Should the information systems department be placed under new leadership? Should the accounting and finance departments be joined under a new vice president of financial services? Should the marketing department be divided along the company's two major product lines? These and similar questions can profoundly affect the overall effectiveness of the organization and should be supported by an ESS.

strategic planning

planning that involves determining long-term objectives by analyzing the strengths and weaknesses of the organization, predicting future trends, and projecting the development of new product lines

Overall direction for staffing decisions and effective communication with labor unions are major decision areas for top-level executives. Middle- and lower-level managers make staffing decisions, but general decisions about the types and number of employees for various departments in the organization are determined at the top level of the organization. In addition, top-level managers are responsible for labor negotiations. Thus, ESSs can be employed to help analyze the impact of staffing decisions, potential pay raises, changes in employee benefits, and new work rules.

Support for strategic control. Another type of executive decision relates to strategic control, which involves monitoring and managing the overall operation of the organization. Goal seeking can be done for each major area to determine what performance these areas need to achieve to reach corporate expectations. Effective ESS approaches can help top-level managers make the most of their existing resources and control all aspects of the organization.

Support for crisis management. Even with careful strategic planning, a crisis can occur. Major disasters, including hurricanes, tornadoes, floods, earthquakes, fires, and sabotage, can totally shut down major parts of the organization. Handling these emergencies is another responsibility for top-level executives. In many cases, strategic emergency plans can be put into place with the help of an ESS. These contingency plans help organizations recover quickly if an emergency or crisis occurs. Read the "Ethical and Societal Issues" box to learn about the use of competitive intelligence in an ESS.

ETHICAL AND SOCIETAL ISSUES

Use of Competitive Intelligence in ESS

Decision making requires input from numerous sources, one of which should be an understanding of the competition and their probable strategy in the marketplace. Competitive intelligence is designed to provide that input. That intelligence might concern the opening of a competitor's new plant, a rival's efforts to raise additional cash for expansion, the design and price of a competitor's upcoming product, the benefits of acquiring a rival business, or the dangers of entering a new market.

The primary goals of competitive intelligence are to avoid surprises, identify threats and opportunities, reduce reaction time, and outwit the competition. Competitive intelligence by itself has little value and should be combined with other data used in strategic planning, market research, and scenario development to improve the decision-making process. Many companies enable senior executives to access competitive intelligence via an executive information system.

The gathering of competitive intelligence is increasing rapidly. The Society of Competitive Intelligence Professionals has grown from a few dozen members in 1986 to more than 6,000 in 1998, doubling in the past two years. Companies such as Microsoft, Motorola, IBM, Procter & Gamble, General Electric, Hewlett-Packard, Coca-Cola, and Intel have become highly skilled at digging up information about competitors. Kellogg Co. hired ex-intelligence officers to beef up their operations. Intel has recruited ex-CIA officers with training in disinformation.

Louis Gerstner, CEO of IBM, set up a squad of competitive intelligence teams, each under a different senior executive. One group collects data about Hewlett-Packard, another about Compaq. Jan Herring, an intelligence consultant who used to work for the CIA, helped IBM establish this system. IBM's intelligence teams target their competitors' consultants, suppliers, customers, and even employees. Young engineers from competing companies are a favorite source of information. While new engineers may not know a great deal, intelligence people are skilled at taking bits of data and assembling them into meaningful patterns. IBM's teams gather bits and pieces of such data and download them into an ESS, accessible to the firm's top executives. Much of the advantage gained is tactical. When a competitor is launching a new product, IBM wants to learn about it in advance. IBM's sales force calls an 800 number to hear details about the rival's product, discounts, and sales pitches.

Larry Hamilton, Dow Corning's manager of global business intelligence, has developed an "early warning" system to detect moves by its competitors in the United States, Japan, Korea, France, and Germany. By tracking competitors' environmental filings, Dow obtains advance knowledge of their expansion plans. "You're not doing anything illegal," says Hamilton. "You're just being smart."

The Society of Competitive Intelligence Professionals promotes strict ethical standards. For example, they recommend that intelligence collectors identify their employers to every source. Most firms do warn their intelligence officers not to misrepresent themselves. But the rules are not always strictly followed. Investigators frequently pose as EPA or OSHA officials to trick companies into giving them information.

Many believe that there is enough information available from above board sources to make unethical methods unnecessary. Eighty percent of the information your company needs is available publicly or from the people in your own organization. The real value comes from the synthesis and analysis you do with the information.

The most important component of a competitive intelligence program is the analysts who read competitors' newspaper ads, prowl trade shows, and work the phones. Competitive intelligence information systems have been built around technology as simple as Internet access and e-mail. However, these applications simply deliver and organize information; they do not analyze the data and hence, do not create intelligence. There is a growing demand for analysis packages that really work.

One such tool is called KnowledgeX. IBM acquired this product from a company called Integrated Technology and will integrate it into future versions of its DB2 Universal Database and business-intelligence products. KnowledgeX lets users enter words or concepts they want to research. It then gathers and analyzes information from sources such as databases, news feeds, and the Internet. It displays the documents it retrieves as icons and draws lines between documents with related content.

Atlantic Richfield Co. is conducting a pilot project using KnowledgeX as a tool to analyze data about its competitors in the petroleum industry. KnowledgeX enables its users to go into that data, organize it, and see how it all relates. Professional services firm Arthur Andersen, electronics maker Texas Instruments, and software vendor SAP America are also evaluating KnowledgeX. Other products that offer visual maps of associated information are CBR Content Navigator from Inference Corp. in Novato, California, and Vineyard from Data Fellows Inc. in San Jose, California.

Discussion Questions
1. Your firm has decided to start a competitive intelligence group and has named you as its head. As your first task, you are to draft a "code of ethics" by which the group is to abide. State a few of the key elements of the team's charter.
2. Outline some of the approaches you will encourage your group to use to gather competitive intelligence.

Sources: Adapted from "Why Competitive Intelligence?" found at the J Thomas Group Web site at http://www.mindspring.com, accessed August 23, 1998; Jason Hibbard, "Net Info Gets Organized," *InformationWeek Online*, News in Review, July 28, 1998; Justin Hibbard, "IBM Acquires Knowledge-Management Technology," *InformationWeek*, July 24, 1998; Gary H. Anthes, "Competitive Intelligence," *Computerworld*, July 6, 1998, pp. 62-63; and William Green, "I Spy," *Forbes*, April 20, 1998.

Decision making is a vital part of managing businesses strategically. IS systems such as decision support, group decision support, and executive support systems help employees by tapping existing databases and providing them with current, accurate information. The increasing integration of all business information systems—from TPSs to MISs to DSSs—can help organizations monitor their competitive environment and make better-informed decisions.

Information Systems Principles

Problem solving can be divided into five stages: intelligence, design, choice, implementation, and monitoring. These stages can apply to a wide range of problems, including the development of an information system.

Management information systems are used to address structured problems; decision support systems are used when the problems are more unstructured.

While an information system cannot prevent a person from making a mistake in decision making or problem solving, it can help identify certain potential mistakes and provide a structure that makes it more difficult for a person to make a mistake.

Various factors (decision objectives, increased alternatives, competition, creativity, social and political actions, international aspects, technology, and time compression) are important in solving problems. An awareness of these factors increases a manager's ability to properly analyze a problem and make a good decision.

With the use of decision support systems, employees risk losing touch with the underlying principles that guide the enterprise.

The responsibility given to top-level executives and decision makers brings unique problems and pressures to their jobs. An effective ESS should have the capability to support executive decisions in the areas of strategic planning and organizing, crisis management, and more.

■ SUMMARY

1. Outline and briefly describe the stages of a problem-solving process.

Problem solving begins with decision making. A well-known model developed by Herbert Simon divides the decision-making phase of the problem-solving process into three stages: intelligence, design, and choice. The three phases of decision making are augmented by implementation and monitoring to result in problem solving.

The first stage in the decision-making phase of the problem-solving process is the intelligence stage. During this stage, potential problems and/or opportunities are identified and defined. Information is gathered that relates to the cause and scope of the problem. Constraints on the possible solution and the problem environment are investigated. In the design stage, alternative solutions to the problem are developed. In addition, the feasibility and implications of these alternatives are evaluated. The last stage of the decision-making phase, the choice stage, requires selecting a course of action.

Problem solving includes and goes beyond decision making. It also includes the implementation stage, when the solution is put into effect. The final stage of the problem-solving process is the monitoring stage. In this stage, the decision makers evaluate the implementation of the solution to determine whether the anticipated results were achieved and to modify the process in light of new information learned during the implementation stage.

2. List and discuss important characteristics of decision support systems (DSSs) that give them the potential to be effective management support tools.

A decision support system (DSS) is an organized collection of people, procedures, software, databases, and devices working to support managerial decision making. DSS characteristics include the ability to handle large amounts of data and obtain and process data from different sources; provide report and presentation flexibility; support drill down analysis; perform complex statistical analysis; offer textual and graphical orientations; support optimization, satisficing, and heuristic approaches; and perform "what-if," simulation, and goal-seeking analyses.

DSSs provide assistance through all phases of the decision-making process. The degree of problem structure and scope contributes to the complexity of the decision support system. Problems may be highly structured or unstructured, infrequent in occurrence or routine and repetitive. An ad hoc DSS addresses unique, infrequent decision situations; an institutional DSS handles routine decisions. A common database is often the link that ties together a company's TPS, MIS, and DSS.

3. Identify and describe the basic components of a DSS.

The components of a DSS are the database, model base, dialogue manager, and a link to external databases, the Internet, the corporate intranet, extranets, networks, and other systems. The model base contains the models used by the decision maker, such as financial, statistical, graphical, and project management models. The dialogue manager provides a dialogue management facility to assist in communications between the system and the user. Access to other computer-based systems permits the DSS to tie into other powerful systems, including the TPS or function-specific subsystems.

4. State the goal of a group decision support system (GDSS) and identify the characteristics that distinguish it from a DSS.

A group decision support system (GDSS), also called a computerized collaborative work system, consists of most of the elements in a DSS, plus GDSS software needed to provide effective support in group decision-making settings. GDSSs are typically easy to learn and use and can offer specific or general decision-making support. GDSS software, also called groupware, is specially designed to help generate lists of decision alternatives and perform data analysis. These packages let people work on joint documents and files over a network.

The frequency of GDSS use and the location of the decision makers will influence the GDSS alternative chosen. The decision room alternative supports users in a single location that meet infrequently. Local area networks can be used when group members are located in the same geographic area and users meet regularly. Teleconferencing is used when decision frequency is low and the location of group members is distant. A wide area network is used for situations where the decision frequency is high and the location of group members is distant.

5. *Identify fundamental uses of an ESS and list the characteristics of such a system.*

Executive support systems (ESSs) are specialized decision support systems designed to meet the needs of senior management. They serve to indicate issues of importance to the organization and new directions the company may take, and help executives monitor the company's progress. ESSs are typically easy to use, offer a wide range of computer resources, and handle a variety of internal and external data. In addition, the ESS performs sophisticated data analysis, offers a high degree of specialization, and provides flexibility and comprehensive communications abilities. An ESS also supports individual decision-making styles. Some of the major decision-making areas that can be supported through an ESS are providing an overall vision, strategic planning and organizing, staffing and labor relations, crisis management, and strategic control.

■ KEY TERMS

ad hoc DSS 449

brainstorming 459

choice stage 439

compound documents 461

computerized collaborative work system 458

decision-making phase 438

decision room 462

delphi approach 459

design stage 439

dialogue manager 453

executive support system (ESS) or executive information system (EIS) 464

financial mode 455

goal-seeking analysis 448

graphical modeling program 456

group consensus approach 460

group decision support system (GDSS) 458

heuristics 441

highly structured problems 450

implementation stage 439

intelligence stage 438

model base 453

model management software (MMS) 453

monitoring stage 439

nominal group technique 460

nonprogrammed decisions 440

optimization model 441

problem solving 439

programmed decisions 440

project management models 456

satisficing model 441

semistructured or unstructured problems 450

simulation 448

statistical analysis model 455

strategic planning 466

time compression 444

virtual workgroup 464

"what-if" analysis 447

■ REVIEW QUESTIONS

1. What are the five stages of problem solving?
2. Describe the difference between a structured and an unstructured problem and give an example of each.
3. Define a decision support system. What are its characteristics?
4. What is the difference between "what-if" analysis and goal-seeking analysis?
5. What are some of the advantages and disadvantages of modeling?
6. How are transaction processing, management information, and decision support systems integrated?
7. What are the components of a decision support system?
8. Describe four models used in decision support systems.
9. What is meant by groupthink?
10. Identify eight factors that increase a manager's ability to properly analyze a problem.
11. State the objective of a group decision support system (GDSS) and identify three characteristics that distinguish it from a DSS.
12. Identify three of the group decision-making approaches often supported by a GDSS.
13. What is an executive support system? Identify three fundamental uses for such a system.

■ DISCUSSION QUESTIONS

1. What functions do decision support systems support in business organizations? You have looked at the functions supported by TPSs and MISs in Chapters 8 and 9. How does a DSS differ from these two types of systems?

2. Why are organizations providing Web-based access to their decision support systems?

3. How is decision making in a group environment different from individual decision making, and why are information systems that assist in the group environment different? What are the advantages and disadvantages of making decisions as a group?

4. Discuss the following issue: if not designed and managed properly, the "high technology" associated with GDSS can actually hinder the decision making process.

5. The use of ESSs should not be limited to the executives of the company. Do you agree or disagree? Why?

6. Imagine that you are the vice president of manufacturing for a Fortune 1000 manufacturing company. Describe the features and capabilities of your ideal ESS.

7. You are the sales forecaster for a manufacturing firm that employs a sophisticated DSS to develop product demand forecasts. What do you need to know about the DSS employed?

■ PROBLEM-SOLVING EXERCISES

1. Review the summarized consolidated statement of income for the manufacturing company shown below. Use graphics software to prepare a set of bar charts that shows data for this year compared with the data for last year.

a. Operating revenues increase by 3.5 percent while operating expenses increase 2.5 percent.

b. Other income and expenses decrease to $13,000.

c. Interest and other charges increase to $265,000.

Operating Results (in thousands)	
Operating Revenues	$2,924,177
Operating Expenses (including taxes)	2,483,687
Operating Income	440,490
Other Income and Expenses	13,497
Income before Interest and Other Charges	453,987
Interest and Other Charges	262,845
Net Income	191,142
Average Common Shares Outstanding	147,426
Earnings per share	$1.30

If you were a financial analyst tracking this company, what detail data might you need to perform a more complete analysis? Write a brief memo summarizing your data needs.

2. As the head buyer for a major supermarket chain, John is constantly being asked by manufacturers and distributors to stock their new products. Over 50 new items are introduced each week. Many times these products are launched with national advertising campaigns and special promotional allowances to both retailers, such as John's firm, and consumers. The store has only a limited amount of shelf and floor space to stock items. Thus, to add new products, the amount of shelf space allocated to existing products must be reduced, or items must be eliminated altogether.

 Develop a spreadsheet that John can use to estimate the change in profits from adding or deleting an item from inventory. The spreadsheet will include input such as estimated weekly sales in units, shelf space allocated to stock item (measured in units), total cost per unit, sales price per unit, and promotional allowance earned per unit. The spreadsheet should calculate total profit by item and then sort the rows in descending order based on total profit. Because of the limited amount of shelf space, the spreadsheet should also calculate the accumulated shelf space based on the items stocked.

■ TEAM ACTIVITY

Imagine that you and your team have decided to develop an ESS software product to support senior executives in the music recording industry. What are some of the key decisions these executives must make? Make a list of the capabilities that such a system must provide to be useful. Identify at least six sources of external information that will be useful to its users.

■ WEB EXERCISE

Group decision support systems (GDSSs) permit collaboration and joint work. Groupware, an important software component, allows people to communicate and work on joint projects. Using an Internet search engine, find one or more companies that make groupware. Describe the features of these groupware products. You may be asked to develop a report or send an e-mail message to your instructor about what you found.

■ CASES

1 Bank Uses Intranet to Support DSS

Jefferies Group, through its principal operating subsidiary Jefferies & Company, a global investment bank, is focused on capital raising, research, mergers and acquisitions, advisory and restructuring services for small- to medium-sized companies, and trading in equity and taxable fixed income securities, convertible bonds, options, futures, and international securities for institutional clients. Total revenue for 1997 was $703 million. Jefferies Group has more than 1,000 investment and technology professionals with offices in Los Angeles, New York, Chicago, Dallas, Boston, Atlanta, New Orleans, Houston, Jersey City, San Francisco, Stamford, London, Hong Kong, Zurich, and Tokyo.

Jefferies Group, Inc. also owns over 80 percent of Investment Technology Group, which through its wholly-owned, broker-dealer subsidiary, ITG Inc., is a leading provider of technology-based equity trading services and transaction research to institutional investors and brokers. ITG services help clients to sell stocks, execute trades more efficiently, and make better trading decisions. With an emphasis on ongoing research, ITG offers the following services: ITG POSIT®, an electronic stock crossing system; ITG QuantEX®, a decision support system; Electronic Trading Desk Services, offering customers trading capabilities through the ITG trading desk, which utilizes multiple sources of liquidity; ITG Platform, a PC-based routing system; and ITG ISIS, a set of analytical tools for systematically lowering transactions costs. Although Jefferies continues to reap benefits from the Investment Technology Group Inc., the two announced their intention to form separate entities. Jefferies founded New York–based ITG in 1987 and holds an 82 percent stake.

Five years ago, Jefferies sold stocks. Today, it provides corporate finance, research, and a slew of other services to institutional customers. It was not easy for the investment firm to broaden its image. The Los Angeles–based brokerage—once known exclusively for institutional or third market block stock trading—recently built a sizable, complementary corporate finance unit. This means that Jefferies plays in two hot markets: high-yield bond underwriting and the trading of blocks of stocks to investors.

To support its movement into new services, Jefferies & Co. recently implemented several intranet-accessible decision support systems. The consolidation of data from nine different data sources into a unified view of customer accounts strongly supports the company's shift to becoming a full investment bank. The system pulls data from the transaction processing systems, cleans it, and places it into one or more of the databases. The databases are broken down by business function such as trading information and financial data. The system is used by 75 salespeople and executives to determine who Jefferies's biggest customers were yesterday and whether they hit their financial targets. Currently the system only supports standard queries. This limitation was accepted to enable a faster rollout of the system. Before the databases were rolled out, it would have taken as much as a month for Jefferies's

information systems staff to generate a report for a financial consultant, so it did so only occasionally. The ultimate goal is to provide data for decision support and enable financial consultants to make buy, sell, or hold recommendations consistent with their customers' long-term goals.

1. What sort of models will be required to implement this decision support system? What kind of data will be required?
2. Although Jefferies has benefited from its association with Investment Technology Group Inc., the two firms are now splitting and going their separate ways. Can you think of some good business reasons why this is happening?

Sources: Adapted from Thomas Hoffman, "Intranet Helps in Market Shift," *Computerworld*, May 25, 1998, pp. 45-46; Ira Breskin, "Jefferies Group Is Broadening Portfolio of Financial Services," *Investor's Business Daily*, January 2, 1998; the "About Our Company" section on the Jefferies Group's Web site at http://www.jefco.com, accessed May 26, 1998; and the "Our Products" section on ITG's Web site at http://www.itginc.com, accessed May 26, 1998.

2 Decision Support for Individual Investors

The last five years has seen a volatile stock market with numerous record highs. With the huge number of investment alternatives, including mutual funds, it can be difficult for individual investors to get the decision support they need. With the increasing popularity of 401ks, IRAs, and related accounts, it is even more important for individuals to make sound financial decisions. Retirement and financial security depend on it.

With all of the choices, more individual investors are looking for help in making critical investment decisions. One source is a Web site, called Motley Fools. (In old theatre productions, the fool was the one telling the king the truth.) Started by two brothers, Motley Fools began with

an investment newsletter in 1994. The newsletter turned into a discussion board on America Online. Today, in addition to America Online, Motley Fools can also be reached directly on the Internet. The America Online site has up to 9 million hits. The Internet site has about 6 million. While the discussion board on America Online and the Web site offers market reviews, daily analysis, and recommended portfolios, the many message boards are one of the most popular features.

A message board is a location on a Web site where people can post messages that anyone visiting the discussion board can read. The message boards are where the action is. On average, close to 5,000 messages are posted to America Online and the Internet daily. The message boards are divided into a vast array of interest areas, including specific stocks, mutual funds, and other investment topics. People can post investment questions or problems and wait for answers from other people visiting the Motley Fools site. People who ask questions one day might provide answers or advice on another day. Because three years of message boards are available, it is possible to read everything that was posted by an individual for several years. This gives people visiting the site the ability to judge who is giving the best advice over the long run. It is also possible to search for investment advice on a specific stock or topic. But what if the information posted by someone is wrong? Can Motley Fools be held accountable? Because Motley Fools is a newsletter and not acting as an advisor, the company is not responsible for what other people post on their site.

1. How can the Motley Fools Web site be used to support individual investor decisions? What are the disadvantages of this type of advice?
2. If you had $10,000 to invest, how would you get decision support to help you make the best investments?

Sources: Jeff Ubois, "Fool's Logic," *Internet World*, February, 1998, p. 69; Mark Hulbert, "Practicing Versus Preaching," *Forbes*, February 23, 1998, p. 146; Robert Barker, "A New Gold Medalist in Fund Software," *Business Week*, February 23, 1998, p. 130; Toddi Gutner, "The View From the Pit," *Business Week*, March 2, 1998, p. 144.

3 Getting Decision Support for Medical Problems

Talking to a doctor about a medical problem is the traditional way to get medical advice and treatment. While there is no replacement for this type of one-on-one visit, it does not always give patients what they want. In some cases, a family doctor may not have all of the answers. If the patient needs a referral, it can take weeks or months before a specialist becomes available. In these cases, patients need a better decision support system. One approach that is increasing in popularity is seeking additional medical advice electronically.

Today, a number of services offer medical advice or diagnosis. This type of decision support can save money and lives. Some medical decision support systems are free. Many are based on the Internet. In 1996, Dr. Michel Bazinet, a Canadian doctor specializing in cancer treatment, developed a decision support Web site to help patients. The original site, which was developed to help prostate patients, has turned into a business providing medical advice on a number of topics. The decision support system is called Mediconsult. To keep the information free, Mediconsult receives fees from sponsors and sells medical books and supplies on the site.

Mediconsult was an immediate success. With about 1.5 million page views and 300,000 visits per month, it was clear that people wanted to use this source to get medical information. Since the start of the site, there have been a number of enhancements. For example, people can ask a medical librarian about medical reports and articles concerning specific medical problems or conditions. Next, MediXpert was started. This site allows people to ask medical experts questions. Answers are sent back in 5 days or fewer using e-mail. There is, however, a cost of $195 for this service. Compared with the expense of traveling to and paying for a visit with a medical expert, many believe the fee is justified. The MediXpert decision support system is not intended to replace a doctor's advice but to support and enhance it. This feature is especially useful for people living in rural areas that might require travel of hundreds of miles to see a specialist.

With Mediconsult and MediXpert, all communications are encoded to protect the privacy of the patient. To protect MediXpert and Mediconsult, patient names are hidden from the specialists and doctor-patient relationships are not formed. Company founders are hoping these measures will protect them from any potential malpractice claims.

1. What are the benefits of Mediconsult and MediXpert? What are the disadvantages?

2. Would you use this type of service? Under what circumstances would you recommend it to others?

Sources: James Evans, "Doctor's on E-mail," *Internet World,* February 1998, p. 48; David Stipp, "Health Help on the Net," *Fortune,* January 12, 1998, p. 135; Staff, "Doctors Take to the Web to Earn Course Credits," *PC Week,* January 26, 1998, p. 83.

Project Management Models at Bank of America

Project management models are an important part of model management in most decision support systems. These models allow decision makers to monitor and control projects. Project management was one of the first applications implemented on personal computers. Project management software is being used for simple tasks, such as coordinating job schedules, to very complex and long projects, including the construction of a new manufacturing plant or the coordination of a major military engagement.

Today, a number of software packages are available to implement the models used by project management. The total market for project management software is close to $900 million annually and growing at about 20 percent per year according to some reports. Project management software is available for mainframe computers, client server systems, and personal computers. Computer Associates' CA-SuperProject/Net is a project management tool for client server systems. Microsoft is the leader in project management software for personal computers, with a majority market share of about two-thirds. Project 98 is Microsoft's most recent entry in the project management market.

Most organizations have to manage a number of large and often complex projects. It is not uncommon for these organizations to have hundreds of projects, from developing new computer programs to starting a completely new business line, that are active at the same time. Some of them are critical to the success or survival of the organization. But large projects often fail to be delivered on time and within budget. Some never get implemented, after millions of dollars of expenditures. The number of unsuccessful or failed computer-related projects may be as high as 50 percent according to some experts.

Like many other software modeling tools, how the software is used can be as important as, or more important than, the sophistication of the software. This was the case with Bank of America. To get its projects on track, Bank of America hired Chris Higgins, a new project manager with a tough military background. Higgins used good software, Microsoft's Project 98, and good leadership skills. He also used the Microsoft Excel spreadsheet program for financial calculations. Higgins instituted a training program that stressed discipline, structure, tools, training, and leadership. He had a team of 140 people and a budget of $100,000. Just a few problems with a few projects could result in millions of dollars of loss. More important than the financial losses are lost opportunities because of failed projects. Higgins knew that the stakes were high. In addition to an extensive training program, Higgins managed by example, showing his project managers how things should be done. To date, his performance has been promising.

Higgins has been able to reduce project times by about 20 percent on average. Personnel expenses have been reduced by about 10 percent. This is impressive when so many projects fail, resulting in millions of dollars of losses and lost opportunities. Higgins has not only shown Bank of America how to correctly use project management models, but he shows all of us that success requires both good software and good leadership skills.

1. What is project management software? Give some examples of how this type of software can be used.
2. How was Bank of America able to manage its project management models? What are the keys to success for using software to manage projects?

Sources: Kathleen Melymuka, "Spit and Polish," *Computerworld*, February 16, 1998, p. 65; Brendan Coffey, "The Buy-Side Elite," *Wall Street and Technology*, January, 1998, p. 50; Susan Garland, "Bank of America's Back Is Against the Wall," *BusinessWeek*, March 9, 1998, p. 64; Karen Carrillo, "Is It All a Project," *InformationWeek*, February 23, 1998; Chris Vandersluis, "Fighting for Project Control," *Computing Canada*, January 26, 1998, p. 18.

CHAPTER 11

Artificial Intelligence and Expert Systems

"The mystery grows more acute. The more we think about computers, the more we realize how strange consciousness is."

—David Chalmers, philosopher and author of the book *The Conscious Mind*

Chapter Outline

An Overview of Artificial Intelligence
Artificial Intelligence in Perspective
The Nature of Intelligence
The Difference between Natural and Artificial Intelligence
The Major Branches of Artificial Intelligence

An Overview of Expert Systems
Characteristics of an Expert System
Capabilities of Expert Systems
When to Use Expert Systems

Components of Expert Systems
The Knowledge Base
The Inference Engine
The Explanation Facility
The Knowledge Acquisition Facility
The User Interface

Expert Systems Development
The Development Process
Participants in Developing and Using Expert Systems
Expert Systems Development Tools and Techniques
Advantages of Expert System Shells and Products
Expert Systems Development Alternatives

Applications of Expert Systems and Artificial Intelligence

Learning Objectives

After completing Chapter 11, you will be able to:

1. Define the term *artificial intelligence* and state the objective of developing artificial intelligence systems.
2. List the characteristics of intelligent behavior and compare the performance of natural and artificial intelligence systems for each of these characteristics.
3. Identify the major components of the artificial intelligence field and provide one example of each type of system.
4. List the characteristics and basic components of expert systems.
5. Identify at least three factors to consider in evaluating the development of an expert system.
6. Outline and briefly explain the steps for developing an expert system.
7. Identify the benefits associated with the use of expert systems.

European Media Buying
Computers Help Select TV Commercial Slots

Media-buying agencies (MBAs) in the European TV commercials market purchase TV commercial slots wholesale from TV stations for subsequent distribution. In purchasing TV slots, the MBAs choose time slots from multiple TV stations, with a single MBA purchasing up to 5,000 TV slots a year. Good buying decisions consider several factors, among them the expected number of viewers for the commercial, the normalized price per viewer (price of the commercial $\times$ length $\div$ total number of viewers), the viewer makeup (age distribution, demographics, professions, and income groups), program type during which the commercial will run (show, soap opera, news, sports, etc.), and the distribution of the MBA's clientele.

One European MBA uses computer software to evaluate the attractiveness of each TV commercial slot. The design of the system was completed in just eight weeks by software consultants. The prototype was tested and refined by comparing the buyer team's decisions for the prior three months with the system's outputs. Testing continued until the management of the MBA was convinced that the performance of the system was as good as the buyer team's decisions. The system is now in continuous use. It is capable of restructuring all buying decisions in only five minutes, which allows adaptation of the portfolio of commercial slots offered by the MBA in case of policy changes, new clients, or any other changes in the TV media market.

As you read this chapter, consider the following questions:

- What does it take to design and implement an effective artificial intelligence system?

- What business needs can be addressed by an expert system?

Sources: Adapted from Business and Financial Application Papers section of the Inform Software Web site at http://www.fuzzytech.com, accessed June 10, 1998 and Science: Buying Terms section of the Alles Media Services Web site at http://www.software.de, accessed September 13, 1998.

artificial intelligence (AI)

computers with the ability to mimic or duplicate the functions of the human brain

At a Dartmouth College conference in 1956, John McCarthy proposed the use of the term **artificial intelligence (AI)** to describe computers with the ability to mimic or duplicate the functions of the human brain. Many AI pioneers attended this first conference; a few predicted that computers would be as "smart" as people by the 1960s. The prediction has not yet been realized, but the benefits of artificial intelligence in business and research can be seen today. Advances in AI have led to many practical applications of systems, like the one used at European MBA, which are capable of making complex decisions.

AN OVERVIEW OF ARTIFICIAL INTELLIGENCE

Science fiction novels and popular movies have featured scenarios of computer systems and intelligent machines taking over the world. Computer systems such as Hal in the movie *2001: A Space Odyssey* are futuristic glimpses of what might be. These accounts are fictional, but we see the real application of many computer systems that use the notion of AI. These systems help to make medical diagnoses, explore for natural resources, determine what is wrong with mechanical devices, and assist in designing and developing other computer systems. In this chapter we explore the exciting applications of artificial intelligence and look at what the future really might hold.

Artificial Intelligence in Perspective

artificial intelligence systems

the people, procedures, hardware, software, data, and knowledge needed to develop computer systems and machines that demonstrate the characteristics of intelligence

Artificial intelligence systems include the people, procedures, hardware, software, data, and knowledge needed to develop computer systems and machines that demonstrate characteristics of intelligence. Researchers, scientists, and experts on how humans think are often involved in developing these systems. The objective in developing contemporary AI systems is not to replace human decision making completely but to replicate it for certain types of well-defined problems. As with other information systems, the overall purpose of artificial intelligence applications in business is to help the organization achieve its goals.

Science fiction movies give us a glimpse of the future, but many practical applications of artificial intelligence exist today, among them medical diagnostics, mechanical diagnostics, and development of computer systems.
(Source: Photofest.)

The Nature of Intelligence

intelligent behavior

the ability to learn from experience and apply knowledge acquired from experience, handle complex situations, solve problems when important information is missing, determine what is important, react quickly and correctly to a new situation, understand visual images, process and manipulate symbols, be creative, imaginative and use heuristics

From the early AI pioneering stage, the research emphasis has been on developing machines with **intelligent behavior**. Some of the specific characteristics of intelligent behavior include the ability to do the following:

Learn from experience and apply the knowledge acquired from experience. Being able to learn from past situations and events is a key component of intelligent behavior and is a natural ability for humans who learn by trial and error. However, learning from experience is not natural for computer systems. This ability must be carefully programmed into the system. Today, researchers are developing systems that have this ability. For instance, computerized AI chess games can learn to improve their game while they play human competitors (Figure 11.1).

In addition to learning from experience, people apply what they have learned to new settings and circumstances. In a number of cases, individuals have taken what they have learned and succeeded with in one endeavor and applied it to another. For example, a company that developed a dish washing product effective for cleaning greasy dishes developed a variation of the product for use in cleaning up messy highway spills using the same overall approach. Although humans have the ability to apply what they have learned to new settings, this characteristic is not automatic with computer systems. Developing computer programs to allow computers to apply what they have learned can be difficult.

Handle complex situations. Humans are involved in complex situations. World leaders face difficult political decisions regarding conflict, global economic conditions, hunger, and poverty. In a business setting, top-level managers and executives are faced with a complex market, difficult and challenging competitors, intricate government regulations, and a demanding workforce. Even human experts make mistakes in dealing with these situations. Developing computer systems that can handle perplexing situations requires careful planning and elaborate computer programming.

Solve problems when important information is missing. The essence of decision making is dealing with uncertainty. Quite often, decisions must be made even when we lack information or have inaccurate information, because obtaining complete information is too costly or impossible. You have probably seen movies in which computers have responded to human commands with statements like "Does not compute" and "Insufficient information." Today, AI systems can make important calculations, comparisons, and decisions even when missing information.

FIGURE 11.1

Computers like Deep Blue attempt to learn from past chess moves. The powerful supercomputer's logic system was able to calculate the ramifications of up to 100 billion chess maneuvers within the allotted time for each move.
(Source: Photo courtesy of the Association for Computing Machinery.)

Determine what is important. Knowing what is truly important is the mark of a good decision maker. Every day we are bombarded with facts and must process large amounts of data, filtering out what is unnecessary. Determining which items are crucial can make the difference between good decisions and those that ultimately lead to problems or failures. Computers, on the other hand, do not have this natural ability. Developing programs and approaches to allow computer systems and machines to identify important information is not a simple task.

React quickly and correctly to a new situation. A small child, for example, can look over a ledge or a drop-off and know not to venture too close. The child reacts quickly and correctly to a new situation. Computers, on the other hand, do not have this ability without complex programming.

perceptive system

a system that approximates the way a human sees, hears, and feels objects

Understand visual images. Interpreting visual images can be extremely difficult, even for sophisticated computers. People and animals can look at objects interacting in our environment and understand exactly what is going on. For instance, we can see a man sitting at a table and know that he has legs and feet that we cannot see. Being able to understand and correctly interpret visual images is an extremely complex process for computer systems. Moving through a room of chairs, tables, and other objects can be trivial for people but extremely complex for machines, robots, and computers. Such machines require an extension of understanding visual images, called a **perceptive system**. Having a perceptive system allows a machine to approximate the way a human sees, hears, and feels objects.

Process and manipulate symbols. People see, manipulate, and process symbols every day. Visual images provide a constant stream of information to our brains. By contrast, computers have difficulty handling symbolic processing and reasoning. Although computers excel at numerical calculations, they aren't as good at dealing with symbols and three-dimensional objects. Recent developments in machine-vision hardware and software, however, allow some computers to process and manipulate symbols on a limited basis.

Be creative and imaginative. Throughout history, some people have turned difficult situations into advantages by being creative and imaginative. For instance, when shipped a lot of defective mints with holes in the middle, an enterprising entrepreneur decided to market these new mints as Lifesavers instead of returning them to the manufacturer. Ice cream cones were invented at the St. Louis World's Fair when an imaginative store owner decided to wrap ice cream with a waffle from his grill for portability. Developing new and exciting products and services from an existing (perhaps negative) situation is a human characteristic. Few computers have the ability to be truly imaginative or creative in this way, although software has been developed to enable a computer to write short stories.

Use heuristics. With some decisions, people use heuristics (rules of thumb arising from experience) or even guesses. In searching for a job, we may decide to rank companies we are considering according to profits per employee. Companies making more profits might pay their employees more. In a manufacturing setting, a corporate president may decide to look at only certain locations for a new plant. We make these types of decisions using general rules of thumb, without completely searching all alternatives and possibilities. Today, some computer systems also have this ability. They can, given the right programs, obtain good solutions that use approximations instead of trying to search for an optimal solution, which would be technically difficult or too time-consuming.

This list of traits only partially defines intelligence. Unlike virtually every other field of information systems research in which the objectives can be clearly defined, the term *intelligence* is a formidable stumbling block. One of the problems in artificial intelligence is arriving at a working definition of real intelligence against which to compare the performance of an artificial intelligence system.

The Difference between Natural and Artificial Intelligence

Since the term *artificial intelligence* was defined in the 1950s, experts have disagreed about the difference between natural and artificial intelligence. For instance, is there a difference between carbon life (human or animal life) and silicon life (a computer chip) in terms of behavior? Profound differences exist, but they are declining in number (Table 11.1). One of the driving forces behind AI research is an attempt to understand how humans actually reason and think. It is believed that the ability to create machines that can reason will be possible only once we truly understand our own processes for doing so.

The Major Branches of Artificial Intelligence

AI is a broad field that includes several specialty areas, such as expert systems, robotics, vision systems, natural language processing, learning systems, and neural networks (Figure 11.2). Many of these areas are related; advances in one can occur simultaneously with or result in advances in others.

TABLE 11.1

A Comparison of Natural and Artificial Intelligence

Attributes	Natural Intelligence (Human)	Artificial Intelligence (Machine)
The ability to use sensors (eyes, ears, touch, smell)	High	Low
The ability to be creative and imaginative	High	Low
The ability to learn from experience	High	Low
The ability to be adaptive	High	Low
The ability to afford the cost of acquiring intelligence	High	Low
The ability to use a variety of information sources	High	High
The ability to acquire a large amount of external information	High	High
The ability to make complex calculations	Low	High
The ability to transfer information	Low	High
The ability to make a series of calculations rapidly and accurately	Low	High

FIGURE 11.2

A Conceptual Model of Artificial Intelligence

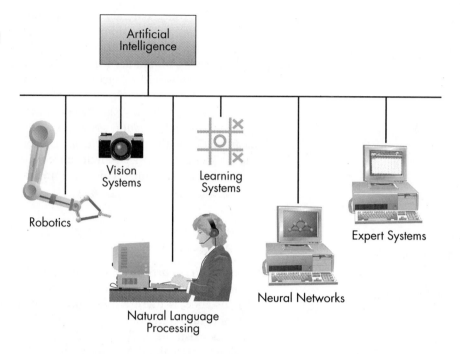

Expert systems. An **expert system** consists of hardware and software that stores knowledge and makes inferences, similar to a human expert. Because of their many business applications, expert systems are discussed in more detail in the next several sections of the chapter.

Robotics. Robotics involves developing mechanical or computer devices that can paint cars, make precision welds, and perform other tasks that require a high degree of precision or are tedious or hazardous for humans. Contemporary robotics combines both high-precision machine capabilities with sophisticated controlling software. The controlling software in robots is what is most important in terms of AI.

Many applications of robotics exist, and research into these unique devices continues (Figure 11.3). Manufacturers use robots to assemble and paint products. Raytheon Premier uses robots to cut production costs for its business jets. Robots are used on the manufacturing floor at Maxwell Products in Cerritos, California. Welding robots have enabled the firm to manufacture top-quality products and reduce labor costs while shortening delivery time to their customers. In addition, the company, which employs 160 workers, reduced the risk of employee injury when it switched to robotic welding.[1]

In addition to their use in supporting manufacturing operations, robots have been applied in more unusual situations. "Standardized Teleoperation System" (STS) is a modular kit of robotic controls for converting a vehicle to unmanned operation. STS controls permit normal (or manual) vehicle operation and unmanned operation with the flick of a switch. The strap-on kit feature converts existing vehicles. Vehicles can also be delivered with the STS installed. STS controls are used in many types and classes of vehicles including bulldozers, tanks, and trucks used for bomb and landmine detection, neutralization, detonation, clearing, and route proofing. Some of the

expert system

the hardware and software that stores knowledge and makes inferences, similar to a human expert

robotics

mechanical or computer devices that perform tasks requiring a high degree of precision or that are tedious or hazardous for humans

more unusual applications of robotics include a ping-pong ball server for the U.S. Olympic team, a toy soldier marching machine, and a bug sorter for museum display.[2]

Although robots are essential components of today's automated manufacturing systems, future robots will find wider applications outside the factory in banks, restaurants, homes, and hazardous working environments such as nuclear stations. A robot must not only execute tasks programmed by the user but must also be able to interact with its environment through its sensors and actuators, sense and avoid unforeseen obstacles, and perform its duties much the same way as humans.

FIGURE 11.3

Robots can be used in situations that are hazardous, repetitive, or difficult to do in other ways. (Source: Courtesy of ABB Flexible Automation.)

vision systems

the hardware and software that permit computers to capture, store, and manipulate visual images and pictures

Vision systems. Another area of AI involves vision systems. **Vision systems** include hardware and software that permit computers to capture, store, and manipulate visual images and pictures. The U.S. Justice Department uses vision systems to perform fingerprint analysis, with almost the same level of precision as human experts. The speed with which the system can search through the huge database of fingerprints has brought a quick resolution to many long-standing mysteries. Vision systems are also effective at identifying people based on facial features. Just three days after installing their newest crime-fighting weapon, facial recognition software, Los Angeles County Sheriff's Department detectives used the technology to identify and arrest a car jacker.[3]

Vision systems can be used in conjunction with robots to give these machines "sight." Robots such as those used in factory automation typically perform mechanical tasks with little or no visual stimuli. Robotic vision extends the capability of these systems allowing the robot to make decisions based on visual input. Generally, robots with vision systems can recognize black and white and some gray-level shades but do not have good color or three-dimensional vision. Other systems concentrate on only a few key features in an image, ignoring the rest. It may take years before a robot or other computer system can "see" in full color and draw conclusions from what it sees, the way humans do.

natural language processing

processing that allows the computer to understand and react to statements and commands made in a "natural" language, such as English

Natural language processing. As discussed in Chapter 4, **natural language processing** allows the computer to understand and react to statements and commands made in a "natural" language, such as English. There are really three levels of voice recognition: command (recognizes dozens to hundreds of words), discrete (recognizes dictated speech with pauses between words), and continuous (recognizes natural speech).[4] For example, a natural language processing system can be used to retrieve important information without typing in commands or

Dragon Systems'
NaturallySpeaking Preferred
edition uses continuous speech,
or natural speech, allowing the
user to speak to the computer
at a normal pace without
pausing between words. The
spoken words are transcribed
immediately onto the computer
screen.
(Source: Courtesy of Dragon
Systems, Inc.)

learning system

a combination of software and
hardware that allows the computer
to change how it functions or
reacts to situations based on
feedback it receives

neural network

a computer system that can act like
or simulate the functioning of a
human brain

searching for key words. With natural language processing, it is possible to speak into a microphone connected to a computer and have the computer convert the electrical impulses generated from the voice into text files or program commands. With some simple natural language processors, you say a word into a microphone and type the same word on the keyboard. The computer then matches the sound with the typed word. With more advanced natural language processors, the recording and typing of words is not necessary. The voice recognition system at *The Boston Globe* is used by employees to reach colleagues at the newspaper by dialing one telephone number and speaking the colleague's name. The system matches the name with a number in its database to connect the call.[5] In financial services, Charles Schwab & Co. implemented Voice Broker, an application that lets users access stock prices and other information by dialing a number and speaking the name of the company, stock symbol, mutual fund, or market indicator.[6] Some natural language processors, however, can have problems understanding context and words that sound alike but have different meanings (e.g., the words *time* and *thyme*). Read the "FYI" box to learn more about the use of natural language processing.

Learning systems. Another part of AI deals with **learning systems**, a combination of software and hardware that allows the computer to change how it functions or reacts to situations based on feedback it receives. For example, some computerized games have learning abilities. If the computer does not win a particular game, it remembers not to make the same moves under the same conditions. Learning systems software requires feedback on the results of its actions or decisions. At a minimum, the feedback needs to indicate whether the results are desirable (winning a game) or undesirable (losing a game). The feedback is then used to alter what the system will do in the future.

Neural networks. An increasingly important aspect of AI involves neural networks.[7] A **neural network** is a computer system that can act like or simulate the functioning of a human brain. The systems use massively parallel processors in an architecture that is based on the human brain's own meshlike structure. In addition, neural network software can be used to simulate a neural network using standard computers. Neural networks can process many pieces of data at once and learn to recognize patterns. The systems then program themselves to solve related problems on their own. Some of the specific features of neural networks include the following:

- The ability to retrieve information even if some of the neural nodes fail
- Fast modification of stored data as a result of new information
- The ability to discover relationships and trends in large databases
- The ability to solve complex problems for which all the information is not present

Neural networks excel at pattern recognition. For example, neural network computers can be used to read bank check bar codes despite smears

FYI

Listening to Our Language

Natural language processing has been a dream of artificial intelligence enthusiasts for decades. Classic movies, such as *2001: A Space Odyssey*, had people talking to and commanding computers. Speech recognition, once the topic of science fiction films and books, is now rapidly becoming a reality. In the last few years, tremendous progress has been made in software that gives a computer the ability to accept and act on human speech. Prices for this advanced software are tumbling, giving many people the advantages of speech recognition technology. No other group of people appreciates these advances more than users with physical disabilities.

A vice president of a manufacturing company has greatly benefited from speech recognition. A debilitating stroke in her late 40s paralyzed the right side of her body and abruptly stopped her career. She could not write, speak, or construct simple sentences in her head. A bright future seemed over, replaced with frustration and disappointment. But help, in terms of natural language processing, made a huge difference. Using a natural language processing software product by Unisys, she was able to get back on the road to recovery. The Unisys system presented her with a vast array of pictures. Using a microphone, she would describe pictures in simple words. Rapid feedback that compares pictures to spoken words is the key to the Unisys system. In less than a year, she has been able to make great progress. She is now able to speak in simple sentences. According to her husband, the help she received from the Unisys system has turned frustration and disappointment into enthusiasm and hope.

The manufacturing vice president is not alone. Thousands of disabled users are also replacing their frustration with hope. A dyslexic student who could not keep up with his course work can now be assisted by a natural language reading machine, making a college education possible. A new Internet browser that uses natural language processing is now able to surf the net. The new Internet browser, now available in Japan, uses speech recognition to allow the blind to access the Internet. A deaf student can learn to speak using a 3-D modeling system that displays a talking head that shows the muscles and jaw structure needed to form sounds. The blind, deaf, hearing impaired, and the dyslexic are just a few groups that are benefiting. In the United States, there are over 20 million people who receive some form of disability assistance, which costs the federal government about $200 billion annually. Some systems that cost $5,000 a few years ago can now be acquired for $200. The total potential cost savings, which could be in the billions of dollars for the U.S. government, is just one benefit. A bigger benefit is how this technology is dramatically changing the lives of so many people.

DISCUSSION QUESTIONS

1. How was natural language processing technology able to help disabled users?
2. What breakthroughs would you expect in the next five years? What impact would these breakthroughs have on you?

Sources: Paul Judge, "For Some a Necessity," *BusinessWeek*, February 23, 1998, p. 74; Michael Caton, "IBM Takes Dictation," *PC Week*, January 9, 1998, p. 49; Neil Gross, "Let's Talk," *BusinessWeek*, February 23, 1998, p. 61; Neil Fawcett, "Speech Recognition Is Talk of the Technologists' Town," *Computer Weekly*, February 2, 1998, p. 25.

or poor-quality printing. Anheuser-Busch uses BrainMaker to identify the organic content of its own and its competitors' beer vapors with 96 percent accuracy. This allows the company to ensure consistent quality for its customers and keep track of any changes made by its competitors.[8] Some hospitals use neural networks to determine a patient's likelihood of contracting cancer or other diseases. Neural nets work particularly well when it comes to analyzing down-to-the-minute trends. Walt Disney and several large banks use neural networks to figure out staffing needs based on customer traffic—a task that requires precise analysis, down to the half-hour. Increasingly, businesses are firing up neural nets to help them navigate ever-thicker forests of data and make sense of myriad customer traits and buying habits. Other applications include prediction of the survival of a business, recognizing a spoken single word from a small vocabulary, prediction of intensive-care survival, paint blending, and predicting imminent failure of jet engines. Like expert systems, neural networks are finding more and more use in businesses today.[9] Read the "Ethical and Societal Issues" box to learn about issues that can arise in applying such systems.

ETHICAL AND SOCIETAL ISSUES

Chicago Police Department Uses Neural Net to Screen Officers

The Chicago Police Department is the third largest metropolitan police force in the United States. It recently experimented with using a neural network to predict which officers on the force are potential candidates for misbehavior. The department's Internal Affairs Division developed a database of patternlike characteristics, behaviors, and demographics found among 200 police officers who had been terminated for disciplinary reasons. The data included sex, race, education, age, citations, performance reports, number of traffic accidents, incidents of lost weapons or badges, marital status, and frequency of sick leave. The neural network used this data to define a pattern from the 200-member control group.

After analysis of the data of the 12,500 current officers, the neural network produced a list of 91 at-risk men and women who, by virtue of matching the pattern or sharing questionable characteristics to some degree, were deemed to be "at risk." Of those 91 people, nearly half were found to be already enrolled in a counseling program to help officers guilty of misconduct. Internal Affairs now intends to make the neural network a supplement to the counseling program because, as Deputy Superintendent Raymond Risely says, the sheer size of the Chicago police force makes it "pretty much impossible for all at-risk individuals to be identified [by supervisors]."

Obviously, this particular application of neural networks has been highly controversial, drawing criticism from several quarters—the most vocal being Chicago's Fraternal Order of Police. William Nolan, the order's president, has made Orwellian references, saying the department's program seems like "Big Brother." The Internal Affairs Division, however, is pleased with the results. Despite arguments raging over whether computers should monitor humans, its supporters argue that the neural network is subjective and is not personally biased as "manned" programs often are. Clearly, the software can hold no personal grudges and seeks only to dispassionately identify patterns and characteristics that could spell trouble.

Discussion Questions

1. If you were a member of Chicago's Fraternal Order of Police, would you prefer such a system or would you prefer a human-based system for identifying "at risk" officers? Why?

2. What changes or improvements might be made to the process of identifying "at risk" officers that might combine the objectivity of a neural network with the wisdom and experience of senior officers and counselors?

Sources: Adapted from "Neural Network Red-Flags Police Officers with Potential for Misconduct," in the Applications Section of the BrainMaker Web site at http://www.calsci.com, accessed September 12, 1998 and "High-Tech Tool to Weed Out Bad Cops Proved a Bust," *The Chicago Tribune*, October 15, 1997.

Many companies and individuals develop neural network software. In addition, some companies sell this type of software. SPSS, a software company known for statistical programs, sells Neural Connection for under $1,000.[10] The neural network software can be used for difficult forecasting and classification problems where traditional linear algorithms are difficult or impossible to apply. Neural Connection is a stand-alone software package that can be integrated with SPSS for Windows, a powerful statistical software package. Other neural network software packages include NEUframe, BrainMaker, Neuralyst, Braincel, and Predict. These software packages are designed to be used with an existing spreadsheet program. Read the "Technology Advantage" box to learn about how a software vendor is incorporating AI techniques into its software to better meet customers' needs.

AN OVERVIEW OF EXPERT SYSTEMS

As discussed, an expert system acts or behaves like a human expert in a particular field. Computerized expert systems have been developed to diagnose problems, predict future events, and solve energy problems. They have also been used to design new products and systems, determine the best use of lumber, and increase the quality of healthcare. Like human experts, computerized expert systems use heuristics, or rules of thumb, to arrive at conclusions or make suggestions. Expert systems have also been used to determine credit limits for credit cards (Figure 11.4). The research conducted in AI during the past two decades is resulting in expert systems that explore new business possibilities, increase overall profitability, reduce costs, and provide superior service to customers and clients.

Characteristics of an Expert System

Expert systems have a number of characteristics and capabilities, including the following:

Difficult forecasting problems can be solved by neural network software. The red line in this screen is the one-step-ahead forecast function. The time series is the annual count of the number of sunspots. (Source: Courtesy of SPSS.)

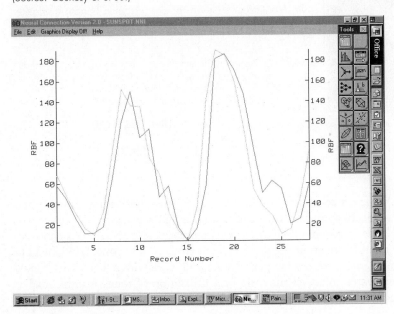

Can explain their reasoning or suggested decisions. A valuable characteristic of an expert system is the capability to explain how and why a decision or solution was reached. For example, the expert system can explain the reasoning behind the conclusion to approve a particular loan application. The ability to explain its reasoning processes can be the most valuable feature of a computerized expert system. The user of the expert system thus gains access to the reasoning behind the conclusion.
Can display "intelligent" behavior. Considering a collection of data, an expert system can propose new ideas or approaches to problem solving. A few of the applications of expert systems are an imaginative medical diagnosis based

TECHNOLOGY ADVANTAGE
Software Vendor Uses Neural Network to Enhance Products

Computer Associates International, Inc. with headquarters in Islandia, New York, is a manufacturer of business software. The company develops, licenses and supports more than 500 integrated products that include enterprise computing and information management, application development, and manufacturing and financial applications. CAI has over 11,000 people in 160 offices in 43 countries and had revenue of $4.5 billion in calendar year 1997.

CAI's software product, Unicenter TNG, provides advanced network management capabilities including monitoring and management of the network performance. Unicenter TNG continually monitors important network connections and their service levels based on predefined policies. In the event of problems, it notifies appropriate personnel and takes corrective actions. Unicenter TNG provides support for neural network agents (neugents). These are software programs that learn, predict, and resolve problems before they arise. The first neugent,

which deals with performance and availability, is available as an add-on option to Unicenter TNG.

The use of neural network technology enables Unicenter TNG to learn how to manage a customer's network. It has had great potential in client/server management because there are so many components that can fail or go wrong. By observing causes and effects, not just individual statistics, the software trains itself with every event. While it is impossible for humans to track what happens within a busy network, for the neugent, every error is meaningful. For example, a neugent watching an e-mail server in CAI's lab recognized a pattern in message buffers and queues that could cause a failure. Left alone, the server soon failed. Other neugents try to detect emerging situations such as a service slowdown or virus activity. Still other agents could enable security policies to take over manual tasks such as scanning a set of servers every week looking for expired user IDs.

Such self-learning capability should help Unicenter TNG users greatly. Currently, users need a lot of expertise to write management policies that tell Unicenter TNG what to do in response to changes in each variable. It's easier to simply identify patterns and trends as good or bad.

DISCUSSION QUESTIONS

1. Why does the monitoring of complex corporate networks represent an area of opportunity to apply neural network technology?
2. What other software packages could benefit from use of this technology?

Sources: Adapted from "Computer Associates Expands Enterprise Management with New Versions of Unicenter TNG," April 27, 1998 press release found on Computer Associates home page at http://www.cai.com, accessed June 7, 1998 and Patrick Dryden, "Neural Agents Spy Network Traffic Errors," *Computerworld*, February 9, 1998, pp. 49–50.

on a patient's condition, a suggestion to explore for natural gas at a particular location, and providing job counseling for workers.[11]

Can draw conclusions from complex relationships. Expert systems can evaluate complex relationships to reach conclusions and solve problems. For example, one proposed expert system will work with a flexible manufacturing system to determine the best utilization of tools. Another expert system can suggest ways to improve quality control procedures.

FIGURE 11.4

Credit card companies often use expert systems to determine credit limits for credit cards.
(Source: David Young Wolff/ Tony Stone Images.)

Can provide portable knowledge. One unique capability of expert systems is that they can be used to capture human expertise that might otherwise be lost. A classic example of this is the expert system called DELTA (Diesel Electric Locomotive Troubleshooting Aid), which was developed to preserve the expertise of the retiring David Smith, the only engineer competent to handle many highly technical repairs of such machines.
Can deal with uncertainty. One of an expert system's most important features is its ability to deal with knowledge that is incomplete or not completely accurate. The system deals with this problem through the use of probability, statistics, and heuristics.

Even though these characteristics of expert systems are impressive, other characteristics limit their current usefulness. Many of these limiting characteristics are related to concerns of cost, control, and complexity. Some of these characteristics are as follows:

Not widely used or tested. Even though successes occur, expert systems are not used in a large number of organizations. In other words, they have not been widely tested in corporate settings.
Difficult to use. Some expert systems are difficult to control and use. In some cases, the assistance of computer personnel or individuals trained in the use of expert systems is required to help the user get the most from these systems. Today's challenge is to make expert systems easier to use by decision makers who have limited computer programming experience.
Limited to relatively narrow problems. Whereas some expert systems can perform complex data analysis, others are limited to simple problems. Furthermore, many problems solved by expert systems are not that beneficial in business settings. An expert system designed to provide advice on how to repair a machine, for example, is unable to assist in decisions about when or whether to repair it. In general, the narrower the scope of the problem, the easier it is to implement an expert system to solve it.
Cannot readily deal with "mixed" knowledge. Expert systems cannot easily handle a knowledge base that has a mixed representation. Knowledge can

be represented through defined rules, through comparison to similar cases, and in various other ways. An expert system in one application might not be able to deal with knowledge that combined both rules and cases.

Possibility of error. Although some expert systems have limited abilities to learn from experience, the primary source of knowledge is a human expert. If this knowledge is incorrect or incomplete, it will affect the system negatively. Other development errors involve programming. Because expert systems are more complex than other information systems, the potential for such errors is greater.

Cannot refine own knowledge base. Expert systems are not capable of acquiring knowledge directly. A programmer must provide instructions to the system that determine how the system is to learn from experience. Also, some expert systems cannot refine their own knowledge bases—such as eliminating redundant or contradictory rules.

Difficult to maintain. Related to the preceding point is the fact that expert systems can be difficult to update. Some are not responsive or adaptive to changing conditions. Adding new knowledge and changing complex relationships may require sophisticated programming skills. In some cases, a spreadsheet used in conjunction with an expert system shell can be used to modify the system. In others, upgrading an expert system can be too difficult for the typical manager or executive. Future expert systems are likely to be easier to maintain and update.

May have high development costs. Expert systems can be expensive to develop when using traditional programming languages and approaches. Development costs can be greatly reduced through the use of software for expert system development. **Expert system shells**, a collection of software packages and tools used to develop expert systems, can be implemented on most popular PC platforms to reduce development time and costs.

Raise legal and ethical concerns. People who make decisions and take action are legally and ethically responsible for their behavior. A person, for example, can be taken to court and punished for a crime. When expert systems are used to make decisions or help in the decision-making process, who is legally and ethically responsible—the human experts used to develop the knowledge base, the expert system developer, the user, or someone else? For example, if a doctor uses an expert system to make a diagnosis and the diagnosis is wrong, who is responsible? These legal and ethical issues have not been completely resolved.

expert system shell

a collection of software packages and tools used to develop expert systems

FIGURE 11.5

Solutions Offered by Expert Systems

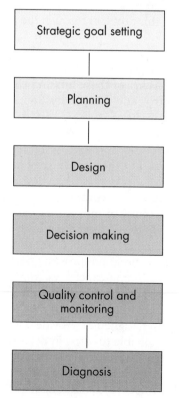

Capabilities of Expert Systems

Compared with other types of information systems, expert systems offer a number of powerful capabilities and benefits. For example, one expert system, called XCON, is often used in designing computer system configurations because it consistently does a better job than human beings.

Expert systems can be used to solve problems in every field and discipline and can assist in all stages of the problem-solving process. Past successes have shown that expert systems are good at strategic goal setting, planning, design, decision making, quality control and monitoring, and diagnosis (Figure 11.5). Read the "E-commerce" box to learn how intelligent agents are being used to help people find what they need on the Internet.

E-COMMERCE
Intelligent Software Agents Help Tame the Internet

Business organizations and workers suffer from information overload—they are being overwhelmed with the flood of electronic mail, business reports, and information necessary to conduct business and make decisions. Even with all the data they receive on a daily basis, there is a need to retrieve even more information from the Internet. Artificial intelligence and agent technology may offer a solution to all this.

Intelligent agents are software entities that assist people and act on their behalf. Intelligent software agents with artificial intelligence capabilities are being used to find useful information on the Internet where an enormous amount of information is available from a wide variety of sources. Intelligent agents can filter this flood of information and pass on to the user only those items in which the user is most likely to be interested. Agents even have the ability to keep track of what information you liked and what information you discarded. This tracking allows intelligent agents to learn your preferences and improve their performance. In this way, intelligent agents reduce the information complexity of the networked world.

Investors looking for hot new stock offerings, economic trends, and other financial news now have a new search engine—FinanceWise (http://www.financewise.com)—that delivers fast, focused and up-to-date financial information free of charge. FinanceWise is specifically designed to retrieve, analyze, index, and present only finance-related information from the Internet. Although chiefly intended for the banking and financial professional, FinanceWise can be a very useful tool for the individual investor.

Intelligent software agents are also useful to mobile computing users. Mobile users, however, are not always connected to a network, and, when they are connected, it can be over a variety of media, with different bandwidth, reliability, and security characteristics. Intelligent agent technology can be used to create "surrogates" within the network that represent mobile users. These surrogates do work on behalf of the mobile users they represent, even when the mobile user is disconnected. When the mobile user reconnects, agent technology can tailor how data is transferred to the mobile user to match the characteristics of the connection. Intelligent agents can thereby reduce the complexity associated with mobility.

DISCUSSION QUESTIONS

1. What are intelligent agents? How do intelligent agents help with the right information, right person, right format, and right time aspects of information delivery?
2. What are the advantages to organizations of implementing intelligent agents?

Source: Don Barker, "A Word to the FinanceWise," at the Best of the Bots Web site at http://botspot.com/best_of_the_bots, created June 18, 1998, accessed October 25, 1998, and Robert Scheier, "Agents May No Longer Be So Secret," *Computerworld*, March 17, 1997, p. 86.

Strategic goal setting. Developing strategic goals for an organization is one of the most important functions of top-level decision makers. Strategic goals provide a framework for all other activities throughout the organization. Expert systems can suggest strategic goals and explore the impact of adopting them. Strategic goals can include identifying opportunities in the marketplace, analyzing the strengths of the existing organization, determining the power and position of competitors, and understanding the existing labor force. For example, say a California wine maker is currently perceived as a low-cost/low-quality producer. An expert system can help the company's top-level management determine costs and benefits involved in producing higher-quality wines and changing its image in the marketplace.

Planning. Expert systems have been employed to assist in the planning process. The ability to reach overall corporate objectives, the impact of plans on organizational resources, and the ways specific plans will help an organization compete in the marketplace can be investigated via expert systems. A manufacturing company, for example, might be exploring the possibility of building a new plant. An expert system can assist with this planning process by suggesting factors that should be considered in making the final decision, based on facts supplied by management.

Design. Designing new products and services requires experience, judgment, and an understanding of the marketplace. Some expert systems have been developed to assist in designing a variety of products, such as computer chips and systems. These types of expert systems use general design principles, an understanding of manufacturing procedures, and a collection of design rules.

Decision making. Wouldn't it be nice to have an expert help us make our day-to-day decisions? Expert systems have provided this type of support for many individuals and organizations. Acting as advisors or counselors, these systems can suggest possible alternatives, ways of looking at problems, and logical approaches to the decision-making process. In addition, expert systems can improve the learning process for those who are not as experienced in decision making.

Quality control and monitoring. Measuring the quality of products and services, determining whether an existing computer system is operating as intended, analyzing the efficiency of a manufacturing plant, and determining the overall effectiveness of a hospital or nursing home are some of the abilities of monitoring systems. Computerized expert systems can assist in monitoring various systems and proposing solutions to system problems. Expert systems can also be used to monitor product quality. When machines are malfunctioning, the expert system can assist in determining possible causes.

Diagnosis. Monitoring and diagnosis go hand in hand. Monitoring determines the current state of a system; diagnosis looks at the causes and proposes solutions. In medicine, expert systems have been employed to diagnose difficult patient conditions. An expert system can analyze test results and patient symptoms. Some systems put probability estimates on potential diseases, given the data and analysis performed. An expert system can provide the doctor with the probable cause of the medical problem and propose treatments or interventions. In a business setting, an expert system can diagnose potential problems of, for example, a chemical distillation facility that is not operating as expected or desired (Figure 11.6).

When to Use Expert Systems

Sophisticated expert systems can be difficult, expensive, and time-consuming to develop. This is especially true for large expert systems implemented on mainframes. Thus, it is important to make sure that the potential benefits are worth the effort and that various expert system characteristics are balanced in terms of cost, control, and complexity.

FIGURE 11.6

In a chemical plant, an expert system can be used to monitor machinery and predict potential problems.
(Source: Image © Copyright PhotoDisc, 1998.)

Following is a list of factors that normally make expert systems worth the expenditure of time and money:

- Provide a high potential payoff or significantly reduced downside risk
- Can capture and preserve irreplaceable human expertise
- Can develop a system more consistent than human experts
- Can provide expertise needed at a number of locations at the same time or in a hostile environment that is dangerous to human health
- Can provide expertise that is expensive or rare
- Can develop a solution faster than human experts can
- Can provide expertise needed for training and development to share the wisdom and experience of human experts with a large number of people

COMPONENTS OF EXPERT SYSTEMS

An expert system consists of a collection of integrated and related components, including a knowledge base, an inference engine, an explanation facility, a knowledge base acquisition facility, and a user interface. A diagram of a typical expert system is shown in Figure 11.7. In this figure, the user interacts with the user interface, which interacts with the inference engine. The inference engine interacts with the other expert system components. These components must work together in providing expertise.

knowledge base

that component of the expert system that stores all relevant information, data, rules, cases, and relationships used by the expert system

if-then statements

rules that suggest certain conclusions

FIGURE 11.7

Components of an Expert System

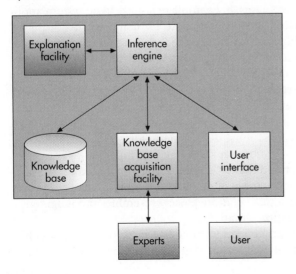

The Knowledge Base

The **knowledge base** stores all relevant information, data, rules, cases, and relationships used by the expert system. A knowledge base must be developed for each unique application. For example, a medical expert system will contain facts about diseases and symptoms. The knowledge base can include generic knowledge from general theories that have been established over time and specific knowledge that comes from more recent experiences and rules of thumb. Knowledge bases, however, go far beyond simple facts, also storing relationships, rules or frames, and cases. For example, certain telecommunications network problems may be related or linked; one problem may cause another. In other cases, rules suggest certain conclusions, based on a set of given facts. In many instances, these rules are stored as **if-then statements**, such as "If a certain set of network conditions exists, then a certain network problem diagnosis is appropriate." Cases can also be used. This technique involves finding instances, or cases, that are similar to the current problem and modifying the solutions to these cases to take into account any differences between the previously solved cases stored in the computer and the current situation or problem.

Purpose of a knowledge base. The overall purpose of the knowledge base is to hold the relevant facts and information for the specific expert system. A knowledge base is similar to the total sum of a human expert's knowledge and experience gained through years of work in a specific area or discipline. The goal of the system is to capture as much experience and knowledge as possible.

Consider an expert system that locates hardware problems for a large mainframe computer system. A human expert will know a large number of facts about the system. The human expert will also look for specific problems that have occurred frequently in the past. Human experts use a number of rules to help locate problems. If the malfunctioning computer displays certain types of behaviors, then the human inspects certain parts of the hardware for potential problems. A knowledge base developed to identify hardware problems also contains important facts on the hardware, information on frequent problems, and relationships between computer performance and what may be wrong.

Assembling human experts. One challenge in developing a knowledge base is to assemble the knowledge of multiple human experts. Typically, the objective in building a knowledge base is to integrate the knowledge of individuals with similar expertise (e.g., many doctors may contribute to a medical diagnosis knowledge base). A knowledge base that contains information from numerous experts can be extremely powerful and accurate in terms of its predictions and suggestions. Unfortunately, human experts can disagree on important relationships and interpretations of data. This presents a dilemma for designers and developers of knowledge bases and expert systems in general. Some human experts are more expert than others; their knowledge, experience, and information are better developed and more accurately represent reality. When human experts disagree on important points, it can be difficult for expert systems developers to determine which rules and relationships to place in the knowledge base.

The use of fuzzy logic. Another challenge for expert system designers and developers is capturing knowledge and relationships that are not precise or exact. Computers typically work with numerical certainty; certain input values will always result in the same output. In the real world, as you know from experience, this is not always the case. To handle this dilemma, a specialty research area in computer science, called **fuzzy logic**, has been developed. Research into fuzzy logic has been going on for decades, but application to expert systems is just beginning to show results in a variety of areas. The opening vignette was an example of a fuzzy logic system application to help media-buying agencies purchase TV commercial slots.

Instead of the usual black and white, yes/no, or true/false conditions of typical computer decisions, fuzzy logic allows shades of gray, or what are known as "fuzzy sets." The criteria on whether a subject or situation fits into a set are given in percentages or probabilities. For example, a weather forecaster might state that "if it is very hot with high humidity, the likelihood of rain is 75 percent." The imprecise terms of "very hot" and "high humidity" are what fuzzy logic must determine to formulate the chance of rain. Fuzzy logic rules help computers evaluate the imperfect or imprecise conditions they encounter and make "educated guesses" based on the likelihood or probability of correctness of the decision. This ability to estimate whether a condition fits a situation more closely resembles the judgment a person makes when evaluating situations.

Fuzzy logic is used in embedded computer technology—for example, autofocus cameras, medical equipment that monitors patients' vital signs

fuzzy logic

a specialty research area in computer science that allows shades of gray and does not require everything to be simple black or white, yes/no, or true/false

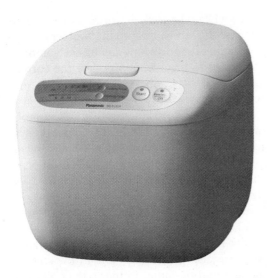

This rice cooker/slow cooker from Panasonic employs fuzzy logic technology. A built-in thermostat maintains heat at a precise and uniform level, virtually eliminating burning or overcooking the rice.
(Source: Courtesy of Panasonic Consumer Electronics Company.)

rule

a conditional statement that links given conditions to actions or outcomes

and makes automatic corrections, and temperature sensors attached to furnace controls.

Fuzzy logic was first applied in Japan. Seiji Yasunobu used it in an automatic control system for the city of Sendai's subway system. Even in a country famed for the precision of its underground railways, Sendai's is impressive. Each train stops to within 7 cm. (3 in.) of the right spot on the platform. In addition, the trains travel more smoothly and use about 10 percent less energy than their human-controlled equivalents. The person in the driver's compartment is there for little more than reassurance. Mercedes Benz Corporation of Germany used fuzzy logic to optimize the design process of truck components, such as gear boxes, axles, or steering. For this optimization, it was necessary to measure the "maturity" of the design process with a single parameter. Fuzzy logic was used to assess this single parameter from the numerous sources of information that describe various aspects of the design process. In a pilot study, a U.S. hospital used fuzzy logic to estimate the length of hospital stay of patients accepted to the hospital. The system uses the information that is provided by the doctor who admits the patient to the hospital. The fuzzy logic system takes into account the diagnosis, the patient's general condition, the likelihood of complications, the patient's previous medical history (if available), and other information. In yet another application of fuzzy logic, data analysis, the decision has to be made whether two similar entries in a database really represent the same person. This decision is harder than it appears at first glance because entries can vary in many ways—minor differences in spelling, order and grouping of words, and typos. While most humans, based on their experience comparing addresses, can come up with a pretty good guess whether two addresses in a database belong to the same person, a mathematical model for the similarity of two addresses is hard to define.[12]

The use of rules. A **rule** is a conditional statement that links given conditions to actions or outcomes. As we saw earlier, a rule is constructed using if-then constructs. If certain conditions exist, then specific actions are taken or certain conclusions are reached. In an expert system for a weather forecasting operation, for example, the rules could state that if certain temperature patterns exist with a given barometric pressure and certain previous weather patterns over the last 24 hours, then a specific forecast will be made, including temperatures, cloud coverage, and the wind-chill factor. Rules are often combined with probabilities, such as if the weather has a particular pattern of trends, then there is a 65 percent probability that it will rain tomorrow. Likewise, rules relating data and conclusions can be developed for any knowledge base. Most expert systems prevent users from entering contradictory rules. Figure 11.8 shows the use of expert system rules in helping to determine whether a person should receive a mortgage loan from a bank. In general, as the number of rules an expert system knows increases, the precision of the expert system increases.

The use of cases. As mentioned previously, an expert system can use cases in developing a solution to a current problem or situation. This process

Mortgage Application for Loans from $100,000 to $200,000

If there are no previous credit problems and

If monthly net income is greater than 4 times monthly loan payment and

If down payment is 15% of the total value of the property and

If net assets of borrower are greater than $25,000 and

If employment is greater than three years at the same company

Then accept loan application

Else check other credit rules

FIGURE 11.8

Rules for a Credit Application

inference engine

a part of the expert system that seeks information and relationships from the knowledge base and provides answers, predictions, and suggestions the way a human expert would

backward chaining

a method of reasoning that starts with conclusions and works backward to the supporting facts

involves (1) finding cases stored in the knowledge base that are similar to the problem or situation at hand and (2) modifying the solutions to the cases to fit or accommodate the current problem or situation. Cases stored in the knowledge base can be identified and selected by comparing the parameters of the new problem with the cases stored in the computer system. For example, a company may be using an expert system to determine the best location of a new service facility in the state of New Mexico. Labor and transportation costs may be the most important factors. The expert system may identify two cases involving the location of a service facility where labor and transportation costs were also important—one in the state of Colorado and the other in the state of Nevada. The expert system will modify the solution to these two cases to determine the best location for a new facility in New Mexico. The result might be to locate the new service facility in the city of Santa Fe.

The Inference Engine

The overall purpose of an **inference engine** is to seek information and relationships from the knowledge base and to provide answers, predictions, and suggestions the way a human expert would. In other words, the inference engine is the component that delivers the expert advice.

The process of retrieving relevant information and relationships from the knowledge base is not simple. As you have seen, the knowledge base is a collection of facts, interpretations, and rules. The inference engine must find the right facts, interpretations, and rules and assemble them correctly. In other words, the inference engine must make logical sense out of the information contained in the knowledge base, the way the human mind does when sorting out a complex situation. The inference engine has a number of ways of accomplishing its tasks, including backward and forward chaining.

Backward chaining. Backward chaining is the process of starting with conclusions and working backward to the supporting facts. If the facts do not support the conclusion, another conclusion is selected and tested. This process is continued until the correct conclusion is identified.

Consider an expert system that forecasts product sales for next month. With backward chaining, we start with a conclusion, such as "Sales next month will be 25,000 units." Given this conclusion, the expert system searches for rules in the knowledge base that support the conclusion, such as "IF sales last month were 21,000 units and sales for competing products were 12,000 units, THEN sales next month should be 25,000 units or greater." The expert system verifies the rule by checking sales last month for the company and its competitors. If the facts are not true—in this case, if last month's sales were not 21,000 units or 12,000 units for competitors—the expert system would start with another conclusion and proceed until rules, facts, and conclusions match.

forward chaining

a method or reasoning that starts with the facts and works forward to the conclusions

Forward chaining. Forward chaining starts with the facts and works forward to the conclusions. Consider the expert system that forecasts future sales for a product. With forward chaining, we start with a fact, like "The demand for the product last month was 20,000 units." With the forward-chaining approach, the expert system searches for rules that contain a reference to product demand. For example, "IF product demand is over 15,000 units, THEN check the demand for competing products." As a result of this process, the expert system might use information on the demand for competitive products. Next, after searching additional rules, the expert system might use information on personal income or inflation on a national basis. This process continues until the expert system can reach a conclusion using the data supplied by the user and the rules that apply in the knowledge base.

A comparison of backward and forward chaining. Forward chaining can reach conclusions and yield more information with fewer queries to the user than backward chaining, but this approach requires more processing and a greater degree of sophistication. Forward chaining is often used by more expensive expert systems. It is also possible to use mixed chaining, which is a combination of backward and forward chaining.

The Explanation Facility

explanation facility

a part of the expert system that allows a user or decision maker to understand how the expert system arrived at certain conclusions or results

An important part of an expert system is the **explanation facility**, which allows a user or decision maker to understand how the expert system arrived at certain conclusions or results. A medical expert system, for example, may have reached the conclusion that a patient has a defective heart valve given certain symptoms and the results of tests on the patient. The explanation facility allows a doctor to find out the logic or rationale of the diagnosis made by the expert system. The expert system, using the explanation facility, can indicate all the facts and rules that were used in reaching the conclusion. This facility allows doctors to determine whether the expert system is processing the data and information in a correct and logical fashion.

The Knowledge Acquisition Facility

A difficult task in developing an expert system is the process of creating and updating the knowledge base. In the past, when more traditional programming languages were used, developing a knowledge base was tedious and time-consuming. Each fact, relationship, and rule had to be programmed into the knowledge base. In most cases, an experienced programmer was required to create and update the knowledge base.

Today, specialized software exists that allows users and decision makers to create and modify their own knowledge bases. This is done through the knowledge acquisition facility (Figure 11.9). The overall purpose of the

FIGURE 11.9

The knowledge acquisition facility acts as an interface between experts and the knowledge base.

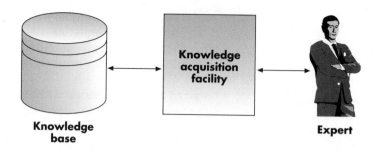

Knowledge base   **Knowledge acquisition facility**   **Expert**

knowledge acquisition facility

the part of the expert system that provides a convenient and efficient means of capturing and storing all components of the knowledge base

knowledge acquisition facility is to provide a convenient and efficient means for capturing and storing all components of the knowledge base. Knowledge acquisition software can present users and decision makers with easy-to-use menus. After filling in the appropriate attributes, the knowledge acquisition facility correctly stores information and relationships in the knowledge base. This makes the knowledge base easier and less expensive to set up and maintain. Knowledge acquisition can be a manual process or a mixture of manual and automated procedures. Regardless of how the knowledge is acquired, it is important to validate and update the knowledge base frequently to make sure it is still accurate.

The User Interface

Specialized user interface software is employed for designing, creating, updating, and using expert systems. The overall purpose of the user interface is to make the development and use of an expert system easier for users and decision makers. At one time, skilled computer personnel were needed to create and operate most expert systems; today, the user interface permits decision makers to develop and use their own expert systems. Because expert systems place more emphasis on directing user activities than do other types of systems, text-oriented user interfaces (using menus, forms, and scripts) may be more common in expert systems than the graphical interfaces often used with DSSs.

Read the "Making a Difference" box for a discussion of the use of knowledge databases and expert systems to provide technical support.

EXPERT SYSTEMS DEVELOPMENT

Like other computer systems, expert systems require a systematic development approach for best results (Figure 11.10). This approach includes determining the requirements for the expert system, identifying one or more experts in the area or discipline under investigation, constructing the components of the expert system, implementing the results, and maintaining and reviewing the complete system.

The Development Process

Specifying the requirements for an expert system begins with identifying the system's objectives and its potential use. Identifying experts can be difficult.

MAKING A DIFFERENCE
Lotus Corporation Builds Knowledge Base to Support Customers

The term "knowledge base" refers to a database of expert information and answers to common questions used by expert systems. An expert system processes its knowledge base using rules called heuristics and responds to a series of user questions and choices to diagnose a problem. The user seems to be talking to a human expert in a particular field.

The term knowledge base has developed another meaning in the context of technical support. In the world of technical support, a knowledge base is a database of technical information and solutions related to a particular product or system and is provided as a customer service on a corporate Web site. Some of these Web knowledge bases use an expert system or other artificial intelligence logic to diagnose problems, but most rely on simple search engines, like those generally used to search the Web as a whole. Compared with more traditional means of providing documentation, they offer many advantages, both to the organizations that build them and to their customers. Information on the Web is cheaper to distribute and easier to keep current than books or CDs. Thus, customers always have the most up-to-date information, available to them anytime from anywhere in the world. Also, the information is easily searched, making for faster access. In addition, a knowledge base can lower the cost of technical support by reducing customer support calls.

After developing a fax-on-demand service, bulletin board system, and a forum on CompuServe, Lotus Development Corp. began a migration to Web-based technical support in 1994. Lotus customers use the site's knowledge base search engine to retrieve technical documentation, scan documentation on a specific product, or enter questions and obtain answers automatically. Their support server now receives 10 million hits per month. Lotus's fax-on-demand and bulletin board services helped develop an editorial staff capable of documenting and transferring information into Lotus's on-line knowledge database. The company's technical support phone analysts write up solutions to customer problems in the form of tech notes. These notes are then submitted to review by 11 technical editors. The reviewed information is fed into the fax-on-demand and bulletin board systems, as well as the knowledge base, where it can be searched. Recently, a knowledge engineer has begun to scrutinize the types and volume of calls for opportunities to reduce call volume through specific tech notes.

DISCUSSION QUESTIONS

1. Is the technical support knowledge base the same knowledge base that is needed for developing an expert system? Why or why not?
2. Imagine that you are developing a Web-based expert system to provide technical support for Lotus Notes (a client/server e-mail package) end users. What additional components are needed besides a knowledge base?

Sources: Adapted from Joyce Flory, "Web Self Support Works! 6 Surprising New Case Studies," *Customer Support Management*, November/December 1997; "What Is a Web Knowledge Base?" at http://www.guidance.com, accessed September 21, 1998.

In some cases, a company will have human experts on hand; in other cases, experts outside the organization will be required. Developing the expert system components requires special skills. Implementing the expert system involves placing it into action and making sure it operates as intended. Like other computer systems, expert systems should be periodically reviewed and maintained to make sure they are delivering the best support to decision makers and users.

Many companies are only now entering the area of expert system use and development. Expert system development is a team effort, but experienced personnel and users may be in short supply. Because development can take from months to years, the cost of bringing in consultants for development can be high. It is critical, therefore, to find and assemble the right people to assist with development.

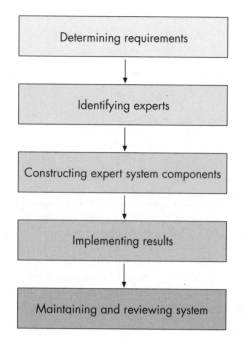

FIGURE 11.10

Steps in the Expert System
Development Process

domain

the area of knowledge addressed
by the expert system

domain expert

the individual or group whose
expertise and knowledge is
captured for use in an expert
system

Participants in Developing and Using Expert Systems

Typically, several people are involved in developing and using an expert system (Figure 11.11).

The domain expert. Because of the time and effort involved in the task, an expert system is developed to address only a specific area of knowledge. This area of knowledge is called the **domain**. A careful evaluation of the domain of the expert system is needed to determine its stability and longevity, which should be weighed against its cost to implement. Many domains, such as the design of microcomputer chips, change quickly in their content and structure. Rapid changes in the knowledge or rules used to make decisions will quickly invalidate the system. On the other hand, the expert system should be built to be flexible so that new rules and knowledge can be added to the system—in effect, permitting the system to learn. The **domain expert** is the individual or group who has the expertise or knowledge one is trying to capture in the expert system. In most cases, the domain expert is a group of human experts. The domain expert (individual or group) usually has the ability to do the following:

- Recognize the real problem
- Develop a general framework for problem solving
- Formulate theories about the situation
- Develop and use general rules to solve a problem
- Know when to break the rules or general principles
- Solve problems quickly and efficiently
- Learn from past experience
- Know what is and is not important in solving a problem
- Explain the situation and solutions of problems to others

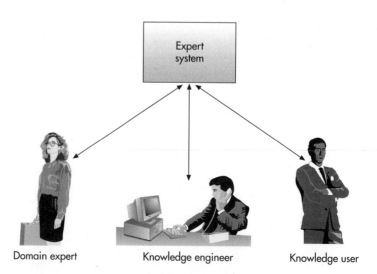

FIGURE 11.11

Participants in Expert Systems
Development and Use

knowledge engineer

an individual who has training and/or experience in the design, development, implementation, and maintenance of an expert system

knowledge user

the individual or group who uses and benefits from the expert system

The knowledge engineer and knowledge users. A **knowledge engineer** is an individual who has training and/or experience in the design, development, implementation, and maintenance of an expert system, including training and/or experience with expert system shells. The **knowledge user** is the individual or group who uses and benefits from the expert system. Knowledge users do not need any previous training in computers or expert systems.

Expert Systems Development Tools and Techniques

Theoretically, expert systems can be developed from any programming language. Since the introduction of computer systems, programming languages have become easier to use, more powerful, and increasingly able to handle specialized requirements. In the early days of expert systems development, traditional high-level languages, including Pascal, FORTRAN, and COBOL, were used (Figure 11.12). LISP was one of the first special languages developed and used for artificial intelligence applications. PROLOG, a more recent language, was also developed for AI applications. Today, however, other expert system products (such as shells) are available that remove the burden of programming, allowing nonprogrammers to develop and benefit from the use of expert systems.

Expert system shells and products. As discussed, an expert system shell is a collection of software packages and tools used to design, develop, implement, and maintain expert systems. Expert system shells exist for both personal computers and mainframe systems. Some shells are inexpensive,

FIGURE 11.12

Software for expert systems development has evolved greatly since 1980, from traditional programming languages to expert system shells.

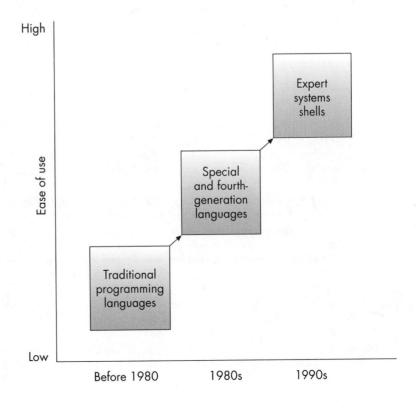

costing less than $500. In addition, off-the-shelf expert system shells are available that are complete and ready to run. The user enters the appropriate data or parameters, and the expert system provides output to the problem or situation. For example, CLIPS is an expert system shell that supports the construction of rule and/or object based expert systems. It is used by all NASA sites and branches of the military, numerous federal bureaus, government contractors, universities, and many companies. CLIPS supports three different programming paradigms: rule-based, object-oriented, and procedural. Rule-based programming allows knowledge to be represented as heuristics, or "rules of thumb," which specify a set of actions to be performed for a given situation. Object-oriented programming allows complex systems to be modeled as modular components (which can be easily reused to model other systems or to create new components). The procedural programming capabilities provided by CLIPS are similar to capabilities found in languages such as C, Pascal, Ada, and LISP. CLIPS can be embedded within procedural code, called as a subroutine, and integrated with languages such as C, FORTRAN, and Ada.[13] Other expert system shells in use today include Financial Advisor, 1st-Class Fusion, Knowledgepro, Leonardo, Personal Consultant, and MindWizard. These are summarized in Table 11.2.

In addition to expert system shells, other expert system development tools make the development of expert systems easier and faster. These products help capture if-then rules for the rule base, assist in using tools such as spreadsheets and programming languages, interface with traditional database packages, generate the inference engine, and perform other functions.

Once developed, an expert system can be run by people with little or no computer experience. The expert system asks the user a series of questions. Subsequent questions are often based on answers to previous questions. After the user answers the system-generated questions, the expert system generates conclusions. Some expert systems with word processing capabilities can generate letters to users requesting additional information, if

TABLE 11.2

Popular Expert System Shells

Financial Advisor can analyze financial investments in new equipment, facilities, and the like. The expert system requests the appropriate data and performs a complete financial analysis.

1st-Class Fusion offers a direct, easy-to-use link to the knowledge base. It also offers a visual rule tree, which graphically shows how rules are related. Because of its features, it has been used to develop models for large, sophisticated expert systems.

Knowledgepro, by Knowledge Garden, is a high-level language that combines functions of expert systems and hypertext. It allows the construction of classic if-then rules and can read dBase and Lotus 1-2-3 files.

Leonardo, an expert system shell that uses an object-oriented language, was used to develop an expert system called COMSTRAT, which can be used to help marketing managers analyze the position of their companies and products relative to their competition. COMSTRAT uses data on the company's operation and competitors' operations to create a marketing knowledge base that is used to give advice to marketing managers.

Personal Consultant (PC) Easy, an expert system shell developed by Texas Instruments, was used to develop an automated guidance expert system to route vehicles in warehouses and manufacturing plants. The expert system developed from PC Easy asks several questions and then determines the best use of automated guidance vehicles in warehouses and manufacturing facilities.

MindWizard, an inexpensive PC-based software package, enables development of compact expert systems ranging from simple models that incorporate their business decision rules to highly sophisticated models.

needed. The expert system can also access needed data and information from files and databases, instead of asking the user for this information.

Advantages of Expert System Shells and Products

Expert system shells and products are used to a greater extent than ever before. As discussed next, these newer programming approaches offer many advantages over traditional programming tools and techniques for developing expert systems.

> *Easy to develop and modify.* As new facts and rules become available and as existing facts and rules need modification, it is desirable to make changes to the knowledge base. Systems developed using PROLOG and LISP are more difficult to modify than expert systems developed with shells. Shells have an editing facility that makes modification relatively easy and cost effective.
>
> *The use of satisficing.* The traditional approach to problem solving attempts to find the optimal, or best, solution; advanced and symbolic languages can deal with more complex problems and return very good— but not necessarily optimal—decisions. As discussed earlier, this is called the satisficing approach. Good decisions that satisfy the requirements of the decision maker are found, whereas the optimal or the best solution would be too difficult or time-consuming to obtain.
>
> *The use of heuristics.* As mentioned previously, expert systems must be able to handle imprecise relationships. Heuristics can help handle these situations and will often return a good solution that satisfies the decision maker. Heuristic rules are often easier to implement on an expert system shell than to code directly.
>
> *Development by knowledge engineers and users.* With expert system shells, knowledge engineers and knowledge users can complete the development process. When developing expert systems with traditional programming languages, systems analysis and computer programming, which are expensive and usually require more time, are often required. In addition, it can be difficult to communicate the exact needs of the expert system to traditional programmers and computer personnel. Using expert system shells can result in a system that is less expensive to develop, requires less time to implement, and more accurately captures the needs of its intended decision makers and users. As expert system shells become easier to use, the role of the knowledge engineer will shift from system developer to more of a consultant.

Expert Systems Development Alternatives

Expert systems can be developed from scratch by using an expert system shell or by purchasing an existing expert system package. The approach selected depends on the system benefits compared with the cost, control, and complexity of each alternative. A graph of the general cost and time of development are shown in Figure 11.13. It is usually faster and less expensive to develop an expert system using an existing package or an expert system shell. Note that there will be an additional cost of developing an existing package or acquiring an expert system shell if the organization does not already have this type of software.

FIGURE 11.13

Some Expert Systems
Development Alternatives
and Their Relative Cost
and Time Values

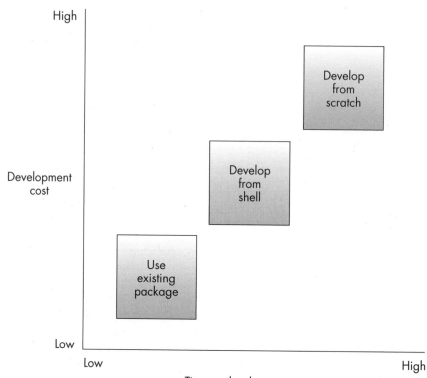

In-house development: develop from scratch. This approach is usually more costly than the other alternatives, but the organization has more control over the features and components of the resulting expert system. Developing an expert system from scratch can also result in a more complex system, resulting in higher maintenance and updating costs.

In-house development: develop from a shell. As you have seen, an expert system shell consists of one or more software products that assist in the development of an expert system. In some instances, the same shell can be used to develop many expert systems. Developing an expert system from a shell can be less complex and easier to maintain than developing an expert system from scratch. However, the resulting expert system may need to be modified to make it appropriate for specific applications. In addition, the capabilities and features of the expert system can be more difficult to control.

Off-the-shelf purchase: use existing packages. Using an existing expert system package is the least expensive and fastest approach, in most cases. An existing expert system package is one that has been developed by a software or consulting company for a specific field or area, such as the design of a new computer chip or a weather forecasting and prediction system. The advantages of using an existing package can go beyond development time and cost. These systems can also be easy to maintain and update over time. A disadvantage of using an off-the-shelf package is that it may not be able to satisfy the unique needs or requirements of the organization.

APPLICATIONS OF EXPERT SYSTEMS AND ARTIFICIAL INTELLIGENCE

Expert systems and artificial intelligence are being used in a variety of ways. Some of the applications of these systems are summarized next.

Credit granting. Many banks employ expert systems to review an individual's credit application and credit history data from credit bureaus to make the credit granting decision.

Information management and retrieval. The explosive growth of information available to decision makers has created a demand for devices to help manage the information. Expert systems can aid this process through the use of agents. Businesses might use an agent to retrieve information from large distributed databases or a vast network like the Internet. Expert system agents help managers find the right data and information while filtering out irrelevant facts that might impede timely decision making.

AI and expert systems embedded in products. The antilock braking system on modern automobiles is an example of a rudimentary expert system. A processor senses when the tires are beginning to skid and releases the brakes for a fraction of a second to prevent the skid. AI researchers are also finding ways to use neural networks and robotics in everyday devices, like toasters, alarm clocks, and televisions.

Plant layout. FLEXPERT is an expert system that uses fuzzy logic to perform plant layout. The software helps companies determine the best placement for equipment and manufacturing facilities.

Hospitals and medical facilities. Some hospitals use expert systems to determine a patient's likelihood of contracting cancer or other diseases. MYCIN is an expert system started at Stanford University to analyze blood infections. A medical expert system used by the Harvard Community Health Plan allows members of the HMO to get medical diagnoses via home personal computers. For minor problems, the system gives uncomplicated treatments; for more serious conditions, the system schedules appointments. The system is highly accurate, diagnosing 97 percent of the patients correctly (compared with the doctors' 78 percent accuracy rating). In order to help doctors in the diagnosis of thoracic pain MatheMEDics® has developed THORASK™, a straightforward, easy-to-use program, requiring only the input of carefully obtained clinical information. The program helps the less experienced to distinguish the three principal categories of chest pain from each other. It does what a true medical expert system should do without the need for complicated user input. You answer basic questions about the patient's history and

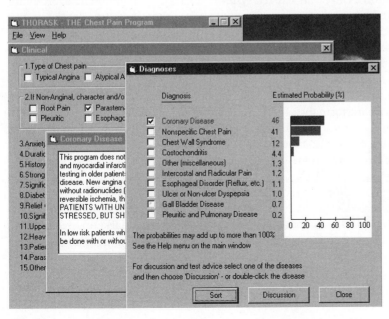

MatheMEDics THORASK™ is a medical expert system used to diagnose chest pain.
(Source: Reprinted by permission from MatheMEDics, Inc. (www.mathemedics.com).)

directed physical findings, and the program immediately displays a list of diagnoses. The diagnoses are presented in decreasing order of likelihood, together with their estimated probabilities. The program also provides concise descriptions of relevant clinical conditions and their presentations as well as brief suggestions for diagnostic approaches. For purposes of record-keeping, documentation, and data analysis, there are options for saving and printing cases.[14]

Help desks and assistance. Expert systems are used by customer service help desks to provide timely and accurate assistance. Kaiser Permanente, a large HMO, uses an expert system and voice response to automate its help desk function. The automated help desk frees up staff to handle more complex needs, while still providing more timely assistance for routine calls.

Employee performance evaluation. An expert system by Austin-Hayne, called Employee Appraiser, provides managers with expert advice for use in employee performance reviews and career development.

Loan analysis. KPMG Peat Marwick uses an expert system called Loan Probe to review loan loss reserves to determine whether sufficient funds have been set aside to cover the risk that some loans will become uncollectible.

Virus detection. IBM is using neural network technology to help create more advanced software for eradicating computer viruses, a major problem in American businesses. IBM's neural network software deals with "boot sector" viruses, the most prevalent type, using a form of artificial intelligence that mimics the human brain and generalizes by looking at examples. It requires a vast number of training samples, which in the case of antivirus software are three-byte virus fragments.

Repair and maintenance. ACE is an expert system used by AT&T to analyze the maintenance of telephone networks. Nynex (New York and New England Telephone Exchange) has expert systems to help its workers locate and solve customer-related phone problems. IET-Intelligent Electronics uses an expert system to diagnose maintenance problems related to aerospace equipment. In the airline industry, prognosis can help reduce the high costs of unscheduled major component removals (such as engines) or failures in flight (such as in-flight engine shutdowns). Assessment of equipment health will also support early and better identification and planning of optimal maintenance actions. General Electric Aircraft Engine Group uses an expert system to enhance maintenance performance levels at all sites, improve diagnostic accuracy and reduce ambiguity in fault isolation, advise on real-time repair action, provide clues to incipient failures, and access and display relevant maintenance information.[15]

Shipping. CARGEX-Cargo Expert System is used by Lufthansa, a German airline, to help determine the best shipping routes.

Marketing. CoverStory is an expert system that extracts marketing information from a database and automatically writes marketing reports.

Warehouse optimization. United Distillers uses an expert system to determine the best combinations of stocks to produce its blends of Scottish whiskey. This information is then supplemented with information about location of the casks for each blend. The system then optimizes the selection of required casks, keeping to a minimum the number of "doors" (warehouse sections) from which the casks must be taken and the number of casks that need to be moved to clear the way. Other constraints must be satisfied, such as the current working capacity of each warehouse and the maintenance and re-stocking work that may be in progress.[16]

Integrating expert systems. As with the other information systems, an expert system can be integrated with other systems in an organization through a common database. An expert system that identifies late-paying customers who should not receive additional credit may draw data from the same database as an invoicing MIS that produces weekly reports on overdue bills. The same database—a by-product of the invoicing transaction processing system—might also be used by a decision support system to perform "what-if" analysis to determine the impact of late payments on cash flows, revenues, and overall profit levels.

In many organizations, these systems overlap. A TPS might be expanded to provide management information, which in turn may provide some DSS functions, and so on. In each progressive phase of this overlap, the information system assists with the decision-making process to a greater extent. Of all these information systems, expert systems display this characteristic most obviously, proposing decisions based on specific problem data and a knowledge base. Understanding the capabilities and characteristics of expert systems is the first step in applying these systems to support managerial decision making and organizational goals.

Information Systems Principles

As with other information systems, the overall purpose of artificial intelligence applications in business is to help the organization achieve its goals.

The ability to create machines that can reason will be possible only once we truly understand our own processes for doing so.

Sophisticated expert systems can be difficult, expensive, and time-consuming to develop. It is important to make sure that the potential benefits are worth the effort and that various expert system characteristics are balanced in terms of cost, control, and complexity.

When human experts disagree on important points, it can be difficult for expert system developers to determine which rules and relationships to place in the knowledge base.

Specifying the requirements for an expert system begins with identifying the system's objectives and its potential use.

■ SUMMARY

1. *Define the term* artificial intelligence *and state the objective of developing artificial intelligence systems.*

Artificial intelligence is used to describe computers with the ability to mimic or duplicate the functions of the human brain. The objective of building AI systems is not to replace human decision making completely but to replicate it for certain types of well-defined problems.

2. *List the characteristics of intelligent behavior and compare the performance of natural and artificial intelligence systems for each of these characteristics.*

Intelligent behavior encompasses several characteristics including the abilities to: learn from experience and apply this knowledge to new experiences; handle complex situations and solve problems for which pieces of information may be missing; determine relevant information in a given situation, think in a logical and rational manner, and give a quick and correct response; and understand visual images and processing symbols. The computer is better than humans at transferring information, making a series of calculations rapidly and accurately, and making complex calculations, but a human is better than a computer at all other attributes of intelligence.

3. *Identify the major components of the artificial intelligence field and provide one example of each type of system.*

Artificial intelligence is a broad field that includes several key components, such as expert systems, robotics, vision systems, natural language processing, learning systems, and neural networks. An expert system consists of the hardware and software to produce systems that act or behave like a human expert in a field or area (e.g., credit analysis). Robotics involves developing mechanical or computer devices to perform tasks that require a high degree of precision or are tedious or hazardous for humans (e.g., stacking cartons on a pallet). Vision systems include hardware and software that permit computers to capture, store, and manipulate images and pictures (face-recognition software). Natural language processing allows the computer to understand and react to statements and commands made in a "natural" language, such as English. Learning systems use a combination of

software and hardware that allows the computer to change how it functions or reacts to situations based on feedback it receives (e.g., computerized chess game). A neural network is a computer system that can act like or simulate the functioning of a human brain (e.g., disease diagnostics system).

4. *List the characteristics and basic components of expert systems.*

Expert systems can explain their reasoning or suggested decisions; display intelligent behavior; manipulate symbolic information and draw conclusions from complex relationships; provide portable knowledge; and can deal with uncertainty. They are not yet widely used; some are difficult to use; are limited to relatively narrow problems; cannot readily deal with mixed knowledge; present the possibility for error; cannot refine their own knowledge base; are difficult to maintain; may have high development costs; and their use raises legal and ethical concerns.

An expert system consists of a collection of integrated and related components, including a knowledge base, an inference engine, an explanation facility, a knowledge acquisition facility, and a user interface. The knowledge base contains all the relevant data, rules, and relationships used in the expert system. The rules are often composed of if-then statements, which are used for drawing conclusions. Fuzzy logic allows expert systems to incorporate facts and relationships into expert system knowledge bases that may be imprecise or unknown.

The inference engine performs the processing of the rules, data, and relationships stored in the knowledge base to provide answers, predictions, and suggestions the way a human expert would. Two common methods for processing include forward and backward chaining. Backward chaining starts with a conclusion, then searches for facts to support it; forward chaining starts with a fact, then searches for a conclusion to support it. Mixed chaining is a combination of backward and forward chaining.

The explanation facility of an expert system allows the user to understand what rules were used in arriving at a decision. The knowledge acquisition facility helps the user add or update knowledge in the knowledge base. The user interface makes it easier to develop and use the expert system.

The individuals involved in the development of an expert system include the domain expert, the knowledge engineer, and the knowledge users. The domain expert is the individual or group who has the expertise or knowledge being captured for the system. The knowledge engineer is the developer whose job is the extraction of the expertise from the domain expert. The knowledge user is the individual who benefits from the use of the developed system.

5. *Identify at least three factors to consider in evaluating the development of an expert system.*

The following is a list of factors that normally make expert systems worth the expenditure of time and money: a high potential payoff or significantly reduced downside risk, the ability to capture and preserve irreplaceable human expertise, the ability to develop a system more consistent than human experts, expertise needed at a number of locations at the same time, and expertise needed in a hostile environment that is dangerous to human health. The expert system solution can be developed faster than the solution from human experts. An ES also provides expertise needed for training and development to share the wisdom and experience of human experts with a large number of people.

6. *Outline and briefly explain the steps for developing an expert system.*

The steps involved in the development of an expert system include determining requirements, identifying experts, constructing expert system components, implementing results, and maintaining and reviewing the system.

Expert systems can be implemented in several ways. Previously, traditional high-level languages, including Pascal, FORTRAN, and COBOL, were used. LISP and PROLOG are two languages specifically developed for creating expert systems from scratch. A faster and less expensive way to acquire an expert system is to purchase an expert system shell or existing package. The shell program is a collection of software packages and tools used to design, develop, implement, and maintain expert systems. Advantages of expert system shells include ease of development and modification, use of satisficing, use of heuristics, and development by knowledge engineers and end users. The approach selected depends on the benefits compared with cost, control, and complexity considerations.

7. *Identify the benefits associated with the use of expert systems.*

The benefits of using an expert system go beyond the typical reasons for using a computerized processing solution. Expert systems display "intelligent" behavior, manipulate symbolic information and draw conclusions, provide portable knowledge, and can deal with uncertainty. Expert systems can be used to solve problems in many fields or disciplines and can assist in all stages of the problem-solving process. Past successes have shown that expert systems are good at strategic goal setting, planning, design, decision making, quality control and monitoring, and diagnosis.

▪ KEY TERMS

artificial intelligence (AI) 480
artificial intelligence systems 480
backward chaining 498
domain 502
domain expert 502
expert system 484
expert system shell 492
explanation facility 499

forward chaining 499
fuzzy logic 496
if-then statements 495
inference engine 498
intelligent behavior 481
knowledge acquisition facility 500
knowledge base 495
knowledge engineer 503

knowledge user 503
learning systems 486
natural language processing 485
neural network 486
perceptive system 482
robotics 484
rule 497
vision systems 485

■ REVIEW QUESTIONS

1. Define the term *artificial intelligence*. What is the difference between natural and artificial intelligence?
2. Define and identify six subfields of artificial intelligence.
3. What is an intelligent agent? Provide three examples of the use of an intelligent agent.
4. Identify three examples of the use of robotics.
5. What is a neural network? Describe two applications of neural networks.
6. What are the capabilities of an expert system?
7. What are some of the limiting characteristics of expert systems?
8. Identify the basic components of an expert system and describe the role of each.
9. What are fuzzy sets and fuzzy logic?
10. How are rules used in expert systems?
11. Expert systems can be built based on rules or cases. What is the difference between the two?
12. Describe the domain expert, the knowledge engineer, and the knowledge user.
13. What are the primary benefits derived from the use of expert systems?
14. Identify three approaches for developing an expert system.

■ DISCUSSION QUESTIONS

1. Can computers think? Will they ever be able to? Explain why or why not.
2. Accuracy slip occurs when there are significant changes in the real world making an expert system less accurate. Identify at least three expert system applications where accuracy slip could lead to the loss of human life. What process can be put in place to safeguard against accuracy slip for these critical expert system applications?
3. You have been hired to capture the knowledge of a brilliant financial adviser who has an outstanding track record for picking growth stocks for a mutual fund before she retires from the firm in six months. This knowledge will be used as the basis for an expert system to help other financial advisers in making decisions on which stocks to add or drop from the mutual fund. Is this system a good candidate for an expert system? Why or why not?
4. Briefly explain why human decision making often does not lead to optimal solutions to problems.
5. Explain the difference between a database and a knowledge base.
6. Describe an application that requires the concurrent use of more than one of the subfields of artificial intelligence.
7. Imagine that you are developing the rules for an expert system to select the strongest candidates for a medical school. What rules and/or heuristics would you include?
8. What skills does it take to be a good knowledge engineer? Would knowledge of the domain help or hinder the knowledge engineer in capturing knowledge from the domain expert?

■ PROBLEM-SOLVING EXERCISES

1. Imagine that you are a knowledge engineer and are developing an expert system to help consumers choose the camera that best meets their needs and their budget. You are going to your first interview with the owner of a camera shop who is the designated expert for this system. Use your word processing software and develop a list of questions that you would ask to begin to capture this individual's knowledge.
2. Assume you live in an area where there is a wide variation in weather from day to day. Develop a simple expert system to provide advice on what sort of clothes to wear based on the weather. The system needs to help you decide which clothes and accessories (umbrella, boots, etc.) to wear for sunny, snowy, rainy, hot, mild or cold days, etc. Key inputs to the system include last night's weather forecast, your observation of the morning temperature and cloud situation, and yesterday's weather. Using your word processing program, create seven or more rules that could be used in such an expert system. Create five cases and use the rules you developed to determine the best course of action.

■ TEAM ACTIVITY

1. With two or three of your classmates, identify three real examples of expert systems in use. Discuss the problems solved by each of these systems. Identify any issues that may arise due to "knowledge creep." Choose which of the three systems provides the most benefit and state why you selected that system.

2. With members of your team, develop an expert system that makes suggestions about what to do if your car does not start. Develop a simple "dialogue" between the expert system and you, the end user. The expert system should suggest different actions to take in an attempt to diagnose and correct whatever may be wrong.

■ WEB EXERCISE

Artificial intelligence includes a number of related fields, including robotics, vision systems, learning systems, neural networks, and expert systems. Using an Internet search engine, explore one of these areas.

Did you find any companies that specialized in one of these areas? You may be asked to develop a report or send an e-mail message to your instructor about what you found.

■ CASES

1 Using an Expert System to Improve Net Presence

Brightware is an expert system that helps companies turn their Internet presence into a round-the-clock selling tool by actively soliciting questions from Web visitors to engage them as sales leads. Brightware receives incoming messages from Web forms or e-mail servers. It reads and accurately interprets each message, then rates it to determine the appropriate actions. It responds to all messages and routes selected ones to the appropriate people according to company business policy. It also sends priority messages of special interest to marketing and sales for review and reports on the types of requests received and how they were handled.

The Brightware expert system recognizes typical words and phrases customers use in messages and it knows common business policies for handling certain messages. Companies can customize and extend knowledge about their business using point and click editors designed for use by business people. Unlike neural networks, Brightware uses an information extraction technique based on the use of cases stored in its knowledge base.

It combines the use of cases with fuzzy logic to improve the accuracy of matching messages to known request types.

Wells Fargo & Company is headquartered in San Francisco and has been serving the financial needs of consumers since 1852. Today, it is the tenth largest bank holding company in the country, with $100 billion in assets as of June 30, 1997. *SmartMoney* magazine named Wells Fargo the "Best Online Bank" in 1996. The Wells Fargo Internet site was also awarded "Best Overall Site by a U.S. Financial Institution" by the Online Banking Association that year. Yahoo!, the Web site locator, gave the bank's home page a four-star rating (its highest).

However, Wells Fargo is not willing to rest on its laurels, and the company is always looking for ways to better serve customers. It is using Brightware to answer consumer inquiries as a way to bring customers quicker and more accurate responses to their on-line requests, while cutting business operation costs. "When customers shop for financial service products on the Net, they expect banks to offer them the means to ask questions on-line and a reasonable likelihood that their questions will be answered correctly and in a timely fashion," explains Chuck Williams, president and CEO of Brightware. "We believe Wells Fargo and its customers will benefit significantly from Brightware and its capability to automatically fulfill electronic requests and turn hits into qualified leads, low-cost sales, and satisfied customers."

Brightware is in use at several companies including American Finance and Investment (AFI) supporting its Cybersmart Instant Mortgage System (http://www.loanshop.com). Other companies who have licensed Brightware include Amway Corporation, E-Care, Fannie Mae, and Swiss Bank.

1. Visit the Web site of one of the companies using Brightware and request information from the company. What sort of response do you receive? Does it address your issue? Is it timely? Write a short paragraph about your experience.

2. Do you think there are any risks associated with the use of an expert system to interact directly with customers? If so, what are those risks? How might such risks be minimized?

Sources: Adapted from Product Description section of the Web site of Brightware, Inc. at http:www.brightware.com, last modified April 8, 1998, accessed June 13, 1998; Web site of the Horn Group, Inc. at http://www.horngroup.com/news, accessed June 13,1998; press release "Leading Online Bank to Handle Electronic Customer Requests on the Net with Brightware 1.0," August 4, 1997; Barb Cole-Gomolski, "E-mail with AI Offers Service with a :-)," *Computerworld*, August 11, 1997, pp. 45, 48.

2   Use of Fuzzy Logic to Predict Length of Patient Stay

Hospitals are under extreme pressure to reduce costs. One way to do this is to increase their efficiency by optimizing the utilization of all resources. Hospitals have several types of scarce resources that must be carefully managed including staff, hospital beds, intensive care unit beds, and the number of patients the different teams can handle.

Hospitals are beginning to experiment with the use of fuzzy logic expert systems to improve their operating efficiency. Some hospitals actually have defined a strategy for admitting patients that helps address this issue. During times when a small fraction of the hospital beds are in use, a hospital's policy is to select and schedule patients in a way that makes best use of the staff. During times when most hospital beds are in use, the policy changes to select and organize work based on more outpatients and patients that only require a short stay.

To assess optimal capacity, it is necessary to develop a good estimate of how long each patient will stay in the hospital—at the time the patient is admitted to the hospital. In a pilot study, a U.S. hospital, which wishes to remain anonymous, used fuzzy logic to estimate the length of hospital stay of patients accepted to the hospital. The system uses the information that is provided by the doctor who admits the patient to the hospital.

1. What specific information do you think is used to estimate the length of the patient's stay?

2. Why would the hospital wish to remain anonymous?

Sources: Adapted from "Hospital Stay Prediction," in the Business and Finance Applications section of the Inform Software Web site at http://www.fuzzytech.com, accessed June 12, 1998; "Introduction to Fuzzy Logic" found at the Center for Fuzzy Logic, Robotics, and Intelligent Systems at Texas A & M University Web site at http://www.cs.tamu.edu, accessed September 19, 1998.

3 Artificial Intelligence: An Intelligent Way to Schedule Jobs at Volvo

Evolutionary change has only recently been applied to the world of computers and software. Once a program is written, it acts or performs the same each time it is run under the same conditions. An exciting development in the field of artificial intelligence is changing how programs operate. The field of changing software is called genetic algorithms. With this branch of artificial intelligence, how a program responds and its output evolves or changes over time is studied. Once a computer science research project, genetic algorithms are being applied in corporations, including Swedish car and truck manufacturer Volvo.

Volvo, known for promoting safety in cars and trucks, has been a pioneer in the auto industry for decades. It was one of the first companies to experiment with worker teams, when the other auto manufacturers still viewed workers as replaceable cogs on the assembly line. So it was no surprise when the innovative auto and truck manufacturer turned to genetic algorithms to help it schedule jobs at one of its manufacturing facilities to do custom work.

Although known primarily for its luxury cars, Volvo is also a successful truck manufacturer. The average car driver typically spends less than a few hours a day in the vehicle. Professional truck drivers hauling loads from one location to another typically spend most of their days or nights behind the wheel. For years, car buyers have demanded options and choice in selecting the interior and exterior features. The same is true for professional truck drivers. Spending large amounts of time in the cab, these drivers want custom interiors to make driving long hours more comfortable and enjoyable. Putting this flexibility into a truck is a joy for the drivers but a headache for manufacturing companies. One problem is scheduling jobs and manufacturing facilities.

Volvo's truck plant in Dublin, Virginia, has over a million square feet of manufacturing facilities to make commercial trailer cabs. The restrictions on what can be produced and in what quantities is exceedingly complex. Adding the flexibility to make custom cabs makes the problem worse. How do you schedule jobs and manufacturing facilities to build flexibility into the manufacturing process? For Volvo, the answer was to use genetic algorithms.

To solve its scheduling problems, Volvo decided to try a new technology. It purchased a program, called OptiFlex, that uses the theory of genetic algorithms. The program allows the schedule to "evolve." It continually improves a series of so-so schedules. It combines and varies schedules and selects the best one. It is like Darwin's theory of evolution that relies on variability and natural selection. The primary difference is that the computer varies the schedules and selects the best one. The result is a good schedule that has evolved from past schedules. The process is automatic. According to one manager, you tell the program what you want to produce and then go get a cup of coffee. When you return, OptiFlex has evolved a schedule for you.

1. What is a genetic algorithm, and how can it be used in scheduling jobs and manufacturing facilities?
2. In what other ways could a genetic algorithm be used to solve business problems?

Sources: Srikumar Rao, "Evolution at Warp Speed," *Forbes*, January 12, 1998, p. 82; David Furlonger, "Swede Smell of Success," *Financial Mail*, January 10, 1998, p. 42; Colin Johnson, "Technique Accelerates Evolution of Genetic Programs," *Electronic Engineering Times*, February 2, 1998; Staff, "Volvo/NFC," *Corporate Money*, January 21, 1998, p. 7.

4 Immigration and Naturalization Service Applies High Technology

The Immigration and Naturalization Service (INS) has long recognized the role technology can play in carrying out the nation's immigration laws. The INS is equipping its officers with the best technology available including the Secure Electronic Network for Travelers' Rapid Inspection, or SENTRI. The SENTRI system significantly improves border inspection capabilities enabling inspectors to rapidly screen vehicles that cross the border frequently. The system minimizes delays without compromising border security. SENTRI benefits businesses and individuals who frequently cross the border by improving the efficiency of border inspection processes. The program was designed to speed the cross-border commute of workers in companies that have facilities in the United States and Mexico. There are more than 1,200 such companies—called *maquiladoras*—along the border.

Currently the SENTRI system is used in two commuter lanes at the Otay Mesa crossing just south of San Diego, which serves about 3,000 border crossers per day. For drivers who can't use these lanes, it can take two hours to cross the border during rush hour. To use the new lanes, applicants pay a $129 annual fee and undergo an extensive background check, which includes running their fingerprints through criminal databases at the FBI and other law enforcement agencies. The background check helps screen out applicants with a criminal history, those most likely to smuggle drugs or other contraband, and gives others quicker access to the border.

SENTRI is composed of two key components: the Global Enrollment System (GES) and a validation system. The GES stores information about applicants, including fingerprints, photos, and biographical data, and screens it against INS databases and the National Crime Information Center. Once an individual is authorized, a SENTRI card is issued, and a transponder is installed on the car. When a SENTRI user approaches the border crossing, the system scans the driver's license plate and identifies the car using the transponder. When the driver stops at the inspection booth, the pertinent data (e.g., license number, digitized photographs of the driver and passengers, and make, model, and color of the vehicle) are displayed on a computer screen in the inspector's booth. The information is used by the inspector to quickly identify the vehicle and its passengers.

The federal INS is considering technology that can identify frequent border crossers by matching images of their faces or snippets of their voices to samples they supply when applying to use the system. If the use of image and voice recognition technology works accurately, it could mean minimal delays for commuters. Ideally the system would identify them without having them stop at all at the border crossing. It also would help the INS to better identify the person behind the wheel.

1. Do you think there are any individual privacy issues with the use of technology to enforce our nation's immigration laws? Why or why not?
2. Can you envision other related potential applications for this technology?

Sources: Adapted from INS Fact Sheet dated May 28, 1998 found at the INS Web site at http://www.ins.usdoj.gov, accessed June 7, 1998 and Barb Cole-Gomolski, "Your Face is Your Ticket," *Computerworld*, March 9, 1998, pp. 61, 64.

PART IV

Systems
Development

CHAPTER 12

Systems Investigation and Analysis

"If you can't do it better, why do it?"

—Herbert H. Dow, founder of the Dow Chemical Company

Chapter Outline

An Overview of Systems Development
Participants in Systems Development
Initiating Systems Development
Information Systems Planning
Establishing Objectives for Systems Development
Systems Development and the Internet
Trends in Systems Development and Enterprise Resource
  Planning

Systems Development Life Cycles
The Traditional Systems Development Life Cycle
Prototyping
Rapid Application Development and Joint Application
  Development
The End-User Systems Development Life Cycle

Factors Affecting Systems Development Success
Degree of Change
Quality of Project Planning
Use of Project Management Tools
Use of Formal Quality Assurance Processes
Use of Computer-Aided Software Engineering (CASE) Tools

Systems Investigation
Initiating Systems Investigation
Participants in Systems Investigation
Feasibility Analysis
The Systems Investigation Report

Systems Analysis
General Considerations
Participants in Systems Analysis
Data Collection
Data Analysis
Requirements Analysis
The Systems Analysis Report

Learning Objectives

After completing Chapter 12, you will be able to:

1. Identify the key participants in the systems development process and discuss their roles.
2. Define the term *information systems planning* and list several reasons for initiating a systems project.
3. Identify important system performance requirements of transaction processing business applications that run on the Internet or a corporate intranet.
4. Discuss three trends that illustrate the impact that the use of enterprise resource planning software packages is having on systems development.
5. Discuss the key features, advantages, and disadvantages of the traditional, prototyping, rapid application development, and end-user systems development life cycles.
6. Identify several factors that influence the success or failure of a systems development project.
7. State the purpose of systems investigation.
8. State the purpose of systems analysis and discuss some of the tools and techniques used in this phase of systems development.

Gerber
Develops System to Manage Customers' Inventory

The Gerber story started in the kitchen of Daniel Gerber in the summer of 1927. Following the advice of a pediatrician, his wife had been hand-straining solid food for their seven-month-old daughter. After many evenings of repeating this chore, Dorothy Gerber suggested that her husband try it. After watching him make several attempts, she pointed out that the work could be easily done at the Fremont Canning Company, where the Gerber family produced a line of canned fruits and vegetables. Experiments with strained baby foods began shortly. Soon workers in the plant requested samples for their babies. By late 1928, strained peas, prunes, carrots, and spinach, not to mention beef vegetable soup, were ready for the national market. Gerber has continued to grow throughout the years. Today nearly 190 food products are labeled in 16 languages and distributed to 80 countries.

Tightening inventory management is a top priority for cost-conscious retailers and their suppliers. Retailers want their products to be available for customers to buy, but they do not want too much. Gerber has convinced 40 major grocery chains to allow the company to manage their inventory of Gerber products. During systems analysis, project participants defined the system objectives—reduce both Gerber's and the customers' inventory costs and provide a strong incentive for store managers to buy from Gerber. The company also decided not to charge for the inventory management service, treating it instead as a way to build customer loyalty and get sales data that can be used to fine-tune baby food production plans. Gerber believes it can gain a competitive advantage with superior forecasting and planning.

An electronic data interchange (EDI) setup was designed and implemented to feed information on sales of Gerber products from the grocery stores to the Fremont, Michigan, company. The data is input to Manugistics software to schedule new deliveries. Manugistics, Inc. is a manufacturer of software for supply-chain management. The company's solutions improve the flow of product within and among companies from raw materials or parts through manufacturing to delivery of product to the end customer. With Manugistics software, Gerber makes informed operational decisions, resulting in increased revenues, reduced inventories, improved customer service, better relationships among trading partners, greater speed to market, and lower overall costs throughout the supply chain.

While the Gerber-Manugistics system worked fairly well, the need to translate all of the EDI messages from various grocery chains into a common format slowed down Gerber's ability to add grocers. And custom-built software for sending alerts and other messages to Gerber's inventory planners gave them only the minimum data they needed.

Planners often had to resort to searching through raw EDI transmissions to find important data. Thus, Gerber requested Manugistics to perform a systems analysis and to design a simpler and improved EDI process.

Manugistics formed a partnership with Frontec AMT, a company that specializes in integrating applications. The alliance created the Intelligent Messenger for Vendor Managed Inventory, a software product to format and prepare customer product activity data for input into Manugistics. The software was designed to present data to Manugistics that is consistent and complete in terms of product identification, unit of measure conversion, data validation, and sequence checking.

Additional features were identified based on user requirements—intelligent routing of messages, event-driven notifications, and predefined trading partner business processes.

Gerber plans to be the first company to implement the new data transformation and messaging software. It has established objectives for this system to dramatically increase the amount of inventory it manages for grocery stores. Gerber sells nearly $700 million worth of baby food in the United States each year; however, it only manages about 27 percent of base sales. Its goal is to manage inventory for 80 percent of sales within two years.

As you read this chapter, consider the following questions:

- What are the stages of an information systems project and what are the objectives of each stage?

- Who are the various players that need to be involved in an information systems development project and what are their roles?

Sources: Adapted from Craig Stedman, "Gerber Tightens Inventory Control," *Computerworld*, June 8, 1998, pp. 57–58; Gerber Story and Products sections of the Gerber Web site at http://www.gerber.com, accessed June 19, 1998; and Press release "Frontec Announces Intelligent Messenger For Vendor Managed Inventory," June 1, 1998 found at Manugistics Web site at http://www.manugistics.com, accessed June 19, 1998.

When organizations such as Gerber need to accomplish a new task or change a work process, how do they do it? They develop a new system. Systems development is the activity of creating or modifying existing business systems. It refers to all aspects of the process—from identifying problems to be solved or opportunities to be exploited to the implementation and refinement of the chosen solution. At some point in your career, you will likely be involved in a systems development project—as a user, as a manager of a business area or project team, as a member of the IS department, maybe even as a CIO. Understanding and being able to apply a systems development life cycle, tools, and techniques discussed in this chapter, and in the next chapter, will help ensure the success of the development projects on which you participate.

AN OVERVIEW OF SYSTEMS DEVELOPMENT

Understanding systems development is important to all professionals, not just those in the field of information systems. In today's businesses, managers and employees in all functional areas work together and use business information systems. As a result, users of all types are helping with development and, in many cases, leading the way. Take the example of Gerber from our opening vignette. The role of the IS department was to invest in the necessary hardware and software and to support the staff as they developed a proprietary application with the help of an outside software firm. By developing their own system employing the Manugistics software, employees at Gerber hoped to get a system that met their needs, supported company goals, and delivered value to their customers. The initial system was close to what was needed, but it had to be refined to improve data capture and simplify adding new grocery chains. Major information system projects often need adjustment following their initial implementation. Indeed, based on interviews with 365 information systems executives at large, medium, and small companies, it is estimated that for every 100 application development projects, there are 94 restarts—where developers actually stop the project and start it all over again.[1] More than 60 percent of 50 companies that took part in a survey rated themselves as successful at meeting the data analysis expectations of end users and senior management.[2] This chapter, and the next one, will provide you with a deeper appreciation of the systems development process and help you avoid costly failures.

Participants in Systems Development

Effective systems development requires a team effort. The team usually consists of stakeholders, users, managers, systems development specialists, and various support personnel. This team, called the development team, is responsible for determining the objectives of the information system and delivering a system that meets these objectives to the organization. **Stakeholders** are individuals who, either themselves or through the area of the organization they represent, ultimately benefit from the systems development project. **Users** are individuals who will interact with the system regularly. They can be employees, managers, customers, or suppliers. For large-scale systems development projects, where the investment in and value of a system can be quite high, it is common to have senior-level managers, including the company president and functional vice presidents (of finance, marketing, and so on), be part of the development team.

Depending on the nature of the systems project, the development team might include systems analysts and programmers, among others. A **systems analyst** is a professional who specializes in analyzing and designing business systems. Systems analysts play various roles while interacting with the stakeholders and users, management, vendors and suppliers, software programmers, and other IS support personnel (Figure 12.1). Like an architect developing blueprints for a new building, a systems analyst develops detailed plans for the new or modified system. The **programmer** is responsible for modifying or developing programs to satisfy user requirements. Like a contractor constructing a new building or renovating an existing one, the programmer takes the plans from the systems analyst and builds or modifies the necessary software.

The other support personnel on the development team are mostly technical specialists, including database and telecommunications experts, hardware engineers, and supplier representatives. One or more of these roles may be outsourced to nonemployees or consultants. Depending on the magnitude of the systems development project and the number of IS systems development specialists on the team, the team may also include one or more IS managers. The composition of a development team may vary over time and from project to project. For small businesses, the development team may consist of a systems analyst and the business owner as the primary stakeholder. For larger organizations, formal IS staff can include hundreds of people involved in a variety of IS activities, including systems development. Every development team should have a team leader. As shown in Table 12.1, different types of team leaders should also be considered for different projects and development teams.

stakeholders

individuals who either themselves or through the organization they represent, ultimately benefit from the systems development project

users

individuals who will interact with the system regularly

systems analyst

professional who specializes in analyzing and designing business systems

programmer

individual responsible for modifying or developing programs to satisfy user requirements

FIGURE 12.1

The systems analyst plays an important role in the development team and is often the only person who sees the system in its totality. The one-way arrows in this figure do not mean that there is no direct communication between other team members. Instead, these arrows just indicate the pivotal role of the systems analyst—an individual who is often called upon to be a facilitator, moderator, negotiator, and interpreter for development activities.

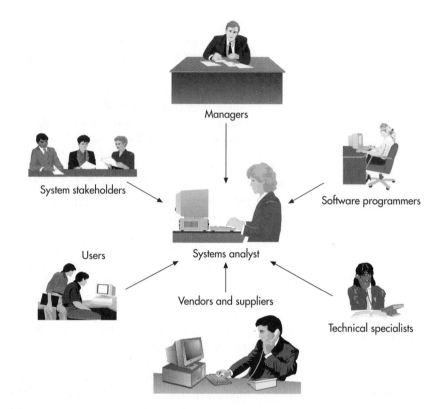

Managers

System stakeholders

Software programmers

Users

Systems analyst

Vendors and suppliers

Technical specialists

Regardless of the specific nature of the project, systems development involves new or modified systems, and this means change. Managing this change effectively requires development team members to communicate well. It is important that you learn communication skills, as you probably will participate in systems development during your career. You may even be the individual who initiates systems development.

Initiating Systems Development

Systems development efforts begin when an individual or group capable of initiating organizational change perceives a potential benefit from a new or modified system. Such individuals have a stake in the development of the system. The systems development effort undertaken by Gerber, for example,

TABLE 12.1

Team Leaders for Different Systems Development Projects

Characteristics of Systems Development Project	Examples of Appropriate Team Leaders
Project involves new and advanced technology	Individual from IS department
Impact will force critical changes in a functional area of the business	Manager from that functional area
Project is extremely large and complex	Specialist in project management
Project will have dramatic impact on personnel	Individual from human resource department
Project will share a combination of the preceding characteristics	Senior management; leader should build a development team that includes personnel skilled in all affected areas

was initiated to improve inventory management. Gerber managers recognized an opportunity to provide superior customer service that could distinguish Gerber from its competition. These perceived benefits resulted in the initiation of systems development.

Equally important, however, is the ability of the individual or group to initiate organizational change. Many individuals cannot initiate change. This may not be because they do not want to but rather because they are constrained by hierarchical status, power, authority, and political standing within the organization. Managers are most often empowered by an organization to initiate change and thus most often initiate systems development projects. In addition, the degree of overall managerial support, especially senior-level managerial support, greatly influences the probable success or failure of a systems development effort.

Systems development initiatives arise from all levels of an organization and are both planned and unplanned. Solid planning and managerial involvement helps ensure that these initiatives support broader organizational goals. Systems development projects may be initiated for a number of reasons, as shown in Figure 12.2.

Information Systems Planning

Because an organization's strategic plan contains both organizational goals and a broad outline of steps required to reach them, the strategic plan affects the type of system an organization needs. For example, a strategic plan may identify as organizational goals a doubling of sales revenue within five years, a 20 percent reduction of administrative expenses over three years, acquisition

FIGURE 12.2

Typical Reasons to Initiate a Systems Development Project

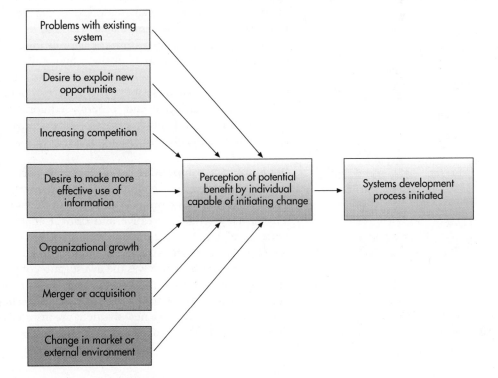

of at least two competing companies within a year, or the capture of market leadership in a given product category. Organizational commitments to policies such as continuous improvement are also reflected in the strategic plan. Such goals and commitments set broad outlines of system performance.

Often, a section of the strategic plan lists guidelines about how to meet organizational goals. Examples of these guidelines might be improving customer service for luxury car buyers, exploring the purchase of existing international distributors, and using a specific amount of profits to buy back company stock. The strategic plan also provides direction to the functional areas within an organization, including marketing, production, finance, accounting, and human resources. For the information systems department, these directions are encompassed in the information systems plan.

The term **information systems planning** refers to the translation of strategic and organizational goals into systems development initiatives (Figure 12.3). Part of the IS plan for the luxury car company might be to build a new product tracking system to improve service. Proper IS planning ensures that specific systems development objectives support organizational goals.

One of the primary benefits of IS planning is that it provides a long-range view of information technology use in the organization. While specific systems development initiatives may spring from the IS plan, the IS plan must also provide a broad framework for future success. The IS plan should provide guidance on how the information systems infrastructure of the organization should be developed over time. Another benefit of IS planning is that it ensures better use of information systems resources, including funds, IS personnel, and time for scheduling specific projects. The steps of IS planning are shown in Figure 12.4.

Overall IS objectives are usually distilled from the relevant aspects of the organization's strategic plan. IS projects can be identified either directly from the objectives determined in the first step or may be identified by others, such as managers within the various functional areas. Setting priorities and selecting projects typically requires the involvement and approval of senior management. When objectives are set, planners consider the resources necessary to complete the projects including employees (systems analysts, programmers, and others), equipment (computers, network servers, printers, and other devices), expert advice (specialists and other consultants), and software among others.

Developing a competitive advantage. As part of translating the corporate strategic plan into the information systems plan, many companies seek systems development projects that will provide a competitive advantage. This usually requires creative and critical analysis. Read the "Making a Difference" box to see how one manufacturer is offering dealer and customer services via an extranet to gain a competitive advantage.

Creative analysis involves the investigation of new approaches to existing problems. By looking at problems in new or different ways and by introducing innovative methods to solve them, many firms have gained a competitive advantage. Typically, these new solutions are inspired by people and things not directly related to the problem.

information systems planning

the translation of strategic and organizational goals into systems development initiatives

creative analysis

the investigation of new approaches to existing problems

FIGURE 12.3

Information systems planning transforms organizational goals outlined in the strategic plan into specific systems development activities.

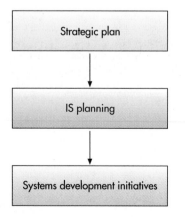

Strategic plan

↓

IS planning

↓

Systems development initiatives

FIGURE 12.4

The Steps of IS Planning

Some projects are identified through overall IS objectives, whereas additional projects, called unplanned projects, are identified from other sources. All identified projects are then evaluated in terms of their organizational priority.

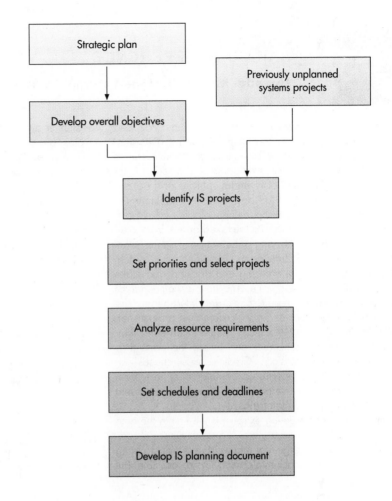

critical analysis

the unbiased and careful questioning of whether system elements are related in the most effective or efficient ways

Critical analysis requires unbiased and careful questioning of whether system elements are related in the most effective or efficient ways. It involves considering the establishment of new or different relationships among system elements and perhaps introducing new elements into the system. Critical analysis in systems development involves the following actions:

• *Going beyond automating manual systems.* Many organizations use systems development to simply automate existing manual systems, which may result in relatively faster, more efficient systems. However, if the underlying manual system is flawed, automating it might just magnify its impact. In addition, automating existing manual systems might cause many opportunities to be lost by continuing to do things in the same old way. For example, rather than automate current customer service systems, many companies are implementing customer service systems based on the use of the Internet and Web technology so that customers can provide a high degree of self-service. Critical analysis in systems development involves asking why things are done a certain way and considering alternative approaches.

MAKING A DIFFERENCE
Hyundai Motors Designs Extranet to Serve Dealers and Customers

Hyundai Motor Company was established in 1967 and is the leader of the Korean automobile industry. Its goal is to become one of the top ten automobile manufacturers by the turn of the century. Hyundai Motor America is an importer and distributor of Hyundai vehicles and parts.

Hyundai Motor America designed and implemented an extranet to eliminate the deluge of phoned-in parts orders from dealers as well as the faxed orders from international distributors. During information systems planning, key project participants identified the system goal—streamline parts ordering and tracking and make it more efficient for everyone. The new system enables dealers, independent repair shops, and international distributors to order parts and track orders over an extranet. With Hyundai's new Web site, customers can enter their vehicle identification number and locate the parts they need using on-line catalogs. Credit card purchases can be made using an electronic commerce system from Chicago-based Click Interactive, which enters orders into Hyundai's IBM 3090 mainframe.

Hyundai's parts-ordering extranet is the first in the auto industry open to consumers, which gives Hyundai a competitive advantage. Larger car manufacturers want to bring as many consumers back into dealerships as possible. But for smaller automakers like Hyundai, with far fewer dealerships, direct ordering capabilities become an important convenience. Hyundai has only 500 U.S. dealerships, compared with about 8,000 for General Motors.

Opening up the parts process globally represents another industry first. This should help Hyundai reduce parts inventories, which means savings, because it is expensive to have excess parts, especially in global operations. In the future, Hyundai plans to extend the ability to check order status to its mobile field representatives.

DISCUSSION QUESTIONS

1. Since this project was identified as part of its information systems plan, what strategic and organizational goals will this system help Hyundai achieve?
2. Is the fact that this system is the first to enable consumers to order parts likely to give the company a competitive advantage? Why or why not?

Source: Adapted from Bob Wallace, "Hyundai Builds Extranet to Streamline Parts Ordering," *Computerworld,* June 15, 1998, p. 6, and the Company Profile section of the Hyundai Web site at http://www.hyundai.com, accessed September 19, 1998.

- *Questioning statements and assumptions.* Questioning users about their needs and clarifying their initial responses can result in better systems and more accurate predictions. Too often, stakeholders and users specify certain system requirements because they assume their needs can only be met that way. Often, an alternative approach would be better. For example, a stakeholder may be concerned because there is always too much of some items in stock and not enough of other items. So, the stakeholder might request a totally new and improved inventory control system. An alternative approach is to identify the root cause for poor inventory management. This latter approach might determine that sales forecasting is inaccurate and needs improvement or that production is not capable of meeting the set production schedule. All too often solutions are selected before a complete understanding of the nature of the problem itself is obtained.
- *Identifying and resolving objectives and orientations that conflict.* Different departments in an organization can have different objectives and orientations. The buying department may want to minimize the cost of spare parts by always buying from the lowest cost supplier, while engineering might want to buy more expensive, higher quality spare parts to reduce the frequency of replacement. These differences must be identified and resolved before a new purchasing system is developed or an existing one modified.

Establishing Objectives for Systems Development

The impact a particular system has on an organization's ability to meet its goals determines the true value of that system to the organization. While all systems should support business goals, some systems are more pivotal in continued operations and goal attainment than others. These systems are called **mission-critical systems**. An order processing TPS, for example, is usually considered mission-critical. Without it, few organizations could continue daily activities, and they clearly would not meet set goals.

The goals defined for an organization will in turn define the objectives set for a system. A manufacturing plant, for example, might determine that minimizing the total cost of owning and operating its equipment is a critical success factor (CSF) in meeting a production volume and profit goals. This CSF would be converted into specific objectives for a proposed plant equipment maintenance system. One specific objective might be to alert maintenance planners when a piece of equipment is due for routine preventive maintenance (e.g., cleaning and lubrication). Another objective might be to alert the maintenance planners when the necessary cleaning materials, lubrication oils, or spare parts inventory levels are below specified limits. These objectives could be accomplished either through automatic stock replenishment via electronic data interchange or through the use of exception reports.

Regardless of the particular systems development effort, the development process should define a system with specific performance and cost objectives. The success or failure of the systems development effort will be measured against these objectives.

Performance objectives. The extent to which a system performs as desired can be measured through its performance objectives. System performance is usually determined by such factors as the following:

- *The quality or usefulness of the output.* Is the system generating the right information for a value-added business process or for use by a goal-oriented decision maker?
- *The quality or usefulness of the format of the output.* Is the output generated in a form that is usable and easily understood? For example, objectives often concern the legibility of screen displays, the appearance of documents, and the adherence to certain naming conventions.
- *The speed at which output is generated.* Is the system generating output in time to meet organizational goals and operational objectives? Objectives such as customer response time, the time to determine product availability, and throughput time are examples.

In some cases, the achievement of performance objectives can be easily measured (e.g., by tracking the time it takes to determine product availability). The achievement of performance objectives is sometimes more difficult to ascertain in the short term. For example, it may be difficult to determine how many customers are lost because of slow responses to customer inquiries regarding product availability. These outcomes, however, are often closely associated with corporate goals and are vital to the long-term success of the organization. Their attainment is usually dictated by senior management.

mission-critical systems

systems that play a pivotal role in an organization's continued operations and goal attainment

Cost objectives. The benefits of achieving performance goals should be balanced with all costs associated with the system, including the following:

* *Development costs.* All costs required to get the system up and running should be included.
* *Costs related to the uniqueness of the system application.* A system's uniqueness has a profound effect on its cost. An expensive but reusable system may be preferable to a less costly system with limited use.
* *Fixed investments in hardware and related equipment.* Developers should consider costs of such items as computers, network-related equipment, and environmentally controlled data centers in which to operate the equipment.
* *Ongoing operating costs of the system.* Operating costs include costs for personnel, software, supplies, and such things as the electricity required to run the system.

The costs of the current system or problem-solving method are subtracted from the costs associated with a given level of performance, control, and complexity in the new or modified system. The resulting figure is an incremental cost differential, for which an objective is set.

Balancing performance and cost objectives within the overall framework of organizational goals can be challenging. Systems development objectives are important, however, in that they allow an organization to effectively and efficiently allocate resources and measure the success of a systems development effort.

Systems Development and the Internet

In an effort to reach more customers, Barnes and Noble, a retail bookstore, has converted a portion of its business to run over the Internet.

Increasingly, companies are converting at least some portion of their business to run over the Internet, intranets, or extranets. Applications that are being moved to the Internet include those that support selling products to customers, placing orders with suppliers, and letting customers access information about production, inventory, orders, or accounts receivable. Internet technology provides a platform for applications that enables companies to extend their transaction processing systems beyond the boundaries of the organization to their customers, suppliers, and partners. This enables companies to conduct business much faster, interact with more people, and try to keep one step ahead of the competition.

Building a static Web site to display simple text and graphics is fairly straightforward. However, implementing a dynamic core business application that runs over the Web is much more complicated. Such applications must meet special business needs. They must be able to scale up to support highly variable transaction throughput from

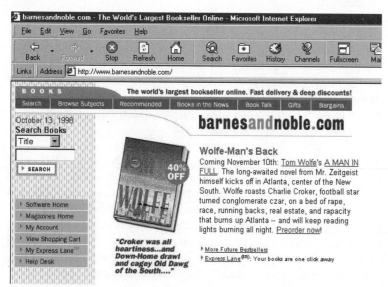

Thomson Financial Services used an application server to build applications to track job candidates and monitor consultants' work hours.

potentially thousands of users. Ideally, they can scale up instantly when needed. They must be reliable and fault tolerant, providing continuous availability while processing all transactions accurately. They must also integrate with existing infrastructure, including customer and order databases, existing applications, and enterprise resource planning systems. Development and maintenance must be quick and easy, as business needs may require changing applications on the fly.

There are many tools available for building and running Web applications.[3] The best tools provide components to support applications on an enterprise scale while speeding development. Several vendors provide what is known as an applications server to provide remote access to databases via a corporate intranet. These include NetDynamics, SilverStream, WebLogic, Novera Software, Netscape Communications, and IBM. Thomson Financial Services used an application server to build two applications—one tracks job candidates and the second monitors consultants' work hours.[4]

Trends in Systems Development and Enterprise Resource Planning

Enterprise resource planning software has reached beyond business processes and is increasingly affecting systems development. Not only are planners considering different types of systems that include ERP software, but the ERP software that is already in place is driving planners to explore and develop different types of systems. In 1996, more than half the enterprise application sales were driven by the basic processing needs of departments such as finance and human resources. But by 2000, those purchases will account for less than a third of sales as buyers switch their focus to uses such as supply-chain management and customer care.[5] Other ERP users are moving from just using the software to run the business to trying to use it to make business decisions.

An important trend in systems development and the use of ERP systems is that companies wish to stay with their primary ERP vendor (SAP, Oracle, PeopleSoft, etc.) instead of looking elsewhere for answers to their data warehousing and production planning needs or developing in-house solutions. Thus, they look for their original ERP vendor to provide these solutions.

A second trend is that many software vendors are building software that integrates with the ERP vendor's package. For example, Aspect Development, a leading supplier of electronic catalogs for high-tech components, has now entered the market for on-line catalogs of maintenance, repair, and operations (MRO) supplies. Its product Morocco is an MRO supplies catalog for users of SAP's ERP software. It contains more than 1.5 million items from 3,000 suppliers. Morocco works in conjunction with the purchasing module of SAP and directs buyers to preferred suppliers specified by the purchasing department.[6] Thus, again there is less in-house development and more dependence on ERP

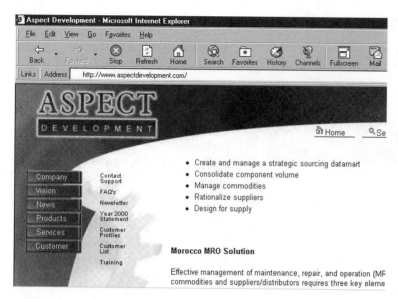

vendors and their strategic partners to provide enhancements and add-ons to the original ERP package.

A third interesting trend is the increase in the number of companies who, once they have successfully implemented their own company's ERP project, are branching out to provide consulting to other companies. See the "Technology Advantage" box to read about how companies are becoming consultants to other companies in the area of implementing ERP systems.

Customers such as Ashland Oil and Occidental Chemical expect to save millions of dollars by using Aspect Development's Morocco system to consolidate MRO purchasing enterprise-wide, analyze buying patterns to identify strategic suppliers, and create an electronic catalog of preferred supplier items that can be used to quickly research the availability of needed items from in-house inventory as well as strategic suppliers.

SYSTEMS DEVELOPMENT LIFE CYCLES

The systems development process is also called a systems development life cycle (SDLC) because the activities associated with it are ongoing. As each system is being built, the project has timelines and deadlines, until at last the system is installed and accepted. The life of the system continues as it is maintained and reviewed. If the system needs significant improvement beyond the scope of maintenance, if it needs to be replaced because of a new generation of technology, or if the IS needs of the organization change significantly, a new project will be initiated and the cycle will start over.

A key fact of systems development is that the later in the SDLC an error is detected, the more expensive it is to correct (Figure 12.5) (this analysis is documented in a classic work by Barry Boehm).[7] One reason for this is that if an error is found in a later phase of the SDLC, the previous phases must

FIGURE 12.5

The later that system changes are made in the SDLC, the more expensive these changes become.

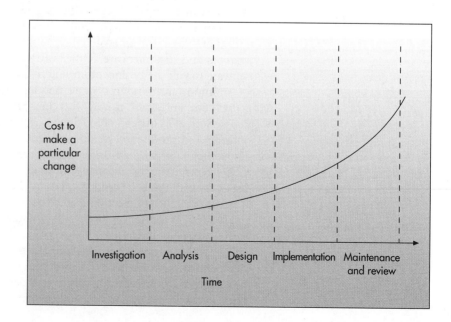

TECHNOLOGY ADVANTAGE

After Learning Systems Development Techniques, Companies Become Consultants

Developing new systems or modifying old ones is expensive. It is also expensive to retain a complete staff of IS professionals, computers, and equipment to complete systems development projects. Although many companies are now outsourcing much or all of their systems development projects to consulting companies, many still believe an internal staff of systems development personnel is the best way to go. Being able to get exactly what you want and having total control over systems development projects are key motivators. These organizations employ a complete staff of systems developers to keep systems development projects inside. Through years of experience, these internal systems developers acquire expertise in systems development. To get the most from their investment in people and experience, some companies are turning to consulting to generate profits. These companies sell their expertise to others. This is especially true for the exceedingly complex enterprise resource planning (ERP) software from companies like SAP, Baan, PeopleSoft, Oracle, and others.

Integrating a company's core business activities, ERP offers the potential to save a company millions of dollars, while improving customer service and satisfaction. Companies from Microsoft to Monsanto are now starting to use ERP software. Using ERP, however, is not easy or inexpensive. Companies have spent hundreds of millions of dollars and years of development effort to get these software packages running. So it is not surprising that a number of companies that have successfully implemented ERP software are now trying to cash in on their hard-won expertise.

GATX is a $5 billion company specializing in asset management and leasing. To better integrate its operations, GATX installed SAP's R/3 enterprise resource planning software. The ERP installation took years and millions of dollars. Because no other asset management and leasing company had yet implemented R/3, GATX had to carefully analyze what R/3 was able to deliver and what functionality was needed. The next step was to customize R/3 for its particular business needs. This required approximately 114,000 hours of work. GATX competitors, including First Chicago NBD corporation and Pitney

Bowes, later asked GATX for help in installing their own ERP software.

GATX is not alone in selling its systems development expertise. Monsanto Corporation is also looking into marketing its ERP expertise. It also took Monsanto over two years and millions of dollars to implement ERP. Like GATX, Monsanto also used R/3 from SAP as its platform to implement enterprise resource planning software. With assistance from IBM, Monsanto is now ready to get an additional return from its experience with R/3 by becoming a consultant to other companies that are starting to implement their own R/3 software. Instead of becoming a consulting company offering a full range of services, however, the IBM/Monsanto Solution Center will concentrate on helping other companies operate R/3 and link it with factory floor systems.

Selling ERP expertise appears to be an excellent way to get a greater return on a huge investment. To some, however, this is a foolish venture. Companies selling their expertise may become liable if something goes wrong. More importantly, a company may be selling its competitive advantage, destroying its future revenue stream for additional dollars today. Still, some companies are becoming consultants to their competitors. They argue that a competitor today will be a partner tomorrow, and providing expertise to other companies will only improve the total market.

DISCUSSION QUESTIONS

1. Why would a company want to sell its systems development expertise?
2. What are the advantages and disadvantages of selling systems development expertise?

Sources: Randy Weston, "R/3 User Markets Customized Code," *Computerworld*, January 5, 1998, p. 61; "Services to the Industry," *Chemical Week*, January 14, 1998, p. 38; Gregory Johnson, "AgriLink Forges Tentative Pact wth GATX," *Journal of Commerce*, February 2, 1998, p. 11A; Craig Stedman, "Monsanto Plans to Market Its Hard-won R/3 Expertise," *Computerworld*, January 12, 1998, p. 53

be reworked to some extent. Another reason is that the errors found late in the SDLC have an impact on more people. For example, an error found after a system is installed may require retraining users once a "workaround" to the problem has been found. Thus, experienced system developers prefer an approach that will catch errors early in the project life cycle.

Four common systems development life cycles exist: traditional, prototyping, rapid application development (RAD), and end-user development. In some companies, these approaches are formalized and documented so that system developers have a well-defined process to follow; in other companies, less formalized approaches are used. Keep Figure 12.5 in mind as you are introduced to alternative SDLCs in the next section.

The Traditional Systems Development Life Cycle

Traditional systems development efforts can range from a small project, such as purchasing an inexpensive computer program, to a major undertaking, such as Gerber's installing a multimillion dollar system. The steps of traditional systems development may vary from one company to the next, but most approaches have five common phases: investigation, analysis, design, implementation, and maintenance and review (Figure 12.6).

In the **systems investigation** phase, potential problems and opportunities are identified and considered in light of the goals of the business. Systems investigation attempts to answer the question "What is the problem, and is it worth solving?" The primary result of this phase is a defined information system project for which business problem or opportunity statements have been created, to which some amount of organizational resources have been committed, and for which systems analysis is recommended. **Systems analysis** attempts to answer the question "What must the information system do to solve the problem?" This phase involves the study of existing systems and work processes to identify strengths, weaknesses, and opportunities for improvement. The major outcome of systems analysis is a list of requirements and priorities. **Systems design** seeks to answer the question "How will the information system do what it must do to solve the problem?" The primary result of this phase is a technical design that either describes the new system or describes how existing systems will be modified. The system design details system outputs, inputs, and user interfaces; specifies hardware, software, database, telecommunications, personnel, and procedure components; and shows how these components are related. **Systems implementation** involves creating or acquiring the various system components detailed in the systems design, assembling them, and placing the new or modified system into operation. An important task during this phase is to train the users. Systems

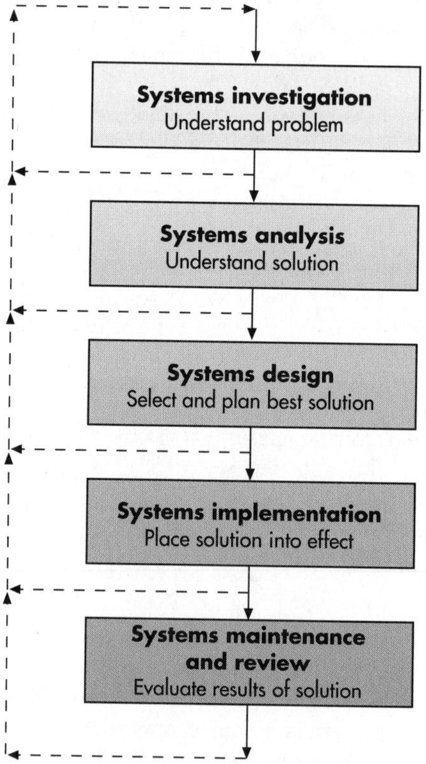

systems investigation

the system development phase during which problems and opportunities are identified and considered in light of the goals of the business

systems analysis

the system development phase during which the existing systems and work processes are studied to identify strengths, weaknesses, and opportunities for improvement

systems design

the system development phase that defines how the information system will do what it must do to solve the problem

systems implementation

the phase during which the various system components detailed in the systems design are assembled and the new or modified system is placed into operation

FIGURE 12.6

The Traditional Systems Development Life Cycle

Sometimes, information learned in a particular phase requires cycling back to a previous phase.

Systems investigation
Understand problem

Systems analysis
Understand solution

Systems design
Select and plan best solution

Systems implementation
Place solution into effect

Systems maintenance and review
Evaluate results of solution

systems maintenance and review

the development phase that ensures the system operates and modifies the system so that it continues to meet changing business needs

prototyping

an iterative approach to the systems development process

TABLE 12.2

Advantages and Disadvantages of Traditional SDLC

implementation results in an installed, operational information system that meets the business needs for which it was developed. The purpose of **systems maintenance and review** is to ensure the system operates and to modify the system so that it continues to meet changing business needs. As shown in Figure 12.6, a system under development moves from one phase of the traditional SDLC to the next.

The traditional SDLC allows for a large degree of management control. At the end of each phase, a formal review is performed and a decision is made whether to continue with the project, terminate the project, or perhaps repeat some of the tasks of the current phase. Use of the traditional SDLC also creates much documentation, such as entity relationship diagrams. This documentation, if kept current, can be useful when it is time to modify the system. The traditional SDLC also ensures that every system requirement can be related to a business need. In addition, resulting products can be reviewed to verify that they satisfy the system requirements and conform to organizational standards.

A major problem with the traditional SDLC is that the user does not use the solution until the system is nearly complete. Quite often, users get a system that does not meet their real needs because its development was based on the development team's understanding of the needs. The traditional approach is also inflexible—changes in user requirements cannot be accommodated during development. In spite of its limitations, however, the traditional SDLC is still used for large, complex systems that affect entire businesses, such as TPS and MIS systems. For example, Aetna U.S. Healthcare, the nation's largest for-profit health maintenance organization followed the traditional SDLC to implement an electronic claims program to speed payments to member physicians and reduce claims-processing errors.[8] It is also frequently employed on government projects because of the strengths mentioned previously. Table 12.2 lists advantages and disadvantages of the traditional SDLC.

Prototyping

Prototyping takes an iterative approach to the systems development process. During each iteration, requirements and alternative solutions to the problem are identified and analyzed, new solutions are designed, and a

Advantages	Disadvantages
Formal review at the end of each phase allows maximum management control.	Users get a system that meets the needs as understood by the developers; this may not be what was really needed.
This approach creates considerable system documentation.	Documentation is expensive and time-consuming to create. It is also difficult to keep current.
Formal documentation ensures that system requirements can be traced back to stated business needs.	Often, user needs go unstated or are misunderstood.
It produces many intermediate products that can be reviewed to see whether they meet the users' needs and conform to standards.	Users cannot easily review intermediate products and evaluate whether a particular product (e.g., data flow diagram) meets their business requirements.

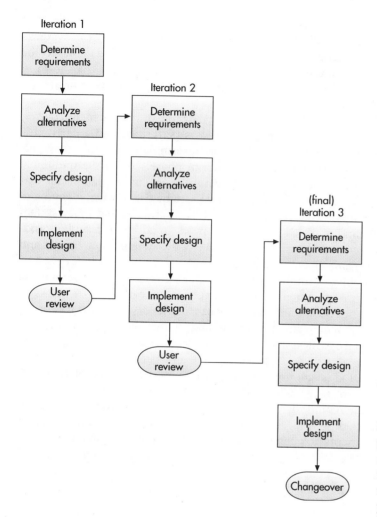

Iteration 1

Determine requirements

Analyze alternatives

Specify design

Implement design

User review

Iteration 2

Determine requirements

Analyze alternatives

Specify design

Implement design

User review

(final) Iteration 3

Determine requirements

Analyze alternatives

Specify design

Implement design

Changeover

FIGURE 12.7

An Iterative Approach to Systems Development

operational prototype

a prototype that works—accesses real data files, edits input data, makes necessary computations and comparisons, and produces real output

nonoperational prototype

a mockup or model that includes output and input specifications and formats

rapid application development (RAD)

a technique that employs tools, techniques, and methodologies designed to speed application development

portion of the system is implemented. Users are then encouraged to try the prototype and provide feedback (Figure 12.7). Prototyping begins with the creation of a preliminary model of a major subsystem or a scaled-down version of the entire system. For example, a prototype might be developed to show sample report formats and input screens. Once developed and refined, the prototypical reports and input screens are used as models for the actual system, which may be developed using an end-user programming language such as SAS, Focus, or Visual Basic. The first preliminary model is refined to form the second- and third-generation models, and so on until the complete system is developed (Figure 12.8).

Types of prototypes. Prototypes can be classified as operational or nonoperational. An **operational prototype** is a prototype that works—accesses real data files, edits input data, makes necessary computations and comparisons, and produces real output. Fully developed financial reports are examples. The operational prototype may access real files but perhaps does no editing of input. A **nonoperational prototype** is a mockup, or model. It typically includes output and input specifications and formats. The outputs include printed reports to managers and the screen layout of reports displayed on personal computers or terminals. The inputs reveal how data is captured, what commands users must enter, and how the system accesses other data files. The primary advantage of a nonoperational prototype is that it can be developed much faster than an operational prototype. Nonoperational prototypes can be discarded, and a fully operational system can be built based on what was learned from the prototypes. The advantages and disadvantages of prototyping are summarized in Table 12.3.

Rapid Application Development and Joint Application Development

Rapid application development (RAD) employs tools, techniques, and methodologies designed to speed application development. RAD reduces paper-based documentation, automates program source code generation, and facilitates user participation in design and development activities. With RAD, entire systems are developed in less than six months. The ultimate goal is to accelerate the process so that applications can go into production much sooner than when using other approaches.

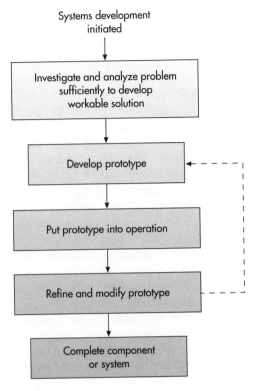

Systems development
initiated

Investigate and analyze problem
sufficiently to develop
workable solution

Develop prototype

Put prototype into operation

Refine and modify prototype

Complete component
or system

FIGURE 12.8

Prototyping is a popular
technique in systems
development. Each generation
of prototype is a refinement of
the previous generation based
on user feedback.

TABLE 12.3

Advantages and Disadvantages
of Prototyping

The use of software tools to support RAD has flourished. PowerBuilder, a RAD tool from Sybase's Powersoft subsidiary, is popular in the federal sector. This tool is used on contracts such as the U.S. Army's Small Multiuser Computer II and the Justice Department's Justice Consolidated Office Network. Microsoft's Visual Basic was used on such contracts as the U.S. Air Force's Non-Appropriated Funds project. The U.S. Postal Service has tapped Visual Basic as its primary RAD development tool.[9] In addition, such database vendors as Computer Associates International, Informix Software, and Oracle market fourth-generation languages and other products targeting the RAD market. Tools from such traditional vendors as Texas Instruments and Sterling Software are also being deployed as RAD products.

Throughout the RAD project, users and developers work together as one team. This promotes healthy risk taking and team-based decision making, resulting in better systems with shorter delivery dates. If the entire system is too large to be completed in less than six months, it is broken into subsystems and is delivered subsystem by subsystem. The first subsystem may be delivered in three to four months, with no delivery date more than six months after the last one. This leads to less waste, because even if there is a serious error in the system, only one subsystem has to be rebuilt.

RAD should not be used on every software development project. In general, it is best suited for decision support and management information systems and less suited for transaction processing applications. During a RAD project, the level of participation on the part of stakeholders and users is much higher than in other approaches. They become working members of the team and can be expected to spend more than 50 percent of their time producing project outcomes. This can be a problem if the users are also needed to perform their normal business role. For this reason, RAD team participants are often taken off their normal assignments and work full-time on the RAD project. Because of the full-time commitment and intense schedule deadlines, RAD is a high-pressure development approach that can

Advantages	Disadvantages
Users can try the system and provide constructive feedback during development.	Each iteration builds on the previous iteration and further refines the solution. This makes it difficult to reject the initial solution as inappropriate and start over. Thus, the final solution will be only incrementally better than the first.
An operational prototype can be produced in weeks.	Formal end-of-phase reviews do not occur. Thus, it is very difficult to contain the scope of the prototype, and the project never seems to end.
Users become more positive about implementing the system as they see a solution emerging that will meet their needs.	System documentation is often absent or incomplete, since the primary focus is on development of the prototype.
Prototyping enables early detection of errors and omissions.	System backup and recovery, performance, and security issues can be overlooked in the haste to develop a prototype.

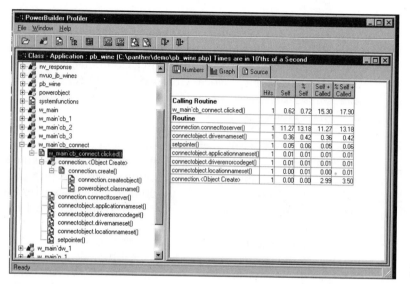

PowerBuilder, a RAD tool from Sybase's Powersoft subsidiary, is used in both the public and private sectors.

joint application development (JAD)

process for data collection and requirements analysis

easily result in employee burnout. Table 12.4 lists advantages and disadvantages of RAD.

RAD makes extensive use of the **joint application development (JAD)** process for data collection and requirements analysis. Originally developed by IBM Canada in the 1970s, JAD involves group meetings in which users, stakeholders, and IS professionals work together to analyze existing systems, propose possible solutions, and define the requirements of a new or modified system. JAD groups consist of both problem holders and solution providers. A group normally requires one or more top-level executives who initiate the JAD process, a group leader for the meetings, potential users, and one or more individuals who act as secretaries and clerks to record what is accomplished and to provide general support for the sessions. Many companies have found that groups can develop better requirements than individuals working independently and have assessed JAD as a very successful development technique.

The End-User Systems Development Life Cycle

As we saw in our opening vignette, systems development initiatives arise from a wide variety of individuals and organizational areas, including users. The proliferation of general-purpose information technology and the flexibility of many packaged software programs have allowed non-IS employees to independently develop information systems that meet their needs. Such employees have believed that, by bypassing the formal requisitioning of resources from the IS department, they can develop systems more quickly. In addition, these individuals often believe that they have better insight into their own needs and can develop systems better suited for their purposes.

End-user-developed systems range from the very small (e.g., a software routine to merge form letters) to those of significant organizational value (such as customer contact databases). Like all projects, some

TABLE 12.4

Advantages and Disadvantages of RAD

Advantages	Disadvantages
For appropriate projects, this approach puts an application into production sooner than any other approach.	This intense SDLC can burn out systems developers and other project participants.
Documentation is produced as a by-product of completing project tasks.	This approach requires systems analysts and users to be skilled in RAD system development tools and RAD techniques.
RAD forces teamwork and lots of interaction between users and stakeholders.	RAD requires a larger percentage of stakeholders' and users' time than other approaches.

end-user-developed systems fail, and others are successful. Initially, IS professionals discounted the value of these projects. As the number and magnitude of these projects increased, however, IS professionals began to realize that for the good of the entire organization, their involvement with the projects needed to increase.

end-user systems development

any systems development project in which the primary effort is undertaken by a combination of business managers and users

Today, the term **end-user systems development** describes any systems development project in which the primary effort is undertaken by a combination of business managers and users. Rather than ignoring these initiatives, astute IS professionals encourage them by offering guidance and support. Technical assistance, communication of standards, and the sharing of "best practices" throughout the organization are just some of the ways IS professionals work with motivated managers and employees undertaking their own systems development. In this way, end-user-developed systems can be structured as complementary to, rather than in conflict with, existing and emerging information systems. In addition, this open communication among IS professionals, managers of the affected business area, and users allows the IS professionals to identify specific initiatives so that additional organizational resources, beyond the prerogative of the initiating business manager or user, are provided for its development.

Many end users are already demonstrating their systems development capability by designing and implementing their own PC-based systems. Sophisticated end users with a general knowledge of technology and the requisite analytical and communication skills may be ideal candidates for systems analyst positions.

FACTORS AFFECTING SYSTEMS DEVELOPMENT SUCCESS

Successful systems development means delivering a system that meets user and organizational needs—on time and within budget. Years of experience in completing systems development projects have resulted in the identification of factors that contribute to the success or failure of systems development. These factors include the degree of change involved with the project, the quality of the project planning, the use of project management tools, the use of formal quality assurance processes, and the use of CASE tools.

Degree of Change

A major factor that affects the quality of systems development is the degree of change associated with the project. This can vary from implementing minor enhancements to an existing system to major reengineering. The project team needs to recognize where they are on this spectrum of change.

As discussed in Chapter 2, continuous improvement projects do not require significant business process or information system changes or retraining of individuals; thus, they have a high degree of success. Typically, because continuous improvements involve minor improvements, they also have relatively modest benefits. On the other hand, reengineering involves fundamental changes in how the organization conducts business and completes tasks. The factors associated with successful reengineering are similar to those of any development effort, including top management support, clearly defined corporate goals and systems development objectives, and careful management of change. Major reengineering projects tend to have a high degree of risk but also a high potential for major business benefits (see Figure 12.9). For example, Thomas Cook Group spent over $40 million on a major reengineering effort to implement a single toll-free number and a Web site to funnel all service requests, such as trip reservations and traveler's check orders, through one customer contact center. The goal is to provide improved customer service at a lower cost and enable more rapid changes in service offerings.[10]

Managing change. The ability to manage change is critical to the success of systems development. The systems created during systems development will inevitably cause change. For example, the work environment and habits of users are invariably affected by the development of a new information system. Unfortunately, not everyone adapts easily. Managing change requires the ability to recognize existing or potential problems (particularly the concerns of users) and deal with them before they become a serious threat to the success of the new or modified system. Although many problems can result from initiating new or modified systems, here are several of the most common:

- Fear that the employee will lose his or her job, power, or influence within the organization
- Belief that the proposed system will create more work than it eliminates
- Reluctance to work with "computer people"
- Anxiety that the proposed system will negatively alter the structure of the organization
- Belief that other problems are more pressing than those solved by the proposed system or that the system is being developed by people unfamiliar with "the way things need to get done"
- Unwillingness to learn new procedures or approaches

FIGURE 12.9

Degree of change can greatly affect the probability of a project's success.

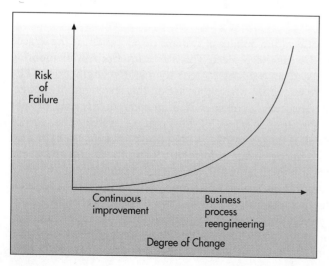

Preventing or dealing with these types of problems requires a coordinated effort involving stakeholders and users, managers, and information systems personnel. One positive step is simply to talk with all people concerned and learn what their biggest concerns are. Management can then deal with those concerns and try to eliminate them. Once the immediate concerns are addressed, these people can often become part of the project team.

Quality of Project Planning

Another key success factor is the quality of project planning. The bigger the project, the more likely that poor planning will lead to significant problems. For example, the IRS wasted somewhere between $3 billion and $4 billion on its ongoing, failed effort to modernize its computer systems to process over 200 million tax returns.[11] Project management failures are also undermining the Federal Aviation Agency's $20 billion effort to modernize the nation's air traffic control systems.[12]

runaways

projects that are far over budget and past delivery dates

Systems development projects like these, which are far over budget and past delivery dates, are termed **runaways**.

These examples are not atypical. Many companies find that large systems projects fall behind schedule, go over budget, and do not meet expectations. Although proper planning cannot guarantee that these types of problems will be avoided, it can minimize the likelihood of their occurrence. Certain factors contribute to the failure of systems development projects. These factors and countermeasures to eliminate or alleviate the problem are summarized in Table 12.5. Read the "Ethical and Societal Issues" box to learn of a potential runaway project of significance to us all.

ETHICAL AND SOCIETAL ISSUES

The IRS Modernization Project

The Internal Revenue Service (IRS) is seeking a prime systems contractor to oversee subcontractors during a 15-year modernization project. The changes to be implemented are defined in a 1,700 page document. The cost of the modernization project is difficult to pin down; however, it is expected to exceed $8 billion.

The need for modernization is clear. The IRS failed its own audit in 1995. It misses up to $50 billion in revenue every year because of its obsolete systems. In addition, the IRS has announced that it will improve its ability to handle electronic tax filings, replace its chief information officer, and fix its year 2000 system problems.

Although modernization is not the primary focus of the Senate taxpayer bill of rights proposal, the current draft of the legislation sets 2007 as a deadline for the IRS to process 80 percent of its annual tax return filings electronically. During 1997, about 15 percent of the nearly 119 million individual filings were done electronically or over the phone. When taxpayers file by computer or phone, they cut out the need for data input of paper returns by IRS staff. Error rates with computer filings are less than 1 percent and well below the 18 percent error rate for paper returns.

Most observers do not think the IRS project has much hope of success, if for no other reason than the project is simply too big and big projects tend to fail. There is even a chance of major changes in the tax structure (e.g., a flat tax or national sales tax). In addition, critics say that any information systems project with a 15-year planning horizon is doomed to failure. The pace of change in information systems is such that every four or five years there is a major technology shift.

Discussion Questions

1. Do you agree that the information system plan for the IRS is appropriate? Why or why not?
2. How could the modernization effort be redefined to be more manageable and have a greater probability of success?

Sources: Adapted from Frank Hayes, "The IRS's Doomed Cure-All," *Computerworld*, April 6, 1998, p. 12 and Matt Hamblen, "IRS Takes Another Stab at Modernization," *Computerworld*, March 30, 1998, p. 8.

Good systems development is not automatic. Companies have lost millions of dollars and wasted years of effort because of faulty systems development. Because many development projects are large and expensive, a substantial amount of time and money is often on the line. Unfortunately,

Factor	Countermeasure
Solving the wrong problem	Establish a clear connection between the project and organizational goals.
Poor problem definition and analysis	Follow a standard systems development approach.
Poor communication	Communicate, communicate, communicate.
Project is too ambitious	Narrow the project focus to address only the most important business opportunities.
Lack of top management support	Identify the senior manager who has most to gain from the success of the project, and recruit this individual to champion the project.
Lack of management and user involvement	Identify and recruit key stakeholders to be active participants in the project.
Inadequate or improper system design	Follow a standard systems development approach.
Poor testing and implementation	Plan sufficient time for this activity.
Users are unable to use the system effectively	Develop a rigorous user training program and budget sufficient time in the schedule to execute it.
Lack of concern for maintenance	Include an estimate of people effort and costs for maintenance in the original project justification.

TABLE 12.5

Project Planning Issues Frequently Contributing to Project Failure

corporate America's track record for delivering these projects is not very good. According to a Standish Group International survey, 73 percent of IS software projects have been canceled, over budget, or late.[13] As a result, most organizations have adopted formal systems development approaches. NCR, for example, has adopted a standard set of project management techniques for IS managers to follow. The company has also established a formal career path for IS project managers, who are also expected to earn a graduate degree and certification in the discipline.[14] An important part of following these methodologies is to consider the objectives of a proposed information system in terms of broader organizational goals.

Use of Project Management Tools

Project management involves planning, scheduling, directing, and controlling human, financial, and technological resources for a fixed-term task that will result in the achievement of specific goals and objectives. A **project schedule** is a detailed description of what is to be done. Each project activity, the use of personnel and other resources, and expected completion dates are described. A **project milestone** is a critical date for the completion of a major part of the project. The completion of program design, coding, testing, and release are examples of milestones for a programming project. The **project deadline** is the date the entire project is to be completed and operational—when the organization can expect to begin to reap the benefits of the project.

Project management tools. Although the steps of systems development seem straightforward, larger projects can become complex, requiring literally hundreds or thousands of separate activities. For these types of systems development efforts, formal project management methods and tools become essential.

In systems development, each activity has an earliest start time, earliest finish time, and slack time, which is the amount of time an activity can be delayed

project schedule

a detailed description of what is to be done

project milestone

a critical date for the completion of a major part of the project

project deadline

the date that the entire project is due to be completed and operational

critical path

the path that consists of all activities that, if delayed, would delay the entire project

Program Evaluation and Review Technique (PERT)

a formalized approach for developing a project schedule

Gantt chart

a graphical tool used for planning, monitoring, and coordinating projects

without delaying the entire project. The **critical path** consists of all activities that, if delayed, would delay the entire project. These activities have zero slack. Any problems with critical-path activities will cause problems for the entire project. To ensure that critical-path activities are completed in a timely fashion, formalized project management approaches have been developed.

A formalized approach called **Program Evaluation and Review Technique (PERT)** involves creating three time estimates for an activity: shortest possible time, most likely time, and longest possible time. A formula is then applied to come up with a single PERT time estimate. A **Gantt chart** is a graphical tool used for planning, monitoring, and coordinating projects; it is essentially a grid that lists activities and deadlines. Each time a task is completed, a darkened line (or bar) is placed in the proper grid cell to indicate the completion of a task (Figure 12.10).

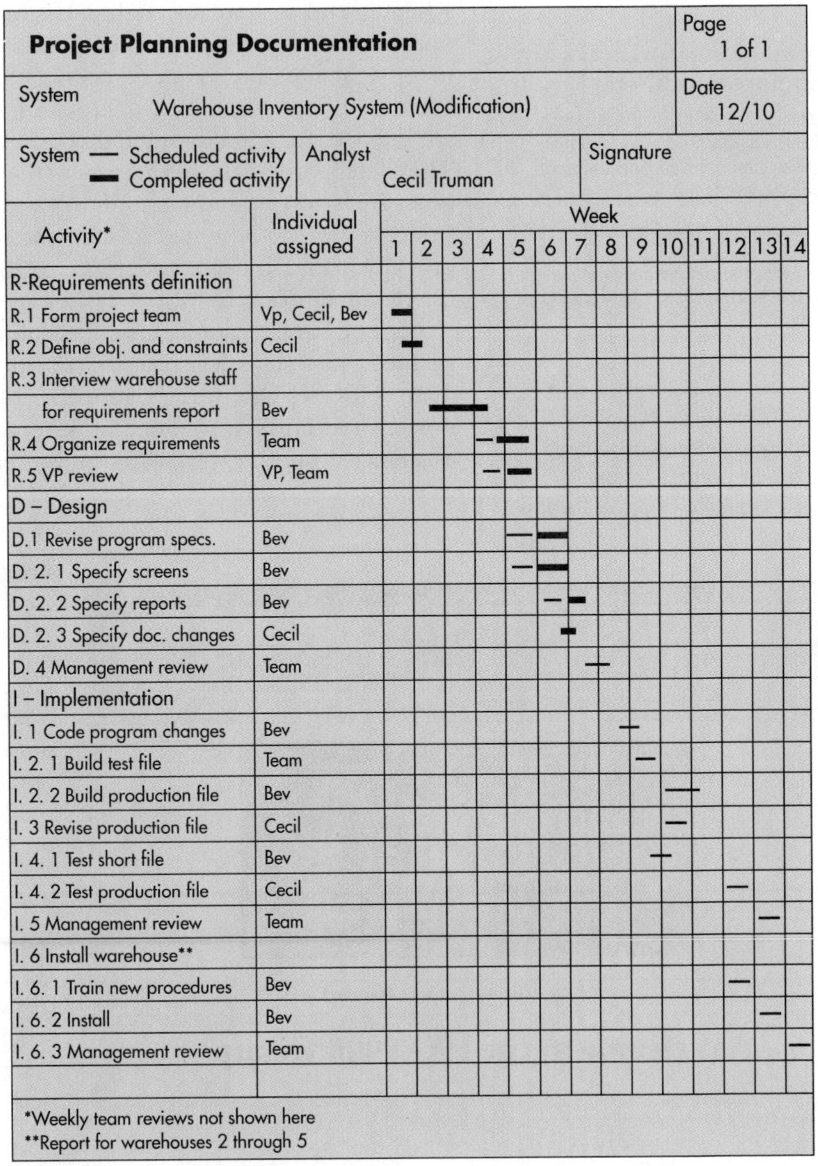

Project Planning Documentation		Page 1 of 1
System	Warehouse Inventory System (Modification)	Date 12/10
System — Scheduled activity ▬ Completed activity	Analyst Cecil Truman	Signature

Activity*	Individual assigned	Week
		1 2 3 4 5 6 7 8 9 10 11 12 13 14
R-Requirements definition		
R.1 Form project team	Vp, Cecil, Bev	▬
R.2 Define obj. and constraints	Cecil	▬
R.3 Interview warehouse staff		
for requirements report	Bev	▬
R.4 Organize requirements	Team	─ ▬
R.5 VP review	VP, Team	─ ▬
D – Design		
D.1 Revise program specs.	Bev	─ ▬
D. 2. 1 Specify screens	Bev	─ ▬
D. 2. 2 Specify reports	Bev	─ ▬
D. 2. 3 Specify doc. changes	Cecil	▬
D. 4 Management review	Team	─
I – Implementation		
I. 1 Code program changes	Bev	─
I. 2. 1 Build test file	Team	─
I. 2. 2 Build production file	Bev	─
I. 3 Revise production file	Cecil	─
I. 4. 1 Test short file	Bev	─
I. 4. 2 Test production file	Cecil	─
I. 5 Management review	Team	─
I. 6 Install warehouse**		
I. 6. 1 Train new procedures	Bev	─
I. 6. 2 Install	Bev	─
I. 6. 3 Management review	Team	─

*Weekly team reviews not shown here
**Report for warehouses 2 through 5

FIGURE 12.10

Sample Gantt Chart

A Gantt chart shows progress through systems development activities by putting a bar through appropriate cells.

Software	Vendor
BeachBox '98	NetSQL Partners
Job Order	Management Software Inc.
OpenPlan	Welcom
Project	Microsoft
Project Scheduler	Scitor Corporation
Super Project	Computer Associates

TABLE 12.6

Selected Project Management Software Packages

computer-aided software engineering (CASE)

technology that automates many of the tasks required in a systems development effort and enforces adherence to the SDLC

upper-CASE tools

tools that focus on activities associated with the early stages of systems development

ISO 9000 are international quality standards used by IS and other organizations to ensure quality of products and services.

Both PERT and Gantt techniques can be automated using project management software. This type of software monitors all project activities and determines whether activities and the entire project are on time and within budget. Project management software also has workgroup capabilities to handle multiple projects and to allow a team of people to interact with the same software. Project management software helps managers determine the best way to reduce project completion time at the least cost. Several project management software packages are identified in Table 12.6.

Use of Formal Quality Assurance Processes

The development of information systems requires a constant trade-off of schedule and cost versus quality. Historically, the development of application software has put an overemphasis on schedule and cost to the detriment of quality. Techniques, such as use of the ISO 9000 standards, have been developed to improve the quality of information systems. ISO 9000 are international quality standards originally developed in Europe in 1987. These standards help businesses define and document their own quality procedures for production and services. They can be used in any type of business and are accepted around the world as proof that a business can provide quality. Indeed, adherence to ISO 9000 is a requirement in many international markets.[15]

Many IS organizations have incorporated ISO 9000, total quality management, and statistical process control principles into the way they produce software. Often, to ensure the quality of the systems development process and finished product, an IS organization will form its own quality assurance groups to work with project teams and encourage them to follow established standards.

Use of Computer-Aided Software Engineering (CASE) Tools

Computer-aided software engineering (CASE) tools automate many of the tasks required in a systems development effort and enforce adherence to the SDLC, thus instilling a high degree of rigor and standardization to the entire systems development process.

CASE packages that focus on activities associated with the early stages of systems development are known as **upper-CASE tools**. These packages provide automated tools to assist with systems investigation, analysis, and design activities. Other CASE packages,

(An affiliate of Internet Commerce Associates)

Welcome to the ISO 9000 Group

lower-CASE tools

tools that focus on the later implementation stage of systems development

integrated-CASE tools (I-CASE)

tools that provide links between upper- and lower-CASE packages, allowing lower-CASE packages to generate program code from upper-CASE package designs

called **lower-CASE tools**, focus on the later implementation stage of systems development and are capable of automatically generating structured program code. Some CASE tools provide links between upper- and lower-CASE packages, thus allowing lower-CASE packages to generate program code from upper-CASE package designs. These are called **integrated-CASE tools (I-CASE)**. Selected I-CASE tools are listed in Table 12.7. Total-CASE tools have also been developed for object-oriented programs and applications.[16]

As with any team, coordinating the efforts of members of a systems development team can be a problem. So, many CASE tools allow more than one person to work on the same system at the same time via a multi-user interface, which coordinates and integrates the work performed by all members of the same design team. With this facility, a person working on one aspect of systems development can automatically share his or her results with someone working on another aspect of the same system. Advantages and disadvantages of CASE tools are listed in Table 12.8.

TABLE 12.7

Selected I-CASE Tools

Product	Vendor
ADW	KnowledgeWare
Bachman	Bachman Information Systems
Composer	TI Information Engineering
CorVision	Cortex Corporation
Developer's Studio	Microsoft
Envision	Future Tech Systems Inc.
Foundation	Andersen Consulting
HPS	Seer Technologies
Implementor	Implementors
Intelligent OOA	Kennedy Carter
Intersolv	Intersolv
KnowledgeWare	KnowledgeWare
Maestro II	Softlab
Methods Factory	VSF Ltd
Oracle CASE	Oracle
PACBASE	CGI
Paradigm Plus	ProtoSoft Inc.
Predict CASE	Software AG
Software through Pictures	Interactive Development Environments
System Architect	Popkin Software Systems Inc.
Systems Engineer	LBMS
Teamwork	ObjectTeam Cadre Technologies
ToolBuilder	IPSYS Software
TopWindows/TopCASE	TopSystems International
Westmount I-CASE	Westmount Technology

TABLE 12.8

Advantages and Disadvantages
of CASE Tools

Advantages	Disadvantages
Produce systems with a longer effective operational life	Produce initial systems that are more expensive to build and maintain
Produce systems that more closely meet user needs and requirements	Require more extensive and accurate definition of user needs and requirements
Produce systems with excellent documentation	May be difficult to customize
Produce systems that need less systems support	Require training of maintenance staff
Produce more flexible systems	May be difficult to use with existing systems

SYSTEMS INVESTIGATION

As discussed earlier in the chapter, systems investigation is the first phase in the traditional SDLC of a new or modified business information system. The purpose is to identify potential problems and opportunities and consider them in light of the goals of the company. For example, for Gerber, an opportunity existed to improve customer service. By identifying this during systems investigation, Gerber was able to develop a system to meet this corporate goal. In general, systems investigation attempts to uncover answers to the following questions:

- What primary problems might a new or enhanced system solve?
- What opportunities might a new or enhanced system provide?
- What new hardware, software, databases, telecommunications, personnel, or procedures will improve an existing system or are required in a new system?
- What are the potential costs (variable and fixed)?
- What are the associated risks?

Initiating Systems Investigation

Because systems development requests can require considerable time and effort to implement, many organizations have adopted a formal procedure for initiating systems development, beginning with systems investigation. The **systems request form** is a document that is filled out by someone who wants the IS department to initiate systems investigation. This form typically includes the following information:

systems request form

a document filled out by someone who wants the IS department to initiate systems investigation

- Problems in or opportunities for the system
- Objectives of systems investigation
- Overview of the proposed system
- Expected costs and benefits of the proposed system

The information in the systems request form helps to rationalize and prioritize the activities of the IS department. Based on the overall IS plan, the organization's needs and goals, and the estimated value and priority of

The Investigation Team

Managers and stakeholders

IS personnel

- Undertakes feasibility analysis
- Establishes system development goals
- Selects system development methodology
- Prepares system investigation report

FIGURE 12.11

The Systems Investigation Team

The team is made up of upper- and middle-level managers, IS personnel, users, and stakeholders.

feasibility analysis

an assessment of the technical, operational, schedule, and economic feasibility of a project

technical feasibility

an assessment of whether the hardware, software, and other system components can be acquired or developed to solve the problem

operational feasibility

a measure of whether the project can be put into action or operation

schedule feasibility

an assessement of whether the project can be completed in a reasonable amount of time

economic feasibility

an assessment of whether the project makes financial sense and whether predicted benefits offset the cost and time needed to obtain them

net present value

the preferred approach for ranking competing projects and determining economic feasibility

the proposed projects, managers make decisions regarding the initiation of each systems investigation for such projects.

Participants in Systems Investigation

Once a decision has been made to initiate systems investigation, the first step is to determine what members of the development team should participate in the investigation phase of the project. Members of the development team change from phase to phase (Figure 12.11).

Ideally, functional managers are heavily involved during the investigation phase. Other members could include users or stakeholders outside management, such as an employee who helped initiate systems development. The technical and financial expertise of others participating in investigation would help the team determine whether the problem is worth solving. The members of the development team who participate in investigation are then responsible for gathering and analyzing data, preparing a report justifying systems development, and presenting the results to top-level managers.

Feasibility Analysis

A key step of the systems investigation phase is **feasibility analysis**, which assesses technical, operational, schedule, and economic feasibility (Table 12.9). **Technical feasibility** is concerned with whether the hardware, software, and other system components can be acquired or developed to solve the problem. **Operational feasibility** is a measure of whether the project can be put into action or operation. It includes both logistical and motivational (acceptance of change) considerations. Motivational considerations are very important because new systems affect people and data flows and may have unintended consequences. As a result, power and politics may come into play, and some people may resist the new system. **Schedule feasibility** determines whether the project can be completed in a reasonable amount of time—a process that involves balancing the time and resource requirements of the project with other projects. **Economic feasibility** determines whether the project makes financial sense and whether predicted benefits offset the cost and time needed to obtain them. Economic feasibility can involve cash flow analysis such as in net present value.

Net present value. **Net present value** is the preferred approach for ranking competing projects and for determining economic feasibility. The net present value represents the net amount by which project savings exceed project expenses, after allowing for the cost of capital and the passage of time. The cost of capital is the average cost of funds used to finance the operations of the business. It represents the minimum desired rate of return on an investment; thus, it is also called the hurdle rate. Net present value takes into account that a dollar returned at a later date is not worth as much as one received today, since the dollar in hand can be invested to

TABLE 12.9

Types of Feasibility

Technical feasibility	Can the hardware, software, and other system components be acquired or developed to solve the problem?
Operational feasibility	Can the project be put into action or operation? What are the logistical and motivational (acceptance of change) considerations?
Schedule feasibility	Can the project be completed in a reasonable amount of time?
Economic feasibility	Does the project make financial sense? Do the predicted benefits offset the cost and time needed to obtain them?

earn profits or interest in the interim. Spreadsheet programs, such as Lotus and Excel, have built-in functions to compute the net present value. The net present value is the sum of the expected cash flows from each year of the project and can be expressed as follows:

$$\text{Net present value} = \sum_{t=1}^{n} (\text{CF}_t)/(1+k)^t$$

where

CF_t is the expected cash flow in period t and k is the project's cost of capital.

Since income tax payments represent a disbursement of cash, all cash flows affecting taxable income are credited to the project on an after-tax basis (i.e., computed by multiplying the tax complement by the pretax cash flow), with the notable exception of depreciation. Since depreciation expense does not involve actual cash disbursement, it is excluded from cash flows. However, depreciation is a deductible expense for federal tax purposes. Therefore, the amount of tax relief from depreciation is included in cash flows. It is computed by multiplying the federal tax rate by the depreciation expense. Table 12.10 illustrates these concepts for a sample project in a firm with a tax rate of 36 percent and a cost of capital of 20 percent.

TABLE 12.10

Sample Net Present Value Calculation

Cash Flow (in thousands of dollars)	Year 1	Year 2	Year 3	Year 4
1. Cash inflow (gross savings)	25	105	125	200
2. Cash outflow (expenses)	-135	-25	-30	-35
3. Pretax cash flow (line 1 + 2)	-110	80	95	165
4. After-tax cash flow [line 3 x (1 − tax rate)]	-70	51	61	106
5. Depreciation	60	50	40	30
6. Tax relief from depreciation (line 5 x tax rate)	22	18	14	11
7. Net after-tax cash flow (line 4 + line 6)	-48	69	75	117
8. Discounted cash flow [line 7 ÷ (1 + cost of capital)^{Year}]	-40	48	43	56
9. Net present value (sum of amounts in row 8)	107			

The key to making accurate estimates of the cash flows associated with a project is to involve the business managers responsible for the business functions served. Another key resource is your organization's financial manager, who should be very familiar with net present value analysis. If a systems development project is determined to be feasible, systems investigation will formally begin.

The Systems Investigation Report

systems investigation report

a report that summarizes the results of the systems investigation and the process of feasibility analysis and recommends a course of action

The primary outcome of systems investigation is a **systems investigation report**. This report summarizes the results of systems investigation and the process of feasibility analysis and recommends a course of action: continue on into systems analysis, modify the project in some manner, or drop it. A typical Table of Contents for the systems investigation report is shown in Figure 12.12.

The systems investigation report is reviewed by senior management, often organized as an advisory or **steering committee** consisting of senior management and users from the IS department and other functional areas. These individuals help IS personnel with their decisions about the use of information systems in the business and give authorization to pursue further systems development activities. After review, the steering committee might agree with the recommendation of the systems development team or suggest a change in project focus to concentrate more directly on meeting a specific company objective. Another alternative is that everyone may decide that the project is not feasible for one reason or another and cancel the project.

steering committee

an advisory group consisting of senior management and users from the IS department and other functional areas

SYSTEMS ANALYSIS

After a project has been approved for further study, the next step is to answer the question "What must the information system do to solve the problem?" The process needs to go beyond mere computerization of existing systems. The entire system, and the business process with which it is associated, should be evaluated. Often, a firm can make great gains if it restructures both business activities and the related information system simultaneously. The overall emphasis of analysis is gathering data on the existing system, determining the requirements for the new system, considering alternatives within these constraints, and investigating the feasibility of the solutions. The primary outcome of systems analysis is a prioritized list of systems requirements.

FIGURE 12.12

A Typical Table of Contents for a Systems Investigation Report

Johnson & Florin, Inc.
Systems Investigation Report

Table of Contents

EXECUTIVE SUMMARY
REVIEW of GOALS and OBJECTIVES
SYSTEM PROBLEMS and OPPORTUNITIES
PROJECT FEASIBILITY
PROJECT COSTS
PROJECT BENEFITS
RECOMMENDATIONS

General Considerations

Systems analysis starts by clarifying the overall goals of the organization and determining how the existing or proposed information system helps meet them. A manufacturing company, for example, might want to reduce the number of equipment breakdowns. This goal can be translated into one or more informational needs. One need might be to create and maintain an accurate list of each piece of equipment and a schedule for preventive maintenance. Another need might be a list of equipment failures and their causes.

Analysis of a small company's information system can be fairly straight-forward. On the other hand, evaluating an existing information system for a large company can be a long, tedious process. As a result, large organizations evaluating a major information system normally follow a formalized analysis procedure, involving these steps:

1. Assembling the participants for systems analysis
2. Collecting appropriate data and requirements
3. Analyzing the data and requirements
4. Preparing a report on the existing system, new system requirements, and project priorities

Participants in Systems Analysis

The first step in formal analysis is to assemble a team to study the existing system. This group includes members of the original development team—from users and stakeholders to IS personnel and management. Most organizations usually allow key members of the development team not only to analyze the condition of the existing system but also to perform other aspects of systems development, such as design and implementation.

Once the participants in systems analysis are assembled, this group develops a list of specific objectives and activities. A schedule for meeting the objectives and completing the specific activities is also developed, along with deadlines for each stage and a statement of the resources required at each stage, such as clerical personnel, supplies, and so forth. Major milestones are normally established to help the team monitor progress and determine whether problems or delays occur in performing systems analysis.

Data Collection

The purpose of data collection is to seek additional information about the problems or needs identified in the systems investigation report. During this process, the strengths and weaknesses of the existing system are emphasized.

Identifying sources of data. Data collection begins by identifying and locating the various sources of data, including both internal and external sources (Figure 12.13).

Performing data collection. Once data sources have been identified, data collection begins. Figure 12.14 shows the steps involved. Data collection may require a number of tools and techniques, such as interviews, direct observation, and questionnaires.

In a **structured interview**, the questions are written in advance. In an **unstructured interview**, the questions are not written in advance; the interviewer relies on experience in asking the best questions to uncover the inherent problems of the existing system. An advantage of the unstructured interview is that it allows the interviewer to ask follow-up or clarifying questions immediately.

With **direct observation**, one or more members of the analysis team directly observe the existing system in action. One of the best ways to understand how the existing system functions is to work with the users to

structured interview

an interview in which the questions are written in advance

unstructured interview

an interview in which the questions are not written in advance

direct observation

analysis conducted by one or more members of the analysis team who observes the existing system in action

FIGURE 12.13

Internal and External Sources of Data for Systems Analysts

Internal Sources
Users, stakeholders, and managers
Organization charts
Forms and documents
Procedure manuals and policies
Financial reports
IS manuals
Other measures of business process

External Sources
Customers
Suppliers
Stockholders
Government agencies
Competitors
Outside groups
Journals, etc.
Consultants

questionnaires

a method of gathering data when the data sources are spread over a wide geographic area

statistical sampling

a method of data gathering that involves taking a random sample of data and applying the characteristics of the sample to the whole group

FIGURE 12.14

The Steps in Data Collection

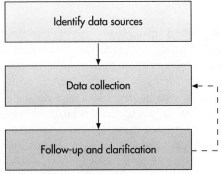

discover how data flows in certain business tasks. This entails direct observation of users' work procedures, their reports, current screens (if automated already), and so on. From this, members of the analysis team determine which forms and procedures are adequate and which are inadequate and need improvement. Direct observation requires a certain amount of skill. The observer must be able to see what is really happening and not be influenced by his or her own attitudes or feelings. This approach can reveal important problems and opportunities that would be difficult to obtain using other data collection methods. An example would be to observe the work procedures, reports, and computer screens associated with an accounts payable system being considered for replacement.

When many data sources are spread over a wide geographic area, **questionnaires** may be the best approach. Like interviews, questionnaires can be either structured or unstructured. In most cases, a pilot study is conducted to fine-tune the questionnaire. A follow-up questionnaire can also capture the opinions of those who do not respond to the original questionnaire.

A number of other data collection techniques can be employed. In some cases, telephone calls are an excellent method. In other cases, activities may be simulated to see how the existing system reacts. Thus, fake sales orders, stockouts, customer complaints, and data-flow bottlenecks may be created to see how the existing system responds.

Statistical sampling, which involves taking a random sample of data, is another technique. For example, suppose we want to collect data that describes 10,000 sales orders received over the last few years. Because it is too time-consuming to analyze each of the 10,000 sales orders, a random sample of 100–200 sales orders from the entire batch can be collected. The characteristics of this sample are then assumed to apply to the 10,000 orders.

Direct observation is a method of data collection. One or more members of the analysis team directly observe the existing system in action.
(Source: Image copyright ©1998 PhotoDisc, Inc.)

data analysis

manipulating the collected data so that it is usable for the development team members who are participating in systems analysis

data modeling

a commonly accepted approach to modeling organizational objects and associations that employ both text and graphics

activity modeling

a method to describe related objects, associations, and activities

data-flow diagram (DFD)

a diagram that models objects, associations, and activities by describing how data can flow between and around them

Data Analysis

The data collected in its raw form is usually not adequate to determine the effectiveness and efficiency of the existing system or the requirements for the new system. The next step is to manipulate the collected data so that it is usable for the development team who are participating in systems analysis. This manipulation is called **data analysis**. Data and activity modeling, using data-flow diagrams and entity-relationship diagrams, are useful during data analysis to show data flows and the relationships among various objects, associations, and activities. Other common tools and techniques for data analysis include application flowcharts, grid charts, and CASE tools.

Data modeling. Data modeling, first introduced in Chapter 5, is a commonly accepted approach to modeling organizational objects and associations that employ both text and graphics. The exact way data modeling is employed, however, is governed by the specific systems development methodology.

Data modeling is most often accomplished through the use of entity-relationship (ER) diagrams. Recall from Chapter 5 that an entity is a generalized representation of an object type—such as a class of people (employee), events (sales), things (desks), or places (Philadelphia)—and that entities possess certain attributes. Objects can be related to other objects in numerous ways. An entity-relationship diagram, such as the one shown in Figure 12.15a, describes a number of objects and the ways they are associated. An ER diagram is not capable in and of itself of fully describing a business problem or solution, because it lacks descriptions of the related activities. It is, however, a good place to start, since it describes object types and attributes about which data may need to be collected for processing.

Activity modeling. To fully describe a business problem or solution, it is necessary to describe the related objects, associations, and activities. Activities in this sense are events or items that are necessary to fulfill the business relationship or that can be associated with the business relationship in a meaningful way.

Activity modeling is often accomplished through the use of data-flow diagrams. A **data-flow diagram (DFD)** models objects, associations, and activities by describing how data can flow between and around various objects. DFDs work on the premise that for every activity there is some communication, transference, or flow that can be described as a data element. DFDs describe what activities are occurring to fulfill a business relationship or accomplish a business task, not how these activities are to be performed. That is, DFDs show the logical sequence of associations and

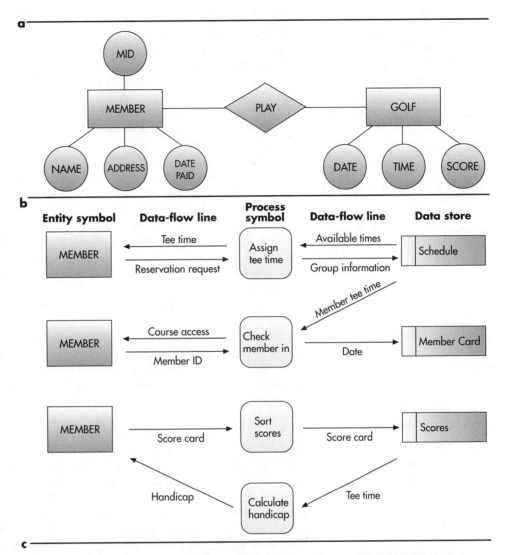

To play golf at the course, you must first pay a fee to become a member of the golf club. Members are issued member cards and are assigned member ID numbers. To reserve a tee time (a time to play golf), a member calls the club house at the golf course and arranges an available time slot with the reception clerk. The reception clerk reserves the tee time by writing the member's name and number of players in the group on the course schedule. When a member arrives at the course, he or she checks in at the reception desk where the reception clerk checks the course schedule and notes the date on the member's card. After a round of golf has been completed, the members leave their score card with the reception clerk. Member scores are tracked and member handicaps are updated on a monthly basis.

FIGURE 12.15

(a) An entity-relationship diagram. (b) A data-flow diagram. (c) A semantic description of the business process.
(Source: G. Lawrence Sanders, *Data Modeling*, Danvers, MA: Boyd & Fraser Publishing, 1995. Reprinted with permission from Course Technology.)

activities, not the physical processes. A system modeled with a DFD could operate manually or could be computer based; if computer based, the system could operate with a variety of technologies.

DFDs are easy to develop and easily understood by nontechnical people. Data-flow diagrams use four primary symbols, as illustrated in Figure 12.15b.

data-flow line

the path that shows the direction of data element movement in a data-flow diagram

process symbol

the symbol that reveals a function that is performed in a data-flow diagram

entity symbol

the symbol that shows either a source or destination of a data element in a data-flow diagram

data store

the symbol that shows a storage location for data in a data-flow diagram

application flowcharts

charts that show relationships among applications or systems

grid chart

a table that shows relationships among the various aspects of a systems development effort

FIGURE 12.16

An application flowchart shows the relationships among various applications.

- *Data flow.* The **data-flow line** includes arrows that show the direction of data element movement.
- *Process symbol.* The **process symbol** reveals a function that is performed. Computing gross pay, entering a sales order, delivering merchandise, and printing a report are examples of functions that can be represented with a process symbol.
- *Entity symbol.* The **entity symbol** shows either the source or destination of the data element. An entity can be, for example, a customer who initiates a sales order, an employee who receives a paycheck, or a manager who gets a financial report.
- *Data store.* A **data store** reveals a storage location for data. A data store is any computerized or manual data storage location, including magnetic tape, disks, a filing cabinet, or a desk.

Comparing entity-relationship diagrams with data-flow diagrams provides insight into the concept of top-down design. Figure 12.15a and b show an entity-relationship diagram and a data-flow diagram for the same business relationship—namely, a member of a golf course playing golf. Figure 12.15c provides a brief description of the business relationship for clarification.

Application flowcharts. Application flowcharts show the relationships among applications or systems. Assume that a small business has collected data about its order processing, inventory control, invoicing, and marketing analysis applications. Management is thinking of modifying the inventory control application. The raw facts collected, however, do not help in determining how the applications are related to each other and the databases required for each. This is done through data analysis with an application flowchart (Figure 12.16). Using this tool for data analysis makes clear the relationships among the order processing, inventory control, invoicing, and marketing analysis applications.

In the simplified application flowchart in Figure 12.16, you can see that the order processing application provides important data to the inventory control and marketing analysis applications. The inventory control application provides data to the invoicing application. Any changes made to any one of these applications must take into account the other applications, which may provide data to or receive data from the others.

Grid charts. A **grid chart** is a table that shows relationships among various aspects of a systems development effort. For example, a grid chart can be used to reveal the databases used by the various applications (Figure 12.17).

The simplified grid chart shown in Figure 12.17 shows that the customer database is used by the order processing, invoicing, and marketing analysis applications. The inventory database is used by the order processing, inventory control, and marketing analysis applications. The supplier database is used by the inventory control application, and the accounts receivable database is used by the invoicing application. This grid chart shows which applications use common databases and

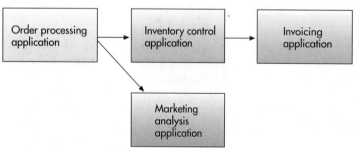

FIGURE 12.17

A grid chart shows the relationships among applications and databases.

Databases → Applications ↓	Customer database	Inventory database	Supplier database	Accounts receivable database
Order processing application	X	X		
Inventory control application		X	X	
Marketing analysis application	X	X		
Invoicing application	X			X

reveals that, for example, any changes to the inventory control application must investigate the inventory and supplier databases.

CASE tools. Many systems development projects use upper-CASE tools to complete analysis tasks. For example, most computer-aided software engineering tools have generalized graphics programs that can generate a variety of diagrams and figures. Entity-relationship diagrams, data-flow diagrams, application flowcharts, and other diagrams can be developed using CASE graphics programs to help describe the existing system. During the analysis phase, a **CASE repository**—a database of system descriptions, parameters, and objectives—will begin to be developed.

CASE repository

a database of system descriptions, parameters, and objectives

Requirements Analysis

requirements analysis

an assessment used to determine user, stakeholder, and organizational needs

The overall purpose of **requirements analysis** is to determine user, stakeholder, and organizational needs. For an accounts payable application, the stakeholders could include suppliers and members of the purchasing department. Questions that should be asked during requirements analysis include the following:

Are these stakeholders satisfied with the current accounts payable application?

What improvements could be made to satisfy suppliers and help the purchasing department?

One of the most difficult procedures in systems analysis is eliciting user or systems requirements. In some cases, communications problems can interfere with the determination of these requirements. For example, an accounts payable manager may want a better procedure for tracking the amount owed by customers. Specifically, the manager would like to have a weekly report that shows all customers who owe more than $1000 and are more than 90 days past due on their account. A financial manager might need a report that summarizes total amount owed by customers to look at the need to loosen or tighten credit limits. A sales manager might want to review the amount owed by a key customer relative to sales to that same customer. The purpose of requirements analysis is to capture these requests in detail. Numerous tools and techniques can be used to capture systems requirements. Often, various techniques are used in the context of a JAD session.

Asking directly. One the most basic techniques used in requirements analysis is asking directly. **Asking directly** is an approach that asks users, stakeholders, and other managers about what they want and expect from the new or modified system. This approach works best for stable systems in which stakeholders and users clearly understand the system's functions. Unfortunately, many individuals do not know exactly or are unable to adequately articulate what they want or need. The role of the systems analyst during the analysis phase is to critically and creatively evaluate needs and define them clearly so that the systems can best meet them.

Critical success factors. Another approach uses **critical success factors (CSFs)**. Managers and decision makers are asked to list only those factors that are critical to the success of their area of the organization. A CSF for a production manager might be adequate raw materials from suppliers; a CSF for a sales representative could be a list of customers currently buying a certain type of product. Starting from these CSFs, the system inputs, outputs, performance, and other specific requirements can be determined.

The IS plan. As we have seen, the IS plan translates strategic and organizational goals into systems development initiatives. The IS planning process often generates strategic planning documents that can be used to define system requirements. Working from these documents ensures that requirements analysis will address the goals set by top-level managers and decision makers (Figure 12.18). There are unique benefits to applying the IS plan to define systems requirements. Because the IS plan takes a long-range approach to using information technology within the organization, the requirements for a system analyzed in terms of the IS plan are more likely to be compatible with future systems development initiatives. Read the "E-Commerce" box to learn how an entrepreneur defined user requirements for a new information service.

Screen and report layout. Developing formats for printed reports and screens to capture data and display information are some of the common tasks associated with developing systems. Screens and reports relating to systems output are specified first to verify that the desired solution is being delivered. Manual or computerized screen and report layout facilities are used to capture both output and input requirements.

 Screen layout is a technique that allows a designer to quickly and efficiently design the features, layout, and format of a display screen. In general, users who interact with the screen frequently can be presented with more data and less descriptive information (Figure 12.19a); infrequent users should have more descriptive information presented to explain the data that they are viewing (Figure 12.19b).

---

asking directly

an approach to gathering data that asks users, stakeholders, and other managers about what they want and expect from the new or modified system

critical success factors (CSFs)

those factors that are critical to the success of a manager's area of the organization

screen layout

a technique that allows a designer to quickly and efficiently design the features, layout, and format of a display screen

FIGURE 12.18

Converting Organizational Goals into Systems Requirements

E-COMMERCE
Defining Requirements for an On-Line Service

Back in 1994, when Michael Levy envisioned the idea of an on-line information service that would deliver sports information, there was not much interest from potential investors or partners. TV, radio, and the print media were so full of sports that another source of information seemed unnecessary. Today, Levy's SportsLine USA has emerged as one of the leading Internet media sites dedicated to sports information and entertainment. The company's Web site, cbs.sportsline.com, provides real-time information on everything related to sports, from the major U.S. team sports to World Cup soccer to even cricket. SportsLine is not only one of the most successful sports Web sites, but one of the most popular content providers on the Internet.

"To be successful, new media businesses must be able to establish a brand, appeal to a designated user group, and amortize fixed costs across multiple revenue streams," says Chris Dixon, a new media analyst for PaineWebber. SportsLine's Levy realized that, to make money, a startup has to have compelling content. While other content sites, such as Cnet, Slate, and CNNSI, pride themselves on originating news and programming, SportsLine made a deliberate decision not to follow their lead. Instead, SportsLine began by buying its sports content from third-party suppliers. As SportsLine has grown, it has added its own

reporters, who attend major sports events and provide live feeds to the editing facilities in New York.

In completing the systems analysis of the information system required to support the information service, SportsLine decided it must employ a stable technology platform to ensure that viewers do not suffer a loss of service. It set up a communications infrastructure that supports different pathways to the Internet with two data centers, and each uses three carriers for Internet transmission.

DISCUSSION QUESTIONS

1. Imagine that you are the information systems project manager responsible for establishing the infrastructure for SportsLine. How would you define the technical requirements for the infrastructure to support this new venture?
2. Identify two basic infrastructure requirements in addition to the need to ensure that viewers do not suffer a loss of service.

Sources: Adapted from Udayan Gupta, "Scoring with Content," *InformationWeek*, June 15, 1998, pp. 115–120 and the SportsLine homepage at http://www.cbs.sportsline.com.

---

report layout
a technique that allows designers to diagram and format printed reports

Report layout allows designers to diagram and format printed reports. These reports can be produced on either standard line printers (using 132 columns or positions) or personal computer printers (with 80 columns or positions). Reports can contain data, graphs, or both. Graphic presentations allow managers and executives to quickly view trends and take appropriate action, if necessary.

Screen layout diagrams can document the screens users desire for the new or modified application. Report layout charts reveal the format and content of various reports that the application will prepare. Other diagrams and charts can be developed to reveal the relationship between the application and outputs from the application.

Requirements analysis tools. A number of tools can be used to document requirements analysis. Again, CASE tools are often employed. As requirements are developed and agreed upon, entity-relationship diagrams, data-flow diagrams, screen and report layout forms, and other types of documentation will be stored in the CASE repository. These requirements might also be used later as a reference during the rest of systems development or for a different systems development project.

FIGURE 12.19

Screen Layouts

(a) A screen layout chart for frequent users who require little descriptive information.
(b) A screen layout chart for infrequent users who require more descriptive information.

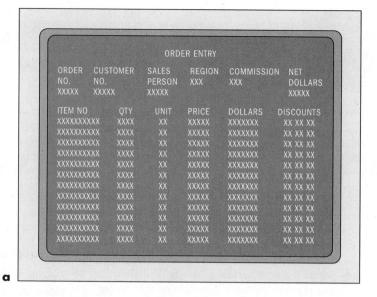

a

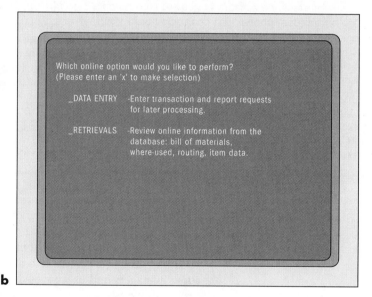

b

The Systems Analysis Report

Systems analysis concludes with a formal systems analysis report. It should cover the following elements:

* The strengths and weaknesses of the existing system from a stakeholder's perspective
* The user/stakeholder requirements for the new system (also called the functional requirements)
* The organizational requirements for the new system
* A description of what the new information system should do to solve the problem

Suppose analysis reveals that a marketing manager thinks a weakness of the existing system is its inability to provide accurate reports on product availability. These requirements and a preliminary list of the corporate objectives for the new system—including performance, cost, control, and complexity—will be in the systems analysis report. Particular attention is placed on areas of the existing system that could be improved to meet user requirements. The Table of Contents for a typical report is shown in Figure 12.20.

The systems analysis report gives managers a good understanding of the problems and strengths of the existing system. If the existing system is operating better than expected or the necessary changes are too expensive relative to the benefits of a new or modified system, the systems development process can be stopped at this stage. If the report shows that changes to another part of the system might be the best solution, the development process might start over, beginning again with systems investigation. Or, if the systems analysis report shows that it will be beneficial to develop one or more new systems or to make changes to existing ones, systems design begins. Read the "FYI" box to learn how utility companies are designing billing systems that provide useful information to customers and are also flexible enough to be used in multiple billing situations.

As the examples and discussion of this chapter have demonstrated, systems development can profoundly affect the current and future business activities of an organization. Certainly, information systems professionals are critical members of the development team, but people from all functional areas of an organization are involved, whether through providing input for data collection, testing of prototypes, or participating in any of the other steps. The goal of development activities is to create and maintain business systems that help an organization meet its strategic goals and remain competitive.

FIGURE 12.20

A Typical Table of Contents for a Report on an Existing System

FYI

Utilities Perform Systems Analysis to Take Advantage of Deregulation

Alliant Utilities has its headquarters in Madison, Wisconsin, and has assets of $1.9 billion. It provides electric energy, natural gas, and water to nearly 400,000 customers in south-central Wisconsin. Alliant recently implemented a new billing and customer service system to help it succeed in the more competitive environment of electric utilities. The bill, once considered merely a demand notice for money, is now turning into a marketing tool for utilities, Internet service providers, and telecommunications carriers, all of whom now face increased competition as a result of deregulation in their industries.

Upon completing a systems investigation, Alliant realized that its 17-year-old COBOL billing application was not flexible enough to accommodate the new billing practices required for the deregulated utility industry. It could not support consolidated electricity, gas, and water bills or provide billing services for new entrants. During the subsequent systems analysis, Alliant determined that it must convert the application into an in-house developed, Unix-based client/server system that connects the billing application to 30 other systems in the organization, including the customer service call center. With the help of a systems consulting firm, Alliant converted 3.5 million lines of COBOL code into an Oracle database application running on Hewlett-Packard servers.

Alliant hopes to improve customer loyalty through this major investment. The new system will consolidate billing to include both phone and electricity charges. It also offers the added dimension of electronic commerce, enabling customers to pay their bills over the Internet. The ultimate vision is to give customers the ability to decide when they want their meter read, when they want their bill, and how often and how they pay. Related to customers, they should have easy access to any information about themselves and their energy usage. Listed are a few of the customer billing services that were defined during systems analysis.

- Budget Billing is a convenient payment plan that eliminates the highs and lows of a customer's energy bill throughout the year. Customers pay the same amount each month, which helps make budgeting easier.
- AutoPay enables customers to stop writing checks and buying stamps and to pay automatically. Alliant works with the cusomter's financial institution to deduct the payment right from the customer's account.
- Customer Assistance Plus provides affordable payment plans and referral services to customers experiencing financial difficulty. This service takes a personal approach to helping customers solve their financial problems.

What carriers and utility companies realize is that a flexible billing solution allows them to complete far more transactions than in the past for a bigger piece of the revenue. The new-found versatility will enable telephone companies, utilities, and ISPs to offer billing services.

DISCUSSION QUESTIONS

1. Imagine that you are leading systems analysis and must define the customer billing requirements for a utility. How are these requirements similar regardless of the type of utility—phone service, water, electric, gas, and Internet access? Can you identify any major differences?
2. If one of the goals of implementing a new billing system is to increase customer loyalty, what specific system requirements would support this? How would you define these requirements?

Sources: Adapted from Stephanie Neil, "Powering Up Customer Relationships," *PC Week*, May 11, 1998, pp. 69, 72 and the "Products and Services" pages of the Alliant Web site at http://www.alliant-energy.com, accessed September 21, 1998.

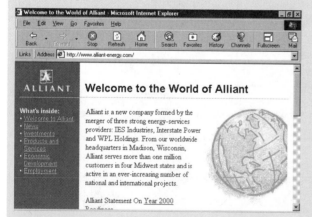

Information Systems Principles

For project success, it is essential to choose the appropriate systems development life cycle and follow it rigorously.

Effective systems development requires a team effort with stakeholders, users, managers, systems development specialists, and various support personnel.

The composition of the development team may vary over time and from project to project.

Systems development initiatives arise from all levels of an organization and are both planned and unplanned. Solid planning and managerial involvement are needed to ensure that these initiatives support broader organization goals.

An IS plan ensures a long-range approach to using information technology and better utilization of information systems resources. The use of the Internet, intranets, and extranets for the delivery of information should be strongly considered.

The extent to which a formal systems development process is followed, the degree of change associated with a project, and the level of top-management support are major determinants of project success.

Lack of top management support is often the cause of project failure. It is critical to identify a senior manager who has the most to gain from success of the project and recruit this individual to champion the project.

Because it accounts for the time value of money, net present value is often used for ranking competing projects and choosing among mutually exclusive alternatives.

■ SUMMARY

1. *Identify the key participants in the systems development process and discuss their roles.*

The systems development team consists of stakeholders, users, managers, systems development specialists, and various support personnel. The development team is responsible for determining the objectives of the information system and delivering to the organization a system that meets its objectives.

Stakeholders are individuals who, either themselves or through the area of the organization they represent, ultimately benefit from the systems development project. Users are individuals who will interact with the system regularly. They can be employees, managers, customers, or suppliers. Managers on development teams are typically representative of stakeholders or may be stakeholders themselves. In addition, managers are most capable of initiating and maintaining change. For large-scale systems development projects, where the investment in and value of a system can be quite high, it is common to have senior-level managers be part of the development team.

A systems analyst is a professional who specializes in analyzing and designing business systems. The programmer is responsible for modifying or developing programs to satisfy user requirements. Other support personnel on the development team include technical specialists, either IS department employees or outside consultants. Depending on the magnitude of the systems development project and the number of IS systems development specialists on the team, the team may also include one or more IS managers.

2. *Define the term* information systems planning *and list several reasons for initiating a systems project.*

Information systems planning refers to the translation of strategic and organizational goals into systems development initiatives. Benefits of IS planning include a long-range view of information technology use and better use of information systems resources. Planning requires developing overall IS objectives; identifying IS projects; setting priorities and selecting projects; analyzing resource requirements; setting schedules, milestones, and deadlines; and developing the IS planning document.

Systems development projects may be initiated for a number of reasons, including the need to solve problems with an existing system to exploit opportunities to gain competitive advantage, to increase competition, to make use of effective information, to create organizational growth, to settle a merger or corporate acquisition, and to address a change in the market or external environment.

3. *Identify important system performance requirements of transaction processing business applications that run on the Internet or a corporate intranet.*

Such applications must be designed to meet special business needs. They must be able to scale up to support highly variable transaction throughput from potentially thousands of users. Ideally, they can scale up dynamically when needed. They must be reliable and fault-tolerant, providing 24 hours a day/7 days a week availability with extremely high transaction integrity. They must be able to integrate with your existing infrastructure including customer and order databases, legacy applications, and enterprise resource planning systems. Development and maintenance must be quick and easy, as business needs may require you to change applications on the fly.

4. *Discuss three trends that illustrate the impact that the use of enterprise resource planning software packages is having on systems development.*

The first trend is that companies wish to stay with their primary ERP vendor (SAP, Oracle, PeopleSoft, etc.) instead of looking elsewhere for answers to their data warehousing and production planning needs or developing in-house solutions. Thus, they look for their original ERP vendor to provide these solutions.

A second trend is that there is less in-house development and more dependence on ERP vendors and their strategic partners to provide enhancements and add-ons to the original ERP package.

A third interesting trend is the increase in the number of companies who, once they have successfully implemented their own company's ERP project, are branching out to provide consulting to other companies.

5. *Discuss the key features, advantages, and disadvantages of the traditional, prototyping, rapid application development, and end-user systems development life cycles.*

The five phases of the traditional SDLC are investigation, analysis, design, implementation, and maintenance and review. Systems investigation involves identifying potential problems and opportunities and considering them in light of organizational goals. Systems analysis seeks a general understanding of the solution required to solve the problem; the existing system is studied in detail and weaknesses are identified. Systems design involves creating new or modified system requirements. Systems implementation encompasses programming, testing, training, conversion, and operation of the system. Systems maintenance and review entails monitoring the system and performing enhancements or repairs.

Advantages of the traditional SDLC include the following: it provides for maximum management control, creates considerable system documentation, ensures that system requirements can be traced back to stated business needs, and produces many intermediate products for review. Its disadvantages include the following: users may get a system that meets the needs as understood by the developers; the documentation is expensive and difficult to maintain; users' needs go unstated or may not be met; and users cannot easily review the many intermediate products produced.

Prototyping is an iterative approach that involves defining the problem, building the initial version, having users utilize and evaluate the initial version, providing feedback, and incorporating suggestions into the second version. Prototypes can be fully operational or nonoperational, depending on how critical the actual system under development is and how much time and money the organization has to spend on prototyping.

Advantages of the prototyping approach include the following: users get an opportunity to try the system before it is completed; useful prototypes can be produced in weeks; users become positive about the evolving system; and errors and omissions can be detected early. Disadvantages include these: the approach makes it difficult to start over if the initial solution misses the mark widely; it is difficult to contain the scope of the project; system documentation is often absent; and key operational considerations are often overlooked.

Rapid application development (RAD) uses tools and techniques designed to speed application development. Its use reduces paper-based documentation, automates program source code generation, and facilitates user participation in development activities. RAD makes extensive use of the joint application development (JAD) process to gather data and perform requirements analysis. JAD involves group meetings in which users, stakeholders, and IS professionals work together to analyze existing systems, propose possible solutions, and define the requirements for a new or modified system.

RAD has the following advantages: it puts an application into production quickly; documentation is produced as a by-product; and it forces good teamwork among users, stakeholders, and developers. Its disadvantages are the following: it is an intense process that can burn out the participants; it requires participants to be skilled in advanced tools and techniques; and a large percentage of time is required from stakeholders and users.

The end-user SDLC is used to support projects where the primary effort is undertaken by a combination of business managers and users.

6. *Identify several factors that influence the success or failure of a systems development project.*

The degree of change introduced by the project, the quality of project planning, the use of project management tools, the use of formal quality assurance processes, and the use of CASE tools are all factors that affect the success of a project. The greater the amount of change, the greater the degree of risk and also frequently the amount of reward. The quality of project planning involves such factors as support from top management, strong user involvement, use of a proven methodology, clear project goals and objectives, concentration on key problems and straightforward designs, staying on schedule and within budget, good user training, and solid review and maintenance programs. The use of automated project management tools enables detailed development, tracking, and control of the project schedule. Effective use of a quality assurance process enables the project manager to deliver a quality system and to make intelligent trade-offs among cost, schedule, and quality. CASE tools automate many of the systems development tasks, thus reducing the time and effort required to complete them while ensuring good documentation.

7. *State the purpose of systems investigation.*

The purpose of systems investigation is to identify potential problems and opportunities and consider them in light of the organization's goals. In most organizations, a systems request form initiates the investigation process. The systems investigation is designed to assess the feasibility of implementing solutions for business problems. An investigation team follows up on the request and performs a feasibility analysis that addresses technical, economic, operational, and schedule feasibility.

If the project under investigation is feasible, major goals are set for the system's development, including performance, cost, managerial, and procedural goals. Many companies choose a popular methodology so that new IS employees, outside specialists, and vendors will be familiar with the systems development tasks set forth in the approach. Selection of a systems development methodology must be made. As a final step in the investigation process, a systems investigation report should be prepared to document relevant findings.

8. *State the purpose of systems analysis and discuss some of the tools and techniques used in this phase of systems development.*

Systems analysis is the examination of existing systems, which begins once approval for further study is received from management. Additional study of a selected system allows those involved to further understand the systems' weaknesses and potential improvement areas. An analysis team is assembled to collect and analyze data on the existing system.

Data collection methods include observation, interviews, questionnaires, and statistical sampling. Data analysis manipulates the collected data. The analysis includes grid charts, application flow-charts, and CASE tools. The overall purpose of requirements analysis is to determine user and organizational needs.

Data modeling is used to model organizational objects and associations using text and graphical diagrams. It is most often accomplished through the use of entity-relationship (ER) diagrams. Activity modeling is often accomplished through the use of data-flow diagrams (DFDs), which model objects, associations, and activities by describing how data can flow between and around various objects. DFDs use symbols for data flows, processing, entities, and data stores.

■ KEY TERMS

activity modeling 550
application flowcharts 552
asking directly 554
CASE repository 553
computer-aided software engineering (CASE) 542
creative analysis 524
critical analysis 525
critical path 541
critical success factors (CSFs) 554
data analysis 550
data modeling 550
data store 552
data-flow diagram (DFD) 550
data-flow line 552
direct observation 548
economic feasibility 545
end-user systems development 537
entity symbol 552
feasibility analysis 545
Gantt chart 541
grid chart 552

information systems planning 524
integrated-CASE (I-CASE) tools 543
joint application development (JAD) 536
lower-CASE tools 543
mission-critical systems 527
net present value 545
nonoperational prototype 534
operational feasibility 545
operational prototype 534
process symbol 552
Program Evaluation and Review Technique (PERT) 541
programmer 521
project deadline 540
project milestone 540
project schedule 540
prototyping 533
questionnaires 549
rapid application development (RAD) 534
report layout 555

requirements analysis 553
runaways 539
schedule feasibility 545
screen layout 554
stakeholders 521
statistical sampling 549
steering committee 547
structured interview 548
systems analysis 532
systems analyst 521
systems design 532
systems implementation 532
systems investigation 532
systems investigation report 547
systems maintenance and review 533
systems request form 544
technical feasibility 545
unstructured interview 548
upper-CASE tools 542
users 521

■ REVIEW QUESTIONS

1. What is an information system stakeholder?
2. What is the goal of information systems planning? What steps are involved in IS planning?
3. What are some of the key system performance requirements of business transaction processing applications that run on the Internet or corporate intranet?
4. Identify three trends that illustrate the impact that the use of enterprise resource planning software packages is having on systems development.

5. Identify each of the four systems development life cycles and summarize their strengths and weaknesses.
6. How is a Gantt chart developed? How is it used?
7. What sort of information system applications are being built to run over the Internet, intranets, or extranets?
8. Why is it important to identify and remove errors early in the systems development life cycle?
9. Identify four reasons a systems development project may be initiated.

10. List factors that have a strong influence on project success.
11. What is a runaway project?
12. What is the purpose of systems investigation?
13. Define the four types of feasibility.
14. What is the net present value of a project?
15. What is the purpose of systems analysis?
16. How does the JAD technique support the RAD systems development life cycle?

■ DISCUSSION QUESTIONS

1. Why is it important for business managers to have a basic understanding of the systems development process?
2. Briefly describe the role of a system user in the systems investigation and systems analysis stages of a project.
3. Discuss the following statement: The adoption of ERP systems is reducing the amount of systems development activity in business organizations. Yes or no? Why?
4. During the systems investigation phase, how important is it to think creatively? What are some approaches to increase creativity?
5. For what types of systems development projects might prototyping be especially useful? What are the characteristics of a system developed with a prototyping technique?
6. Imagine that your firm has never developed an information systems plan. What sort of issues between the business functions and IS organization might exist?

7. Why are companies building applications that run over the Internet, intranets, or extranets? What are some of the technical challenges for such applications?
8. How important are communications skills to IS personnel? Consider this statement: "IS personnel need a combination of skills—one-third technical skills, one-third business skills, and one-third communications skills." Do you think this is true? How would this affect the training of IS personnel?
9. Imagine that you are a highly paid consultant who has been retained to evaluate an organization's systems development processes. With whom would you meet? How would you make your assessment?
10. You are a senior manager of a functional area in which a mission-critical system is being developed. How can you safeguard this project from becoming a runaway?

■ PROBLEM-SOLVING EXERCISES

 1. Develop a spreadsheet program to determine net present value using the form outlined on page 564.

Use the spreadsheet to select among two projects with the following cash flows. Project number one has a gross savings of $100,000 per year and expenses of $25,000 per year plus $15,000 per year in depreciation. Project two has a gross savings of $75,000 the first year with $125,000 per year thereafter. Expenses are $35,000 per year plus $18,000 per year in depreciation. Assume a capital cost rate of 15 percent and tax rate of 35 percent.

	Cash Flow (in thousands of dollars)			
	Year 1	**Year 2**	**Year 3**	**Year 4**
1. Cash inflow (gross savings)				
2. Cash outflow (expenses)				
3. Pretax cash flow (line 1 + 2)				
4. After-tax cash flow [line 3 x (1 − tax rate)]				
5. Depreciation				
6. Tax relief from depreciation (line 5 x tax rate)				
7. Net after-tax cash flow (line 4 + line 6)				
8. Discounted cash flow [line 7 ÷ (1 + cost of capital)^{Year}]				
9. Net present value (sum of amounts in row 8)				

2. You are developing a new information system for The Fitness Center, a company that has five fitness centers in your metropolitan area, with about 650 members and 30 employees in each location. This system will be used by both members and fitness consultants to keep track of participation in various fitness activities, such as free weights, volleyball, swimming, stair climbers, and aerobic classes. One of the performance objectives of the system is that it help members plan a fitness program to meet their particular needs. The primary purpose of this system, as envisioned by the director of marketing, is to assist The Fitness Center in obtaining a competitive advantage over other fitness clubs.

Use word processing software to prepare a brief memo to the required participants in the development team for this systems development project. Be sure to specify what roles these individuals will play and what types of information you hope to obtain from them. Assume that the relational database model will be the basis for building this system. Use a database management system to define the various tables that will make up the database.

3. Develop a list of the phases, steps, and tasks associated with planning the construction of a new home. Estimate the duration of each task. Use your presentation graphics program to develop a Gantt chart to show the schedule for this project.

■ TEAM ACTIVITY

Systems development is more of an art and less of a science with a wide variety of approaches in how companies perform this activity. You and the members of your team are to interview members of an information systems organization's development group. Prepare a list of questions to determine if they follow the approach outlined in this chapter. During the course of your interview, when you find discrepancies between the approach they follow and the process suggested in the text, find out why there is a difference. Also learn what tools and techniques are most frequently employed. Prepare a short report on your findings.

■ WEB EXERCISE

A number of companies were discussed in this chapter. Locate the Web site of one of these companies. What are the goods and/or services that this company produces? After visiting the company Web site, describe how systems development could be used to improve the goods and/or services it produces. You may be asked to develop a report or send an e-mail message to your instructor about what you found.

■ CASES

1 FAA Systems Development Project Slows

The desire to modernize the Federal Aviation Administration (FAA) started as early as 1981, when the nation's air traffic controllers decided to strike. In an unexpected move, then President Reagan decided to fire the air traffic controllers. Although there were some rough times, the air transportation system survived. During this time the government decided that air traffic control should be modernized to save money and reduce any impact of a strike. It was a challenging but exciting time. The Federal Aviation Administration (FAA) also had a mandate to improve flight safety. There were too many close calls, and the airways were getting more crowded. The primary objectives of the new system were to make air travel safer and to reduce total costs of air traffic control. These objectives were seen as critical to the long-term success of the FAA.

To solve the problem, the FAA decided an aggressive systems development effort was needed. The goal was to develop a network of ground stations that worked in conjunction with existing military satellite systems. The systems development cost was projected to range from $400 million to $500 million. The systems development effort called for the project to be completed by 1997. In the early stages of the project, there was tremendous enthusiasm. The project was called revolutionary. Unfortunately, the project cost and completion date were too aggressive.

Early in the systems development project, it was clear there would be problems. To make the situation worse, a number of planners wanted new features and enhancements, which added a significant amount of effort to the project. The result was higher costs and a longer development and implementation schedule. Furthermore, a new FAA director was appointed in the middle of the systems development effort. At this time, some of the systems development staff revealed that they thought the original cost and schedule projections were too optimistic and politically motivated. The original cost estimates did not include testing and development, which would likely cost an additional $300 million. The estimates may not have included other costs, such as additional hardware, software, and upgrades. All of these factors placed a stress on the systems development effort. After four years, the project was still not near completion.

Another negative aspect of the systems development effort related to the airlines. The large airline companies did not know what equipment to order for their new planes. Depending on which system is eventually implemented, different sets of equipment would be required. Airline companies faced a dilemma. The equipment that had to be purchased with today's airplanes depended on the future success or failure of the FAA systems development project.

1. Describe the problems with this FAA systems development effort. If you were the FAA director, how would you prevent these problems from happening in the future?
2. What impact, if any, did the systems development project have on airline companies?

Sources: Jeff Cole, "How Major Overhaul of Air-Traffic Control Lost Its Momentum," *The Wall Street Journal,* March 2, 1998, p. A1; Edward Cone, "Crash Landing Ahead," *Information Week,* January 12, 1998, p. 38; Andy Chuter, "White Paper Aims for US Airspace Reform," *Flight International,* February 25, 1998, p. 26; Bruce Nordwall, "Air Traffic Control Outlook," *Aviation Week and Space Technology,* February 2, 1998, p. 43.

2 GATX Capital Corp.

GATX Capital Corp. is a diversified international financial services company that provides asset-based financing for transportation, industrial, and information technology equipment. GATX Capital arranges full-payout finance leases, secured loans, operating leases, and other structured financing both as an investing principal and alongside institutional investors. Earned income for 1997 was $613 million, with a net income of $55 million and $2.3 billion in assets.

GATX Capital is part of the GATX family of companies that provides services to help its customers transport, store, or distribute their products. GATX's assets include railcars and locomotives, bulk liquid terminals and pipelines, ships, warehouses, commercial aircraft, technology equipment, and other capital assets and related services worldwide. GATX also offers a variety of financial services focusing on owning or managing transportation and distribution assets.

Information technology represents a growing segment of GATX Capital's core business. Through the Technology Group, GATX provides large corporate customers with full life-cycle asset management and information services. Investments in information technology totaled $238 million by year-end 1996, representing 12 percent of the GATX Capital portfolio. The Technology Group meets customers' equipment and support needs through financing of client/server assets and development of network equipment and integration solutions.

Michael Cromar, the chief financial officer of GATX Corp., is a Certified Professional Accountant (CPA) by profession. He started in accounting management in 1977 at a company whose accounting systems were entirely manual. However, he learned early in his career that he needed to be a technology advocate. He wanted a better way to get things done and to make his job easier. Automation was the best way he could see to do that. So, Cromar became an unlikely champion of technology. He often fought the negative perceptions his superiors had of computers. For example, he persuaded his company to hire a computer systems person and to buy hardware and software. His persistence paid off and his career has advanced down two career paths—a CPA path and an information systems one.

At GATX, Cromar is leading a major SAP implementation effort. The project requires rewriting much of the SAP code to meet the needs of the leasing and asset industry. Upon completion of this effort, the SAP system should be a more attractive software solution for members of the industry.

1. What traits and experience does Michael Cromar possess that make him an excellent sponsor for his company's SAP effort?
2. What specific systems development actions would you expect Michael Cromar to take in sponsoring the SAP conversion project?

Sources: Adapted from the Corporate Information section of the GATX Capital Corporation Web site at http://www.gatxcap.com, accessed June 22, 1998; Randy Weston, "New Breed Manages Projects," *Computerworld*, April 13, 1998, pp. 37, 40; and "GATX Partnerships Order B737-800s and B757-200s Valued at $1.2 Billion Including Options," *PRNewswire*, July 31, 1997.

3 CompUSA

CompUSA is a large reseller of personal computers and related products and services. The company currently operates 160 stores in 73 major metropolitan areas across the United States, which serve retail, corporate, government, and education customers and include technical service departments and classroom training facilities. CompUSA also offers its own build-to-order personal computer series, the CompUSA PC trademark, and operates an Internet Web site located at http://www.compusa.com.

Customer service is a priority at CompUSA, and every CompUSA Superstore offers a full menu of services. CompUSA Technical Services include repairs, upgrades, custom configuration, and installation. In addition, the Superstores provide computer and software training for businesses and end users in over 100 standard applications and in advanced technologies such as Certified Novell Engineer and Microsoft Certified Systems Engineer. CompUSA has a dedicated, trained corporate sales team who work hand-in-hand with business customers, as well as governmental and educational institutions.

Honorio Padron was recently appointed Chief Information Officer (CIO) of CompUSA. His goal is to help the Dallas-based computer retailer rethink its business processes and to lay the technological groundwork to support one-stop shopping for customers of its retail, technical training, and other products and services. Padron took four months to complete a new systems architecture that will replace 80 percent of the retailer's existing systems. Key to this plan is the replacement of these systems with a multimillion dollar enterprise resource planning system. The effort is expected to take 18 months.

In addition, CompUSA has been experiencing a turnover rate of 50 percent in its 300-person information systems staff. Padron has worked with the human resources director to implement pay raises for most of his staff ranging from 2 to 20 percent.

1. What factors affecting systems development success will have the greatest impact on CompUSA's implementing its new systems architecture?

2. Which systems development life cycle is CompUSA likely to follow in implementing the new enterprise resource planning system? Why?

Sources: Adapted from "CompUSA to Relocate Cincinnati-Area Computer Superstore to the Tri-County Marketplace Shopping Center in Springdale," *PRNewswire*, May 29, 1998; Thomas Hoffman, "Business Savvy is CIO's Best Weapon," *Computerworld*, April 13, 1998, pp. 37, 40; and 1997 CompUSA Annual Report.

4   Walgreens Pharmacies' Successful Completion of Strategic Project

Walgreens is a leader in sales and profits in the drugstore industry. Sales were $13.4 billion during fiscal 1997, making Walgreens the nation's largest drugstore chain. Earnings have increased from $56 million in 1982 to $436 million in 1997. Walgreens is also a leader in the use of sophisticated technology. It was the first chain to link all its pharmacy computers, called Intercom, via satellite. Today, Intercom Plus, an updated version of the original system, is operating throughout the chain.

Handling prescriptions quickly and accurately is extremely important in Walgreens' business strategy. It foresees a significant upswing in prescriptions as baby boomers move into middle and old age. Intercom Plus provides more time for the pharmacist to counsel patients by automating many tasks. Other patient benefits include tax/insurance records and patient profiles. Where state law allows, Intercom Plus also permits "prescription transferability," so you can obtain refills at Walgreens stores other than the one where you first bought your medication, even if it's in another state. Intercom Plus has many enhanced customer service features, including automatic refill requests via touchtone phone and patient information provided with every prescription. Pharmacy patients are even able to request automatic notification from the pharmacy when a medication is due for a refill.

This new system revolutionizes the way prescriptions are filled and raises Walgreens' service and productivity to a new level. While providing increased patient access to its pharmacists, it also substantially raises the number of prescriptions each store can efficiently dispense. The company already fills more prescriptions than any other American retailer (more than 200 million in 1997).

It expects its satellite-based Intercom Plus system to help boost the average number of prescriptions handled daily per store from 230 now to more than 400 by the end of the decade. By then, Walgreens expects pharmacy sales to account for half of total company revenue. It also sees continuing strong corporate growth fueled by the intelligent use of innovation to better serve its customers.

Walgreens spent $150 million over five years developing the proprietary Intercom Plus software with Andersen Consulting. IS staff, consultants, and Walgreens workers spent months defining system requirements and designing how the system would work. The system's installation came at a critical time for Walgreens; two recent megamergers have transformed the competitive landscape. Rite Aid Corp. recently completed its merger with Thrifty PayLess Holdings to form a 3,500-store chain. JC Penney Co. has completed its $3.5 billion acquisition of Clearwater, Florida–based Eckerd Corp. and has begun melding Eckerd and its Thrift Drug chain into a 2,800-store combination. Walgreens operates over 2,400 drugstores in 34 states and Puerto Rico.

1. How important is the Intercom Plus system to Walgreens as it fights against growing competition?
2. What must Walgreens' IS and business managers do to keep the Intercom Plus system more useful to its pharmacists and customers than similar systems offered by its competitors?

Source: Adapted from "Walgreens Co. Reports Record Sales, Earnings for Second Quarter, First Half 1998," Company Press Release, March 30, 1998; Letter to Shareholders as published in Walgreens Annual Report, November 18, 1997; Thomas Hoffman, "Walgreens Seeks Network Rx," *Computerworld*, January 27, 1997, pp. 63–64; "Fleet-Footed Pharmacy," *Uplink*, Fall 1996, p. 7, published by Hughes Communications, Inc.

CHAPTER 13

Systems Design, Implementation, Maintenance, and Review

"If we can be four times more productive building applications by flip-flopping the 80/20 rule (80 percent of IS effort goes to maintenance, 20 percent to new development), the benefits will be enormous."

— Office Depot Chairman and CEO David Fuentes

Chapter Outline

Systems Design
Logical and Physical Design
Special Systems Design Considerations
Emergency Alternate Procedures and Disaster Recovery
Systems Controls
The Importance of Vendor Support
Generating Systems Design Alternatives
Evaluating and Selecting a System Design
Evaluation Techniques
Freezing Design Specifications
The Contract
The Design Report

Systems Implementation
Acquiring Hardware from an Information Systems Vendor
Acquiring Software: Make or Buy?
Externally Developed Software
In-House-Developed Software
Tools and Techniques for Software Development
Acquiring Database and Telecommunications Systems
User Preparation
IS Personnel: Hiring and Training
Site Preparation
Data Preparation
Installation
Testing
Start-Up
User Acceptance

Systems Maintenance
Reasons for Maintenance
Types of Maintenance
The Request for Maintenance Form
Performing Maintenance
The Financial Implications of Maintenance
The Relationship between Maintenance and Design

Systems Review
Types of Review Procedures
Factors to Consider During Systems Review
System Performance Measurement

Learning Objectives

After completing Chapter 13, you will be able to:

1. State the purpose of systems design and discuss the differences between logical and physical systems design.
2. Outline key steps taken during the design phase.
3. Define the term *RFP* and discuss how this document is used to drive the acquisition of hardware and software.
4. Describe the techniques used to make systems selection evaluations.
5. List the advantages and disadvantages of purchasing versus developing software.
6. Discuss the software development process and some of the tools used in this process.
7. State the purpose of systems implementation and discuss the various activities associated with this phase of systems development.
8. State the importance of systems and software maintenance and discuss the activities involved.
9. Describe the systems review process.

Empire District Electric Company
New Customer Information System Produces Flexibility in Changing Industry

Empire District Electric Company is an investor-owned electric utility company that provides electric service to a 10,000-square-mile area covering parts of southwest Missouri, southeast Kansas, northwest Arkansas, and northeast Oklahoma. It also provides water service to three incorporated Missouri communities. Annual revenue exceeds $200 million, with earnings around $24 million.

The deregulation of the utility industry is forcing utilities such as Empire to become more sophisticated marketers. The days of one bill with one rate for one service are gone. Utility companies must be able to bill customers for multiple products, from multiple vendors, and at multiple prices. As a result, Empire is developing a new customer information system called Centurion. This system will replace an old mainframe-based system written in COBOL and accessed via IBM 3270 terminals. Centurion is based on the client/server architecture. The server portion of Centurion runs on an inexpensive Windows NT server, with Windows 95 workstations for the client side. The server hardware for Centurion

will cost around $90,000, and Empire will spend around $125,000 for new client computers.

Centurion was built using the Java programming language and Java development tools. Java's modular structure and rapid debugging cycle enabled three programmers to write the application in one year. As a result, the development cost is about half what it might have cost using other development approaches. This estimate is based on an industry benchmark of $30 to $70 per customer to reengineer a utility's customer information system, or somewhere between $400,000 and $1 million based on Empire's 142,000 customers. Centurion's actual development cost was around $300,000.

Its developers did a good job of defining the data model to meet a wide range of customer requirements—the application is extremely flexible. In addition, the business process model used for the system is a basic retail business process model—a radical departure from the formerly regulated utility business but a close match to the new, unregulated environment. The

billing modules of the system include advanced features that enable it to calculate fractional billing rates including consumption charges, taxes, and other variable costs. Another plus is that the user interface is far simpler for customers than the company's existing COBOL mainframe system. The goal is to make the Centurion system as adaptable as possible so that small and mid-size utilities may consider purchasing it to update their billing systems.

As you read this chapter, consider the following questions:

- What are the various tasks associated with design and implementation of a new information system?

- What tools and techniques are available to assist in the design and implementation activities?

Sources: Adapted from "Empire Facts and Stats" in the "Information" section of the Empire Electric Web site at http://www.empiredistrict.com, accessed July 5, 1998 and Charles Waltner, "Java-Charged Power," *InformationWeek*, June 22, 1998, pp. 88–92.

The way an information system is designed, implemented, and maintained profoundly affects the daily functioning of an organization. As we saw with the Empire District Electric Co., information systems must be replaced as the needs of a business change. But, it is important that the new systems be flexible to meet evolving needs. Otherwise, companies repeat a costly cycle of building and replacing information systems. This chapter presents the basics of systems design, implementation, maintenance, and review. Both users and IS personnel need to be aware of these stages so that they can participate in good systems development in their organizations.

SYSTEMS DESIGN

systems design

a phase in the development of an IS system that answers the question "How will the information system do what it must do to obtain a solution to a problem?"

The purpose of **systems design** is to answer the question "How will the information system solve a problem?" The primary result of the systems design phase is a technical design that details system outputs, inputs, and user interfaces; specifies hardware, software, databases, telecommunications, personnel, and procedures; and shows how these components are related. The new system should overcome shortcomings of the existing system and help the organization achieve its goals. Of course, the system must also meet certain guidelines, including user and stakeholder requirements and the objectives defined during previous development phases.

Systems design is typically accomplished using the tools and techniques discussed in previous chapters. Depending on the specific application, these methods can be used to support and document all aspects of systems design. Two key aspects of system design are logical and physical design.

Logical and Physical Design

Information systems must be designed along two dimensions: logical and physical. The **logical systems design** refers to what the system will do. The **physical systems design** refers to how the tasks are accomplished, including how the components work together and what each component does.

logical systems design

the dimension of the design phase that describes the functional requirements of a system

physical systems design

the dimension of the design phase that specifies the characteristics of the system components necessary to put the logical design into action

Logical design. Logical design describes the functional requirements of a system. That is, it conceptualizes what the system will do to solve the problems identified through earlier analysis. Without this step, the technical aspects of the system (such as which hardware devices should be acquired) often obscure the solution. Logical design involves planning the purpose of each system element, independent of hardware and software considerations. The logical design specifications that are determined and documented include the following:

Output design. Output design describes all outputs from the system and includes the types, format, content, and frequency of outputs. For example, a requirement that all company invoices include the customers' original invoice number is a logical design specification. Screen and report layout tools can be used during output design to capture the output requirements for the system.

Input design. Once output design has been completed, input design can begin. Input design specifies the types, format, content, and frequency of input data. For example, the requirement that the system capture customers' phone numbers from their incoming calls and use that to automatically look up account information is a logical design specification. A variety of diagrams and screen and report layouts can be used to reveal the type, format, and content of input data.

Processing design. The types of calculations, comparisons, and general data manipulations required of the system are determined during processing design. For example, a payroll program will require gross and net pay computations, state and federal tax withholding, and various deductions and savings plans.

File and database design. Most information systems require files and database systems. The capabilities of these systems are specified during the logical design phase. For example, the ability to obtain instant updating of customer records is a logical design specification. In many cases, a database administrator is involved with this aspect of logical design. Data-flow and entity-relationship diagrams are typically used during file and database design.

Telecommunications design. During logical design, the network and telecommunications systems need to be specified. For example, a hotel might specify a client/server system with a certain number of workstations that are linked to a server. From these requirements, a hybrid topology might be chosen. Graphics programs and CASE tools can be used to facilitate logical network design.

Procedures design. All information systems require procedures to run applications and handle problems if they occur. These important policies are captured during procedures design. Once designed, procedures can be described by using text and word processing programs. For example, the steps to add a new customer account may involve a series of both manual and computerized tasks. Written procedures would be developed to provide an efficient process for all to follow.

Controls and security design. Another important part of logical design is to determine the required frequency and characteristics of backup systems. In general, everything should have a backup, including all hardware, software, data, personnel, supplies, and facilities. Planning how to avoid or recover from a computer-related disaster should also be considered in this stage of the logical design phase.

Personnel and job design. Some systems require additional employees; others may need modification of the tasks associated with one or more existing IS positions. The job titles and descriptions of these positions are specified during personnel and job design. Organization charts are useful during personnel design to diagram various positions and job titles. Word processing programs are also used to describe job duties and responsibilities.

Physical design. Physical design specifies the characteristics of the system components necessary to put the logical design into action. In this phase, the characteristics of each of the following components must be specified:

Hardware design. All computer equipment, including input, processing, and output devices, must be specified by performance characteristics. For example, if the logical design specified that the database must hold large amounts of historical data, then the system storage devices must have large capacity.

Software design. All software must be specified by capabilities. For example, if the ability for dozens of users to update the database concurrently is specified in the logical design, then the physical design must specify a database management system that allows this to occur. In some cases, software can be purchased; in other situations it will be developed internally. Logical design specifications for program outputs, data inputs, and processing requirements are also considered during the physical design of the software. For example, the ability to access data stored on certain disk files that the program will use is specified.

Database design. The type, structure, and function of the databases must be specified. The relationships between data elements established in the logical design must be mirrored in the physical design as well. These relationships include such things as access paths and file structure organization. Fortunately, many excellent database management systems exist to assist with this activity.

Telecommunications design. The necessary characteristics of the communications software, media, and devices must be specified. For example, if the logical design specifies that all members of a department must be able to share data and common software, then the local area network configuration and the communications software that are specified in the physical design must possess this capability.

Personnel design. This step involves specifying the background and experience of individuals most likely to meet the job descriptions specified in the logical design.

Procedures and control design. How each application is to run, as well as what is to be done to minimize the potential for crime and fraud, must be specified. These specifications include auditing, backup, and output distribution methods.

Special System Design Considerations

A number of special system characteristics should be considered during both logical and physical design. These characteristics include sign-on procedures, interactive processing, interactive dialogue, and error prevention and detection.

Procedures for signing on. System control methods are established during systems design. With almost any system, control problems may exist, such as criminal hackers breaking into the system or an employee mistakenly accessing confidential data. Sign-on procedures are the first line of defense against these problems. A **sign-on procedure** consists of identification numbers, passwords, and other safeguards needed for an individual to gain access to computer resources. A **system sign-on** allows the user to gain access to the computer system; an **application sign-on** permits the user to start and use a particular application, such as payroll or inventory control. These sign-on procedures help ensure that only authorized individuals can access a particular system or application.

The sign-on, also called a "logon," can identify, verify, and authorize access and usage (Figure 13.1). Identification means that the computer

sign-on procedure

the identification numbers, passwords, and other safeguards needed for an individual to gain access to computer resources

system sign-on

the procedure that allows the user to gain access to the computer system

application sign-on

the procedure that permits the user to start and use a particular application

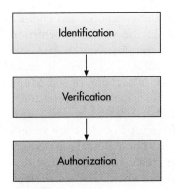

FIGURE 13.1

The Levels of the Sign-On Procedure

menu-driven system

a system that allows users to simply pick what they want to do from a list of alternatives

help facility

a tool that aids a user who is having difficulty understanding what is happening or what type of response is expected

lookup tables

tables within an application that simplify and shorten data entry

restart procedures

procedures that make it very simple for an individual to restart an application from where it left off

dialogue

the series of messages and prompts communicated between the system and the user

identifies the user as valid. If you must enter an identification number and password when logging on to a mainframe, you have gone through the identification process. For systems and applications that are more sensitive or secure, verification is used. Verification involves entering an additional code before access is given. Finally, authorization allows the user to gain access to restricted parts of a system or application. Consider a credit-checking application for a major credit card company. To grant credit, a clerk may be given only basic credit information on the screen. A credit manager, however, will have an authorization code to get additional credit information about a client.

Interactive processing. Today, most computer systems allow interactive processing. With this type of system, people directly interact with the processing component of the system through terminals or networked PCs. The system and the user respond to each other in a real-time mode, which means within a matter of seconds. Interactive real-time processing requires special design features for ease of use such as menu-driven systems, help commands, table lookup facilities, and restart procedures. With a **menu-driven system** (Figure 13.2), users simply pick what they want to do from a list of alternatives. Most people can easily operate these types of systems. They select their choice or respond to questions (or prompts) from the system, and the system does the rest.

Many designers incorporate a **help facility** into the system or applications program. When a user is having difficulty understanding what is happening or what type of response is expected, he or she can activate the help facility. The help screen relates directly to the problem the user is having with the software. The program responds with information on the program status, what possible commands or selections the user can give, and what is expected in terms of data entry.

Incorporating tables within an application is another very useful design technique. **Lookup tables** can be developed and used by computer programs to simplify and shorten data entry. For example, if you are entering a sales order for a company, you simply type in its abbreviation, such as ABCO. The program will then go to the customer table, normally stored on a disk, and look up all the information pertaining to the company abbreviated ABCO that is required to complete the sales order. This information is then displayed on a terminal screen for confirmation. The use of these tables can prevent wasting a tremendous amount of time entering the same data over and over again into the system.

If a problem occurs in the middle of an application—such as a temporary interruption of power or a printer running out of paper—the application currently being run is typically shut down. As a result, easy-to-use **restart procedures** are developed and incorporated into the design phase. With a restart procedure, it is very simple for an individual to restart an application where it left off.

Designing good interactive dialogue. **Dialogue** refers to the series of messages and prompts communicated between the system and the user. From a user's point of view, good interactive dialogue from the computer system is essential, making data entry faster, easier, and more accurate.

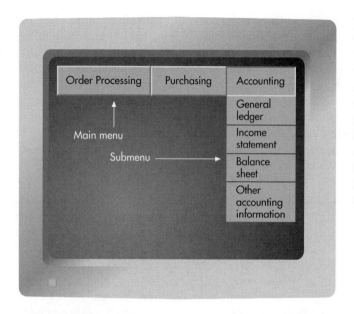

FIGURE 13.2

A menu-driven system allows you to choose what you want from a list of alternatives.

Poor dialogue from the computer can confuse the user and result in the wrong information being entered. If the computer system prompts the user to enter ACCOUNT, does it mean account number, type of account, or something else? Prototyping and an iterative approach are essential to good dialogue design, as users must have a great deal of input to ensure the system can be used properly. The following list covers some elements that create good interactive dialogue. These elements should be considered during systems design.

Clarity. The computer system should ask for information using language easily understood by users. Whenever possible, the users themselves should be involved in selecting the words and phrases used for dialogue with the computer system.

Response time. Ideally, response from the computer system should approximate a normal response time from a human being carrying on the same sort of dialogue.

Consistency. The system should use the same commands, phrases, words, and function keys for all applications. After a user learns one application, all others will then be easier to use.

Format. The system should use an attractive format and layout for all screens. The use of color, highlighting, and the position of information on the screen should be considered carefully and consistently.

Jargon. All dialogue should be written in easy-to-understand terms. Avoid jargon known only to IS specialists.

Respect. All dialogue should be developed professionally and with respect. Dialogue should not talk down to or insult the user. Avoid statements such as "You have made a fatal error."

Preventing, detecting, and correcting errors. The best and least expensive time to deal with potential errors is early in the design phase. During installation or after the system is operating, it is much more expensive and time-consuming to handle errors and related problems. Good systems design attempts to prevent errors before they occur, which involves recognizing what can happen and developing steps and procedures that can prevent, detect, and correct errors. This process includes developing a good backup system that can recover from an error. Table 13.1 lists the major causes of system errors.

Emergency Alternate Procedures and Disaster Recovery

As part of ongoing information systems assessment, organizations need to identify key information systems that control cash flow (e.g., invoicing, accounts receivable, payroll) and support other key business operations

TABLE 13.1

Major Causes of System Errors

Human	Natural	Technical
IS personnel	Wind	Hardware
Authorized users	Fire	Application software
Nonauthorized users	Earthquakes	Systems software
	Extreme temperatures	Communications
	Floods	Electricity
	Hurricanes	

(e.g., inventory control, shipping customer products). Organizations typically develop a set of emergency procedures and disaster recovery plans for these systems.

In the event that a key information system becomes unusable, the end users need a set of emergency alternate procedures to follow to meet the needs of the business. End users should work with IS personnel to develop these procedures. They may be manual procedures to work around the unavailable automated work process. For example, if the order entry transaction processing system is unavailable, users would resort to the use of special preprinted forms to capture basic order data for later entry when the system becomes available. In some cases, the emergency alternate procedures involve accessing a remote computer for computing resources.

Disaster planning is the process of anticipating and providing for disasters. A disaster can be an act of nature (a flood, fire, or earthquake), a human act, error, labor unrest, or erasure of an important file. Disaster planning focuses primarily on two issues: maintaining the integrity of corporate information and keeping the information system running until normal operations can be resumed.

One of the first steps of disaster planning is to identify potential threats or problems, such as natural disasters, employee errors, and poor internal control procedures. Disaster planning also involves disaster preparedness. IS managers should occasionally hold an unannounced "test disaster"—similar to a fire drill—to ensure that the disaster plan is effective.

Disaster recovery is defined as the implementation of the disaster plan. Although companies have known about the importance of disaster planning and recovery for decades, many do not adequately prepare. A few examples of recent computer disasters are provided in the "Ethical and Societal Issues" box to illustrate the wide range of potential problems for which organizations must prepare.

disaster recovery

the implementation of the disaster plan

Companies that rely on computer systems for their business may lose revenue if the system fails and they do not have a recovery plan.

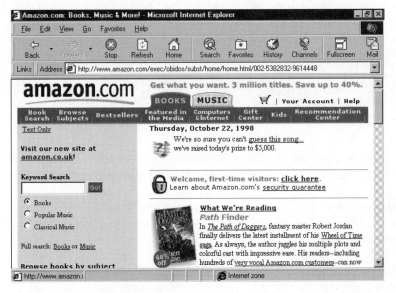

ETHICAL AND SOCIETAL ISSUES

Business Disruptions from Computer Problems

For three hours on June 24, 1998, three failed circuit breakers at the Sabre Group Holding Inc.'s main data center were blamed for an outage that left on-line consumers, 40,000 travel agencies, and more than 50 airlines—including carriers such as American Airlines and Hawaiian Airlines—without access to the passenger reservation system that manages customer booking and ticketing information.

Barnes & Noble's on-line site was busier than usual on June 5, 1997, possibly from the problems Amazon.com had on its site. Amazon.com. had been down most of the day. As of Friday evening, it was still not taking orders. The bookseller posted this message on its site: "We're working on the system. We expect to be back in a while. Please try us again later." Quite a few people checked out the competition evidently, as the BarnesandNoble.com customer service department handled heavier than usual traffic. Amazon.com, based in Seattle, said it took down the system early Friday for routine maintenance. But when engineers put the system back online, they discovered a bug. Rather than patching it up, they decided not to risk "the integrity of the system."

In June 1997, Web-related failures shut down stock market trading on the E★Trade Web site for about an hour. A week later, a software glitch in Charles Schwab & Co.'s electronic brokerage service left customers unable to get information about the status of trades or account balances. Then, a combination of computer error and operator carelessness caused a Network Solutions Domain Naming Service (DNS) server to send out corrupted Internet address information. Because Network Solutions is responsible for managing and assigning Internet addresses worldwide, the error had far-reaching consequences: Some sites remained inaccessible for hours. Worst of all, some Internet service providers (ISP) said at the time, DNS problems of this kind happen regularly, and Network Solutions still has no fail-safe mechanism to stop them.

Discussion Questions

1. What do you think was the impact of the outages mentioned above on the companies whose services were disrupted? Is there a way to quantify this impact?
2. How serious is the potential for widespread computer problems to cause a major disruption in our society?

Sources: Adapted from Connie Guglielmo, "Power Outage Downs Sabre Group's Reservation System," *Inter@ctive Week*, June 24, 1998; Renee Deger, "Amazon.com Site Hits Snag," *ZD Net News*, June 6, 1997; and Elisabeth Horwitt, "When Things Go Wrong," *Computerworld*, December 15, 1997.

The primary tools used in disaster planning and recovery are hardware, software and database, telecommunications, and personnel backup.

Hardware backup. It is common for a company to form an arrangement with its hardware vendor or a disaster recovery company to provide access to a compatible computer hardware system in the event of a loss of hardware. A duplicate, operational hardware system (or having immediate access to one through a specialized vendor) is an example of a **hot site**. A hot site is a compatible computer system that is operational and ready to

hot site

a duplicate computer system that is operational and ready for use

cold site

also called a *shell*, a computer environment that includes rooms, electrical service, telecommunications links, data storage devices, etc.

use. If the primary computer has problems, the hot site can be used immediately as a backup. Another approach is to use a **cold site**, also called a *shell*, which is a computer environment that includes rooms, electrical service, telecommunications links, data storage devices, and the like. If a problem with the primary computer occurs, backup computer hardware is brought into the cold site, and the complete system is made operational. For both hot and cold sites, telecommunications media and devices are used to provide fast and efficient transfer of processing jobs to the disaster facility.

There are a number of firms that offer disaster recovery services. For example, Business Recovery Management offers a Disaster Recovery Center in Pittsburgh that provides the facilities, computer systems, and other equipment needed to recover from unplanned business interruptions. In order to provide an effective, fully-equipped workplace in time of disaster or special need, the center uses technology to ensure that a business stays connected both locally and globally including: fully-equipped computer room facilities for local and remote processing; wiring and PBX facilities in place to support voice, data, facsimile, and video telecommunications requirements; complete office facilities and services including conference rooms, private offices, and workstation areas to accommodate more than 250 people; and operational and technical support to assist in recovery.[1] Guardian Computer Support is an international computer support company providing a variety of computer services including on-site hardware and software maintenance, contract staffing, disaster recovery services, and outsourcing services.[2] SunGard Recovery Services provides hot sites and cold sites that are available for extended recovery, strategically located around the country, and ready to respond quickly to multiple regional disasters. The company's consulting arm analyzes a disaster's potential impact on a business and identifies the resources needed for recovery. It also structures an effective disaster recovery plan, which is constantly monitored and revised, so a company will have the equipment, information, communications, records, supplies, work space, and personnel needed in the event of emergency. SunGard's expertise is available on disk, in the form of a Windows-based planning software, which can be use for disaster recovery planning.[3]

Software and database backup. Software and databases can be backed up by making duplicate copies of all programs, files, and data. At least two backup copies should be made. One backup copy can be kept in the information systems department in case of accidental destruction of the software; another backup copy should be kept off-site in a safe, secure, fireproof, and temperature and humidity-controlled environment. A number of service companies provide this type of backup environment.

Backup is also essential for the data and programs on users' desktop computers. The advent of more distributed systems, like client/server systems, means that many users now have important, and perhaps mission-critical, data and applications on their desktop computers. Utility packages inexpensively provide backup features for desktop computers by copying data onto magnetic disk or tape.

Software and database backup can be very difficult if an organization has a large amount of data. For some companies, making a backup of the entire database could take hours. A tight budget may also prohibit backing up significant quantities of data. As a result, some companies use **selective backup**, which involves creating backup copies of only certain

selective backup

backup copies of only certain files

files. For example, only critical files might be copied every night for backup purposes.

Another backup approach is to make a backup copy of all files changed during the last few days or the last week, a technique called **incremental backup**. This approach to backup uses an **image log**, which is a separate file that contains only changes to applications. Whenever an application is run, an image log is created that contains all changes made to all files. If a problem with a database occurs, an old database with the last full backup of the data, along with the image log, can be used to re-create the current database. Many disaster plans also include recovery of on-line hardware, software, and databases, which entails having another computer perform real-time backup of data at all times.

Telecommunications backup. Some disaster recovery plans call for the backup of vital telephone communications. Complex plans might call for recovering whole networks. In other plans, the most critical nodes on the network are backed up by duplicate components. Using such fault-tolerant networks, which will not break down when one node or part of the network malfunctions, can be a more cost-effective approach to telecommunications backup.

Personnel backup. Information systems personnel must also have backup. This can be accomplished in a number of ways. One of the best approaches is to provide cross-training for IS and other personnel so that each individual can perform an alternate job if required. For example, a company might train employees in accounting, finance, or other IS departments to operate the system if a disaster strikes. The company could also make an agreement with another information systems department or an outsourcing company to supply IS personnel if necessary.

Security, fraud, and the invasion of privacy. Security, fraud, and the invasion of privacy can present disastrous problems. For example, because of an inadequate security and control system, a futures and options trader for a British bank lost about $1 billion. A simple system might have prevented a problem that caused the 200-year-old bank to collapse. In addition, IRS employees were caught looking at the tax returns of celebrities and other people. Preventing and detecting these problems is an important part of systems design. Prevention includes the following:

• Determining potential problems
• Determining the importance of these problems
• Determining the best place and approach to prevent problems
• Determining the best way to handle problems if they occur

Every effort should be made to prevent problems, but companies must establish procedures to handle problems if they occur.

Systems Controls

Most IS departments establish tight **systems controls** to maintain data security. Systems controls can help prevent computer misuse, crime, and fraud by employees and others. Most IS departments have a set of general

---

incremental backup
backup copies of all files changed during the last few days or the last week

image log
a separate file that contains only changes to applications

systems controls
procedures to maintain data security

closed shops

IS department in which only authorized operators can run the computers

open shops

IS department in which other people, such as programmers and systems analysts, are also authorized to run the computers

deterrence controls

rules that prevent problems before they occur

operating rules that help protect the system. Some information systems departments are **closed shops**, in which only authorized operators can run the computers. Other IS departments are **open shops**, in which other people, such as programmers and systems analysts, are also authorized to run the computers. Other rules specify the conduct of the IS department.

These rules are examples of **deterrence controls**, which involve preventing problems before they occur. Making a computer more secure and less vulnerable to a break-in is another example. Good control techniques should help an organization contain and recover from problems. The objective of containment control is to minimize the impact of a problem while it is occurring, and recovery control involves responding to a problem that has already occurred.

Many types of system controls may be developed, documented, implemented, and reviewed. These controls touch all aspects of the organization, including the following:

Input controls. Input controls maintain input integrity and security. Some input controls involve the people who use the system; others relate to the data. The overall purpose is to reduce errors while protecting the computer system against improper or fraudulent input. Input controls range from using standardized input forms to eliminating data entry errors and using tight password and identification controls. For example, based on their logon identification, users are provided with access to a subset of the system and its capabilities. Some users can only view data, other users can update and view data. In addition, input controls can involve more sophisticated hardware and software that can use voice, fingerprints, and related techniques to identify and permit access to sensitive computer systems.

Processing controls. Processing controls deal with all aspects of processing and storage. In many cases, hardware and software are duplicated to provide procedures that ensure processing is as error-free as possible. In addition, storage controls prevent users from gaining access to or accidentally destroying data. The use of passwords and identification numbers, backup copies of data, and storage rooms that have tight security systems are examples of storage controls.

Output controls. Output controls are developed to ensure that output is handled correctly. In many cases, output generated from the computer system is recorded in a file that indicates the reports and documents generated, the time they were generated, and their final destinations.

Database controls. Database controls deal with ensuring an efficient and effective database system. These controls include the use of subschemas, identification numbers, and passwords, without which a user is denied access to certain data and information. Many of these controls are provided by database management systems.

Telecommunications controls. Telecommunications controls are designed to provide accurate and reliable data and information transfer among systems. Some telecommunications controls include hardware and software and other devices developed to ensure correct communications while eliminating the potential for fraud and crime. Examples are encryption devices and expert systems that can be used to protect a network from unauthorized access.

Many companies use ID badges to prevent unauthorized access to sensitive areas in the information systems facility. (Source: Sensormatic Electronics Corporation.)

Personnel controls. Various personnel controls can be developed and implemented to make sure only authorized personnel have access to certain systems to help prevent computer-related mistakes and crime. Personnel controls can involve the use of identification numbers and passwords that allow only certain people access to certain data and information. ID badges and other security devices (such as "smart cards") can prevent unauthorized people from entering strategic areas in the information systems facility.

Once controls are developed, they should be documented in various standards manuals that indicate how the controls are to be implemented. They should then be implemented and frequently reviewed. It is common practice to measure the extent to which control techniques are used and to take action if the controls have not been implemented.

The Importance of Vendor Support

Whether an individual is purchasing a personal computer or an experienced company is acquiring an expensive mainframe computer, the system could be obtained from one or more vendors. Some of the factors to consider in selecting a vendor are the following:

- The vendor's reliability and financial stability
- The type of service offered after the sale
- The goods and services the vendor offers and keeps in stock
- The vendor's willingness to demonstrate its products
- The vendor's ability to repair hardware
- The vendor's ability to modify its software
- The availability of vendor-offered training of IS personnel and system users
- Evaluations of the vendor by independent organizations

Read the "E-Commerce" box to learn how one software vendor is providing customer support via the Web.

Generating Systems Design Alternatives

When additional hardware and software are not required, alternative designs are often generated without input from vendors. If the new system is complex, the original development team may want to involve other personnel to generate alternative designs. If new hardware and software are to be acquired from an outside vendor, a formal request for proposal (RFP) should be made.

Request for Proposal (RFP)

a document that specifies in detail required resources such as hardware and software

The **request for proposal (RFP)** is one of the most important documents generated during systems development. It often results in a formal bid that is used to determine who gets a contract for new or modified systems. The RFP specifies in detail required resources such as hardware and software. In some cases, separate RFPs are developed for different needs. For example, a company might develop separate RFPs for hardware, software, and database systems. The RFP also communicates these needs to one or more vendors, and it provides a way to evaluate whether the vendor

E-COMMERCE

Geac Designs and Implements New Customer Support System

Geac Computer Corp. is an international software company, employing approximately 3,000 professionals in more than 70 offices in 18 countries. Its headquarters are in Markham, Ontario. Founded in 1971, Geac supplies software for library automation, hotel management, property management, construction, restaurant management, publishing, public safety, manufacturing and distribution, and banking. In addition, Geac also provides financial and human resource applications. Recent annual sales have been around $300 million.

In October 1997, the company launched AnswerLink, a Web-based, customer-support system that provides messaging, case handling, and database utilities designed specifically for intranet and Internet deployment. The project cost about $500,000 for the software, consulting, staff time, and server hardware. Geac estimates that it will save about $1 million a year on reduced 800-number costs alone.

Geac designed the system so that customers can access AnswerLink via the Web by pointing their browser to the URL of the AnswerLink Web page. The design includes a requirement for them to enter their AnswerLink user ID and password supplied by Geac. Once successfully logged on to the application, customers can send e-mail to the support staff to automatically open a case. The system routes the customer message based on product, customer, or customer class to send the e-mail to the proper support personnel. A key design feature of the system is that an automatic response is generated immediately, citing the case number and support rep to whom the case has been assigned. If the support rep is busy and the case is not picked up within the time allowed, the system alerts a support manager. Support staff can author new solutions for new problems—though more often than not, they reply with Web-page citations or pages they get themselves from the Web site. The customer can search solutions archived at the site in a variety of ways. Customers can also pick up e-mail, browse posted solutions, or follow the progress of their case.

DISCUSSION QUESTIONS

1. Imagine that you were the project manager in charge of the design and implementation of the customer support system for Geac. How would you ensure that the system provides a good interactive dialogue for the customer?
2. How important are disaster recovery considerations for this type of system? What safeguards and measures would you put in place?

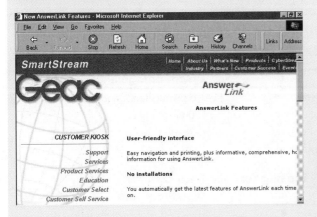

Sources: Adapted from the "About Geac" section of the Geac Web site at http://www.geac.com, accessed July 4, 1998; "AnswerLink Features" at http://www.smartstream.geac.com, accessed September 22, 1998; and Peter Cassidy, "At Your eService," *InformationWeek*, June 1, 1998, pp. 91–98.

---

has delivered what was expected. In some cases, the RFP is made part of the vendor contract. The Table of Contents for a typical RFP is shown in Figure 13.3.

Financial options. When it comes to acquiring computer systems, three choices are available: purchase, lease, or rent. Cost objectives and constraints set for the system play a significant role in the alternative chosen, as do the advantages and disadvantages of each. Table 13.2 summarizes the advantages and disadvantages of these financial options.

Determining which option is best for a particular company in a given situation can be difficult. Financial considerations, tax laws, the organization's

Johnson & Florin, Inc.
Systems Investigation Report

Table of Contents

COVER PAGE (with company name and contact person)
BRIEF DESCRIPTION of the COMPANY
OVERVIEW of the EXISITING COMPUTER SYSTEM
SUMMARY of COMPUTER-RELATED NEEDS and/or PROBLEMS
OBJECTIVES of the PROJECT
DESCRIPTION of WHAT IS NEEDED
HARDWARE REQUIREMENTS
PERSONNEL REQUIREMENTS
COMMUNICATIONS REQUIREMENTS
PROCEDURES to BE DEVELOPED
TRAINING REQUIREMENTS
MAINTENANCE REQUIREMENTS
EVALUATION PROCEDURES (how venders will be judged)
PROPOSAL FORMAT (how venders should respond)
IMPORTANT DATES (when tasks are to be completed)
SUMMARY

FIGURE 13.3

A Typical Table of Contents for
a Request for Proposal

policies, its sales and transaction growth, marketplace dynamics, and the organization's financial resources are all important factors. In some cases, lease or rental fees can amount to more than the original purchase price after a few years. As a result, most companies prefer to purchase their equipment.

On the other hand, constant advances in technology can make purchasing risky. A company would not want to purchase a new multimillion-dollar computer only to have newer and more powerful computers available a few months later at a lower price. Some companies employ several people to determine the best option based on all the factors. This staff can also help negotiate purchase, lease, or rental contracts.

Evaluating and Selecting a System Design

The final step in systems design is to evaluate the various alternatives and select the one that will offer the best solution for organizational goals. Performance, cost, control, and complexity objectives must be considered and balanced during design evaluation. Depending on their weight, any one of these objectives may result in the selection of one design over another. For example, financial

TABLE 13.2

Advantages and Disadvantages
of Acquisition Options

Renting (Short-Term Option)	
Advantages	**Disadvantages**
No risk of obsolescence	No ownership of equipment
No long-term financial investment	High monthly costs
No initial investment of funds	Restrictive rental agreements
Maintenance usually included	

Leasing (Longer-Term Option)	
Advantages	**Disadvantages**
No risk of obsolescence	High cost of canceling lease
No long-term financial investment	Longer time commitment than renting
No initial investment of funds	No ownership of equipment
Less expensive than renting	

Purchasing	
Advantages	**Disadvantages**
Total control over equipment	High initial investment
Can sell equipment at any time	Additional cost of maintenance
Can depreciate equipment	Possibility of obsolescence
Low cost if owned for a number of years	Other expenses, including taxes and insurance

concerns might make a company choose rental over equipment purchase. Specific performance objectives—say, that the new system performs on-line data processing—may result in a complex network design for which control procedures must be established. Evaluating and selecting the best design involves a balance of system objectives that will best support organizational goals. Normally, evaluation and selection involves both a preliminary and a final evaluation before a design is selected.

preliminary evaluation

evaluation that begins after all proposals have been submitted for the purpose of dismissing unwanted proposals

The preliminary evaluation. A **preliminary evaluation** begins after all proposals have been submitted. The purpose of this evaluation is to dismiss unwanted proposals. Several vendors can usually be eliminated by investigating their proposals and comparing them with the original criteria. Those that compare favorably are asked to make a formal presentation to the analysis team. The vendors should also be asked to supply a list of companies that use their equipment for a similar purpose. The organization then contacts these references and asks them to evaluate their hardware, software, and the vendor.

final evaluation

evaluation that involves detailed investigation of the proposals offered by the remaining vendors

The final evaluation. The **final evaluation** begins with a detailed investigation of the proposals offered by the remaining vendors. The vendors should be asked to make a final presentation and to fully demonstrate the system. The demonstration should be as close to actual operating conditions as possible. Such applications as payroll, inventory control, and billing should be conducted using a large amount of test data.

After the final presentations and demonstrations have been given, the organization makes the final evaluation and selection. Cost comparisons, hardware performance, delivery dates, price, flexibility, backup facilities, available software training, and maintenance factors are considered. Although it is good to compare computer speeds, storage capacities, and other similar characteristics, it is also necessary to carefully analyze whether the characteristics of the proposed systems meet the company's objectives. In most cases, the RFP captures these objectives and goals. Figure 13.4 illustrates the evaluation process.

FIGURE 13.4

The Stages in Preliminary and Final Evaluations

Note that the number of possible alternatives decreases as the firm gets closer to making a final decision.

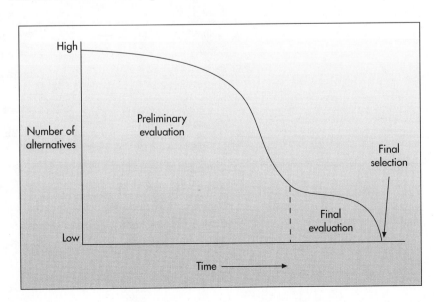

Evaluation Techniques

Although the exact procedure used to make the final evaluation and selection varies from one organization to the next, four approaches are commonly considered: group consensus, cost/benefit analysis, benchmark tests, and point evaluation.

group consensus

an approach to evaluation in which a decision-making group is appointed and given the responsibility of making the final evaluation and selection

Group consensus. In **group consensus**, a decision-making group is appointed and given the responsibility of making the final evaluation and selection. Usually, this group includes the members of the development team who participated in either systems analysis or systems design. This approach might be used to evaluate which of several screen layouts or report formats is best.

cost/benefit analysis

an approach that lists the costs and benefits of each proposed system

Cost/benefit analysis. **Cost/benefit analysis** is an approach that lists the costs and benefits of each proposed system. Once expressed in monetary terms, all the costs are compared with all the benefits. Table 13.3 lists some of the typical costs and benefits associated with the evaluation and selection procedure. This approach is used to evaluate options whose costs can be quantified, such as which hardware or software vendor to select.

TABLE 13.3

Cost/Benefit Analysis Table

Costs	Benefits
Development costs	**Reduced costs**
Personnel	Fewer personnel
Computer resources	Reduced manufacturing costs
	Reduced inventory costs
	More efficient use of equipment
	Faster response time
	Reduced down- or crash time
	Less spoilage
Fixed costs	**Increased revenues**
Computer equipment	New products and services
Software	New customers
One-time license fees for software and maintenance	More business from existing customers
	Higher price as a result of better products and services
Operating costs	**Intangible benefits**
Equipment lease and/or rental fees	Better public image for the organization
Computer personnel (including salaries, benefits, etc.)	Higher employee morale
Electric and other utilities	Better service for new and existing customers
Computer paper, tape, and disks	The ability to recruit better employees
Other computer supplies	Position as a leader in the industry
Maintenance costs	System easier for programmers and users
Insurance	

benchmark test

an examination that compares computer systems operating under the same conditions

point evaluation system

an evaluation process in which each evaluation factor is assigned a weight, in percentage points, based on importance; the system with the greatest total score is selected

FIGURE 13.5

An Illustration of the Point Evaluation System

In this example, software has been given the most weight (40 percent), compared with hardware (35 percent) and vendor support (25 percent). When system A is evaluated, the total of the three factors amounts to 82.5 percent. Systems B's rating, on the other hand, totals 86.7 percent, which is closer to 100 percent. Therefore, the firm chooses System B.

Benchmark tests. A **benchmark test** is an examination that compares computer systems operating under the same conditions. Although most computer companies publish their own benchmark tests, the best approach is for an organization to develop its own tests, then use them to compare the equipment it is considering. Several independent companies also rate computer systems. Computerworld, Datamation, PC Week, and DataPro, for example, not only summarize various systems but also evaluate and compare computer systems and manufacturers according to a number of criteria. This approach might be used to compare the end user system response time on two similar systems.

Point evaluation. One of the disadvantages of cost/benefit analysis is the difficulty of determining the monetary values for all the benefits. An approach that does not employ monetary values is a **point evaluation system**. Each evaluation factor is assigned a weight, in percentage points, based on importance. Then each proposed information system is evaluated in terms of this factor and given a score that might range from 0 to 100, where 0 means the alternative does not address the feature at all and 100 means the alternative addresses that feature perfectly. The scores are totaled, and the system with the greatest total score is selected.[4] When using point evaluation, literally hundreds of factors can be listed and evaluated. Figure 13.5 shows a simplified version of this process. This approach is used when there are many factors on which options are to be evaluated, such as which software best matches a particular business's needs.

Because many elements must be considered before making a final selection, point evaluation can include a large number of factors. Performance concerns might include speed, storage capacity, and processing capabilities. Costs might include the deposit required on contract signing, payment schedules, lease and rental arrangements, maintenance costs, and availability of leasing companies. Complexity factors could include compatibility and ease of use, while control might include considerations such as training and maintenance offered by vendors, as well as system reliability and backup. When all these factors are added to the point evaluation system, a very large grid can result. The rows of the grid list the various factors important to the client company, and the columns of the grid represent the various vendors that responded to the request for proposal. Even if weights are not used, this type of chart can be very helpful. Some companies just use check marks to indicate which vendors have satisfied certain factors.

		System A			System B		
Factor's importance		Evaluation		Weighted evaluation	Evaluation		Weighted evaluation
Hardware	35%	95	35%	33.25	75	35%	26.25
Software	40%	70	40%	28.00	95	40%	38.00
Vendor support	25%	85	25%	21.25	90	25%	22.50
Totals	100%			82.5			86.75

Freezing Design Specifications

Near the end of the design stage, an organization prohibits further changes in the design of the system. The design specifications are then said to be frozen. Freezing systems design specifications means that the user agrees in writing that the design is acceptable. Most system consulting companies insist on this step to avoid cost overruns and missed user expectations.

A good analogy is the construction industry. When building a home, it is much less expensive and time-consuming to make changes at the blueprint stage than during construction. Likewise, it is less expensive and faster to make changes during systems design than during systems implementation. Changes made during systems implementation can send the entire project over budget and past the project completion date.

For organizations, freezing systems design specifications is one of the biggest problems with formal systems development. It assumes that users are able to correctly state their requirements to analysts and that these requirements do not change during implementation. However, formal systems development can take more than a year. During that time, users can experience changing business conditions and may need to change their system requirements. If the design has been locked in, these changes cannot occur. Because companies want to build the most effective systems possible, they often spend the extra money to make necessary changes during implementation. Thus, development projects are often not frozen in the real world.

A problem that often arises during the implementation of any major project is that of "scope creep." As users more clearly understand how the system will work and what their needs are, they begin to request changes to the original design. Each change may be relatively minor, so the project team is strongly tempted to expand the scope of the project and incorporate the requested changes. However, the aggregate impact of many small changes can be significant; implementation of all minor changes can delay the project and/or increase the cost significantly. A common example is the request for a new report that was not included in the original design. If not managed carefully, the implementation of the first new report will lead to the request for "just one more report" and then another and another.

Prior to implementation, experienced project managers place formal controls on the project scope. A key component of the process is to assess the cost and schedule impact of each requested change, no matter how small, and to decide whether to include the change. Often the users and the project team decide to hold all changes until the original effort is completed and then prioritize the entire set of requested changes.

The Contract

One of the most important steps in systems implementation is to develop a good contract if new computer facilities are being acquired. Most computer vendors provide standard contracts; however, such contracts are designed to protect the vendor, not necessarily the organization buying the computer equipment.

More and more organizations are using outside consultants and legal firms to help them develop their own contracts. Such contracts stipulate exactly what they expect from the system vendor and what interaction will occur between the vendor and the organization. All equipment specifications, software, training, installation, maintenance, and so on are clearly stated. Furthermore, the contract stipulates deadlines for the various stages or milestones of installation and implementation, as well as actions the vendor will take in case of delays or problems. Some organizations include penalty clauses in the contract, in case the vendor is unable to meet its obligation by the specified date. Typically, the request for proposal becomes part of the contract. This saves a considerable amount of time in developing the contract, because the RFP specifies in detail what is expected from the vendors.

The Design Report

design report

a report that reflects the decisions made for system design and prepares the way for systems implementation

System specifications are the final results of systems design. These include a technical description that details system outputs, inputs, and user interfaces, as well as all hardware, software, databases, telecommunications, personnel, and procedure components and the way these components are related. The specifications are contained in a **design report**, which is the primary result of systems design. The design report reflects the decisions made for system design and prepares the way for systems implementation. The contents of the design report are summarized in Figure 13.6.

When developing a system, it is important to understand and thoroughly complete the systems development activities covered in this chapter. These phases provide the blueprints and groundwork for the rest of systems development. The activities of the next phases will be easier, faster, and more accurate and will result in a more efficient, effective system if the design is complete and well thought out.

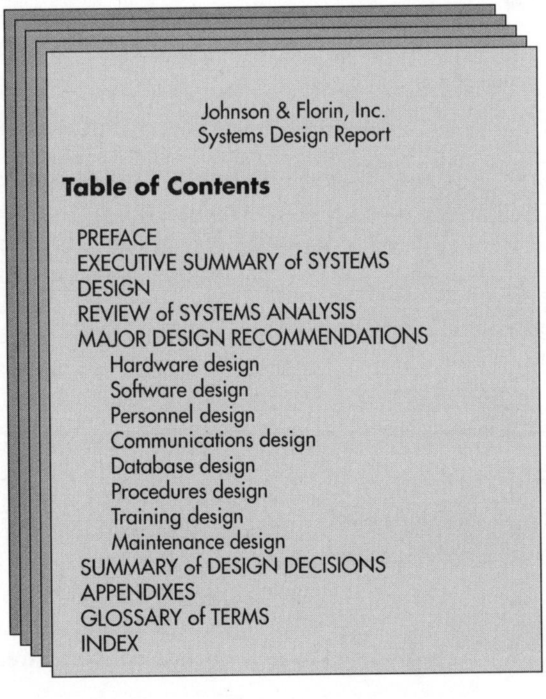

FIGURE 13.6

A Typical Table of Contents for a Systems Design Report

Johnson & Florin, Inc.
Systems Design Report

Table of Contents

PREFACE
EXECUTIVE SUMMARY of SYSTEMS DESIGN
REVIEW of SYSTEMS ANALYSIS
MAJOR DESIGN RECOMMENDATIONS
    Hardware design
    Software design
    Personnel design
    Communications design
    Database design
    Procedures design
    Training design
    Maintenance design
SUMMARY of DESIGN DECISIONS
APPENDIXES
GLOSSARY of TERMS
INDEX

SYSTEMS IMPLEMENTATION

systems implementation

the phase that includes hardware acquisition, software acquisition or development, user preparation, hiring and training of personnel, site and data preparation, installation, testing, start-up, and user acceptance

make-or-buy decision

the decision regarding whether to obtain the necessary software from internal or external sources

FIGURE 13.7

Typical Steps in Systems Implementation

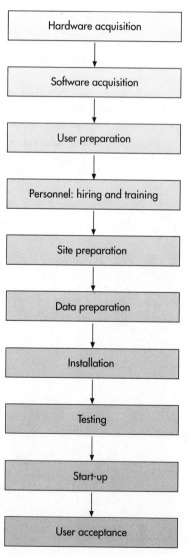

After the information system has been designed, a number of tasks must be completed before the system is installed and ready to operate. This process, called **systems implementation**, includes hardware acquisition, software acquisition or development, user preparation, hiring and training of personnel, site and data preparation, installation, testing, start-up, and user acceptance. The typical sequence of these activities is shown in Figure 13.7.

Choices and trade-offs are made at each step of systems implementation, which involve analyzing the benefits in terms of performance, cost, control, and complexity. Unfortunately, many organizations do not take full advantage of these steps or carefully analyze the trade-offs and hence never realize the full potential of new or modified systems. The hassles and carelessness must be avoided if organizations are to maximize their return on their information systems investment.

Acquiring Hardware from an Information Systems Vendor

To obtain the components for an information system, organizations can purchase, lease, or rent computer hardware and other resources. During implementation, organizations must identify and select one or more information systems vendors. An information systems vendor is a company that offers hardware, software, telecommunications systems, databases, information systems personnel, and/or other computer-related resources. Types of information systems vendors include general computer manufacturers (e.g., IBM, Compaq, and Hewlett-Packard), small computer manufacturers (e.g., Compaq, Dell, and Gateway), peripheral equipment manufacturers (e.g., Epson, Canon, and Piiceon), computer dealers and distributors (e.g., Canadian Communication Products Inc., Compaq, Dell, and Gateway) and leasing companies (e.g., National Computer Leasing, ECONOCOM-US, and Paramount Computer Rentals, plc).

Acquiring Software: Make or Buy?

As with hardware, software can be acquired several ways. As previously mentioned, it can be purchased from external developers or developed in-house. This decision is often called the **make-or-buy decision**. In some cases, companies use a blend of external and internal software development. That is, off-the-shelf or proprietary software programs are modified or customized by in-house personnel. The advantages and disadvantages of these approaches were discussed in Chapter 4.

Externally Developed Software

Some of the reasons a company might purchase or lease externally developed software include lower costs, less risk regarding the features and performance of the package, and ease of installation. The cost of the software package is known and there is little doubt that it will meet the company's accounting needs. The amount of development effort is also less when software is purchased, compared with in-house development.

Computer dealers such as CompUSA manufacture build-to-order computer systems and sell computers and supplies from other vendors.
(Source: Courtesy of CompUSA, Inc.)

For example, a company may decide to purchase a general ledger software package developed by a major international consulting firm such as Ernst and Young that is used widely throughout their industry. If the company were to decide to build the software, it could take many months (or even years) and there is a high degree of risk that when the system is implemented it may not meet the business needs as well as a purchased software package.

Should a company choose off-the-shelf or contract software in its new systems, it must take the following steps:

Rather than develop its own catering reservation software system, Sheraton hotels contracted with Newmarket Software Systems. DelphiPower BEO is a software package that tracks every facet of function planning—from video equipment availability to menus to automated banquet checks.
(Source: Courtesy of Newmarket Software Systems.)

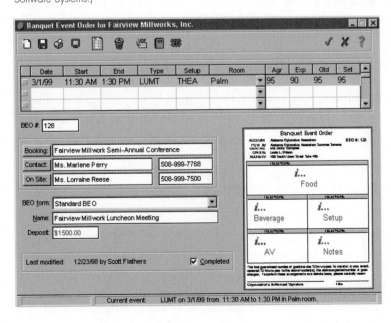

Review needs and requirements. It is important to analyze the program's ability to satisfy user and organizational needs.

Acquire software. Many of the approaches discussed in previous sections, including the development of requests for proposals, performing financial analysis, and negotiating the software contract, should be undertaken.

Modify or customize software. Externally developed software seldom does everything the organization requires. Thus, it is likely that externally developed software will have to be modified to satisfy user and organizational needs. Some software vendors will assist with the modification, but others may not allow their software to be modified at all.

software interface

the programs or program modifications that allow proprietary software to work with other software used in the organization

Acquire software interfaces. Usually, proprietary software requires a **software interface**, which consists of programs or program modifications that allow proprietary software to work with other software used in the organization. For example, if an organization purchases a proprietary inventory software package, software interfaces must allow the new software to work with other programs, such as sales ordering and billing programs.

Test and accept the software. Externally developed software should be completely tested by users in the environment in which it is to run before it is accepted.

Maintain the software and make necessary modifications. With many software applications, changes will likely have to be made over time. This aspect should be considered in advance because, as mentioned before, some software vendors do not allow their software to be modified.

In-House-Developed Software

Another option is to make or develop software internally. This requires the company's IS personnel to be responsible for all aspects of software development. Some advantages inherent with in-house-developed software include meeting user and organizational requirements and having more features and increased flexibility in terms of customization and changes. Software programs developed within a company also have greater potential for providing a competitive advantage because they are not easily duplicated by competitors in the short term.

chief programmer team

a group of skilled IS professionals with the task of designing and implementing a set of programs and has total responsibility for building the best software possible

Chief programmer teams. On software programming projects, the emphasis is on results—the finished package of computer programs. To get a smooth and efficient set of programs operating, the programming team must strive for the same overall objective. The **chief programmer team** is a group of skilled IS professionals with the task of designing and implementing a set of programs. This team has total responsibility for building the best software possible. Although the makeup of the chief programmer team varies with the size and complexity of the computer programs to be developed, a number of functions are common for all teams. A typical team has a chief programmer, a backup programmer, one or more other programmers, a librarian, and one or more clerks or secretaries (Figure 13.8).

The programming life cycle. Developing in-house software requires a substantial amount of detailed planning. A series of steps and planned

FIGURE 13.8

A Hierarchy Chart Showing the Typical Structure of a Chief Programmer Team

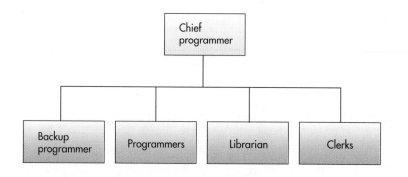

activities can maximize the likelihood of developing good software. These phases make up the **programming life cycle**, as illustrated in Figure 13.9 and described next.

Investigation, analysis, and design activities have already been completed. So the programmer already has a detailed set of documents that describe what the system should do and how it should operate. An experienced programmer will begin with a thorough review of these documents before any code is written.

Language selection involves determining the best programming language for the application. Important characteristics to be considered are (1) the difficulty of the problem, (2) the type of processing to be used (batch or on-line), (3) the ease with which the program can be changed later, and (4) the type of problem, such as business or scientific. Often, a trade-off will need to be made between the ease of use of a language and the efficiency with which programs execute. Older, hard-to-write machine and assembly language programs are more efficient than easier-to-use, high-level language programs.

Program coding is the process of writing instructions in the language selected to solve the problem. Like a contractor building a house, the computer programmer follows the plans and documents developed in the previous steps. This ensures that the software actually accomplishes the desired result.

Testing and debugging are vital steps in developing computer programs. In general, testing is the process of making sure the program performs as intended; debugging is the process of locating and eliminating errors.

Documentation is the next step. This can include technical and user documentation. **Technical documentation** is used by computer operators

programming life cycle

a series of steps and planned activities developed to maximize the likelihood of developing good software

technical documentation

documentation used by computer operators to execute the program and by analysts and programmers in case there are problems with the program or if the program needs modification

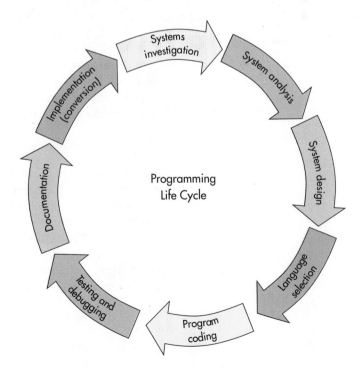

FIGURE 13.9

Steps in the Programming Life Cycle

user documentation

documentation developed for the individuals who use the program. This type of documentation shows users, in easy-to-understand terms, how the program can and should be used.

to execute the program and by analysts and programmers in case there are problems with the program or if the program needs modification. In technical documentation, each line of computer code is written out and explained with an English statement describing the function of the statement. Each variable is also described. **User documentation** is developed for the individuals who use the program. This type of documentation shows users, in easy-to-understand terms, how the program can and should be used. Incorporating a description of the benefits of the new application into user documentation may help stakeholders understand the reasons for the program and speed user acceptance. The software vendor often provides such documentation or it may be obtained from a technical publishing firm. For example, Microsoft provides technical documentation for its spreadsheet package Excel, but there are literally hundreds of books and manuals for this software from other sources.

Implementation, or *conversion,* is the last step in developing new computer software. It involves installing the software and making it operational. Several approaches are discussed later in the chapter when we discuss installation.

The same basic steps can be followed for programming in both fourth-generation languages (4GLs) and traditional high-level programming languages. 4GLs, however, may be easier and faster to use. They are appropriate for iterative and prototyping development techniques, because prototypes can be developed quickly. The ease of coding with fourth-generation languages also allows more emphasis on creating programs to meet user and organizational needs. Read the "FYI" box to learn about the importance of high-quality software development.

object-oriented software development

development of a collection of existing modules of code or objects that can be used across any number of applications without being rewritten

Companies can also use **object-oriented software development** to develop programs. With this approach, a collection of existing modules of code, or objects, can be used across a number of applications. In most cases, minimal coding changes will be required to mesh with the use of pre-developed objects or modules of code. While object-oriented software development does not require using object-oriented languages, most developers use them for the structure and ease they provide. Read the "Technology Advantage" box to see how one company developed an effective application using object-oriented software development.

Tools and Techniques for Software Development

If software will be developed in-house, the chief programmer team can use a number of tools, techniques, and approaches. Options include structured design, structured programming, CASE tools, cross-platform development, integrated development environments, and structured walkthroughs. Table 13.4 on page 595 compares the usage of these various techniques.

Structured design. Structured design is one well-known approach to designing and developing application software. The overall objective of

FYI

The Cost of Software Defects

Poorly written software code causes much rework for the system development staff, keeping them mired in maintenance duties and exacerbating the shortage of system development resources. But, the impact of software errors goes far beyond the cost of rework for computer programmers. Software errors can lead to a loss of revenue, missed market opportunities, and incorrectly processed information—all reducing the profits of a business. For example, faulty baggage-handling software caused baggage to be literally chewed up and clothing and other personal belongings to be lost or destroyed. As a result, the opening of the Denver International Airport was delayed by 16 months at a cost of $1 million per day. Then there's the cost of customer dissatisfaction, intangible but real, to be considered. How many customers have been lost because an information system caused an order to be filled incorrectly, failed to credit payment to an account, or sent an incorrect bill? And consider the cost of maintaining a help desk. Commercial software vendors field over 200 million calls for technical support each year at an average cost of around $23 for a total in excess of $4.6 billion.

The problem is that U.S. companies are such successful software innovators that they do not focus on good "process discipline," which includes program testing and project management. That can lead to code that does not always work as advertised and runaway projects that require additional staff to manage. Organizations need to put standard system development processes in place and measure the processes to identify projects that are running into potential problems. However, strengthening software development practices is a long-term solution that will take time and money up front, and not everyone is willing to make that investment.

Many companies have had their development practices rated by the Software Engineering Institute at Carnegie Mellon University in Pittsburgh. A rating of 1 to 5 shows companies the strength of their software development organization. The institute looks at whether development shops have a defined process, integrated training, coordination among different groups, quality assurance

plans, and ways to track their progress, among other variables. Here are specific steps to improve software quality:

- Set up a measurement system to see how you're doing now. Pick a few projects and determine time, effort, errors, and time to fix.
- Document your current development process.
- Have the IS organization focus on development of software for strategic projects versus less important efforts.
- Focus on eliminating errors and omissions in definition of system requirements; the sooner in the system development process errors are caught, the easier and less expensive they are to fix.
- Have one programmer review another programmer's code for errors and ease of maintenance.
- Test individual program modules before whole programs are completed.
- Keep working at process improvement.

DISCUSSION QUESTIONS

1. You have been hired as a consultant to review the software practices of a software manufacturer who produces financial applications. How would you begin to evaluate the firm's current software development practices?

2. Some organizations follow a software development approach where programmers begin coding while the systems analysts talk to the users to find out what they need. What is wrong with this approach?

Sources: Adapted from Tim Ouellete, "Poor Coding Hampers Developer Productivity," *Computerworld*, February 2, 1998, pp. 37, 40; Miryam Williamson, "Quality Pays," Special Report in *Computerworld*, August 18, 1997, pp. 78–81; and Steve Devinney and John W. Horch, "Sound Off: Should We Give Up on Quality Vendor-Supplied Software?" *Computerworld*, August 18, 1997, pp. 82–84.

TECHNOLOGY ADVANTAGE
Travelers Designs Object-Oriented System to Improve Productivity

To boost productivity in handling workers' compensation claims, Travelers Property Casualty has designed numerous object-oriented software applications and linked them to more than 100 database servers distributed throughout its offices. In the past four years, the company has designed and implemented new applications for handling claims for workers' compensation. And recently, the company followed an object-oriented design approach to tie workers' compensation claims to medical management services. The goal is to get injured workers back to work faster and at less cost. The combined system, called TravComp 2000, links the in-house developed claims application with packaged medical case management software. This system uses technology to set Travelers apart from rival insurers.

Focused on helping people get well and back to work as quickly as possible, TravComp 2000 brings workers' compensation claim and medical management specialists together in 52 field offices throughout the country. The system has met its design goal to reduce workers' compensation costs significantly for customers. Employers and employees benefit from a faster, more effective system. The appropriate level of medical expertise is applied in each case, resulting in a healthier work force and returning employees to work as quickly as possible.

When a claim is reported via a Travelers' 800 number, the information is sent instantaneously to a local Travelers' team of medical and claim professionals. Supported by the state-of-the-art software system, they obtain expert information on appropriate medical care procedures and return-to-work time frames. Based on this information, the medical and claim management team gets the right professionals involved early in the case. Case management nurses direct all aspects of the return-to-work process, from the medical diagnosis and treatment plan through arranging for appropriate services. Along with improving medical care for injured workers,

the program provides a new opportunity to reduce costs and speed the delivery of services to customers and their employees.

Travelers spent more than $26 million to design and implement the C++ applications, acquire hardware, and set up the databases that make up TravComp 2000. System benefits include a 15 percent reduction in the cost of handling insurance claims, largely because of reduced personnel needs. The system has been so successful that it is being reapplied with some minor modifications to the personal and commercial property lines of insurance. As a result, the systems effort in these areas is proceeding rapidly and with fewer initial problems.

Travelers took a very aggressive approach to the implementation of the system. It began using the workers' compensation software early—when it was still being tested. As a result, the start-up was a little rocky; one out of eight end-user PCs crashed daily. In addition to completing the roll out of the object-based applications to its 6,500 claims handlers, Travelers is simultaneously moving the software from IBM's OS/2 platform to Windows NT and switching from Sybase databases to Microsoft's SQL Server.

DISCUSSION QUESTIONS

1. Was the use of object-oriented programming appropriate for this application? Why or why not?
2. Why is Travelers in such a hurry to introduce and complete the roll out of the TravComp 2000 system? Do you agree with this rush even though it is likely to cause some start-up problems?

Sources: Adapted from Craig Stedman, "Insurance Company Claims Object Gain," *Computerworld*, March 16, 1998, pp. 55–56 and The Workers Compensation/Managed Care Claim section under "Commercial Insurance" at the Travelers Web site at http://www.travelers.com, accessed July 4, 1998.

structured design

an approach to designing and developing application software that breaks large problems into smaller, simpler problems

structured design is to develop better software by reusing procedures and approaches that solve a variety of problems. **Structured design** breaks a large, difficult problem into smaller problems, each simple enough to manage and solve independently. These modules can then be reused in new and different programs. This building-block, or modular, software usually costs less to develop and maintain and is easier to modify and update over time. For example, a module to generate a certain type of report can be developed independently, then plugged into numerous programs that require this type of report. One way to implement structured design is structured programming.

TABLE 13.4

Usage of Tools by Systems Analysts

Tool	Percentage of Systems Analysts Using
Structured analysis	70 %
Prototyping	50 %
CASE tools	25 %
No tool, approach, or method	16 %

Source: Mark Misic, "The Skills Needed by Today's Systems Analysts," *Journal of Systems Management*, vol. 47, no. 3, May–June 1996, p. 34.

Structured programming. For many programming projects, maintaining and updating the programs can take more time and effort than the original development process. Because structured programming makes program maintenance easier and faster, it continues to be an important programming technique.

A critical goal of structured programming is to untangle and reduce the complexity of computer programs. For example, consider the simple GOTO statement of the BASIC programming language. This statement instructs the CPU to move, or "go to," another part of the program to read a different program statement. If a human analyst or programmer were attempting to follow the logic of such a program, too many uses of the GOTO statement would create confusion because of all the movements in the program from one location to another, often distant, location.

The basic idea behind structured programming is to improve the logical program flow by breaking the program into groups of statements. These groups are formed according to their function. One group may read data, while another group does a certain processing task. When using structured programming, statements in a group must conform to a standard structure. To begin with, there can be only one entering point into the block of statements and only one exit point from it. Therefore, you cannot branch to or from the middle of the structured group of statements. This restriction eliminates many programming errors and makes debugging much easier. Furthermore, there should be no groups of statements that cannot be reached or executed. Using the structured programming approach, each group can be tested separately.

sequence structure

a structure in programming in which statements are executed one after another until all the statements in the sequence have been executed

decision structure

a structure in programming that allows the computer to branch, depending on certain conditions

loop structure

a structure in programming that uses loops; in the do-until structure, the loop is done until a certain condition is met. For the do-while structure, the loop is done while a certain condition exists.

As shown in Figure 13.10, only three types of structures are allowed when using structured programming. In the **sequence structure**, there must be definite starting and ending points. After starting the sequence, programming statements are executed one after another until all the statements in the sequence have been executed. Then the program either ends or continues onto another sequence. The **decision structure** allows the computer to branch, depending on certain conditions. Normally, there are only two possible branches. The final structure is the **loop structure**. Actually, there are two commonly used structures for loops. One is the do-until structure, and the other is the do-while structure. Both accomplish the same thing. In the do-until structure, the loop is done until a certain condition is met. For the do-while structure, the loop is done while a certain condition exists. Structured program code development can be the key to developing good program code. Some of the characteristics of structured programming are shown in Table 13.5.

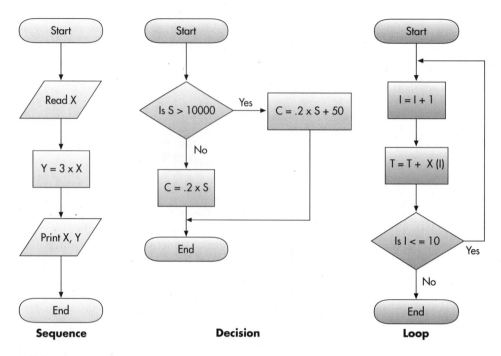

Sequence   **Decision**   **Loop**

FIGURE 13.10

The Three Structures Used in
Structured Programming

top-down approach

a programming approach in which
the main module is written before
the other modules

Structured programming: The top-down approach. In general, a good
approach to writing a large program is to start with the main module and
work down to the other modules. This is called the **top-down approach**
to programming and is used in structured program design and structured
programming. Although the concept of top-down programming is simple,
its use is beneficial in untangling or avoiding coding and debugging problems. The process begins by writing the main module. Then the modules at
the next level are written. This procedure continues until all the modules
have been written. Figure 13.11 can be used to visualize the top-down
approach. In addition to program coding, the top-down approach should
be used in testing and debugging. Thus, after the first, or main, module is
written, it is tested and debugged. But the main module sends the computer to modules at the second level, which have not been written yet.
Thus, simple modules at the second level are written to send the computer
back to the main module so it can be fully and completely tested. If errors
are found in the main module, they are corrected immediately.

CASE tools. CASE tools are often used during software development to automate some of the techniques. For example, source code can be automatically

TABLE 13.5

Characteristics of Structured
Programming

Program code is broken into modules.

Each module has one and only one function. Such modules are said to have
tight internal cohesion.

There is one and only one logic path into each module and one logic exit
from each module.

The modules are loosely coupled.

GOTO statements are not allowed.

FIGURE 13.11

The Top-Down Approach to Writing, Testing, and Debugging a Modular Program

Level 1 (The main module)

a. Write the main module.
b. Write any necessary dummy modules at the second level.
c. Test the main module.
d. Debug the main module.

Level 2 (This procedure is done for each module one at a time.)

a. Write the module.
b. Write any necessary dummy modules for next lower level.
c. Test the module (this will automatically test all the modules that are above this one in the structure chart.)
d. Debug the module.

Level N (This procedure is repeated for all levels.)

generated using CASE tools. Of the types of CASE tools previously discussed, lower-CASE tools are most likely to be used for software programming. Lower-CASE tools can provide a graphical programming environment and include compilers, syntax checkers, and software modules that generate the actual program code. CASE tools may also have interfaces to the code generators of other vendors' CASE tools, a situation that allows a programmer to mix and match the program code generated. Using CASE can help increase programmer accuracy and productivity, particularly in terms of time spent on maintenance.

Cross-platform development. In the past, most applications were developed and implemented using mainframe computers. Today, many applications are developed on personal computers by users of the system. In response to the growth of end-user development, software vendors now offer more tools and techniques to PC users. One software development technique, called **cross-platform development**, allows programmers to develop programs that can run on computer systems having different hardware and operating systems, or platforms. With cross-platform development, the same program might be able to run on both a personal computer and a mainframe, or on two different types of PCs. Cross-platform development can be done on all sizes of computers, including PCs. One benefit of cross-platform development is that programs can run on both small and large systems, and users can perform software development activities on their PCs.

cross-platform development

a software development technique that allows programmers to develop programs that can run on computer systems having different hardware and operating systems, or platforms

Integrated development environments. Software vendors also offer integrated development environments to assist with programming on personal computers. **Integrated development environments (IDEs)** combine the tools needed for programming with a programming language into one integrated package. IDE allows programmers to use simple screens, customized pull-down menus, and graphical user interfaces. Some even use different color text to allow a programmer to quickly locate sections, verbs, or errors in program code. In general, IDE can make programming software more intuitive as well as make personal computer programmers more productive. Having these tools combined with the language itself makes it

integrated development environments (IDEs)

software that combines the tools needed for programming with a programming language into one integrated package easier for programmers to develop sophisticated programs on personal computers.

Turbo Pascal includes an IDE with a code editor, a program debugger, and a compiler. Visual C++ is another example of a programming language with an IDE. The MULTI Software Development Environment provides a framework of interacting tools to support program development by small or large workgroups. MULTI includes all of the tools needed to support major programming projects, including a program builder, version control, program editor, debugger, execution profiler, run-time error checker, and source-code control. JIG is a Java integrated development environment that helps programmers write, manage and debug applets and applications written in Java. JIG is written entirely in Java.

Structured walkthroughs. Regardless of the tools or techniques used, companies should review software throughout the development process. Companies often use a structured walkthrough technique, which is typically performed by chief programmer teams. As shown in Figure 13.12, a **structured walkthrough** is a planned and preannounced review of the progress of a program module, a structure chart, or a human procedure. The walkthrough helps team members review and evaluate the progress of components of a structured project. The structured walkthrough approach is also useful for programming projects that do not use the structured design approach.

structured walkthrough

a planned and preannounced review of the progress of a program module, a structure chart, or a human procedure

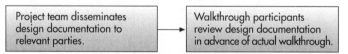

Walkthrough Planning and Preparation

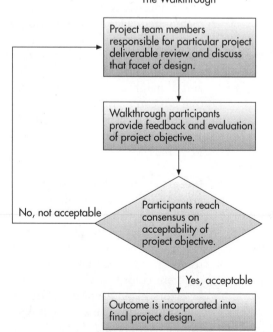

FIGURE 13.12

A structured walkthrough is a planned, preannounced review of the progress of a particular project objective.

Acquiring Database and Telecommunications Systems

Acquiring or upgrading database systems can be one of the most important steps of a systems development effort. Because databases are a blend of hardware and software, many of the approaches discussed earlier for acquiring hardware and software also apply to database systems. For example, an upgraded inventory control system may require database capabilities, including more hard disk storage or a new DBMS. Additional storage hardware will have to be acquired from an information systems vendor. New or upgraded software might also be purchased or developed in-house.

Telecommunications is one of the fastest growing applications for today's businesses and individuals. Like database systems, telecommunications systems require a blend of hardware and software. For personal computer systems, the primary piece of hardware is a modem. For client/server and mainframe systems, the hardware can include multiplexers, concentrators, communications processors, and a variety of network equipment. Communications software will also have to be acquired from a software company or developed in-house. Again, the earlier discussion on acquiring hardware and software also applies to the acquisition of telecommunications hardware and software.

User Preparation

user preparation

the process of readying managers, decision makers, employees, other users, and stakeholders for the new systems

User preparation is the process of readying managers, decision makers, employees, other users, and stakeholders for the new system. With the growing trend to employee empowerment, system developers need to provide users with the proper training to make sure they use the information system correctly, efficiently, and effectively. User preparation can include active participation, marketing, training, documentation, and support. Top-management support in ensuring that sufficient time and resources are allocated to user preparation is absolutely essential to a successful system start-up.

Informing and preparing users for new or modified systems can be done in a variety of ways. User preparation actually begins with user participation in system analysis. Some organizations also actively market new systems to users via brochures, newsletters, and seminars. The idea is to promote the new computer system the way they would a new product or service.

Stakeholders, who benefit from the system but do not directly use or interact with it, need to be made aware of the results of the systems analy-. sis and design effort. For example, suppose a toy manufacturer integrates a new inventory control system with the order processing system and production planning system so that it can quickly respond to changes in product demand. If a successful product promotion dramatically increases customer demand, the new application would alert managers to schedule larger or additional production runs to meet customer demand. Informed stakeholders—in this instance, customers—will likely shop at that chain, knowing that they can rely on finding shovels in stock. Some companies even go so far as to advertise their use of information systems to add product and service value.

Without question, training users is an essential part of user preparation, whether they are trained by internal personnel or by external training firms. In some cases, companies that provide software will provide user training at no charge or at a reasonable price. The cost of training can be negotiated during the selection of new software. Other companies conduct user training throughout the systems development process. Concerns and apprehensions about the new system must be eliminated through these training programs. Employees should be acquainted with the system's capabilities and limitations.

Continuing support provides assistance to users after a new or modified application has been installed. The overall purpose of support is to make sure users understand and benefit from the new or modified system. This support can be additional hardware, software, and service. Continuing support is provided by most vendors for a fee. Seminars, training programs, and consulting personnel are also popular. The preparation and distribution of user documentation is another important aspect of continuing support.

IS Personnel: Hiring and Training

Depending on the size of the new system, an organization may have to hire and, in some cases, train new IS personnel. An information systems manager, systems analysts, computer programmers, data entry operators, and similar personnel may be needed for the new system.

As with users, the eventual success of any system depends on how it is used by the personnel within the organization. Training programs should be conducted for the IS personnel who will be using the computer system. These programs will be similar to those for the users, although they may be more detailed in the technical aspects of the system. Effective training will help IS personnel use the new system to perform their jobs and will help them support other users in the organization.

Site Preparation

site preparation

the process of preparing the actual location of the new system

The location of the new system needs to be prepared in a process called **site preparation**. For a small system, this may simply mean rearranging the furniture in an office to make room for a computer. With a larger system, this process is not so easy. Larger systems may require special wiring and air-conditioning. One or two rooms may have to be completely renovated. Additional furniture may have to be purchased. A special floor may have to be built, under which the cables connecting the various computer components are placed, and a new security system may have to be installed to protect the equipment. For larger systems, additional power circuits may also be required.

Data Preparation

data preparation, or **data conversion**

the process of converting all manual files into computer files

If the organization is computerizing its work processes, all manual files must be converted into computer files in a process called **data preparation**, or **data conversion**. All the permanent data must be placed on a permanent storage device, such as magnetic tape or disk. Usually the organization hires some temporary, part-time data-entry operators or a service company to convert the manual data. Once the data has been converted into computer files, the temporary workers are no longer needed. A computerized database system or other software will be used to maintain and update these computer files.

Installation

Installation is the process of physically placing the computer equipment on the site and making it operational. Although normally the manufacturer is responsible for installing the computer equipment, someone from the organization (usually the IS manager) should oversee the process, making sure that all the equipment specified in the contract is installed at the proper location. After the system is installed, the manufacturer performs several tests to ensure that the equipment is operating as it should.

Testing

Several forms of testing are used, including testing each of the individual programs (**unit testing**), testing the entire system of programs (**system testing**), testing the application with a large amount of data (**volume testing**), and testing all related systems together (**integration testing**), as well as conducting any tests required by the user (**acceptance testing**). The sequence in which these testing activities normally occur is shown in Figure 13.13.

Unit testing is accomplished by developing test data that will force the computer to execute every statement in the program. In addition, each program is tested with abnormal data to determine how it will handle problems. System testing requires the testing of all the programs together. It is not uncommon for the output from one program to become the input for another. In these cases, system testing ensures that the output from one program can be used as input for another program within the system. Volume testing ensures that the entire system can handle a large amount of data under normal operating conditions. Integration testing ensures that the new programs can interact with other major applications. It also ensures that data flows efficiently and without error to other applications. For example, a new inventory control application may require data input from an older order processing application. Integration testing would be done to ensure smooth data flow between the new and existing applications. Integration testing is typically done after unit and system testing. Finally, acceptance testing makes sure that the new or modified system is operating as intended. Run times, the amount of memory required, disk access methods, and more can be tested during this phase. Acceptance testing ensures that all performance objectives defined for the system or application are satisfied. Involving users in acceptance testing may help them better understand and effectively interact with the new system. Acceptance testing is the final check of the system before start-up.

Most organizations testing their computer systems for year 2000 readiness either set up a test environment with duplicate equipment or partition off part of their mainframes for testing. These approaches keep businesses from shutting down production systems while they do testing. But buying or leasing duplicate equipment can cost millions of dollars, and no matter how much money a company spends, test environments can never duplicate all the quirks of the production system. Information systems professionals at the Philadelphia Stock Exchange found another test solution. The exchange has been testing production data that is mirrored in real time to a remote Sungard Data Systems disaster recovery facility. The stock exchange can test the systems for procedures such as

FIGURE 13.13

Types of Testing

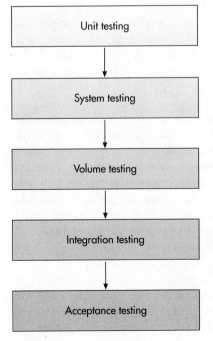

executions of stock trades as they are being processed and does not have to simulate such tests at night or on weekends. Because the exchange already owns the duplicate IBM OS/390 environment housed at Sungard, it is saving millions of dollars because it does not need to buy duplicate test equipment. Part of the exchange's cost savings stems from the fact that it does not have to pay year 2000 team members to work nights or weekends. And the stock exchange does not pay extra for the test time because testing uses system time that is already paid for.[5]

Start-Up

start-up

procedures that begin with the final tested information system; when start-up is finished, the system will be fully operational

direct conversion (also called plunge or direct cutover)

start-up procedure that involves stopping the old system and starting the new system on a given date

phase-in approach

sometimes called a piecemeal approach, the approach to start-up in which components of the new system are slowly phased in while components of the old one are slowly phased out

pilot start-up

the start-up approach that involves running the new system for one group of users rather than all users

parallel start-up

the start-up approach that involves running both the old and new systems for a period of time

user acceptance document

a formal agreement signed by the user that states that a phase of the installation or the complete system is approved

systems maintenance

the checking, changing, and enhancing of the system to make it more useful in achieving user and organizational goals

Start-up begins with the final tested information system. When start-up is finished, the system will be fully operational. Various start-up approaches are available (Figure 13.14). **Direct conversion** (also called plunge or direct cutover) involves stopping the old system and starting the new system on a given date. This is usually the least desirable approach because of the potential for problems and errors when the old system is completely shut off and the new system is turned on at the same instant. The **phase-in approach** is a popular technique preferred by many organizations. In this approach, sometimes called a piecemeal approach, components of the new system are slowly phased in while components of the old one are slowly phased out. When everyone is confident that the new system is performing as expected, the old system is completely phased out. This process is repeated for each application until the new system is running every application. **Pilot start-up** involves running the new system for one group of users rather than all users. For example, a manufacturing company with a number of retail outlets throughout the country could use the pilot start-up approach and install a new inventory control system at one of the retail outlets. When this pilot retail outlet runs without problems, the new inventory control system can be implemented at other retail outlets. **Parallel start-up** involves running both the old and new systems for a period of time. The output of the new system is compared closely with the output of the old system, and any differences are reconciled. When users are comfortable that the new system is working correctly, the old system is eliminated.

User Acceptance

Most mainframe computer manufacturers use a formal **user acceptance document**—a formal agreement signed by the user that states that a phase of the installation or the complete system is approved. This is a legal document that usually removes or reduces the information systems vendor from liability for problems that occur after the user acceptance document has been signed. Because this document is so important, many companies get legal assistance before they sign the acceptance document. Stakeholders may also be involved in acceptance to make sure that the benefits to them are indeed realized.

SYSTEMS MAINTENANCE

Systems maintenance involves checking, changing, and enhancing the system to make it more useful in achieving user and organizational goals. Software maintenance is a major concern for organizations.[6] During the

FIGURE 13.14

Start-Up Approaches

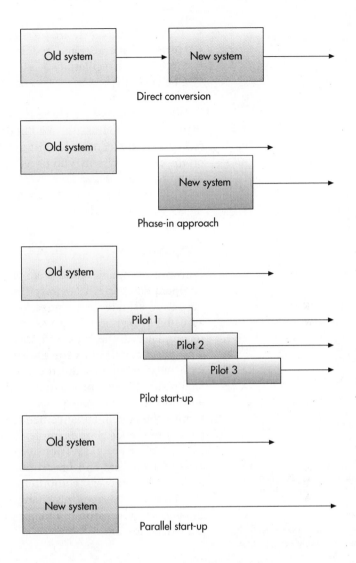

Direct conversion

Phase-in approach

Pilot start-up

Parallel start-up

late 1990s, the need to modify systems to properly account for the year 2000 was a major cause for system maintenance, with companies spending millions on this activity. In some cases, an organization will encounter major problems that involve recycling the entire systems development process. In other situations, minor modifications are sufficient.

Reasons for Maintenance

Once a program is written, it is likely to need ongoing maintenance. To some extent, a program is like a car that needs oil changes, tune-ups, and repairs at certain times. Experience shows that frequent, minor maintenance to a program, if properly done, can prevent major system failures later. Some of the major reasons for program maintenance are the following:

- Changes in business processes
- New requests from stakeholders, users, and managers
- Bugs or errors in the program

- Technical and hardware problems
- Corporate mergers and acquisitions
- Government regulations
- Change in the operating system or hardware on which the application runs

When it comes to making necessary changes, most companies modify their existing programs instead of developing new ones. That is, as new systems needs are identified, most often the burden of fulfilling the needs falls on the existing system. Old programs are repeatedly modified to meet ever-changing needs. Over time, these modifications tend to interfere with the system's overall structure, reducing its efficiency and making further modifications more burdensome.

Types of Maintenance

Software companies and many other organizations use four generally accepted categories to signify the amount of change involved in maintenance. A **slipstream upgrade** is a minor upgrade—typically a code adjustment or minor bug fix—not worth announcing. It usually requires recompiling all the code and, in so doing, it can create entirely new bugs. This practice accounts for the exact same computers working differently with what is supposedly the exact same software. A **patch** is a minor change to correct a problem or make a small enhancement. It is usually an addition to an existing program. That is, the programming code representing the system enhancement fix is usually "patched into," or added to, the existing code. For example, Microsoft Corp. released patches to remedy a glitch in several versions of Internet Explorer, a problem that opened a way for hackers and unscrupulous Web site operators to read the contents of files on users' computers.[7] A new **release** is a significant program change that often requires changes in the documentation of the software. Finally, a new **version** is a major program change, typically encompassing many new features.

The Request for Maintenance Form

Because of the amount of effort that can be spent on maintenance, many organizations require a **request for maintenance form** to authorize modification of programs. This form is usually signed by a business manager who documents the business case for the need for the change and identifies the priority of this change relative to other work that has been requested. The IS group reviews the form and identifies the programs to be changed, determines the programmer assigned to the project, estimates the expected completion date, and develops a technical description of the change. A cost/benefit analysis may be required if the change requires substantial resources.

Performing Maintenance

Depending on organizational policies, the people who perform systems maintenance vary. In some cases, the team that designs and builds the system also performs maintenance. This gives the designers and programmers an

slipstream upgrade

a type of maintenance that involves a minor upgrade—typically a code adjustment or minor bug fix—not worth announcing

patch

a minor change to correct a problem or make a small enhancement

release

a significant program change that often requires changes in the documentation of the software

version

a major program change, typically encompassing many new features

request for maintenance form

a form that is used to authorize modification of programs

maintenance team

the team responsible for modifying, fixing, and updating existing software

incentive to build systems well from the outset: if there are problems, the designers and programmers will have to fix them. In other cases, organizations have a separate **maintenance team**. This team is responsible for modifying, fixing, and updating existing software. Because experience and skills are important in maintenance, some organizations utilize a specialized maintenance team or department. Java and the object-oriented programming languages hold the promise of reducing the program maintenance effort.

Regardless of who performs maintenance, the same tools and techniques used for earlier phases of systems development—CASE tools, flowcharts, structured programming, and so on—should be used. In addition, all maintenance should be fully documented. Unfortunately, this is often not done. For example, if an order processing application crashes during peak hours, the main objective is to get the application running as soon as possible. Documenting the problem or any changes made to the application may be overlooked in the rush to restart the application. This can cause future problems if programmers reference outdated data-flow diagrams, layout charts, and so on when they perform maintenance for the system. Thus, it is essential that the tools used in maintenance allow easy documentation to accurately reflect the changes.

A number of vendors have developed tools to ease the software maintenance burden. Relativity Technologies recently unveiled RescueWare, a product that converts third-generation code such as COBOL to highly maintainable C++, Java, or Visual Basic object-oriented code. Using RescueWare, developers download mainframe code to Windows NT or Windows 95 workstations. They then use the product's graphical tools to analyze the original system's inner workings. RescueWare lets the programmer see the original system as a set of object views, which visually illustrate module and program structures. Developers choose one of three available levels of transformation: revamping the user interface, converting the database access, and transforming procedure logic.[8]

RescueWare provides companies with automated software that accelerates the process of transforming third-generation code such as COBOL to Internet or client/server platforms. (Source: Courtesy of Relativity Technologies, Inc.)

The Financial Implications of Maintenance

The cost of maintenance is staggering. For older programs, the total cost of maintenance can be up to five times greater than the total cost of development. In other words, a program that originally cost $25,000 to develop may cost $125,000 to maintain over its lifetime. The average programmer can spend over 50 percent of his or her time on maintaining existing programs instead of developing new ones. Furthermore, as programs get older, total maintenance expenditures in time and money increase, as illustrated in Figure 13.15. With the

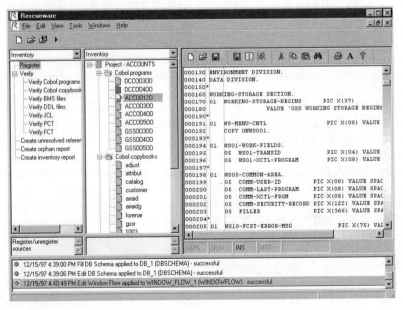

FIGURE 13.15

Maintenance Costs as a
Function of Age

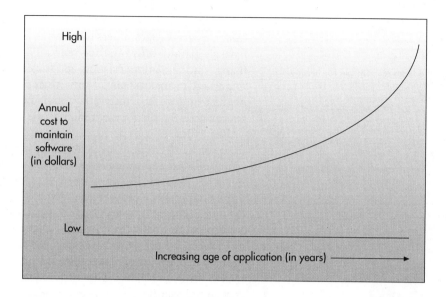

use of newer programming languages and approaches, including object-oriented programming, maintenance costs are expected to decline. Even so, many organizations have literally millions of dollars invested in applications written in older languages (such as COBOL), which are both expensive and time-consuming to maintain.

The financial implications of maintenance make it important to keep track of why systems are maintained, instead of simply keeping cost figures. This is another reason documentation of maintenance tasks is so crucial. A determining factor in the decision to replace a system is the point at which it is costing more to fix it than to enhance it. Read the "Making a Difference" box to learn about a company that decided to reduce the amount of maintenance to enable more flexible pricing and increase revenues.

The Relationship between Maintenance and Design

Programs are expensive to develop, but they are even more expensive to maintain. Programs that are well designed and documented to be efficient, structured, and flexible are less expensive to maintain in later years. Thus, there is a direct relationship between design and maintenance. More time spent on design up front can mean less time spent on maintenance later.

In most cases, it is worth the extra time and expense to design a good system. Consider a system that costs $250,000 to develop. Spending 10 percent more on design would cost an additional $25,000, bringing the total design cost to $275,000. Maintenance costs over the life of the program could be $1 million. If this additional design expense can reduce maintenance costs by 10 percent, the savings in maintenance costs would be $100,000. Over the life of the program, the net savings would be $75,000 ($100,000 − $25,000). This relationship between investment in design and long-term maintenance savings is graphically displayed in Figure 13.16.

The need for good design goes beyond mere costs. There is a real risk in ignoring small system problems when they arise, as these small problems may become large in the future. As mentioned earlier, because maintenance

MAKING A DIFFERENCE

Sanofi Reduces Maintenance with New Contract Management System

Sanofi specializes in the manufacture and sale of prescription drugs for people suffering from diseases of the central nervous system and cardiovascular problems. DMR Consulting Group is an international provider of information systems services to businesses and public enterprises, with offices in the United States, Canada, Asia-Pacific region, and Europe and a global base of nearly 8,000 professionals worldwide. Recently, Sanofi and DMR teamed up to implement a state-of-the-art contract management system.

Thousands of hospitals and healthcare networks provide Sanofi's drugs. The company constructs contracts to meet the prescription needs of its customers and to increase total revenue. What each group pays for medicine is negotiated into a contract and varies widely. One group may receive a 2 percent discount on each drug it prescribes, and another group may receive a 2 percent discount on the first 3,000 doses and a 4 percent discount thereafter. Keeping track of these different contracts was complex and time consuming.

With the help of DMR Consulting, Sanofi recently implemented a client/server contract management system to keep track of these one-of-a-kind deals. The company expects to save more than $.5 million per year on its $1.5 million systems investment. The system also enables Sanofi to execute more creative and competitive contracts to help it achieve its business goals.

The old system Sanofi used was written in COBOL, and it was extremely difficult to make the programming changes required to track and manage new contract terms. For example, if Sanofi wanted the contract to tie a customer's discount to the percentage of the customer's total drug sales that were represented by Sanofi's products, extensive reprogramming was required. With the new system, contract administrators can quickly configure contracts and discounts by using a set of built-in tools that comes with the application.

DMR Consulting managed the overall project. It began with a prototype of the system so that test users could experiment with the system as it was being configured. This way changes could be identified before it was rolled out for broad use. Users also made many suggestions about the system's ease of use and screen layouts. Following this approach, the implementation was completed in just six months.

DISCUSSION QUESTIONS

1. What other types of information systems are likely to have a heavy, ongoing maintenance effort to meet rapidly changing business needs?
2. Does a more flexible contracting and pricing system provide Sanofi with a competitive advantage?

Sources: Adapted from Julia King, "Drug Maker Simplifies Deals," *Computerworld*, September 28, 1998, pp. 55–56 and the About DMR section of the DMR Consulting Group Web page at http://www.dmr.com, accessed October 9, 1998.

programmers spend an estimated 50 percent or more of their time deciphering poorly written, undocumented program code, there is little time to spend on developing new, more effective systems. If put to good use, the tools and techniques discussed in this chapter will allow organizations to build longer-lasting, more reliable systems.

SYSTEMS REVIEW

systems review
the final step of systems development; the process of analyzing systems to make sure they are operating as intended

Systems review, the final step of systems development, is the process of analyzing systems to make sure they are operating as intended. This often involves comparing the performance and benefits of the system as it was designed with the actual performance and benefits of the system in operation. Cost, performance, control, and complexity factors investigated during design are revisited after the system has been operating. Problems and opportunities uncovered during systems review will trigger systems development and begin the process anew. For example, as the number of users of an interactive system increases, it is not unusual for system response time to

FIGURE 13.16

The Value of Investment in
Design

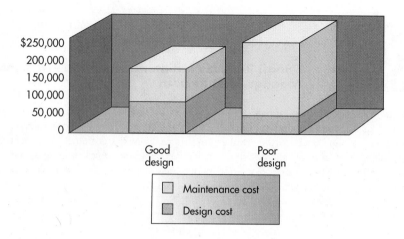

increase. If the degradation in response time is too great, it may be necessary to do some redesign of the system, modify databases, or increase the power of the computer hardware.

Types of Review Procedures

There are two types of review procedures: event driven and time driven. An **event-driven review** is triggered by a problem or opportunity such as an error, a corporate merger, or a new market for products. In some cases, a company will wait until a large problem or opportunity occurs before a change is made. In this case, minor problems may be ignored. Some companies use a continuous improvement approach to systems development. With this approach, an organization makes changes to a system even when small problems or opportunities occur. While this approach can keep the system current and responsive, doing the repeated design and implementation can be both time-consuming and expensive.

A **time-driven review** is performed after a specified amount of time. Many application programs are reviewed every six months to a year. With this approach, an existing system is monitored on a schedule. If problems or opportunities are uncovered, a new systems development cycle may be initiated. A payroll application, for example, may be reviewed once a year to make sure it is still operating as expected. If it is not, changes are made.

Many companies use both approaches. A billing application, for example, might be reviewed once a year for errors, inefficiencies, and opportunities to reduce operating costs. This is a time-driven approach. In addition, the billing application might be redone if there is a corporate merger, if one or more new managers require different information or reports, or if federal laws on bill collecting and privacy change. This is an event-driven approach.

event-driven review

review of the system that is triggered by a problem or opportunity such as an error, a corporate merger, or a new market for products

time-driven review

review of the system that is performed after a specified amount of time

Factors to Consider During Systems Review

Systems review should investigate a number of important factors, such as the following:

Mission. Is the computer system helping the organization toward its overall mission? Are stakeholder needs and desires satisfied or exceeded with the new or modified system?

Organizational goals. Does the computer system support the specific goals of the various areas and departments of the organization?

Hardware and software. Are hardware and software up to date and adequate to handle current and future processing needs?

Database. Is the current database up to date and accurate? Is database storage space adequate to handle current and future needs?

Telecommunications. Is the current telecommunications system fast enough, and does it allow managers and workers to send and receive timely messages? Does it allow for fast order processing and effective customer service?

Information systems personnel. Are there sufficient IS personnel to perform current and projected processing tasks?

Control. Are rules and procedures for system use and access acceptable? Are the existing control procedures adequate to protect against errors, invasion of privacy, fraud, and other potential problems?

Training. Are there adequate training programs and provisions for both users and IS personnel?

Costs. Are development and operating costs in line with what is expected? Is there an adequate information systems budget to support the organization?

Complexity. Is the system overly complex and difficult to operate and maintain?

Reliability. Is the system reliable? What is the mean-time between failures (MTBF)?

Efficiency. Is the computer system efficient? Are system outputs generated by the right amount of inputs, including personnel, hardware, software, budget, and others?

Response time. How long does it take the system to respond to users during peak processing times?

Documentation. Is the documentation still valid? Are changes in documentation needed to reflect the current situation?

System Performance Measurement

system performance measurement

systems review that involves monitoring the system—the number of errors encountered, the amount of memory required, the amount of processing or CPU time needed, and other problems

system performance products

products developed to measure all components of the computer-based information system, including hardware, software, database, telecommunications, and network systems

Systems review often involves monitoring the system. This is called **system performance measurement**. The number of errors encountered, the amount of memory required, the amount of processing or CPU time needed, and other problems should be closely observed. If a particular system is not performing as expected, it should be modified, or a new system should be developed or acquired. In some cases, specialized software products have been developed that do nothing but monitor system performance. **System performance products** have been developed to measure all components of the computer-based information system, including hardware, software, database, telecommunications, and network systems. When properly used, system performance products can quickly and efficiently locate actual or potential problems.

Candle is a leading provider of mainframe performance monitoring and enterprise management solutions. Its products include Candle Command Center, advanced systems management tools for optimizing an organization's computing resources and maximizing business application availability, and Omegamon II performance monitors, for realtime and historical analysis of performance on a variety of systems.[9] Precise/Pulse is a product from Precise Software Solutions that provides around-the-clock performance monitoring for Oracle database applications. It detects and reports potential problems through leading systems management consoles. Precise/Pulse monitors the performance of critical database applications and issues alerts about database inefficiencies before they turn into application performance problems.[10]

Measuring a system is, in effect, the final task of systems development. The results of this process may bring the development team back to the beginning of the development life cycle, where the process begins again.

Information Systems Principles

Information systems must be designed along two dimensions: logical and physical. Logical design refers to the way various components of an information system work together. Physical design refers to the specification of the actual physical components.

Organizations need to define emergency alternate and disaster recovery procedures for key information systems that control cash flow and support key business operations.

Tight systems controls are necessary to maintain data security and to prevent computer misuse, crime, and fraud by employees and others.

Near the end of the design stage, the design specifications need to be frozen. Formal change management procedures should be established to avoid "scope creep."

Take the time to develop your own contract with computer facilities, hardware, and software vendors—their standard contracts were designed to protect them, not you.

Users need proper training to make sure they use the information system correctly.

Maintenance of software adds to its useful life but can consume large amounts of resources. This activity can benefit from the same rigorous methods and project management techniques applied to the system development process.

Organizations should conduct an ongoing system review of their information systems.

■ SUMMARY

1. *State the purpose of systems design and discuss the differences between logical and physical systems design.*

The purpose of systems design is to prepare the detailed design needs for a new system or modifications to the existing system. The logical systems design dimension refers to the way the various components of an information system will work together. The logical design includes data specifications for output and input, processing, files and databases, telecommunications, procedures, personnel and job design, and controls and security design. The physical systems design dimension refers to the specification of the actual physical components. The physical design must specify characteristics for hardware and software design, database and telecommunications, and personnel and procedures design.

2. *Outline key steps taken during the design phase.*

A number of special design considerations should be taken into account during both logical and physical system design. A sign-on procedure consists of identification numbers, passwords, and other safeguards needed for individuals to gain access to computer resources. A system sign-on allows the user to gain access to the computer; an application sign-on permits the user to start and use a particular application.

If the system under development is interactive, the design approach must consider using menus, help facilities, table lookup facilities, and restart procedures. A good interactive dialogue will ask for information in a clear manner, respond rapidly, contain consistency between applications, and use an attractive format. Also, it will avoid use of computer jargon and will treat the user with respect.

Error prevention, detection, and correction should be part of the system design process. Causes of errors include humans, natural phenomena, and technical problems. Designers should be alert to prevention of fraud and invasion of privacy.

At the end of the systems design step, the final specifications are frozen and no changes are allowed so that implementation can proceed. A final design report is developed, containing the products from the design process.

3. *Define the term RFP and discuss how this document is used to drive the acquisition of hardware and software.*

If new hardware or software will be purchased from a vendor, a formal request for proposal (RFP) is needed. The RFP outlines the company's needs; in response, the vendor provides a written reply. In addition to responding to the company's stated needs, the vendor provides data on its operations. This data might include the vendor's reliability and stability, the type of post-sale service offered, the vendor's ability to perform repairs and fix problems, vendor training, and the vendor's reputation.

RFPs from various vendors are reviewed and narrowed to the few most likely candidates. In the final evaluation, a variety of techniques—including group consensus, cost/benefit analysis, point evaluation, and benchmark tests—can be used. After the vendor is chosen, contract negotiations can begin.

If externally developed software is to be used, a company should review software needs and requirements, acquire and customize software, and develop appropriate interfaces between the purchased software and existing software. Software should also be tested prior to its acceptance for use, and modifications and/or maintenance agreements should be reached with the vendor. Problems to avoid include paying for extras the organization will not use; underestimating the cost of the modifications, interfaces, implementation, and installation; and being one of the first to try out a new package.

4. *Describe the techniques used to make system selection evaluations.*

There are four commonly used approaches to making a final system evaluation and selection: group consensus, cost/benefit analysis, benchmark tests, and point evaluation.

In group consensus, a decision-making group is appointed and given responsibility for making the final evaluation and selection. With cost/benefit analysis, all costs and benefits of the alternatives are expressed in monetary terms. Benchmarking involves comparing computer systems operating under the same condition. Point evaluation assigns weights to evaluation factors and each alternative is evaluated in terms of each factor and given a score from 0 to 100.

5. *List the advantages and disadvantages of purchasing versus developing software.*

Software can be purchased from external vendors or developed in-house—a decision termed the make-or-buy decision. A purchased software package usually has a lower cost, less risk regarding the features and performance, and easy installation. The amount of development effort is also less when software is purchased. Developing software can result in a system that more closely meets the business needs and has increased flexibility in terms of customization and changes. Developing software also has greater potential for providing a competitive advantage.

6. *Discuss the software development process and some of the tools used in this process.*

Software development is often performed by a chief programmer team—a group of IS professionals who design, develop, and implement a software program. Programming using traditional programming languages follows a life cycle that includes investigation, analysis, design, language selection, program coding, testing and debugging, documentation, and implementation (conversion). Documentation includes technical and user documentation. Implementation of the new computer program may be done in parallel, in small pilot tests, phased in slowly, or direct conversion.

There are many tools and techniques for software development. Structured design is a philosophy of designing and developing application software. Structured programming is not a new programming language; it is a way to standardize computer programming using existing languages. The top-down approach starts with programming a main module and works down to the other modules. Other tools, like cross-platform development and integrated development environments, make software development easier and more thorough. CASE tools are often used to automate some of these techniques.

Fourth-generation languages (4GLs) and object-oriented languages offer another alternative to in-house development. Development using these fast and easy-to-use languages requires several steps, much like the programming life cycle. The main difference is that, with object-oriented languages, one must identify and select objects and integrate them into an application, instead of using step-by-step coding.

7. *State the purpose of systems implementation and discuss the various activities associated with this phase of systems development.*

The purpose of systems implementation is to install the system and make everything, including users, ready for its operation. Systems implementation includes hardware acquisition, software acquisition or development, user preparation, hiring and training of personnel, site and data preparation, installation, testing, start-up, and user acceptance. Hardware acquisition requires purchasing, leasing, or renting computer resources from a vendor. Types of vendors include small and general computer manufacturers, peripheral equipment manufacturers, leasing companies, time-sharing companies, software companies, dealers, distributors, service companies, and others.

Implementation must address personnel requirements. User preparation involves readying managers, employees, and other users for the new system. New IS personnel may need to be hired, and users must be well trained in the system's functions. Preparation of the physical site of the system must be done, and any existing data to be used in the new system will require conversion to the new format. Hardware installation is done during the implementation step, as is testing. Testing includes program (unit) testing, systems testing, volume testing, integration testing, and acceptance testing.

Start-up begins with the final tested information system. When start-up is finished, the system will be fully operational. There are a number of different start-up approaches. Direct conversion (also called plunge or direct cutover) involves stopping the old system and starting the new system on a given date. With the phase-in approach, sometimes called a piecemeal approach, components of the new system are slowly phased in while components of the old one are slowly phased out. When everyone is confident that the new system is performing as expected, the old system is completely phased out. Pilot start-up involves running the new system for one group of users rather than all users. Parallel start-up involves running both the old and new systems for a period of time. The output of the new system is compared closely with the output of the old system, and any differences are reconciled. When users are comfortable that the new system is working correctly, the old system is eliminated.

8. *State the importance of systems and software maintenance and discuss the activities involved.*

Systems maintenance involves checking, changing, and enhancing the system to make it more useful in obtaining user and organizational goals. Maintenance is critical for the continued smooth operation of the system. The costs of performing maintenance can well exceed the original cost of acquiring the system. Some major causes of maintenance are new requests from stakeholders and managers, enhancement requests from users, bugs or errors, technical or hardware problems, newly added equipment, changes in organization structure, and government regulations.

Maintenance can be as simple as a program patch to correct a small problem to the more complex upgrading of software with a new release from a vendor. Requests for maintenance should be documented with a request for maintenance form, a document that formally authorizes modification of programs. The development team or a specialized maintenance team may then make approved changes.

9. *Describe the systems review process.*

Systems review is the process of analyzing systems to make sure that they are operating as intended. It involves monitoring systems to be sure they are operating as designed. The two types of review procedures are event-driven review and time-driven review. An event-driven review is triggered by a problem or opportunity. A time-driven review is started after a specified amount of time.

Systems review involves measuring how well the system is supporting the mission and goals of the organization. System performance measurement monitors the system for number of errors, amount of memory and processing time required, and so on.

■ KEY TERMS

acceptance testing 601
application sign-on 572
benchmark test 585
chief programmer team 590
closed shops 579
cold site 577
cost/benefit analysis 584
cross-platform development 597
data conversion 600
data preparation 600
decision structure 595
design report 587
deterrence controls 579
dialogue 573
direct conversion 602
disaster recovery 575
event-driven review 608
final evaluation 583
group consensus 584
help facility 573
hot site 576
image log 578
incremental backup 578
installation 601
integrated development environments (IDEs) 597

integration testing 601
logical systems design 570
lookup tables 573
loop structure 595
maintenance team 605
make-or-buy decision 588
menu-driven system 573
object-oriented software development 592
open shops 579
parallel start-up 602
patch 604
phase-in approach 602
physical systems design 570
pilot start-up 602
point evaluation system 585
preliminary evaluation 583
programming life cycle 591
release 604
request for maintenance form 604
request for proposal (RFP) 580
restart procedures 573
selective backup 577
sequence structure 595
sign-on procedure 572
site preparation 600

slipstream upgrade 604
software interface 590
start-up 602
structured design 594
structured walkthrough 598
system performance measurement 609
system performance products 609
system sign-on 572
system testing 601
systems controls 578
systems design 570
systems implementation 588
systems maintenance 602
systems review 607
technical documentation 591
time-driven review 608
top-down approach 596
unit testing 601
user acceptance document 602
user documentation 592
user preparation 599
version 604
volume testing 601

■ REVIEW QUESTIONS

1. What is the purpose of systems design?
2. What are some of the special design considerations that should be taken into account during both the logical and physical design?
3. What are some differences between emergency alternate procedures and a disaster recovery plan?
4. Identify specific controls that are used to maintain input integrity and security.
5. What is meant by cross-platform development?
6. What is an RFP? What is typically included in one? How is it used?
7. What is the difference between system interface and end user interface?
8. What activities go on during the user preparation phase of system implementation?
9. What are the major steps of systems implementation?
10. What are some tools and techniques for software development?
11. Explain the three types of structures allowed in structured programming.
12. What are the steps involved in testing the information system?
13. What are some of the reasons for program maintenance? Explain the four types of maintenance.
14. Describe the point evaluation system for selection of the optimum system alternative.
15. How is systems performance measurement related to the systems review?

■ DISCUSSION QUESTIONS

1. Identify some of the advantages and disadvantages of purchasing versus leasing hardware.
2. Discuss the relationship between maintenance and system design.
3. Is it equally important for all systems to have a disaster recovery plan? Why or why not?
4. Four approaches were discussed to evaluate a number of systems alternatives. No one approach is always the best. How would you decide which approach to use in a particular instance?
5. Identify the various forms of testing used. Why are there so many different types of tests?
6. Discuss the key responsibilities of end users and general management during software coding, data preparation, installation, and start-up of a new system.
7. What is the goal of conducting a systems review? What factors need to be considered during systems review?
8. What features and terms would you insist on in a software package contract?
9. What issues might you expect to arise if you initiate the use of a request for maintenance form where none had been required? How would you deal with these issues?
10. How would you go about evaluating a software vendor?

■ PROBLEM-SOLVING EXERCISES

 1. Make a list of the tasks associated with the systems design and implementation phases of an information systems project. Develop a Gantt chart using a project management software package such as Microsoft Project or a spreadsheet program. Which of these tasks can be done concurrently and which must be done sequentially? Which of these tasks are likely to be relatively minor and take a lesser amount of effort? Which tasks are more involved and will take longer? Assume that each minor task takes four elapsed days and each major task takes twelve elapsed days. How long will it take to complete the project?

2. A project team has estimated the costs associated with the development and maintenance of a new system. One approach requires a more complete

design and will result in a slightly higher design and implementation cost but a lower maintenance cost over the life of the system. The second approach cuts the design effort short, saving some dollars but with a likely increase in maintenance cost.

 a. Enter the following data in the spreadsheet. Print the result.

The Benefits of Good Design

	Good Design	Poor Design
Design Costs	$14,000	$10,000
Implementation Cost	$42,000	$35,000
Annual Maintenance Cost	$32,000	$40,000

 b. Create a stacked bar graph that shows the total cost, including design, implementation, and maintenance costs. Be sure that the chart has a title and that the costs are labeled on the chart.

 c. Use your word processing software to write a paragraph that recommends which approach to take and why.

3. To get a better understanding of the value of a good interactive dialogue, go to the Web site of Amazon.com at http://www.amazon.com. Search for several books based on different criteria. In searching for these materials, use the author as one criteria (e.g., R. M. Stair), the subject (e.g., information systems), and the title (e.g., *Principles of Information Systems*). If you have access to another on-line library catalog, conduct the same searches using that system. Now evaluate the interactive dialogue of the systems, based on the elements discussed in the text. Write up your observations using your word processor. If you can, also send this evaluation to your instructor via e-mail.

■ TEAM ACTIVITIES

1. Assume your project team has been working three months to complete the systems design of a new Web-based customer ordering system. They are two possible options that seem to meet all users' needs. The project team must make a final decision on which option to implement. The table that follows summarizes some of the key facts about each option.
   a. What process would you follow to make this important decision?
   b. Who needs to be involved?
   c. What additional questions need to be answered to make a good decision?
   d. Based on the data, which option would you recommend and why?
   e. How would you account for project risk in your decision making?

Factor	Option #1 (Millions)	Option #2 (Millions)
Annual gross savings	$1.5	$3.0
Total development cost	$1.5	$2.2
Annual operating cost	$0.5	$1.0
Time required to implement	9 months	15 months
Risk associated with project (expressed in probabilities)		
Benefits will be 50% less than expected	20%	35%
Cost will be 50% greater than expected	25%	30%
Organization will not/cannot make changes necessary for system to operate as expected	20%	25%
Does system meet all mandatory requirements?	Yes	Yes

2. Assume that your team is working for a medium-size consulting firm. You have been engaged to develop a database for tracking information about each project currently in progress at the firm. The database must provide the information necessary to create a quick status report for the senior consultants and partners of the firm. It is especially important to be able to identify participants from the firm involved in the project and the key customer contacts. In addition, the controller wants to be able to produce an expected cash flow analysis report that shows the consulting dollars billed to each project.

a. Develop a project proposal for this effort including an estimate of the time required and a statement of potential project benefits.
b. Develop a mock-up report showing what will be displayed to the senior consultants and partners if they want a project status report.
c. Develop a mock-up report showing the expected cash flow analysis report.
d. Use a database management system to document the key data elements that must be stored in the database.
e. Use a database management system to define the various tables and associated primary keys and data elements.

▪ WEB EXERCISE

EDS is just one of the many systems development outsourcing companies. Find the Web site of this company or another outsourcing company by entering the Internet address or using a search engine. If you used a search engine, did you discover any irrelevant information? Describe the company, its services, and any investor or employment information. You may be asked to develop a report or send an e-mail message to your instructor about what you found.

▪ CASES

1   Outsource It All

Ten years ago, companies acquired a variety of different businesses to diversify. A manufacturing company, for example, would purchase a printing business or a lumber operation. The idea was to reduce risk by getting into different businesses. The results were mixed. For some, acquiring different companies increased costs and reduced profitability. More recently, companies have concentrated on their core business or expertise, merging only with companies in the same industry. Getting more focused on core business activities has also led to the large number of outsourcing deals, where a company will hire an outside company to perform systems development or run the entire IS operation. This allows a company to concentrate on what it does best, while spinning off computer operations to an expert outsourcing firm.

While outsourcing IS services can increase a company's competitive position, it can also be very expensive. High costs, which can climb into the millions of dollars, can remove the advantage of outsourcing. Outsourcing can be beneficial, but cost savings are not guaranteed. As a result, many companies considering outsourcing are now carefully structuring the outsourcing deal to make sure it is cost-effective. The U.S. Chamber of Commerce and Ryder Truck are just two companies that are controlling costs while entering into outsourcing deals.

The U.S. Chamber of Commerce had a large number of computer programs that would malfunction due to the year 2000 problem. Because the U.S. Chamber of Commerce did not have the staff or expertise to solve the problem and wanted to concentrate on core operations, it recently entered into a large outsourcing deal with Cap Gemini America. The deal was worth $75 million over a ten-year period. In the deal, Cap Gemini America agreed to perform basically all of the IS functions required by the U.S. Chamber of Commerce. The deal is to be self-funding, meaning that the cost savings are supposed to equal the cost of the outsourcing deal. Before outsourcing, the U.S. Chamber of Commerce had an annual IS budget of about $7 million, with a staff of 60 IS personnel. With the outsourcing deal, the entire IS budget will be saved. Only one person of the original 60-person staff was retained.

Ryder Truck also decided to hire an outsourcing company to handle its entire IS operation. Ryder, however, had a different approach. It used a two-stage process. First, Cambridge Technology Partners was hired to develop and implement 18 business systems. Perot Systems was then hired to run the daily IS operations. Ryder also used a different approach to make sure that the costs of the outsourcing deal were balanced by the benefits. To achieve this, Ryder made a deal with the outsourcing companies where 20 percent of the fees would be withheld if the new systems failed to perform as expected and the additional revenues were not achieved. With this approach, the outsourcing companies had a financial interest in making sure the systems they built returned a profit.

1. What are the advantages of outsourcing? What are the disadvantages?
2. If you were in charge of structuring an outsourcing deal, what provisions would you put in the outsourcing agreement?

Sources: Julia King, "Ryder Trucks Out Entire IS Operation," *Computerworld*, February 9, 1998, p. 1; Matthew Lake, "BAAN Moves into Outsourcing," *Computer Weekly*, February 5, 1998, p. 14; Bob Wallace, et al., "Rare Deal: Self-Funded Outsourcing," *Computerworld*, February 23, 1998, p. 1; Tom Stein, "JD Edwards Debuts Outsourcing," *Information Week*, February 16, 1998.

2   Measuring Return: A Systems Development Success Story

Business managers and executives often look at any investment in similar terms. The investment could be a new plant, an updated warehouse, a short-term investment of cash in the stock market or bonds, a new marketing campaign, or a new computer system. Common measures of the success of an investment include payback period, rate of return, and other measures. Payback period is the number of years it takes to pay back the investment in returns or profits. A return of $5,000 on an investment of $50,000 would have a payback period of 10 years. Rate of return is the percentage return the investment is able to achieve. Getting an $8,000 profit or return on an investment of $100,000 would have an 8 percent rate of return.

Using payback period, rate of return, or a similar measure allows corporate executives to compare different investments using the same framework. A new warehouse with a rate of return of 9 percent can be compared with a new marketing effort with a rate of return of 11 percent. Many corporate executives try to achieve 20 percent returns or greater.

Measuring return on systems development projects can be difficult. For example, it can be very difficult to determine the value of a systems development effort that resulted in a new payroll system. What is the return? How do you compute all of the costs? In addition to the measuring difficulties, many systems development efforts are behind schedule and over budget. Some experts believe that 50 percent or more of all systems development efforts are delivered late, cost more than they should, or are not able to perform as expected. But there are incredible systems development successes.

Enron, a Houston energy company, was able to turn a systems development project into a huge financial success. When the California Public Utility Commission required power companies to develop meter data management systems for commercial customers, many power companies were not in a position to bid for the new deregulated business. Enron turned the requirement into an opportunity. Using rapid application development techniques, the company was able to satisfy all of the requirements and bid on providing electric service to some of California's universities and state colleges. The systems development effort took four months and about $1 million to complete. What was the result of this investment?

Enron was able to sign a deal worth $500 million. This represented a 500 percent return on the four-month systems development effort.

1. How would you measure the return of a systems development effort?
2. Why do you think Enron used rapid application development techniques?

Sources: Thomas Hoffman, "Power Play," *Computerworld*, March 2, 1998, p. 1; "Enron, Electric Avenues," *The Economist*, February 28, 1998, p. 69; "Enron on Blast Charge," *The Engineer*, January 8, 1998, p. 3; Barbara Shook, "Core Business Boost Enron Profits," *The Oil Daily*, January 21, 1998, p. 30.

3 Slowing Systems Integration at Aetna

Even with advanced systems development tools and techniques, getting a successful system within budget and on time is challenging. While there have been successes, there have also been systems development failures that have cost companies millions of dollars and lost opportunities. In some cases, a company's survival can be threatened when systems development fails. When a large number of systems have to be modified or created from scratch concurrently, getting good systems development results can seem impossible. This is what Aetna faced when it merged with New York Life.

A corporate merger can present the most difficult systems development problems, especially when the systems of the two companies are not similar or the companies are large. In 1998, Aetna completed a $105 billion acquisition of New York Life. The challenge was to integrate the two corporate information systems. In addition to the transaction processing systems, including payroll and claims processing, management information and decision support systems that gave managers mission-critical information also had to be integrated. To help the systems development effort, Aetna could look to past mergers, including the U.S. Healthcare merger.

In 1996, Aetna acquired U.S. Healthcare for $8.3 billion. This merger caused many systems development integration problems. Merging the claims processing transaction-processing system, for example, was difficult. Many doctors were not paid on time. The inadequate integration also caused problems for many customers and clients. Some experts on Wall Street realized the problem and did not report favorably about the merger or the integration of the two information systems. Even a few years after the merger, Aetna was still struggling with integration problems. Based on this merger experience, Aetna decided to undertake a slow and deliberate systems development effort with the New York Life merger.

Aetna's overall approach for the systems development effort is to gradually merge the two information systems. Currently, New York Life has 2.2 million members. These members must be added to the 14 million Aetna members. To help ensure high quality, Aetna plans to integrate smaller groups of customers first. This pilot-program approach will help Aetna get bugs out of the process before full-scale integration is attempted.

Slowing the integration process, however, presents its own problems. With different customer and client codes and procedures, Aetna must run different systems during the integration period, which could be years. What does the future hold for Aetna and other healthcare corporations? More mergers are likely, causing more development efforts to integrate information systems. Travelers Insurance, which owns Aetna, has announced plans to merge with Citicorp. The resulting company will be huge. The information systems departments of Travelers and Citicorp are hoping that the integration problems will not loom as large.

1. What is Aetna's approach to implementing the information systems for the merger with New York Life?
2. What systems development issues and problems are caused by mergers?

Sources: Thomas Hoffman, "Aetna Learns from Previous Integration Snafus," *Computerworld*, March 23, 1998, p. 14; Michael Slezak, "Aetna Slapped with Lawsuits," *American Druggist*, January 1998, p. 10; "For the Record: Aetna to Buy NYLCare Plans," *Modern Healthcare*, March 23, 1998, p. 18; Gerald Meuchner, "Merger Could Create Clash of the Titans," *Final Chaser*, April 7, 1998, p. E1.

4   Company Frees IS Staff for Future Application Development

For most companies, 80 percent of their IS organization's time is spent on the maintenance of old systems and only 20 percent on developing new business applications. In 1995, Office Depot made a deliberate move to increase the amount of resources available for future new application development. Its strategy—forfeit short-term business gains by delaying new application development. The company stabilized its application portfolio by converting old applications to new technology over a 14-month period.

One example of the maintenance work that was done occurred in late 1995 and early 1996 when the company was having trouble absorbing eight stationers it had acquired in 1994. Profit margins for Office Depot's business services division (sales to corporate customers) sank from 8 percent to less than 3 percent during that time. The IS organization replaced more than 100 obsolete IBM AS/400 applications with new IBM ES/9000 applications to improve management of this part of the business. Management gives this project strong credit for helping return profit margins to 8 percent over the next year.

This strategy is beginning to show results with Office Depot's information systems staff, which now has more time available to work on new applications. For example, by completing key maintenance efforts, the IS organization had more resources available when they turned their attention to developing inventory control and management systems in mid-1996. Those systems helped tighten and reduce inventory levels by $160 million in 1997 even though Office Depot added 42 stores that year. Another new application that was staffed by the freed-up IS resources is a "consultative sales" system. Furniture sales will be a top priority for the company this year. Office Depot expects this furniture system pilot to go live soon among a select group of stores in south Florida.

Senior management is confident that the maintenance turnaround has aided the businesses. Once the company finishes converting its California warehouses to the new order-entry and warehousing systems later this year, the IS organization can stop converting systems and focus instead on driving new business—where the fun is. "If we can be four times more productive building applications by flip-flopping the 80/20 rule, the benefits will be enormous," says Office Depot Chairman and CEO David Fuentes.

1. Do you think other companies should follow Office Depot's strategy to free up their IS resources to work on new business applications? Why or why not?
2. What should Office Depot do if it sees the scales beginning to tip back where maintenance effort begins to exceed the effort on new application development?

Sources: Adapted from "OfficeDepot.com Named Premier E-Commerce Site in Sweep of CIO Web Business 50/50 and Retail Network Innovation Awards," July 1, 1998, *Business Wire;* "Office Depot Ratings Raised by Standard & Poor's," *PRNewswire,* June 16, 1998; Thomas Hoffman, "Short Term Sacrifices," *Computerworld,* March 16, 1998, pp. 1, 97.

PART

Information
Systems in
Business and
Society

CHAPTER 14

Security, Privacy, and Ethical Issues in Information Systems and the Internet

"Risk is what companies must live with when they only allocate limited monies and resources for network security. And corporations that do that have to hope they can live with the threat."

—John Davis, Director,
National Computer
Security Center[1]

Chapter Outline

Computer Waste and Mistakes
  Computer Waste
  Computer-Related Mistakes
  Preventing Computer-Related Waste and Mistakes

Computer Crime
  The Computer as a Tool to Commit Crime
  The Computer as the Object of Crime
  Preventing Computer-Related Crime

Privacy
  Privacy Issues
  Fairness in Information Use
  Federal Privacy Laws and Regulations
  State Privacy Laws and Regulations
  Corporate Privacy Policies
  Protecting Individual Privacy

The Work Environment
  Health Concerns
  Avoiding Health and Environmental Problems

Ethical Issues in Information Systems

Learning Objectives

After completing Chapter 14, you will be able to:

1. Describe some examples of waste and mistakes in an IS environment, their causes, and possible solutions.
2. Explain the types and effects of computer crime, along with measures for prevention.
3. Discuss the principles and limits of an individual's right to privacy.
4. List the important effects of computers on the work environment.
5. Outline the criteria for the ethical use of information systems.

CERT
Policing the Internet

Unauthorized computer access is one of the most common, troublesome, and potentially destructive behavioral problems facing society. It is the high-tech equivalent of breaking and entering a home or business. Once the unlawful entry has been accomplished, what happens next depends on the level of the intruder's destructive intent, computer skills, and the value of the property available for destruction or theft.

The Computer Emergency Response Team (CERT) was established to monitor Internet security and coordinate the response of law enforcement agencies to incidents of unauthorized computer access on the Internet. In fulfilling this role, CERT also studies Internet security vulnerabilities, provides incident response services to sites that have been the victims of attack, publishes a variety of security alerts, researches security and survivability in wide-area-network computing, and develops information to help you improve security at your site.

CERT was formed in 1988 with federal funding as part of the Software Engineering Institute at Carnegie Mellon University. Incident reports have increased from roughly 400 that first year to 2,300 in recent years, and some 50 other CERTs have emerged worldwide. These numbers, however, do not tell the whole story. The incidents reported to CERT are typically only the most serious. Furthermore, it is likely that many other serious break-ins are never revealed by the victims because they do not wish to advertise that their security systems can be breached. Also some proportion of break-ins are simply never discovered. A sufficiently skillful intruder is difficult to detect, particularly if no damage was done or alteration was made to the invaded system.

CERT comprises a small group of highly-skilled individuals. To relieve the stress, members rotate through periods of administrative, research, and education responsibilities to avoid the burnout that would inevitably come from constantly handling calls. More often than not, each incident involves hundreds (occasionally thousands) of sites. Some incidents may also involve ongoing activity for long periods of time, sometimes more than a year. Some attacks are on the Internet infrastructure itself and receive top priority along with life-threatening incidents such as attacks on hospital, fire, and police computers. Fewer intruders are succeeding, but the number of attempts is increasing.

The conflict between the responsibility to report known security hazards and the reluctance to place information in the wrong hands is a constant challenge. The CERT Coordination Center Vulnerability unit carefully weighs which advisories to release and when. The team painstakingly crafts each advisory to be useful without revealing clues that would aid attackers. Despite all of CERT's deliberation, caution, and thoroughness, its warnings often go unheeded. Software with known security problems remains unchanged by IS organizations and vendors alike. Today's systems administrators are novices in an arena that's expanding faster and in more directions simultaneously than any technology in history. Under pressure from senior management to get on the Web, systems administrators routinely add software and hardware to their networks that create ideal entry points for intruders.

As you read this chapter, consider the following questions:

- What motivates an individual to commit a computer crime?

- What can be done to detect and avoid computer crime?

Sources: Adapted from Web site for the Computer Emergency Response Team at http://www.cert.org, last updated June 26, 1998, accessed July 6, 1998 and Leslie Goff, "Search and Rescue," *Computerworld*, April 7, 1997, pp. 87–88.

Earlier chapters have detailed the amazing benefits of computer-based information systems in business, including increased profits, superior goods and services, and higher quality of work life. Computers have become such valuable tools that today's businesspeople would have difficulty imagining work without them. Yet, the Information Age has also brought some potential problems for workers, companies, and society in general (see Table 14.1).

To a large extent, this book has focused on the solutions—not the problems—presented by information systems. In this chapter we discuss details regarding the problems as a reminder of the social and ethical considerations underlying the use of computer-based information systems. No business organization, and hence no information system, operates in a vacuum. All IS professionals, managers, and users have a responsibility to see that the potential consequences of IS use are fully considered.

Managers and users at all levels play a major role in helping organizations achieve the positive benefits of IS. These individuals must also play a major role in helping to minimize or eliminate the negative consequences of poorly designed and improperly utilized information systems.

For managers and users to have such an influence, they must be properly educated. Many of the problems presented in this chapter, for example, should cause you to think back to some of the systems design and systems control issues we have already discussed. They should also help you look forward to how these issues and your choices might affect your future IS management considerations.

COMPUTER WASTE AND MISTAKES

Computer-related waste and mistakes are major causes of computer problems, contributing as they do to unnecessarily high costs and lost profits. Computer waste involves the inappropriate use of computer technology and resources. Computer-related mistakes refer to errors, failures, and other computer problems that make computer output incorrect or not useful, caused mostly by human error. In this section we explore the damage that can be done as a result of computer waste and mistakes.

Computer Waste

The U.S. government is the largest single user of information systems in the world. It should come as no surprise that it is also perhaps the largest misuser. The government is not unique in this regard. The same type of waste and misuse found in the public sector also exists in the private sector. Some companies discard old diskettes and even complete computer systems when they still have value. Others waste corporate resources to build and maintain complex systems never used to their fullest extent. A less dramatic, yet still

TABLE 14.1

Social Issues in Information Systems

• Year 2000 issues	• Privacy
• Computer waste and mistakes	• Health concerns
• Computer crime	• Ethical issues

relevant, example of waste is the amount of company time and money employees may waste playing computer games, sending unimportant e-mail, or accessing the Internet. Junk e-mail, also called spam, and junk faxes also cause waste.[2] People receive hundreds of e-mail messages and fax documents advertising products and services not wanted or requested by users. This not only wastes time, but it also wastes paper and computer resources. When waste is identified, it typically points to one common cause: the improper management of information systems and resources.

Computer-Related Mistakes

Despite many people's distrust, computers themselves rarely make mistakes. Even the most sophisticated hardware cannot produce meaningful output if users do not follow proper procedures. Mistakes can be caused by unclear expectations and a lack of feedback. Or a programmer might develop a program that contains errors. In other cases, a data-entry clerk might enter the wrong data. Unless errors are caught early and prevented, the speed of computers can intensify mistakes by an order of magnitude. As information technology becomes faster, more complex, and more powerful, organizations and individuals face greater risk of experiencing the results of computer-related mistakes. Take, for example, these cases from recent news:

- The fire brigade in Grampian, Scotland, said a combination of human and computer errors resulted in a budget shortfall of £688,000, or approximately $1.2 million. Significant human error and erroneous practices, as well as technological shortcomings, contributed to this matter. The financial information on which the senior officers and members of the joint fire board were reliant was fundamentally flawed in its construction and presentation.
- In Bernalillo County, New Mexico, a computer glitch turned portions of 50,000 absentee ballots into an illegible jumble. The ballots contained Spanish text for a constitutional amendment and state and county bond issues to be voted on in a November election. Words were misspelled, missing consonants or vowels, or they ran together. It looked like a language no human had ever seen.
- The wireless communications venture Globalstar Telecommunications lost 12 of its satellites when its rocket crashed in Kazakhstan. The Ukranian-built Zenit-2 rocket crashed just minutes after liftoff from the Baikonur Cosmodrome. NPO Yuzhnoye, the company that manufactured the rocket, said that two computer glitches occurred in rapid succession. A company statement said that the computer "as a result sent an order to cut the engines." The satellites were valued at $185 million.
- A Utah retiree was ordered to pay back nearly $14,000 in excess retirement benefits he received. A retirement fund employee testified in the trial that a portion of the retiree's benefits were calculated by hand and another portion by computer, and the two figures were accidentally combined, resulting in double payments. The company overpaid in two ways. It paid a lump-sum distribution that was $14,000 too high. In addition, the company overpaid the retiree's monthly benefits.

- A recall of Toyota and Lexus cars, due to faulty on-board computers, may cost the company as much as $82.5 million. The California Air Resources Board ordered the recall of 330,000 1996-1998 Toyota and Lexus models. The ARB said that the computers fail to properly detect gasoline vapors. ARB estimates the cost to Toyota of fixing the computers will be approximately $250 each.
- Students in Chippewa Falls, Wisconsin, received free school lunches as a result of a computer glitch. New computer software that was supposed to run the lunch program did not communicate properly with the computer hardware. During tests, duplicate student identification numbers were issued. On the first day of the school year, students were sent home with notices saying that they were eligible for free lunches that week.

Preventing Computer-Related Waste and Mistakes

To remain profitable in a competitive environment, organizations must use all resources wisely. Preventing computer-related waste and mistakes like those just described should therefore be a goal. Today, most organizations use some type of CBIS. To employ IS resources efficiently and effectively, employees and managers alike should strive to minimize waste and mistakes. Preventing waste and mistakes involves (1) establishing, (2) implementing, (3) monitoring, and (4) reviewing effective policies and procedures.

Establishing policies and procedures. The first step to prevent computer-related waste is to establish procedures and policies regarding efficient acquisition, use, and disposal of systems and devices. Computers permeate organizations today, but it is critical for organizations to ensure that systems are used to their full potential. As a result, most companies have implemented stringent policies on the acquisition of computer systems and equipment. These include requiring a formal justification statement before computer equipment is purchased, definition of standard computing platforms (operating system, type of computer chip, minimum amount of RAM, etc.), and the use of a preferred vendor list for all acquisitions.

Prevention of computer-related mistakes begins by identifying the most common types of errors. There are surprisingly few (see Table 14.2). To control and prevent potential problems caused by computer-related mistakes, companies have developed preventive policies and procedures that cover the following:

- Acquisition and use of computers, with a goal of avoiding waste and mistakes
- Training programs for individuals and workgroups
- Manuals and documents on how computer systems are to be maintained and used
- Approval of certain systems and applications before they are implemented and used to ensure compatibility and cost-effectiveness
- Requirement that documentation and descriptions of certain applications be filed or submitted to a central office, including all cell formulas for spreadsheets and a description of all data elements and relationships in a database system; such standardization can ease access and use for all personnel

TABLE 14.2

Types of Computer-Related Mistakes

- Data-entry or capture errors
- Errors in computer programs
- Errors in handling files, including formatting a disk by mistake, copying an old file over a newer one, and deleting a file by mistake
- Mishandling of computer output
- Inadequate planning for and control of equipment malfunctions
- Inadequate planning for and control of environmental difficulties (electrical problems, humidity problems, etc.)

Once companies have planned and developed policies and procedures, they must consider how best to implement them.

Implementing policies and procedures. Implementing policies and procedures to minimize waste and mistakes varies according to the business conducted. Most companies develop such policies and procedures, often in conjunction with advice provided by the firm's internal auditing group or its external auditing firm. The policies often focus on the implementation of source data automation, the use of data editing to ensure data accuracy and completeness, and assigning clear responsibility for data accuracy within each information system. Table 14.3 lists some useful policies to minimize waste and mistakes.

Training is another key aspect of implementation. Many users are not properly trained in developing and implementing applications. Their mistakes can be very costly. Since more and more people use computers in their daily work, it is important that they understand how to use them. Training is often the key to acceptance and implementation of policies and procedures. Because of the importance of accurate data and people understanding their responsibilities, companies converting to ERP systems invest weeks of training for key users of the system's various modules.

Monitoring. To ensure that users throughout the organization are following established procedures, the next step is to monitor routine practices and take corrective action if necessary. By understanding what is happening in day-to-day activities, organizations can make adjustments or develop new

TABLE 14.3

Useful Policies to Eliminate Waste and Mistakes

- Changes to critical tables should be tightly controlled, with all changes authorized by responsible owners and documented.
- A user manual should be available that covers operating procedures and that documents the management and control of the application.
- Each system report should indicate its general content in its title and specify the time period it covers.
- The system should have controls to prevent invalid and unreasonable data entry.
- Controls should exist to ensure that data input is valid, applicable, and posted in the right time period.
- Users should implement proper procedures to ensure correct input data.

procedures. Many organizations implement internal audits to measure actual results against established goals for things such as percent of end-user reports produced on time, percent of data-input errors rejected, number of input transactions entered per eight-hour shift, etc.

Reviewing policies and procedures. The final step is to review existing policies and procedures and determine whether they are adequate. During review, people should ask the following questions:

- Do current policies cover existing practices adequately? Were any problems or opportunities uncovered during monitoring?
- Does the organization plan any new activities in the future? If so, does it need new policies or procedures on who will handle them and what must be done?
- Are contingencies and disasters covered?

This review and planning allows companies to take a proactive approach to problem solving, which can avert disasters. During such a review, companies are alerted to upcoming crises in information systems that could have a profound effect on many business activities. As mentioned previously, one such problem is the year 2000 crisis. Today, many companies are trying to prevent this crisis, which is caused by how older computer programs store dates.[3] For example, interest on a loan in the year 2003 could be computed wrong—starting in the year 1903—because the program would think that "03" was "1903" instead of "2003." The cost to prevent the 2000 crisis is staggering. Read the "FYI" box to learn more about the year 2000 problem. GTE, for example, estimates that solving the 2000 crisis for its information systems might cost more than $200 million and require about 1,000 people. The Gartner Group, a research company, estimates that the total cost to prevent the 2000 crisis for all companies could reach $600 billion.

Information systems professionals and users need to be aware of the misuse of resources throughout an organization. Preventing errors and mistakes is one way to do so. Another is implementing in-house security measures and legal protections to detect and prevent a dangerous type of misuse: computer crime.

COMPUTER CRIME

Even good IS policies may not be able to predict or prevent computer crime. A computer's ability to process millions of pieces of data in less than a second can help a thief steal data worth thousands or millions of dollars. Compared with the physical dangers of robbing a bank or retail store with a gun, a computer criminal with the right equipment and know-how can steal large amounts of money from the privacy of a home. Computer crime often defies detection, the amount stolen or diverted can be substantial, and the crime is "clean" and nonviolent. According to the Computer Security Institute (CSI), the annual cost of computer crime is rising rapidly. In a recent study ("1998 Computer Crime and Security Survey" conducted with participation by the FBI International Computer Crime Squad), CSI received responses from 520 companies. This data is summarized in Table 14.4.

FYI

Y2K: This One Is Costing Billions and Taking Years to Fix

Mistakes can be minor, such as a misspelling, or major, such as a decimal place that is wrong in data. When it comes to the year 2000 (Y2K) fix, the problem is huge. The Y2K fix is costing companies billions of dollars and may take years to completely solve. The problem has been embedded in computer code used for key operations in corporations for decades.

Any company or governmental agency with older software, called legacy applications, is having to carefully check each line of code to determine whether and where this problem is located. With millions of lines of older program code for some companies, this process is tedious, time consuming, and very expensive. Today, companies are scrambling to find and fix their Y2K problems. Here are some examples:

- The securities industry makes heavy use of data—tracking of trade, dividend, stock split information, and the way funds are posted to client accounts. Realizing the potential for disaster, the Securities Industry Association decided to coordinate the testing and act as a conduit for the Y2K teams at the different member companies. Without such leadership, there was the potential for jumbling the billions in daily stock market sales.
- Union Pacific believes it is ahead of schedule and under budget for fixing its year 2000 problem. It is so confident that the slogan "January 1, 2000—Just Like Any Other Day" can be seen in offices and hallways. Its current budget is $46 million, and it has more than 100 people assigned to fix the problem. Thousands of older programs are either being fixed or replaced with newer ones that don't have the problem. Like other companies, Union Pacific had to hire additional

programmers and systems analysts to get the Y2K problem solved before the mistake became a disaster.

- Norstan attacked the Y2K problem beginning back in 1993 when it realized that it could be used to justify replacing its aging inventory and financial systems. Not only does Norstan end up with systems that are year 2000 compliant, but it gets new systems that are more aligned to its current business.

While companies like these are ahead of schedule and under budget in solving their Y2K problem, many companies are not. The problem is so serious for some companies that they are putting themselves up for sale. Others are looking to be merged or acquired. Their Y2K problem is simply too big to be solved alone. Resources required to fix the Y2K problem means that other critical programming projects are being put on hold. In addition, there is a programmer shortage as a result of the Y2K mistake.

The U.S. government has been cited for being far behind on its work to avoid the millennium bug and prepare critical computers for the year 2000 and beyond. The Department of Defense, with more than 600,000 potential problem computer chips in computers and weapons systems, is so far behind schedule that at its current rate of progress it will be 2002 before all necessary changes are made. The Federal Aviation Administration has failed to make adequate progress, and without dramatic improvements, the nation's air traffic could face serious disruptions for an extended period of time after December 31, 1999. The Treasury Department's Financial Management Service must make improvements to be able to continue issuing social security checks.

DISCUSSION QUESTIONS

1. Why is the Y2K problem so serious?
2. What are the social implications of the Y2K problem? What impact will it have on creating new applications and programmer salaries?

Sources: Kevin Burden, "Spinning Gold from Flax," Computerworld, September 28, 1998, pp. 75–78; Joanne Morrison, "U.S. Gets Failing Grade for Millennium Bug," Yahoo News at Web site http://dailynews.yahoo.com, June 2, 1998; Debra Sparks, "Will Your Bank Live to See the Millennium?" Business Week, January 26, 1998, p. 74; William Ulrich, "Package Vendors Better Unbundle Year 2000 Fixes," Computerworld, January 12, 1998, p. 37; Brian Jaffe, "01/01/2000: Where Will You Be?" Computerworld, January 19, 1998, p. 93; Julia King, "Union Pacific on Fast Track," Computerworld, January 12, 1998, p. 28.

TABLE 14.4

Summary of Key Data from 1998 Computer Crime and Security Survey

• Companies reporting security breaches in the last 12 months	64%
• Respondents who suffered financial losses from security breaches	72%
• Respondents who could quantify financial losses from security breaches	46%
• Total financial loss	$136 million
• Respondents citing their Internet connection as a frequent point of attack	54%

Today, computer criminals are a new breed—bolder and more creative than ever. With the increased use of the Internet, computer crime is becoming global. A criminal in Japan could steal money from someone in Europe doing business on the Internet. Regardless of its nonviolent image, computer crime is only different because a computer is used. It is still a crime. Part of what makes computer crime so unique and hard to combat is its dual nature—it can be both the tool used to commit a crime and the object of that crime.

The Computer as a Tool to Commit Crime

A computer can be used as a tool to gain access to valuable information and as the means to steal thousands or millions of dollars. It is, perhaps, a question of motivation: many individuals who commit computer-related crime claim they do it for the challenge, not for the money. Credit card fraud, where a criminal illegally gains access to another's line of credit with credit card numbers, is a major concern for today's banks and financial institutions. In general, two capabilities are required to commit most computer crimes. First, the criminal needs to know how to gain access to the computer system. Sometimes this requires knowledge of an identification number and a password. Second, the criminal must know how to manipulate the system to produce the desired result.

In one instance of computer crime, a 30-year-old hacker working from Russia used his laptop computer to hack into Citibank's computer system in the United States and stole $10 million from the bank, remotely. Also, with desktop publishing programs and high-quality printers, crimes involving counterfeit money, bank checks, traveler's checks, and stock and bond certificates are on the rise. As a result, the Department of Treasury redesigned and printed new currency that is much more difficult to counterfeit.

The Computer as the Object of Crime

A computer can also be the object of the crime, rather than the tool for committing it. Tens of millions of dollars of computer time and resources are stolen every year. Each time system access is illegally obtained, data or computer equipment is stolen or destroyed, or software is illegally copied, the computer becomes the object of crime. These crimes fall into several categories: (1) illegal access and use, (2) data alteration and destruction, (3) information and equipment theft, (4) software and Internet piracy, (5) computer scams, and (6) international computer crime.

Illegal access and use. Crimes involving illegal system access and use of computer services are a concern to both government and business. Federal, state, and local government computers are sometimes left unattended over weekends without proper security, and university computers are often used for commercial purposes under the pretense of research or other legitimate academic pursuits. A 28-year old computer expert allegedly tied up thousands of US West computers in an attempt to solve a classic math problem. The individual allegedly obtained the passwords to 2,585 computers and diverted them to search for a new prime number. This person ran up 10 years of computer processing time. The alleged hacking was discovered by a US West Intrusion Response Team after company officials noticed that computers were taking up to five minutes to retrieve telephone numbers, when normally they only require three to five seconds. At one point, customer calls had to be rerouted to other states, and the delays threatened to close down the Phoenix Service Delivery Center.[4]

Since the outset of information technology, computers have been plagued by criminal hackers. A **hacker** is a person who enjoys computer technology and spends time learning and using computer systems. A **criminal hacker**, also called a **cracker**, is a computer-savvy person who attempts to gain unauthorized or illegal access to computer systems. In many cases, criminal hackers are people who are looking for fun and excitement—the challenge of beating the system. In other cases, they are looking to steal passwords, files and programs, or even money.

Catching and convicting criminal hackers remains a difficult task.[5] The method behind these crimes is often hard to determine. Even if the method behind the crime is known, tracking down the criminals can take a lot of time. It took years for the FBI to arrest one criminal hacker for the alleged "theft" of almost 20,000 credit card numbers that had been sent over the Internet. Table 14.5 provides some guidelines to follow in the event of a computer security incident.

hacker

a person who enjoys computer technology and spends time learning and using computer systems

criminal hacker

a computer-savvy person who attempts to gain unauthorized or illegal access to computer systems

cracker

another term for a criminal hacker

TABLE 14.5

How to Respond to a Security Incident

- Follow your site's policies and procedures for a computer security incident. (They are documented, aren't they?)
- Contact the incident response group responsible for your site as soon as possible.
- Inform others, following the appropriate chain of command.
- Further communications about the incident should be guarded to ensure intruders do not intercept information.
- Document all follow-up actions (phone calls made, files modified, system jobs that were stopped, etc.)
- Make backups of damaged or altered files.
- Designate one person to secure potential evidence.
- Make copies of possible intruder files (malicious code, log files, etc.) and store them off-line.
- Evidence, such as tape backups and printouts, should be secured in a locked cabinet, with access limited to one person.
- Get the National Computer Emergency Response Team involved if necessary.
- If you are unsure of what actions to take, seek additional help and guidance before removing files or halting system processes.

Data alteration and destruction. Data and information are valuable corporate assets. The intentional use of illegal and destructive programs to alter or destroy data is as much a crime as destroying tangible goods. Most common of these types of programs are viruses and worms, which are software programs that, when loaded into a computer system, will destroy, interrupt, or cause errors in processing. There are more than 20,000 known computer viruses today with more than 6,000 new viruses and worms being discovered each year.[6] A **virus** is a program that attaches itself to other programs. A **worm** functions as an independent program, replicating its own program files until it interrupts the operation of networks and computer systems. In perhaps the most famous case involving a worm, an individual inserted a worm into the ARPANET that infected more than 6,000 computers on the network. In some cases, a virus or a worm can completely halt the operation of a computer system or network for days or longer until the problem is found and repaired. In other cases, a virus or a worm can destroy important data and programs. If backups are inadequate, the data and programs may never be fully functional again.

Some viruses and worms attack personal computers, while others attack network and client/server systems. A personal computer can get a virus from an infected disk, an application, or opening e-mail attachments that were received from the Internet. A virus or worm that attacks a network or client/server system is usually more severe because it can affect hundreds or thousands of personal computers and other devices attached to the network. Workplace computer virus infections are increasing rapidly due to the increased spread of macro viruses usually found in files attached to e-mail. The number of infections per 1,000 PCs was 21.45 in 1997; in 1998 it had grown to 31.85, according to the International Computer Security Association, Inc.[7] The primary ways to avoid viruses and worms are to install virus scanning software on all systems, update it routinely, and abstain from using disks or files from unknown or unreliable sources.

The two most common kinds of viruses are application viruses and system viruses. **Application viruses** infect executable application files such as word processing programs. When the application is executed, the virus infects the computer system. Because these types of viruses normally attach themselves to application files, they can often be detected by checking the length or size of the file. If the file is larger than it should be, a virus may be attached. A **system virus** typically infects operating systems programs or other system files. These types of viruses usually infect the system as soon as the computer is started. The well-publicized Michelangelo virus (which first appeared March 6, 1992) is an example of a system virus. Other system viruses include Die Hard 2, which struck 100 networked computers in San Diego County, and Stealth, which caused system failures at Florida State University. A small security hole in Microsoft Corp.'s Internet Explorer 4.0 browser allowed a particular virus to flood a user's desktop with white pixels. This type of "ghosting attack" takes place when a malicious Java applet on a Web page transfers to a user's desktop and, in effect, starts throwing virtual snowballs at the screen. The applet can be stopped easily with the standard "control-alt-delete" program shutdown key sequence. The number and types of viruses are almost limitless.[8]

Another type of program that can destroy a system is a **logic bomb**, an application or system virus designed to "explode" or execute at a specified time and date. Logic bombs are often disguised as a Trojan horse, a program

---

virus

a program that attaches itself to other programs

worm

an independent program that replicates its own program files until it destroys other systems and programs or interrupts the operation of networks and computer systems

application viruses

types of viruses that infect executable application files such as word processing programs

system virus

a type of virus that typically infects operating systems programs or other system files

logic bomb

an application or system virus designed to "explode" or execute at a specified time and date that appears to be useful but actually masks the destructive program. Some of these programs execute randomly; others are designed to remain inert in software until a certain code is given. In this fashion, the bomb will explode months, or even years, after being "planted." A former programmer at Omega Engineering Corp. activated a "logic bomb" that permanently deleted all of the company's software. The resulting loss to Omega was at least $10 million in lost sales and contracts.[9]

macro virus

a virus that uses an application's own macro programming language to distribute itself

A **macro virus** is a virus that uses an application's own macro programming language to distribute itself. Unlike previous viruses, macro viruses do not infect programs; they infect documents. The document could be a letter created using a word processing application, a graphics file developed for a presentation, or a database file. WW6macro, for example, is a macro virus that attaches itself to Microsoft Word for Windows documents. Macro viruses that are hidden in a document file can be difficult to detect. As with other viruses, however, virus detection and correction programs can be used to find and remove macro viruses.

A malicious macro virus that targeted Microsoft Corp.'s Word 97 wreaked havoc on July 12, 1998, the date the World Cup Soccer games ended in France. If the user's word processor was opened on July 12, or if the seconds of the internal clock were at 12, the WorldCup98 macro executed. A dialog box appeared containing the names of the nine teams competing in the soccer championship. The user was then prompted to type in his or her favorite team. If the choice coincided with the one the virus program liked, a congratulations screen popped up. Otherwise, a message appeared expressing sympathy. The virus executed regardless of the answer and whether or not a response was given.[10]

Most macro viruses are written for Microsoft's Word for Windows and Excel for Windows. However, there are also macro viruses for Lotus AmiPro (APM/Greenstripe). By the end of May 1997 the total number of macro viruses had reached many hundreds. If you count every single-bit difference as a virus variant, the total number is well above 1,800. This figure is growing fast, with more than five new macro viruses every day.[11]

The Macro Virus Protection tool is a free tool that installs a set of protective macros which detect suspicious Word files and alert customers to the potential risk of opening files with macros. Upon being alerted, users are given the choice of opening the file without executing the macros, thereby ensuring that no viruses are transmitted. The tool works with Word 6.0 for Windows 3.1, Word 6.0.1 for the Macintosh, Word 6.0 for Windows NT, Word for Windows 95, and Windows NT. It can be downloaded from the following on-line services: The Microsoft World Wide Web site at http://www.microsoft.com/msoffice; The Microsoft Network, using go word *macrovirustool;* and The Word forums on other on-line services such as America Online.[12]

Hoax, or false, viruses are another problem. Criminal hackers will warn the public of some new and devastating virus that doesn't exist. Companies can spend hundreds of hours warning employees and taking preventive action against a nonexistent virus. Take, for example, the vice president of a major footwear manufacturer in the Northeast who forwarded a virus hoax called Join the Crew to other top managers at his company. "WARN-ING," reads Join the Crew. "If you receive an e-mail titled 'Join the Crew,' DO NOT open it! It will erase EVERYTHING on your hard drive! Send this letter out to as many people as you can. This is a new virus that is not yet detectable." By the time the information systems liaison in the footwear maker's legal department received the message, it had spread throughout the company—maybe faster than any real virus would have.[13]

Security specialists recommend that IS personnel establish a formal paranoia policy to thwart virus panic among gullible end users. Stress that before users forward an e-mail alert to colleagues and higher-ups, they should send it to the help desk or the security team. Use the corporate intranet to explain the difference between real viruses and fakes, and provide links to Web sites that can set the record straight. Table 14.6 lists some of the most informative sites to learn about viruses.

Information and equipment theft. Data and information represent assets or goods that can also be stolen. Individuals who illegally access systems often do so to steal data and information. This requires identification numbers and passwords. Some criminals try different identification numbers and passwords until they find ones that work. Using **password sniffers** is another approach. A password sniffer is a small program hidden in a network or a computer system that records identification numbers and passwords. In a few days, a password sniffer can record hundreds or thousands of identification numbers and passwords. Using a password sniffer, a criminal hacker can gain access to computers and networks to steal data and information, invade privacy, plant viruses, and disrupt computer operations. Read the "E-Commerce" box on page 636 to learn how one hacker employed a sniffer program to gain credit card accounts.

All types of computer systems and equipment have been stolen. Computer theft is now second only to automobile theft according to recent U.S. crime statistics. In fact, in 1995 the theft of portable computers accounted for $640 million in losses, a 50 percent increase over 1994! These figures represent the United States alone. European statistics are just as staggering. In the United Kingdom more than 30 percent of all reported thefts are computer related. Printers, desktop computers, and scanners are often targets. Portable computers such as laptops (and the data and information stored in them) are especially easy for thieves to take. In some cases, the data and information stored in these systems are more valuable than the equipment. Without adequate protection and security measures, equipment can easily be stolen.

password sniffer

a small program hidden in a network or a computer system that records identification numbers and passwords

TABLE 14.6

Sources of Information about Viruses
Source: Leslie Goff, "Resources," *Computerworld*, May 25, 1998.

Web Page	Address
CIAC Internet Hoaxes Page	http://ciac.llnl.gov/ciac/CIACHoaxes.html
Computer Virus Myths Homepage	http://www.kumite.com/myths/
ICSA Virus Myths Page	http://www.icsa.net/services/consortia/alerthoax.htm
The Truth About E-mail Viruses	http://www.gerlitz.com/virushoax/
Virus Hoax FAQ	http://chekware.simplenet.com/hoaxfaq.htm
Dr. Solomon's Software	http://www.drsolomon.com
Symantec Corp.	http://www.symantec.com
Network Associates, Inc.	http://www.networkassociates.com

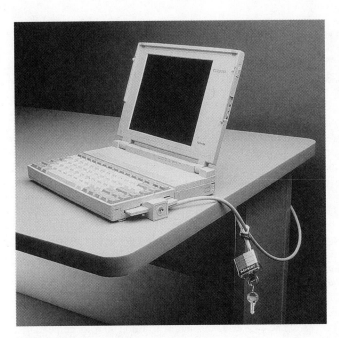

To fight computer crime, many companies use devices such as BookLock (shown here), which disables the disk drive and locks the computer to the desk. (Source: Courtesy of Secure-It Inc.)

software piracy

the act of illegally duplicating software

Internet piracy

the act of illegally gaining access to and using the Internet

Software and Internet piracy. Each time you use a word processing program or access software on a network, you are taking advantage of someone else's intellectual property. Like books and movies—other intellectual properties—software is protected by copyright laws. Often, people who would never think of plagiarizing another author's written work have no qualms about using and copying software programs they have not paid for. Such illegal duplicators are called pirates; the act of illegally duplicating software is called **software piracy**. Technically, software purchasers are granted the right only to use the software under certain conditions; they don't really own the software.

Internet piracy involves illegally gaining access to and using the Internet. While not as prevalent as software piracy, the amount of Internet piracy is growing rapidly.[14] Many companies on the Internet receive fees from customers for information, services, and even products. Some investment firms, for example, offer market analysis and investment information for a monthly or annual fee. Other companies offer information on sports or provide research information on a variety of topics. For some services, the fees can be thousands of dollars annually. Typically, Internet companies give customers identification numbers or passwords. Like customers illegally copying software, some customers illegally share their identification numbers and passwords with others. In other cases, criminal hackers obtain these numbers illegally on the Internet or from other sources. In these cases, Internet firms lose valuable revenues. Internet piracy can also be directed against individuals. While users are surfing the Web, outsiders can download an applet to the browser's machine, use its processor to perform computations and send the results back to a host. This technique is called MIPs-sucking.[15]

Computer-related scams. People have lost hundreds of thousands of dollars on real estate, stock, and other business scams. Today, many of these same types of scams are being performed using computers. Using the Internet, scam artists offer real estate deals, bank fraud, telephone lotteries, penny stocks, tax avoidance, and other get-rich-quick schemes. In one Internet scam, a telephone lottery investment of $129 was reported to guarantee a return that could range from $600 per week to $10,000 per week. The scam raised over $3 million in less than a year. In most cases, only the scam artists get rich. Here are some general tips to help you avoid becoming a victim:

• Don't agree to anything in a high-pressure meeting or seminar. Insist on having time to think it over and to discuss things with a spouse, partner, or even your lawyer. If a company won't give you the time you need to check it out and think things over, you don't want to do business with it. A good deal now will be a good deal tomorrow; the only reason for rushing you is if the company has something to hide.

E-COMMERCE

The Dark Side of Electronic Commerce

Money attracts criminals. And many hundreds of millions of dollars are flowing into the Internet each month. According to the Forrester Group (Cambridge, Massachusetts), the total value of goods and services traded between companies over the Internet will reach $8 billion this year and soar to $327 billion by 2002. Such large amounts of money are bound to draw the criminal element.

Exploiting known operating system flaws and utilizing commonly available hacking tools, an individual gained unauthorized access to a major Internet provider and gathered 100,000 credit card numbers, along with enough information to use them. The hacker allegedly inserted a program that gathered the credit information from a dozen companies selling products over the Internet. Had he succeeded, at a minimum, 100,000 customers accounts would have been compromised, and they would not have known it until they got their bill at the end of the month. The arrest thwarted potential losses of up to $1 billion in credit card fraud.

The scheme was discovered by the unidentified Internet service provider during routine maintenance. Technicians found an intruder had placed a program in their server called a "packet sniffer," which locates specified blocks of information such as credit card numbers. The FBI traced the intruder program to the individual who was using an account with the University of California–San Francisco. With the cooperation of another computer

user, who communicated with the perpetrator, the FBI arranged to have an undercover agent buy the stolen credit card information. After making two small buys, the FBI agents arranged to meet the perpetrator to pay $260,000 for 100,000 credit card numbers with credit limits that ranged up to $25,000 each. After decrypting and checking that the information was valid, the individual was taken into custody.

He waived his rights and acknowledged breaking into computers, including the Internet service provider. The FBI has not found any evidence that the individual made any purchases with the numbers himself, but the investigation is continuing. He appeared before a federal magistrate and was released on a $100,000 personal bond. As a condition of bail, the judge forbade him to come anywhere near a computer.

DISCUSSION QUESTIONS

1. What preventive measures could the ISP vendor have taken to avoid this problem altogether?
2. What impact might such incidents have on the growth of E-commerce?

Sources: Adapted from "Hacker Pleads Guilty to Stealing Credit Cards," *USA Today*, September 30, 1997; Richard Cole, "Hacker Sold Credit Card Nos.," at http://www.infowar.com/hacker, accessed July 10, 1998; and Emily Gurnon, "Credit Card Theft on Net Is Not Rocket Science," *San Francisco Examiner*, 1997.

- Don't judge a company based on appearances. Flashy Web sites can be created and up on the Net in a matter of days. After a few weeks of taking money, a site can vanish without a trace in just a few minutes. You may find that the perfect money-making opportunity Web site was a money maker for the crook and a money loser for you.
- Avoid any plan that pays commissions simply for recruiting additional distributors. Your primary source of income should be your own product sales. If the earnings are not made primarily by sales of goods or services to consumers or sales by distributors under you, you may be dealing with an illegal pyramid.
- Beware of shills, people paid by the company to lie about how much they've earned and how easy the plan was to operate. Check with an independent source to make sure that you aren't having the wool pulled over your eyes.
- Beware of the company's claim that it can set you up in a profitable home-based business, but you must first pay up front to attend a seminar

and buy expensive materials. Frequently, seminars are high-pressure sales pitches, and the material is so general that it is worthless.

- If interested in starting a home-based business, get a complete description of the work involved before you send any money. You may find that what you are asked to do after you pay is far different from what was stated in the ad. You should never have to pay for a job description or for needed materials.
- Get in writing the refund, buy-back, and cancellation policies of any company you deal with. Do not depend on oral promises.
- Do your homework. Check with your state attorney general and the National Fraud Information Center before getting involved, especially when the claims about the product or potential earnings seem too good to be true.

If you need advice about an Internet or on-line solicitation, or you want to report a possible scam, use the Online Reporting Form or Online Question & Suggestion Form features on the Web site for the National Fraud Information Center at http://fraud.org, or call the NFIC hotline at 1-800-876-7060.

International computer crime.   Computer crime is also an international issue, and it becomes more complex when it crosses borders. Estimates on software piracy in the global marketplace indicate that more than 27 percent of software is pirated, adding up to an estimated $2.8 billion in lost revenue. In China, 96 percent of software is pirated, with lost revenue totaling $1.4 billion. Indeed, the U.S. software industry lost $11.4 billion of revenue in 1997 due to illegal copying of programs such as Microsoft's Excel and Adobe Systems Illustrator.[16] Hardware is also affected. In Scotland, knife-wielding, masked criminals stole nearly $4 million worth of microchips. Another issue in international computer crime relates to illegally obtaining and selling restricted information in countries with less stringent laws. To avoid these problems, many countries require that computer equipment and software be registered with appropriate authorities before it can be brought into the country. With cash and funds being transferred electronically, some are concerned that international drug dealers and criminals are using information systems to launder illegally obtained funds.[17] Computer terrorism is another aspect of computer crime. Read the "Making a Difference" box on page 639 to learn how terrorists are going high-tech.

Preventing Computer-Related Crime

Because of increased computer use today, greater emphasis is placed on the prevention and detection of computer crime. Although more than 45 states have passed computer crime bills, some believe that they are not effective because (1) companies do not always actively detect and pursue computer crime, (2) security is inadequate, and (3) convicted criminals are not severely punished. However, all over the United States, private users, companies, employees, and public officials are making individual and group efforts to curb computer crime, and recent efforts have met with some success.

Crime prevention by state and federal agencies.   State and federal agencies have begun aggressive attacks on computer criminals, including criminal hackers of all ages. In 1986, Congress enacted the Computer Fraud and Abuse Act, which mandates punishment based on the victim's dollar loss. As discussed in the opening vignette, the Department of Defense also supports the Computer Emergency Response Team (CERT) that responds to network security breaches and monitors systems for emerging threats. Law enforcement agencies are also increasing their efforts to stop criminal hackers, and many states are now passing new, comprehensive computer crime bills to help eliminate computer crimes. Recent court cases and police reports involving computer crime show that lawmakers are ready to introduce newer and tougher computer crime legislation.

Several states have passed laws in an attempt to outlaw spam, the practice of sending large amounts of unsolicited e-mail to overwhelm users' e-mail boxes or the e-mail servers on a network. A California bill prohibits companies from sending e-mail in violation of an ISP's stated policies. The bill allows California-based ISPs to sue the company that sent the e-mail and to seek damages of $50 per e-mail sent for up to $25,000 plus attorney's fees.[18]

Crime prevention by corporations. Companies are also taking crime-fighting efforts seriously. Many businesses have designed procedures and specialized hardware and software to protect their corporate data and systems. Specialized hardware and software, such as encryption devices, can be used to encode data and information to help prevent unauthorized use.[19] **Biometrics** is another a way to protect important data and information systems. Biometrics simply means the measurement of a living trait, whether physical or behavioral.[20] Biometric techniques compare a person's unique characteristics against a stored set for the purpose of detecting differences between them. Biometric systems can scan fingerprints, faces, handprints, and retinal images to prevent unauthorized access to important data and computer resources. Most of the interest among corporate users is in fingerprint technology, followed by face recognition. Fingerprints hit the middle ground between price and usability. Iris and retina scans are more accurate, but they are more expensive, and they involve more equipment. Compaq has begun shipment of a fingerprint identification system for PCs called Fingerprint Identification Technology that will replace passwords with unique fingerprints.[21]

biometrics

the measurement of a living trait, whether physical or behavioral, for the purpose of protecting important data and information systems

EyeDentify's Icam 2000 is a retinal recognition device. These devices are used to grant or deny access by mapping the vascular pattern of the eye, thereby eliminating the need for cards, codes, or PIN numbers.
(Source: Courtesy of EyeDentify, Inc.)

Many of the early adopters of biometric technologies are healthcare organizations, because new federal legislation requires healthcare organizations to protect the privacy of patients.[22] Instead of typing in a password, a doctor or hospital employee puts his or her finger on a scanner connected to a personal computer and the image is matched against a set of authorized fingerprints. With some systems, instead of an image of a fingerprint, software uses an algorithm to create a personal identification number that equates to a person's fingerprint.

MAKING A DIFFERENCE

New Center to Combat Terrorists Trading Explosives for Computers

Not long ago, if a terrorist wanted to cause a blackout in, say, Los Angeles, it would have taken some work. He might have gained a job as a utility worker so he could sabotage the electrical system. Or, he might have loaded a truck with explosives and directed it into a power plant. Today, intelligence experts say, it's possible for trained computer hackers to darken the City of Angels from the comfort of their home, even if their home is half a world away. Worse yet, warned CIA Director George Tenet recently, they may enjoy the full backing and technical support of a foreign government.

According to the CIA chief, at least a dozen countries, some hostile to America, are developing programs to attack other nations' information and computer systems. China, Libya, Russia, Iraq, and Iran are among those considered a threat. Indeed, the People's Liberation Daily in China notes that a foe of the United States "only has to mess up the computer systems of its banks by high tech means. This would disrupt and destroy the U.S. economy."

Officials are worried because so much of America's infrastructure is either driven or connected by computers. Computers run financial networks, regulate the flow of oil and gas through pipelines, control water reservoirs and sewage treatment plants, power air traffic control systems, and sustain telecommunications networks, emergency services, and power grids. All are vulnerable. A computer terrorist capable of implanting the right virus or accessing the right terminal can cause massive damage. "Today, because of the networked nature of our critical infrastructures, our enemies needn't risk attacking our strong military if they can much more easily attack our soft digital underbelly," said Senator Jon Kyl of Arizona, chairman of the technology, terrorism, and government information subcommittee.

To help industries fend off hacker attacks, both foreign and domestic, the government has created the National Infrastructure Protection Center, to be staffed by 125 people from the FBI, other agencies, and industry.

Recent events clearly demonstrate that tighter defenses are needed. In 1996, a Swedish hacker wormed his way through cyberspace from London to Atlanta to Florida, where he rerouted and tied up telephone lines to 11 counties, put 911 emergency service systems out of commission, and impeded the emergency responses of police, fire, and ambulance services.

In 1997, several serious incidents occurred. A 14-year-old boy with a home computer disabled control-tower communications at a Worcester, Massachusetts, airport for six hours. A secret war game was run with a set of written scenarios in which energy and telecommunications utilities were disrupted by computer attacks. During these same war games, three two-person "red teams" from the National Security Agency actually used hacker techniques that can be learned on the Internet to penetrate Department of Defense computers. After gaining access to the military's electronic message systems, the teams were poised to intercept, delete, and modify all messages on the networks. Ultimately, the hackers achieved access to the DOD's classified network and, if they had wished, could have denied the Pentagon the ability to deploy forces. In yet another computer incident, teenagers in the United States and Israel were able to break into Pentagon computers. The intrusion was quickly detected, but it took several days before the military and law enforcement authorities determined that teenagers, and not a foreign government, were behind the attack. "It is often impossible to determine at the outset if an intrusion is an act of vandalism, computer crime, terrorism, foreign intelligence activity, or some form of strategic attack," says Michael Vatis of the FBI's newly formed National Infrastructure Protection Center. The Department of Defense found that 63 percent of test attacks on its own systems went undetected.

DISCUSSION QUESTIONS

1. Develop three scenarios of how computer terrorists could disrupt our daily lives by attacking civilian computer systems.

2. Search the Internet for information on the newly formed National Infrastructure Protection Center. What evidence can you find of specific actions that this agency has taken to improve U.S. computer security?

Sources: Adapted from Douglas Pasternak and Bruce B. Auster, "Terrorism at the Touch of a Keyboard," *US News and World Report*, July 13, 1998; "Lawmakers: Cyber Terrorism is a Worry," *USA Today*, June 11, 1998; Matt Hamblen, "Security Pros: Hacking Penalties Too Lenient," *Computerworld*, March 23, 1998, p. 2.

TABLE 14.7

Common Methods Used to Commit Computer Crimes

Even though the number of potential computer crimes appears to be limitless, the actual methods used to commit crime are limited.

Methods	Examples
Add, delete, or change inputs to the computer system.	Delete records of absences from class in a student's school records.
Modify or develop computer programs that commit the crime.	Change a bank's program for calculating interest to make it deposit rounded amounts in the criminal's account.
Alter or modify the data files used by the computer system.	Change a student's grade from C to A.
Operate the computer system in such a way as to commit computer crime.	Access a restricted government computer system.
Divert or misuse valid output from the computer system.	Steal discarded printouts of customer records from a company trash bin.
Steal computer resources, including hardware, software, and time on computer equipment.	Make illegal copies of a software program without paying for its use.

This saves data storage and avoids having actual fingerprints stored on file. Some companies actually hire former criminals to thwart other criminals.[23]

Crime-fighting procedures usually require additional controls on the information system. Before designing and implementing controls, organizations must consider the types of computer-related crime that might occur, the consequences of these crimes, and the cost and complexity of needed controls. In most cases, organizations conclude that the trade-off between crime and the additional cost and complexity weighs in favor of better system controls. Having knowledge of some of the methods used to commit crime is also helpful in preventing, detecting, and developing systems resistant to computer crime (see Table 14.7). Table 14.8 provides a set of useful guidelines to protect your computer from hackers.

Companies are also joining to fight crime. The **Software Publishers Association (SPA)**, which was formed by a number of leading software companies, audits companies and checks for software licenses. Organizations

Software Publishers Association (SPA)

an organization formed by a number of leading software companies to audit and check for software licenses

TABLE 14.8

How to Protect Your Corporate Data From Hackers

- Install strong user authentication and encryption capabilities on your firewall
- Install the latest security patches, which are often available at the vendor's Internet site
- Disable guest accounts and null user accounts that let intruders access the network without a password
- Do not provide overfriendly log-in procedures for remote users (e.g., an organization that used the word *welcome* on their initial log-on screen found they had difficulty prosecuting a hacker)
- Give an application (e-mail, file transfer protocol, and domain name server) its own dedicated server
- Restrict physical access to the server and configure it so that breaking into one server won't compromise the whole network
- Turn audit trails on
- Consider installing caller ID

that are found to be illegally using software can be fined or sued. Depending on the violation, the fine can be hundreds of thousands of dollars. While the SPA is an effective deterrent in the United States, efforts to curb software abuse in other countries is much more difficult.

Using antivirus programs. Companies and individuals must protect their computer systems and networks from viruses. Most people use **antivirus programs** or utilities to prevent viruses and recover from them if they infect a computer. These programs range in cost from free (shareware) to a few hundred dollars. Antivirus programs are developed for different operating systems. Norton Antivirus for Windows 95/98 and NT workstations, Norton Antivirus for Macintosh, Quarterdeck Utility Pack for Windows 95/98, Dr. Solomon's Anti-Virus Toolkit for Windows 95/98, and Network Associate's Virex for Macintosh are just a few examples of antivirus programs. Proper use of antivirus software requires the following steps:

antivirus programs

programs or utilities that prevent viruses or help recover from them if they infect a computer

1. *Install a virus scanner and run it often.* Many of these programs automatically check for viruses each time you boot up your computer or insert a diskette, and some even monitor all transmissions and copying operations.
2. *Update the virus scanner often.* Old programs may fail to detect new viruses.
3. *Scan all diskettes before copying or running programs from them.* Hiding on diskettes, viruses often move between systems. If you carry document or program files on diskettes between computers at school or work and your home system, always scan them.
4. *Install software only from a sealed package produced by a known software company.* Even software publishers can unknowingly distribute viruses on their program diskettes. Most scan their own systems, but viruses may still remain.
5. *Follow careful downloading practices.* If you download software from the Internet or a bulletin board, check your computer for viruses immediately after completing the transmission.
6. *If you detect a virus, take immediate action.* Early detection often allows you to remove a virus before it does any serious damage.

Despite careful precautions, viruses can still cause problems. They can elude virus-scanning software by lurking almost anywhere in a system. Future antivirus programs may incorporate "nature-based models" that check for unusual or unfamiliar computer code. The advantage of this type of virus program is the ability to detect new viruses that are not part of an antivirus database.

Internet laws and protection for libel and decency. On February 8, 1996, President Clinton signed the Telecommunications Act of 1996 into law. This bill also included an act called the Communications Decency

Antivirus software should be used and updated often.
(Source: Courtesy of Symantec Corporation.)

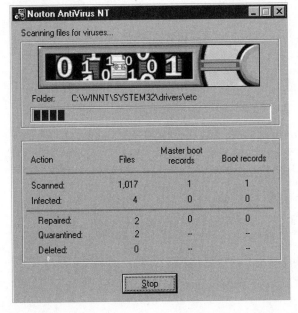

Net Nanny is a filtering
software that helps block
unwanted Internet content from
children and young adults.
(Source: Courtesy of Net Nanny
Software International, Inc.)

Act. One of the original provisions of this act is the ability of the government to jail or fine anyone up to $100,000 for sending indecent materials to a minor electronically. People and companies were very concerned about the free speech implications of this provision and turned their Web page to black in protest. Subsequently, the courts limited the Communications Decency Act. In an unrelated case, the German government forced CompuServe to suspend about 200 newsgroups on its service that Germany claimed violated German law. The law was Section 184 of the German Criminal Code, which deals with the distribution of pornographic materials to minors.

To help parents control what their children see on the Internet, some companies are developing software—called *censorware*—to help screen Internet content. Many of these screening programs also prevent children from sending personal information over e-mail or through chat groups. This stops children from broadcasting their name, address, phone number, or other personal information over the Internet. The two approaches used are filtering, which blocks certain Web sites, and rating, which places a rating on Web sites. Examples of censorware include Border Manager, Choice Net, Click & Browse Junior, Cybersitter, Cyber Patrol, Net Nanny, Specs for Kids, Surf Guard, and Surf Watch.

With the increased popularity of networks and the Internet, libel and decency become important legal issues. A publisher, such as a newspaper, can be sued for libel, which involves publishing a written statement that is damaging to a person's reputation. Generally, a bookstore cannot be held liable for statements made in a newspaper or other publications it sells. On-line services, such as CompuServe and America Online, have control over who puts information on their service but may not have direct control over the content of what is published by others on their service. Can on-line services be sued for libel for content that someone else publishes on their service? Are on-line services more like a newspaper or a bookstore? While this legal issue has not been completely resolved, some court cases have been decided. The *Cubby, Inc. v. CompuServe* case ruled that CompuServe was more like a bookstore and not liable for content put on its service by others. In this case, Judge Leisure stated, "While CompuServe may decline to carry a given publication altogether, in reality, once it does decide to carry a given publication, it will have little or no editorial control over that publication's content."

Preventing crime on the Internet. As mentioned in Chapter 7, Internet security can include firewalls and a number of methods to secure financial transactions. A firewall can include hardware and software combinations that act as a barrier between an organization's information system and the outside world. A number of systems have been developed to safeguard financial transactions on the Internet.

To help prevent crime on the Internet, the following steps can be taken.

1. Develop effective Internet and security policies for all employees.
2. Use a stand-alone firewall (hardware and software) with network monitoring capabilities.

TECHNOLOGY ADVANTAGE
Becton Dickinson Implements Intranet

Becton Dickinson manufactures and sells a broad range of medical supplies, devices, and diagnostic systems for healthcare professionals, medical research institutions, and the general public. Recent annual sales exceeded $3 billion.

Much of the value of a high-tech company like Becton Dickinson lies in the collective knowledge of its employees. Recognizing an opportunity to harness this knowledge, the company implemented an intranet to enable sharing of the combined wisdom of its 19,000 employees. The intranet provides access to a database of the best practices developed by employees and also serves as a contact resource and corporate technical encyclopedia. Using this database of expertise, employees can find an in-house expert in key topics, such as clinical microbiology, in a few keystrokes. Likewise, information on intravenous catheterization, another core competency of the company, is readily available on-line.

Department committees made up of Web-site development staff, IS employees, and marketing managers meet once a month. A worldwide quarterly meeting of selected participants decides what new information to make accessible over the intranet. Under development is intranet Web-site data to expedite the budget process.

Prior to the intranet, there was no way for Becton Dickinson employees to share knowledge. The intranet is currently available to about 70 percent of the employees. The goal is to give intranet access to all employees with a desktop PC within a year.

DISCUSSION QUESTIONS

1. How important is it to limit access to this intranet to employees? What sort of measures should be implemented to ensure this?
2. What additional data and Web sites do you think would be of value to the employees?

Sources: Adapted from Han Greenberg, "Knowledge Sharing Via Intranet," October 5, 1998, *InformationWeek*, p. 14SS and "HT Medical Systems Announces Marketing Alliance with Becton Dickinson," *PRNewswire*, September 22, 1998.

---

3. Monitor managers and employees to make sure they are using the Internet for business purposes only.
4. Use Internet security specialists to perform audits of all Internet and network activities.

Even with these precautions, computers and networks can never be completely protected against crime. One of the biggest threats is from employees. According to Rich Ayers, vice president of Network Information Management at Chase Manhattan Bank Corporation, "Just blocking everyone externally doesn't mean you are safe." Some believe that 80 percent of all computer attacks come from employees or managers inside the company. Although firewalls provide good perimeter control to prevent crime from the outside, procedures and protection measures are needed for personnel. Passwords, identification numbers, and tighter control of employees and managers also help prevent Internet-related crime. Read the Technology Advantage box to gain an appreciation of the need to provide safeguards over the use of corporate intranets.

PRIVACY

Another important social issue in information systems involves privacy. In 1890, U.S. Supreme Court Justice Louis Brandeis stated that the "right to be left alone" is one of the most "comprehensive of rights and the most valued by civilized man." Basically, the issue of privacy deals with this right to be left alone or to be withdrawn from public view. With information systems, privacy deals with the collection and use or misuse of data. Data is constantly being collected and stored on each of us. This data is often distributed over easily accessed networks and without our knowledge or consent. Concerns of privacy regarding this data must be addressed.

With today's computers, the right to privacy is an especially challenging problem. More data and information is produced and used today than ever before. A difficult question to be answered is, "Who owns this information and knowledge?" If a public or private organization spends time and resources in obtaining data on you, does the organization own the data and can it use the data in any way it desires? Government legislation answers these questions to some extent for federal agencies, but the questions remain unanswered for private organizations.

Privacy Issues

The issue of privacy is important because data on an individual can be collected, stored, and used without that person's knowledge or consent. When someone is born, takes certain high school exams, starts working, enrolls in a college course, applies for a driver's license, purchases a car, serves in the military, gets married, buys insurance, gets a library card, applies for a charge card or loan, buys a house, or merely purchases certain products, data is collected and stored somewhere in computer databases.

Privacy and the federal government. The federal government is perhaps the largest collector of data. Close to four billion records exist on individuals, collected by about 100 federal agencies, ranging from the Bureau of Alcohol, Tobacco, and Firearms to the Veterans Administration. Other data collectors include state and local governments and profit and nonprofit organizations of all types and sizes.

In recent years, there have been a number of examples to cause concern for privacy of on-line data. In 1997, the Social Security Administration suspended a service that let people look up personal earnings, disability information, and benefits estimates when the public realized that the data could be widely accessed. In July 1998, America Online raised consumers' ire after it proposed giving member information to partners that could then telemarket to AOL's 11 million subscribers. AOL backed down. In August 1998, Experian, a credit bureau site, dropped a service that let consumers check their credit history after a nasty glitch in which one person after another was mistakenly given someone else's report.

As a result, the Federal Trade Commission is considering action, and there are about 70 laws before Congress regarding various aspects of Internet privacy. The European Union has already passed a data-protection directive that requires firms transporting data across national boundaries to have certain privacy procedures in place. This affects virtually any company doing business in Europe; the directive is driving much of the attention being given to privacy in the United States.

Most user companies and computer vendors are wary of the federal government dictating Internet privacy standards. A group called the Online Privacy Alliance is developing a voluntary code of conduct. It is backed by companies such as AT&T, IBM, Dun & Bradstreet, Time-Warner, Walt Disney, the Lexis-Nexis division of Reed Elsevier Inc., Microsoft, and Netscape. The alliance's guidelines will call on companies to notify users when they are collecting data at Web sites to gain consent for all uses of that data, to provide for the enforcement of privacy policies, and to have a clear process in place for receiving and addressing user complaints. The alliance's policy can be found at http://www.privacyalliance.org.

Privacy at work. The right to privacy at work is also an important issue. Some experts believe that there will be a collision between workers that want their privacy and companies that demand to know more about their employees. Recently, issues of companies monitoring their employees have raised concerns. Workers may find that they are being closely monitored via computer technology. These computer monitoring systems tie directly into computerized workstations; specialized computer programs can track every keystroke made by a user. This type of system can determine what workers are doing while at the keyboard. The system also knows when the worker is not using the keyboard or computer system. These systems can estimate what a person is doing and how many breaks he or she is taking. Needless to say, many workers consider this type of supervision very dehumanizing.

E-mail privacy issues. E-mail also raises some interesting issues about work privacy. Federal law permits employers to monitor e-mail sent and received by employees. Furthermore, e-mail messages that have been erased from hard disks may be retrieved and used in lawsuits because the laws of discovery demand that companies produce all relevant business documents (see Figure 14.1). On the other hand, the use of e-mail among public officials may violate "open meeting" laws. These laws, which apply to many local, state, and federal agencies, prevent public officials from meeting in private about matters that affect the state or local area.

FIGURE 14.1

E-mail has changed how workers and managers communicate. With e-mail, people can communicate in the same building or around the world. E-mail, however, can be monitored and intercepted. As with other services—like cellular phones—the convenience of e-mail must be balanced with the potential of privacy invasion.
(Source: Image copyright © 1998 PhotoDisc, Inc.)

There has been significant commentary from the legal community regarding the applicability of privileges—specifically, the attorney-client privilege—to communications made electronically, including electronic mail. Perhaps in response to this commentary, but more likely as a reflection of a technologically-advanced society, the state of New York amended its Civil Practice Law and Rules and clarified the matter: A privileged communication does not lose its privileged character by virtue of its communication or transmission electronically. Over a dozen more states have either put into effect a similar law or are debating this matter.[24] Navigating these paradoxical privacy concerns will be a challenge for organizations in the Information Age.

Privacy and the Internet. Some people assume that there is no privacy on the Internet and that you use the Internet at your own risk. Others believe that companies with sites on the Internet should have strict privacy procedures and should be accountable for privacy invasion. With either view, the potential of privacy invasion on the Internet is huge. People invading your privacy could be anyone from criminal hackers to marketing companies to corporate bosses. Personal and professional information on you can be seized on the Internet without your knowledge or consent. E-mail is a prime target, as discussed previously.[25] Sending an e-mail message is like having an open conversation in a large room. People can listen to your messages. When you visit a Web site on the Internet, information about you and your computer can be captured. When this information is combined with other information, companies can know what you read, what products you buy, and what your interests are. According to an executive of an Internet software monitoring company, "It's a marketing person's dream."

In a recent survey conducted by Louis Harris, 72 percent of people who buy products on the Web say it's very important to have a policy explaining how personal information is used. Companies that use Web sites need to have a policy statement that people feel comfortable with and that is extremely clear in terms of what information is collected and what will and will not be done with the information. However, the Federal Trade Commission found privacy policies were posted at only about 14 percent of 1,400 Web sites it recently surveyed. And an FTC sweep in October 1997 of 126 children's commercial sites found that 86 percent of those sites collected data about children, including e-mail addresses and phone numbers, most without seeking parental consent. A *Business Week* check of the top 100 Web sites found that 43 percent displayed privacy policies. Of the notices posted, some were difficult to find and inconsistent in explaining how data is tracked and used. Moreover, TRUSTE, a nonprofit organization that provides a "trustmark" to put on Web sites signifying disclosure of privacy policies and outside auditing practices, has only 75 sites signed up after nine months of operation, only 10 percent of its goal of 750. The real issue is over what content providers want with registration information. If a site requests that you provide your name and address, you have every right to know why and what will be done with it. If you buy something and provide a shipping address, will it be sold to other retailers? Will your e-mail address be sold on a list of active Internet shoppers? And if so, you should realize that it's no different than the lists compiled from the orders you place with catalog retailers. You have the right to be taken off any mailing address. Read the "Ethical and Societal Issues" box to learn more about how personal data is captured and used.

ETHICAL AND SOCIETAL ISSUES

Personalization

Today more and more companies are seeking to gain a closer relationship with their customers using the Internet. On the other hand, heightened sensitivity to privacy has even the most ambitious businesses treading lightly. "Privacy is clearly at the forefront of our minds in the on-line arena," says Sandy Herdon, manager of on-line marketing at American Airlines, which recently unveiled a major overhaul of its Web site for frequent fliers. American is using software that generates personalized

Web pages by monitoring and responding to customers' activities while on the Web site. This process is known as personalization.

The American Airlines site represents one of the most aggressive uses of personalization to date. Each time a person visits the site, he or she will see Web pages driven by data from three sources: American's existing database, a detailed questionnaire submitted by customers, and observation of what the person does while on the site. American's cross-marketing partners—Citibank, Hertz, and Hilton—will also participate in the site. A person who requests a rate for a flight to Tuscon will receive additional information, such as rates for a Hertz car rental and a Hilton hotel room. Such sites may be the wave of the future. Advertisers and Website managers stand to profit from this additional exposure but are concerned with censure from privacy advocates.

How do Web sites collect data? Sites gather information with and without consumer's knowledge. The most common way is through "clickstream" data—information about where people go within a site and the ads and content they see. Clickstream data is collected by so-called cookies, or small data files placed on users' hard drives when they first visit a Web site. When the individual goes back to that Web site, the site's computer server can read the usage data from the cookies. This data is then stored in a database and can be used to target ads or content, based on preferences tracked. Personal information—such as e-mail address, name, street address, age, or sex—is gathered through registration at such sites as Time Warner's Pathfinder or Amazon.com. The same sort of data can also be obtained from so called "swebstakes," promotional giveaway programs run by companies such as Excite's MatchLogic, a subsidiary that posts ads and marketing campaigns for 65 clients across a host of Web sites.

Marketers say the ability to gather data and target consumers is what makes the net unique and is key to attracting advertisers and spurring E-commerce. Many say they are taking steps to protect their consumers, but they use information to deliver customized services and ads rather than something random. However, companies simply can't ignore the fact that people on the Web have strong privacy concerns. If they are a little too explicit about what they know about a person, and that comes across, they can offend that person and he or she may never come back to that site.

Discussion Questions

1. Do you think that personalization is good for the consumer? Why or why not?
2. What can you do to protect your privacy when you surf the Web?

Sources: Adapted from Gregory Dalton, et al., "Pressure for Better Privacy," *Information Week*, June 22, 1998, pp. 36–38; Rebecca Quick, "On-Line Groups Are Offering Up Privacy Plans, *The Wall Street Journal*, June 22, 1998, pp. B1, B3; Heather Green, et al., "A Little Privacy, Please," *Business Week*, March 16, 1998, pp. 98–100; and Bill Machrone, "The Cookie Monsters Are After Our Computers," *PC Week*, June 23, 1997, p. 89.

Fairness in Information Use

Selling information to other companies can be so lucrative that many companies will probably continue to store and sell the data they collect on customers, employees, and others. When is this information storage and use fair and reasonable to the individuals whose data is stored and sold? Do individuals have a right to know about data stored about them and to

TABLE 14.9

The Right to Know and the
Ability to Decide

Fairness Issues	Database Storage	Database Usage
The right to know	Knowledge	Notice
The ability to decide	Control	Consent

Knowledge. Should individuals have knowledge of what data is stored on them? In some cases, individuals are informed that information on them is stored in a corporate database. In others, individuals do not know that their personal information is stored in corporate databases.

Control. Should individuals have the ability to correct errors in corporate database systems? This is possible with most organizations, although it can be difficult in some cases.

Notice. Should an organization that uses personal data for a purpose other than the original purpose notify individuals in advance? Most companies don't do this.

Consent. If information on individuals is to be used for other purposes, should these individuals be asked to give their consent before data on them is used? Many companies do not give individuals the ability to decide if information on them will be sold or used for other purposes.

decide what data is stored and used? As shown in Table 14.9, these questions can be broken down into four issues that should be addressed: knowledge, control, notice, and consent.

Federal Privacy Laws and Regulations

In the past few decades, significant laws have been passed regarding an individual's right to privacy. Others relate to business privacy rights and the fair use of data and information.

The Privacy Act of 1974. The major piece of legislation on privacy is the Privacy Act of 1974 (PA74), enacted by Congress during Gerald Ford's presidency. PA74 applies only to certain federal agencies. The act, which is about 15 pages long, is straightforward and easy to understand.

The purpose of this act is to provide certain safeguards for individuals against an invasion of personal privacy by requiring federal agencies (except as otherwise provided by law) to do the following:

- Permit individuals to determine what records pertaining to them are collected, maintained, used, or disseminated by such agencies
- Permit individuals to prevent records pertaining to them obtained by such agencies for a particular purpose from being used or made available for another purpose without their consent
- Permit individuals to gain access to information pertaining to them in federal agency records, to have a copy of all or any portion thereof, and to correct or amend such records
- Collect, maintain, use, or disseminate any record of identifiable personal information in a manner that ensures that such action is for a necessary and lawful purpose, that the information is current and accurate for its intended use, and that adequate safeguards are provided to prevent misuse of such information

- Permit exemptions from the requirements with respect to records provided for in this act only in those cases where there is an important public need for such exemption as has been determined by specific statutory authority
- Be subject to civil suit for any damages that occur as a result of willful or intentional action that violates any individual's rights under this act

PA74, which applies to all federal agencies except the CIA and law enforcement agencies, also established a Privacy Study Commission to study existing data banks and to recommend rules and legislation for consideration by Congress. PA74 also requires training for all federal employees who interact with a "system of records" under the act. Most of the training is conducted by the Civil Service Commission and the Department of Defense. Another interesting aspect of PA74 concerns the use of social security numbers—federal, state, and local governments and agencies cannot discriminate against any individual for not disclosing or reporting his or her social security number.

Other federal privacy laws. In addition to PA74, other pieces of federal legislation relate to privacy. A federal law that was passed in 1992 bans unsolicited fax advertisements. This law was upheld in a 1995 ruling by the Ninth U.S. Circuit Court of Appeals that concluded that the law is a reasonable way to prevent the shifting of advertising costs to customers. Table 14.10 lists additional laws related to privacy.

TABLE 14.10

Federal Privacy Laws and their Provisions

Law	Provisions
Fair Credit Reporting Act of 1970 (FCRA)	Regulates operations of credit-reporting bureaus, including how they collect, store, and use credit information
Tax Reform Act of 1976	Restricts collection and use of certain information by the Internal Revenue Service
Electronic Funds Transfer Act of 1979	Outlines the responsibilities of companies that use electronic funds transfer systems, including consumer rights and liability for bank debit cards
Right to Financial Privacy Act of 1978	Restricts government access to certain records held by financial institutions
Freedom of Information Act of 1970	Guarantees access for individuals to personal data collected about them and about government activities in federal agency files
Education Privacy Act	Restricts collection and use of data by federally funded educational institutions, including specifications for the type of data collected, access by parents and students to the data, and limitations on disclosure
Computer Matching and Privacy Act of 1988	Regulates cross-references between federal agencies' computer files (e.g., to verify eligibility for federal programs)
Video Privacy Act of 1988	Prevents retail stores from disclosing video rental records without a court order
Telephone Consumer Protection Act of 1991	Limits telemarketers' practices
Cable Act of 1992	Regulates companies and organizations that provide wireless communications services, including cellular phones
Computer Abuse Amendments Act of 1994	Prohibits transmissions of harmful computer programs and code, including viruses

State Privacy Laws and Regulations

Some states either have or are proposing their own privacy legislation. The use of social security numbers and medical records, the disclosure of unlisted telephone numbers by telephone companies and credit reports by credit bureaus, the disclosure of bank and personal financial information, and the use of criminal files are some of the issues being considered by state legislators. Furthermore, many of these proposed legislative actions apply to both public and private organizations.

Corporate Privacy Policies

Even though privacy laws for private organizations are not very restrictive, most organizations are very sensitive to privacy issues and fairness. They realize that invasion of privacy problems can hurt their business, turn away customers, and dramatically reduce revenues and profits. Consider a major international credit or charge card company. If it were reported that the company sold confidential financial information on millions of customers to other companies, the results could be disastrous. In a matter of days, the firm's business and revenues could be reduced dramatically. Thus, most organizations maintain privacy policies, even though they are not required by law. Some companies even have a privacy bill of rights that specifies how the privacy of employees, clients, and customers is to be protected. Corporate privacy policies should address a customer's knowledge, control, notice, and consent over the storage and use of information. They may also cover who has access to private data and when it may be used.

Protecting Individual Privacy

While numerous state and federal laws deal with privacy, privacy laws do not completely protect individual privacy. In addition, not all companies have privacy policies. As a result, many people are taking steps to increase their own privacy protection. Some of the steps that individuals can take to protect personal privacy include the following:

- *Find out what is stored about you in existing databases.* Call the major credit bureaus to get a copy of your credit report for $8 (you can obtain one free if you have been denied credit in the last 60 days). The major companies are Equilan (800-392-1122), Trans Union (312-258-1717), and Equifax (800-685-1111). You can also submit a Freedom of Information Act request to a federal agency that you suspect may have information stored on you.
- *Be careful when you share information about yourself.* Don't share information unless it is absolutely necessary. Every time you give information about yourself through an 800, 888, or 900 call, your privacy is at risk. You can ask your doctor, bank, or financial institution not to share information about you with others without your written consent.
- *Be proactive to protect your privacy.* You can get an unlisted phone number and ask the phone company to block caller ID systems from reading your phone number. If you change your address, don't fill out a change-of-address form with the U.S. Postal Service; you can notify the people

and companies that you want to have your new address. Be careful about sending personal e-mail messages over a corporate e-mail system. You can also avoid junk mail and telemarketing calls by contacting Direct Marketing Association at P.O. Box 3861, New York, NY 10163.

THE WORK ENVIRONMENT

The use of computer-based information systems has changed the makeup of the workforce. Jobs that require IS literacy have increased, while many less-skilled positions have been eliminated. Corporate programs, such as reengineering and continuous improvement, bring with them the concern that, as business processes are restructured and information systems are integrated within them, the people involved in these processes will be removed.

However, the growing field of computer technology and IS has opened up numerous avenues to professionals and nonprofessionals of all backgrounds. Enhanced telecommunications has been the impetus for new types of business and has created global markets in industries once limited to domestic markets. Even the simplest tasks have been aided by computers, making cash registers faster, causing order processing to be smoother, and allowing people with disabilities to participate more actively in the workforce. As computers and other IS components drop in cost and become easier to use, more workers will benefit from the increased productivity and efficiency provided by computers.

Information systems, while increasing productivity and efficiency, can raise other concerns.

Health Concerns

Organizations can increase employee effectiveness by paying attention to the health concerns in today's work environment. For some people, working with computers can cause occupational stress. Workers' anxieties about job insecurity, loss of control, incompetence, and demotion are just a few of the fears they might experience. In some cases, the stress may become so severe that workers may sabotage computer systems and equipment. Monitoring of employee stress may alert companies to potential problems. Training and counseling can often help the employee and deter problems.

Computer use may affect physical health as well. Strains, sprains, tendinitis, and other problems account for more than 60 percent of all occupational illnesses and about a third of workers' compensation claims, according to the Joyce Institute in Seattle. The costs to U.S. corporations for these types of health problems is as high as $27 billion annually. Claims relating to **repetitive motion disorder**, which can be caused by working with computer keyboards and other equipment, have increased greatly. Also called **repetitive stress injury (RSI)**, the problems can include tendinitis, tennis elbow, the inability to hold objects, and sharp pain in the fingers. Also common is **carpal tunnel syndrome (CTS)**, which is the aggravation of the pathway for nerves that travel through the wrist (the carpal tunnel). CTS involves wrist pain, a feeling of tingling and numbness, and difficulty grasping and holding objects. It may be caused by a number of factors, such as stress, lack of exercise, and the repetitive motion of typing on a computer keyboard.

repetitive motion disorder
health problems caused by working with computer keyboards and other equipment

repetitive stress injury (RSI)
such problems as tendinitis, tennis elbow, the inability to hold objects, and sharp pain in the fingers; same as *repetitive motion disorder*

carpal tunnel syndrome (CTS)
the aggravation of the pathway for nerves that travel through the wrist (the carpal tunnel)

Decisions on workers' compensation related to repetitive stress syndrome can go either way. A federal judge threw out a $5.3 million verdict against Digital Equipment Corporation in a keyboard-injury case after submission of new evidence suggested the plaintiff's debilitating wrist injury was not work related. In another verdict, a different judge upheld a $274,000 award by the jury for repetitive stress syndrome.[26]

Other work-related health hazards involve emissions from improperly maintained and used equipment. Some studies show that poorly maintained laser printers may release ozone into the air; others dispute the claim. Numerous studies on the impact of emissions from display screens have also resulted in conflicting theories. Although some medical authorities believe that long-term exposure can cause cancer, studies are not conclusive at this time. Regardless, many organizations are developing conservative and cautious policies.

Most computer manufacturers publish technical information on radiative emissions from their screens, and many companies pay close attention to this information. San Francisco was one of the first cities to propose a video display terminal (VDT) bill. The bill requires companies with 15 or more employees who spend at least four hours a day working with computer screens to give 15-minute breaks every two hours. In addition, adjustable chairs and workstations are required if requested by employees.

Avoiding Health and Environmental Problems

ergonomics

the study of designing and positioning computer equipment to reduce health problems

Many computer-related health problems are minor and caused by a poorly designed work environment. The computer screen may be hard to read, with problems of glare and poor contrast. Desks and chairs may also be uncomfortable. Keyboards and computer screens may be fixed in place or difficult to move. The hazardous activities associated with these unfavorable conditions are collectively referred to as *work stressors*. Although these problems may not be of major concern to casual users of computer systems, continued stressors such as repetitive motion, awkward posture, and eyestrain may cause more serious and long-term injuries. If nothing else, these problems can severely limit productivity and performance.

The study of designing and positioning computer equipment, called **ergonomics**, has suggested a number of approaches to reducing these health problems. The objective is to have "no pain" computing. The slope of the keyboard, the positioning and design of display screens, and the placement and design of computer tables and chairs have been carefully studied. Flexibility is a major component of

Prolonged use of keyboards and computer equipment may cause RSI and other problems. (Photo source: Steve Kahn.)

ergonomics and an important feature of computer devices. People of differing sizes and preferences require different positioning of equipment for best results. Some people, for example, want to have the keyboard in their laps; others prefer to place the keyboard on a solid table. Because of these individual differences, computer designers are attempting to develop systems that provide a great deal of flexibility.

In addition to steps taken by companies, individuals can also reduce RSI and develop a better work environment. Here is what can be done.

- Maintain good posture and positioning. In addition to good equipment, good posture and work habits can eliminate or reduce the potential of RSI.
- Don't ignore pain or discomfort. Many workers ignore early signs of RSI and as a result the problem becomes much worse and more difficult to treat.
- Use stretching and strengthening exercises. Often, such exercises can prevent RSI.
- Find a good physician. Some doctors are not familiar with RSI and how it can be treated.
- After treatment, start back slowly and pace yourself. Many people who are treated for RSI start back to work too soon and injure themselves again.

While we have investigated how computers may be harmful to your health, the computer can also be used to help prevent and treat general health problems. As discussed in Part III on business information systems, we have seen how computers can be used to assist doctors and other medical professionals by diagnosing medical problems and suggesting potential treatments. People can also use computers to get medical information. Special medical software for personal computers can help people get medical information and determine whether they need to consult a physician. In addition, a wealth of information is available on the Internet on a variety of medical topics. See Table 14.11 for a few examples.

TABLE 14.11

Medical Topics on the Internet

Internet Address	Description
http://www.neoforma.com	Enables users to conduct electronic commerce with healthcare vendors, automatically send out requests-for-proposals via broadcast e-mail, establish free e-mail accounts, post classified advertisements, obtain career information and job postings, and participate in topical discussion groups.
http://dialspace.dial.pipex.com	Anatomy and Medical Graphics Visible Human Project is a complete, anatomically detailed, three-dimensional representations of the male and female human body. The current phase of the project is collecting transverse CAT, MRI, and frozen section images of representative male and female cadavers at one-millimeter intervals.
http://www.WebMD.com	Provides access to reference material and on-line professional publications from Thomson Healthcare Information Group, Stamford, Connecticut.
http://www.cancer.org	Web site of the American Cancer Society.
http:/www.mayo.edu	A tour of the Mayo clinic.
gopher://gopher.med.harvard.edu	A Web site that contains medical information from the Harvard Medical school.
http://oncolink.upenn.edu	A University of Pennsylvania site that deals with cancer information.

ETHICAL ISSUES IN INFORMATION SYSTEMS

Ethical issues deal with what is generally considered right or wrong. Some information systems professionals believe that many opportunities for unethical behavior exist in their field. They also believe that unethical behavior can be reduced by top-level managers developing, discussing, and enforcing codes of ethics. Information systems professionals are usually more satisfied with their jobs when top management stresses ethical behavior.

According to one view of business ethics, the "old contract" of business, the only responsibility of business is to its stockholders and owners. According to another view, the "social contract" of business, businesses are responsible to society. At one point or another in their operations, businesses may have employed one or both philosophies.

Various organizations and associations promote ethically responsible use of information systems and have developed codes of ethics. These organizations include the following:

- The Association of Information Technology Professionals (AITP), formerly the Data Processing Management Association (DPMA)
- The Association for Computing Machinery (ACM)
- The Institute of Electrical and Electronics Engineers (IEEE)
- Computer Professionals for Social Responsibility (CPSR)

The AITP code of ethics. The AITP has developed a code of ethics, standards of conduct, and enforcement procedures that give broad responsibilities to AITP members (Figure 14.2).[27] In general, the code of ethics is an obligation of every AITP member in the following areas:

- Obligation to management
- Obligation to fellow AITP members
- Obligation to society
- Obligation to college or university
- Obligation to the employer
- Obligation to country

FIGURE 14.2

AITP Code of Ethics

CODE OF ETHICS

I acknowledge:
    That I have an obligation to management, therefore, I shall promate the understanding of information processing methods and procedures to management using every resource at my command.
    That I have an obligation to my fellow members, therefore, I shall uphold the high ideals of AITP as outlined in its International Bylaws. Further, I shall cooperate with my fellow members and treat them with honesty and respect at all times.
    That I have an obligation to society and will participate to the best of my ability in the dissemination of knowledge pertaining to the general development and understanding of information processing. Further, I shall not violate the privacy and confidentiality of information entrusted to my employer whose trust I hold, therefore, I shall endeavor to discharge this obligation to the best of my ability, to guard my employer's interests, and to advise him or her wisely and honestly.
    That I have an obligation as a personal representative and as a member of this association. I shall actively discharge these obligations and I dedicate myself to that end.

For each area of obligation, standards of conduct describe the specific duties and responsibilities of AITP members. In addition, enforcement procedures stipulate that any complaint against an AITP member must be in writing, signed by the individual making the complaint, properly notarized, and submitted by certified or registered mail. Charges and complaints may be initiated by any AITP member in good standing.

The ACM code of professional conduct. The ACM has developed a number of specific professional responsibilities.[28] These responsibilities include the following:

- Strive to achieve the highest quality, effectiveness, and dignity in both the process and products of professional work
- Acquire and maintain professional competence
- Know and respect existing laws pertaining to professional work
- Accept and provide appropriate professional review
- Give comprehensive and thorough evaluations of computer systems and their impacts, including analysis of possible risks
- Honor contracts, agreements, and assigned responsibilities
- Improve public understanding of computing and its consequences
- Access computing and communication resources only when authorized to do so

The mishandling of the social issues discussed in this chapter—including waste and mistakes, crime, privacy, health, and ethics—can devastate an organization. The prevention of these problems and recovery from them are important aspects of managing information and information systems as critical corporate assets. Increasingly, organizations are recognizing that people are the most important component of a computer-based information system and that long-term competitive advantage can be found in a well-trained, motivated, and knowledgeable workforce.

Information Systems Principles

Control standards, policies and procedures, training programs, and manuals should be developed to help eliminate computer waste and mistakes.

Knowing how a computer can be used as the object of crime, as well as the tool to commit crime, is the first step in crime prevention.

Privacy can include knowledge, notice, control, and consent. Know your rights and be active to protect your privacy.

Good posture, stretching and strengthening programs, and good treatment can help reduce RSI. Don't ignore pain, and don't try to do too much too soon when recovering from RSI.

▪ SUMMARY

1. *Describe some examples of waste and mistakes in an IS environment, their causes, and possible solutions.*

Computer waste is the inappropriate use of computer technology and resources in both the public and private sectors. Computer mistakes relate to errors, failures, and other problems that result in output that is incorrect and without value. Waste and mistakes occur in government agencies as well as corporations. At the corporate level, computer waste and mistakes impose unnecessarily high costs for an information system and drag down profits. Waste often results from poor integration of IS components, leading to duplication of efforts and overcapacity. Inefficient procedures also waste IS resources, as do thoughtless disposal of useful resources and misuse of computer time for games and personal processing jobs. Inappropriate processing instructions, inaccurate data entry, mishandling of IS output, and poor systems design all cause computer mistakes. Careful programming practices, thorough testing, flexible network interconnections, and rigorous backup procedures can help an information system prevent and recover from many kinds of mistakes. Companies should develop manuals and training programs to avoid waste and mistakes. Company policies should specify criteria for new resource purchases and user-developed processing tools to help guard against waste and mistakes.

2. *Explain the types and effects of computer crime, along with measures for prevention.*

Some crimes use computers as tools (e.g., to manipulate records, counterfeit money and documents, commit fraud via telecommunications links, and make unauthorized electronic transfers of money). Other crimes target computer systems, including illegal access to computer systems by criminal hackers, alteration and destruction of data and programs by viruses (system, application, and document), and simple theft of computer resources. A virus is a program that attaches itself to other programs. A worm functions as an independent program, replicating its own program files until it destroys other systems and programs or interrupts the operation of computer systems and networks. Application viruses infect executable application files, and a system virus infects operating system programs. A macro virus uses an application's own macro programming language to distribute itself. Unlike previous viruses, macro viruses do not infect programs; they infect documents. A logic bomb is designed to "explode" or execute at a specified time and date. Because of increased computer use, greater emphasis is placed on the prevention and detection of computer crime.

Software and Internet piracy may represent the most common computer crime. Computer scams have cost individuals and companies thousands of dollars. Computer crime is also an international issue. Preventing computer crime is done by state and federal agencies, corporations, and individuals. Security measures, such as using passwords, identification numbers, and data encryption, help to guard against illegal access, especially when supported by effective control procedures. Virus scanning software identifies and removes damaging computer programs. Law enforcement agencies armed with new legal tools enacted by Congress now actively pursue computer criminals.

3. *Discuss the principles and limits of an individual's right to privacy.*

Although most companies use data files for legitimate, justifiable purposes, opportunities for invasion of privacy abound. Privacy issues are a concern with government agencies, e-mail use, corporations, and the Internet. The Privacy Act of 1974, with the support of other federal laws, establishes straightforward and easily understandable requirements for data collection, use, and distribution by federal agencies; federal law also serves as a nationwide moral guideline for privacy rights and activities by private organizations. Some states supplement federal protections and limit activities within their jurisdictions by private organizations. A business should develop a clear and thorough policy about privacy rights for customers, including database access. That policy should also address the rights of employees, including electronic monitoring systems and e-mail. Fairness in information use for privacy rights emphasizes knowledge, control, notice, and consent for people profiled in databases. Individuals should have knowledge of the data that is stored about them and have the ability to correct errors in corporate database systems. If information on individuals is to be used for other purposes, these individuals should be asked to give their consent beforehand. Each individual has the right to know and the ability to decide.

4. *List the important effects of computers on the work environment.*

Computers have changed the makeup of the workforce and even eliminated some jobs, but they have also expanded and enriched employment opportunities in many ways. Computers and related devices affect employees' emotional and physical health, especially by causing repetitive stress injury (RSI). Some critics blame computer systems for emissions of ozone and electromagnetic radiation. Ergonomic design principles help to reduce harmful effects and increase the efficiency of an information system. RSI prevention includes keeping good posture, not ignoring pain or problems, performing stretching and strengthening exercises, and seeking proper treatment. In addition to these negative health con-

sequences, information systems can be used to provide a wealth of information on health topics through the Internet and other sources.

5. *Outline the criteria for the ethical use of information systems.*

Ethics determine generally accepted and discouraged activities within a company and the larger society. Ethical computer users define acceptable practices more strictly than just refraining from committing crimes; they also consider the effects of their IS activities, including Internet usage, on other people and organizations. The Association for Computing Machinery and the Association of Information Technology Professionals have developed a code of ethics that provides useful guidance. To help prevent ethical problems, the Telecommunications Act of 1996 set a standard for decency issues on the Internet. Software products have been developed to help filter and rate Internet content. Many IS professionals join computer-related associations and agree to abide by detailed ethical codes.

■ KEY TERMS

antivirus program 641
application virus 632
biometrics 638
carpal tunnel syndrome (CTS) 651
cracker 631
criminal hacker 631
ergonomics 652

hacker 631
Internet piracy 635
logic bomb 632
macro virus 633
password sniffer 634
repetitive motion disorder 651
repetitive stress injury (RSI) 651

software piracy 635
Software Publishers Association (SPA) 640
system virus 632
virus 632
worm 632

■ REVIEW QUESTIONS

1. Provide several examples of computer waste and computer mistakes.
2. How can organizations prevent computer-related waste and mistakes?
3. What are some of the ways the computer can be used as a tool to commit crimes?
4. How might the computer be the object of crime?
5. What is the difference between a hacker and a criminal hacker? What are the major problems caused by criminal hackers?

6. What is the difference between a worm and a virus?
7. What are application viruses, system viruses, and macro viruses?
8. What is software piracy, and why is it so common?
9. What is computer terrorism? How serious a threat is it?
10. What is Internet piracy, and how can companies avoid it?
11. What four issues should be addressed when considering the individual's right to privacy?

12. What are the provisions of the Privacy Act of 1974?
13. Describe other federal privacy laws.
14. What is personalization?

15. What are repetitive motion disorders, and how can they be prevented?
16. Describe the traditional views of ethics in business.
17. What is a code of ethics? Give an example.

■ DISCUSSION QUESTIONS

1. You have just been appointed director of an information systems department. Discuss what steps you would take to assess your organization's ability to prevent computer crime.
2. You are surprised to see that your paycheck is one hundred times its usual amount. How could this have happened? What would you do?
3. What can you do as a computer user to detect and prevent computer crimes?
4. Your marketing department has just opened a Web site and is requesting visitors to register at the site to enter a promotional contest where the chances of winning a prize are better than one in three. Visitors must provide the information necessary to contact them plus fill out a brief survey about the use of your company's products. What data privacy issues may arise?
5. If you were a U.S. senator, what laws would you propose to protect people from computer crime and scams?
6. Based on a number of recent workers' compensation cases, your employer has pledged to spend additional money for office furniture, equipment, and computers that are ergonomically designed so that employees avoid repetitive stress injuries. Will this solve the problem?
7. How could you use the Internet to help improve your health?

8. Consider this quote: "There's barely a piece of information about people that isn't used for far different purposes than it was initially gathered for, and always without approval." (Janlori Goldman, an attorney at the American Civil Liberties Union, quoted in Jeffrey Rothfeder, *Privacy for Sale*, New York: Simon & Schuster, 1992, p. 25.) Do you think this statement is true? Which of the four issues associated with privacy does the quote address? Which of the four is most important to you?
9. Using information presented in this chapter on federal privacy legislation, identify which federal law regulates the following areas and situations: cross-checking IRS and social security files to verify the accuracy of information; credit bureaus processing home loans; customer liability for debit cards; individuals' right to access data contained in federal agency files; IRS obtaining personal information; government obtaining financial records; and employer's access to university transcripts.
10. How can organizations ensure that their information systems are used ethically and morally? Who should audit the organization, departments, and employees to ensure that an ethical code of conduct is established and followed?

■ PROBLEM-SOLVING EXERCISES

1. Using the Internet, search for information about computer crime. Try to get statistics to quantify the number of incidents and their dollar impact. Find recent examples of computer crime. Use your word processor to summarize your findings. Discuss your view of ethics and codes of conduct.

 2. Using a spreadsheet application's random number generator capability, create passwords for the employees listed in the table on page 659. Passwords can be from five to eight characters in length and can consist of letters and/or numbers—no spaces, special characters, or punctuation. Finally, the passwords should be based on the employee's ID number and name.

Name	ID #	Name	ID #	Name	ID #
John	13506	Mike	34186	Alice	18637
Sandy	12987	Wilma	31742	Fred	21901
June	37264	Barney	14773	Betty	36471

Using your spreadsheet, enter the names (Column A) and ID numbers (Column B). For cells C2 through C10, use the random number generator to create random numbers for each employee. In Column D, choose some constant (e.g., the Julian date, which would be 32 for February 2, 1999*). In Column E, multiply the ID number by the random number and the day. Then, create passwords by combining name and number.

Your spreadsheet may look like this:

Name	ID #	Random #	Day*	Password
Mike	34186	0.629710	32	
Wilma	31742	0.039855	32	
Barney	14773	0.579455	32	
Alice	18637	0.418630	32	
Fred	21901	0.008777	32	
Betty	36471	0.116566	32	

■ TEAM ACTIVITIES

1. Your Web site "Hot Spots in the Desert Southwest" has done extremely well, with the number of visitors exceeding five hundred per day. The site provides useful information on places to go and things to do for travelers to the desert portions of New Mexico, Arizona, and California. Based on the site's success, you are considering a joint venture with a car rental agency, a hotel chain, and a clothier that specializes in Western clothing. As part of the venture, you will capture visitor registration information and clickstream data to provide to your new partners at a rate of $1 per new visitor and $.10 per returning visitor. You have called a meeting of your partners to develop a Web-site policy statement. You want to inform visitors what will and will not be done with the information. Your goal is to make visitors feel comfortable with the collection of data and to avoid driving them away. Conduct this meeting with two or three of your classmates. Create a policy statement for the Web site that will be viewed by each visitor. Address the issues discussed in the chapter and what responsibilities Web-site operators should have.

2. Your company, Smart Designs, specializes in computer systems and office equipment. The company has manufacturing facilities in Portland, Oregon, that make desks, chairs, and other office equipment. Your team is assigned the task of developing policies, acquiring computers, and designing office equipment to provide the best ergonomic office environment. What are your policies, acquired computer systems, and office equipment?

▪ WEB EXERCISE

Several computer-related organizations were discussed in this chapter, including AITP, ACM, IEEE, and CPSR. Locate the Web pages for two of these associations. What did you find at the sites? If you were an information systems professional, which organization would you join given what you found at the Web sites? Explain your answer. You may be asked to develop a report or send an e-mail message to your instructor about what you found.

▪ CASES

The GAO Finds Waste and Mistakes in Federal Agencies

The U.S. government is one of the largest users of computer systems, and some believe it is one of the biggest wasters of computer systems and resources. Federal agencies have spent a staggering $145 billion on computer systems and equipment during the last 5 or 6 years. With a budget this large, it is no wonder that there is evidence of computer waste and mistakes. Who is responsible for overseeing these large expenditures? It is the General Accounting Office (GAO). This agency is charged with investigating computer use and reporting problems or potential problems. In the recent past, it has uncovered computer-related waste and mistakes in a number of federal agencies.

The U.S. Department of Agriculture (USDA) has a systems development project estimated to cost $2.6 billion. The objective of the project is to develop a network of 2,500 service centers. The large expenditures, however, may be aimed at acquiring more technology instead of improving service to farmers and other customers.

The U.S. Army developed a logistics system to improve the accuracy of their procurement and inventory control procedures. The system, called the Continuing Balance System-Expanded (CBSX), reports the types, quantities, and location of military equipment. The system also determines the readiness of the equipment in warfighting units. The CBSX system, however, does not adequately track the sources of any adjustments or identify how potential problems can be resolved. Furthermore, the Army lacks a methodology for determining the performance and effectiveness of the CBSX system.

The Federal Deposit Insurance Corporation (FDIC) was having difficulties upgrading its computer systems for the year 2000. Without adequate upgrades, the FDIC system will make mistakes in processing fund transfers. During senate hearings, Jack Brock of the GAO reported that these mistakes in the computer programs of the FDIC could have serious consequences worth trillions of dollars. The conclusion of the hearings was to do more and do it faster to get rid of the mistakes.

In addition to these agencies, the GAO is looking into waste and mistakes in other agencies, including the IRS, the Federal Aviation Administration, the Department of Defense, and the National Weather Service. To help prevent further problems with federal waste and mistakes, the federal Chief Information Officers Council (CIOC) recently issued a report on best information systems practices in the federal government. While some view the report as a contradiction, most believe that the efforts by the council will help federal agencies better manage their information systems investments by reducing computer-related waste and mistakes.

1. Why are computer-related waste and mistakes such a big problem with federal agencies?
2. If you were in the GAO, what steps would you take to try to reduce federal waste and mistakes?

Sources: Patrick Thibodeau, "GAO Report Slams Fed's IT Spending," *Computerworld*, March 2, 1998, p. 12; Staff, "Army Needs to Address Software Problems," *Federal Computer Market Report*, February 6, 1998, p. 9, Rory Thompson, "Starting to Add Up," *Information Week*, February 1, 1998; Staff, "FDIC Lagging in Conversion," *PC Week*, March 2, 1998, p. 87; Edward Cone, "Federal CIOs Look Past Failures," *Information Week*, January 12, 1998, p. 44.

2   Computer Terrorism

The United States is not the only government concerned about computer terrorism. Australian security experts have warned that computer terrorists could target the Sydney 2000 Olympic Games. "Techniques of 'information warfare' may be employed by terrorist organizations with no less effect than the traditional bomb," security researchers Russell Smith and Peter Grabosky said in a study released at an Australian Institute of Criminology computer crime conference. They warned that authorities around the world had failed to grasp the potential for computer terrorism, which includes hijacking air traffic control systems to crash planes and cutting power supplies. Terrorists could also take hostage computerized services such as telecommunications and power supplies. Many of these things depend on software and the convergence of computing and communications, and they could be vulnerable to disruption by pranksters, extortionists, or terrorists. Attacks on computers could be launched via the Internet. "Some people regard their information systems with a degree of nonchalance," Grabosky said. "It's the contemporary equivalent of leaving your home with the door unlocked."

Imagine that you have been retained as a computer security consultant to the Sydney 2000 Olympics.

1. Identify the top three or four potential targets for a computer terrorist attack.
2. How would you begin to assess the adequacy of these computer systems?

Sources: Adapted from Maria Moscaritolo, "Will Terrorists Target Games Via Computer?" *Deseret News*, February 16, 1998 and the "National Security/Defense" section of the Web page for the National Center for Policy Analysis at http://www.ncpa.org, accessed October 6, 1998.

3 AOL Works to Improve Security

America Online (AOL), the world's largest Internet service provider and provider of unique content and services, has been under constant attack from hackers bent on finding ways to exploit the system and steal user information. Attacks on AOL go back to March 1995, when it was discovered that staff-only areas could be accessed just by knowing the secret links to them. Additional attempts to compromise AOL security have led to the defacing of AOL content areas, password crackers, and other troublemakers causing problems on the system. The majority of these security problems, said Tatiana Gau, the company's security czar, are caused by end users having their password compromised—usually because they were duped by a malicious user.

In a recent attack, a database containing sensitive account information about America Online community leaders was hacked and the data circulated via e-mail. The list contained the screen names, true names, and account numbers of more than 1,300 AOL community leaders. Community leaders are AOL members who volunteer their time as guides and chat room monitors in exchange for free membership. They discover and deal with prohibited behavior and try to generate goodwill from members of the on-line community. They dispense handy tips on password security and how to find AOL staff. They also seek out those who have violated the Terms of Service (TOS) procedures outlined in the Guide Policies.

The list was obtained when a hacker broke into the account of an AOL employee who oversees community leaders. The perpetrator sifted through the employee's e-mail to find the document and mass-e-mailed the list. As a result of the security breach, some community leaders say they have been subjected to harassing phone calls, and some have been threatened with violence. (A community leader's phone number can be obtained via Web phone directories using the information provided on the list).

"The good news is that these volunteers' accounts are not in jeopardy because of this file, but it's obviously not information we'd want out there," AOL spokeswoman Tricia Primrose said. "We are in the process of investigating what happened." Many community leaders fear "social-engineering" hacks of their accounts, with hackers persuading or tricking someone into willingly handing over information. With the community leader's account number, a hacker can call up AOL and say that he or she is the community leader.

Primrose maintained that community leader accounts would be secure, noting that the service validates all volunteer and community leader account information. AOL has started a community leader help desk and has defined an expedited process for community leaders to change their volunteer screen names. Community leader passwords were not contained in the list.

During a Commerce Department summit examining self-regulatory policies to shield Net users' privacy, AOL Senior Vice President and General Counsel George Vrandenberg said, "We are working our tails off" to improve privacy protection for AOL members. For example, he said that every AOL employee had to review the company's privacy policies and sign a statement pledging compliance with them. Vrandenberg also said the company was boosting security measures. "We're doing internal audits to test our own system to ensure it is as good as it can get," he said.

1. What additional steps could be taken to minimize the impact of this type of hack?
2. How might you conduct an internal audit to test AOL security?

Source: Adapted from Jim Hu, "AOL Volunteer List Hacked," *CNET NEWS.COM,* June 30, 1998 and Michael Stutz, "America Online Under Attack from Hackers, *Wired,* January 29, 1998.

4 The Politics of Technology

Influential lawmakers up for re-election in Congress say technology issues are relatively unimportant with the public. Although executives of high-tech firms may worry about issues such as on-line privacy, Internet content controls, and E-commerce taxes, technology is not even on most people's radar screens. Technology issues such as data privacy aren't as worrisome as bottom-line concerns such as education, social security, taxes, and the economy. This is perhaps to be expected since only recently have technology companies established significant Washington lobbying efforts.

Appeals for policies that will limit regulation and help the Internet industry grow are coming from within technology companies, not from the public. In meeting with the lawmakers, voters tend to express opinions about technology only when the subject comes up in conversations about other issues. For example, taxpayers do want more money spent on education, including dollars to put the best technology in the schools. There's also a growing discomfort with pornography on the Internet, particularly among parents who are anxious to shield their children. Once more people become familiar with and use new technology, they are likely to become more organized and demand solutions to Internet privacy and on-line pornography from the technology industry, possibly including stronger filtering software.

Lawmakers, meanwhile, are beginning to sense the importance of issues such as on-line content controls. For example, the inclusion of an Internet content control measure and a bill to increase immigration opportunities for foreign high-tech workers were among the controversies holding up the 1999 federal budget agreement.

1. Make a "top ten" list of what you think are the biggest computer security and privacy issues (e.g., on-line pornography, free speech, money for technology in the schools, data privacy, etc.). Rank them from one through ten in order of their importance to you.
2. Talk with your friends, fellow students, and family members about the importance of these issues in their life. Do any themes emerge?

Sources: Adapted from Maria Seminerio, "Political Paradox: Technology is Everywhere, Except in Voters' Hearts," *ZDNet News,* October 16, 1998 and Maria Seminerio, "Free Speech Advocates Ready to Fight Child Protection Law," *ZDNet News,* October 15, 1998.

acceptance testing conducting any tests required by the user

accounting MIS an MIS that provides aggregate information on accounts payable, accounts receivable, payroll, and many other applications

accounting systems systems that include budget, accounts receivable, payroll, asset management, and general ledger

accounts payable system a system that increases an organization's control over purchasing, improves cash flow, increases profitability, and provides more effective management of current liabilities

accounts receivable system a system that manages the cash flow of the company by keeping track of the money owed the company on charges for goods sold and services performed

activity modeling a method to describe related objects, associations, and activities

ad hoc DSS a DSS concerned with situations or decisions that come up only a few times during the life of the organization

analog signal a continuous, curving signal

antivirus programs programs or utilities that prevent viruses or help recover from them if they infect a computer

applet a small program embedded in Web pages

application flowcharts charts that show relationships among applications or systems

application sign-on the procedure that permits the user to start and use a particular application

application software programs that help users solve particular computing problems

application viruses types of viruses that infect executable application files such as word processing programs

arithmetic/logic unit (ALU) the part of the CPU that performs mathematical calculations and makes logical comparisons

ARPANET a project started by the U.S. Department of Defense (DOD) in 1969 as both an experiment in reliable networking and a means to link DOD and military research contractors, including a large number of universities doing military-funded research

artificial intelligence (AI) a field that involves computer systems taking on the characteristics of human intelligence

artificial intelligence systems the people, procedures, hardware, software, data, and knowledge needed to develop computer systems and machines that demonstrate the characteristics of intelligence

asking directly an approach to gathering data that asks users, stakeholders, and other managers about what they want and expect from the new or modified system

assembly language second-generation language that replaced binary digits with symbols programmers could more easily understand

asset management transaction processing system a system that controls investments in capital equipment and manages depreciation for maximum tax benefits

attribute a characteristic of an entity

audit trail the trace to any output from the computer system back to the source documents

auditing the process that involves analyzing the financial condition of an organization and determining whether financial statements and reports produced by the financial MIS are accurate

backbone one of the Internet's high-speed, long-distance communications links

backward chaining a method of reasoning that starts with conclusions and works backward to the supporting facts

batch processing system a system whereby business transactions are accumulated over a period of time and prepared for processing as a single unit or batch

benchmark test an examination that compares computer systems operating under the same conditions

best practices the most efficient and effective ways to complete a business process

bill of materials the description of the parts needed to make final products

biometrics the measurement of a living trait, whether physical or behavioral, for the purpose of protecting important data and information systems

bit BInary digiT—0 or 1

brainstorming a decision-making approach that fosters creativity and free thinking by allowing members of the group to offer ideas "off the top of their heads"

bridge a device that connects two or more networks at the media access control portion of the data link layer; the two networks must use the same communications protocol

budget transaction processing system a system that automates many of the tasks required to amass budget data, distribute it to users, and consolidate the prepared budgets

bus line the physical wiring that connects the computer system components

bus network a type of topology that consists of computers and computer devices on a single line. Each device is connected directly to the bus and can communicate directly with all other devices on the network. The bus network is one of the most popular types of personal computer networks.

business resumption planning the process of anticipating and providing for disasters

byte eight bits together that represent a single character of data

C++ a popular programming language that is an enhancement of the original C programming language

cache memory a type of high-speed memory that a processor can access more rapidly than main memory

carpal tunnel syndrome (CTS) the aggravation of the pathway for nerves that travel through the wrist (the carpal tunnel)

CASE repository a database of system descriptions, parameters, and objectives

CD-rewritable (CD-RW) a common form of optical disk that allows personal computer users to replace their diskettes with high-capacity CDs that can be written upon and edited over

central processing unit (CPU) the part of the computer that consists of three associated elements: the arithmetic/logic unit, the control unit, and the register areas

centralized processing data processing that occurs in a single location or facility

certification process for testing skills and knowledge that results in a statement by the certifying authority that says an individual is capable of performing a particular kind of job

change model a representation of change theories that identifies the phases of change and the best way to implement them

character the basic building block of information, represented by a byte

chat room a facility that enables two or more people to engage in interactive "conversations" over the Internet

chief programmer team a group of skilled IS professionals with the task of designing and implementing a set of programs and has total responsibility for building the best software possible

choice stage the last stage of the decision-making phase during which a course of action is selected

client application the application that accepts objects from other applications

client/server an architecture in which multiple computer platforms are dedicated to special functions such as database management, printing, communications, and program execution

clock speed the predetermined rate at which the CPU produces a series of electronic pulses

closed shops IS department in which only authorized operators can run the computers

cold site also called a shell, a computer environment that includes rooms, electrical service, telecommunications links, data storage devices, etc.

collaborative computing software software that helps teams of people work together toward a common goal

command-based user interface a part of the operating system that requires text commands be given to the computer to perform basic activities

common carriers long-distance telephone companies

communications software software that provides error checking, message formatting, communications logs, data security and privacy, and translation capabilities for networks

compact disk read-only memory (CD-ROM) a common form of optical disk on which data, once recorded, cannot be modified

competitive advantage a significant and (ideally) long-term benefit to a company over its competition

compiler a language translator that converts a complete program into a machine language to produce a program that the computer can process in its entirety

complementary metal oxide semiconductor (CMOS) a semiconductor fabrication technology that uses special material to achieve low-power dissipation

complex instruction set computing (CISC) a computer chip design that places as many microcode instructions into the central processor as possible

compound documents components of GDSS software that include information from spreadsheet programs, database packages, word processors, and other applications

computer literacy a knowledge of computer systems and equipment and the ways they function

computer network the communications media, devices, and software needed to connect two or more computer systems and/or devices

computer programs sequences of instructions for the computer

computer system architecture the structure, or configuration, of the hardware components of a computer system

computer system platform the combination of a particular hardware configuration and systems software package

computer-aided software engineering (CASE) technology that automates many of the tasks required in a systems development effort and enforces adherence to the SDLC

computer-assisted manufacturing (CAM) a system that directly controls manufacturing equipment

computer-integrated manufacturing (CIM) systems that use computers to link the components of the production process into an effective system

computerized collaborative work system another term for a group decision support system

concurrency control a method of dealing with a situation in which two or more people need to access the same record in a database at the same time

content streaming a method for transferring multimedia files over the Internet so that the data stream of voice and pictures plays continuously, without a break, or very few of them. It also enables users to browse large files in real time.

continuous improvement constantly seeking ways to improve the business processes to add value to products and services

contract software software developed for a particular company

control unit the part of the CPU that sequentially accesses program instructions, decodes them, and coordinates the flow of data in and out of the ALU, the registers, primary storage, and the secondary storage and various output devices

coprocessor a part of the computer that speeds processing by executing specific types of instructions while the CPU works on another processing activity

cost centers divisions within the company that do not directly generate revenue

cost/benefit analysis an approach that lists the costs and benefits of each proposed system

cracker another term for a criminal hacker

creative analysis the investigation of new approaches to existing problems

criminal hacker a computer-savvy person who attempts to gain unauthorized or illegal access to computer systems

critical analysis the unbiased and careful questioning of whether system elements are related in the most effective or efficient ways

critical path the path that consists of all activities that, if delayed, would delay the entire project

critical success factors (CSFs) those factors that are critical to the success of a manager's area of the organization

cross-platform development a software development technique that allows programmers to develop programs that can run on computer systems having different hardware and operating systems, or platforms

cryptography the process of converting a message into a secret code and changing the encoded message back to regular text

culture a set of major understandings and assumptions shared by a group

customer interaction system a system that monitors and tracks each customer interaction with the company

data raw facts

data analysis manipulating the collected data so that it is usable for the development team members who are participating in systems analysis

data cleanup the process of looking for and fixing inconsistencies to ensure that data is accurate, complete, economical, flexible, reliable, relevant, simple, timely, verifiable, accessible, and secure

data collection the process of capturing and gathering all data necessary to complete transactions

data communications a specialized subset of telecommunications that refers to the electronic collection, processing, and distribution of data—typically between computer system hardware devices

data correction the process of reentering miskeyed or misscanned data that was found during the data editing

data definition language (DDL) a collection of instructions and commands used to define and describe data and data relationships in a specific database

data dictionary a detailed description of all data used in the database

data editing the process of checking data for validity and completeness

data entry the process by which human-readable data is converted into a machine-readable form

data input the process of transferring machine-readable data into the computer system

data integrity the degree to which the data in any one file is accurate

data item the specific value of an attribute

data manipulation the process of performing calculations and other data transformations related to business transactions

data manipulation language (DML) the commands that are used to manipulate the data in a database

data mart a subset of a data warehouse for small and medium-size businesses or departments within larger companies

data mining the automated discovery of patterns and relationships in a data warehouse

data model a map or diagram of entities and their relationships

data modeling a commonly accepted approach to modeling organizational objects and associations that employ both text and graphics

data preparation, or data conversion the process of converting all manual files into computer files

data redundancy the duplication of data in separate files

data storage the process of updating one or more databases with new transactions

data store the symbol that shows a storage location for data in a data-flow diagram

data warehouse a relational database management system designed specifically to support management decision making

data-flow diagram (DFD) a diagram that models objects, associations, and activities by describing how data can flow between and around them

data-flow line the path that shows the direction of data element movement in a data-flow diagram

database an organized collection of facts and information in integrated and related files

database approach an approach to data management in which a pool of related data is shared by multiple application programs. Rather than having separate data files, each application uses a collection of data that is either joined or related in the database.

database management system (DBMS) a group of programs that manipulate the database and provide an interface between the database and the user of the database or other application programs

decentralized processing data processing that occurs when devices are placed at various remote locations

decision room a facility that supports decision making when decision makers are in the same building, combining face-to-face verbal interaction

with technology-aided formalization to make the meeting more effective and efficient

decision structure a structure in programming that allows the computer to branch, depending on certain conditions

decision support system (DSS) an organized collection of people, procedures, software, databases, and devices used to support problem-specific decision making

decision-making phase the phase of the problem-solving process that includes the intelligence, design, and choice stages

dedicated line a line that provides a constant connection between two points. No switching or dialing is needed; the two devices are always connected.

delphi approach a decision-making approach in which group decision makers are geographically dispersed throughout the country or the world

demand reports reports developed to give certain information at a manager's request

design report a report that reflects the decisions made for system design and prepares the way for systems implementation

design stage the second stage of the problem-solving process during which alternative solutions to the problem are developed

deterrence controls rules that prevent problems before they occur

dialogue the series of messages and prompts communicated between the system and the user

dialogue manager a component of the DSS that allows decision makers to easily access and manipulate the DSS and to use common business terms and phrases

digital computer cameras input devices that record and store images and video in digital form

digital signal a signal represented by bits

digital signal processor (DSP) a chip that improves the analog-to-digital-to-analog conversion process

digital signature an encryption technique used to meet the critical need for processing on-line financial transactions

Digital Subscriber Line (DSL) a line that uses existing phone wires going into today's homes and businesses to provide transmission speeds exceeding 500 Kbps at a cost of $100–$300 per month

digital video disk (DVD) storage format used to store digital video or computer data

direct access the process by which data can be retrieved without the need to read or pass by other data in sequence

direct access storage devices (DASDs) devices used for direct access of secondary storage data

direct conversion (also called plunge or direct cutover) start-up procedure that involves stopping the old system and starting the new system on a given date

direct observation analysis conducted by one or more members of the analysis team who observes the existing system in action

disaster recovery the implementation of the business resumption plan

disk mirroring a process of storing data that provides an exact copy that fully protects users in the event of data loss

distance learning the use of telecommunications to extend the classroom

distributed database a database in which the actual data may be spread across several smaller databases connected via telecommunications devices

distributed processing data processing that occurs when computers are placed at remote locations but are connected to each other via telecommunications devices

document production the process of generating output records and reports

documentation description of the program functions to help the user operate the computer system

domain the allowable values for an attribute; the area of knowledge addressed by the expert system

domain expert the individual or group whose expertise and knowledge is captured for use in an expert system

downsizing reducing the number of employees to cut costs

drill down reports reports that provide detailed data about a situation

E-commerce involves any business transaction executed electronically between parties such as companies (business-to-business), companies and consumers (business-to-consumer), business and the public sector, and consumers and the public sector

e-mail technology that enables a sender to connect his or her computer to a network, type in a message, and send it to another person on the network

economic feasibility an assessment of whether the project makes financial sense and whether predicted benefits offset the cost and time needed to obtain them

economic order quantity (EOQ) a method used to determine the quantity of inventory that should be reordered to minimize total inventory costs

effectiveness a measure of the extent to which a system achieves its goals

efficiency a measure of what is produced divided by what is consumed

electronic data interchange (EDI) an intercompany, application-to-application communication of data in standard format, permitting the recipient to perform the functions of a standard business transaction

electronic document distribution a process that involves transporting documents—such as sales reports, policy manuals, and advertising brochures—over communications lines and networks

electronic software distribution a process that involves installing software on a file server for users to share by signing onto the network and requesting that the software be downloaded onto their computers over a network

embed procedure used when you want an object to become part of the client document

empowerment giving employees and their managers more power, responsibility, and authority to make decisions, take certain actions, and have more control over their jobs

encapsulation the process of grouping items into an object

encryption the original conversion of a message into a secret code

end-user systems development any systems development project in which the primary effort is undertaken by a combination of business managers and users

enterprise data modeling data modeling done at the level of the entire organization

enterprise resource planning (ERP) a set of integrated programs that manage a company's vital business operations for an entire multisite, global organization

enterprise sphere of influence information systems that support the firm in its interaction with its environment

entity a generalized class of people, places, or things (objects) for which data is collected, stored, and maintained

entity symbol the symbol that shows either a source or destination of a data element in a data-flow diagram

entity-relationship (ER) diagrams a data model that uses basic graphical symbols to show the organization of and relationships between data

ergonomics the study of designing and positioning computer equipment to reduce health problems

event-driven review review of the system that is triggered by a problem or opportunity such as an error, a corporate merger, or a new market for products

exception reports reports that are automatically produced when a situation is unusual or requires management action

execution time (E-time) the time it takes to complete the execution phase of the execution of an instruction

executive support system (ESS) or executive information system (EIS) a specialized DSS that includes all hardware, software, data, procedures, and people used to assist senior-level executives within the organization

expert system the hardware and software that stores knowledge and makes inferences, similar to a human expert

expert system shell a collection of software packages and tools used to develop expert systems

explanation facility a part of the expert system that allows a user or decision maker to understand how the expert system arrived at certain conclusions or results

external auditing auditing performed by an outside group

extranet a network based on Web technologies that links selected resources of the intranet of a company with its customers, suppliers, or other business partners

feasibility analysis an assessment of the technical, operational, schedule, and economic feasibility of a project

feedback output that is used to make changes to input or processing activities

field a group of characters

file a collection of related records

file server an architecture in which the application and database reside on one host computer, called the file server

File Transfer Protocol (FTP) a protocol that describes a file transfer process between a host and a remote computer. FTP allows users to copy a file from one computer to another.

final evaluation evaluation that involves detailed investigation of the proposals offered by the remaining vendors

financial MIS an MIS that provides financial information to all financial managers within an organization

financial models DSS models that provide cash flow, internal rate of return, and other investment analysis

firewall a device that sits between your internal network and the outside Internet and limits access into and out of your network based on your organization's access policy

five-force model a widely accepted model that identifies five key factors that can lead to attainment of competitive advantage including (1) rivalry among

existing competitors, (2) the threat of new entrants, (3) the threat of substitute products and services, (4) the bargaining power of buyers, and (5) the bargaining power of suppliers

flash memory a silicon computer chip that, unlike RAM, is nonvolatile and keeps its memory when the power is shut off

flat organizational structure organizational structure with a reduced number of management layers

flexible manufacturing system (FMS) an approach that allows manufacturing facilities to rapidly and efficiently change from making one product to another

forecasting a proactive approach to feedback

forward chaining a method or reasoning that starts with the facts and works forward to the conclusions

fourth-generation languages (4GLs) programming languages that are less procedural and even more English-like than third-generation languages

front-end processor a special-purpose computer that manages communications to and from a computer system

function points standard measures used to gauge IS developers' productivity

fuzzy logic a specialty research area in computer science that allows shades of gray and does not require everything to be simple black or white, yes/no, or true/false

Gantt chart a graphical tool used for planning, monitoring, and coordinating projects

gateway a device that operates at or above the OSI transport layer and links LANs or networks that employ different, higher-level protocols, thus allowing networks with very different architectures and using dissimilar protocols to communicate

general ledger system a system that produces a detailed list of business transactions designed to automate financial reporting and data entry

general-purpose computers computers that are used for a wide variety of applications

geographic information system (GIS) a computer system capable of assembling, storing, manipulating, and displaying geographically referenced information, i.e., data identified according to its location

goal-seeking analysis the process of determining the problem data required for a given result

graphical modeling programs software that assists decision makers in designing, developing, and using graphic displays of data and information

graphical user interface (GUI) a part of the operating system that uses pictures (icons) and menus displayed on the screen to send commands to the computer system

grid chart a table that shows relationships among the various aspects of a systems development effort

group consensus an approach to evaluation in which a decision-making group is appointed and given the responsibility of making the final evaluation and selection

group consensus approach a decision-making approach that forces members in the group to reach a unanimous decision

group decision support system (GDSS) a system that consists of most of the elements in a DSS, plus GDSS software needed to provide effective support in group decision-making settings

groupware software that helps groups of people work together more efficiently and effectively

hacker a person who enjoys computer technology and spends time learning and using computer systems

hardware any machinery (most of which uses digital circuits) that assists in the input, processing, storage, and output activities of an information system

help facility a tool that aids a user who is having difficulty understanding what is happening or what type of response is expected

hertz one cycle or pulse per second

heuristics commonly accepted guidelines or procedures that usually find a good solution

hierarchical database model a data model in which the data is organized in a top-down, or inverted tree, structure

hierarchical network a type of topology that uses a treelike structure with messages passed along the branches of the hierarchy until they reach their destination

hierarchy of data bit, characters, fields, records, files, and databases

highly structured problems problems that are straightforward, requiring known facts and relationships

home page the cover page for a Web site that has graphics, titles, and black and blue text

hot site a duplicate computer system that is operational and ready for use

HTML tags codes that let the Web browser know how to format text: as a heading, as a list, or as body text and whether images, sound, and other elements should be inserted

human resource MIS an MIS that is concerned with all of the activities related to employees and potential employees of the organization

hypermedia tools that connect the data on Web pages, allowing users to access topics in whatever order they wish

Hypertext Markup Language (HTML) the standard page description language for Web pages

icon picture

if-then statements rules that suggest certain conclusions

image log a separate file that contains only changes to applications

implementation stage the stage where action is taken to put a solution into effect

in-house development development of application software using the company's resources

incremental backup backup copies of all files changed during the last few days or the last week

inference engine a part of the expert system that seeks information and relationships from the knowledge base and provides answers, predictions, and suggestions the way a human expert would

information a collection of facts organized in such a way that they have additional value beyond the value of the facts themselves

information center provides users with assistance, training, application development, documentation, equipment selection and setup, standards, technical assistance, and troubleshooting

information service unit a miniature IS department

information system (IS) a set of interrelated elements or components that collect (input), manipulate (process), and disseminate (output) data and information and provide a feedback mechanism to meet an objective

information systems literacy a knowledge of how data and information are used by individuals, groups, and organizations

information systems planning the translation of strategic and organizational goals into systems development initiatives

inheritance property used to describe objects in a group of objects taking on characteristics of other objects in the same group or class of objects

input the activity of gathering and capturing raw data

installation the process of physically placing the computer equipment on the site and making it operational

instruction time (I-time) the time it takes to perform the instruction phase of the execution of an instruction

integrated development environments (IDEs) software that combines the tools needed for programming with a programming language into one integrated package

Integrated Services Digital Network (ISDN) technology that uses existing common-carrier lines to simultaneously transmit voice, video, and image data in digital form

integrated-CASE tools (I-CASE) tools that provide links between upper- and lower-CASE packages, allowing lower-CASE packages to generate program code from upper-CASE package designs

integration testing testing all related systems together

intelligence stage the first stage of the problem-solving process during which potential problems and/or opportunities are identified and defined

intelligent behavior the ability to learn from experience and apply knowledge acquired from experience, handle complex situations, solve problems when important information is missing, determine what is important, react quickly and correctly to a new situation, understand visual images, process and manipulate symbols, be creative, imaginative and use heuristics

internal auditing auditing performed by individuals within the organization

international network a network that links systems between countries

Internet a collection of interconnected networks, all freely exchanging information

Internet piracy the act of illegally gaining access to and using the Internet

Internet Protocol (IP) conventions that enable traffic to be routed from one network to another as needed

Internet service provider (ISP) any company that provides individuals or organizations with access to the Internet

interpreter a language translator that translates one program statement at a time into machine code

intranet an internal corporate network built using Internet and World Wide Web standards and products that allows employees of an organization to gain access to corporate information

inventory control system a system that updates the computerized inventory records to reflect the exact quantity on hand of each stock keeping unit

Java an object-oriented programming language from Sun Microsystems based on C++ that allows small programs—applets—to be embedded within an HTML document

joining data manipulation that combines two or more tables

joint application development (JAD) process for data collection and requirements analysis

just-in-time (JIT) inventory approach an approach by which inventory and materials are delivered just before they are used in a product

key a field or set of fields in a record that is used to identify the record

key-indicator report a report that summarizes the previous day's critical activities and is typically available at the beginning of each workday

knowledge an awareness and understanding of a set of information and how that information can be made useful to support a specific task

knowledge acquisition facility the part of the expert system that provides a convenient and efficient means of capturing and storing all components of the knowledge base

knowledge base that component of the expert system that stores all relevant information, data, rules, cases, and relationships used by the expert system

knowledge engineer an individual who has training and/or experience in the design, development, implementation, and maintenance of an expert system

knowledge user the individual or group who uses and benefits from the expert system

knowledge-based programming an approach to development of computer programs in which you do not tell a computer how to do a job, but what you want it to do

LAN administrator person who sets up and manages the Local Area Network hardware, software, and security processes

language translator systems software that converts a programmer's source code into its equivalent in machine language

learning system a combination of software and hardware that allows the computer to change how it functions or reacts to situations based on feedback it receives

link procedure used when you want any changes made to the server object to automatically appear in all linked client objects

linked related tables in a relational database together

local area network (LAN) a network that connects computer systems and devices within the same geographic area

logic bomb an application or system virus designed to "explode" or execute at a specified time and date

logical systems design the dimension of the design phase that describes the functional requirements of a system

lookup tables tables within an application that simplify and shorten data entry

loop structure a structure in programming that uses loops; in the do-until structure, the loop is done until a certain condition is met. For the do-while structure, the loop is done while a certain condition exists.

low-level language basic coding scheme using the binary symbols 1 and 0

lower-CASE tools tools that focus on the later implementation stage of systems development

machine cycle the instruction phase and the execution phase of an instruction

machine language the first-generation programming language

macro virus a virus that uses an application's own macro programming language to distribute itself

magnetic disks common secondary storage medium, with bits represented by magnetized areas

magnetic tape a common secondary storage medium that consists of Mylar film coated with iron oxide with magnetized portions to represent bits

magneto-optical disk a hybrid that combines magnetic disk technologies and CD-ROM technologies

mainframe computers large, powerful computers that are

often shared by hundreds of concurrent users connected to the machine via terminals

maintenance team the team responsible for modifying, fixing, and updating existing software

make-or-buy decision the decision regarding whether to obtain the necessary software from internal or external sources

management information system (MIS) an organized collection of people, procedures, software, databases, and devices used to provide routine information to managers and decision makers

manufacturing resource planning (MRPII) an integrated, companywide system based on network scheduling that enables people to run their business with a high level of customer service and productivity, while lowering costs and inventories

marketing MIS an MIS that supports managerial activities in product development, distribution, pricing decisions, and promotional effectiveness

material requirements planning (MRP) a set of inventory control techniques that help coordinate thousands of inventory items when the demand for one item is dependent on the demand for another

megahertz (MHz) millions of cycles per second

menu-driven system a system that allows users to simply pick what they want to do from a list of alternatives

microcode predefined, elementary circuits and logical operations that the processor performs when it executes an instruction

midrange computers computer systems that are about the size of a small three-drawer file cabinet and can accommodate several users at one time

MIPS millions of instructions per second

mission-critical systems systems that play a pivotal role in an organization's continued operations and goal attainment

model an abstraction or an approximation that is used to represent reality

model base a component of the DSS that provides decision makers with access to a variety of models and assists them in decision making

model management software (MMS) software that coordinates the use of models in a DSS

modems devices that translate data from digital to analog and analog to digital

monitoring stage the final stage of the problem-solving process where decision makers evaluate the implementation

Moore's Law a hypothesis that states transistor densities on a single chip will double every 18 months

multidimensional organizational structure structure that may incorporate several structures at the same time

multifunction device a device that combines a printer, fax machine, scanner, and copy machine into one unit

multiplexer a device that allows several telecommunications signals to be transmitted over a single communications medium at the same time

multiprocessing the simultaneous execution of two or more instructions at the same time

multitasking a processing activity that allows a user to run more than one application at the same time

multithreading a processing activity that is basically multitasking within a single application

natural language processing processing that allows the computer to understand and react to statements and commands made

in a "natural" language, such as English

net present value the preferred approach for ranking competing projects and determining economic feasibility

network computer a cheaper-to-buy and cheaper-to-run version of the personal computer that is used primarily for accessing networks and the Internet

network management software software that enables a manager on a networked desktop to monitor the use of individual computers and shared hardware (like printers), scan for viruses, and ensure compliance with software licenses

network model an expansion of the hierarchical database model with an owner-member relationship in which a member may have many owners

network operating system (NOS) systems software that controls the computer systems and devices on a network and allows them to communicate with each other

network topology a logical model that describes how networks are structured or configured

networks used to connect computers and computer equipment in a building, around the country, or across the world to enable electronic communications

neural network a computer system that can act like or simulate the functioning of a human brain

newsgroups on-line discussion groups that focus on a specific topic

nominal group technique a decision-making approach that encourages feedback from individual group members, and the final decision is made by a voting approach similar to the one used to elect public officials

nonoperational prototype a mockup or model that includes output and input specifications and formats

nonprogrammed decisions decisions that deal with unusual or exceptional situations

object a collection of data and the programs to process this data

object code another name for machine language code

object linking and embedding (OLE) a software feature that allows you to copy text from one document to another or embed graphics from one program into another program or document

object-oriented languages languages that allow interaction of programming objects, including data elements and the actions that will be performed on them

object-oriented software development development of a collection of existing modules of code or objects that can be used across any number of applications without being rewritten

object-relational database management system (ORDBMS) a DBMS capable of manipulating audio, video, and graphical data

off-the-shelf software an existing software program

on-line analytical processing (OLAP) programs used to store and deliver data warehouse information

on-line transaction processing (OLTP) a system whereby each transaction is processed immediately, without the delay of accumulating transactions into a batch

open database connectivity (ODBC) a set of standards that ensures software written to comply with these standards can be used with any ODBC-compliant database

open shops IS department in which other people, such as programmers and systems analysts,

are also authorized to run the computers

Open Systems Interconnection (OSI) model a standard model for network architectures that divides data communications functions into seven distinct layers to promote the development of modular networks that simplify the development, operation, and maintenance of complex telecommunications networks

operating system (OS) a set of computer programs that control the computer hardware and act as an interface with applications programs

operational feasibility a measure of whether the project can be put into action or operation

operational prototype a prototype that works—accesses real data files, edits input data, makes necessary computations and comparisons, and produces real output

optical disk a rigid disk of plastic onto which data is recorded by special lasers that physically burn pits in the disk

optical processors computer chips that use light waves instead of electrical current to represent bits

optimization model an approach to decision support that involves finding the best solution, usually the one that will best help the organization meet its goals

order entry system a system that captures the basic data needed to process a customer order

order processing systems systems that process order entry, sales configuration, shipment planning, shipment execution, inventory control, invoicing, customer interaction, and routing and scheduling in an organization

organization a formal collection of people and other

resources established to accomplish a set of goals

organizational change deals with how for-profit and nonprofit organizations plan for, implement, and handle change

organizational culture the major understandings and assumptions for a business, a corporation, or an organization

organizational learning organizations adapt to new conditions or alter their practices over time

organizational structure organizational subunits and the way they are related to the overall organization

output useful information, usually in the form of documents and/or reports

outsourcing contracting with outside professional services to meet specific business needs

paging a function of virtual memory that allows the computer to store currently needed pages of a number of programs in RAM while the rest of these programs wait on the disk

parallel processing a form of multiprocessing that speeds processing by linking several processors to operate at the same time or in parallel

parallel start-up the start-up approach that involves running both the old and new systems for a period of time

password sniffer a small program hidden in a network or a computer system that records identification numbers and passwords

patch a minor change to correct a problem or make a small enhancement

payroll journal a report that contains employees' names, the area where employees worked during the week, hours worked, the pay rate, a premium factor for overtime pay, earnings, earnings type, various deductions, and net pay calculations

perceptive system a system that approximates the way a human sees, hears, and feels objects

personal computers (PCs) relatively small, inexpensive computer systems, sometimes called microcomputers

personal productivity software information systems that enable their users to improve their personal effectiveness, increasing the amount of work and its quality

personal sphere of influence information systems that serve the needs of an individual user

phase-in approach sometimes called a piecemeal approach, the approach to start-up in which components of the new system are slowly phased in while components of the old one are slowly phased out

physical systems design the dimension of the design phase that specifies the characteristics of the system components necessary to put the logical design into action

pilot start-up the start-up approach that involves running the new system for one group of users rather than all users

pipelining a CPU operation in which multiple execution phases are performed in a single machine cycle

pixel a point of light on a display screen

planned data redundancy a way of organizing data in which the logical database design is altered so that certain data entities are combined, summary totals are carried in the data records rather than calculated from elemental data, and some data attributes are repeated in more than one data entity to improve database performance

plotters a type of hard-copy output device used for general design work

point evaluation system an evaluation process in which each evaluation factor is assigned a weight, in percentage points, based on importance; the system with the greatest total score is selected

Point to Point Protocol (PPP) communications protocol software that transmits packets over telephone lines, allowing dial-up access to the Internet

point-of-sale devices (POS) terminals used in retail operations to enter sales information into the computer system

polymorphism a process allowing the programmer to develop one routine or set of activities that will operate on multiple objects

preliminary evaluation evaluation that begins after all proposals have been submitted for the purpose of dismissing unwanted proposals

primary key a field or set of fields that uniquely identifies the record

primary storage the part of the computer that holds program instructions and data, also called main memory or just memory

private branch exchange (PBX) a communications system that can manage both voice and data transfer within a building and to outside lines

problem solving the process of combining intelligence, design, and choice stages along with implementation and monitoring to reach a solution

process a set of logically related tasks performed to achieve a defined outcome

process symbol the symbol that reveals a function that is performed in a data-flow diagram

processing converting or transforming data into useful outputs

productivity a measure of the output achieved divided by the input required

profit centers departments within an organization that track total expenses and net profits

Program Evaluation and Review Technique (PERT) a formalized approach for developing a project schedule

program-data dependence a situation in which programs and data organized for one application are incompatible with programs and data organized differently for another application

programmed decisions decisions that are made using a rule, procedure, or quantitative method

programmer individual responsible for modifying or developing programs to satisfy user requirements

programming languages coding schemes used to write both systems and application software

programming life cycle a series of steps and planned activities developed to maximize the likelihood of developing good software

project deadline the date that the entire project is due to be completed and operational

project management models DSS models that are used to handle and coordinate large projects; they are also used to identify critical activities and tasks that could delay or jeopardize an entire project

project milestone a critical date for the completion of a major part of the project

project organizational structure structure centered on major products or services

project schedule a detailed description of what is to be done

projecting data manipulation that eliminates columns in a table

proprietary software software designed to solve a unique and specific problem

protocol rules that ensure communications among computers of different types and from different manufacturers

prototyping an iterative approach to the systems development process

public network services services that give personal computer users access to vast databases and other services, usually for an initial fee plus usage fees

purchase order processing system a system that helps purchasing departments complete their transactions quickly and efficiently

purchasing transaction processing systems processing systems that include inventory control, purchase order processing, receiving, and accounts payable

push technology technology that enables users to automatically receive information over the Internet rather than searching for it using a browser

quality the ability of a product (including services) to meet or exceed customer expectations

quality control a process that ensures that the finished product meets the customers' needs

query languages used to ask the computer questions in English-like sentences

questionnaires a method of gathering data when the data sources are spread over a wide geographic area

random access memory (RAM) a form of memory where instructions or data can be temporarily stored

rapid application development (RAD) a technique that employs tools, techniques, and methodologies designed to speed application development

read-only memory (ROM) a form of memory that provides permanent storage for data and instructions

receiving system a system that creates a record of expected and actual receipts

record a collection of related fields

reduced instruction set computing (RISC) a computer chip design based on reducing the number of microcode instructions built into a chip to an essential set of common microcode instructions

redundant array of independent/inexpensive disks (RAID) a method of storing data that generates extra bits of data from existing data, allowing the system to create a "reconstruction map" so that if a hard drive fails, it can rebuild lost data

reengineering (process redesign) the radical redesign of business processes, organizational structures, information systems, and values of the organization to achieve a breakthrough in business results

registers high-speed storage areas used to temporarily hold small units of program instructions and data immediately before, during, and after execution by the CPU

relational model a data model in which all data elements are placed in two-dimensional tables, called relations, that are the logical equivalent of files

release a significant program change that often requires changes in the documentation of the software

removable storage devices internal or external devices that provide additional storage capacity in the form of removable disks or cartridges

reorder point (ROP) the critical inventory quantity level

repetitive motion disorder health problems caused by working with computer keyboards and other equipment

repetitive stress injury (RSI) such problems as tendinitis, tennis elbow, the inability to hold objects, and sharp pain in the fingers; same as repetitive motion disorder

replicated database a database that holds a duplicate set of frequently used data

report layout a technique that allows designers to diagram and format printed reports

request for maintenance form a form that is used to authorize modification of programs

Request for Proposal (RFP) a document that specifies in detail required resources such as hardware and software

requirements analysis an assessment used to determine user, stakeholder, and organizational needs

restart procedures procedures that make it very simple for an individual to restart an application from where it left off

return on investment (ROI) one measure of IS value that investigates the additional profits or benefits that are generated as a percentage of the investment in information systems technology

reusable code the instruction code within an object that can be reused in different programs for a variety of applications

revenue centers divisions within the company that track sales or revenues

ring network a type of topology that contains computers and computer devices placed in a ring, or circle. With a ring network, there is no central coordinating computer. Messages are routed around the ring from one device or computer to another.

robotics mechanical or computer devices that perform tasks requiring a high degree of precision or that are tedious or hazardous for humans

router a device that operates at the network level of the OSI model and features more sophisticated addressing software than bridges. Whereas bridges simply pass along everything that comes to them, routers can determine preferred paths to a final destination.

routing system a system that determines the best way to get goods and products from one location to another

rule a conditional statement that links given conditions to actions or outcomes

runaways projects that are far over budget and past delivery dates

sales configuration system a system that ensures that the products and services ordered are sufficient to accomplish the customer's objectives and will work well together

satisficing model an approach to decision support that involves finding a good—but not necessarily the best—solution to a problem

scalability the ability of the computer to smoothly handle an increasing number of concurrent users

schedule feasibility an assessement of whether the project can be completed in a reasonable amount of time

scheduled reports reports produced periodically, or on a schedule, such as daily, weekly, or monthly

scheduling system a system that determines the best time to deliver goods and services

schema a description of the entire database

screen layout a technique that allows a designer to quickly and efficiently design the features, layout, and format of a display screen

search engine a Web search tool

secondary storage the part of the computer that stores large amounts of data, instructions, and information more permanently than main memory, also called permanent storage

selecting data manipulation that eliminates rows according to certain criteria

selective backup backup copies of only certain files

semistructured or unstructured problems more complex problems wherein the relationships among the data are not always clear, the data may be in a variety of formats, and the data is often difficult to manipulate or obtain

sensitivity analysis analysis that allows a manager to determine how the production schedule would change with different assumptions concerning demand forecasts or costs figures

sequence structure a structure in programming in which statements are executed one after another until all the statements in the sequence have been executed

sequential access the process by which data must be retrieved in the order in which it is stored

sequential access storage devices (SASDs) devices used to sequentially access secondary storage data

Serial Line Internet Protocol (SLIP) communications protocol software that transmits packets over telephone lines, allowing dial-up access to the Internet

server application the application that supplies objects you place into other applications

shipment execution system a system that coordinates the outflow of all products and goods from the organization, with the objective of delivering quality products on time to customers

shipment planning system a system that determines which open orders will be filled and from which location they will be shipped

sign-on procedure the identification numbers, passwords, and other safeguards needed for an individual to gain access to computer resources

simulation the ability of the DSS to act like or duplicate the features of a real system

site preparation the process of preparing the actual location of the new system

slipstream upgrade a type of maintenance that involves a minor upgrade—typically a code adjustment or minor bug fix—not worth announcing

Smalltalk a popular object-oriented programming language

software interface the programs or program modifications that allow proprietary software to work with other software used in the organization

software piracy the act of illegally duplicating software

Software Publishers Association (SPA) an organization formed by a number of leading software companies to audit and check for software licenses

software suite a collection of single application software packages in a bundle

source code high-level program code written by the programmer

source data automation the process of capturing data at its source, recording it accurately, in a timely fashion, with minimum manual effort, and in a form that can be directly entered to the computer rather than keying the data from some type of document

special-purpose computers computers that are used for limited applications by military and scientific research groups

sphere of influence the scope of the problems and opportunities addressed by a particular organization

stakeholders individuals who either themselves or through the organization they represent, ultimately benefit from the systems development project

star network a type of topology that has a central hub or computer system, and other computers or computer devices are located at the end of communications lines that originate from the central hub or computer

start-up procedures that begin with the final tested information system; when start-up is finished, the system will be fully operational

statistical analysis models DSS models that provide such data as summary statistics, trend projections, and hypothesis testing

statistical sampling a method of data gathering that involves taking a random sample of data and applying the characteristics of the sample to the whole group

steering committee an advisory group consisting of senior management and users from the IS department and other functional areas

strategic alliance (strategic partnership) an agreement between two or more companies that involves the joint production and distribution of goods and services

strategic planning planning that involves determining long-term objectives by analyzing the strengths and weaknesses of the organization, predicting future trends, and projecting the development of new product lines

structured design an approach to designing and developing application software that breaks large problems into smaller, simpler problems

structured interview an interview in which the questions are written in advance

structured query language (SQL) a standardized language often used to perform database queries and manipulations

structured walkthrough a planned and preannounced review of the progress of a program module, a structure chart, or a human procedure

subschema a file that contains a description of a subset of the database and identifies which users can perform modifications on the data items in that subset

supercomputers the most powerful computers, with the fastest processing speeds

superconductivity a property of certain metals that allows current to flow with minimal electrical resistance

switch a device that routes or switches data to its destination

switched lines lines that use switching equipment to allow one transmission device (e.g., your telephone) to be connected to other transmission devices (e.g., the telephones of your friends and relatives)

syntax a set of rules associated with a programming language

system a set of elements or components that interact to accomplish goals

system boundary defines the system and distinguishes it from everything else

system parameter a value or quantity that cannot be controlled by the decision maker

system performance measurement systems review that involves monitoring the system—the number of errors encountered, the amount of memory required, the amount of processing or CPU time needed, and other problems

system performance products products developed to measure all components of the computer-based information system, including hardware, software, database, telecommunications, and network systems

system performance standard a specific objective of the system

system sign-on the procedure that allows the user to gain access to the computer system

system testing testing the entire system of programs

system variable a quantity or item that can be controlled by the decision maker

system virus a type of virus that typically infects operating systems programs or other system files

systems analysis the system development phase during which the existing systems and work processes are studied to identify strengths, weaknesses, and opportunities for improvement

systems analyst professional who specializes in analyzing and designing business systems

systems controls procedures to maintain data security

systems design a phase in the development of an IS system that answers the question "How will the information system do what it must do to obtain a solution to a problem?"

systems development the activity of creating or modifying existing business systems

systems implementation involves creating or acquiring the various system components (hardware, software, databases, etc.) defined in the design step, assembling them, and putting the new system into operation

systems investigation the system development phase during which problems and opportunities are identified and considered in light of the goals of the business

systems investigation report a report that summarizes the results of the systems investigation and the process of feasibility analysis and recommends a course of action

systems maintenance the checking, changing, and enhancing of the system to make it more useful in achieving user and organizational goals

systems maintenance and review the development phase that ensures the system operates and modifies the system so that it continues to meet changing business needs

systems request form a document filled out by someone who wants the IS department to initiate systems investigation

systems review the final step of systems development; the process of analyzing systems to make sure they are operating as intended

systems software the set of programs designed to coordinate the activities and functions of the hardware and various programs throughout the computer system

team organizational structure structure centered on work teams or groups

technical documentation documentation used by computer operators to execute the program and by analysts and programmers in case there are problems with the program or if the program needs modification

technical feasibility an assessment of whether the hardware, software, and other system components can be acquired or developed to solve the problem

technology infrastructure a computer-based information system that consists of the shared IS resources that form the foundation of the information system

telecommunications the electronic transmission of signals for communications, including such means as telephone, radio, and television

telecommunications medium anything that carries an electronic signal and interfaces between a sending device and a receiving device

telecommuting enables employees to work away from the office using personal computers and networks to communicate via electronic mail with other workers and to pick up and deliver results

Telnet a terminal emulation protocol that enables users to log on to other computers on the Internet to gain access to public files

terminal-to-host an architecture in which the application and database reside on one host computer, and the user interacts with the application and data using a "dumb" terminal

time compression a phenomenon whereby activities occur in a shorter time frame than was previously possible

time-driven review review of the system that is performed after a specified amount of time

time-sharing a processing activity that allows more than one person to use a computer system at the same time

top-down approach a programming approach in which the main module is written before the other modules

total quality management (TQM) a collection of approaches, tools, and techniques that offers a commitment to quality throughout the organization

traditional approach an approach to data management in which separate files are created and stored for each application program

traditional organizational structure organizational structure in which major department heads report to a president or top-level manager

transaction any business-related exchange

transaction processing cycle the process of data collection, data editing, data correction, data

manipulation, data storage, and document production

transaction processing system audit an examination of the TPS in an attempt to answer three basic questions: Does the system meet the business need for which it was implemented? What procedures and controls have been established? Are these procedures and controls being properly used?

transaction processing systems (TPSs) systems that process the detailed data necessary to update records about the fundamental business operations of the organization

transactions the basic business operations such as customer orders, purchase orders, receipts, time cards, invoices, and payroll checks in an organization

Transmission Control Protocol/Internet Protocol (TCP/IP) standard originally developed by the U.S. government to link defense research agencies; it is the primary communications protocol of the Internet

Transport Control Protocol (TCP) a protocol that operates at the transport layer and is used in combination with IP by most Internet applications

tunneling the process by which VPNs transfer information by encapsulating traffic in IP packets and sending the packets over the Internet

uniform resource locator (URL) an assigned address on the Internet for each computer

unit testing testing of individual programs

unstructured interview an interview in which the questions are not written in advance

upper-CASE tools tools that focus on activities associated with the early stages of systems development

Usenet a system closely allied with the Internet that uses e-mail to provide a centralized news service. It is actually a protocol that describes how groups of messages can be stored on and sent between computers.

user acceptance document a formal agreement signed by the user that states that a phase of the installation or the complete system is approved

user documentation documentation developed for the individuals who use the program. This type of documentation shows users, in easy-to-understand terms, how the program can and should be used.

user interface a part of the operating system that allows individuals to access and command the computer system

user preparation the process of readying managers, decision makers, employees, other users, and stakeholders for the new systems

users individuals who will interact with the system regularly

utility programs programs used to merge and sort sets of data, keep track of computer jobs being run, compress files of data before they are stored or transmitted over a network, and perform other important tasks

value chain a series (chain) of activities that includes inbound logistics, warehouse and storage, production, finished product storage, outbound logistics, marketing and sales, and customer service

value-added carriers companies that have developed private telecommunications systems and offer their services for a fee

version a major program change, typically encompassing many new features

very long instruction word (VLIW) a computer chip design based on further reductions in the number of instructions in a chip by lengthening each instruction

video compression a process that reduces the number of bits required to represent a single video frame by using mathematical formulas

videoconferencing systems that combine video and phone call capabilities with data or document conferencing

virtual memory memory that allocates space on the hard disk to supplement the immediate, functional memory capacity of RAM

virtual private network (VPN) a secure connection between two points across the Internet

virtual workgroups teams of people located around the world working on common problems

virus a program that attaches itself to other programs

vision systems the hardware and software that permit computers to capture, store, and manipulate visual images and pictures

visual programming languages languages that use a mouse, icons, or symbols on the screen and pull-down menus to develop programs

voice mail technology that enables users to leave, receive, and store verbal messages for and from other people around the world

voice-over-IP (VOIP) technology that enables network managers to route phone calls and fax transmissions over the same network they use for data

voice-recognition devices input devices that recognize human speech

volume testing testing the application with a large amount of data

Web browser software that creates a unique, hypermedia-based menu on your computer screen and provides a graphical interface to the Web

webmaster person who sets up and manages a company's Internet site

"what-if" analysis the process of making hypothetical changes to problem data and observing the impact on the results

wide area network (WAN) a network that ties together large geographic regions using microwave and satellite transmission or telephone lines

wordlength the number of bits the CPU can process at any one time

workgroup two or more people who work together to achieve a common goal

workgroup sphere of influence information systems that support a workgroup in attainment of a common goal

workstations computers that fit between high-end microcomputers and low-end midrange computers in terms of cost and processing power

World Wide Web a collection of tens of thousands of independently-owned computers that work together as one in an Internet service

worm an independent program that replicates its own program files until it destroys other systems and programs or interrupts the operation of networks and computer systems

write-once, read-many (WORM) an optical disk that allows businesses to record customized data and information; once data and information has been recorded onto a WORM disk, it can be repeatedly accessed, but it cannot be altered

CHAPTER 1

1. Justin Hibbard, "Spreading Knowledge," *Computerworld*, April 7, 1997, pp. 63-64.
2. "Business Commerce," I/PRO CyberAtlas at http://www.economist.com accessed May 6, 1998.
3. "Survey: E-Commerce to Double by End of 1998," www.internetnews.com, March 27, 1998.
4. "Price Waterhouse Predicts Explosive E-Commerce Growth," www.internetnews.com, March 26, 1998.
5. E-commerce tutorial site located at http://www.netvendor.com accessed on April 8, 1998.
6. Randy Weston, "Whirlpool to Try Pricing Systems," *Computerworld*, March 23, 1998, pp. 1, 14.
7. Kim Girard, "Want to See That Desk in 3-D," *Computerworld*, April 6, 1998, pp. 55–56.
8. Stephanie Neil, "Guiding Airlines To Greater Efficiency," *PC Week*, January 19, 1998, pp. 79–80.
9. "Charles Schwab & Co., Inc." *PRNewswire*, March 20, 1998.
10. Thomas Hoffman and Kim S. Nash, "Titanic Tangle," *Computerworld*, April 13, 1998, pp. 1, 94.
11. Monua Jonah and Clinton Wilder, "Special Delivery," *InformationWeek*, October 27, 1997, pp. 42–60.
12. Don Peppers, "Knocking on Healthcare's Door," *Healthcare Forum Journal*, January/February 1998, p. 29.
13. Justin Hibbard, "Spreading Knowledge," *Computerworld*, April 7, 1997, pp. 63–64.

CHAPTER 2

1. Atrium Web site at http://www.atrium-hr.com accessed on June 30, 1998.
2. E. H Schein, *Process Consultation: Its Role in Organizational Development* (Reading, Mass.: Addison-Wesley, 1969). See also Peter G.W. Keen, "Information Systems and Organizational Change," *Communications of the ACM*, vol. 24, no. 1, January 1981, pp. 24–33.
3. Sharon Machlis, "A Battery of Benefits Online," *Computerworld*, June 1, 1998, p. 35.
4. Bronwyn Fryer, "Decision Support Aiding Health Care," *InformationWeek*, June 15, 1998, pp. 159–164.
5. Mike Boyer, "GEAE Mantra: Six Sigma," *Cincinnati Enquirer*, April 19, 1998, p. 1 of Business Section.
6. Marianne Kolbasuk McGee with Gregory Dalton, "Lilly Outsources to EDS," *InformationWeek*, February 23, 1998, p. 34.
7. Bruce Caldwell, "Why Outsource?" *InformationWeek*, December 1, 1997, p. 115.

8. Ann Monroe, "Getting Rid of the Gray, Will Age Discrimination be the Downfall of Downsizing?" http://bsd.mojones.com, accessed April 14, 1998.
9. M. Porter and V. Millar, "How Information Systems Give You Competitive Advantage," *Journal of Business Strategy*, Winter 1985. See Also M. Porter, *Competitive Advantage* (New York: Free Press, 1985).
10. Bill Vlasic, et al., "The First Global Car Colossus," *BusinessWeek*, May 16, 1998, pp. 40–42.
11. Peter Coy, "Trustbusters: Why They Do What They Do," *BusinessWeek*, June 1, 1998, pp. 42–44.
12. Rich Levin, "Kmart To Expand Data Warehouse," *InformationWeek*, June 15, 1998, p. 128.
13. Thomas Hoffman, "ROI Aimed at Costs, Not Competitiveness," *ComputerWorld*, March 30, 1998, p. 4.
14. Jeffrey M. Kaplan, "Focus On IT's Return," *InformationWeek*, June 22, 1998, p. 186.
15. Marianne Kolbasuk McGee, "School Daze," *InformationWeek*, February 2, 1998, pp. 44–52.

CHAPTER 3

1. "Federal Aviation Administration Year 2000 Project Plan," at http://faay2k.com/text/projplan.html, last updated April 1998, accessed July 25, 1998.
2. Lisa Digarlo, "Digital Revs Alpha to 600 MHz," *PC Week*, February 9, 1998, p. 47.
3. Intel Web site at http://www.intel.com, accessed July 25, 1998.
4. Andy Reinhart, Ira Sager, Peter Burrows, "Intel—Can Andy Grove Keep Profits Up in an Era of Cheap PCs?," *BusinessWeek*, December 22, 1997, pp. 70–77.
5. Dean Takahashi, "Membrane May Help Man Meet Machine," *The Wall Street Journal*, January 31, 1997, p. 81.
6. "Rambus Completes Direct RDRAM Interface Design," press release on Rambus, Inc. Web site at http://www.rambus.com, February 17, 1998, accessed July 26, 1998.
7. Paul Massiglia, "A Buyer's Guide to Storage for the Enterprise," Digital Equipment Corporation, December 5, 1997.
8. Mary E. Thyfault, "Voice Recognition Enters The Mainstream," *InformationWeek*, July 14, 1997, p. 146.
9. Jon Pepper, "Get the Picture," *InformationWeek*, January 26, 1998, pp. 120–123.
10. Unisys Web site at http://www.unisys.com, accessed July 26, 1998.

11. "Compaq Brings PC & TV Together," Intel Web site at http://www.developer.intel.com/ design/idf, accessed July 25, 1998.
12. Larry Armstrong, "For This Printer, Scanning's a Snap," *BusinessWeek*, August 11, 1997, p. 16.
13. Reinhardt, Sager, and Burrows, "Intel, Can Andy Grove Keep Profits Up in an Era of Cheap PCs?"
14. Gordon Mah Ung, "Handheld PCs," *Computerworld*, March 2, 1998, p. 26.
15. Lisa Digarlo, "JavaStation to Finally Debut," *PC Week*, March 9, 1998, pp. 51, 54.
16. Bob Francis and Bruce Caldwell, "PC Ownership Cost Control," *InformationWeek*, January 27, 1997, p. 150.
17. Randy Weston and April Jacobs, "Early Users Tout NC," *Computerworld*, February 10, 1997, p. 1.
18. April Jacobs, "PC Vendors Get Management Fever," *Computerworld*, March 2, 1998, p. 20.
19. Karen J. Bannan, "HP Kayaks Join 333 MHz Pentium II Fleet," *PC Week*, March 2, 1998, pp. 43, 49.
20. Karen J. Bannan, " Hardware Scaling Higher," *PC Week*, February 16, 1998, pp. 55, 59.
21. Jikumar Vijayan, "IBM Mainframe Gains Appeal," *Computerworld*, March 2, 1998, p. 6.
22. Gary Anthes, "Supercomputer Tops 1 Teraflop," *Computerworld*, January 2, 1997, p. 124.

CHAPTER 4

1. Vance McCarthy, "IT Staff Shortage Nears 350,000," *HP World*, February 1998, p. 5.
2. Kathleen Gow, "The Support Burden," *Computerworld Global Innovators*, June 9, 1997, p. 8.
3. Sam Kennedy, "Intel and Microsoft to Decrease Application Start-up Time for Win 98," Gaming Age Web site at http://www.gamingage.com/news, accessed July 20, 1998.
4. "Upgrade to Windows 98 . . . Before Microsoft Does Get the OSR2 Downloadable Components," *PC Magazine*, March 10, 1998.
5. WavePhone homepage at http://www.wavephore.com.
6. Steve Rigney, "What to Expect from Memphis Beta 2," *PC Magazine*, September 9, 1997.
7. Larry Seltzer, "System Plumbing," *PC Magazine* August 4, 1997.
8. Andrew Seybold's Outlook, CE Watch, "Perspectives on Window CE," at http://www.cewatch.com, accessed October 3, 1998.
9. Apple Computer Inc. homepage at http://www.apple.com, accessed January 19, 1998.
10. Ed Bott, "NT 5.0 Exposed," *PC Magazine*, February 2, 1998.

11. Lisa Dicarlo, "No Go on Windows '00'," *PC Week,* July 28, 1997, p. 8.
12. Jaikumar Vijayan, "Unix Grows at High End; NT Takes Over at Low End," *Computerworld,* August 4, 1997, p. 14.
13. Peter Coffee, "Workgroup Products That Work," *PC Week,* December 22/29, 1997, p. 49.
14. Peter Coffee, "Do It Yourself Microsoft Office," *PC Week,* February 9, 1998, p. 51.
15. Patrick Dryden, "Caution Advised on Management Suites," *Computerworld,* January 13, 1997, p. 4.
16. Barb Cole-Gomolski, "Insurer Uses Groupware to Cut Back Claims Work," *Computerworld,* September 15, 1997, pp. 41–42.
17. Christy Walker, "cc:Mail Winds Down," *PC Week,* February 2, 1998, p. 13.
18. John Taschek, "Groupware: 2 Decades Old and It's Still Not Ready," *PC Week,* January 12, 1998, p. 64.
19. Christy Walker, "Domino.Doc Opens Up," *PC Week,* February 9, 1998, p. 10.
20. Tim Stevens, "ERP Explodes," *Industry Week,* vol. 245, no. 13, July 1, 1996, p. 37; and Jack Neff, "Grand Plans: ERP Software Aims to Tame Planning Beast," *Food Processing,* vol. 58, no. 3, March 1997, p. 109.
21. Julia King, "Smalltalk Boot Camp Pays Off," *Computerworld,* vol. 31, no. 10, March 10, 1997, p. 65.
22. Peter Coffee, "Why C++ Is Not Yet Ready for Retirement," *PC Week,* February 16, 1998, p. 39.
23. Michael Moeller, "Sun Sows Java Seeds," *PC Week,* February 25, 1998, p. 39.
24. Gordon Mah Ung, "Nabisco, MediaSolv Cook Up App.," *Computerworld,* vol. 32, no. 1, December 29, 1997, p. 61.
25. "Beyond Visual Basic," *PC Week,* vol. 14, no. 9, March 3, 1997, p. 69.
26. "The Next Generation: 5GL Programming," *PC Week,* vol. 14, no. 6, February 10, 1997, p. 119.
27. Randy Weston, "The Fifth Generation," *PC Week,* vol. 14, no. 1, January 6, 1997, p. 103.
28. David Watson, "Programmers Show Dedication to C++," *Computer Dealer News,* vol. 13, no. 21, October 20, 1997, p. 102.

CHAPTER 5

1. John W. Verity, "Coaxing Meaning Out of Raw Data," *BusinessWeek,* February 3, 1997, pp. 134–138.
2. Thomas Hoffman, "Data Warehouse, The Sequel," *Computerworld,* June 2, 1997, pp. 69–72.
3. Business Research Paper, "On Doing Well by Doing Good," *Datamation,* March 1997, p. S11.
4. Craig Stedman, "Vendors Ready to Harvest Database Crop," *Computerworld,* vol. 31, no. 35, September 8, 1997, p. 16.

5. Nelson King, "Dealing with Dynamic Data," *InternetWorld,* July 1997, pp. 71–83.
6. Sedar Yegulalp, "The Programmer's Database Gets Even Better," *Windows* magazine, vol. 8, no. 6, June 1997, p. 14.
7. Chris O'Malley, "Do-It-All Databases," *PC Computing,* vol. 10, no. 3, March 1997, p. 171.
8. Timothy Dyck, "Borland Revitalizes dBASE," *PC Week,* vol. 14, no. 44, October 20, 1997 p. 65.
9. John Foles, "Data Warehouse Pitfalls," *InformationWeek,* May 19, 1997, pp. 93–96.
10. Charles B. Darling, "Ease Implementation Woes with Packaged Datamarts," *Datamation,* March 1997, pp. 94–98.
11. Teresa Wingfield, "Data Management Strategies: Meeting the Needs of Departments and the Enterprise," Analyst Briefing at 1997 Giga Information Group, Phoenix, Arizona.
12. Craig Stedman, "Going Virtual Eases Pain of Managing Multiple Data Marts," *Computerworld,* August 4, 1997, p. 8.
13. Craig Stedman, "Data Vaults Unlocked," *Computerworld,* June 2, 1997, pp. 1, 17.
14. Verity, "Coaxing Meaning Out of Raw Data."
15. John Foley and Joy D. Russell, "OLAP Goes Mainstream," *InformationWeek,* January 19, 1998, p. 18.
16. Ibid.
17. Theresa Rigney, "Clean Data at the End of the Tunnel," *Data Quality Maze,* Special Supplement to *Software* magazine, October 1997, p. S3.
18. John Taschek, "The Midyear State of the Database Industry," *PC Week,* .vol. 14, no. 33, August 4, 1997, p. 6.
19. Stacy Lavilla, "DB2 Universal Database Supports More Data Types," *PC Week,* vol. 14, no. 39, September 15, 1997, p. 18.
20. Theresa Rigney, "Clean Data at the End of the Tunnel."

CHAPTER 6

1. Marianne Kolbasuk McGee, "Ford Plans Supply-Chain Integration," *InformationWeek,* July 27, 1998, p. 126.
2. Eric Schine, Peter Elstrom, Amy Barrett, Gail Edmondson, and Michael Shari, "The Satellite Biz Blasts Off," *BusinessWeek,* January 27, 1997, pp. 62–70.
3. Jack Lyon, "Blast onto the Net by Satellite," *PC Computing,* March 1997, p. 71.
4. Frederick Rose, "New Modems are Fast, Cheap," *The Wall Street Journal,* February 11, 1997, p. 81.
5. Scott Wooley and Nikhil Hutheesing, "Pushing Phone Lines to the Limit," *Forbes,* March 23, 1998, pp. 160–162.

6. David Sullivan, *The New Computer User,* 2nd ed., Orlando, Florida: Harcourt Brace & Company, 1997.
7. Les Freed, "Digital Dragsters," *PC Magazine,* February 4, 1997, p. 137.
8. Kim Gerard, "Faster Phone Links Move Toward Reality," *Computerworld,* February 17, 1997, p. 57.
9. Tim Greene, "US West Full Speed Ahead with DSL," *Network World,* February 2, 1998, p. 6.
10. UPS Web site at http://www.upsnet.com, accessed March 15, 1998.
11. Charles Waltner, "Unilever's Direct Approach," *InformationWeek,* February, 16, 1998, pp. 118–119.
12. David Kosiur, "Routers: The Past or the Future?" *PC Week,* August 4, 1997, p. 96.
13. Michael Surkan, "Is It a Switch or Router?" *PC Week,* January 12, 1998, p. 106.
14. Web site for Telecommuting Success, Inc. at http://www.telsuccess.com, "The Why's of Telecommuting," accessed August 4, 1998.
15. Ad for NuVision Technologies, Inc., *Forbes,* March 23, 1998, p. 155.
16. "America Online Inc. Completes Acquisition of Compuserve Worldwide Online Services and Sale of ANS Communications," press release found at the CompuServe Web site at: http://www.compuserve.com, accessed June 1, 1998.
17. Marc Gunther, "The Internet is Mr. Case's Neighborhood," *Fortune,* March 30, 1998, pp. 69–80.

CHAPTER 7

1. Scott Berinato, "The Net's Next Frontiers," *PC Week,* March 2, 1998, p. 21.
2. Erick Von-Schweber, "Projects Promise IS Plenty," *PC Week,* February 9, 1998, p. 93.
3. John Borland, "Network Solutions Girds For Loss Of Monopoly," TechWeb, *Internet News,* August 6, 1998, at http://www.techweb.com accessed August 12, 1998.
4. Kevin Burden, "Choosing an Internet Service Provider, What Your Peers Say," *Computerworld,* January 26, 1998, p. 79.
5. Jeff Sweat, "ISPs Step Out," *InformationWeek,* January 19, 1998, pp. 73–80.
6. Ibid.
7. Carmen Nobel, "Despite Hiccups, Satellite Web Access Looms," *PC Week,* April 13, 1998, p. 105.
8. John Rendleman and Scott Berinato, "Voice Over IP Alters Telephony Market," *PC Week,* March 16, 1998, p. 108.
9. Andrew Cray, "Voice Over IP, Hear's How," *Data Communications,* April 1998, pp. 44–56.

10. Larry Armstrong, Neal Sandler, and Peter Elstrom, "You're Coming Over Loud—and Almost Clear," *BusinessWeek,* October 27, 1997, pp. 116–118.
11. David Sullivan, *The New Computer Users,* 2nd ed. Orlando, Florida: Harcourt Brace & Company, 1997, pp. 246–277.
12. Michael Csenger, "Stream On," *Intranet,* January 1998, pp. 17–20.
13. Les Freed, "Digital Dragsters," *PC Magazine,* February 4, 1997, p. 137.
14. "Bell Atlantic to Launch High-speed Internet Access," *Reuters News Service,* at http://www.reuters.com, accessed June 4, 1998.
15. Clare Haney, "Micrososft Moves to Allay ActiveX Security Worries," *TechWire,* at http://www.ezwire.com, accessed February 20, 1997.
16. Scott Mace, Udo Flohr, Rick Dobson, and Tony Graham, "Weaving a Better Web," *Byte,* March 1998, pp. 58–68.
17. Lawrence Aragon, "Go With the Flow," *PC Week,* June 2, 1997, p. 135.
18. Rick Whiting, "How IT Works—Push Technology," *Client/Server Computing,* June 1997, p. 48.
19. Press Release "EFF Builds DES Cracker that Proves that Data Encryption Standard is Insecure," July 17, 1998, Electronic Frontier Foundation Web site at http://www.eff.org, accessed August 13, 1998.
20. Rutrell Yasin, "Hackers Break Pentagon Defense," *Internet News,* at http://www.internetnews.com, accessed March 2, 1998.
21. Trusted Information Systems Web site at http://www.tis.com, (updated 2/4/1997), accessed June 1, 1998.

CHAPTER 8

1. Bernard Wysocki Jr. "Internet is Opening Up a New Era of Pricing," *The Wall Street Journal,* June 8, 1998, p. 1.
2. Tim Ouellette and Thomas Hoffman, "Ice Storm Freezes Operations," *Computerworld,* January 19, 1998, pp. 1, 116.
3. Joseph R. Gerber, "Know Your Customer," *Forbes,* February 10, 1997, p. 128.
4. Maura Cleveland, "It's Time for Internet Commerce," *Software Strategies,* March 1998, p. 25.
5. Bernard Wysocki Jr., " Internet is Opening Up a New Era of Pricing," *The Wall Street Journal*, p.A1, June 8,1998.
6. Ibid.
7. Erin Callaway, "ERP Choices," *Managing Automation,* August 1998, pp. 28–39.
8. "Baan In Business," *PC Week,* October 20, 1997, p. 43.
9. Craig Stedman, "ERP More than a 2000 Fix," *Computerworld,* August 3, 1998, pp. 1, 84.

10. Randy Weston and Craig Stedman, "Forget ROI—Just Install It," *Computerworld,* April 13, 1998, pp. 1, 14.
11. Callaway, "ERP Choices."
12. Randy Weston, "ERP Vendors Going with the Flow," *Computerworld,* January 26, 1998, p. 53.
13. Nancy H. Bancroft, Henning Seip, and Andrea Sprengel, *Implementing SAP R/3,* 2nd ed., p. 14, copyright 1998, Greenwich, CT: Manning Publications Co.

CHAPTER 9

1. Carolyn T. Geer, "New Software Is Helping Business Cut Legal Costs-Haggle No More," *Forbes,* January 27, 1997.
2. "Hyperion Essbase in Action: Nabisco," under the Customers section of the Hyperion Web site at http://www.hyperion.com, accessed September 6, 1998.
3. Doug Bartholomew, "Successful? Try, Try Again," *IndustryWeek,* February 2, 1998.
4. Tim McCollum, "Taking Account of Software," *Nation's Business,* vol. 85, no. 1, January 1997, p. 41.
5. Thomas Hoffman, "Green-Light Special: IS Reviving Kmart," *Computerworld,* December 16, 1996, p. 8.
6. Alex Taylor III, "The Gentlemen at Ford are Kicking Butt," *Fortune,* June 22, 1998.
7. Hoffman, "Green-Light Special."
8. Electronic Commerce World-Marketing on the Internet at http://ecworld.utexas.edu/ejou/mkt/mkt.html, accessed May 28, 1998.
9. "Boise and Hyperion Solutions Create High-Performance Business Analysis Applications," under the Customers section of the Hyperion Web site at http://www.hyperion.com, accessed September 6, 1998.
10. Esther Shein, "Show 'Em What You Got," *Inside Technology Training,* May 1998, p. 39.
11. April Jacobs, "Mapping Software Puts an End to Paper Chase," *Computerworld,* March 3, 1997, pp. 45, 48.
12. The Application of GIS—Emergency Response Planning section of the U.S. Geological Service Web site at http://www.usgs.gov, accessed September 3, 1998.

CHAPTER 10

1. John W. Verity, "Coaxing Meaning Out of Raw Data," *BusinessWeek,* February 3, 1997, pp. 134-138.
2. "Softkit Technologies Decision Grid 3.1," *Computer Shopper Product,* June 1998.
3. Patrick Thibodeau, "Need Info? Help Yourself," *Computerworld,* April 27, 1998, p. 24.

4. Thomas Hoffman, "IT Plays Key Role in Acquisition," *Computerworld,* August 10, 1998, p. 8.
5. Elliott Gold and Stephen Shaw, " Enterprisewide Conferencing: The Era of Networked Teams," Special Advertising Supplement, *Fortune,* April 1997.
6. Daniel Levine, "Notes Takes on the Web," *PC Computing,* March 1997, p. 86.
7. David Brown, "Collaboration Tools Link Far-Flung Office Mates," *Network Computing Online* at http:www.idg.net updated Monday, June 1, 1998, accessed August 26, 1998.
8. Gold and Shaw, "Enterprisewide Conferencing: The Era of Networked Teams."
9. Craig Stedman, "Information Tools for CEOs See Renaissance," *Computerworld,* July 6, 1998.

CHAPTER 11

1. "Robotic Welding Offers Cost-Effective Solution," *MetalWorking Digest Online* found at http://www.manufacturing.net/magazine/metal/features, accessed September 12, 1998.
2. Homepage of Arrick Robotics at http://www.arrick.com, accessed June 10, 1998.
3. "Los Angeles Sheriff's Department Identifies Suspect with New Crime Fighting Software" from the "News About Viisage" Web site at http://www.viisage.com, accessed September 13, 1998.
4. Peter Coffee, "Telling PCs Where to Go," *PC Week,* June 2, 1997, p. 14.
5. Matt Hamblen, "Boston Globe: The Power of Your Voice," *Computerworld,* March 30, 1998, p. 49.
6. Mary E. Thyfault, "Nuance Aims to Boost Voice Recognition in Apps," *InformationWeek,* May 25, 1998, p. 168.
7. Suleiman Kassicieh et al., "NeuralWorks," *OR/MS Today,* February 1997, p. 68.
8. "BrainMaker Tracks Beer Quality" in the Applications Section of the BrainMaker Web site at http://www.calsci.com, accessed September 12, 1998.
9. Suleiman K. Kassicieh, et al., "Neuralworks Predict", *OR/MS Today,* February 1997, pp. 68–70.
10. Product Information at the Web site for SPSS at http://www.spss.com/software, accessed September 12, 1998.
11. Home page of Acquired Intelligence Inc. at http://www.aii.com, accessed June 8, 1998.
12. Inform Software Web page at http://www.fuzzytech.com, accessed June 12, 1998.
13. Web page of GH Corp at http://www.ghgcorp.com, accessed September 12, 1998.

14. Web page of MatheMedics at http:www.mathemedics.com, accessed June 11, 1998.
15. Web page of the Institute for Information Technology in Ottawa, Canada (part of the National Research Council of Canada) at http://ai.iit.nrc.ca, last revised March 3, 1998 by Michael Lehane, accessed June 11, 1998.
16. Web page of Attar Software Limited at http://www.attar.com, accessed June 11, 1998.

CHAPTER 12

1. The Standish Group International, Inc. and Dennis Mass, "Project Management Survey," *Computerworld,* June 8, 1998, p. 57.
2. Craig Stedman, "Warehousing Projects Hard to Finish," *Computerworld,* March 23, 1998, p. 29.
3. Jeetu Patel, "Make the Web Work for You," *InformationWeek,* June 22, 1998, pp. 63–80.
4. Carol Silva, "Tools Provide Web Access to Databases," *Computerworld,* June 22, 1998, p. 41.
5. Craig Stedman, "SAP Shoots Beyond Back-Office Realm," *Computerworld,* June 22, 1998, p. 8.
6. Clinton Wilder, "MRO Catalog for R/3," *InformationWeek,* June 22, 1998, p. 83.
7. Barry W. Boehm, *Software Engineering Economics* (Englewood Cliffs, NJ: Prentice-Hall, 1981).
8. Sharon Machlis, "HMO Giant Launches E-payment Program," *Computerworld,* June 22, 1998, p. 6.
9. Charolette Adams, "Agencies Take a RAD Approach to Development," Tech Briefing, April 15, 1997 at http://www.fcw.com, accessed June 20, 1998.
10. Kim Girard, "Call Center Reorg Unites Travel Info," *Computerworld,* July 13, 1998, pp. 1, 85.
11. Michael Goldberg, "IRS Warning," *Computerworld,* February 10, 1997, p. 32.
12. "Air Traffic Control: Weak Computer Security Practices Jeopardize Flight Safety," Abstracts of GAO Reports and Testimony, FY98, AIMD-98-155, May 18, 1998, found at http://www.gao.gov, accessed September 19, 1998.
13. Julia King, "IS Reins in Runaway Projects," *Computerworld,* February 24, 1997, p. 1.
14. Ibid.
15. ISO 9000 Group Web page at http://www.isogroup.simplenet.com, accessed June 19, 1998.
16. Kevin Burden "Buyer's Guide to Application Development Tools: Simplicity Sells," *Computerworld,* March 24, 1998, pp. 89–92.

CHAPTER 13

1. Disaster Recovery Services at the Web site of Business Recovery Management located at http://www.businessrecords.com, accessed September 19, 1998.
2. Disaster Recovery at the Guardian Computer Support Web site at http://www.guardian-computer.com, accessed September 19, 1998.
3. Products section of the SunGard Recovery Services Web site at http://www.sungard-recovery.com, accessed September 19, 1998.
4. Matthew Schwartz, "A Methodology to Evaluation Madness," *Software Magazine,* June 1998, p. 76.
5. Thomas Hoffman, "Testing Y2K on Mirror Site Saves for Exchange," *Computerworld,* September 21, 1998.
6. Rajiv D. Banker and Sandra A. Slaughter, "A Field Study of Scale Economies in Software Maintenance," *Management Science,* vol. 43, no. 12, December 1997, pp. 1709–1724.
7. Tom Diederich, "Microsoft Patches Browser Glitch," *Online News,* September 8, 1998.
8. "New Life for Legacy Code with Component Mining," *InformationWeek,* February 16, 1998, p. 18A.
9. Products and Services Section of the Candle Corporation Web site at http://www.candle.com, accessed September 25, 1998.
10. Press Release, "Precise Software Solutions Announces Precise/Pulse! Application Performance Monitoring Solution," July 28, 1998 in the news section of the Precise Software Web page at http://www.precisesoft.com, accessed September 25, 1998.

CHAPTER 14

1. Laura DiDio, "Computer Crime Costs on the Rise," *Computerworld,* April 20, 1998, pp. 55–56.
2. S. Machlis and B. Cole, "E-mail Floods Cause Network Havoc," *Computerworld,* March 10, 1997, p. 10.
3. Ann Coffou and Stephanie Moore, "Year 2000 Contingency Plans: What Can Be Done in the Time Remaining?" *GigaWorld IT Forum 98,* June 1–4, 1998, Scottsdale, Arizona.
4. Leonard Lee, "FBI Investigates a 'Prime' Hacking Suspect," *Technology: GLITCHES Updated:* September 26, 1998.
5. Sharon Machlis, "Hacker Detection IS Key," *Computerworld,* March 3, 1997, p. 59.
6. Virus Statistics at Dr. Solomon Web page at http://www.drsolomon.com, accessed October 3, 1998.
7. Patrick Thibodeau, "Workplace Computer Virus Infections on the Rise," *Computerworld Online News,* September 14, 1998.

8. Stewart Deck, "Browser Bug Blanks Out PC Desktops," *Software News,* April 23, 1998.
9. "U.S. Programmer Charged with Computer Sabotage," *Yahoo! News,* February 17, 1998.
10. Tom Diederich, "World Cup Virus Gives Word Users a Boot," *Computerworld Online News,* July 2, 1998.
11. "What's a Macro Virus?" Dr. Solomon's Web site at http:www.drsolomon.com, accessed October 5, 1998.
12. "Macro Virus" section of the Animal Health Diagnostic Lab at Michigan State University Web page at http://www.ahdl.com.msu.edu, accessed October 5, 1998.
13. Leslie Goff, "In-Depth," *Computerworld,* May 25, 1998.
14. Rebecca Quick, "Web Sites Find Members Don't Keep Secrets," *The Wall Street Journal,* February 21, 1997, p. B1.
15. Sharon Machlis, "MIPS-Snatching Apps: Threat May Be Benefit," *Computerworld,* March 17, 1997, pp. 67, 71.
16. "U.S. Software Firms Lost Billion to Pirates in 1997," Yahoo News at Web site http://dailynews.yahoo.com, accessed June 17, 1998.
17. Ann Davis, "Concerns Rise on Laundering Money On Line," *The Wall Street Journal,* March 17, 1997, p. B1.
18. Margaret Kane, "California Passes Spam Bill," August 29, 1998, Tech News from ZD Net.
19. Sheryl Canter, "E-mail Encryption," *PC Magazine,* April 8, 1997, p. 243.
20. Ken Phillips, "Unforgettable Biometrics," *PC Week,* October 27, 1997, pp. 95, 111.
21. Chris Oakes, "Fingerprint Computer IDs Enter the Mainstream," Yahoo News Web site at http://dailynews.com, accessed July 8, 1998.
22. Lawrence Aragon, "Show Me Some ID," *PC Week,* January 12, 1998, pp. 77–78.
23. Jennifer Bresnahan, "To Catch a Thief," *CIO,* March 1, 1997, p. 69.
24. James M. Wicks and Eric W. Penzer, "Is it Safe? New CPLR Section Says E-Mail Communications Retain Evidentiary Privilege," *New York Law Journal,* August 17, 1998.
25. Bill Mann, "Stop Watching Me," *Internet World,* April 1997, p. 46.
26. Jon G. Auerbach, "Judge Throws Out $5.3 Million Verdict Against Digital on Keyboard Injury," *The Wall Street Journal,* April 30, 1997, p. 85.
27. Web site of the Association of Information Technology Professionals at http:// www.aitp.org, accessed October 7, 1998.
28. Association for Computing Machinery Web site at http://www.acm.org, accessed October 7, 1998.

■ INDEX

A boldface page number indicates a key term and the location where its definition can be found.

A

ABAP/4, 376, 378
Acceptance testing, **601**
Access 98 (Microsoft), 214
Accounting, information system in, 30
Accounting management information systems, **424**
Accounting systems, 348, 360–366
　accounts receivable system, **263**
　asset management, 364–365
　budget transaction processing system, 360–361
　general ledger system, 365–366
　payroll system, 363–364
Accounts payable system, **360**
Accounts receivable system, **263**
Accuracy of TPS, 339
ACE, 508
ACM code of professional conduct, 655
Acrobat (Adobe), 271–272
Active-matrix display, 110
Activity modeling, 550–552
Ad hoc DSS, **449**–450
Administrators, LAN, 71
ADSTAR Distributed Storage Manager (ADSM), 134–135
Advanced Configuration and Power Interface (ACPI), 150
Advanced Micro Devices, Inc. (AMD), 131–132
Advanced Power Management (APM) 1.2, 150
Advantage, competitive. *See* Competitive advantage
Advertising, 416–417
Aetna, 533, 618–619
Age discrimination, 56
Agents, intelligent, 359, 495
Airline industry, information systems in, 31
AITP code of ethics, 654
ALADIN system, 283–284
Alameda Alliance for Health, 53
Algorithms, genetic, 515
Allchin, Jim, 152
Alliance, strategic (strategic partnership), 59–60
Alliant Utilities, 558
Allina Health System, 198–199
Alpha 21264 processors, 83, 88
Alphanumeric data, 5
Alta Vista, 310–311
Amazon.com, 24, 317, 576
Amdahl, 119, 120
American Airlines, 186–187
American Civil Liberties Union, 196
American Express, 273
American Standard Code for Information Interchange (ASCII), 170
America Online (AOL), 277, 644, 662
AmiPro (Lotus), 633
Analog signal, **248**
Analysis, systems, **29**
Analytical processing, on-line (OLAP), **223**–225
Anchor Glass, 351
Anheuser-Busch, 487
Anonymous input, 460
AnswerLink, 581
Antivirus programs, **641**
Apple operating systems, 148, 150–152
Applets, **308**–309, 312–313
Application flowcharts, **552**
Application Launch Accelerator, 149
Application servers in SAP system, 376, 377

Application sign-on, **572**
Application software, **138**–139, 155–169
　enterprise, 167–169
　object linking and embedding (OLE), **163**–164
　personal, 158–163
　　creativity, 158
　　database, 158, 159–160, 161
　　desktop publishing (DTP), 158
　　financial management, 158, 159
　　graphics, 158, 160, 161
　　on-line information services, 158, 160–161, 162
　　project management, 158
　　spreadsheet, 158, 159, 160
　　word processing, 158, 159, 160
　types of, 156–157
　workgroup, 165–166
Application viruses, **632**
Approach 98 (Lotus), 214–215
Ariba software, 131–132
Arithmetic/logic unit (ALU), **86**
Arizona Telemedicine Program, 275
ARPANET, **291**, 632
Arthur Andersen & Co., 19, 20
Artificial intelligence (AI), 27–28, **480**–489. *See also* Expert systems
　applications of, 407–409
　for job scheduling, 515
　major branches of, 483–489
　natural intelligence vs., 483
Artificial intelligence systems, **480**
Ashcroft, John, 323
Asking directly, **554**
Aspect Development, 530
Assembly language, **171**
Asset management transaction processing system, 364–365
Asynchronous transfer mode (ATM), 321
Athleta, 314
Atlantic Richfield Co., 468
Atrium Empowerment, 47
AT&T, 251, 254
Attribute, **191**–192
Audio components, in multimedia systems, 122–123
Audio data, 5
Auditing, **403**–404
Audit trail, **347**
Auto-By-Tel, 307
Automated record keeping, 460
Automatic teller machines (ATMs), 108, 342
Auto PC, 151
Ayers, Rich, 643

B

Backbone, **292**
Backup, 345–347, 577–578
　incremental, **578**
　selective, **577**–578
Backward chaining, **498**–499
Banana Republic, 184
BankBoston, 371
Bank of America, 476–477
Bank of Montreal, 433–434
Banks, information systems in, 31
Barclays Merchant Services, 52
Bar code scanners, 108–109
Bargaining power of buyers and suppliers, competitive advantage and, 58, 62
Barnes and Noble Web site, 528
Barriers to E-commerce, 58
BASIC programming language, 595

Batch processing system, **338**
Becton Dickinson, 643
Behavior, intelligent, **481**
Benchmark test, **585**
Bernalillo County, New Mexico, 625
Berners-Lee, Tim, 303
Bernoff, Josh, 229
Best practices, 375
Beverly Hills travel shop Web site, 306
Bib Net, 289
Bill of materials, **409**
Biomedical science, ethical and societal issues in, 120–121
Biometrics, **638**
BioNOME project, 121
Bit, **89**
BJC-4304 inkjet printer, 111
Black & Veatch, 78–79
Boise Cascade Office Products, 416–417
Bonzelaar, Doug, 437
Boscov, 78
Boston Convention Center and Visitor's Bureau (BCVB), 431–432
BrainMaker, 487
Brainstorming, **459**
Bridges, **268**–269
Brightware, 513–514
British Petroleum Company (BP), 335–336, 337
Browsers, web, **308**–309, 312
"Browsing" (improper prying), 196
Budget transaction processing system, **360**–361
Business
　disruptions from computer problems, 576
　information systems. *See under* Information systems (IS)
　Web usage by, 314–316, 317
Business application programming interfaces (BAPIs), 376–377, 378
Business Recovery Management, 577
Business resumption planning, **345**
Business Travel Solutions, 187
Bus line, **89**
Bus line width, **89**–90
Bus network, **258**, 259, 261
Buyers, bargaining power of, 58, 62
Buyout package, 56
Byte, **92**

C

C++, **174**, 178
C3P, 411
Cable modem, 255
Cables, telecommunications, 245–246
Cache memory, **94**–95, 125
Cameras, digital computer, **105**–106
Canadian Air Traffic Control Systems (CATCS), 284–285
Candle, **609**
Car buying on Internet, 307
Careers in IS, 67–72
CARGEX-Cargo Expert System, 508
CARL system, 260
Carpal tunnel syndrome (CTS), **651**
CarPoint, 307
Carriers, telecommunications, 251–256
Cascading style sheets (CSS), 310
CASE. *See* Computer-aided software engineering (CASE)
CASE repository, **553**
Cases, in expert systems, 497–498
The Case Web site, 306
CA-SuperProject/Net (Computer Associates), 476
CD-audio, 122
CD-rewritable (CD-RW) disk, **100**
CD-ROM, **99**–100

Cellular transmission, 247–248
Censorware, 642
Centralized processing, **256**
Central processing unit (CPU), **86**. *See also under* Hardware
Centurion, 569
Cerf, Vinton, 320
Certification, **71**–72
Chalmers, David, 478
Chamberlain, D.D., 211
Change
　managing, 538
　organizational, **49**
Change model, **49**, 51
Character, **190**
Character recognition devices, 106–107
Charles Schwab & Co., 50
Chat rooms, **300**
Chief executive officer (CEO), 71
Chief financial officer (CFO), 71
Chief information officer (CIO), 71, 72
Chief Information Officers Council (CIOC), 661
Chippewa Falls, Wisconsin, 626
Choice stage, **439**
Chrysler Corporation, 59, 435
Cinergy Corporation, 449
Citibank, 241
Citicorp, 63
Citigroup, 63
Cleanup, data, **204**–205
Clickstream data, 647
Client application, 163
Client/server systems, **263**–265
CLIPS, 504
Clock speed, **88**
Closed shops, **579**
Coaxial cable, 245
Cocke, John, 91
Code, reusable, **173**
Codec (compression-decompression) device, 303
Code of ethics, AITP, 654
Cold site, 346, **577**
Collaborative computing software, **165**
Color graphics adapter (CGA), 100
Command-based user interface, **143**
Commercial Internet Exchange (CIX), 314
Common carriers, **251**
Communication(s), 242–243. *See also* Telecommunications
　cellular, 247–248
　infrared, 248
　microwave, 246–247
　parallel, 460
　satellite, 247
Communications Decency Act, 641–642
Communications protocols, 266–268
Communications software, **265**–266
Community Playthings, 432–433
Compact disk read-only memory (CD-ROM), **99**–100
Compaq Computer, 318
Comparative Microprocessor Performance index (iCOMP index), 89–90
Competition
　for local telephone services, 252–253
　as problem-solving factor, 443
Competitive advantage, **57**–63
　factors for seeking, 57–58, 62
　IS planning to develop, 524
　MIS for, 396

strategic planning for, 58–62
TPS and, 341
Competitive intelligence, 467–468
Compiler, **179**
Complementary metal oxide semiconductor (CMOS), **118**
Complex instruction set computing (CISC), **91**–92
Compound documents, **461**
Comp-U-Card (CUC), 272
CompUSA, 566–567, 589
Computer-aided software engineering (CASE), 542–544
  for requirements analysis, 555
  for software development, 596–597
  for systems analysis, 553
Computer-assisted design (CAD), 31, 407–408
Computer-assisted manufacturing (CAM), 31, **411**
Computer Associates International, Inc., 490
Computer-based information systems (CBIS), 17, 19
Computer crime, 628–643
  computer as target, 630
    data alteration and destruction, 632–634
    illegal access and use, 631
    information and equipment theft, 634
    international crime, 637
    scams, 635–637
    software and Internet piracy, **635**
  computer as tool for, 630
  prevention of, 637–643
    with antivirus programs, **641**
    by corporations, 638–641
    on Internet, 642–643
    with Internet laws, 641–642
    by state and federal agencies, 637–638
Computer downsizing, 118
Computer Emergency Response Team (CERT), 623
Computer-integrated manufacturing (CIM), 31, **411**
Computerized Collaborative Work System, **458**. See also Group DSS (GDSS)
Computer literacy, **30**
Computer network, **244**. See also Network(s)
Computer output microfilm (COM) devices, 111
Computer programs, **138**
Computer-related mistakes, 625–628
Computer scientists, 68
Computer system architecture, **123**–124
Computer system platform, **138**
Computer systems. See under Hardware
Computer terrorism, 639, 661
Computer waste, 624–625, 626–628
Concurrency control, **211**
Concurrent users, 216
Consumer Packaged Goods (CPG) software, 184
Content streaming, **303**
Continuing Balance System—Expanded (CBSX), 660
Continuous improvement, **53**–54, 412
Continuous replenishment program (CRP), 79
Contract management system, 607
Contract software, **156**

Controls design, 571, 572
Control unit, **86**
Conversion, data, **600**
Cookies, 647
CoolTalk, 309
Coprocessor, **95**
Copying with OLE, 163
Corbis Corp., 229
Corporate mergers, system development and, 618–619
Corporate policies
  on financial MIS, 401–402
  on human resource MIS, 419
  on manufacturing MIS, 405
  on marketing MIS, 413
  on privacy, 650
Correction, data, **344**
Cost/benefit analysis, **584**
Cost centers, **403**
Cost objectives, 528
Cost of DBMS packages, 217
Cost systems, 403
CoverStory, 508
Cracker, **631**
Creative analysis, **524**
Creativity, need for, 443
Creativity software, 158
Credit card companies, 491
Credit granting, expert systems and AI in, 507
Credit reporting, privacy in, 195–196
Credit risks, identifying bad, 362
Crime. See Computer crime
Criminal hacker, **631**
Crisis management, ESS support for, 467
Critical analysis, **525**–526
Critical path, **541**
Critical success factors (CSFs), 527, **554**
Cromar, Michael, 566
Cross-platform development, **597**
CRT, 109
Cryptography, **321**, 322
Cryptosystem, 321
Culture, organizational, **48**–49
Customer awareness and satisfaction, 66
Customer interaction system, 353–355, **354**
Customer loyalty, TPS and, 341
Customer relationship databases, 213
Customized software package, 157
CyberCash, 371
Cyberstar service, 296

D
Daimler-Benz, 59
Daimler-Chrysler AG, 59
Darigold Inc., 137
Data, **5**–6
  alteration and destruction of, 632–634
  clickstream, 647
  external, 407, 414, 420, 465
  identifying sources of, 548
  in manufacturing MIS, 406
  patient-care, 420–421
  payroll, 419–420
  personnel, 420
  types of, 5
Data analysis, 550–553
Database(s), 18, 190, **191**
  acquisition of, 599
  backup, 577–578
  customer relationship, 213
  design of, 571, 572
  digital image, 229
  distributed, **218**–219
  replicated, 219
  size of, 215

small, 216
very large (VLDBs), 220
Database controls, 579
Database management system (DBMS), **190**, 206–217
  creating and modifying database, 207–209
  for end users, 212–215
  manipulating data, 211–212
  object-relational (ORDBMS), **226**–229
  report generator, 211–212
  selecting, 215–217
  storing and retrieving data, 209–211
  user view of, 206–207
Database models, 201–206
  comparison of, 205–206
  hierarchical, **201**–202
  network, **202**
  relational, 203–204, 206
Database server in SAP system, 377
Database software, 158, 159–160, 161
Data cleanup, 204–205
Data collection, 343, 548–549
Data communications, 243
Data correction, **344**
Data definition language (DDL), **207**
Data dictionary, 208–209
Data editing, **344**
Data Encryption Standard (DES), 322
Data entry, **104**
Data-flow diagram (DFD), 14, **550**–552
Data-flow line, **552**
Data input, **104**
Data integrity, **194**
Data item, **192**
Data management, 190–199
  database approach to, **195**–199
  data entities, attributes, and keys, 191–192
  hierarchy of data and, 190–**191**
  traditional approach to, 192–195, **194**
Data manipulation, **344**
Data manipulation language (DML), **211**
Data marts, **221**–222
Data mining, **222**–223
Data model, **200**
Data modeling, 199–201, 550
  enterprise, **200**
Data preparation or data conversion, **600**
Data readers, optical, 106–107
Data redundancy, **194**, 209
  planned, **200**
Data reliability, 209
Data storage, **344**
Data store, **552**
Data warehouse, **219**–221, 225, 237, 239
Davis, John, 622
Deadline, project, **540**
Debugging, 591
Decency, Internet laws and protection for, 641–642
Decentralized processing, **256**
Decision(s)
  make-or-buy, **588**
  nonprogrammed, **440**
  programmed, **440**
  repetitive, 450
Decision Grid, 447–448
Decision making
  executive, 465
  expert systems to assist in, 493
  operational, 375
  problem solving and, 438–444

Decision-making phase, **438**
Decision room, **462**–463
Decision structure, 595
Decision support systems (DSS), **26**–27, 436–477
  ad hoc, **449**–450
  capabilities of, 448–451
  characteristics of, 445–448
  components of, 453–458
    dialogue manager, **453**, 458
    model base, **453**–458
  executive support system (ESS) or executive information system (EIS), **464**–469
  group (GDSS), **458**–464
  for individual investors, 474–475
  integration of TPS, MIS, and, 451
  intranet support of, 473–474
  for medical problems, 475–476
  MIS compared to, 452
  project management models, 476–477
  web-based, 452–453
Dedicated line, **251**–252
Dell, Michael, 114
Dellagen, Kate, 272
Dell Computer, 114
Delphi approach, **459**
DelphiPower BEO, 589
Demand reports, 26, **392**, 393
Dependence, program-data, **194**
Deregulation of utilities industry, 558, 569
Design. See also Systems design
  expert systems to assist in, 493
  in manufacturing MIS, 407–408
Design report, **587**
Design stage, **439**
Desktop computers, 113
Desktop publishing (DTP) software, 158
Deterrence controls, **579**
Development, software. See under Software
Development, systems. See Systems development
Diagnosis, expert systems to assist in, 493–494
Dialing services, 253–254
Dialog manager, **453**
Dialogue, **573**–574
Dialogue manager, **453**, 458
Dictionary, data, **208**–209
Diesel Electric Locomotive Troubleshooting Aid (DELTA), 491
Diggins, John, 446
Digital computer cameras, **105**–106
Digital Equipment Corporation (DEC), 83
Digital images, on demand, 229
Digital signal, **248**
Digital signal processor (DSP), **122**–123
Digital signature, **322**
Digital subscriber line (DSL), 255, **256**, 257
Digital video disks (DVD), 100, **101**
Digital video interactive (DVI), 123
DirecPC, 295
Direct access storage devices (DASDs), **97**
Direct assess, **97**
Direct conversion (plunge or direct cutover), **602**
Direct observation, **548**–549, 550
Directories, subject, 311–312
Direct Rambus (Direct RDRAM) technology, 93
Disaster recovery, **345**–347, 574–578, **575**
Disaster Recovery Center, 577
Discrimination, age, 56

Disk(s)
 DVD, 100, **101**
 magnetic, **98**–99
 magneto-optical, **100**
 optical, **99**–100
Disk mirroring, **99**
Display monitors, 109–110
Distance learning, 277–**278**
Distributed databases, 218–219
Distributed processing, **256**–257
Dixon, Chris, 555
DMR Consulting Group, 607
Document(s)
 compound, **461**
 electronic distribution of, 270–272, **271**
 user acceptance, **602**
Documentation, **138**
 technical, **591**–592
 TPS, 340
 user, **592**
Document production, **345**
Domain, **203**, **502**
Domain affiliations, 292–293
Domain expert, **502**
Domino's Pizza, 396
Dow, Herbert H., 518
Downsizing, **56**
 computer, 118
Downstream management, 45
Dreamworks, 330
Drill down abilities of ESS, 465
Drill down analysis, 446
Drill down reports, **394**–395
Drug reactions, adverse, 53
Dupont & Company, 55–56
Dynamic HTML (DHTML), 310
Dynamic RAM (DRAM), 93

E
Early retirement, 56
Earnings growth, 66
Easy Router, 356
E-commerce, **22**–24, 50, 367–373
 Barnes and Noble, 528
 barriers to, 58
 business on the Web, 314–316, 317
 business-to-business, 367–368
 car buying, 307
 Comp-U-Card, 272
 for consumers, 367
 dark side of, 636
 by Eddie Bauer, 117
 electronic check payment, 371–372
 five-stage model for, 368–373
  after-sales support, 369, 373
  product or service delivery, 369, 372–373
  purchasing, 369, 371
  search and identification, 368–371
  selection and negotiation, 369, 371
 for food service industry, 370
 forecasted volume of, 368
 MasterCard International and, 193
 on-line service requirements, 555
 in perspective, 368
 reengineering for, 53
 security issues in, 371
 Ticketmaster's use of, 39
Economic feasibility, **545**, 546
Economic order quantity (EOQ), **408**
Eddie Bauer, 117
Editing, data, **344**
Edit Paste command, 164
EDS, 55
EFF DES Cracker, 322
Effectiveness, **11**

Efficiency, **11**
Efficient consumer response (ECR), 79
Electronic commerce. *See* E-commerce
Electronic Data Interchange (EDI), **274**–276, 349–350
Electronic document distribution, **271**–272
Electronic Frontier Foundation (EFF), 322
Electronic markets, 368
Electronic software distribution, 270–272, **271**
Eli Lilly & Company, 55
E-mail, **270**, 271, 296–298, 646
Embed, **164**
Embedded computers, 114
Empire District Electric Company, 569
Employee performance evaluation, expert systems and AI in, 508
Empowerment, **46**–47
Encapsulation, **173**
Encryption, **321**–323, 323–324
End-user systems development, 536–**537**
ENEN, 304
Energy Plan (EPLAN), 449
Engineer, knowledge, **503**, 505
Engineering, in manufacturing MIS, 407–408
Enron, 617–618
Enterprise data modeling, **200**
Enterprise marketing automation (EMA), 418
Enterprise resource planning (ERP), **168**–169, 373–379, 404–405
 advantages and disadvantages of, 374–376
 core business activities for, 379
 example of, 376–378
 leading vendors, 373
 overview of, 373–374
 selling expertise on, 531
 trends in, 529–530
Enterprise software, 167–169
Enterprise sphere of influence, **140**
Entity, **191**
Entity-relationship (ER) diagrams, **200**–201, 550
Entity symbol, **552**
E-PRIVACY Act (1998), 323–324
Equipment theft, 634
Erasable programmable read-only memory (EPROM), 94
Ergonomics, **652**–653
ERP. *See* Enterprise resource planning (ERP)
Errors, system, 574, 575
eService, 297
Essbase 5 (Hyperion), 223, 224
Essbase (Arbor Software Corp.), 442
Ethernet, 266
Ethical and societal issues, 654–655
 in biomedical science, 120–121
 in business disruptions from computer problems, 576
 in competition for local telephone services, 252–253
 in competitive intelligence in ESS, 467–469
 in electronic checking, 371–372
 in encryption, 323–324
 in enterprise resource planning (ERP), 168–169
 in information technology's impact on society, 4–5
 in IRS modernization project, 539
 in mergers, 63
 in neural nets, 488
 in personalization on Web sites, 646–647

in privacy standards in credit reporting, 195–196
in protecting patient-care data, 420-421
Ethics code, AITP, 654–655
E*Trade Web site, 576
European media-buying agencies (MBAs), 479
European Union, 644
Event-driven review, **608**
Excel (Microsoft), 441, 633
Exception reports, 26, **392**–394
Exchange (Microsoft), 461
Execution time (E-time), 87
Executive decision making, layers of, 465
Executive support system (ESS) or executive information system (EIS), **464**–469
 capabilities of, 466–469
 characteristics of, 465–466
 ethical and societal issues in, 467–469
 future orientation of ESS, 466
Experian, 644
Expert systems, 27–28, **484**, 489–509
 applications of, 507–509
 capabilities of, 492–494
 characteristics of, 489–492
 components of, 494–500
  explanation facility, **499**
  inference engine, **498**–499
  knowledge acquisition facility, 499–**500**
  knowledge base, **494**–498
  user interface, 500
 development of, 500–506
  alternatives to, 505–506
  participants in, 502–503
  process of, 500–501
  shells and products, 503–505
  tools and techniques for, 503–505
 to improve Internet presence, 513–514
 integrating, 509
 usage of, 494
Expert system shell, **492**, 503–505
Explanation facility, **499**
Extended Data Out (EDO RAM), 93
Extended format (Mac OS), 152
Extensible Markup Language (XML), 310
External auditing, **404**
Extranets, **318**–319, 526
EZ 135 drive, 102

F
Facility Check, 331
Fair Credit Reporting Act (1970), 195
Fairness in information use, 647–648
False (hoax) viruses, 633–634
Farley, Gloria, 236
Fax modems, 249
Federal Aviation Administration (FAA), 84–85, 349, 629
Federal Computer Incident Response Capability, 325
Federal Deposit Insurance Corporation (FDIC), 661
Federal Express (FedEx), 41–42, 385–386
Federal Trade Commission, 644
Feedback, **16**
Feedback mechanism, 8–9, 10
Fiber-optic cable, 245–246
Fibre channel, 124
Field, **191**
Fifth-generation languages, 176
File, **191**
File design, 571

File management by operating systems, 147
File servers, **263**
File Transfer Protocol (FTP), **298**
File-Update command, 164
Final evaluation, **583**
Finance, informations systems in, 30
Financial Advisor, 504
Financial management information system, 389, **400**–404
Financial management software, 158, 159
Financial models, **455**
Finder (Mac OS), 152
Fingerprint Identification Technology, 640
Firewall, **316**, 324–325
First Chicago NBD Corporation, 446
First-generation languages, 170
1st-Class Fusion, 504
Fisher, Donald G., 184
Five-force model, 57
Flash memory, **101**–102
Flat organizational structure, **46**
Flexible manufacturing system (FMS), **411**–412
FLEXPERT, 507
Flood damage, 8
Flowcharts, 14
 application, **552**
FocalPt network, 332
Food service industry, E-commerce for, 370
Forbes, Walter A., 272
Ford Motor Company, 332, 411
Forecasting, **16**, 422, 489
Forward chaining, **499**
Fourth-generation languages (4GL), **171**–172, 592
Fraud, 578
Frequently Asked Questions (FAQs), 300
Front-end processor, **249**–250
Fuentes, David, 568
Function points, **174**
Functions in IS department, 68–72
Funds, uses and management of, 404
Future orientation of ESS, 466
Fuzzy logic, **496**–497, 514
Fuzzy sets, 496

G
GAF Materials Corp. (GAFMC), 389
Gallium arsenide (GaAs), 90–91
Gantt chart, **541**, 542
Gap, The, 184–185
Gartner Group, 115
Garvey, Jane, 85
Gate Assignment and Planning System, 457
Gates, Bill, 152
Gateways, **268**–269
GATX Capital Corp., 531, 565–566
Gauld, Bill, 77
Geac Computer Corp., 581
Gehring, Greg, 325
GE Information Services (GEIS), 303
General Accounting Office (GAO), 660
General Electric, 22
General Electric Aircraft Engine Group, 54–55, 508
General ledger system, **365**–366
General-purpose computers, **113**
Genetic algorithms, 515
Geographic information systems (GIS), **425**–427
Georgia-Pacific, 56
Gerber, 519–520
Gerstner, Louis, 40, 468
Glendale Federal, 287
Global positioning system (GPS), 8

Global software support, 141
Globalstar Telecommunications, 625
Goal-seeking analysis, 447, **448**
Goods, creating new, 60
GOTO statement, 595
Grampian, Scotland, 625
Graphical modeling programs, **456**
Graphical process modeling, 437
Graphical user interface
    (GUI), **143**–144
Graphics components, in multimedia
    systems, 122
Graphics software, 158, 160, 161
Grid chart, **552**–553
Group consensus, **584**
Group consensus approach, **460**
Group DSS (GDSS), **458**–464
    alternatives to, 462–464
    characteristics of, 458–460
    components of, 461
    configuration of, 459
Group scheduling software, 165
Groupware, **165**–166. See also
    Group DSS (GDSS)
GTE Corp., 119
Guardian Computer Support, 577
Gulf Canada, 454

H
Hacker, **631**, 640
Hamilton, Larry, 468
Handheld (palmtop) computers, 113
Hard drive, selecting or upgrading,
    124
Hardware, 17, 82–136, **84**
    acquisition from IS vendor, 588
    backup, 576–577
    computer systems, 84–87,
        112–125
        general-purpose, **113**
        mainframes, 112, **118**–120
        midrange computers (mini-
            computers), 112, **117**–118
        multimedia computers,
            121–123
        network computers, 112,
            114–115
        personal computers (PCs),
            112, **113**–114, 115, 116
        prices of, 116
        selecting and upgrading,
            123–125
        special-purpose, **112**–113
        standards for, 123, 124
        supercomputers, 112, **120**
        workstations, 112, **115**–116
    design of, 571
    input devices, 103–109,
        111–112
    memory, 92–95
        cache, **94**–95, 125
        flash, **101**–102
        main, 86, 124–125
        random access (RAM), **92**–94
        ROM, **94**
        selecting or upgrading,
            124–125
        storage capacity, 92
        virtual, **145**, 146
    multiprocessing, **95**–96
    operating systems and,
        142–143, 144
    output devices, 103–104,
        109–112
    processing devices (CPU), 87–92
        clock speed, **88**
        instruction sets, 91–92
        machine cycle time, **88**
        physical characteristics of,
            90–91
        wordlength and bush line
            width, **89**–90

secondary storage, **96**–103
    access methods, 96–97
    comparison of devices,
        97, 103
    digital video disk (DVD),
        100, **101**
    magnetic disks, **98**–99
    magnetic tape, **98**
    magneto-optical disks, **100**
    memory cards, 101–102
    optical disks, **99**–100
    RAID, 99
    removable devices, 102
    stolen, 637
Hawaiian Greenhouse, 177
Haworth, Inc., 36–37
Health Care Financing
    Administration, 421
Healthcare organizations, 19, 31–32
Health concerns, work-related,
    651–653
Health Insurance Portability and
    Accountability Act of 1996
    (HIPAA), 421
Health maintenance organizations
    (HMOs), 19
Heating, venting, and air-conditioning
    (HVAC) equipment, 118
Helix Water District, 427
Help desks and assistance, expert
    systems and AI in, 507–508
Help facility, **573**
Hendry, Bob, 235
Herdon, Sandy, 646
Herring, Jan, 468
Hertz car rental company, 466
Hertz (Hz), **88**
Heuristics, 441–442, 447, 501
    in expert system shells and
        products, 505
    intelligence and, 482–483
Hewlett-Packard, 56, 83, 88, 449
Hierarchical network, **258**, 259
Hierarchical structure, 45
Higgins, Chris, 477
High Availability Cluster
    MultiProcessing (HACMP), 134
Highly structured problems, **450**
Hiring, personnel, 600
Hitachi, 119, 120
Hoax (false) viruses, 633–634
Holiday Inn, 247, 314–315
Home page, 36, **305**
Hospitals
    expert systems and AI in, 507
    fuzzy logic to predict patient
        stay in, 514
    telemedicine use by, 275
HotBot, 310
Hotel Vintage Park, 285–286
Hot site, 346, **576**–577
HTML. See HyperText Markup
    Language (HTML)
HTML tags, **305**–308
Hughes Network Systems Inc., 295
Hukezalie, Mary Lou, 434
Human resource management infor-
    mation systems, 31, **418**–424,
    425, 433–434
    inputs to, 419–420
    subsystems and outputs,
        421–424
Hybrid networks, 258, 259
Hypermedia, 228, **305**
Hypertext, 227–228
HyperText Markup Language (HTML),
    305, 308
    dynamic (DHTML), 310
    standards, 309–310
HyperText Transport Protocol
    (HTTP), 292

Hyundai Motors Company, 526

I
IBM, 83, 119, 120, 468
Icam 2000 (EyeDentify), 638
Icon, **143**
ID badges, 580
Identity theft, 196
IET-Intelligent Electronics, 508
If-then statements, **495**
Illinois Power, 166
Image data, 5
Image log, **578**
Images, digital, 229
Immigration and Naturalization
    Service (INS), 516
Implementation, systems. See Systems
    implementation
Implementation stage, **439**
Improper prying ("browsing"), 196
Improvement, continuous,
    53–54, 412
In-Box Direct Web site, 306
Incremental backup, **578**
Indexes in search engines, 311
Individual Reference Services Group
    (IRSG), 195–196
Industry, information systems in, 31–32
Inference engine, **498**–499
Information, **5**–6
    characteristics of valuable, 6–7
    theft of, 634
    value of, 7
Information center, **70**
Information management and
    retrieval, expert systems and AI
    in, 507
Information services, specialized and
    regional, 277
Information service unit, **70**–71
Information systems (IS), 2–39, **4**
    business, 21–28
        artificial intelligence (AI),
            27–28
        decision support systems
            (DSS), **26**–27
        E-commerce, **22**–24
        expert systems, 27–28
        management information
            systems (MIS), **25**–26
        transaction processing
            systems (TPS), **21**–22
    careers in, 67–72
    computer-based (CBIS), 17, 19
    databases of, 18
    data vs. information, **5**–6
    feedback in, **16**
    in functional areas of
        business, 30–31
    hardware in, 17
    impact on society, 4–5
    in industry, 31–32
    input in, **15**
    Internet and, 18, 19
    intranets and, 19, 20
    justifying, 66–67
    manual, 17
    networks and, 18–19
    output in, **16**
    people in, 20
    performance-based, 64–67
    procedures of, 20
    processing in, **15**–16
    reasons for studying, 29–32
    responsibilities of, 68
    software in, 17–18
    stages in business use of, 64
    for strategic purposes, 60–62
    telecommunications and, 18
    value of, 19
Information systems literacy, **30**
Infrared transmission, 248

Inheritance, **173**
In-house-developed software,
    590–592
In-house development, **156**
Input(s), 8, 9, 10, **15**
    anonymous, 460
    data, **104**
        to financial MIS, 401–402
        to MIS, 391–392, 405–407,
            412–414, 419–420
Input controls, 579
Input design, 570–571
Input devices, 103–109, 111–112
Instill Corporation, 370
Instruction time (I-time), **87**
Integrated-CASE tools (I-CASE), **543**
Integrated development environments
    (IDEs), **597**–598
Integrated services digital network
    (ISDN), **254**, 255, 257
Integrated supply chain management
    software, 167
Integration testing, **601**
Integrity
    of data, **194**
    of TPS, 339–340
Intel Corp., 83, 89
Intelligence. See also Artificial
    intelligence (AI)
    competitive, 467–468
    natural, 483
    nature of, 481–483
Intelligence stage, **438**–439
Intelligent agents, 359, 495
Intelligent behavior, **481**
Interactive processing, 573
Intercom Plus, 567
Interface. See Software interface;
    User interface
InterMind, 314
Internal auditing, **404**
Internal Revenue Service (IRS),
    modernization project of, 539
International aspects of problem
    solving, 444
International computer crime, 637
International networks, **261**–262
International Systems Program (ISP),
    335–336, 337
Internet, 18, 19, 289–316, **290**,
    319. See also World Wide Web
    accessing, 293–295
    car buying on, 307
    crime prevention on, 642–643
    expert systems to improve
        presence in, 513–514
    functioning of, 292–293
    intelligent agents on, 359, 495
    libel and decency laws and
        protection on, 641–642
    market research and, 415–416
    medical topics on, 653
    Next Generation Internet
        (NGI), 291
    piracy on, **635**
    privacy and, 645–646
    sales on, 53
    services available, 296–303
        chat rooms, **300**
        content streaming, **303**
        e-mail, 296–298
        File Transfer Protocol
            (FTP), **298**
        newsgroups, **299**–300
        phone and videoconferencing,
            300–303
        Telnet, **298**
        usenet, **298**–299
    systems development and,
        528–529

travel planning on, 330–331
use of, 290–291
Internet 2 (I2), 291
Internet Activities Board (IAB), 320
Internet-based routing service, 356
Internet Explorer (Microsoft), 309
Internet Protocol (IP), **291**
Internet Relay Chat (IRC), 300
Internet Satellite Systems (ISS), 295–296
Internet Service Provider (ISP), **295**–296
Internet Society, 320
Interpreter, **178**
INTERSOLV, 325
Interviews, 548
Intranet(s), 19, 20, **316**–318, 319
Becton Dickinson, 643
DSS support by, 473–474
Saab, 235
Intranet Retail Information System (IRIS), 235
Inventory control system, **352**–353, 356, 408–409
Inventory data, in manufacturing MIS, 406
Inventory Status Report, 352
Investigation, systems. See Systems investigation
Investment, return on (ROI), **65**–66
Investment firms, information systems in, 31
Invoicing system, 353
Iomega, 102
Iron Mountain, 346
IS plan, 554
iVillage, 50

J

J. Crew, 24
Jango, 359
Java, **174**–175, 178, **312**–313, 605
Jaz drive, 102
Jefferies Group, 473–474
Jetstream 1000 AIT Tape Array, 98
JIG, 598
Job design, 571
Job placement, 423
Job scheduling, AI for, 515
Joining, **204**
Joint application development (JAD), **536**
Just-in-time (JIT) inventory approach, **410**–411

K

Kanban, 410
Karten, Naomi, 2, 38
Kay, Alan, 143, 174
Kayak XA, 115
Kellogg Co., 184
Kennedy Space Center, 453
Key, **192**
Keyboard, ergonomic, 105
KeyCorp, 213
Key-indicator report, **392**, 393
Kinetra, 55
Kmart, 61, 415
Knowledge, **6**
Knowledge acquisition facility, 499–**500**
Knowledge base, **6**, **494**–498
Knowledge-based programming, **176**
Knowledge engineer, **503**, 505
Knowledgepro, 504
KnowledgeSpace, 19, 20
Knowledge user, **503**, 505
Knowledge workers, 4
KnowledgeX, 468

L

Labor efficiency, TPS and, 341

LAN administrators, 71
Land, 99
Language. See also Programming languages
data manipulation (DML), **211**
Structured Query (SQL), 12, **172**, **211**
Language translators, **178**–179
Laptop computers, 113
Laser printers, 111
Law Library Web site, 306
Laws
crime prevention, 637–638
libel and decency, 641–642
privacy, 648–650
Leahy, Patrick J., 323
Learning
distance, 277–**278**
organizational, 49
Learning system, **486**
Leased line, 251–252
Legacy applications, 374, 629
Legal information systems, 31
Leonardo (expert system), 504
Levy, Michael, 555
Lewin, Kurt, 49
Lexus, 626
Libel, Internet laws and protection for, 641–642
Library of Congress Web site, 306
Licenses, software, 140, 185–186
Light pens, 108
Liquid crystal displays (LCDs), 110–111
LizCADalyst, 38
Liz Claiborne, 38
Loan analysis, expert systems and AI in, 508
Loan Probe, 508
Local area decision network, 463
Local area network (LAN), **258**–261, 293
Lockheed Martin, 216
Logic, fuzzy, 496–497, 514
Logical design, **570**–572
Logic bomb, **632**–633
Logon, 572–573
Lookup tables, **573**
Loop structure, **595**
Loral Space and Communications Ltd., 296
Lotus Development Corp., 501
Lower-CASE tools, **543**
Low-level language, **170**

M

Machine cycle, **87**
Machine cycle time, **88**
Machine language, **170**
MCI Communications Corporation, 238
MacManus Group, 319
Mac OS 8.1, 148, 151–152
Macro virus, **633**
Macro Virus Protection tool, 633
Magnetic disks, **98**–99
Magnetic ink character recognition (MICR) device, 107
Magnetic tape, **98**
Magneto-optical disks, **100**
Mainframe-and-network-to-PC links, 269–270
Mainframes, 112, **118**–120
Main memory (primary storage), 86, 124–125
Maintenance, system. See Systems maintenance
Maintenance and review, system, **29**

Maintenance team, **605**
Make-or-buy decision, **588**
Management, downstream and upstream, 45
Management information systems (MIS), **25**–26, 388–435
accounting, **424**
characteristics of, 395–396
for competitive advantage, 396
DSS compared to, 452
financial, 389, **400**–404
functional aspects of, 397–400
geographic information systems (GIS), **425**–427
human resource, 31, **418**–424, 425, 433–434
inputs to, 391–392
integration of DSS, TPS, and, 451
manufacturing, **404**–412
marketing, 30, **412**–418, 431–433, 435
outputs of, 392–395
in perspective, 390–391
Web technology and, 396–397
Manual information systems, 17
Manufacturing management information systems, 31, **404**–412
inputs to, 405–407
subsystems and outputs of, 407–412
design and engineering, 407–408
inventory control, 408–409
just-in-time (JIT) inventory approach, **410**–411
manufacturing resource planning (MRPII), **409**–410
master production scheduling, 408
process control, 411–412
quality control and testing, 412
Manufacturing resource planning (MRPII), **409**–410
Manugistics (software), 519–520
Marketing
expert systems and AI in, 508
on the Web, 50
Marketing management information systems, 30, **412**–418, 431–433
inputs to, 413–414
subsystems and outputs, 414–417
web-based, 435
Marketing research, 415–416
Marketing software, 432
Markets, electronic, 368
Market share, 66
Marshall Industries, 304
MasterCard International, 193
Master production scheduling, 408
Materials requirement planning (MRP), **409**
Mathematical model, 13, 14–15
MatheMEDics, 507, 508
Media, telecommunications, 245–248
Media-buying agencies (MBAs), European, 479
Media Revolution, 330
Medical facilities, expert systems and AI in, 507
Medical problems, decision support for, 475–476
Medical topics on Internet, 653
Mediconsult, 475–476
MediXpert, 475–476
Medscape Web site, 306
Meet Me Bridge, 331
Megahertz (Mhz), **88**
Memory. See under Hardware
Memory cards, 101–102

Memory management by operating system, 144–145
Menu-driven system, **573**
Merced, 83, 88
Mergers, corporate, 63, 618–619
Message board, 474
Metropolitan Regional Information System (MRIS), 314
Michelin Tire, 289
Microcode, **88**
Microseconds, 88
Microsoft Corp., 151, 315, 323, 604
Microsoft Network, 277
Microwave transmission, 246–247
Midrange computers (minicomputers), 112, **117**–118
Milestone, project, **540**
Miller SQA, 437–438
MILNET, 291
MindWizard, 504
Minicomputers (midrange computers), 112, **117**–118
MIPS, **88**
Mission-critical systems, **528**
Mistakes, computer-related, 625–628
MIT Lab for Computer Science Web site, 306
Model(s), **13**–15
change, 49, 51
database, 201–206
financial, **455**
five-force, **57**
network, **202**
open systems interconnection (OSI), **266**, 267
project management, **456**, 476–477
relational, **203**–204
statistical analysis, **455**
tree, 201–202
Model base, **453**–458
Modeling
activity, 550–552
advantages and disadvantages of, 456–458
data, 199–201, 550
Model management software (MMS), **453**
Modems, **248**–249
Moncla, Brenda, 236
Monitoring, 627–628
Monitoring stage, **439**–440
Monitors, display, 109–110
Monsanto Corporation, 531
Moore, Gordon, 90
Moore's Law, **90**, 91
Morocco (MRO supplies catalog), 530
Motif (Open Systems), 154
Motley Fools, 474
Mouse, 105
Moving, 49, 51
MS-DOS, 147–148
Multidimensional organizational structure, **48**
Multifunction device, **111**–112
Multimedia computers, 121–123
MultiMedia Extension (MMX), 124
Multimedia PC Council (MPC), 124
Multiple Virtual Storage/Enterprise System Architecture (MVS/ESA), 154
Multiplexers, **249**, 250
Multiprocessing, **95**–96
MULTI Software Development Environment, 598
Multitasking, **146**
Multithreading, **146**
Musical Instrument Digital Interface (MIDI), 124
MYCIN, 507

N

Nabisco, 175–176
Nanoseconds, 88
Narrative model, 13, 14
NASA, 228
National Infrastructure Protection Center, 639
National Semiconductor Corp., 41
National Technological University, 277–278
Natural language processing, 485–486, 487
NaturallySpeaking preferred (Dragon Systems), 486
Natural Resources Conservation Service, 8
Natural vs. artificial intelligence, 483
Nav Canada, 284–285
Navigator (Netscape), 308–309
NCR Corporation, 189
Neal, Doug, 188
Negative group behavior, reduction of, 460
Nelson, Scott, 388
netMarket, 272, 317
NetMeeting (Microsoft), 461
Net present value, 545–547
NetSeminar, 304
NetWare, 265
Network(s), 18–19, 244, 256–265. *See also* Internet
    client/server systems, **263**–265
    extranets, **318**–319, 526
    file servers, **263**
    international, **261**–262
    intranets, 19, 20, 235, **316**–318, 319, 473–474, 643
    local area decision, 463
    local area (LAN), **258**–261
    management issues, 320
    neural, **486**–489
    privacy and security of, 321–324
        firewalls, **316**, 324–325
    service bottlenecks, 320–321
    terminal-to-host, **263**
    topology of, **258**, 259, 261
    virtual private (VPN), **319**
    wide area decision, 463
    wide area (WAN), **261**, 262
Network computers, 112, **114**–115
Network Driver Interface Specification (NDIS), 293
Networking capability of operating systems, 147
Network management software, **266**
Network model, **202**
Network operating system (NOS), **265**
Network Solutions Inc. (NSI), 293, 576
Neural Connection, 489
Neural network, **486**–489
New entrants, threat of, 57, 62
New Holland NV, 397
Newsgroups, **299**–300
New York Life, 618–619
New York Times Web site, 306
Next Generation Internet (NGI), 291
Nippon Telegraph and Telephone (NTT), 283–284
Nominal group technique, **460**
Nonoperational prototype, **534**
Nonprogrammed decision, **440**
Norstan, 629
Norton Utility, 154
Notebook computers, 113
Notes (Lotus), 140, 165, 166, 461
NTFS file system, 153

O

Object, **377**
Object code, **178**
Object linking and embedding (OLE), **163**–164

Object-oriented design approach, 594
Object-oriented languages, **172**–174
Object-oriented programming, 184–185
Object-oriented software development, **592**
Object-relational database management system (ORDBMS), **226**–229
Observation, direct, **548**–549, 550
Office Depot, 317, 442, 620
Office suite (Microsoft), 162–163
Off-the-shelf software, **156**–157
Old Navy Clothing Co., 185
Omaha Steaks, 61
1-800-Flowers, 386–387
On-line analytical processing (OLAP), **223**–225
Online Career Center Web site, 306
On-line entry with delayed processing, 339
On-line information services software, 158, 160–161, 162
Online Privacy Alliance, 644–645
On-line Privacy Protection Act, 196
On-line services, 294–295, 555
On-line shopping, 22–23, 315. *See also* E-commerce
On-line transaction processing (OLTP), 219, 225, **338**–339
Open database connectivity (ODBC), **225**–226
Open Datalink Interface (ODI), 293
Open Look, 154
Open shops, **579**
Open systems interconnection (OSI) model, **266**, 267
Operating resource management (ORM), 131
Operating system (OS), **141**–154
    access to system resources, 147
    Apple, 148, 150–152
    DOS with Windows, 148
    file management by, 147
    hardware functions and, 142–143
    hardware independence of, 144
    memory management by, 144–145
    MS-DOS, 147–148
    Multiple Virtual Storage/Enterprise System Architecture (MVS/ESA), 154
    networking capability of, 147
    network (NOS), **265**
    OS/2, 148–149
    processing task of, 145–147
    Unix, 153–154
    user interface, **143**–144
    Windows 95, 148, 149
    Windows 98, 148, 149–150
    Windows CE, 148, 150, 151
    Windows NT, 148, 152–153, 265
Operational decision making, 375
Operational feasibility, **545**, 546
Operational prototype, **534**
Operations, 69
Optical character recognition (OCR), 107
Optical data readers, 106–107
Optical disks, **99**–100
Optical mark recognition (OMR), 107
Optical processors, **91**
Optimization approach, 447
Optimization model, **441**
Oracle Consumer Packaged Goods (CPG) software, 137
Oracle Financials, 137
Oracle Lite, 216
Order entry system, **349**–350
Order processing systems, **347**–355, 384–385

customer interaction system, 353–355, **354**
    in human resource MIS, 420
    inventory control system, 352–353
    invoicing, 353
    in manufacturing MIS, 406
    order entry system, **349**–350
    routing system, **355**, 356
    sales configuration system, **350**
    scheduling system, **355**, 356
    shipment execution system, **352**
    shipment planning system, **350**–351
Organizational change, **49**
Organizational culture, **48**–49
Organizational learning, **49**
Organizational structure, **45**–48
    flat, **46**
    multidimensional, **48**
    project, **47**
    team, **47**–48
    traditional, **45**–47
Organizations, 40–80, **42**
    classifying by system type, 11
    competitive advantage of, **57**–63
        factors for seeking, 57–58, 62
        strategic planning for, 58–62
    continuous improvement in, **53**–54
    general model of, 43
    outsourcing and downsizing by, **55**–56
    reengineering (process redesign) by, 51–53, 54
    total quality management (TQM) in, **54**–55
    value-added processes from, 43–44
Ornda Health Corporation, 425
OS/2, 148–149
Output(s), 8, 9, 10, **16**
    of MIS, 392–395, 402–404, 407–412, 414–417, 421–424
Output controls, 579
Output design, 570
Output devices, 103–104, 109–112
Outsourcing, **55**–56, 77–78, 616–617
Owens & minor, 222

P

Pacific Bell, 273, 297
Packaged software, 67
Packets, IP, 301
Packet sniffer, 636
Padron, Honorio, 566
Paging, **145**
Palm PC, 151
Palmtop (handheld) computers, 113
Parallel communication, 460
Parallel processing, **95**–96
Parallel start-up, **602**
Partnership, strategic (strategic alliance), **59**–60
Passive-matrix display, 110
Password sniffer, **634**
Past Link command, 163
Patch, **604**
Patient-care data, 420–421
Payroll data in human resource MIS, 419–420
Payroll journal, 363–**364**
Payroll system, 21–22, 363–364
PC-to-mainframe-and-network links, 269–270
PeerScape (Deliotte & Touche), 402
Peer-to-peer network, 260
Pen input devices, 108
Pens, light, 108
Perceptive system, **482**

Performance-based information systems, 64–67
Performance measurement, system, **609**–610
Performance Now! (software), 423
Performance objectives, 527
Performance of DBMS, 217
Personal application software, 158–163
    creativity, 158
    database, 158, 159–160, 161
    desktop publishing (DTP), 158
    financial management, 158, 159
    graphics, 158, 160, 161
    on-line information services, 158, 160–161, 162
    project management, 158
    spreadsheet, 158, 159, 160
    word processing, 158, 159, 160
Personal Computer Memory Card International Association (PCMCIA), 101, 124
Personal computers (PCs), 112, **113**–114, 115, 116
    input devices for, 105
Personal Consultant (PC) Easy, 504
Personal information managers (PIMs), 165
Personal Information Privacy Act (1997), 196
Personal productivity software, **139**
Personal sphere of influence, **139**
Personnel
    backup, 578
    data on, 406, 420
    hiring and training of, 600
Personnel controls, 580
Personnel design, 571, 572
Personnel selection and recruiting, 422–423
Phase-in approach, **602**
Philadelphia Stock Exchange, 601–602
Phillips Petroleum Company, 134–135
Phoenix Additional Line Modeling System (PALMS), 236–237
Phone services, 253–254
Physical design, **570**–572
Physical model, 13, 14
Picking list, 351
Picoseconds, 88
Pilot Commercial Credit Analysis software, 446
Pilot Internet Publisher (PIP), 453
Pilot start-up, **602**
Pipelining, **87**
Piracy, software and Internet, **635**, 637
Pixel, **109**–110
Plan, IS, 554
Planned data redundancy, **200**
Planning. *See also* Strategic planning
    expert systems to assist in, 493
    human resource, 422
    information systems, 523–526, **524**
    project, 538–540
Plant layout, expert systems and AI in, 507
Platform, computer system, **138**
Plotters, **111**
Plug'n'Play (PnP), 124
Plunge (direct cutover; direct conversion), **602**
PointCast, 306, 314
Point evaluation system, **585**
Point-of-sale (POS) devices, **107**–108
Point-of-sale transaction processing system, 344
Point to Point Protocol (PPP), **294**
Policies *See* Corporate policies

Political and social actions, as problem-solving factor, 443
Politics of technology, 663
Polymorphism, **173**
Porter, Michael, 24, 43–44
PowerBuilder, 535
Power management, information systems in, 32
PowerPC chip, 92
Precise/Pulse (Precise Software Solutions), 609–610
Preliminary evaluation, **583**
Present value, net, **545**–547
Price Waterhouse, 461
Pricing, product, 417
Pricing systems, automated, 3
Primary key, **192**
Primary storage (main memory), **86**, 124–125
Printers, 111
Privacy, 643–651
    corporate policies on, 650
    in credit reporting, 195–196
    e-mail, 645
    fairness in information use, 647–648
    federal government and, 644–645
    federal laws and regulation on, 648–649
    immigration laws and, 516
    individual, 650–651
    Internet and, 645–646
    invasion of, 578
    of networks, 321–324
        firewalls, 324–325
    state laws and regulation on, 650
    at work, 645
Privacy Act of 1974 (PA74), 648–649
Privacy Study Commission, 649
Private branch exchange (PBX), **253**
Problem solving, 438–444, **439**
Problem-solving factors, 442–444
Procedures design, 571, 572
Process, **6**
Process control, in manufacturing MIS, 411–412
Processing, **15**–16
    parallel, **95**–96
Processing controls, 579
Processing design, 571
Processing devices (CPU). *See under* Hardware
Processing mechanisms, 8–9, 10
Processing tasks of operating systems, 145–147
Processors, 83
    optical, **91**
Process symbol, **552**
Procter & Gamble, 31
Product development, 416
Production process, in manufacturing MIS, 406–407
Production scheduling, in manufacturing MIS, 408
Productivity, **64**–65
Product lines and services, improving existing, 60
Product pricing, 417
Products, substitute, 57–58, 62
Professional conduct, ACM code of, 655
Professional services, information systems in, 32
Profit centers, **403**
Profit/loss systems, 403
Program-data dependence, **194**
Program Evaluation and Review Technique (PERT), **541**, 542
Program flowcharts, 14
Programmable read-only memory (PROM), 94

Programmed decision, **440**
Programmer, **521**, 522
Programming
    life cycle, 590–591
    rule-based, 176, 504
Programming languages, 169–179. *See also specific languages*
    fifth-generation, 176
    first-generation, 170
    fourth-generation (4GL), **171**–172, 592
    object-oriented, **172**–174, 184–185
    second-generation, 170–171
    selecting, 176–178, 591
    standards and characteristics of, 169–170
    third-generation, 171
    translators, 178–179
    visual, **175**–176
Project 98 (Microsoft), 476
Project deadline, 540
Project Gutenberg Web site, 306
Projecting, **204**
Project management models, **456**, 476–477
Project management software, 158, 422
Project management tools, 540–542
Project milestone, **540**
Project organizational structure, **47**
Project planning, 538–540
Project schedule, **540**
Promotion, 416–417
Proprietary software, **156**, 157
ProShare Video System 500 (Intel), 303
Protocol, **266**
Prototyping, **533**–534, 535
Prying, improper ("browsing"), 196
Public network services, **276**–277
Publishing companies, information systems in, 31
Purchase order processing system, 356–358, **357**
Purchasing, web-based, 385. *See also* E-commerce
Purchasing transaction processing systems, 348, **355**–360
    accounts payable system, **360**
    inventory control system, 356
    purchase order processing system, 356–358, **357**
    receiving system, 358–359
Push technology (webcasting), **314**

Q
Quality, **54**
Quality control, **412**, 493
Quality Decision Management, 449
Query languages, **172**
Questionnaires, **549**
Quicken, 159
Quick Response Methods (QRM), 437

R
R/3 (SAP), 374, 376–378, 385–386, 531
Rambus Inc., 93
Random access memory (RAM), **92**–94
Rapid application development (RAD), **534**–536
RCA, 449
Read-only memory (ROM), **94**
RealAudio, 304
Receiving and inspecting data in manufacturing MIS, 406
Receiving system, 358–359
Record, **191**
Record keeping, automated, 460
Recruiting, personnel, 422–423

Reduced instruction set computing (RISC), **91**–92
Redundancy, data, **194**, 209
    planned, **200**
Redundant array of independent/inexpensive disks (RAID), **99**
Reengineering (process redesign), 51–53, 54
Refreezing, 49, 51
Regional Bell Operating Companies, 252
Regional information services, 277
Registers, **86**
Relational model, **203**–204
Release, **604**
Reliability, data, 209
Removable storage devices, **102**
Reorder point (ROP), **408**
Repair and maintenance, expert systems and AI in, 508
Repetitive decisions, 450
Repetitive motion disorder, **651**
Repetitive stress injury (RSI), **651**, 653–654
Replenishment program, continuous (CRP), 79
Replicated database, 219
Report(s)
    accounts receivable, 362–363
    decision support system and, 445
    demand, 26, **392**, 393
    design, **587**
    developing effective, 395
    drill down, **394**–395
    exception, 26, **392**–394
    general ledger, 365
    Inventory Status, 352
    key-indicator, **392**, 393
    for marketing managers, 417
    sales-related, 417
    scheduled, 25, **392**, 393
    systems analysis, 556–557
    systems investigation, 547
    TPS, 340
Report generator, 211–212
Report layout, **555**
Repository, ABAP/4, 378
Request for maintenance form, **604**
Request for proposal (RFP), **580**–581
Requirements analysis, **553**–556
RescueWare, 605
Research, marketing, 415–416
Resources in Motion Management System (RiMMS), 356
Restart procedures, **573**
Retail companies, information systems in, 32
Retirement, early, 56
Retter, Terry, 22
Return on investment (ROI), **65**–66
Reusable code, **173**
Revenue centers, **403**
Review, systems. *See* Systems review
Rightsizing, 55, 56
Ring network, **258**, 259
Rivalries, competitive advantage and, 57, 62
Robotics, **484**–485
Roles in IS department, 68–72
Routers, **268**–269, 320–321
Routing system, **355**, 356
RSA protocol, 322
Rubric Inc., 418
Rule-based programming, 504
Rules, **497**
Rules-based programming, 176
Rumor Mill, 331
Runaways, **539**
Ryder Truck, 616–617

S
Saab Cars USA, 235
SABRE Business Travel Solutions (SABRE BTS), 330–331
Sabre Group Holding Inc., 576
SABRE (semi-automated Business Research Environment), 60–61, 186–187
Salary administration, 424
Sales
    analysis of, 417
    informations systems in, 30
    on Internet, 53
    reports relating to, 417
Sales Builder Engine, 37
Sales configuration order processing system, **350**
Sales force automation, 37
Salt Lake City, Utah, 426
Sampling, statistical, **549**
San Diego Supercomputer Center (SDSC), 120
Sanofi, 607
SAS, 455
Satellite transmission, 247
Satisficing model, **441**, 447, 505
Saturn, 46
Savings and loan companies, information systems in, 31
Scalability, **146**
Scams, 635–637
Scanning devices, 106, 108–109
Schedule, project, **540**
Scheduled reports, 25, **392**, 393
Schedule feasibility, **545**, 546
Scheduling and job placement, 423
Scheduling system, **355**, 356, 423, 515
Schein, Edgar, 49
Schema, **206**, 208
Schematic model, 13–14
Schmidt, Eric E., 288
Schneider, Carolyn, 425
Scope creep, 586
Scottsdale, Arizona, 426
Screen layout, **554**, 556
Screens, touch-sensitive, 108
Search engines, **310**–312
Sears, Roebuck and Co., 24, 224, 239
Secondary storage, **96**. *See also under* Hardware
Second-generation languages, 170–171
Secure Electronic Network for Travelers' Rapid Inspection (SENTRI), 516
Secure Electronic Transactions (SET), 371
Security, 578
    AOL and, 662
    of client/server architectures, 264
    design of, 571
    in E-commerce, 371
    firewalls, **316**, 324–325
    of Java applets, 313
    of networks, 321–325
Selecting, **204**
Selective backup, **577**–578
Semistructured or unstructured problems, **450**–451
Sensitivity analysis, **408**
Sequence structure, **595**
Sequential access, **97**
Sequential access storage devices (SADS), **97**
Serial Line Internet Protocol (SLIP), **294**
Serial Storage Architecture (SSA), 134

Server(s), 119
file, **263**
LAN, Internet connection via, 293
in SAP system, 376, 377
universal database, 227
web, 305
Server application, **163**
Services
creating new, 60
improving existing, 60
substitute, 57–58, 62
telecommunications, 251–256
Shell, expert system, **492**, 503–505
Sheraton hotels, 589
Shielded twisted-pair wire cable, 245
Shipment execution system, **352**
Shipment planning system, **350**–351
Shipping, expert systems and AI in, 508
Shopping, on-line, 22–23, 315. *See also* E-commerce
Shreiner, Jay, 184
Signature, digital, **322**
Sign-on procedure, **572**–573
Simulation, **448**
Site preparation, **600**
Six Sigma, 54–55
Skills inventory, 423
Slipstream upgrade, **604**
Small Computer System Interface (SCSI), 124
Smalltalk, **174**
Smart cards, 371, 372
Smart modems, 248–249
Smialowski, Joseph, 239
Smith, David, 491
Social and political actions, as problem-solving factor, 443
Societal issues. *See* Ethical and societal issues
Society of Competitive Intelligence Professionals, 467–468
Soft copy, 345
Software, 17–18, 136–188
acquisition of, 588–592
external development, 588–590
in-house development, 590–592
application software, **138**–139, 155–169
backup, 577–578
censorware, 642
communications, **265**–266
cost of defects in, 593
design of, 572
development tools and techniques, 592–598
CASE tools, 596–597
cross-platform development, **597**
integrated development environments (IDEs), **597**–598
structured design, 592–595, **594**
structured programming, 595–596
structured walkthroughs, **598**
electronic distribution of, 270–272, **271**
global support of, 141
intelligent agents, 359, 495
issues and trends, 140–141
marketing, 432
model management (MMS), **453**
network management, **266**
object-oriented development of, **592**
operating systems (OS), 141–154
packaged, 67

pirating of, 635, 637
programming languages, **169**–179
project management, 422
routing, 356
sales force automation, 37
scheduling, 356
for supporting goals, 139–140
upgrades of, 140–141
utility programs, **154**–155
Software interface, **590**
Software licenses, 140, 185–186
Software piracy, **635**, 637
Software Publishers Association (SPA), 185–186, **640**
Software suite, **161**–163
Sony, 60
Source code, **178**
Source data automation, **104**, **343**
Spatial data technology, 228–229
Specialized and regional information services, 277
Special-purpose computers, **112**–113
Special-purpose modems, 249
Speech recognition, 485–486, 487
Sphere of influence, **139**–140
SportsLine USA, 306, 555
Spreadsheet software, 158, 159, 160
SPSS, 455, 489
Stakeholders, **521**, 522, 599
Standardized Teleoperation System (STS), 484–485
Starfire (Sun Enterprise 10000), 119
Star network, **258**, 259
Start-up, **602**
Statistical analysis models, **455**
Statistical sampling, **549**
Steering committee, **547**
Steinhardt, Barry, 196
Stock keeping unit (SKU), 349
Storage
data, **344**
primary (main memory), **86**, 124–125
removable, **102**
secondary, **96**–103
Strategic alliance (strategic partnership), 59–60
Strategic control, ESS support for, 467
Strategic goal setting, expert systems to assist in, 493
Strategic planning, **466**
for MIS, 401–402, 405, 413, 419
Structured interview, **548**
Structured Query Language (SQL), 12, **172**, **211**
Structured walkthroughs, **598**
Subject directories, 311–312
Subnotebook computers, 113
Subschema, **207**
Substitute products and services, threat of, 57–58, 62
Sun Enterprise 10000 (Starfire), 119
Sungard Data Systems, 601–602
SunGard Recovery Services, 577
Sun Microsystems, 119
Supercomputers, 112, **120**
Superconductivity, **91**
Superdisk, 102
Super video graphics array (SVGA), 110
Suppliers, bargaining power of, 58, 62
Supply and demand curve for pricing analysis, 416
Support, vendor, 580
Support component of IS department, 69–70
Switched line, **251**
Switches, **268**–269

Sybase, 216
Symbol(s)
entity, **552**
process, **552**
processing and manipulation of, **482**
Synchronous dynamic RAM (SDRAM), 93
SyQuest drives, 102
System(s), **8**–15
components of, 9–10
modeling, 13–15
performance and standards, 11–12
types of, 10–11
System boundary, **9**
System development, corporate mergers and, 618–619
System ESS, 137
System maintenance and review, **29**
System parameter, **13**
System performance measurement, **609**–610
System performance products, **609**
System performance standard, **11**–12
Systems analysis, **29**, **532**, 547–558
data analysis, **550**–553
data collection for, 548–549
participants in, 548
report on, 556–557
requirements analysis, **553**–556
by utility companies, 558
Systems analyst, 68, **521**, 522
Systems control, **578**–580
Systems design, **29**, **532**, 570–587
contract, 586–587
design report, **587**
emergency alternative procedures and disaster recovery, 574–578
evaluating and selecting, 582–586
freezing design specifications, 586
generating alternatives, 580–582
logical and physical design, **570**–572
special considerations in, 572–574
systems control, **578**–580
systems maintenance and, 606–607
vendor support and, 580
Systems development, **29**, 69, 518–544
information systems planning, 523–526, **524**
initiating, 522–523
Internet and, 528–529
life cycles of, 530–537
end-user, 536–**537**
error correction, 530–531
joint application development (JAD), **536**
prototyping, **533**–534, 535
rapid application development (RAD), **534**–536
traditional, 532–533
measuring return on, 617–618
objectives for, 527–528
participants in, 521–522
success of, 537–544
degree of change and, 537–538
formal quality assurance processes and, 542–544
project management tools and, 540–542
quality of project planning and, 538–540
trends in, 529–530

System sign-on, **572**
Systems implementation, **29**, **532**–533, **588**–602
database and telecommunications systems acquisition, 599
data preparation/data conversion, **600**
hardware acquisition from IS 588
installation, **601**
personnel hiring and training, 600
site preparation, **600**
software. *See under* Software
software acquisition, 588–592
external development, 588–590
in-house development, 590–592
start-up, **602**
testing, 601–602
user acceptance document, **602**
user preparation, 599–600
Systems investigation, **29**, **532**, 544–547
feasibility analysis, 545–547
initiating, 544–545
participants in, 545
Systems investigation report, **547**
Systems maintenance, **602**–607
design and, 606–607
financial implications of, 605–606
performing, 604–605
reasons for, 602–604
request for maintenance form, **604**
types of, 604
Systems maintenance and review, **533**
Systems Network Architecture (SNA), 266
Systems request form, **544**–545
Systems review, **607**–610
factors to consider during, 608–609
system performance measurement, **609**–610
types of procedures, 608
Systems software, **138**, 139, 141–155
operating systems (OS), **141**–154
utility programs, **154**–155
System testing, **601**
System variable, **13**
System virus, **632**

T
T1 carrier, 254–256, 257
Tables
lookup, **573**
R/3, 378
Tape, magnetic, **98**
Team organizational structure, 47–48
Technical documentation, 591–592
Technical feasibility, **545**, 546
Technology
ERP and infrastructure of, 375
as problem solving factor, 444
Telecommunications, 18, 240–288, **243**
acquisition of, 599
applications of, 269–278
distance learning, 277–**278**
Electronic Data Interchange (EDI), 274–276
electronic software and document distribution, 270–272, **271**
e-mail, **270**, 271
PC-to-mainframe-and-network links, 269–270
public network services, **276**–277

specialized and regional information services, 277
telecommuting, **273**
videoconferencing, **273**–274
voice mail, **270**
backup, 578
bridges, **268**–269
carriers and services, 251–256
design of, 571, 572
devices for, 248–250
gateways, **268**–269
general model of, 243–244
media types for, 245–248
networks, 256–265
    client/server systems, **263**–265
    file servers, **263**
    international, **261**–262
    local area (LAN), **258**–261
    terminal-to-host, **263**
    topology of, **258**, 259, 261
    wide area (WAN), **261**, 262
routers, **268**–269
switches, **268**–269
Telecommunications Act of 1996, 252, 641–642
Telecommunications controls, 579
Tele-Communications Inc. (TCI), 151
Telecommunications medium, **243**
Telecommuting, **273**
Teleconferencing alternative, 463
Telephone services, competition for local, 252–253
Telephony, Internet, 300–303
Telnet, **298**
Tenet, George, 639
Ten Eyck, Lynn F., 121
Terminals, 106
Terminal-to-host network, **263**
Terrorism, computer, 639, 661
TestDrive, 155
Testing
    computer programs, 591
    system, 601–602
Textron Inc., 77–78
Third-generation languages, 171
Thomson Financial Services, 529
THORASK, 507, 508
Thread, 300
"Three Cs" rule for groupware, 166
3M Dental, 44
Ticketmaster, 39
Tilling, Mack, 370
Time compression, **444**
Time-driven review, **608**
Time-sharing, **146**, 147
Top-down approach, **596**
Total quality management (TQM), 412
Touch of Lace, 303
Touch-sensitive screens, 108
Toyota, 626
Traditional organizational structure, **45**–47
Training, 599–600, 627
Training and skills inventory, 423
Transaction, **21**
Transaction processing cycle, **343**
Transaction processing system audit, **347**
Transaction processing systems (TPS), **21**–22, 335–366, **337**
    accounting systems, 348, **360**–366
    activities, 342–345
    control and management issues in, 345–347

in financial MIS, 402
in human resource MIS, 419–420
integration of DSS, MIS, and, 451
in manufacturing MIS, 406–407
in marketing MIS, 413–414
objectives of, 339–342
on-line (OLTP), 219, 225, **338**–339
order processing systems, **347**–355, 384–385
point-of-sale, 344
purchasing systems, 348, **355**–360
traditional, 337–339
Transactions, **337**
Translators, language, **178**–179
Transmission Control Protocol/Internet Protocol (TCP/IP), **266**, 321
Transportation Evacuation Decision Support System (TEDSS), 449
Transportation industry, information systems in, 31
Transport Control Protocol (TCP), **292**
TravComp 2000, 594
Travelers Group, 63
Travelers Property Casualty, 594
Travelocity, 186–187
Travel planning, 330–331
Tree models, 201–202
Trigger points, 393
Trilogy application, 37
Tucker, Laurie, 386
Tunneling, **319**
Turbo Pascal, 598
TurboTax, 159
Tuscan-Leigh Dairies, 356
Twisted-pair wire cable, 245

U
UltilPro, 434
Ultimedia Solution, 124
Uncertainty, ESS and, 465
Unfreezing, 49, 51
Unicenter TNG, 490
Uniform Resource Locator (URL), **292**
Union Bank of Switzerland (UBS), 132–133
Union Gas Ltd., 216
Union Pacific, 629
Unisys, 119, 120, 132–133, 487
Unisys clearPath HMP NX4600, 132
United Airlines, 133–134, 457
United Distillers, 508–509
United Parcel Service (UPS), 41–42, 315, 337
U.S. Army, 449, 660
U.S. Chamber of Commerce, 616–617
U.S. Department of Agriculture (USDA), 660
U.S. government
    privacy and, 644–645
    Y2K problem and, 629
Unit testing, **601**
Universal database server, 227
Universal Product Code (UPC), 343
Unix, 153–154
Unshielded twisted-pair (UTP) wire cable, 245
Unstructured interview, **548**
Unstructured problems, **450**–451
Upper-CASE tools, **542**
UPSnet, 262
Upstream management, 45

Usenet, **298**–299
User, knowledge, **503**, 505
User acceptance document, **602**
User documentation, **592**
User interface, **143**–144
    command-based, **143**
    in expert systems, 500
    graphical (GUI), **143**–144
User preparation, **599**–600
Users, **521**, 522
US West, 236–237, 331–332
US West Intrusion Response Team, 631
Utilities industry, deregulation of, 558, 569
Utility companies, 32, 558
Utility programs, **154**–155

V
VAIO 505 notebook computer, 60
Value-added carriers, **251**
Value-added processes, 43–44, 466
Value chain, 43–**44**, 45
Vanguard Group, 318
Vendor(s)
    client/server architecture and, 265
    DBMS, 217
    hardware acquisition from, 588
    support from, 580
Ventana Group Systems, 461
Version, **604**
Vertical blanking interval (VBI), 150
Very large databases (VLDBs), 220
Very long instruction word (VLIW), **92**
Video components, in multimedia systems, 122, 123
Video compression, **123**
Videoconferencing, **273**–274, 300–303
Video data, 5
Video disks, digital (DVD), 100, **101**
Virtual memory, **145**, 146
Virtual private network (VPN), **319**
Virtual workgroups, **464**
Virus, **632**–634
Virus detection, expert systems and AI in, 508
Visa, 323
Vision systems, **485**
Visual Basic, 175–176, 535
Visual Basic for Applications (VBA), 162, 214
Visual C++, 598
Visual dBASE 7 (Inprise), 215
Visual programming languages, **175**–176
Voice Broker, 486
Voice mail, **270**
Voice-over-IP (VOIP), **300**–302
Voice recognition, 485–486
Voice-recognition devices, **105**
Volume testing, **601**
Volvo, 515

W
Wage and salary administration, 424
Wahl, Paul, 385, 386
Walgreens Pharmacies, 567
Walkthroughs, structured, **598**
Wall Street Journal Web site, 306
Wal-Mart, 189, 358, 410–411
Walt Disney, 487
Walton, Sam, 114
Warehouse management system, 358
Warehouse optimization, expert systems and AI in, 508–509
Waste, computer, 624–625, 626–628

WATS service (wide-area telecommunications service), 253
WavePhore's WaveTop technology, 150
Web. *See* World Wide Web
Web-based DSS, 452–453
Web-based purchasing, 385
Web browsers, **308**–309
Webcasting (push technology), **314**
Webmasters, **71**
Web servers, 305
Web sites, 36, 634, 646–647. *See also* World Wide Web
Weinbach, Larry, 82
Welch, Jack, 436
Wells Fargo & Company, 513–514
West Group, 384–385
"What-if" analysis, **447**
Whirlpool Corporation, 3
White House Web site, 306
Wide area decision network, 463
Wide area network (WAN), **261**, 262
Williams, Chuck, 514
Windows 95 (Microsoft), 148, 149
Windows 98 (Microsoft), 148, 149–150
Windows CE (Microsoft), 148, 150, 151
Windows (Microsoft), 143, 148
Windows NT (Microsoft), 148, 152–153, 265
Word for Windows (Microsoft), 633
Wordlength, **89**–90
Word processing software, 158, 159, 160
Work environment, 651–653
Workgroup(s), **139**–140, 318
    virtual, **464**
Workgroup software, 165–166. *See also* Group DSS (GDSS)
Workgroup sphere of influence, **140**
Workplace privacy, 645
Workstations, 112, **115**–116
World Wide Web, 19, 36, 257, **303**–316. *See also* Internet
    businesses on, 314–316, 317
    developing web content, 309–310
    Dreamworks on, 330
    Java, **312**–313
    marketing on the, 50
    MIS and, 396–397
    push technology (webcasting), **314**
    search engines, **310**–312
    structure of, 305–308
    Web browsers, **308**–309, 312
Worm, **632**
Write-once, read-many (WORM) disk, **100**
Wysocki, Bernard, Jr., 334

X
X.400, 266–268
X.500, 266–268
XCON, 492
Xerox PARC, 143

Y
Yankee Group, 331
Yasunobu, Seiji, 497
Year 2000 (Y2K) problem, 84–85, 629

Z
Zip drive, 102